2018 IFIP/IEEE International Conference on Very Large Scale Integration (VLSI-SoC 2018)

Verona, Italy
8-10 October 2018

IEEE Catalog Number: CFP18LSI-POD
ISBN: 978-1-5386-4757-8

**Copyright © 2018 by the Institute of Electrical and Electronics Engineers, Inc.
All Rights Reserved**

Copyright and Reprint Permissions: Abstracting is permitted with credit to the source. Libraries are permitted to photocopy beyond the limit of U.S. copyright law for private use of patrons those articles in this volume that carry a code at the bottom of the first page, provided the per-copy fee indicated in the code is paid through Copyright Clearance Center, 222 Rosewood Drive, Danvers, MA 01923.

For other copying, reprint or republication permission, write to IEEE Copyrights Manager, IEEE Service Center, 445 Hoes Lane, Piscataway, NJ 08854. All rights reserved.

*** This is a print representation of what appears in the IEEE Digital Library. Some format issues inherent in the e-media version may also appear in this print version.

IEEE Catalog Number:	CFP18LSI-POD
ISBN (Print-On-Demand):	978-1-5386-4757-8
ISBN (Online):	978-1-5386-4756-1
ISSN:	2324-8432

Additional Copies of This Publication Are Available From:

Curran Associates, Inc
57 Morehouse Lane
Red Hook, NY 12571 USA
Phone: (845) 758-0400
Fax: (845) 758-2633
E-mail: curran@proceedings.com
Web: www.proceedings.com

Table of Contents

Preface

Message from the General Co-Chairs i
Graziano Pravadelli and Todd Austin

Message from the Program Co-Chairs iii
Nicola Bombieri and Masahiro Fujita

Organization

Organizing Committee iv

Program Committee vi

Keynotes

Cyber-medical systems: requirements, components, design x
Giovanni De Micheli

Re-imagining scalable system design xi
Valeria Bertacco

Computational challenges in the design automation for synthetic biology xii
Heinz Koeppl

Test and reliability challenges in the Internet of Things xiii
Yervant Zorian

Panel Session

Computation-in-memory: Hype or Hope? xiv
Said Hamdioui (Moderator), Ian O'Connor, Mehdi Tahoori, Henri-Pierre
Charles and Lionel Torres

Session 1: Emerging technologies and computing paradigms

ReRAM-based in-memory computation of Galois field arithmetic 1
Swagata Mandal, Debjyoti Bhattacharjee, Yaswanth Tavva and Anupam
Chattopadhyay

Minimizing performance and energy overheads due to fanout in memristor based
logic implementations 7
Md Adnan Zaman and Srinivas Katkoori

Versatile ring-based architecture and synthesis flow for general-purpose digital
microfluidic biochips 13
Juinn-Dar Huang, Chia-Hung Liu and Wei-Hao Yang

Directed graph placement for SNN simulation into a multi-core GALS architecture ... 19
Francesco Barchi, Gianvito Urgese, Andrea Acquaviva and Enrico Macii

Session 2: Digital architectures: NoC, multi- and many-core, hybrid, and reconfigurable

Traffic aware deflection rerouting mechanism for mesh network on chip 25
Simi Zerine Sleeba, John Jose, Maurizio Palesi, Rekha K. James and Mini Nair

A parallel hardware architecture for quantum annealing algorithm acceleration 31
Evelina Forno, Andrea Acquaviva, Yuki Kobayashi, Enrico Macii and Gianvito Urgese

An instruction set architecture for low-power, dynamic IoT communication 37
Shahzad Muzaffar and Ibrahim Elfadel

Session 3: Prototyping, verification, modeling, and simulation: from digital to analog circuits

An open-source verification framework for open-source cores: a RISC-V case study ... 43
Pasquale Davide Schiavone, Ernesto Sanchez, Annachiara Ruospo, Francesco Minervini, Florian Zaruba, Germain Haugou and Luca Benini

On the rectifiability of arithmetic circuits using Craig interpolants in finite fields 49
Utkarsh Gupta, Irina Ilioaea, Vikas Rao, Arpitha Srinath, Priyank Kalla and Florian Enescu

A synthesizable digital low-dropout regulator based on voltage-to-time conversion 55
Naoki Ojima, Toru Nakura, Tetsuya Iizuka and Kunihiro Asada

Session 4: Variability, reliability, and test

An analysis of test solutions for COTS-based systems in space applications 59
Riccardo Cantoro, Sara Carbonara, Andrea Floridia, Ernesto Sanchez, Matteo Sonza Reorda and Jan-Gerd Mess

Robust detection of bridge defects in STT-MRAM cells under process variations 65
Andrés Felipe Gómez, Freddy Forero, Kaushik Roy and Victor Champac

Evaluating the impact of process variability and radiation effects on different transistor arrangements 71
Leonardo H. Brendler, Alexandra L. Zimpeck, Cristina Meinhardt and Ricardo Reis

An accurate novel gate-sizing metric to optimize circuit performance under local intra-die process variations 77
Zahira Perez, Hector Villacorta and Victor Champac

Session 5: Hardware security

On the effectiveness of the satisfiability attack on split manufactured circuits 83
Suyuan Chen and Ranga Vemuri

Secure and compact full NTRU hardware implementation 89
Konstantin Braun, Tim Fritzmann, Georg Maringer, Thomas Schamberger and Johanna Sepúlveda

Lightweight and high performance SHA-256 using architectural folding and 4-2 adder compressor 95
Ming Ming Wong, Vikramkumar Pudi and Anupam Chattopadhyay

Low-budget energy sector cyberattacks via open source exploitation 101
Anastasis Keliris, Charalambos Konstantinou, Marios Sazos and Michail Maniatakos

Session 6: Machine learning and emerging technologies for low-power and energy-efficient SoC design

Differential power analysis mitigation technique using three-independent-gate field effect transistors 107
Edouard Giacomin and Pierre-Emmanuel Gaillardon

Energy-driven precision scaling for fixed-point ConvNets 113
Valentino Peluso and Andrea Calimera

Session 7: Embedded and cyberphysical systems: architecture, design, and software

Enhancing performance of computer vision applications on low-power embedded systems through heterogeneous parallel programming 119
Stefano Aldegheri, Silvia Manzato and Nicola Bombieri

An FPGA-based hardware accelerator for scene text character recognition........... 125
Luiz Antonio De Oliveira Junior and Edna Barros

VLSI-architecture of radix-2/4/8 SISO decoder for Turbo decoding at multiple data-rates... 131
Rahul Shrestha and Ashutosh Sharma

Session 8: CAD: Synthesis and analysis

Two combinatorial problems on the layout of switching lattices 137
Anna Bernasconi, Antonio Boffa, Fabrizio Luccio and Linda Pagli

HLS support for polymorphic parallel memories................................ 143
Luca Stornaiuolo, Marco Rabozzi, Donatella Sciuto, Marco Domenico Santambrogio, Giulio Stramondo, Catalin Bogdan Ciobanu and Ana Lucia Varbanescu

Inferential logic: a machine learning inspired paradigm for combinational circuits 149
Valerio Tenace and Andrea Calimera

Special Session 1: IoT for health, wellness and personal assistance

Implantable IoT system for closed-loop epilepsy control based on electrical neuromodulation 155
Reza Ranjandish and Alexandre Schmid

Mobile phones hematophagous diptera surveillance in the field using deep learning and wing interference patterns 159

Marc Souchaud, Pierre Jacob, Camille Simon-Chane, Aymeric Histace, Olivier Romain, Maurice Tchuenté, and Denis Sereno

Radar for assisted living in the context of Internet of Things for health and beyond .. 163

Julien Le Kernec, Francesco Fioranelli, Shufan Yang, Jordane Lorandel and Olivier Romain

A hybrid bioimpedance spectroscopy architecture for a wide frequency exploration of tissue electrical properties 168

Achraf Lamlih, Philippe Freitas, Mohamed-Moez Belhaj, Vincent Kerzerho, Fabien Soulier, Serge Bernard, Tristan Rouyer, Sylvain Bonhommeau and Jérémie Salles

Special Session 2: Design understanding

Design understanding: from logic to specification 172

Gorschwin Fey, Tara Ghasempouri, Swen Jacobs, Gianluca Martino, Jaan Raik and Heinz Riener

Special Session 3: Neuromorphic computing - from robust hardware architectures to testing strategies

Neuromorphic computing - from robust hardware architectures to testing strategies .. 176

Lorena Anghel, Denys Ly, Giorgio Di Natale, Benoit Miramond, Elena-Ioana Vatajelu and Elisa Vianello

Special Session 4: Non-volatile emerging memories - breaking down the memory wall

Prospects for energy-efficient edge computing with integrated HfO_2-based ferroelectric devices 180

Ian O'Connor, Mayeul Cantan , Cédric Marchand, Bertrand Vilquin, Stefan Slesazeck, Evelyn T. Breyer, Halid Mulaosmanovic, Thomas Mikolajick, Bastien Giraud, Jean-Philippe Noël, Adrian Ionescu and Igor Stolichnov

Multi-bit nonvolatile flip-flop based on NAND-like spin transfer torque MRAM 184

Erya Deng, Zhaohao Wang, Wang Kang, Shaoqian Wei and Weisheng Zhao

From spintronic devices to hybrid CMOS/magnetic system on chip 188

Sophiane Senni, Frederic Ouattara, Jad Modad, Kaan Sevin, Guillaume Patrigeon, Pascal Benoit, Pascal Nouet, Lionel Torres, François Duhem, Gregory Di Pendina and Guillaume Prenat

Reliable ReRAM-based logic operations for computing in memory 192

Mathieu Moreau, Eloi Muhr, Marc Bocquet, Hassen Aziza, Jean-Michel Portal, Bastien Giraud and Jean-Philippe Noël

Poster Session

An ultra-low power active diode using a hysteresis common gate comparator for low-voltage and low-power energy harvesting systems 196
Kaori Matsumoto, Tetsuya Hirose, Hiroki Asano, Yuto Tsuji, Yuichiro Nakazawa, Nobutaka Kuroki and Masahiro Numa

A bandwidth-aware authentication scheme for packet-integrity attack detection on trojan infected NoC 201
Mubashir Hussain and Hui Guo

Upgrading QoSinNoC: efficient routing for mixed- criticality applications and power analysis 207
Serhiy Avramenko, Siavoosh Payandeh Azad, Behrad Niazmand, Massimo Violante, Jaan Raik and Maksim Jenihhin

Testability of switching lattices in the stuck at fault model 213
Anna Bernasconi, Valentina Ciriani and Luca Frontini

An MCTS-based framework for synthesis of approximate circuits 219
Muhammad Awais, Hassan Ghasemzadeh Mohammadi and Marco Platzner

FLUTE-EM: electromigration-optimized net topology considering currents and mechanical stress 225
Steve Bigalke and Jens Lienig

Meta-model based automation of properties for pre-silicon verification 231
Keerthikumara Devarajegowda and Wolfgang Ecker

Cyber-physical systems integration in a production line simulator 237
Stefano Centomo, Marco Panato and Franco Fummi

Key architectural optimizations for hardware efficient JPEG-LS encoder 243
Yakup Murat Mert

A graph-based approach for mobile localization exploiting real and virtual landmarks. 249
Florenc Demrozi, Kevin Costa, Federico Tramarin and Graziano Pravadelli

Understanding the design space of wavelength-routed optical NoC topologies for power-performance optimization 255
Mahdi Tala and Davide Bertozzi

Design of latch based configurable ring oscillator PUF targeting secure FPGA 261
Mahabub Hasan Mahalat, Nikhil Ugale, Rohit Shahare and Bibhash Sen

A low power keyword spotting algortihm for memory constrained embedded systems . 267
Gionata Benelli, Gabriele Meoni and Luca Fanucci

Author Index 273

Message from the General Co-Chairs

The 26th IFIP/IEEE Conference on Very Large Scale Integration (VLSI-SoC 2018), turning around the overarching theme: **Design and Engineering of Electroni Systems Based on New Computing Paradigms**, was held between October 8th to 10th 2018 in Verona (Italy). Verona, the city of love, theater of Shakespear's tragedy *Romeo and Juliet*, is located on the bank of the Adige river in the Veneto region. Awarded World Heritage Site status by UNESCO, Verona is one of the main tourist destinations in northern Italy, owing to its artistic heritage, annual fairs, shows and operas, such as the lyrical season in the *Arena*, the Roman amphitheater. As far as the quality and the preservation of its Roman antiquities are concerned, Verona is second only to Rome, but its historical value is not restricted to the Roman times. The period of Scaligeri and the Venetian domination have left their mark on the city as well. From the naturalistic point of view, Verona is 20 minutes away from the Garda Lake, a real natural gem, and one of the most popular tourist destinations in Italy, which offers wonderful landscapes among vineyards, olive orchards and the crystal-blue water of the lake. Less than 2 hours by car or train from the Dolomites, Venice, Milan and Florence, Verona represents a fantastic base for visiting many destinations in the northern and the central part of Italy. Besides these tourist attractions, today Verona is a dynamic business city with a relevant impact on the agro-industrial, manufacturing and fashion sectors, home to a university that attracts approximately 25,000 students every year. Thanks to VLSI-SOC 2018, during the days of the conference, Verona was visited by 95 scientists and researchers from 20 countries worldwide. In addition to the great technical program, these delegates enjoyed the full social program of the conference, which included in the first day a guided city tour followed by a welcome reception based on Italian pizza, and in the second day a wine tasting tour during which they explored the winery of Villa Mosconi-Bertani, tasted different kinds of wine, including the famous Amarone of Valpolicella, and enjoyed the excellent food of the Principe Amedeo Restaurant in the Custoza hills. Overall, based on the result of the evaluation forms filled by the participants, we can say VLSI-SOC 2018 was a successful event from both the scientific and the cultural points of view, but this would not have been possible without the great effort of many volunteers. As general co-chairs of VLSI-SoC 2018, we would like then to express our gratitude to all the people that cooperated for the organization of the conference, as well as to all the participants. By scrolling through a long list of thanks, our first big appreciation goes to the program co-chairs, Nicola Bombieri and Masahiro Fujita, for their invaluable effort to put together an outstanding technical program. We would also like to thank the Program Committee members that, under the supervision of Nicola and Masahiro, made an impressive review work to select an excellent set of technical papers. As tradition, we have also enriched the program by including special sessions on excellent hot topics, and we gave the possibility to PhD students to discuss their research in a dedicated PhD Forum. This was possible thanks to the great effort of the special session co-chairs, Katell Morin-Allory and Srinivas Katkroori, and of the PhD forum co-chairs, Sara Vinco and Kiyoung Choi. We would also like to extend many thanks to those people who devoted their time to contribute to the support activities of the conference. Dr. Yervant Zoriant, the industrial chair, offered a great industrial keynote and worked to guarantee the sponsorship of Synopsys for the best paper award. Matteo Sonza Reorda and Ricardo Reis, the publicity co-chairs, did a very good job in promoting the conference to attract submissions and participants. Davide Bertozzi and Madhi Tala, the publications co-chairs, managed very efficiently all the activities related to publication of the proceedings. Michele Lora, the registration chair, took care of all the aspects related to the registration of the participants and supported many of them in getting VISA and other necessary documents to allow their attendance to the conference. A big thanks goes to the local arrangement committee composed of Stefano Aldegheri, Stefano Centomo, Alessandro Danese, Florenc Demrozi, and Franco Fummi, that worked very hard for setting up the web site and organizing all the local aspects before and even more during the conference, tirelessly contributing to make VLSI-SoC a great experience.

A special thank is directed also to the VLSI-SoC Steering Committee and the IFIP 10.5 working group for giving us the opportunity of hosting VLSI-SoC in Italy for the first time, and specially to Salvador Mir who has guided us in all the activities related to the conference organization since October 2016. Our huge recognition goes also to Aurora Miorelli, Barbara Zaccheddu and Leonardo Bonfiglio from the administrative staff of the Computer Science Department of the University of Verona, who supported us for contracting, budgeting and managing all the financial aspects of VLSI-SOC. We conclude this list, by thanking also our sponsors: IFIP, IEEE CAS, IEEE CEDA, ACM SIGDA, AICA, Synopsys, who contributed in many relevant ways to support VLSI-SOC 2018, and the Computer Science Department of the University of Verona, which acted as organizing institution, supporting both financially and operatively the concretization of this year's event. VLSI-SoC 2019 will be held in magnificent Cuzco, Peru. We hope to see you all there. Direct questions and comments about this article to Graziano Pravadelli, University of Verona (Italy), through his email *graziano.pravadelli@univr.it*.

Graziano Pravadelli and Todd Austin
Conference General Co-Chairs

Message from the Program Co-Chairs

We are pleased to present the technical program for the 26th IFIP/IEEE International Conference on Very Large Scale Integration 2018 (VLSI-SoC 2018). The high number of submissions (106), which is stable with the last years' conference editions, is a strong testimony of the vitality of the community, and confirms VLSI-SoC stature as the premier technical forum on very large scale integration. The technical program committee was formed first by grouping 22 people to chair the 11 conference tracks (2 co-Chairs per track). They were selected as experts in the fields, and each of them selected additional TPC members for a total of 98 people to carefully review and select the submitted papers. The TPC finally selected 39 papers to appear at the conference through a process that required over 400 reviews. Each submitted paper had more than 4 reviews in average and no paper had less than 3 reviews. Two papers received 8 reviews. Conflicts of interest were managed by the chairing online system and thanks to the double track chairing modality. We also asked the track Chairs to identify best paper candidates, whose presentations at the conference were attended and evaluated by the TPC members to assign the Best Paper Award. In general, beside the traditional VLSI-SoC topics such as digital architectures, CAD synthesis and analysis, variability, reliability and test, emerging technologies and computing paradigms cover a significant fraction of this conference edition program. Embedded and cyberphysical systems and hardware security have emerged as some of the other dominant themes in VLSI-SoC 2018, reflecting expectations of changing demands on future system integration paradigms. We are also glad to have four excellent keynote speakers this year: Giovanni De Micheli (EPFL Switzerland), who addresses requirements, components, and design of cyber-medical systems; Valeria Bertacco (Univ. of Michigan, USA), who talks about re-imaging of scalable system design; Heinz Koeppl (Technische Universitat Darmstadt, Germany) addresses computational challenges in the design automation for synthetic biology. Finally, Yervant Zorian (Synopsys-Armenia) talks about test and reliability challenges in the Internet of Things. Four additional special sessions on emerging technologies and computing paradigms and a discussion panel on computation-in-memory complete an exciting and cutting-edge program. We would like to thank the VLSI-SoC Steering Committee and the IFIP 10.5 working group, who were very helpful in the program committee selection. We would also like to thank the PC members who put enormous effort into reviewing the papers and selecting the final program.

Nicola Bombieri and Masahiro Fujita
Conference Program Co-Chairs

Organization

General Co-Chairs

Graziano Pravadelli .... University of Verona, Italy
Todd Austin .... University of Michigan, USA

Program Co-Chairs

Nicola Bombieri .... University of Verona, Italy
Masahiro Fujita .... University of Tokyo, Japan

Special Session Co-Chairs

Srinivas Katkoori .... University of South Florida, USA
Katell Morin-Allory .... TIMA Laboratory, France

PhD Forum Co-Chairs

Kiyoung Choi .... Seoul National University, Korea
Sara Vinco .... Politecnico di Torino, Italy

Industrial Chair

Yervant Zorian .... Synopsys, USA

Local Chair

Franco Fummi .... University of Verona, Italy

Publicity Co-Chairs

Ricardo Reis .... UFRGS, Brasil
Matteo Sonza Reorda ...... Politecnico di Torino, Italy

Publication Co-Chairs

Davide Bertozzi .... University of Ferrara, Italy
Mahdi Tala .... University of Ferrara, Italy

Registration Chair

Michele Lora .... Singapore University of Technology and Design, Singapore

Steering Committee

Manfred Glesner .... TU Darmstadt, Germany
Matthew Guthaus .... UC Santa Cruz, USA
Luis Miguel Silveira .... INESC ID, Portugal
Fatih Uğurdağ .... Özyeğin University, Turkey
Salvador Mir .... TIMA Laboratory, France
Ricardo Reis .... UFRGS, Brasil
Chi-Ying Tsui .... HKUST, Hong Kong, China
Ian O'Connor .... Lyon Institute of Nanotechnology, France
Masahiro Fujita .... The University of Tokyo, Japan

Program Committee

Track 1: Analog, mixed-signal, and sensor architectures

Piero Malcovati .... University of Pavia, Italy
Tetsuya Iizuka .... University of Tokyo, Japan

Track 2: Digital architectures: NoC, multi- and many-core, hybrid, and reconfigurable

Ian O'Connor .... Lyon Institute of Nanotechnology, France
Michael Huebner .... Ruhr-Universität Bochum, Germany

Track 3: CAD: Synthesis and analysis

Srinivas Katkoori .... University of South Florida, USA
Ibrahim Elfadel .... Masdar Institute, UAE

Track 4: Prototyping, verification, modeling, and simulation

Tiziana Margaria .... Lero, Ireland
Katell Morin-Allory .... TIMA Laboratory, France

Track 5: Circuits and systems for signal processing and communications

Fatih Uğurdağ .... Özyeğin University, Turkey
Luc Claesen .... Hasselt University, Belgium

Track 6: IoT, embedded and cyberphysical systems: Architecture, design, and software

Zebo Peng .... Linköping University, Sweden
Donatella Sciuto .... Politecnico di Milano, Italy

Track 7: Low-power and thermal-aware IC design

Dimitrios Soudris .... National Technical University of Athens NTUA, Greece
Alberto Macii .... Politecnico di Torino, Italy

Track 8: Emerging technologies and computing paradigms

Andrea Calimera .... Politecnico di Torino, Italy
Ricardo Reis .... UFRGS, Brasil

Track 9: Variability, reliability, and test

Salvador Mir .... University of Grenoble Alpes, France
Matteo Sonza Reorda ...... Politecnico di Torino, Italy

Track 10: Hardware security

Mihalis Maniatakos .... New York University Abu Dhabi, UAE
Lilian Bossuet .... University St. Etienne, France

Track 11: Machine learning for SoC design and for electronic design automation

Mehdi Tahoori .... Karlsruhe Institute of Technology, Germany
Manuel Barragan .... TIMA Laboratory, France

Members of the Technical Program Committee

Abdulkadir Akin, ETHZ, Switzerland
Aida Todri-Sanial, LIRMM, France
Alberto Bosio, LIRMM, France
Alberto Gola, AMS, Italy
Andrea Acquaviva, Politecnico di Torino, Italy
Anupam Chattopadhyay, Nanyang Technological University, Singapore
Arun Kanuparthi, Intel, USA
Bei Yu, University of Texas at Austin, USA
Brice Colombier, CEA, France
Carlos Silva Cardenas, Pontificia Universidad Catolica del Peru, Peru
Cecile Braunstein, PMC/LIP6, France
Chengmo Yang, University of Delaware, USA
Chun-Jen Tsai, National Chiao Tung University, Taiwan
Diana Goehringer, TU Dresden, Germany
Diego Barrettino, Ecole Polytechnique Federale de Lausanne, France
Donghwa Shin, Yeungnam University, Korea
Edoardo Bonizzoni, University of Pavia, Italy
Elena Ioana Vatajelu, TIMA Laboratory, France
Federico Tramarin, CNR-IEIIT, Italy
Franck Courbon, University of Cambridge, UK
Fynn Schwiegelshohn, Ruhr University Bochum, Germany
Georg Sigl, Technical University of Munich, Germany
Gildas Leger, Instituto de Microelectronica de Sevilla IMSE-CNM-CSIC, Spain
Giorgio Di Natale, LIRMM, France
Haluk Konuk, Broadcom, USA
Haris Javaid, Xilinx, Australia
Houman Homayoun, George Mason University, USA
Ippei Akita, Toyohashi University of Technology, Japan
Iraklis Anagnostopoulos, National Technical University of Athens, Greece
Jaan Raik, Tallin University of Technology, Estonia
Jones Yudi Mori, University of Brasilia, Brasil
Jinmyoung Kim, Samsung Advaced Institute of Technology, South Korea
Johanna Sepulveda, Technical University of Munich, Germany
Jose Monteiro, INESC-ID, IST Ulisboa, Portugal
Ke Huang, San Diego State University, USA
Kostas Siozios, Aristotle University of Thessaloniki, Greece
Lars Bauer, Karlsruhe Institute of Technology, Germany
Leandro Indrusiak, University Of York, UK
Lionel Torres, LIRMM, France
Luciano Ost, University of Leicester, UK
Maksim Jenihhin, Tallinn University of Technology, Estonia
Maria Michael, University of Cyprus, Cyprus
Massimo Poncino, Politecnico di Torino, Italy

Matthias Sauer, University of Freiburg
Mirko Loghi, Universit di Udine, Italy
Nadine Azemard, LIRMM / CNRS, France
Nele Mentens, Katholieke Universiteit Leuven, Belgium
Nektarios Georgios Tsoutsos, New York University, USA
Ozgur Tasdizen, ARM, UK
Paolo Amato, Micron, Italy
Patri Sreehari, National Institute of Technology, Warangal, India
Peng Liu, Zhejiang University, China
Per Larsson-Edefors, Chalmers University, Sweden
Philippe Coussy, University De Bretagne, France
Pierre-Emmanuel Gaillardon, University of Utah, USA
Po-Hung Chen, National Chiao Tung University, Taiwan
Raik Brinkmann, OneSpin Solutions, Germany
Rani S. Ghaida, GlobalFoundries, USA
Robert Wille, Johannes Kepler University Linz, Austria
Rouwaida Kanj, American University Of Beirut, Libano
Said Hamdioui, Technical University of Delft, The Netherlands
Salvatore Pennisi, University of Catania, Italy
Sezer Goren, Yeditepe University, Turkey
Shahar Kvatinsky, Technion - Israel Institute of Technology, Israel
Sicheng Li, HP, USA
Soheil Samii, General Motors, USA
Sri Parameswaran, University of New South Wales, Australia
Tetsuya Hirose, Kobe University, Japan
Theocharis Theocharides, University of Cyprus, Cyprus
Tolga Yalcin, NXP, UK
Valerio Tenace, Politecnico di Torino, Italy
Victor Champac, National Institute or Astrophysics, Optics and Electronics, Mexico
Victor Kravets, IBM, USA
Virendra Singh, Indian Institute of Technology Bombay, India
Vladimir Zolotov, IBM, USA
Wenjing Rao, University of Illinois at Chicago, USA
Yier Jin, University of Florida, USA

KEYNOTE PRESENTATIONS

Cyber-medical systems: requirements, components, design

Giovanni De Micheli

EPFL (Switzerland)

Abstract

Distributed data acquisition and control systems for health care are playing an important in prevention and cure. Examples include remote patient monitoring, emergency care as well as routine care and they benefit from organized and optimized means to quantify clinical data, handle large data sets as well as controlling and personalizing therapy and drug administration. Current electronic devices and systems incorporate bio-chemical interfaces, such as sensors, to perform data acquisition directly. The fusion of sensing and microelectronic technologies, as well as the ability of volume production of integrated sensing systems that can be personalized in the very back end of the line or after fabrication is an important scientific and commercial goal. Electronic design automation is a key technology to realize cyber-medical systems. Examples of specific EDA tools and methods encompass physical design of integrated sensors and their coupling to electronics, simulation of complex systems with bio-chemical stimuli, synthesis of decision making circuitry based on plurality of inexact inputs, policies design for therapies exploiting on-line data acquisition, and verification of life-critical applications under broadly-varying and unpredictable input conditions. Overall, cyber-medical systems represent an important and large market opportunity as well as a chance to realize the promises of better and less expensive care for everyone.

Re-imagining scalable system design

Valeria Bertacco

University of Michigan (USA)

Abstract

In the past 50 years, silicon dimensional scaling has been delivering the lion's share of advancement and economy of scale in computing. Unfortunately, this source of value is quickly drying out, highlighting the urgent need to provide novel approaches to accelerate innovation in computer design. In this talk, I outline three research domains that promise to provide effective alternatives to the once-plentiful benefits of silicon dimensional scaling. First, we now have the opportunity to move beyond silicon dimensional scaling and explore new silicon structures that boost system value in alternate ways. Second, we need to fully embrace specialized heterogeneous design, tailoring systems to application domains, thus leading to orders-of-magnitude performance and value benefits. Finally, and perhaps most importantly, we must break the cycle of ever-increasing design costs and diminishing innovation. By embracing novel low-cost and agile design technologies, we can engage a broader community of engineers in accelerating innovation in computing system design.

Computational challenges in the design automation for synthetic biology

Heinz Koeppl

Technische Universität Darmstadt (Germany)

Abstract

Design automation in synthetic biology is hampered by insufficient accuracy of computational models of synthetic circuits. More specifically, traditional models ignore the molecular context of the host cell that interacts with a synthetic circuit in numerous ways. For instance, transcription factors of the circuit may get tethered away by unspecific binding sites on the host DNA. Hence, in order to finally realize a predictive in-silico design framework, as done in electronic design automation, computational models need to account for all context effects encountered by the circuits. In this talk I will provide an overview of different context effects and our formal attempts to accommodate some of them in computational models. In particular, we derive a stand-alone equivalent model of a circuit that behave as if the circuit is still embedded in the host cell. Such contextualized models also lead to new calibration algorithms on which I will briefly touch upon.

Test and reliability challenges in the Internet of Things

Yervant Zorian

President of Synopsys Armenia

Abstract

The Internet of Things (IoT) is an extremely fragmented market and can be defined as anything from sensors to small servers. It is estimated that over 30 billion IoT devices will ship by 2020. The ability to sense countless amounts of information that communicates to the cloud is driving innovation into IoT applications, such as in wearable devices (for health, fitness or infotainment applications) and in machine-to-machine applications (in smart appliances, smart cities or commerce). It has become crucial for today?s IoT chips to use a range of new solutions during the design stage to ensure test and reliability. DFT designers need to use new test and reliability solutions to enable power reductions during test, concurrent test, isolated debug and diagnosis, pattern porting, calibration, and uniform access. Moreover, the per unit IoT price remains a key factor in high volume production. Thus, minimizing the test cost while meeting the above technical issues is one of the major challenges of the IoT industry. This presentation, besides discussing the key trends and challenges of IoT, will cover solutions to handle the wide range of above challenges during all periods of the IoT lifecycle from design to in-system operation.

26th IFIP/IEEE International Conference on Very Large Scale Integration

Verona, 8-10 October, Roseo Hotel Leon D'Oro

Panel session

Computation-in-Memory: Hype or Hope?

Abstract

It is well recognized that today's computer architectures and technologies are unable to deal with electronic applications that are extremely demanding in terms of storage and computing power. In order for computing systems to keep delivering sustainable benefits for the foreseeable future society, alternative computing architectures and notions have to be explored in the light of emerging new device technologies.

This panel aims at gathering opinions from researchers working on emerging non-volatile memories to enable computation-in-memory paradigm being able to solve (even partially) the major bottlenecks today's architecture and technologies are facing such as memory and power/energy bottlenecks. Some questions we hope to get answered are:

- What does computation-in-memory mean? What is the main difference between other computation paradigms (near-memory computing, conventional computing, etc.)?
- What makes computation-in-memory feasible or infeasible?
- What kind of emerging memories are suitable for computation-in-memory?
- How will computation-in-memory change the way we used to design our circuits and computers?
- What kind of applications can/cannot benefit from such new architectures?
- What are the major challenges that should be solved before having computation-in-memory become a reality?
- Etc.

Organizer and moderator

- Said Hamdioui – Delft University of Technology (the Netherlands)

Panelists

1. **Ian O'Connor (Lyon Institute of Nanotechnology, France)**
2. **Mehdi Tahoori (Karlsruhe Institute of Technology, Germany)**
3. **Henri-Pierre Charles (CEA, France)**
4. **Lionel Torres (LIRMM, France)**

Proceedings of the 2018 IFIP/IEEE International Conference on Very Large Scale Integration (VLSI-SoC)

October 8-10, 2018, Verona, Italy

26th Edition

Sponsored by

Proceedings of the 2018 IFIP/IEEE International Conference on Very Large Scale Integration (VLSI-SoC)

Proceedings of the 2018 IFIP/IEEE International Conference on Very Large Scale Integration (VLSI-SoC)

October 8-10, 2018, Verona, Italy

Sponsored by

Proceedings of the 2018 IFIP/IEEE International Conference on Very Large Scale Integration (VLSI-SoC)

ReRAM-based In-Memory Computation of Galois Field arithmetic

Swagata Mandal, Debjyoti Bhattacharjee*, Yaswanth Tavva, Anupam, Chattopadhyay

School of Computer Science and Engineering, Nanyang Technological University, Singapore

Email*: *debjyoti001@ntu.edu.sg*

Abstract—Robust data communication is a prime need in the age of Internet-of-things (IoT), where multiple connected devices actively exchange information. To permit robustness of this information exchange, error resilient secure communication is necessary. Security, error detection as well as correction are fundamentally based on Galois Field (GF) arithmetic. In this work, we present a novel method for performing GF arithmetic on a state-of-the art ReRAM-based in-memory computing platform. ReRAM devices offer low leakage power, high endurance and non-volatile storage capabilities, coupled with stateful logic operations. The proposed lightweight library presents the mapping of GF element generation, addition and multiplication. We have experimentally verified the results. For $GF(2^4)$, 3.8 nJ, 0.1 nJ and 3.1 nJ energy are required for element generation, addition and multiplication operations respectively, which demonstrates the efficacy of the mapping.

I. INTRODUCTION

Low power consumption, high circuit density, retention capacity and endurance have lead to the growth of Resistive Random Access Memory (ReRAM) technology as an alternative solution of non-volatile NAND or NOR based flash memory. Some other advantages of ReRAM technology are compatibility with traditional CMOS based design flow and inherent parallelism of its crossbar architecture [1]. Unlike other CMOS or TTL based memory elements, it can also perform logically complete set of Boolean operation, apart from the data storage. ReRAM is mainly a two terminal passive circuit element which is developed using different organic and inorganic dielectric materials. The basic working principle of ReRAM technology is the formation of highly conducting path or channel through the dielectric material applying high voltage across it which arises due to different mechanism like metal defect migration [2]. The low resistive conducting path can be controlled (SET or RESET) by another voltage.

ReRAM has been widely adopted for use in different application specific designs like deep learning [3], implementation of different encryption, compression algorithm, and also general purpose-programmable platforms [4]. The use of ReRAMs in these applications has been made possible by the ability to provide in-memory computation platform, as an alternative solution to classical Von-Neumman computational model. Apart from these applications, ReRAM can also be used for implementation of primitive mathematical operation like addition [5], linear algebra [6], [7],etc.

Finite field or Galois field (GF) acts as the basic building block in multiple domains of mathematics and computer science — cryptography, coding theory, algebraic geometry and number theory, etc. In coding theory, many codes specially block codes like Bose Chaudhuri Hocquenghem (BCH) or Reed Solomon (RS) codes are developed as subspaces of vector spaces over GF. Discrete logarithmic problem of GF is the basics of different cryptographic algorithm like elliptic curve cryptography (ECC) or different cryptographic protocols like Diffie-Hellman protocol [8].

Authors in [9] proposed high throughput finite field multiplier circuit on Application Specific Integrated Circuit (ASIC) and Field Programmable Gate Array (FPGA), which are the most commonly used CMOS-based devices in industry. Apart from these basic modules, efficient hardware implementation of different error correcting codes like RS [10], BCH [11] or cryptographic algorithm like ECC [12] have been proposed. However, there are no exisiting implementations for emerging in-memory computing platforms. In this work, we propose the first in-memory GF operation library. Specifically, our contributions are as follows :-

- This work presents the first in-memory implementation of GF operation using ReRAM crossbar array.

- The proposed mapping harnesses the bit-level parallelism offered by ReRAM crossbar arrays and supports a wide variety of crossbar dimensions.

- The proposed implementation has a very low footprint in terms of devices required as well as energy, which makes it suitable for use as building blocks for other applications (error detection, error correction, etc.).

The rest of the paper is organized as follows. Section II presents the fundamentals of GF arithmetic, along with a succint introduction to ReVAMP, a state-of-the-art ReRAM based in-memory computing platform. Section III presents detailed implementation of various GF aithmetic operations for the ReVAMP plafform. Experimental results are described in section IV, followed by conclusion in Section V.

II. PRELIMINARIES

In this section, we present fundamentals of GF operations — element generation, addition and multiplication along with preliminaries of logic operation using ReVAMP architecture.

A. Galois Field Arithmetic

A field is a set of elements on which basic mathematical operations like addition and multiplication can be performed without leaving the set. Hence, these basic operations must satisfy *distributive*, *associative* and *commutative law* [13]. The number of elements in the field is known as *order* of the field. A field with finite number of elements (field order) is known as GF. The order of the GF is always prime number or power of prime number. If p be a prime number and m be a positive integer then the GF will contain p^m elements and can be represented as $GF(p^m)$.

978-1-5386-4757-8/18 $31.00 © 2018 IEEE

For $m = 1$, $p = 2$, the elements in GF will be $\{0, 1\}$ and this is known as *binary field*. Here, we will consider GF of 2^m elements from binary field GF(2) where $m > 1$. If α be an element of $GF(2^m)$ and S be the set of the elements of the field, then S can be represented by the following equation 1:-

$$S = [0, \alpha^0, \alpha^1, \alpha^2, \alpha^3, \ldots \alpha^{2^m-1}] \quad (1)$$

Let $f(x)$ be a polynomial over $GF(2^m)$ and it is said to be *irreducible* if $f(x)$ is not divisible by any other polynomial in $GF(2^m)$ with degree less than m but greater than zero [14]. The irreducible polynomial is a *primitive polynomial*, if the smallest positive integer q for which $f(x)$ divides $x^q + 1$, where $q = 2^m - 1$. For each value of m, there can be multiple primitive polynomials, but we will use the primitive polynomial with least number of terms for computation over GF.

(a) (b)

$GF(2^m)$	Primitive Polynomial	Power Repr.	Polynomial Repr.	3-Tuple Repr.
2^2	$x^2 + x + 1$	0	0	(0,0,0)
2^3	$x^3 + x + 1$	1	α^0	(0,0,1)
2^4	$x^4 + x + 1$	α	α^1	(0,1,0)
2^5	$x^5 + x^2 + 1$	α^2	α^2	(1,0,0)
2^6	$x^6 + x + 1$	α^3	$\alpha + 1$	(0,1,1)
		α^4	$\alpha^2 + \alpha$	(1,1,0)
		α^5	$\alpha^2 + \alpha + 1$	(1,1,1)
		α^6	$\alpha^2 + 1$	(1,0,1)

Fig. 1: *(a) Primitive polynomial for various order GF. (b) Representation of elements in $GF(2^3)$.*

List of primitive polynomials for different values of m is shown in Fig. 1a. These primitive polynomials are the basis of computation using elements of GF. For the generation of elements of GF, we will start from two basic elements 0, 1 and another new element α.

In this paper, we will use operations on $GF(2^3)$ as running example. Here, we consider generation of elements for $GF(2^3)$. An element α of $GF(2^3)$ must satisfy the primitive polynomial corresponding to $GF(2^3)$, i.e, $\alpha^3 + \alpha + 1 = 0$ and it is used to calculate k^{th} power of α (α^k) when k is greater than two. From the primitive polynomial, α^k can be expressed by the recursive equation 2.

$$\alpha^k = \alpha^{k-2} + \alpha^{k-3} \quad (2)$$

Power, polynomial and 3−Tuple representation of all the elements of $GF(2^3)$ are shown in Fig. 1b.

Consider two polynomials A and B, where $A, B \in GF(2^3)$, their addition and product can be defined by equation 5 and 6 respectively. Using the primitive polynomial for $GF(2^3)$ equation 6 can be reduced to equation 7.

$$A = a_0 + a_1 * \alpha + a_2 * \alpha^2 \quad (3)$$

$$B = b_0 + b_1 * \alpha + b_2 * \alpha^2 \quad (4)$$

$$A + B = (a_0 + b_0) + (a_1 + b_1)\alpha + (a_2 + b_2)\alpha^2 \quad (5)$$

$$A * B = a_0b_0 + a_0b_1\alpha + a_0b_2\alpha^2 + a_1b_0\alpha +$$
$$a_1b_1\alpha^2 + a_1b_2\alpha^3 + a_2b_0\alpha^2 + a_2b_1\alpha^3 + a_2b_2\alpha^4 \quad (6)$$

$$A * B = (a_0b_0 + a_1b_2 + a_2b_1) + (a_0b_1 + a_1b_0 + a_1b_2$$
$$+ a_2b_1 + a_2b_2)\alpha + (a_0b_2 + a_1b_1 + a_2b_0)\alpha^2 \quad (7)$$

In general, if we consider two polynomial A and B, where $A, B \in GF(2^m)$, their addition and multiplication can be represented by equation 8 and 9 respectively.

$$A + B = \sum_{i=0}^{(m-1)} (a_i \oplus b_i)\alpha^i \quad (8)$$

$$A * B = \sum_{i=0}^{(m-1)} \sum_{j=0}^{(m-1)} a_ib_j\alpha^{i+j} \quad (9)$$

a_i, b_i used in equations 3 to 7 are binary, i.e, $a_i, b_i \in \{0, 1\}$. Since we are working on the binary field, addition operation represents modulo-2 add, which is Boolean XOR while multiplication operation a_ib_i is Boolean AND.

B. In-memory computing using ReRAM

In this subsection, we describe the ReRAM-based in-memory computing platform — ReVAMP, introduced in [4]. The architecture, presented in Fig. 2, utilizes two ReRAM crossbar memories with light weight peripheral circuitry. The instruction memory (IM) is used as regular memory, with the program counter (PC) being used to access the next instruction. The second ReRAM crossbar memory is used as data storage and computation memory (DCM). This is where in-memory computation using ReRAM devices takes place. A ReRAM cross-

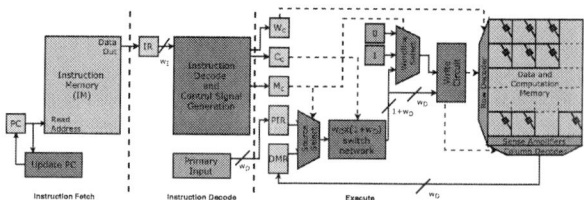

Fig. 2: *ReVAMP architecture. Instruction format:*

Read wl		Apply wl s ws wb (v val_{w_D-1}) … (v val_0)

bar memory consists of multiple 1 Select-1 Resistance (1S1R) ReRAM devices [15], arranged in the form of a crossbar [16]. A V/2 scheme is used for programming the ReRAM array. Unselected lines are kept to ground. In a readout phase, the presence of a 5 μA current is considered as logic '1' while presence of a low current ($< 2 \mu A$) is interpreted as logic '0'. Like conventional RAM arrays, ReRAM memories are accessed as w_D-bit wide words. Each ReRAM device has two input terminals, namely the wordline wl and bitline bl. The internal resistive state Z of the ReRAM acts as a third input and stored bit. The next state of the device Z^n can be expressed as Boolean majority function with three inputs, where the bitline input is inverted.

$$Z^n = M_3(Z, wl, \overline{bl}) \quad (10)$$

This forms the fundamental logic operation that can be realized using ReRAM devices. Using the intrinsic function Z^n, inversion operation can be realized. Since majority and inversion operation form a functionally complete set, any Boolean function can be realized using the Z^n.

The ReVAMP architecture supports two instructions — *Read* and *Apply*, shown in Fig. 2. The architecture has a

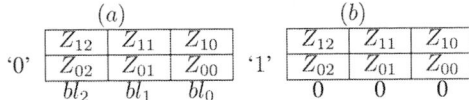

	(a)				(b)		
'0'	Z_{12}	Z_{11}	Z_{10}	'1'	Z_{12}	Z_{11}	Z_{10}
	Z_{02}	Z_{01}	Z_{00}		Z_{02}	Z_{01}	Z_{00}
	bl_2	bl_1	bl_0		0	0	0

Fig. 3: *A* 2×3 *ReRAM crossbar, i.e., a crossbar with two rows and three bitlines.* Z_{ij} *represents the state of device at wordline* i *and bitline* j. *(a) Computation on* 0th *row with '0' and* $\{bl_0, bl_1, bl_2\}$ *as the wordline and bitline inputs respectively. Valid inputs can be either* {'1' (+2.4V) or '0' (−2.4V) }. *(b)* 0th *row is being read out, by setting wordline to '1' (+2.4V) and the bitlines to 0 (0V).*

three-stage pipeline with Instruction Fetch, Instruction Decode and Execution stages. *Read* instruction reads a specified word, wl from the DCM and stores it in the Data Memory Register (DMR). The read out word, available in the DMR, can be used as input by the following instructions. The *Apply* instruction is used for computation in the DCM. The address wl specifies the word in the DCM that will be computed upon. A bit flag s chooses whether the inputs will be from primary input register (PIR) or DMR. Two-bit flag ws is used to select the wordline input – 11 selects '1', 10 selects '0', 00 selects wb bit within the chosen data source for use as wordline input while 01 is an invalid value for ws. Pairs (v, val) are used to specify individual bitline inputs. Bit flag v indicates if the input is NOP or a valid input. Similar to wb, bits val specifies the bit within the chosen data source for use as bitline input. Fig. 3 shows a 2×3 ReRAM crossbar array, which can act as the DCM. The operation in Fig. 3 (a) can be expressed as an *Apply* instruction. Apply 0 00 00 00 1 00 1 01 1 10, with the PIR contents set to bl_0 bl_1 bl_2. The operation in Fig. 3 (b) can be expressed as Read 0. In the rest of the paper, we express the in-memory compute operations in the crossbar representation.

III. METHODOLOGY

In this section, we will discuss mapping of GF arithmetic using ReVAMP array. The DCM dimension required for operations is based on the value of the GF order m.

A. GF Generation

Here, we will illustrate the generation of elements of $GF(2^3)$ as an example. As each element in $GF(2^3)$ is a three tuple, the number of the columns in the DCM should either three or a multiple of three. For this purpose, we need DCM having 8 wordlines and 3 bitlines. Table I presents the intermediate state of the DCM and inputs used for the generation of elements of GF. In Step 1, '1' is applied on 7th wordline and $\overline{a_{00}}$, $\overline{a_{01}}$, $\overline{a_{02}}$ are applied on bitlines. Here, $(\overline{a_{00}}, \overline{a_{01}}, \overline{a_{02}})$ represent 1, 1 and 0 respectively (from the 3-tuple representation shown in Fig. 1b). Similarly, 0, 0 and 1 are loaded into the 7th row which basically represents α^0. In the next two steps, α and α^2 are loaded into the sixth and fifth row applying $(\overline{a_{10}}, \overline{a_{11}}, \overline{a_{12}})$ and $(\overline{a_{20}}, \overline{a_{21}}, \overline{a_{22}})$ in bitlines respectively which represents $(1,0,1)$ and $(0,1,1)$.

Now α^3 will be calculated by modulo-2 addition of elements in 7th and 6th row. Modulo-2 addition between $a_{i,j}$ and $a_{(i+1),j}$ can be broken down into two operations :

$$a_{i,j} \oplus a_{(i+1),j} = a_{i,j} . \overline{a_{(i+1),j}} + (\overline{a_{i,j} + \overline{a_{(i+1),j}}})$$
$$= f_{i,j} + \overline{g_{i+1,j}} \qquad (11)$$

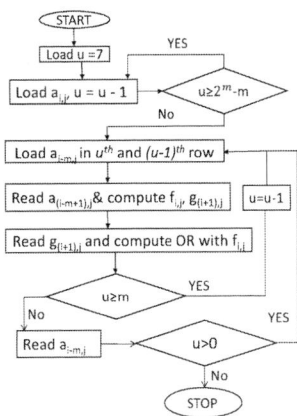

Fig. 4: *Flowchart for the generation of elements of* $GF(2^m)$.

To compute $f_{i,j}$ and $g_{i+1,j}$, we require two copies of $a_{i,j}$. Here i reprsents power of α and j reprsents postion of a bit when α^i is expressed in 3-Tuple format. Hence, in Step 5 and Step 6, we have loaded a_{00}, a_{01} and a_{02} in 4th and 3rd row. $f_{0,j}$ and $g_{1,j}$ are calculated by applying '0' in 4th row, '1' in 3rd row and 0 along all the bitlines in Step 7 and Step 8 respectively. In order to do OR operation between $f_{0,j}$ and $\overline{g_{1,j}}$, $g_{1,j}$ is read from the 3rd row in Step 9 and then in Step 10, apply $g_{1,j}$ along all bitlines and '1' in wordline of 4th row. Finally, 4th row will store the value α^3. Obeying the same procedure α^4, α^5 and α^6 will be calculated and stored in the 3rd, 2nd and 1st row of the crossbar respectively. Flowchart shown in Fig. 4 describes generation of elements for $GF(2^m)$ ReRAM DCM with m bitlinee, where u respresents the number of rows or worldlines in the DCM.

B. GF Addition

Different steps for addition of two element over $GF(2^3)$ using ReRAM DCM is shown in Table II. For addition, we require DCM having 4 wordlines and 3 bitlines. Step 1 involves loading of A into 3rd row of crossbar by applying '1' in 3rd wordline and $(\overline{a_0}, \overline{a_1}, \overline{a_2})$ in bitlines. In Step 2 to Step 6, data from 3rd row is read and loaded into 1st and 0th row. Step 7 loads B into 2nd row of crossbar. In Step 8, (b_0, b_1, b_2) is read out from second row and applied into bitlines with '0' and '1' in first and 0th wordline in Step 9 and Step 10 respectively. This will perform AND and OR operation and generate $a_i . \overline{b_i}$ and $a_i + \overline{b_i}$ $(0 \leq i \leq 2)$, and will be loaded into the 1st and 0th row. In Step 11, content of 0th row will be readout and it is applied in bitlines along with '1' in first wordline in Step 12. This will compute $a_i \oplus b_i$ in first row in Step 13, which is the result of GF addition.

C. GF Multiplication

In this subsection, we present the mapping of the GF field multiplication. Some of the steps of the multiplication operation for elements in $GF(2^3)$ are shown in Table III, on a crossbar with 10 wordlines and 3 bitlines. From the definition of GF, it is clear that number of components in the element generated after the multiplication will be same as the number of components in the elements used in the multiplication process. The result of $A * B$ in equation 7 shows that some terms in the coefficient of α^0 and α^1 are same. Hence, we

978-1-5386-4757-8/18 $31.00 © 2018 IEEE

TABLE I: *Generation operation for elements in $GF(2^3)$ using 8×3 DCM.*

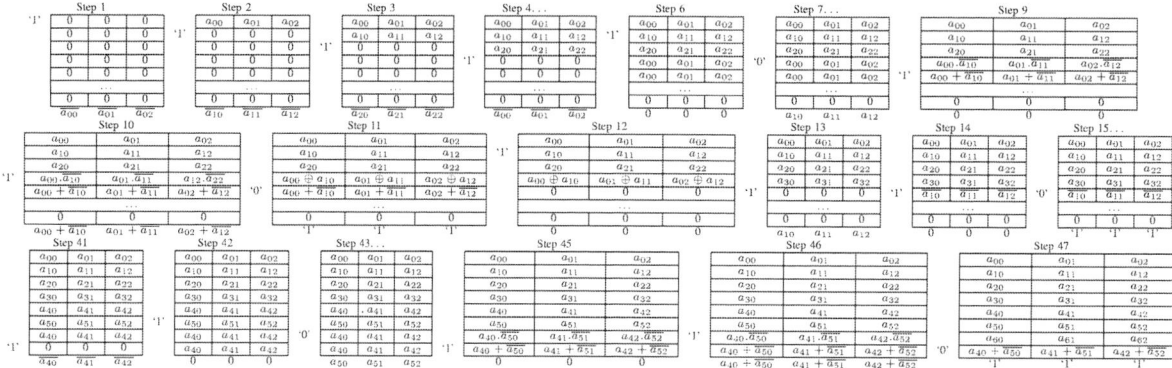

TABLE II: *Mapping of addition of two elements A and B using a 4×3 DCM, $A, B \in GF(2^3)$.*

can reuse the partial result obtained during the generation of coefficient of α^0, during the generation of coefficient of α^1 to reduce the number of steps required for multiplication.

In Step 1 and Step 2, three components of A and B *i.e*, (a_0, a_1, a_2) and (b_0, b_1, b_2) are loaded into the 9th and 8th row of crossbar respectively. Step 3 involves in loading of $(\overline{a_0}, \overline{a_1}, \overline{a_2})$ in 7th row of the crossbar and Step 4 reads the content of 7th row. Step 5 to Step 7 load three copies of (a_0, a_1, a_2) into $6^{th}, 5^{th}$ and 4th row applying '1' in wordline and $(\overline{a_0}, \overline{a_1}, \overline{a_2})$ in the bitlines. Step 8 reset the 7th row while Step 9 reads the content of the 8th row by applying '1' in word line and 0 in all bitlines. In the Step 10, we have applied (b_0, b_1, b_2) in three bitlines and loaded $(\overline{b_0}, \overline{b_1}, \overline{b_2})$ into the 7th row. The 7th row is read out in the next step. In Step 12, '0' is applied in wordline of 6th row and $(\overline{b_0}, \overline{b_0}, \overline{b_0})$ in bitline to perform AND operation. This operation will generate a_0b_0, a_1b_0, a_2b_0 and load them into 6th row. Similarly, Step 13 to Step 16 involves

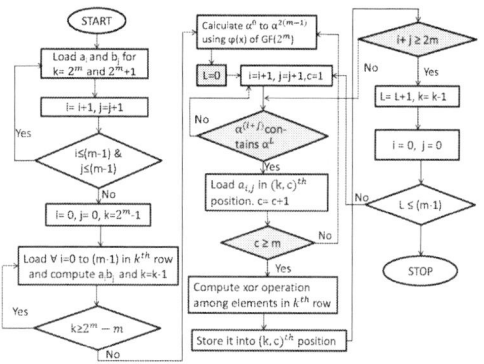

Fig. 6: *Flowchart for the multiplication of A and B over $GF(2^m)$ using ReVAMP.*

generation of (a_0b_1, a_1b_1, a_2b_1) and (a_0b_2, a_1b_2, a_2b_2) and are loaded into the 5th and 4th row. From the equation 7, we have noticed that coefficient of α^0 contains the term a_0b_0, a_1b_2 and a_2b_1. These terms are loaded into the 3rd and 2nd row of the

crossbar in Step 26. In perform the modulo-2 addition using these terms, Step 27 involves in reading the content of 3rd row. Step 28 and Step 29 perform AND and OR operation by applying a_2b_1 in second bitline and '0' and '1' in wordline of 3rd and 2nd row respectively. This process will generate $f_{i,j}$ and $g_{(i+1),j}$ as an intermediate term in Step 30. In the same step, we have read $g_{(i+1),j}$ from 2nd row. '1' is applied along 3rd row and $g_{(i+1),j}$ is applied in the 1st bitline of DCM in Step 31 which generates modulo-2 addition between a_1b_2 and a_2b_1 (indicated by ab^*_{1221} in Table 6) in Step 32. In the next step, '1' is applied in 3rd wordline and 0 in all bitlines of ReRAM crossbar which will read ab^*_{1221} from 3rd row and 1st column of crossbar. ab^*_{1221} is applied in 1st bitline and '0' and '1' are applied along 3rd and 2nd wordline which will perform AND and OR operation respectively in Step 33 and Step 34. Step 35 and 36 involve in the generation of $a_0b_0 \oplus ab^*_{1221}$ which is indicated by C^{α^0} in Table 6. Finally, we will reset 2nd row in Step 37 so that it can be reuse for further computation and 2nd column of 3rd row will hold coefficient of α^0 in A*B.

In the next stage, we will generate the coefficient of α which is the modulo-2 addition of the terms $a_0b_1, a_1b_0, a_1b_2, a_2b_1$ and a_2b_2. As we can see modulo-2 addition between a_1b_2 and a_2b_1 has already performed in Step 27 to Step 32 and stored in 1st column of 3rd row, we can reuse ab^*_{1221} directly here. Initially we will perform modulo-2 addition between a_0b_1, a_1b_0 and a_2b_2 that is stored in the 2nd column of 2nd row. Then, modulo-2 addition operation will be performed between temporary data stored into 2nd column of 2nd row and ab^*_{1221} and store the result again in 2nd column of 2nd row which is indicated by C^{α^1} as shown in Table 6. Similarly we will calculate the coefficient of α^2 (indicated by C^{α^2}) which is generated by modulo-2 addition of a_0b_2, a_1b_1 and a_2b_0 and store into the 2nd column of 1st row.

Multiplication of two elements over other dimensions can be executed using the same strategy. We have provided a flow chart for multiplication of two elements over $GF(2^m)$ using ReVAMP in Fig. 6, where k indicates row in the DCM and c represents a counter.

978-1-5386-4757-8/18 $31.00 © 2018 IEEE

TABLE III: *Mapping of multiplication of two elements A and B over $GF(2^3)$ using 10×3 DCM.*

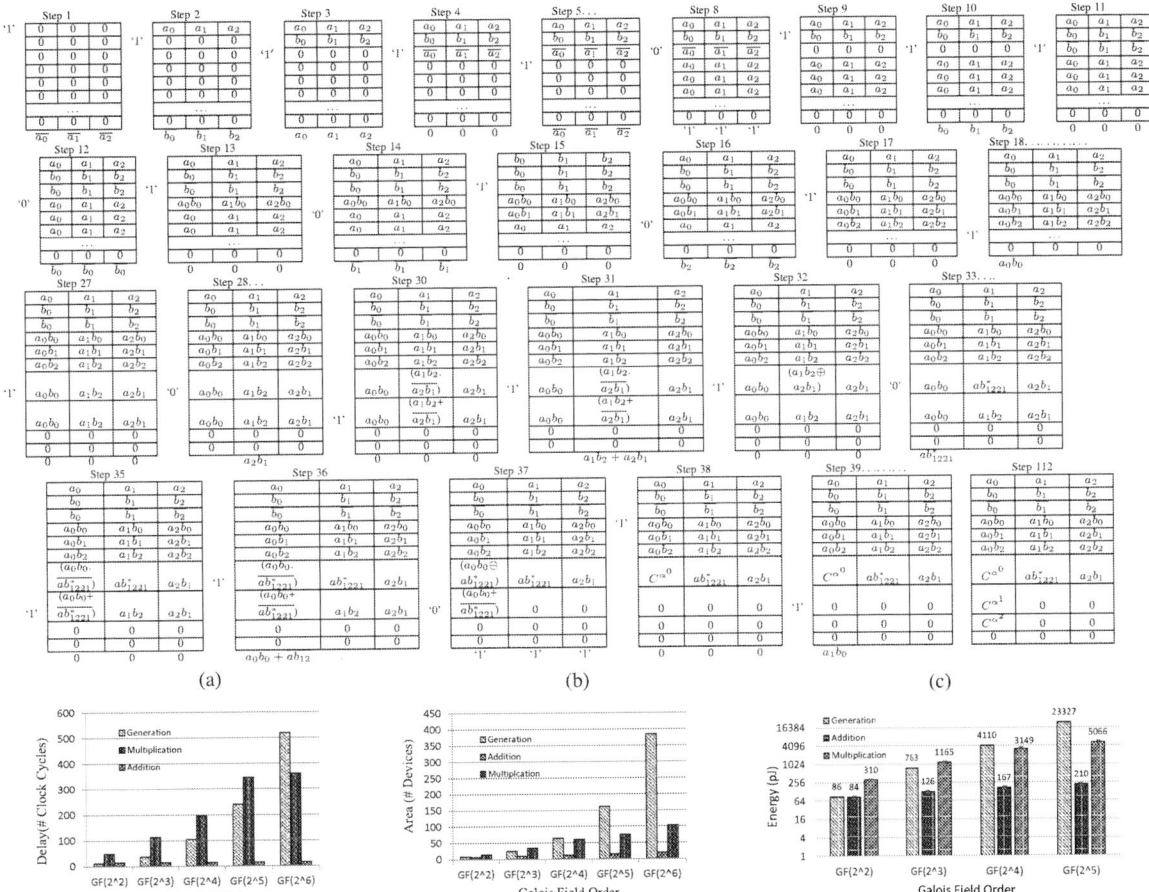

Fig. 5: *Impact of GF order on (a) Computation delay (in cycles) (b) Area (in terms of ReRAM devices) (c) Energy (in pJ).*

IV. EXPERIMENT

In this section, we analyze the performance of in-memory computation of GF for various order m and crossbar dimension. The experiments were performed using Cadence Virtuoso, using device-accurate model of ReRAM devices [15]. Fig. 5a and Fig. 5b shows the delay (#cycles) and area (number of ReRAM devices used in the DCM and size of instruction memory) required for individual operations for various order GF. Fig. 5c presents the mean energy (in log scale) required for each operation. Due to the large runtimes of device-accurate simulations for all possible input combinations, we only report energy number for operations till $GF(2^5)$.

The increase in order m leads to exponential increase in the number of elements in GF, i.e., the delay as well as area requirements for generation operation grow exponentially. Specifically, the number of instructions I_m required for generating all the elements of $GF(2^m)$ can be expressed as :

$$I_m = m + 11(2^m - m - 1) \qquad (12)$$

Addition operation involves XORing the individual bits of the input operands, which is done parallely using the rows $0-3$, as

illustrated in Table II. Therefore, the delay remains constant, even with increase in the value of m. However, the number of bitlines in a row is equal to m for addition of two elements in $GF(2^m)$, hence the number of devices increase with m.

For multiplication of GF elements, we refer to each α^i term in GF element as a component. With increase in m, the number of components increase. For example, number of components in A is 3 for $m = 3$ and 4 for $m = 4$. Also, number of $a_i b_i$ terms in the coefficients of the components increase with increase in the order m, where a_i, b_i are coefficients in the multiplier A and multiplicand B respectively. For example, there are 3, 5 and 4 terms in the coefficient of α^0, α^1 and α^2 in the multiplication result $A * B$, for $A, B \in GF(2^3)$. During the multiplication, terms within each coefficients are read one after another and are placed into a single row in the crossbar as shown in Step 26 in Table III. The individual terms are then added by modulo-2 addition (basically XOR) to generate the final coefficient of each component. As number of terms in each coefficient increases with the increase in order m, the delay increases for computing the resultant coefficient of the components in the result of $A * B$.

978-1-5386-4757-8/18 $31.00 © 2018 IEEE

The number of bit-level parallel operations on ReRAM crossbar arrays is dependent on the number of bitlines present in the crossbar. Fig. 7 demonstrates the impact of number of bitlines on the mapping delay for operations in $GF(2^4)$. With the increase in number of bitlines, the delay of operations reduce, which demonstrate the effectively of the proposed mapping in harnessing the bit-level parallelism offered by the ReRAM crossbar array.

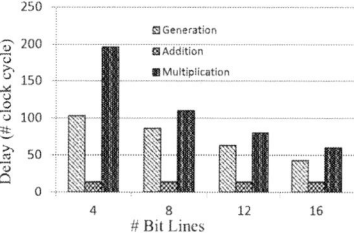

Fig. 7: *Impact of number of bitlines on delay for GF (2^4) operations.*

TABLE IV: *Comparison of Area and Delay for GF implementation on ReVAMP, ASIC and FPGA.*

Op.	GF Order.	ReVAMP Delay ns	DCM $bits$	IM $bits$	ASIC Delay ns	Area GE	FPGA Delay ns	Area $\#LUT$
Generation	$GF(2^2)$	11	8	154	4.975	2283	50	266
	$GF(2^3)$	36	24	918	9.09	3260	90	316
	$GF(2^4)$	103	64	2992	18.19	4619	170	341
	$GF(2^5)$	239	160	10528	36.96	6883	330	390
	$GF(2^6)$	519	384	20498	74.1	11028	650	456
Addition	$GF(2^2)$	14	6	154	1.35	424	30	107
	$GF(2^3)$	14	9	238	1.82	451	40	109
	$GF(2^4)$	14	12	280	2.35	479	50	109
	$GF(2^5)$	14	15	406	2.88	500	60	112
	$GF(2^6)$	14	18	462	3.36	538	70	114
Multiplication	$GF(2^2)$	47	14	564	34.88	2319	320	415
	$GF(2^3)$	112	33	2128	51.52	3480	460	490
	$GF(2^4)$	196	60	4312	81.36	4256	720	560
	$GF(2^5)$	345	75	10695	107.825	5140	950	630
	$GF(2^6)$	360	102	12600	128.8	7170	1120	668

Even though the contemporary technologies such as ASIC and FPGA cannot be directly compared with ReRAM based implementation, we report a coarse comparison for the sake of completeness in Table IV. For ReRAM, we assume mature ReRAM technology with 1 ns read/write times [17]. The DCM column presents the size of the DCM used for in-memory computation, in terms of number of bits. The Instruction Memory (IM) column in the Table IV presents the number of bits required for instruction storage in the IM of the ReVAMP architecture. The FPGA implementation (100 MHz) is on Kintex-7 evaluation board and ASIC implementation ($\approx$ 1GHz) was performed using Synopsys Design Compiler with TSMC 65nm technology library. Since computing using ReRAM is inherently sequential, increase in order m leads to direct increase in delay, along with corresponding increase in area (DCM size). For ASIC and FPGA, the increase in delay is relatively lesser. The key advantage that stands out for ReRAM is *low area* in terms of number of devices required for mapping. For example, generation of all elements in $GF(2^6)$ requires $\approx 11k$ GE for ASIC, 456 LUTs for FPGA but only 384 devices for ReVAMP.

V. CONCLUSION

In this work, efficient mapping of GF operations was proposed on state-of-the-art ReRAM based in-memory computing platform. We have explored multiple configurations for the crossbar dimensions and demonstrated performance trade-offs while varying the GF order. The proposed mapping has a low energy footprint and shows good improvements in terms of area requirements compared to traditional CMOS based design. In future, we plan to extend this work for in-memory implementation of error detection and correction algorithms, as well as cryptographic algorithms.

REFERENCES

[1] D. Bhattacharjee, W. Kim, A. Chattopadhyay, R. Waser, and V. Rana, "Multi-valued and Fuzzy Logic Realization using TaOx Memristive Devices," *Scientific reports*, vol. 8, no. 1, p. 8, 2018.

[2] L. Zhu, J. Zhou, Z. Guo, and Z. Sun, "An overview of materials issues in resistive random access memory," *Journal of Materiomics*, vol. 1, no. 4, pp. 285–295, 2015.

[3] L. Song, X. Qian, H. Li, and Y. Chen, "Pipelayer: A pipelined reram-based accelerator for deep learning," in *2017 IEEE International Symposium on HPCA*, pp. 541–552, Feb 2017.

[4] D. Bhattacharjee, R. Devadoss, and A. Chattopadhyay, "ReVAMP: ReRAM based VLIW architecture for in-memory computing," in *2017 DATE*, pp. 782–787, IEEE, 2017.

[5] A. Siemon, S. Menzel, A. Chattopadhyay, R. Waser, and E. Linn, "In-memory adder functionality in 1S1R arrays," in *2015 IEEE ISCAS*, pp. 1338–1341, IEEE, 2015.

[6] L. Ni, Z. Liu, H. Yu, and R. V. Joshi, "An energy-efficient digital ReRAM-crossbar-based CNN with bitwise parallelism," *IEEE Journal on Exploratory Solid-State Computational Devices and Circuits*, vol. 3, pp. 37–46, 2017.

[7] D. Bhattacharjee and A. Chattopadhyay, "Efficient Binary Basic Linear Algebra Operations on ReRAM Crossbar Arrays," in *2017 30th International Conference on VLSI Design and 2017 16th International Conference on Embedded Systems (VLSID)*, pp. 277–282, Jan 2017.

[8] T. Kerins, E. M. Popovici, and W. P. Marnane, "Fully paramaterisable galois field arithmetic processor over gf(3m) suitable for elliptic curve cryptography," in *2004 24th International Conference on Microelectronics (IEEE Cat. No.04TH8716)*, vol. 2, pp. 739–742 vol.2, May 2004.

[9] J. Xie, P. K. Meher, and Z. H. Mao, "High-throughput finite field multipliers using redundant basis for fpga and asic implementations," *IEEE Transactions on Circuits and Systems I: Regular Papers*, vol. 62, pp. 110–119, Jan 2015.

[10] M. A. Khan, S. Afzal, and R. Manzoor, "Hardware implementation of shortened (48,38) reed solomon forward error correcting code," in *7th International Multi Topic Conference, 2003. INMIC 2003.*, pp. 90–95, Dec 2003.

[11] M. E. Haroussi, I. Chana, and M. Belkasmi, "VHDL design and FPGA implementation of a fully parallel BCH SISO decoder," in *2010 5th International Symposium On I/V Communications and Mobile Network*, pp. 1–4, Sept 2010.

[12] Z. U. A. Khan and M. Benaissa, "High-Speed and Low-Latency ECC Processor Implementation Over GF(2^m) on FPGA," *IEEE Transactions on VLSI Systems*, vol. 25, pp. 165–176, Jan 2017.

[13] J.-M. Couveignes and B. Edixhoven, *Computational aspects of modular forms and Galois representations*. Princeton University Press, 2011.

[14] M. K. Kyuregyan, "Recurrent methods for constructing irreducible polynomials over Fq of odd characteristics," *Finite Fields and Their Applications*, vol. 9, no. 1, pp. 39 – 58, 2003.

[15] A. Siemon, S. Menzel, A. Marchewka, Y. Nishi, R. Waser, and E. Linn, "Simulation of TaOx-based complementary resistive switches by a physics-based memristive model," in *2014 IEEE ISCAS*, pp. 1420–1423, IEEE, 2014.

[16] E. Linn, R. Rosezin, S. Tappertzhofen, U. Böttger, and R. Waser, "Beyond von Neumann-logic operations in passive crossbar arrays alongside memory operations," *Nanotechnology*, vol. 23, no. 30, p. 305205, 2012.

[17] "Emerging Research Devices (ERD) report," in *International Technology Roadmap for Semiconductors (ITRS)*, 2013.

Minimizing Performance and Energy Overheads Due to Fanout In Memristor based Logic Implementations

Md Adnan Zaman and Srinivas Katkoori

Department of Computer Science & Engineering,
University of South Florida, Tampa, FL
Email: {mdadnanz, katkoori}@mail.usf.edu

Abstract—To enable memristor based in-memory computing, crossbar architecture is considered as one of the most preferred structures that provide high density. A given logic function is first synthesized as a netlist of NOR and NOT gates and then it is mapped to the crossbar architecture. In this approach, fanout incurs significant performance and energy overheads. As fanout helps in compact designs, it is essential that we investigate ways to minimize the overheads. In this work, a novel approach has been proposed to address this problem – instead of copying the logic value as inputs to the driven memristors, we propose that the controller reads the logic value and then applies it in parallel to the driven memristors. In comparison to recently published works, experimental evaluation on ISCAS85 benchmarks resulted in average performance improvements of 51.08%, 38.66%, and 63.18% considering three different mapping scenarios (average, best, and worst). In regards to energy dissipation, we have also obtained average improvements of 91.30%, 88.53%, and 74.04% considering the aforementioned scenarios.

I. INTRODUCTION

Von Neumann architecture suffers from *memory wall* problem due to bandwidth mismatch between slower memory and faster CPU [1]. To overcome *memory wall* problem, non von-Neumann architecture is being actively considered where storage and computing can be performed in the same location. This computing inside memory is known as *in-memory computing*. Emerging non-volatile resistive memory technology such as memristor can enable such non von-Neumann computing paradigm. A Memory Processing Unit (MPU) has been proposed [2], where memristive memory is used as storage in conjunction with logical operations.

Due to high speed, low power consumption, scalability, data retention, endurance, and compatibility with conventional CMOS, many memristor based logic families and circuits have been proposed [3]. Based on logic state variable, memristor based logic families can be classified into stateful (logic value represented with memristor resistance) and non-stateful logic families [4]. In this work, we employ a stateful logic family, known as Memristor-Aided loGIC (MAGIC) [5]. In this logic style, for a given logic gate, input values and output value(s) are stored as memristor states. Memristors can be fabricated on a crossbar array, which offers high storage density and low power consumption [6]. With MAGIC, only NOR and NOT gates can be directly mapped to crossbar array.

In recent years, researchers have proposed a few in-memory logic synthesis approaches based on MAGIC logic style, where a given circuit netlist consisting of only NOR and NOT gates is mapped to a memristor crossbar. In [5], a detailed procedure to map NOR/NOT logic gates to crossbar has been discussed and also transpose crossbar concept has been introduced to allow gates to be mapped along the rows as well as columns in a crossbar architecture. In [7], a synthesis tool has been proposed that maps arbitrary logical functions within the memristive memory in an optimal manner. In [8], a scalable design flow for in-memory computing has been proposed that allows a given circuit netlist to be implemented in transpose crossbar. In both of these approaches, a given gate netlist is first converted into a netlist of NOR/NOT gates using an existing logic synthesis tool [9] and mapped to the crossbar architecture. While mapping, we come across fanout where a single output (driving) memristor of a logic gate has to be used as input (driven) memristors of multiple gates connected to it. For a fanout of two or more leaf memristors, current methods (previous approaches) perform the *copy* operation for a number of times equal to the number of driven memristors that are not on the same row or column as of the driving memristor. Such copy style requires two NOT operations which in turn requires two extra cycles. With the increment of fanouts in a given netlist, the number of extra cycles increases hence energy as well. To the best of our knowledge, no other previous works attempted to reduce the additional cycle count of a copy operation inherent to a fanout.

In this paper, we propose a novel approach that will reduce the performance and energy overheads originated from fanout in a given circuit netlist. Instead of copying the value, the proposed controller can read the value and apply in parallel to the driven memristors. It provides an added advantage to decide whether to write the read value or not to the driven memristors. As proposed in MAGIC logic style, the output memristors are initialized to logic-1 prior to logic execution and it allows the controller to skip write cycle if the read value is one. We have compared our work with a recently published work [8] for three different mapping scenarios. We obtain average improvements of 51.08%, 38.66%, and 63.18% and 91.30%, 88.53%, and 74.04% in performance and energy dissipation respectively.

The rest of the paper is organized as follows: Section II presents background and related work. Section III describes the proposed approach to reduce the number of cycles and energy dissipations related to fanout. Section IV reports experimental results. Section V draws conclusions.

978-1-5386-4757-8/18 $31.00 © 2018 IEEE

II. BACKGROUND AND RELATED WORK

In this section, we discuss the basic working principle of a memristor, relevant logic design styles, crossbar architecture, and in-memory computing. We also review some of the recent works on memristor based in-memory logic synthesis for a given gate-level netlist with particular concentration on fanout.

A. Memristor

Memristor is a two terminal device that can remember its previous state and change its resistance based on a given potential across the device. Chua [10] first proposed memristor that links flux (ϕ) and charge (q) to its memristance (M) according to,

$$d\phi = M dq \qquad (1)$$

Structurally, memristor can be thought of as a thin semiconductor film with thickness (D) sandwiched between two contacts. We can change the overall resistance by changing its doped region width, w [11].

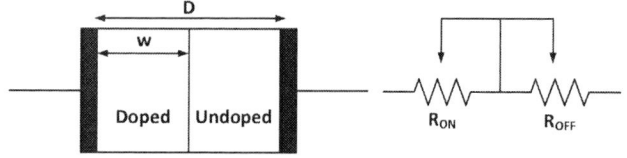

Fig. 1: Physical and circuit model of memristor [2].

B. Memristor Aided LoGIC (MAGIC)

Our proposed methodology is based on MAGIC logic style [5]. As shown in Fig. 2, memristors' resistances of the IN_1 and IN_2 are considered as input values and we can determine the output logic value by measuring OUT memristor resistance. An execution voltage V_G is applied to both input memristors and the output value is stored in the output memristor.

C. In-memory Computation Using Memristor Crossbar

A Memory Processing Unit (MPU) based on *in-memory computing* architecture is as shown in Fig. 3 [2]. It consists of a controller, crossbar memory, and analog multiplexers. The analog multiplexers' outputs are connected to the bitlines and wordlines of the memristive memory and the voltage select lines of the multiplexers are connected to the controller. To carry out a regular read or write operation, controller sends suitable signals to the addressed memristor cells through multiplexers. We have also shown two signals, V_{SET} and V_{RESET} that can enable writing logic-1 and logic-0 to the memristor cells respectively.

The suitability of the crossbar architecture can be better explained by NOR operation as shown in Fig. 2. We, first,

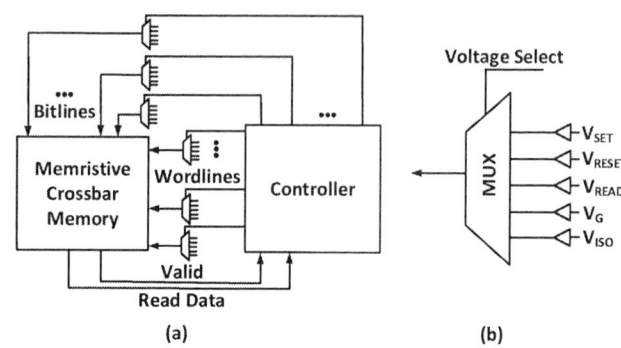

Fig. 3: (a) Memory processing unit (MPU), (b) Analog mux.

initialize the output memristor (OUT) to logic-1 and connect it to the ground (0V) and then we apply execution voltage V_G to the input memristors (IN_1 and IN_2). This voltage may corrupt data on other memristors on the same row/column, to avoid this, an isolation voltage V_{iso} needs to be applied to the columns and rows that we want to unselect. Parallel execution of NOR/NOT gates on the crossbar requires the alignment of the inputs and outputs of the respective gates. Since we are considering transpose memory here, gates can be aligned either by rows or columns.

D. Fanout

Fanout occurs when an output memristor at any circuit depth (excluding primary output) has to drive multiple memristors. It can degrade performance as circuit depth increases and incurs additional energy overheads by introduction of extra cycles. For a given gate-level netlist, a single memristor cell can only be used either as an input or output of a gate. Therefore, the value that is stored in the output memristor if needed can work as input to multiple following memristors. A naïve approach can perform this fanout operation by multiple copies of the logic value as equal to number of the driven memristors (on different rows or columns than driving memristor). This copy operation introduces additional cycles, as proposed in MAGIC logic style where each copy operation requires cascade of two NOT operations. Therefore, it will require two extra memristors as well as two extra cycles for a single copy operation. As the number of fanout increases for a specific gate output, the number of copy operations, hence additional cycles also increases linearly. Moreover, the energy dissipation also increases due to the extra NOT operations. Works [7], [8] based in-memory logic synthesis did not address these performance related issues.

III. PROPOSED APPROACH

A. Overall Approach

In this work, we propose a novel approach that will reduce the performance and energy overheads due to fanout in memristor based in-memory computing. The controller will apply relevant signals in proper sequence to the rows and columns in a crossbar architecture. It will carry out the proposed method according to the following steps:

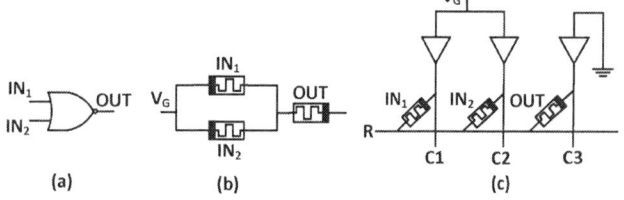

Fig. 2: NOR gate implementation using MAGIC on crossbar.

978-1-5386-4757-8/18 $31.00 © 2018 IEEE

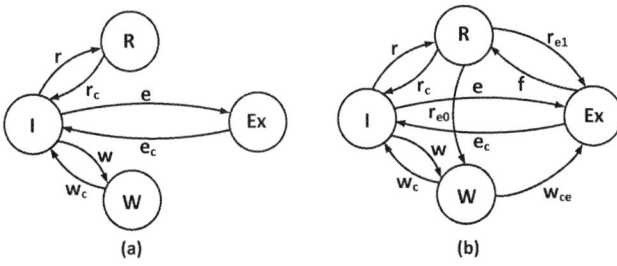

Fig. 4: Controller state diagram. (a) Previous approach and (b) Proposed approach. The states are given as: I = Initial state, R = Read state, W = Write state, and Ex = Execution state; The different inputs denote the followings: r = read, r_c = read complete, w = write, w_c = write complete, e = execution, e_c = execution complete, f = fanout event, r_{e1} = read logic-1, r_{e0} = read logic-0, and w_{ce} = write complete and go to execution state.

- *Cycle 1*: Controller reads the logic value of the output memristor which has fanout of two or more.
- *Cycle 2*: Controller writes the logic value as inputs to the driven memristors only if the sensed value is zero, otherwise, it can skip the cycle 2.

Previous works [7], [8] utilized a controller that produces a similar state diagram as shown in Fig. 4(a). For regular memory read and write operations, it goes to the read and write states respectively and for executing a logic function, it goes to the execution state. Execution state consists of micro operations where the controller executes the logic function in multiple cycles.

With our proposed approach, the number of state transitions will increase as shown in the Fig. 4(b) but the number of states will remain the same. While executing the micro operations in the execution state, whenever there is a fanout event, the controller will go to the read state and depending on the read value, it will decide whether to go to the write state or not. We can achieve this by modifying the controller circuit. The sensed bit can act as selector for a 2:1 multiplexer and depending on the select bit, controller will determine whether to write to the memory or not. If the read value is 1, the multiplexer will pass the input signal r_{e1} to the controller to go the execution state skipping the write state, otherwise, it will go to the write state and then to the execution state. This allows us to avoid writing logic-1 if the sensed logic value is 1.

B. Mapping Scenario Analysis

We have made the following assumptions for three different mapping scenarios of a logic function on a memristor crossbar.

We only assume that a single copy of primary input is stored in a single memristor cell and following a fanout scenario from this cell, one can do the copy operation on-demand. Previous works assume that multiple copies of primary inputs are already available depending on the number of gates they are driving. Hence, their assumptions underestimate the cycles required as they are able to eliminate the copy cycles produced from the fanouts of primary inputs.

Like the previous works [8], [12], we have considered the output memristors are initialized to logic value one at the beginning of the execution process which is a requirement for MAGIC logic style. The work [8] has also initialized the output memristors to logic-1 required for copy operations (which eventually become the first input memristors of the driven gates). With our proposed approach, we are only considering fanout event related metrics not the whole synthesis procedure. Therefore, in our result analyses, we are not accounting the initialization cycles and energy consumptions generated from initializing the output memristors. For fair comparison, we have maintained the same assumption while comparing with work [8]. In our work, the added consideration is, we can avoid writing logic value one. After reading the logic value of the driving memristor, if the controller finds the value as one, it can skip the write operation as the memristors are already initialized to one.

To demonstrate the efficacy of the approach, we have considered three different mapping scenarios (average, best, and worst) depending on the location of the second input memristors of the driven gates. To explain three scenarios, we consider the gate-level netlist shown in Fig. 5(a). We observe that there is a fanout of 4 from the node h. We also consider the logic value of h to be mapped on the location $(1, 1)$ in the memristor crossbar architecture.

1) Scenario 1: Here, we consider that all the second input memristors are aligned along the same row or same column. The mapped variable h in Fig. 5(b) needs to be copied 3 times to use as first input memristors of the driven gates and the variables a, i, j, and g are mapped as second input memristors on the second column of the crossbar. With the previous approach, it requires 4 cycles to copy logic value of h and now with all the input memristors of the driven gates in aligned position, it is possible to execute the gates in one cycle. The following equation can be used to estimate the total cycles required for copy operations with previous approach:

$$\sum_{n=2}^{N}[n * freq(n)] \tag{2}$$

where, n denotes different fanouts (i.e., number of gates it is driving), N is maximum fanout degree, and $freq(n)$ is the number of times, a fanout of n is found in a netlist.

With the proposed approach, we need 1 cycle to read the logic value and one cycle to write multiple copies of h along the same row or column (given that the read value is logic-0). We consider that on average 50% of the time the read value is 0 which allows us to consider only half of the write cycles to the total cycle count. Therefore, we can estimate the total cycles required with the equation given below:

$$1.5 \sum_{n=2}^{N}[freq(n)] \tag{3}$$

It should be noted that with our proposed approach, we require at most two cycles (read and write) for each fanout event and this is true for all scenarios considered here. Therefore, for all three scenarios, we can reuse Equation 3 to estimate the total cycle counts.

978-1-5386-4757-8/18 $31.00 © 2018 IEEE

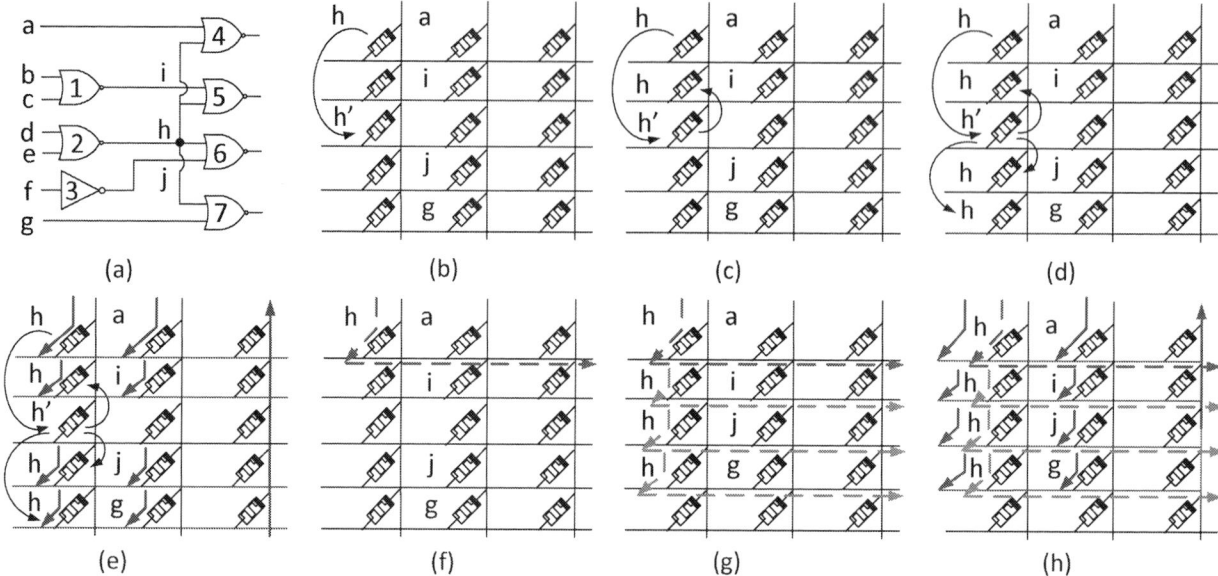

Fig. 5: Scenario 1 with Previous and Proposed Approach. (a) NOR/NOT based synthesized netlist. Regular Approach: (b) 1st cycle: 1st NOT execution to initiate the copy operation, (c) 2nd cycle: 2nd NOT execution to complete the first copy operation, (d) 3rd & 4th cycles: two NOT executions to complete all the copy operations, (e) execution of all the NOR gates in parallel. Proposed Approach: (f) 1st cycle: read operation to sense the logic variable h, (g) 2nd cycle: write operation to write h required number of times in one cycle, (h) 3rd cycle: execution of all NOR gates in parallel. Here, black arrows denote the NOT execution required for copy operation, blue arrow denotes the read operation, green arrows denote the write operations and red arrows denote the execution of NOR gates.

We are providing a general formula to estimate the energy dissipation with both the previous and proposed approach due to the fanout only. According to the work [8], 52.49 fJ is the energy required for one NOT execution. Therefore, with previous approach, we can estimate the total energy dissipation by multiplying 52.49 fJ with the total cycle count (same as total NOT executions required for copy operations). The following equation gives us the total energy dissipation with this approach:

$$52.49 \sum_{n=2}^{N} [n * freq(n)] \qquad (4)$$

Whereas, for the proposed approach, the total energy required for a specific benchmark circuit can be estimated as:

$$\sum_{n=2}^{N} [\{(RE + (n-1) * WE)\} * freq(n)] \qquad (5)$$

Where, RE denotes the total energy needed to read the variable and to decide (using multiplexer) whether to go to the write state or not. For fanout of n, we need to write $(n-1)$ memristors in a single cycle. Here, WE denotes the write energy required for writing a single memristor.

2) Scenario 2: We consider here some of the second input memristors are aligned on the same row as of the first input memristor (h) and some are scattered throughout the crossbar. Here, we divide different fanouts (n) according to the number of second input memristors residing on the same row. For example, for fanout of 3, 4, 5, and 6, we have considered 2 second input memristors are on the same row as first input memristor. From Fig. 6(a), it can be seen that the logic variables a and i that

are mapped to the second input memristors of the gates 4 and 5 are aligned on the same row and variables j and g are in random locations. With the previous approach, h needs to be copied two times (3 NOT executions) to use it as an inputs of gates 6 and 7. We can evaluate the following equation for the previous approach:

$$2 + \sum_{i=1}^{M} \sum_{n=4i-1}^{4i+2} [\{n - (2i-1)\} freq(n)] \qquad (6)$$

where, 2 cycles are required for fanout of 2. Here, i denotes the number of sub-divisions made on the different fanout and M denotes maximum degree of sub-division. For example, fanout of 4 falls under first sub-division ($i=1$) which requires 3 cycles (given by the term $\{n - (2i-1)\}$) to make two copies of the driving memristor. In this way, we find cycle count for each division and add together to get the total cycle counts. As discussed earlier, with the proposed approach, we will get the cycle counts from Equation 3.

With the previous approach, as in Scenario 1, by simply multiplying the total cycle count with energy requirement of one NOT execution, we can estimate the total energy dissipation:

$$52.49 * [2 + \sum_{i=1}^{M} \sum_{n=4i-1}^{4i+2} [\{n - (2i-1)\} freq(n)]] \qquad (7)$$

and for the proposed approach:

$$(RE + WE) + \sum_{i=1}^{M} \sum_{n=4i-1}^{4i+2} [\{RE + (n-2i) * WE\} freq(n)] \qquad (8)$$

978-1-5386-4757-8/18 $31.00 © 2018 IEEE

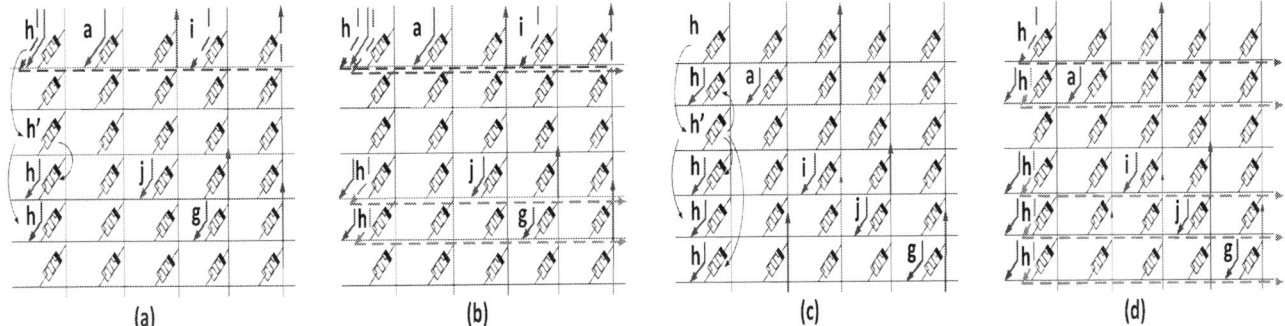

Fig. 6: Scenario 2: (a) previous approach, (b) proposed approach. Scenario 3: (c) regular approach, (d) proposed approach. Here, same color notation as Fig. 5 is maintained, the additional colors, orange, purple, and dark green represent different execution cycles and each composed of all the participating gates in that same cycle.

Where, first term $(RE + WE)$ is the energy required for fanout of 2. The term $(n-2i)$ gives us the number of memristors that needs to be written in a single write cycle.

3) Scenario 3: For the third scenario, we consider that the second input memristors are scattered throughout the crossbar i.e., aligned neither horizontally nor vertically. From Fig. 6(c), we can see that variables a, i, j, and g are mapped on the memristors that are scattered on the crossbar. With the previous approach, it requires $(n + 1)$ copy cycles for a specific fanout of n and the total cycles required can be estimated as:

$$\sum_{n=2}^{N} [(n + 1)freq(n)] \qquad (9)$$

For the proposed approach, we use Equation 3 which gives us the same cycle counts as of scenario 1 and 2.

By maintaining the same procedure as of Scenarios 1 and 2, the total energy dissipation for the previous approach can be obtained. For the proposed approach, we need to write n memristors in a single write cycle for a fanout of n and the term $(n * WE)$ gives us the write energy required for this operation. For the previous and proposed approach, the energy dissipation can be estimated by the following two equations respectively:

$$52.49 * \sum_{n=2}^{N} [(n + 1)freq(n)] \qquad (10)$$

$$\sum_{n=2}^{N} [\{RE + (n * WE)\}freq(n)] \qquad (11)$$

IV. EXPERIMENTAL RESULTS

To validate the proposed approach, we performed a set of experiments on ISCAS'85 benchmark circuits and compared our results with a recently published work [8]. While doing so, we maintained the assumptions as stated in Section III. In a given NOR/NOT based synthesized netlist, total number of gate executions consist of regular execution of gates pertaining only to the netlist and the extra NOT executions required for copy operations produced by fanout. We estimated the total number of extra NOT executions over the total number of gate executions for a specific benchmark circuit and then averaged over all the benchmark circuits. An average of 47.41%, 41.82%,

and 54.40% additional cycle count due to NOT operations has been measured for scenario 1, scenario 2, and scenario 3 respectively. With our proposed approach, we are able to reduce performance and energy overheads produced by these aforementioned additional cycles.

Previous works such as [7], [8] have only considered energy dissipation due to gate executions, but as energy consumptions from read cycle, write cycle, and 2:1 multiplexer are inherent in our proposed approach, we have included these in our analysis. Performance and energy overheads for the multiplexer are obtained for a 45nm CMOS process technology from PTM using HSPICE and our circuit level evaluation shows that the delay introduced by a multiplexer is in the picosecond range. The work [8] has used the VTEAM model [13] to find the maximum latency for MAGIC NOR operations (when either of the inputs is at logic-1) and also to find the read time. We consider these times as the cycle time and read time of the system respectively. According to [14], the latency introduced by MAGIC operations are much higher than that of read and write operations. Hence, the effective delay produced by the sense circuit and the multiplexer still stays below 1 cycle time. Therefore, in our analysis, we will not consider any performance overhead due to this added multiplexer.

A. Experimental Setup

The experimental methodology we follow to validate our approach is as follows:

1) A given gate-level netlist is first synthesized with ABC tool [9] and target library of NOR and NOT gates.
2) For three different scenarios, number of fanout events has been found and cycle counts and energy dissipations estimation have been performed with the proposed method and compared with [8].

B. Results and Analysis

Table I reports the improvements we obtained in comparing our proposed approach with a recently published work with respect to the number of cycles due to fanout. We observe average improvements of 51.08%, 38.66%, and 63.18% in cycle reduction.

Table II reports the improvements in energy consumption. For estimating the energy requirement, we have taken the same

TABLE I: Comparison of Number of Cycles With the Regular and Proposed Approach.

Benchmark	Fanout Events	Number of Cycles Due to Fanout Only								
		Scenario 1 (Average)			Scenario 2 (Best)			Scenario 3 (Worst)		
		Approach [8]	Proposed	Improvement (%)	Approach [8]	Proposed	Improvement (%)	Approach [8]	Proposed	Improvement (%)
c432	80	260	120	53.85	207	120	42.03	340	120	64.71
c499	215	602	323	46.35	499	323	35.27	817	323	60.47
c880	159	448	239	46.65	357	239	33.05	607	239	60.63
c1355	215	602	323	46.35	499	323	35.27	817	323	60.47
c1908	215	610	323	47.05	492	323	34.35	825	323	60.85
c2670	328	1024	492	51.95	832	492	40.87	1352	492	63.61
c3540	352	1343	528	60.69	1015	528	47.98	1695	528	68.85
c5315	596	2193	894	59.23	1703	894	47.50	2789	894	67.95
c6288	1223	3544	1835	48.22	2672	1835	31.32	4767	1835	61.51
c7552	966	2924	1449	50.44	2373	1449	38.94	3890	1449	62.75
Average Improvements				51.08			38.66			63.18

TABLE II: Comparison of Energy Dissipation With the Regular and Proposed Approach.

Benchmark	Fanout Events	Energy Dissipation in pJ Due to Fanout Only								
		Scenario 1 (Average)			Scenario 2 (Best)			Scenario 3 (Worst)		
		Approach [8]	Proposed	Improvement (%)	Approach [8]	Proposed	Improvement (%)	Approach [8]	Proposed	Improvement (%)
c432	80	28.82	2.32	91.95	10.87	1.28	88.22	17.85	5.57	68.80
c499	215	52.60	4.63	91.20	26.19	2.93	88.81	42.88	10.04	76.59
c880	159	43.46	3.62	91.67	18.74	2.06	89.01	31.86	8.36	73.76
c1355	215	52.60	4.63	91.20	26.19	2.93	88.81	42.88	10.04	76.59
c1908	215	54.69	4.76	91.30	25.83	2.87	88.89	43.30	10.46	75.84
c2670	328	106.19	8.77	91.74	43.67	5.09	88.34	70.97	20.44	71.20
c3540	352	133.85	11.82	91.17	53.28	6.54	87.73	88.97	25.54	71.29
c5315	596	219.20	19.19	91.25	89.39	10.93	87.77	146.40	41.88	71.39
c6288	1223	265.02	25.86	90.24	140.25	15.19	89.17	250.22	49.86	80.07
c7552	966	268.38	23.51	91.24	124.56	14.30	88.52	204.19	51.28	74.89
Average Improvements				91.30			88.53			74.04

parameter values as discussed in [8]. The energy required for NOT execution is 52.49 fJ. The energy required to read logic-0 is 0.03 fJ, while that for logic-1 is 3.34 fJ. For our estimation purpose, we have averaged these two values and considered the read energy as 1.685 fJ. The energy required to write logic-1 to a memristor which is termed as SET operation is 66.09 fJ and through RESET operation, we can write logic-0 to a memristor and the energy required is 17.60 fJ. As discussed in Section III, we do not need to write logic-1 with our proposed approach, thus, no energy is consumed. Therefore, the average energy considered for SET and RESET operation is 8.8 fJ. For a specific benchmark circuit, we find all different fanout occurrences and add the energies required to estimate the total energy dissipation. We observe average improvements of 91.30%, 88.53%, and 74.04% by experimental evaluation on ISCAS'85 benchmark circuits. This improvement can be primarily attributed to the fact that we were able to eliminate the need for writing logic-1.

V. CONCLUSIONS

In this work, we outline an effective approach that would significantly reduce the performance and energy overheads due to fanout while mapping a logic function in memristor based crossbar architecture. Comparison has been made with a recently published work and shows significant average improvements in average, best, and worst mapping scenarios in performance and energy dissipation. Future research direction would be to implement the entire logic synthesis process utilizing the fanout optimization discussed here.

REFERENCES

[1] W. A. Wulf and S. A. McKee. Hitting the memory wall: implications of the obvious. *ACM SIGARCH computer architecture news*, 23(1):20–24, 1995.

[2] R. Ben Hur and S. Kvatinsky. Memristive memory processing unit (mpu) controller for in-memory processing. In *2016 IEEE International Conference on the Science of Electrical Engineering (ICSEE)*, pages 1–5, Nov 2016.

[3] S. Kvatinsky, E. G. Friedman, A. Kolodny, and U. C. Weiser. The desired memristor for circuit designers. *IEEE Circuits and Systems Magazine*, 13(2):17–22, Secondquarter 2013.

[4] E. Lehtonen, J. H. Poikonen, and M. Laiho. Memristive stateful logic. In *Memristor Networks*, pages 603–623. Springer, 2014.

[5] N. Talati, S. Gupta, P. Mane, and S. Kvatinsky. Logic design within memristive memories using memristor-aided logic (magic). *IEEE Transactions on Nanotechnology*, 15(4):635–650, July 2016.

[6] R. Nair. Evolution of memory architecture. *Proceedings of the IEEE*, 103(8):1331–1345, Aug 2015.

[7] R. Ben Hur, N. Wald, N. Talati, and S. Kvatinsky. Simple magic: Synthesis and in-memory mapping of logic execution for memristor-aided logic. In *2017 IEEE/ACM International Conference on Computer-Aided Design (ICCAD)*, pages 225–232, Nov 2017.

[8] R. Gharpinde, P. L. Thangkhiew, K. Datta, and I. Sengupta. A scalable in-memory logic synthesis approach using memristor crossbar. *IEEE Transactions on Very Large Scale Integration (VLSI) Systems*, 26(2):355–366, Feb 2018.

[9] Berkeley Logic Synthesis and Verification Group, ABC: A System for Sequential Synthesis and Verification, Release 90215. http://www.eecs.berkeley.edu/~alanmi/abc/.

[10] L. Chua. Memristor-The missing circuit element. *IEEE Transactions on Circuit Theory*, 18(5):507–519, September 1971.

[11] D. B. Strukov, G. S. Snider, D. R. Stewart, and R. S. Williams. The missing memristor found. *nature*, 453(7191):80–83, 2008.

[12] A. Haj-Ali, R. Ben-Hur, N. Wald, and S. Kvatinsky. Efficient algorithms for in-memory fixed point multiplication using magic. In *2018 IEEE International Symposium on Circuits and Systems (ISCAS)*, pages 1–5, May 2018.

[13] S. Kvatinsky, M. Ramadan, E. G. Friedman, and A. Kolodny. Vteam: A general model for voltage-controlled memristors. *IEEE Transactions on Circuits and Systems II: Express Briefs*, 62(8):786–790, Aug 2015.

[14] N. Talati, A. H. Ali, R. Ben Hur, N. Wald, R. Ronen, P. E. Gaillardon, and S. Kvatinsky. Practical challenges in delivering the promises of real processing-in-memory machines. In *2018 Design, Automation Test in Europe Conference Exhibition (DATE)*, pages 1628–1633, March 2018.

978-1-5386-4757-8/18 $31.00 © 2018 IEEE

Versatile Ring-Based Architecture and Synthesis Flow for General-Purpose Digital Microfluidic Biochips

Juinn-Dar Huang, Chia-Hung Liu, and Wei-Hao Yang

Department of Electronics Engineering

National Chiao Tung University, Hsinchu, Taiwan

jdhuang@mail.nctu.edu.tw, terryliu.phd@gmail.com, 19881214gary@gmail.com

Abstract – Digital microfluidic biochip (DMFB) is a tiny device that can carry out a rich set of bioassays without the need of bulky equipment. However, designing a good general-purpose DMFB architecture is still considered a big challenge today. NP-hard synthesis problems make on-line synthesis virtually impossible on exiting array-based architectures. In this paper, we first elaborate on the major concerns in a DMFB design flow, from the aspects of both synthesis and physical design. We then propose a versatile ring-based architecture VERBA and its corresponding fast one-pass synthesis flow. Experimental results show that VERBA incorporated with the proposed synthesis flow is a better solution than existing architectures and synthesis algorithms especially for real-time cyber-physical systems.

I. Introduction

Digital microfluidic biochip (or DMFB) is a new kind of lab-on-a-chip (LoC) device. Through integrating different functions in one small chip, an LoC can serve as a platform for developing various biomedical and chemical applications. Unlike those bulky analysis systems used in laboratories and hospitals, an LoC device is very tiny, only few square centimeters in area. Nevertheless, it is capable of handling complicated biochemical reactions through appropriately manipulating reactants. Hence, LoC is a promising solution in many real-life applications because of its advantages such as portability, less reagent consumption, faster analysis time, higher throughput, and better reliability due to less human interference [1], [2]. A DMFB dispenses reactants as discrete droplets and manipulates them through properly actuating control electrodes. By utilizing different electrode actuation sequences, a DMFB can perform various bioassays. Due to the potentials of DMFBs, numerous studies aiming at related topics have been conducted in the past decade.

A DMFB can be either a *general-purpose* chip or an *application-specific* one [3]. Each type has its own target applications and requires different dedicated design flows. A general-purpose biochip is a platform that can carry out a large set of various biochemical applications. When a target application is set, synthesis algorithms are employed to determine the fluid behavior on the chip surface. The runtime efficiency of algorithms determines whether the synthesis process must be executed off-line or can be executed on-the-fly. On the other hand, for an application-specific chip, the runtime complexity of synthesis algorithms is usually not the major concern since the target assay is pre-determined. Design automation tools are utilized to generate a fully customized chip layout and an optimized fluid behavior (i.e.,

less reactant consumption, lower chip manufacturing cost, and shorter assay completion time). The quality of result is the major consideration for synthesis algorithms targeting application-specific biochips.

Recently, an increasing number of feasible assays on DMFBs raise the demand for general-purpose DMFB designs. Nevertheless, most existing DMFB design flows focus on application-specific architectures. Though some existing studies pay attention to synthesis algorithms for array-based general-purpose DMFBs, the huge gaps between different phases in an array-based DMFB synthesis flow make global optimization extremely difficult [4], [5]. Regarding the growth of chip size and the number of applications, this problem is getting even intractable. Hence, existing synthesis algorithms for general-purpose DMFBs require extensive runtime and thus can only be executed off-line [6]. It limits the applications of general-purpose DMFBs, especially in cyber-physical systems, which demand instant error detection and correction [7], [8]. As a result, a new synthesis-friendly architecture for general-purpose DMFBs and a corresponding fast synthesis algorithm are required.

In this paper, we first highlight known difficulties in different phases of present DMFB design flows and then provide a promising solution. This paper is organized as follows. In Section II, we first introduce a typical architecture of array-based DMFB and several common basic functional units on it. We then give a brief review of the typical DMFB design flow and point out the difficulties at each design step. In Section III, our *versatile ring-based architecture* (VERBA) and a fast one-pass synthesis algorithm is presented. In Section IV, real-life bioassays are demonstrated using VERBA, and the results are compared with other synthesis algorithms on both VERBA and the traditional array-based architecture. Finally, Section V gives the concluding remarks.

II. Preliminaries

A. Array-based DMFB architecture

A typical DMFB architecture is a two-dimensional array of cells as shown in Fig. 1(a). The space sandwiched between two plates is named the *fluid layer*, in which fluidic droplets are transported. Each cell on the array is associated with a dedicated electrode to drive fluidic droplets through the electrowetting-on-dielectric (EWOD) effect [9]. An actuated electrode can pull droplets around onto it. A droplet

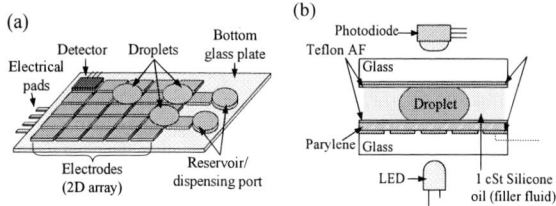

Fig. 1. (a) Top view of an array-based DMFB. (b) Side view of an optical detector.

of most fluidic types can be transported on the chip surface at a frequency of 100Hz (electrode size = 1.5mm) [3], [10]. Compared with the duration of other assay operations like mixing or heating, the time for droplet transportation is small and can generally be ignored during synthesis.

Besides the electrodes for droplet movement, other devices, such as reservoirs, dispensing ports, and detecting cells, are served for specific assay operations. Reservoirs and dispensing ports are used to keep, generate, and inject reactant droplets into the chip, and weed out waste droplets. Different types of detectors are utilized for different analysis demands. For example, optical detectors are required by colorimetric assays [10]. In this case, a photodiode and a light-emitting diode (LED) are placed over the top and under the bottom of the cell respectively, as illustrated in Fig. 1(b).

Besides those mentioned above, a mixer is a "virtual" device comprised of a group (usually a square) of adjacent cells, named *module*. Two droplets are combined into one and then moves rapidly in a module to accomplish a mixing operation. As the operation is completed, all cells associated with this module are set free immediately, and they can be utilized again to form another new module for next mixing operations. The execution duration of a mixing operation depends on the module size in general. A larger module generally leads to shorter mixing duration since droplets can be better accelerated. How to make a good tradeoff between the module size and the duration is important since it also determines the quality of synthesis outcome.

B. Current design flow of DMFBs

Currently, a DMFB design flow can be divided into two phases, *fluidic-level synthesis* and *chip-level design* (Fig. 2), according to their optimization goals [4]. The fluidic-level synthesis determines the fluidic behavior. It includes three major parts: *high-level synthesis*, *module placement*, and *droplet routing*. For an application-specific biochip, this phase is performed before chip fabrication; in contrast, for a general-purpose biochip, this phase must be executed whenever a new target application is loaded. Meanwhile, manufacturing-related issues are the major concerns in the chip-level design phase, such as *pin assignment* and *PCB wiring*. Poor chip-level design affects the chip reliability or even results in whole chip malfunctioned. For both DMFB types, this phase must be done before chip fabrication starts. These two phases are detailed in the following subsections.

C. Fluidic-level synthesis

High-level synthesis is the first part of fluidic-level synthesis, which consists of *operation scheduling*, *resource (module) binding*, and resource *allocation*. These three steps are highly interdependent to each other and dominate the overall assay completion time, (a typical optimization goal). For most existing synthesis methods, the array-based DMFB is regarded as the default architecture. The aforementioned three steps are detailed as follows.

1) Scheduling: A series of biochemical reactions are represented as a sequencing graph, shown in the left part of Fig. 2(a) [11]. The synthesizer takes the sequencing graph as its input, schedules the operations to a definite cycle based on given dependency constraints. Since a sequencing graph is very similar to the data flow graph used in classical VLSI synthesis, many existing methods (e.g., list scheduling) can be applied to DMFB scheduling. However, unlike classical VLSI synthesis, storage is an important concern and needs a special attention here [12]. If two dependent operations do not execute consecutively, a 1-cell space, named *storage unit*, must be reserved to hold the intermediate droplet from the end of the former operation to the start of the latter one. For example, in Fig. 2(b), a storage unit St is required to keep the resultant droplet of dispensing operation B. Since the number of cells on the chip array is limited, reducing the amount of required storage units is crucial for latency minimization.

2) Binding: Binding is another important step in high-level synthesis. For those operations who need specific functional units, an inactive unit will be assigned to each unbound operation during binding. For a mixing operation, the binding process also determines the size of module, which is also called *module selection*. The relation between the module size and the duration of a mixing operation is obtained from experiments on fabricated chips [11]. After binding, the total space occupied by all activated modules, storage units, and detectors in every cycle must be less than the array size.

3) Allocation: Resource allocation determines the required number of functional units. For general-purpose biochips, the count of each type of functional units is given by the specification and regarded as resource constraints. Therefore, the resource allocation step is actually omitted. However, it is a mandatory step for application-specific biochips for cost minimization. Resource allocation is usually carried out in two previous steps, i.e., resource-aware scheduling and resource-aware binding [4]. Through evenly distributing the same kind of operations in every execution cycle, the number of required physical functional units can be well minimized.

As aforementioned, three steps in high-level synthesis are highly interdependent to each other. Hence, most existing algorithms not only cover a single step but several steps for better quality of result [11], [13-15].

The back-end of fluidic-level synthesis includes module placement and routing. Similar to the front-end, the assay latency is typically the major optimization goal. However, the improvement is generally insignificant in this step since the time for droplet routing is quite small. On the other hand, the placement and routing problems are considered more difficult than previous steps due to additional spatial constraints.

978-1-5386-4757-8/18 $31.00 © 2018 IEEE

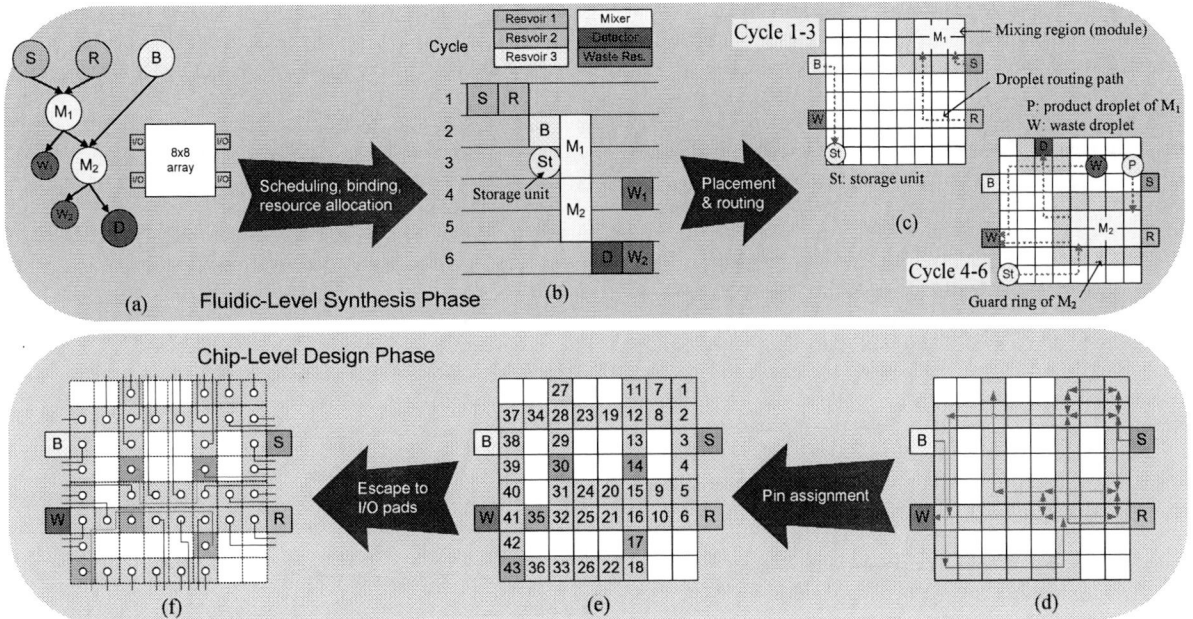

Fig. 2. Design flow for DMFBs. (a) Sequencing graph of a bioassay. (b) High-level synthesis results of (a).
(c) Module placement and droplet routing. (d) Electrode utilization. (e) Electrode addressing. (f) Full chip wiring.

4) Placement: As scheduling and binding are finished, the number and size of modules in each cycle are fixed. The module placement step is then carried out to decide the actual module locations. During placement, any placed module cannot overlap other modules scheduled at the same cycle. For an operation across multiple cycles, the associated module location should be fixed during its life time. Besides, every module is suggested to be wrapped by a guard ring to avoid cross contamination [16]. After placement, all modules must be successfully placed without any overlaps. If not, some operations should be rescheduled to another cycle or rebound with a smaller module to satisfy all the constraints.

5) Routing: Given scheduling and module placement results, this step decides the routing path for all kinds of droplets, including the droplets of reactant, intermediary product, and waste [17]. At any moment, no droplets can appear in the eight neighboring cells of any other droplet. Otherwise, the two droplets may be accidently merged in an unanticipated way and make the entire assay failed.

An example of placement and routing is shown in Fig. 2(c), which takes the result in Fig. 2(b) as its input. Again, module placement, droplet routing, and high-level synthesis parts are highly interdependent. Hence, several algorithms tackle these three issues simultaneously to give a better solution [5], [18-20]. However, the above methods are only suitable for off-line synthesis due to the runtime issue. As a result, the notion of "virtual topology" for general-purpose DMFBs was given in [6] to solve the problem from the architecture point of view. Fig 3(a) shows the proposed design, which includes predefined regions of modules and droplet routing channels. Hence, it guarantees placeable scheduling and deadlock-free routing in terms of simple on-line synthesis algorithms.

D. Chip-level design

Chip-level design determines the fabrication details, including pin-assignment and PCB wiring. In DMFB fabrication, the major cost is determined by the number of required PCB layers. The design with fewer layers is cheaper. However, few layers may lead to serious congestion during wire routing. How to make a prudent tradeoff among fabrication cost, design flexibility, and chip performance is the major challenge in the chip-level design phase.

1) Pin-assignment: A direct-addressing biochip, where every electrode must connect to an individual I/O pad (control pin), may have a serious wiring problem, especially as a small number (< 2) of PCB layers has been utilized. Most designs tackle this dilemma through pin minimization techniques. Unlike direct-addressing biochips, those designs allow a set of electrodes shares a common control pin. These electrodes are connected together at PCB wiring. For example, the electrodes {14, 30} and {17, 35, 43} in Fig. 2(e)

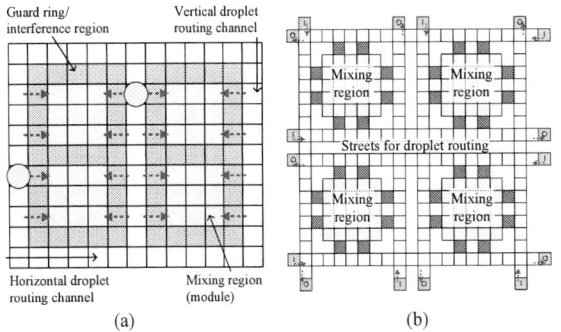

Fig. 3. Virtual topology imposed onto a DMFB.

are assigned to the same control pins, respectively. As a result, not only the wiring problem can be relaxed due to less required escape wires and routing rings on PCBs, but the fabrication cost can also be reduced due to less I/O pads [21]. Nevertheless, on the other hand, pin-sharing inevitably lowers the control flexibility, which further complicates the fluidic-level synthesis. That is also the major reason why the pin-sharing method is mainly adopted by application-specific biochip designs, which has no need for re-synthesis.

2) PCB wiring: PCB wiring is a classical problem in VLSI design. In this step, every electrode is connected to an external control pin at chip boundary via a dedicated or shared wire. For those designs with only one-layer bottom plate, connection wires can only pass through the gap between electrodes or dead space on the chip. For a design with tens of electrodes or more, one routing layer is simply not enough. Hence, most DMFBs utilize two-layer PCBs [21]. Even so, wiring difficulty is still a big concern for those direct-addressing biochips.

Most existing works only focus on the design iteration problem between different steps in fluidic-level synthesis since the chip architecture is unchanged during re-synthesis. However, the design iteration problem also exists during chip-level design; for example, a pin-assignment result may lead to no feasible wiring solutions. Once it happens, an extra PCB layer must be added, and the chip cost is thus increased. This problem may become even messy as the pin-assignment result is affected by the result of fluidic-level synthesis. Chang et al. first target the fluidic-level and chip-level co-design problem for application-specific DMFBs [4]. For general-purpose DMFBs, Grissom et al. provided a solution through further improving the virtual topology with wiring consideration [3]. Fig. 3(b) shows the new design with four tiles, which requires two PCB layers for wire routing [22]. However, this architecture can be further improved by considering more implementation details. Next, we will propose our new versatile ring-based DMFB architecture and the corresponding one-pass synthesis flow.

III. Versatile Ring-Based Architecture and the Corresponding One-Pass Synthesis Flow

Time-efficient fluidic-level synthesis on general-purpose biochips is considered as a big challenge due to the on-line synthesis demand. In contrast, time complexity is generally not a problem on application-specific biochips since the synthesis is only applied once before chip fabrication. Authors of [3] presented a promising fluidic-level synthesis solution for general-purpose biochips in their new chip architecture. Unlike conventional array-based architecture, theirs provides separate work chambers for (mixing) operations as well as predefined channels for droplet routing to fulfill real-time demands. The fluidic-level synthesis is thus claimed to be one-pass through decoupling 1) operation scheduling and module placement, as well as 2) module placement and droplet routing, respectively.

However, even if the method can solve most problems in those interdependent scheduling, placement, and routing phases, there are still issues remaining unsolved; for example,

storage unit management – storage allocation and binding for intermediary droplets. Since the architecture shown in Fig. 3(a) and Fig. 3(b) are lack of storage units, intermediary droplets can only be kept either in work chambers (, which prevents those chambers from operation) or routing channels (, which increases the possibility of routing congestion). Hence, if an excessive number of intermediary droplets are present at the same time, the prior scheduling and binding outcomes are very likely to be voided, which further implies re-synthesis is inevitable. Moreover, it is not easy to do wire routing for the architecture shown in Fig. 3(b). Two dedicated routing layers are required at least. If wires are further shared due to pin minimization (e.g., bus [4]), increased PCB layers definitely lead to a higher fabrication cost. As a result, in this paper, we propose a new versatile ring-based architecture, VERBA, to tackle the above two critical issues.

A. VERBA, *Versatile Ring-Based Architecture*

VERBA is a 13×13 design with 106 cells, as shown in Fig. 4, which can provide similar or richer functionality than a 22×22 design with 284 cells, as shown in Fig. 3(b). VERBA has 16 I/O ports (8 for input and 8 for output), 4 2×4 mixers, 8 storage slots, 4 specialized functional units (detectors), a ring bus, and 24 gate electrodes to properly turn on/off the fluidic channels between the ring bus and specific on-chip functional parts. Note that 4 2×4 mixers are included to support more simultaneous mixing operations [11]. The other detailed features of VERBA are elaborated as follows.

1) I/O ports: One major difference between VERBA and other existing architectures is the I/O design. Conventionally, I/O ports for fluids are always located beside transportation buses. The main reason is that buses can provide temporary storage spaces for input droplets and waste droplets can be weeded out without affecting ongoing operations. However, by observing sequencing graphs, waste droplets are generated after mixing operations, or are those droplets have been examined by detectors. Therefore, allocating output ports near mixers and detectors should be a better choice. It implies waste droplets can be weeded out right after being generated, and the number of droplets on chip can thus be lowered, which effectively reduces the traffic on the ring bus.

2) Ring bus: Unlike Fig. 3(b) where mixers are surrounded by vertical/horizontal channels, VERBA adopts a ring bus and locates at the chip center. It is obvious that a shorter ring

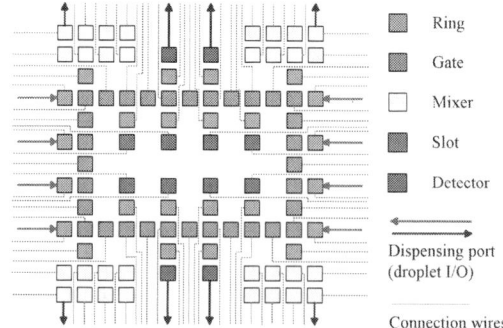

Fig 4. Versatile Ring-Based Architecture (VERBA).

bus provides more efficient droplet transportation, better chip area utilization, and much less wiring complexity than those H/V routing channels shown in Fig. 3(b). All droplets on the ring bus must proceed in one direction, counterclockwise or clockwise, to avoid droplet collisions. The ring bus can transport 8 droplets simultaneously as long as they keep at least a 3-cell distance from each other. Each functional unit is connected to the ring bus via a one-cell gate electrode. The functional unit is isolated from the ring bus if the gate is off, and hence they do not interfere with each other. The gate is on only if a droplet I/O is required by the functional unit.

3) Slot: In VERBA, dedicated storage slots are allocated inside the ring bus. Any intermediary droplets can be temporarily stored in these slots for future use. There are three advantages from dedicated storage slots: a) mixers are no longer used as storage, which is likely to accelerate the assay, b) the ring bus are no longer used as storage, which obviously reduces the traffic, c) storage slots can be regarded as operations and resources during scheduling and binding, which makes true one-pass synthesis possible.

4) PCB wiring: Fig. 4 also illustrates the PCB wiring result. VERBA allocates larger functional units (e.g., mixers) at the chip margin, which significantly reduces the wiring difficulty. As also shown in Fig. 4, merely one PCB layer is enough for allocating both electrodes and wires. Note that only one wire passing through the gap between any two neighboring electrodes. This is another key advantage over other existing architectures due to the extremely low fabrication cost.

B. One-Pass Synthesis Flow

Even if VERBA can provide a low-cost chip-level design solution for general-purpose DMFBs, a reliable one-pass fluidic-level synthesis flow is still demanded. Among the previous works, [4] provided a complete design methodology covering the fluidic-level synthesis and chip-layout generation. However, its synthesis flow cannot be applied on VERBA easily since it targets the application-specific designs. Also, its time-consuming ILP-based scheduling algorithm makes its flow inadequate to those applications requiring real-time on-the-fly synthesis. Therefore, in this paper, we also present a one-pass synthesis flow for VERBA with on-line synthesis ability. The proposed flow incorporates all scheduling, binding (placement), and routing synthesis steps. Allocation is omitted here since the number of available functional units on a VERBA chip is predetermined. With VERBA in use, any step in the proposed flow is mutually independent to each other, which is the key to achieve a true one-pass synthesis. In the following, each individual step of our proposed one-pass synthesis flow is briefly described.

1) Scheduling: Different from [4], we adopt the notion of multi-cycle scheduling for latency minimization. In addition, one cycle time is reserved for every droplet routing operation, which is long enough to complete any droplet routing. The default scheduling algorithm is the storage-aware scheduling (SAS) [15], which concentrates on latency minimization. The most important feature of SAS is that it can minimize the use of storage slots while keeping the latency overhead as small as possible. Other algorithms, such as [11] and [12], can also be selected to achieve different optimization objectives.

2) Binding (Placement): In fact, there is no complicated placement issue here because VERBA provides dedicated mixers and storage slots. As a result, the only task left here is binding. Mixers and storage slots are actually treated as functional units (like detectors and dispensers), and then are bound with mixing and slot access operations, respectively. Consequently, a widely-used left-edge binding algorithm in classical high-level synthesis for VLSI can be adopted to ensure both optimality and runtime efficiency.

3) Routing: Droplet routing on VERBA is extremely simple. As aforementioned, one cycle time is reserved for each end-to-end droplet routing operation during scheduling. Moreover, it is guaranteed that one cycle is definitely enough for any droplet transportation on VERBA based on the following calculation. Assume the cell width is 1mm, and the cycle time is 1 second. According to [10], a droplet can be moved at a speed of up to 15cm/s. It means a droplet can travel at least 150 cells within a cycle, which is a multiple times of the maximum possible distance between any two cells on VERBA. Besides, due to the fixed routing direction on the ring bus (clockwise or counterclockwise), every routing task can have only one possible routing path. These two features help simplify the task of droplet routing and complete the last mile of our one-pass synthesis flow.

IV. Experimental Results

Table I shows the comparison between three DMFB design methodologies including VERBA and two other existing works: virtual platform [3] and co-design methodology [4]. As Table I shows, VERBA and the virtual platform are similar since both of them target general-purpose biochips. They both perform latency minimization and adopt the direct addressing method for better controllability and higher operation parallelism. Nevertheless, VERBA uses the ring bus instead of vertical/horizontal routing channels to significantly increase the area utilization - over 50% of cells in the virtual platform are occupied by channels, whereas the ring bus merely uses 30% of cells in VERBA. Moreover, VERBA is much easier for one-layer PCB wiring due to its simple structure. The dedicated storage slots incorporated with the SAS algorithm facilitate simple droplet routing, and finally a real one-pass synthesis flow is achieved.

TABLE I. Comparison between different design methodologies.

	Virtual Platform [3]	Co-Design Methodology [4]	VERBA
Application	General-purpose	App-specific	General-purpose
Optimization goal	Latency	Resource	Latency
Bus design	Channel	Ring	Ring
Pin assignment	Direct addressing	Pin sharing	Direct addressing
Wiring complexity	Moderate	Moderate	Easy
Storage space	Mixers & channels	Mixers & channels	Slots
Scheduling	List scheduling	Synchronous scheduling	SAS
One pass synthesis	**No**	**No**	**Yes**

Table II shows the synthesis results for a set of biochemical applications on a 13×13 array-based architecture and VERBA. Three different scheduling algorithms are utilized for evaluation; they are list-scheduling (List) [11], path-scheduler (PS) [12], and SAS [15]. All modules are wrapped with a guard ring when they are placed into the cell array. We further assume the droplet routing time is one cycle for both architectures. First, we compare the synthesis results on a 13×13 cell array and VERBA over a set of test cases with different sizes. It is found that there is almost no difference in terms of the assay completion time if list-scheduling (List) or path-scheduler (PS) is in use. Even so, we still want to highlight again that VERBA guarantees the synthesis is always one-pass, whereas the synthesis for a typical array-based architecture is usually iterative and thus time-consuming. Next, we compare SAS with List and PS on VERBA. In small cases (#operations < 100), SAS performs equally well as List, which always guarantees the smallest latency if there are no storage concerns. However, in a large test case (Protein), the performance of List degrades due to an increasing number of storage units, which results in a number of rescheduling runs. Instead, PS and SAS perform better due to their storage-awareness. Furthermore, SAS outperforms PS since SAS can achieve a better balance between the number of storage units and latency minimization.

V. Conclusion

In this paper, we comprehensively elaborate on the design challenges and concerns in the DMFB synthesis flow, from the very beginning of fluidic-level synthesis to the last step of chip-level design. We introduce well-known algorithms at each design step whenever possible. We also point out the most critical challenge in general-purpose biochip design: those interdependent steps in fluidic-level synthesis can drastically increase the time complexity due to iterations, which makes real-time on-line synthesis virtually impossible. We then propose a versatile ring-based architecture VERBA, which not only minimizes the synthesis complexity but also significantly lowers the fabrication cost. We also present a true one-pass synthesis flow for VERBA. The experimental results clearly show that VERBA with its one-pass synthesis flow is indeed a more promising solution than the typical array-based architecture with present synthesis algorithms.

References

[1] R. B. Fair, A. Khlystov, T. D. Tailor, V. Ivanov, R. D. Evans, P. B. Griffin, V. Srinivasan, V. K. Pamula, M. G. Pollack, and J. Zhou, "Chemical and biological applications of digital-microfluidic devices," *IEEE Des. Test of Comput.*, vol. 24, no. 1, pp. 10–24, Jan. 2007.

[2] T.-Y. Ho, J. Zeng, and K. Chakrabarty, "Digital microfluidic biochips: a vision for functional diversity and more than Moore," in *Proc. IEEE/ACM ICCAD*, 2010, pp. 578–585.

[3] D. Grissom, C. Curtis, and P. Brisk, "Interpreting assays with control flow on digital microfluidic biochips," *ACM J. Emerg. Technol. Comput. Syst.*, vol. 10, no. 3, pp. 24:1–24:29, May 2014.

[4] J.-W. Chang, S.-H. Yeh, T.-W. Huang, and T.-Y. Ho, "Integrated fluidic-chip co-design methodology for digital microfluidic biochips," *IEEE Trans. Computer-Aided Design Integr. Circuits Syst.*, vol. 32, no. 2, Feb. 2013.

[5] O. Keszocze, R. Wille, T.-Y. Ho, and R. Drechsler, "Exact one-pass synthesis of digital microfluidic biochips," in *Proc. IEEE/ACM DAC*, 2014, pp. 1–6.

[6] D. Grissom and P. Brisk, "Fast online synthesis of digital microfluidic biochips," *IEEE Trans. Computer-Aided Design Integr. Circuits Syst.*, vol. 33, no. 3, pp. 356–369, Mar. 2014.

[7] Y. Luo, K. Chakrabarty, and T.-Y. Ho, "Error recovery in cyberphysical digital microfluidic biochips," *IEEE Trans. Computer-Aided Design Integr. Circuits Syst.*, vol. 32, no. 1, pp. 59–72, Jan. 2013.

[8] Y.-L. Hsieh, T.-Y. Ho, and K. Chakrabarty, "Biochip synthesis and dynamic error recovery for sample preparation using digital microfluidics," *IEEE Trans. Computer-Aided Design Integr. Circuits Syst.*, vol. 33, no. 2, pp. 183–196, Feb. 2014.

[9] M. G. Pollack, A. D. Shenderov, and R. B. Fair, "Electrowetting-based actuation of droplets for integrated microfluidics," *Lab on a Chip*, vol. 2, no. 2, pp. 96–101, Feb. 2002.

[10] V. Srinivasan, V. K. Pamula, M. G. Pollack, R. B. Fair, "Clinical diagnostics on human whole blood, plasma, serum, urine, saliva, sweat, and tears on a digital microfluidic platform," in *Proc. µTAS*, 2003, pp. 1287–1290.

[11] F. Su and K. Chakrabarty, "High-level synthesis of digital microfluidic biochips," *ACM J. Emerg. Technol. Comput. Syst.*, vol. 3, no. 4, pp. 16:1–16:32, Jan. 2008.

[12] D. Grissom and P. Brisk, "Path scheduling on digital microfluidic biochips," in *Proc. IEEE/ACM DAC*, 2012, pp. 26–35.

[13] A. J. Ricketts, K. Irick, N. Vijaykrishnan, and M. J. Irwin, "Priority scheduling in digital microfluidics-based biochips," in *Proc. IEEE/ACM DATE*, 2006, pp. 329–334.

[14] E. Maftei, P. Pop, and J. Madsen, "Tabu search-based synthesis of dynamically reconfigurable digital microfluidic biochips." in *Proc. CASES*, 2009, pp. 195–203.

[15] C.-H. Liu, K.-C. Liu, and J.-D. Huang, "Latency-optimization synthesis with module selection for digital microfluidic biochips," in *Proc. IEEE SOCC*, 2013, pp.159–164.

[16] T. Xu and K. Chakrabarty, "Integrated droplet routing and defect tolerance in the synthesis of digital microfluidic biochips," *ACM J. Emerg. Technol. Comput. Syst.*, vol. 4, no. 3, pp. 11:1–11:24, Aug. 2008.

[17] T.-W. Huang, C.-H. Lin, and T.-Y. Ho, "A contamination aware droplet routing algorithm for the synthesis of digital microfluidic biochips," *IEEE Trans. Computer-Aided Design Integr. Circuits Syst.*, vol. 29, no. 11, Nov. 2010.

[18] F. Su and K. Chakrabarty, "Unified high-level synthesis and module placement for defect-tolerant microfluidic biochips," in *Proc. IEEE/ACM DAC*, 2005, pp. 825–830.

[19] P.-H. Yuh, C.-L. Yang, and Y.-W. Chang, "Placement of defect-tolerant digital microfluidic biochips using the T-tree formulation," *ACM J. Emerg. Technol. Comput. Syst.*, vol. 3, no. 3, pp. 13:1–13:32, Nov. 2007.

[20] E. Maftei, P. Pop, and J. Madsen, "Module-based synthesis of digital microfluidic biochips with droplet-aware operation execution," *ACM J. Emerg. Technol. Comput. Syst.*, vol. 9, no. 1, pp. 2:1–2:21, Feb. 2013.

[21] Y. Luo, K. Chakrabarty, and T.-Y. Ho, "Pin-limited cyberphysical microfluidic biochip," in *Hardware/Software Codesign and Optimization for Cyberphysical Integration in Digital Microfluidic Biochips.* Springer, 2015, pp 185–194.

[22] D. Grissom, J. McDaniel, and P. Brisk, "Performance and cost analysis of NoC-inspired virtual topologies for digital microfluidic biochips," in *Proc. IEEE ISIC*, 2014, pp. 352–355.

TABLE II. Synthesis results on VERBA and array-based chip.

Application	#op	Assay Latency (cycle)				
		Cell Array (13×13)		VERBA		
		List	PS	List	PS	SAS
in-vitro 1	16	14	14	14	14	14
in-vitro 2	24	18	18	18	19	18
in-vitro 3	36	28	31	28	31	28
in-vitro 4	48	29	33	29	33	29
in-vitro 5	64	37	40	37	40	37
PCR	7	17	17	17	17	17
Protein	103	160	157	160	157	145
One pass synthesis		No		Yes		

Directed Graph Placement for SNN Simulation into a multi-core GALS Architecture

Francesco Barchi, Gianvito Urgese, Andrea Acquaviva, and Enrico Macii
Department of Control and Computer Engineering
Politecnico di Torino
Torino, Italy 10129
francesco.barchi@polito.it

Abstract—In this paper, we present a methodology for efficiently mapping neural networks over a neuromorphic computing architecture. The target architecture is a globally asynchronous locally synchronous (GALS) multi-core designed for simulating spiking neural networks (SNN) in real-time, that is spike timings should be the same as in the human brain. The SNN is implemented as a set of concurrent tasks modelling the behaviour of biological neurons, which are executed on the processing cores and communicate through spikes travelling on a network-on-chip. The problem of neuron-to-core mapping is relevant as a non-efficient allocation may impact real-time and reliability of the neural network execution. We designed a task placement pipeline capable of analysing the network of neurons and producing a placement configuration that enables a reduction of communication between computational nodes. The neuron-to-core mapping problem has been formalised as a problem of minimisation of synaptic elongation. Intuitively, this metric represents the cumulative distance that spikes generated by neurons running on a specific core have to travel to reach their destination core. The proposed placement methodology allows using different techniques to solve the problem. In this work Spectral Analysis, Multilevel Static Mapping, and Simulated Annealing were compared evaluating the overall post-placement synaptic elongation. Results point out that mapping solutions taking into account the directionality of the SNN provide a better placement and quantify this impact. Between all techniques considered only the Simulated Annealing was able to overcome an improvement of 25% compared to a random placement.

I. INTRODUCTION

Finding the best way to map processes to physical cores (PCs) in multi and many-core systems is a relevant optimisation problem, with significant impact on application reliability, performance, and energy consumption. We explored this problem in the case of a globally asynchronous locally synchronous (GALS) multicore architecture designed for neuromorphic applications. Here, tasks to be executed are physical neuron models running in parallel on the platform and communicating through messages. These messages represent signals, called spikes, which biological neurons exchange through their physical (neural) connections inside the brain.

The overall purpose of this application is to execute a Spiking Neural Network (SNN) in real-time. In this case, real-time means that the timings of the spikes generated by the neurons should be compliant with the one of the real human brain. Thus opening the way for the use of neuromorphic

platforms to interface external physical systems and elaborates their signals (e.g. images, sounds) in the same way as the brain does. Being the neurons executed as concurrent tasks by the general purpose cores, there is a problem of efficient mapping of neurons to cores to optimise the communication between them. Indeed, these tasks are characterised by intensive communication activity.

Generalising, the problem we faced concerns the mapping of a large number of light parallel tasks with intensive communication to a many-core architecture. A non-efficient communication, in the specific case of SNN execution, may impact real-time capabilities as well as the reliability of the application. Indeed, spikes can be lost due to congestion problems. In general, a possible approach to face the mapping problem is to model the tasks and their communication as a graph to be mapped over the underlying hardware architecture, represented by another graph.

In this paper, we present a methodology for mapping a task graph representing the SNN computation on a multi-chip many-core architecture with communication awareness. To achieve this target, we designed a task mapping framework capable of analysing the network of neurons to find a configuration with the target of reducing the communication between computational nodes. The neuron-to-core mapping problem has been formalised as a problem of minimisation of synaptic elongation. Intuitively, this metric represents the cumulative distance that spikes generated by neurons running on a specific core have to travel to reach their destination core.

The framework starts by extracting a graph of independent processes from a neural network description. In the case of SNN, the direction of a communication path is also to be represented using a directed graph. On the platform side, the interconnect structure is described as a graph where nodes represent on-chip cores while edges represent physical communication links between them. In this way, we formalised a neuron-to-core mapping as a graph-matching problem solvable through the exploitation of various algorithms available in the literature. The specific formulation we devised for SNN mapping takes into account the typical organisation of these type of neural networks into neuron populations, sharing similar characteristics as well as the neuron model.

The results obtained by comparing four mapping algorithms

978-1-5386-4757-8/18 $31.00 © 2018 IEEE

points out and quantify the relevance of the communication direction information to achieve a better mapping if compared with non-directional algorithms.

The paper is organised as follows. In Section II we give a brief overview of the application and the target board used in this work. In Section III we detailed the problem formulation under the graph theory view-point, while we describe our process placement method in Section IV. In Section V we report on the results obtained during the validation process performed on the cortical microcircuit SNN simulation. Finally in Section VI we reach some conclusions and list some future declinations of this piece of work.

II. BACKGROUND

In this section, we will introduce the application and the MCSoC board selected as a target for demonstrating the advantages of adopting our process-placement communication aware framework.

Spiking Neural Network (SNN) is a particular neural model used by neuroscientist for simulating biologically plausible brain activity. During SNN simulations neurons and their synapses are modelled as differential equations capable of emulating the behaviours observed in biological networks [1]. Two of the most adopted neuron models are the *Leaky Integrate and Fire* (LIF) [2] and *Izhikevich* (IZK) [3], because they are able to ensure a plausible picture of the biological behaviours with reduced computational costs. Van Albada et al. [4] designed an SNN application implementing the cell-type specific *cortical microcircuit* (CM) model created by Potjans et al. [5]. Then they simulated this SNN on a neuromorphic multi-chip many-core platform called SpiNNaker [6] using the standard application partitioning and placement system for setting up the simulation on the board.

For validating our placement methodology framework, we took as target a GALS Neuromorphic many-core architecture and used its native application such as an example case. We used the SpiNNaker architecture [6], which is a general-purpose real-time many-core platform mainly used for simulating neural networks following an event-driven computational approach but the methodology could also be applicable for the new Loihi platform [7] and the future SpiNNaker 2 architecture.

The SpiNNaker chip has 18 ARM 968 cores running at 200MHz, a full-custom router for intra/inter-chip communications, and an SDRAM external to the chip and accessible through the PL340 interface [8]. The SpiNNaker system is built with boards of 48 chips interconnected for forming a toroidal shaped triangular mesh where each chip is connected to six neighbours chips.

Sugiarto et al. [9] presented an approach for improving the overall performance of general-purpose applications running as a task graph on the same many-core neuromorphic supercomputer. Whereas in a recent paper, we have used the cortical microcircuit application as a test case for demonstrating that an enhanced partitioning and placement system studied for the SNN topology can produce a more reliable and stable configuration for the simulation on the SpiNNaker system [10].

III. PROBLEM FORMULATION

The SNN placement into the neuromorphic architecture can be view as an optimisation problem that involves two graphs: $\mathcal{G}_{\mathcal{N}}$ and $\mathcal{G}_{\text{CPU}}$.

A graph $\mathcal{G} = (V, E, \mathcal{W})$ is a mathematical representation for describing a set of elements V and a set of relations $E \subseteq \{(v_i, v_j) : v_i, v_j \in V\}$ among them. The elements are called *nodes* of the graph and the relations are called *edges* of the graph. An edge $e_{ij} \in E$ binds two nodes $v_i, v_j \in V$ to each other. A graph can have a $\mathcal{W} : E \to W$ function that associates an edge $e_{ij} \in E$ to a value $w_{ij} \in W$. The value $w_{ij} = \mathcal{W}(e_{ij})$ is called edge weight. A graph can be categorised according to two properties: i) If the nodes on edges form unordered pairs $e_{ij} : \{v_i, v_j\}$ the graph is said *undirected* otherwise it is said *directed* and the nodes on edges form ordered pairs $e_{ij} : (v_i, v_j)$. ii) If the weight set W is empty the graph is said *unweighted*, otherwise it is said *weighed*.

A Spiking Neural Network (SNN) can be represented using a directed and weighted graph called *neuron graph* $\mathcal{G}_{\mathcal{N}}$. In $\mathcal{G}_{\mathcal{N}}$ the nodes are the SNN neurons and the edges are the SNN synapses. Taking into account a synapse $e_{ij} : (v_i, v_j)$, the neuron v_i is called pre-synaptic neuron and the neuron v_j is called post-synaptic neuron. The edge weight w_{ij} represent the synapse contribution to injected current into the post-synaptic neuron after a stimulus received by the pre-synaptic neuron and is called synaptic weight.

The neuromorphic architecture can be represented using an undirected and weighed graph, called *target graph* $\mathcal{G}_{\mathcal{T}}$. The graph nodes are the ARM processors, and the graph edges are the connections between them. The SpiNNaker board has 48 chip with 16 available processors each, and each chip is connected to other six chip. The edge between two processors of the same chip has a weight of one, the edge between two processors of two neighbour chip has a weight of two.

We can define the placement problem $\Pi : \mathcal{G}_{\mathcal{N}} \to \mathcal{G}_{\mathcal{T}}$ as a minimization problem (1).

$$\underset{f(\pi)}{\text{minimize}} \quad f : \sum_{e_{ij} \in E_{\mathcal{N}}} d(\pi(v_i), \pi(v_j)) \tag{1a}$$

$$\text{subject to} \quad \pi(i) = \pi(j) \to \mathcal{M}(i) = \mathcal{M}(j), \; i, j \in V_{\mathcal{N}} \tag{1b}$$

$$|\pi(i) = p| \leq \mathcal{S}(\mathcal{M}(i)), \; i \in V_{\mathcal{N}}, p \in V_{\mathcal{T}} \tag{1c}$$

The goal of a placement procedure is to minimise the *overall synaptic stretching* (1a) to reduce the communication along the network nodes. The *synaptic stretching* is the distance between the processors where two adjacent neurons are placed. Where $\pi : V_{\mathcal{N}} \to V_{\mathcal{T}}$ is the placement rule, $\mathcal{M} : V_{\mathcal{N}} \to M$ is the neuron-model association rule and, $\mathcal{S} : M \to \mathbb{N}$ is the association rule between a neuron model and the maximum number of neuron per CPU. The constraints of the placement problem are two: i) All neurons mapped into a CPU must be of the same model (1b). ii) Each CPU can simulate only a certain number of neurons, and the quantity depends on the complexity of the neuron model (1c).

A. Problem Relaxation

A SNN is almost never described in $\mathcal{G}_{\mathcal{N}}$ form, due the high complexity in manage all neurons and synapses, but is

978-1-5386-4757-8/18 $31.00 © 2018 IEEE

normally described in terms of Population and Projection. A Population $\mathcal{P}$ is a set of neurons that share the same model and the same properties. A Projection between two Population $\mathcal{P}^{(a)}$ and $\mathcal{P}^{(b)}$ defines a rule for create a set of synapses where the pre-synaptic neurons are in $\mathcal{P}^{(a)}$ and the post-synaptic neurons are in $\mathcal{P}^{(b)}$. We will refer to the *Population-Projection graph* using the notation $G_\mathcal{P}$.

We can eliminate the two constrains (1b, 1c) redefining the problem Π working from the graph $\mathcal{G}_\mathcal{P}$. The first step is splitting each population $\mathcal{P}^{(i)}$ into a set of partial populations $\left\{ \mathcal{P}_1^{(i)}, \mathcal{P}_2^{(i)}, \dots, \mathcal{P}_z^{(i)} \right\}$. All partial populations must contains at most a number of neurons equal to the maximum number of neurons allowed to be simulated in a processor: $|\mathcal{P}_j^{(i)}| \leq n^{(i)} \ \forall j = 1, \dots, z$, with $n^{(i)} = \mathcal{S}(\mathcal{M}(\mathcal{P}^{(i)}))$.

In this way we obtain the *partial population graph* $\mathcal{G}_{\mathrm{pp}}$. The edges of the partial population graph are weighed and ordered. Given an edge $e_{ij} \in E_{\mathrm{pp}}$ between two partial population, its weight w_{ij} is equal to the number of synapses shared between the neurons belonging the two partial populations.

We can redefine (1) using the placement rule $\pi : V_{\mathrm{pp}} \to V_\mathcal{T}$ that map a partial population into a processor (2).

$$\underset{f(\pi)}{\text{minimize}} \qquad \sum_{e_{ij} \in E_{\mathrm{pp}}} d(\pi(v_i), \pi(v_j)) * w_{ij} \qquad (2a)$$

$$\text{subject to} \qquad |\pi(i) = p| \leq 1, \ i \in V_{\mathrm{pp}}, p \in V_\mathcal{T} \qquad (2b)$$

In (2a) we modify the cost function to take into account the number of synapses shared between the processors. The rule in (2b) describes the single constraint of the problem: a processor may contain only one partial population.

B. Graph Partitioning

The partition problem of $\mathcal{G}_\mathcal{P}$ can be solved in different ways. In [10] it was treated as a problem of clustering. The provided solution was divided into three step:

- Graph expansion: $\mathcal{G}_\mathcal{P} \to \mathcal{G}_\mathcal{N}$
- Spectral clustering: $\mathcal{G}_\mathcal{N} \to \mathbb{R}^{|V_\mathcal{N}|}$
- Legalization and clusters fusion: $\mathbb{R}^{|V_\mathcal{N}|} \to \mathcal{G}_{\mathrm{pp}}$.

The first step is to create the neuron graph $\mathcal{G}_\mathcal{N}$ by applying the synaptic generation rules defined into the Population-Projections graph $\mathcal{G}_\mathcal{P}$. In the second step, a spectral clustering procedure is applied to the neuron graph.

The Spectral Clustering involves the eigendecomposition of a representative matrix of the graph. In the case of $\mathcal{G}_\mathcal{N}$, a directed graph, it was used a Laplacian Matrix (3) obtained throught a transition matrix induced by a random walk [11].

$$L = I - \frac{(\Phi^{\frac{1}{2}} P \Phi^{-\frac{1}{2}} + \Phi^{-\frac{1}{2}} P^T \Phi^{\frac{1}{2}})}{2} \qquad (3)$$

The results of the Spectral Clustering is the $\mathcal{G}_\mathcal{N}$ rapresentation into the eigenspace of $\mathbf{L}$, a space belonging to $\mathbb{R}^{|V_\mathcal{N}|}$. The neurons can be clustered into the eigenspace using the KMeans algorithm. After the clustering, a legalisation phase gathers in groups all neurons belonging to the same cluster and the same population. Finally, a second legalisation phase, called Fusion, builds the partial populations putting together the nearby groups of neurons until reach the maximum number of neurons that a processor can simulate.

Other techniques of graph clustering are Multilevel Graph Partitioning and Markov Cluster Algorithm [12], [13]. These techniques, like the Spectral Cluster, was born for undirected graph and their usage should be analysed using different symmetrisation techniques if applied to a directed graph.

IV. PLACEMENT

As seen in section III our goal is placing $\mathcal{G}_\mathcal{N}$ into a set of processors $\mathcal{G}_\mathcal{T}$. In subsection III-A we have relaxed the constraints of the problem separating it into two subproblems: i) Clustering $\mathcal{G}_\mathcal{N}$ (or partitioning if consider $\mathcal{G}_\mathcal{P}$ as a starting point) into the partial population graph. ii) Placement of $\mathcal{G}_{\mathrm{pp}}$ into $\mathcal{G}_\mathcal{T}$. We have briefly described the clustering (or partitioning problem) in the section III-B. In this section, we independently explore the placement problem (2) by comparing different techniques: Naïve, Spectral Embedding, Scotch and Simulated Annealing.

A. Naïve Placement

The Naïve approach is the standard mapping procedure adopted in the SpiNNaker toolchain for assigning populations of neurons to be simulated on the cores available in the SpiNNaker Platform. It is a simple and computationally light method to perform the graph placement without taking into account neither source and target graph connectivity.

The processor graph was ordered following a polar coordinate system (ρ, φ) starting from a chip of choice. The radius $\rho = \max(|x|, |y|, |x - y|)$ has been calculated using the hexagonal distance. The angle $\varphi \in [0, 2\pi)$ is expressed in radians. The procedure starts to place a partial population into each processor and change the chip when all processors inside a chip are used. As the ρ increases, the sub-populations will be distributed along the chip on the circumference and will be separated by a greater and greater distance.

B. Spectral Embedding

The Spectral Embedding placement was partially used in a previous work described in [10]. The procedure involves the spectral analysis of the graph and a dimension reduction procedure to obtain a planar representation of it. By doing so, the target graph can be directly superimposed on the graph of the partial populations. Contrary to previous work, in which a greedy heuristic was used, the association of partial populations with processors was finally described through an *Integer Linear Programming* (ILP) problem.

The procedure starts with the extraction of the first five eigenvalues, and the relative eigenvectors, from the matrix $\mathbf{L}$. The eigenvectors form a matrix Λ that represents the partial populations in a $\mathbb{R}^5$ space. We apply a non-linear dimension reduction procedure using *Sammon Mapping* obtaining a space in $\mathbb{R}^2$.

The Sammon Mapping algorithm minimise the error function in (4) where d_{ij} is the distance in the high-dimensional space (eigenspace) and d_{ij}^* is the distance in the low-dimensional space (placement space) [14].

978-1-5386-4757-8/18 $31.00 © 2018 IEEE

$$E = \frac{1}{\sum_{i<j} d_{ij}} \sum_{i<j} \frac{(d_{ij} - d_{ij}^*)^2}{d_{ij}} \tag{4}$$

Each chip, in the chip mesh, is represented as a point (x, y) in an axial coordinate system. We superimpose the graph $\mathcal{G}_{\mathcal{T}}$ on $\mathcal{G}_{\mathrm{pp}}$ projecting the chip mesh in the placement space (5).

$$\begin{pmatrix} x^* \\ y^* \end{pmatrix} = \sqrt{\frac{2A_h}{3\sqrt{3}}} \begin{pmatrix} \sqrt{3} & -\frac{\sqrt{3}}{2} \\ 0 & \frac{3}{2} \end{pmatrix} \begin{pmatrix} x \\ y \end{pmatrix} \tag{5}$$

Where (x, y) is the chip coordinate in the hex mesh, and (x^*, y^*) is the chip coordinate in the placement space. The side length of the hex is used as a normalising factor and calculated using the area $A_h = \frac{A}{m}$ occupied by each chip. The normalising factor allows scaling the chip mesh concerning the area A occupied by the partial populations. After projecting the points into the placement space, they are translated to centre them on the median of the points representing the partial populations. Now we can describe the placement problem using the ILP formulation (6).

$$\underset{f(\mathbf{X})}{\text{minimize}} \quad f : \sum_{i=1}^{n} \sum_{j=1}^{m} x_{i,j} d_{i,j} \tag{6a}$$

$$\text{subject to} \quad \sum_{i=1}^{n} x_{i,j} \leq k \quad \forall j \in \{1, \ldots, m\} \tag{6b}$$

$$\sum_{j=1}^{m} x_{i,j} = 1 \quad \forall i \in \{1, \ldots, n\} \tag{6c}$$

Where the $\mathbf{X} = (x_{ij})$, $x_{ij} \in \{0, 1\}$ matrix is the placement matrix. An entry $x_{ij} = 1$ means that partial population i is mapped on the target node j. The problem constraints are two: i) Each target node can host at most k partial populations (6b). ii) Each partial population can be associated to only one target node (6c). The ILP problem was modelled using PuLP Python library and solved with *COIN-OR branch and cut* (CBC) solver.

C. Scotch

The Scotch mapping procedure makes use of the programs available in the homonym software suite (SCOTCH). The Dual Recursive Bipartitioning (DRB) is the primary procedure used by this tool [15]. The DRB can use a plethora of other bipartitioning methods according to a strategy defined by the user or deducted by graph properties. The main available methods are: Gibbs-Poole-Stockmeyer [16], Fiduccia-Mattheyses [17], Greedy Graph Growing [12] and Diffusion [18].

The mapping workflow with SCOTCH plans to pre-partition the target graph through the *amk_grf* program. The *amk_grf* program take in input a graph in *grf* format and create a target file (*tgt* format) which contains a decomposition-defined target architecture of same topology as the input graph.

Once a decomposition of the target graph has been obtained, the graph of the partial populations is placed on the target graph using the *gmap* program. The program *gmap* take in input the partial population graph in *grf* format and the target graph in *tgt* format and perform the DRB procedure minimising the communication cost function[1]. The *gmap* output file

[1]The SCOTCH cost function is similar to our Synaptic Stretching

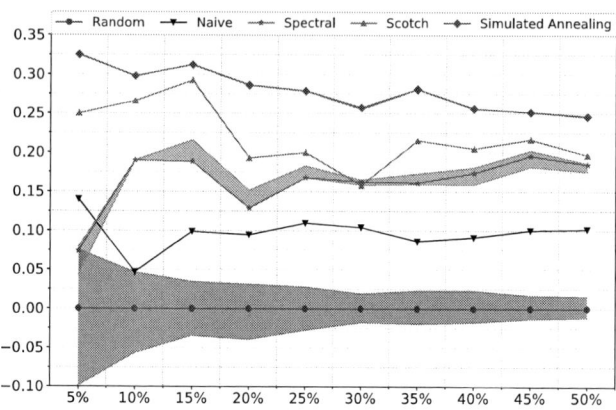

Fig. 1: The graph represents the improvement of a mapping technique with respect to the median of the results obtained with a random placement, y-axis. The x-axis shows the CM scale factor.

is a mapping file (*map* format) that contains the association between the Source and the Target nodes.

We had developed a Python module able to exporting a NetworkX graph to a file according to the *grf* format used by SCOTCH and capable of automating the procedures described above.

D. Simulated Annealing

The Simulated Annealing is a well know procedure used to find a good solution to an optimisation problem [19]. Given the problem in (2a), it is convenient to express the overall synaptic stretching in a matrix form and define a cost function to minimise. Given the partial population graph $\mathcal{G}_{\mathrm{pp}}$ we build its Adjacency matrix $\mathbf{A} = (a_{ij})$ as described in (7).

$$a_{ij} = \begin{cases} w_{ij} & \text{if } \exists (v_i, v_j) \in E_{\mathrm{pp}} \\ 0 & \text{otherwise} \end{cases} \quad \forall i, j \in \{1, \ldots, n\} \tag{7}$$

Given the processor graph $\mathcal{G}_{\mathcal{T}}$ we build its distance matrix $\mathbf{D} = (d_{ij})$ where each entry d_{ij} is the lenght of the mimimum path between two target nodes cpu_i and cpu_j. The distance matrix can be build using the Floyd–Warshall algorithms or repeating Dijkstra's algorithms if $|E_{\mathcal{T}}| \ll |V_{\mathcal{T}}|^2$.

Assuming to have as many subpopulations as processors and a placement rule $\Pi : \{v_1, \ldots, v_n\} \rightarrow \{\mathrm{cpu}_1, \ldots, \mathrm{cpu}_n\}$ we construct the *permutation vector* $\pi : (\Pi(v_1), \ldots, \Pi(v_n))$ and the *permutation matrix* $\mathbf{P}_\pi = (p_{ij})$ in row form (8).

$$p_{ij} = \begin{cases} 1 & \text{if } i = \pi_j \\ 0 & \text{otherwise} \end{cases} \quad \forall i, j \in \{1, \ldots, n\} \tag{8}$$

The permutation matrix is applied to $\mathbf{D}$ to permutate its rows and columns. We obtain the matrix $\mathbf{D}_\pi = \mathbf{P}_\pi \mathbf{D} \mathbf{P}_\pi$. The overall synaptic stretching can be expressed in a matrix form and used as the cost function for the simulated annealing algorithm (9).

$$f : \mathbf{e}^T (\mathbf{A} \odot \mathbf{D}_\pi) \mathbf{e} = \sum_{i,j} a_{ij} * d_{ij}^{(\pi)} \tag{9}$$

Where $\odot$ is an element-wise multiplication and $\mathbf{e}$ is a column vector whose all elements are equal to one.

We used the Simulated Annealing implementation provided in the SciPy ecosystem using the *temperature* to decide how many elements of the permutation vector π to swap.

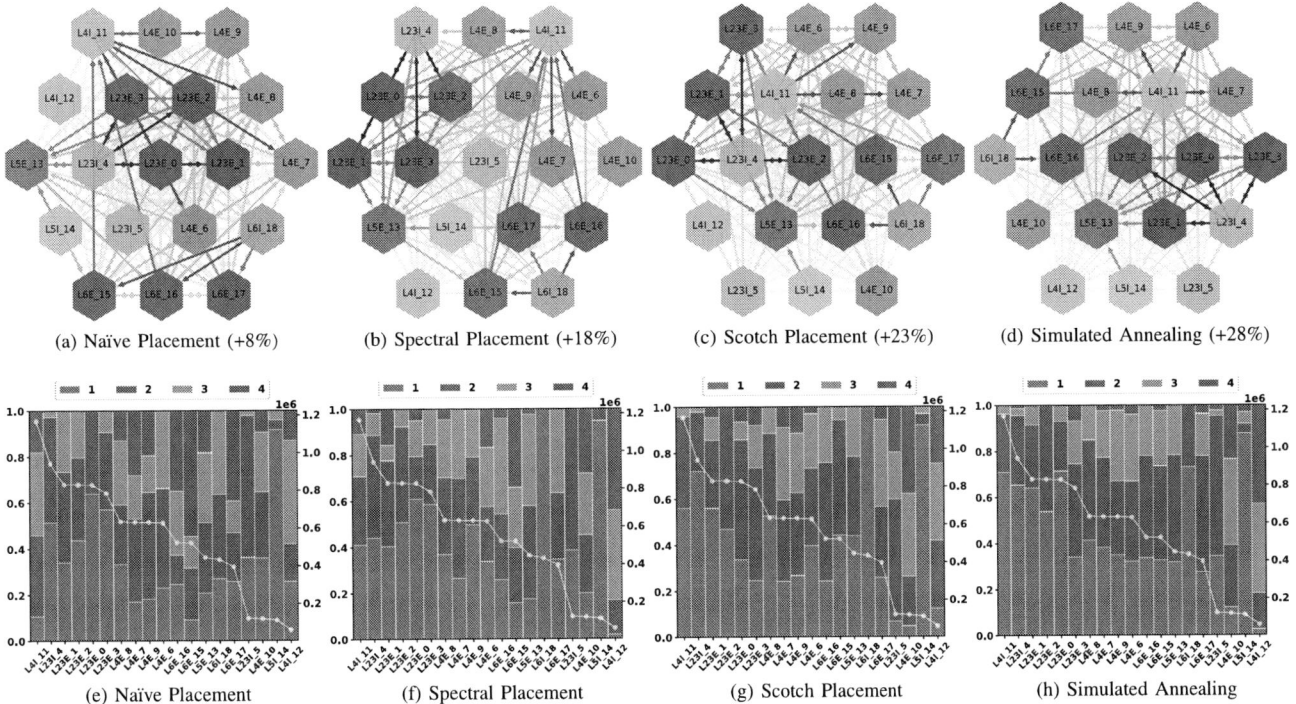

Fig. 2: The figures in the first row represent the placement of the partial population graph build from a $CM_{20\%}$ with 1000 neurons per chip on 19 chip (5 processors per chip). The figures in the second-row represent for each partial population the number of synapses (white line) and the percentage of synapse stretching.

V. RESULTS

In this section, we present the exploration experiments using the methods described in Section IV.

We use the Cortical Microcircuit (CM) as benchmark network, [5]. This network represents the connectivity of neurons inside a slice of the cerebral cortex with an area of $1\ mm^2$. The CM has been chosen because it is a rapresentative biological model with a relativly high global connectivity (5%) and natural clusters defined by the four cerebral cortex layers $\{L_{23}, L_4, L_5, L_6\}$. The CM is described in terms of Population and Projection with two populations for each layer, for a total of 8 Population and 64 Projections.

The network is composed of Integrate and Fire (LIF) and Spike Source (SRC) neuron models. The LIF neurons are models that mimic the biological neurons behaviour. The SRC neurons are simple programmable applications for outputting signals when desired. In this network, the SRC neurons are used to simulate the background activity of cortical neurons not presents in the model. Each SRC neuron is connected to only one LIF neuron, so they can be excluded by the $\mathcal{G}_N$ provided that processors are reserved for their execution.

The CM model has 7.72e+4 LIF neurons and 2.99e+8 synapses. The network can be down-scaled to a percentage CM_p, for example:

- $CM_{5\%}$ has 3.86e+3 neurons and 7.47e+5 synapses.
- $CM_{10\%}$ has 7.72e+3 neurons and 2.99e+6 synapses.
- $CM_{50\%}$ has 3.86e+4 neurons and 7.47e+7 synapses.

For each processor in charge of simulating a LIF partial population, we must reserve two further processors. A processor is reserved for the simulation of paired SRC neurons. A further processor is reserved to host a special application necessary to manage synapses with delays greater than 10 ms, as described in [20]. Taking into account a set of 16 processors belonging to the same chip, we can place 5 partial population per chip for a total of a thousand neurons per chip.

For simplifying the problem we perform a sequential slicing of each population in order to obtain partial populations with at most 1 000 neurons. In this way, we use an entire chip (5 processors with 200 LIF neurons each) for a partial population.

The experiment environment is composed of four different mapping procedures: Naïve, Spectral, Scotch and Simulated Annealing. We had generated 5 CM networks for 10 different scale factors, from 5% to 50%, for a total of 50 networks. For each network, we applied all mapping procedures 20 times. We evaluate the performance of each mapping procedure for each scale factor, using the fitness function (9). As a result, we obtain a distribution of 100 different placement results concerning overall synaptic stretching.

The performance of mapping procedures is compared to the performance of random placement. The median value of the results obtained with the Random procedure is used to compute the percentage improvement of the results obtained with other techniques.

In figure 1 is depicted a chart that summarize all the experiments. On the x-axis, there are the network scale factors, on the y-axis the percentage placement improvements versus random. The data series are represented by polylines of different colours representing the medians of the results set. Each polyline is drawn within an area whose extremes delimit the first and third quartile of the results set.

978-1-5386-4757-8/18 $31.00 © 2018 IEEE

In figure 2 are depicted the mapping results of a $CM_{20\%}$ into a target graph of 19 chip using the four placement techniques. Each hex represents a SpiNNaker chip connected with six neighbours. The colour of the hex area points out the belonging of the neurons, mapped on the chip, to one of the eight populations of the CM. The number of synapses shared between two partial populations is highlighted with the colour intensity of the edge that connects them. The different concentration of the connections with more synapses can be appreciated qualitatively from the figures 2a to 2d and quantitatively from the figures 2e to 2h.

In figure 2a can be seen how the Näive method does not consider the connectivity but place each partial population sequentially following the polar ordering of the chip. Indeed there are many connections with a large number of synapses directed towards distant chips. This not happens in figure 2d where the Simulated Annealing can localise in a defined area all partial population with a high number of shared synapses. In figures 2e and 2h the same information can be appreciated quantitatively. The chart has a bar for each partial population. Each bar represents the overall outgoing synapses of a partial population and shows the percentage of synapses at different levels of elongation. The white line depicts the number of synapses belonging to each partial population. The partial populations are sorted in descending order according to the total number of synapses.

We can see how better methods improve the percentage of synapses at a distance of 1 chip (Green) and decrease the percentage of synapses at a distance of 4 chips (Red).

VI. CONCLUSIONS

In this paper, we described a mapping problem that involves a complex directed graph to be placed in a mesh of processors. We have modelled the mapping problem of SNN into SpiNNaker processor-mesh and split the problem into 3 phases: the expansion, clustering, and mapping. Focusing on the mapping phase, we have identified and test 4 methodologies to solve the problem. The *Naïve* method maintains the proximity of clusters but does not take into account their connectivity. The *Spectral* method uses the graph eigendecomposition to obtain a planar representation of it and perform the node association with the chip mesh through an ILP formulation. The *Scotch* method uses the Dual Recursive Bipartitioning heuristic for fast mapping of a source graph into a target graph. The *Simulated Annealing* method uses the well-known procedure to minimise a cost function.

We are redefining the cost function of the placement problem bringing it into matrix form as a function of a permutation vector. We have chosen the cortical microcircuit at different scale factors as our benchmark network, preferring it for its high connectivity and the presence of clusters. After performing several tests on the chosen benchmark network, the results highlight the superiority of the Simulated Annealing method that works natively on direct graphs. This modelling system for SNN placement problems can be adapted to other architectures such as Intel Lohi and SpiNNaker 2 for investigating new mapping techniques to be adopted for improving the usability of these emerging architectures.

In the next works, we will implement these techniques within the placement pipeline of the SpiNNaker neuromorphic architecture, to offer an alternative to the currently implemented method (Naïve) and evaluating experimentally the reduction of communications between the chips involved.

ACKNOWLEDGMENT

The research leading to these results has received funding from European Union Horizon 2020 Programme [H2020/2014-20] under grant agreements no. 720270 and no. 785907 [HBP].

REFERENCES

[1] W. Maass, "Networks of spiking neurons: the third generation of neural network models," *Neural networks*, vol. 10, no. 9, pp. 1659–1671, 1997.

[2] L. F. Abbott, "Lapicque's introduction of the integrate-and-fire model neuron (1907)," *Brain research bulletin*, vol. 50, no. 5, pp. 303–304, 1999.

[3] E. M. Izhikevich, "Simple model of spiking neurons," *IEEE Transactions on neural networks*, vol. 14, no. 6, pp. 1569–1572, 2003.

[4] S. J. e. a. Van Albada, "Full-scale simulation of a cortical microcircuit on spinnaker," in *Front. Neuroinform. Conference Abstract: Neuroinformatics*, vol. 10, 2016.

[5] T. C. e. a. Potjans, "The cell-type specific cortical microcircuit: relating structure and activity in a full-scale spiking network model," *Cerebral cortex*, vol. 24, no. 3, pp. 785–806, 2014.

[6] S. B. e. a. Furber, "The spinnaker project," *Proceedings of the IEEE*, vol. 102, no. 5, pp. 652–665, 2014.

[7] M. e. a. Davies, "Loihi: A neuromorphic manycore processor with on-chip learning," *IEEE Micro*, vol. 38, no. 1, pp. 82–99, 2018.

[8] S. e. a. Furber, "On-chip and inter-chip networks for modeling large-scale neural systems," in *Circuits and Systems, 2006. ISCAS 2006. Proceedings. 2006 IEEE International Symposium on.* IEEE, 2006, pp. 4–pp.

[9] I. e. a. Sugiarto, "Optimized task graph mapping on a many-core neuromorphic supercomputer," in *High Performance Extreme Computing Conference (HPEC), 2017 IEEE.* IEEE, 2017, pp. 1–7.

[10] G. e. a. Urgese, "Optimizing network traffic for spiking neural network simulations on densely interconnected many-core neuromorphic platforms," *IEEE Transactions on Emerging Topics in Computing*, 2016.

[11] F. Chung, "Laplacians and the cheeger inequality for directed graphs," *Annals of Combinatorics*, vol. 9, no. 1, pp. 1–19, 2005.

[12] G. Karypis and V. Kumar, "A fast and high quality multilevel scheme for partitioning irregular graphs," *SIAM Journal on scientific Computing*, vol. 20, no. 1, pp. 359–392, 1998.

[13] S. Van Dongen, "Graph clustering via a discrete uncoupling process," *SIAM Journal on Matrix Analysis and Applications*, vol. 30, no. 1, pp. 121–141, 2008.

[14] J. W. Sammon, "A nonlinear mapping for data structure analysis," *IEEE Transactions on computers*, vol. 100, no. 5, pp. 401–409, 1969.

[15] F. Pellegrini, "Static mapping by dual recursive bipartitioning of process architecture graphs," in *Scalable High-Performance Computing Conference, 1994., Proceedings of the.* IEEE, 1994, pp. 486–493.

[16] N. E. e. a. Gibbs, "A comparison of several bandwidth and profile reduction algorithms," *ACM Transactions on Mathematical Software (TOMS)*, vol. 2, no. 4, pp. 322–330, 1976.

[17] C. M. Fiduccia and R. M. Mattheyses, "A linear-time heuristic for improving network partitions," in *Papers on Twenty-five years of electronic design automation.* ACM, 1988, pp. 241–247.

[18] F. Pellegrini, "A parallelisable multi-level banded diffusion scheme for computing balanced partitions with smooth boundaries," in *European Conference on Parallel Processing.* Springer, 2007, pp. 195–204.

[19] S. e. a. Kirkpatrick, "Optimization by simulated annealing," *science*, vol. 220, no. 4598, pp. 671–680, 1983.

[20] G. e. a. Urgese, "Top-down profiling of application specific many-core neuromorphic platforms," in *Embedded Multicore/Manycore SoCs (MCSoC), 2015 IEEE 9th International Symposium on.* IEEE, 2015.

Traffic Aware Deflection Rerouting Mechanism for Mesh Network on Chip

Simi Zerine Sleeba*, John Jose†, Maurizio Palesi‡, Rekha K. James§ and M.G. Mini¶

* Dept. of Electronics Engineering, Model Engineering College, Cochin, India
† Dept. of Computer Science and Engineering, Indian Institute of Technology Guwahati, India
‡ Dept. of Electrical, Electronic and Computer Engineering, University of Catania, Italy
§ Division of Electronics Engineering, School of Engineering, CUSAT, India
¶ Dept. of Electronics Engineering, College of Engineering, Cherthala, India

simi.abie@gmail.com, johnjose@iitg.ac.in, maurizio.palesi@dieei.unict.it, rekhajames@cusat.ac.in, minimg@cectl.ac.in

Abstract—In two dimensional mesh Network on Chips (NoC), efficient routing algorithms route majority of the flits through the central routers of the network, whereas routers at the edges and corners experience relatively lesser flit flow. This in turn leads to higher traffic towards central routers than to edge and corner routers. Such uneven traffic distribution causes thermal hot-spots at the center of the chip where the load is high, and reduces the average life-time of the chip. In existing buffer-less deflection routing techniques, load balanced traffic distribution is not considered as a factor during assignment of links to mis-routed flits. Devising deflection routing techniques with greater load balancing capability is a major challenge for efficient thermal management of the chip. This paper proposes an adaptive routing mechanism that can provide a more balanced traffic profile in a deflection router based mesh NoC. Significant number of deflected flits are rerouted towards the edges/corners of the mesh, thereby reducing the load on the central routers. From evaluations, it is seen that the proposed technique reduces traffic variance compared to NoCs using baseline deflection routers. Transient temperature variation studies using Hotspot tool substantiate our findings.

Index Terms—Network on Chip, Deflection routing, Traffic rerouting, Traffic variance, Average latency

I. INTRODUCTION

With the aim of enhancing the performance of processors, multiple computational cores are integrated on a single chip and are termed as Tiled Chip Multiprocessors (TCMP) [1]. Network on Chip (NoC) is widely envisioned as the interconnect of such TCMPs. In a homogeneous TCMP, various tiles are connected using a two dimensional mesh topology where each Processing Element (PE) is connected to a dedicated router and routers are interconnected using links. Data is exchanged between tiles in the form of packets. A packet is further divided into flits (flow control units). Packets generated from a PE make multiple hops through intermediate routers and links and finally reach their destination core. Each router has input/output ports to North, South, East and West directions and also to the local core.

This research is supported in part by Department of Science and Technology(DST), Government of India vide project grant ECR/2016/000212

The first generation NoCs used input buffered routers that use store-and-forward wormhole routing technique [2]. A flit occupies a buffer in the router until it wins arbitration for a productive output port. Buffers play a major role in improving the network performance parameters, but they consume significant amount of chip power. Buffer-less deflection routers are proposed as an alternate method for achieving energy efficient on chip communication [3]. Experiments show that buffer-less routers outperform buffered ones at low to medium injection rate [3]. Due to absence of buffers, flits that fail to occupy productive output ports are deflected through available output ports of the router. In deflection routing, a flit with higher priority is allocated to an output port of its choice. Output ports obtained by other flits are determined by the flit priority, port conflict and port allocation method. Consequently, traffic due to flit deflections may either be towards the center or the edges/corners of the mesh. Majority of the productive flit movements as well as large number of flit deflections occur through the central routers causing traffic imbalance and uneven thermal distribution across the mesh. In this paper, we propose a simple logic unit in the output port allocation stage of deflection routers that reroutes deflected flits away from the center of the mesh and improves the traffic evenness across the network.

II. RELATED WORK

Most of the NoC routers adopt minimal routing techniques that focus on network performance rather than traffic balancing [4]. Due to restrictions imposed by the routing algorithm, certain regions in the network tend to have more concentration of traffic than the rest, creating an uneven traffic profile. Over the past decade, a wide variety of routing techniques for resolving network congestion have been proposed for NoCs with input buffered routers. Beginning with the Free Buffer Priority (FBP) scheme, the count of free input buffers in downstream routers is taken as a measure for adaptive selection of output ports [5]. BOFAR utilizes the history of buffer occupancy time of flits to determine congestion in downstream routers [6]. Another work introduces an aging-aware adaptive routing algorithm that routes packets along

978-1-5386-4757-8/18 $31.00 © 2018 IEEE

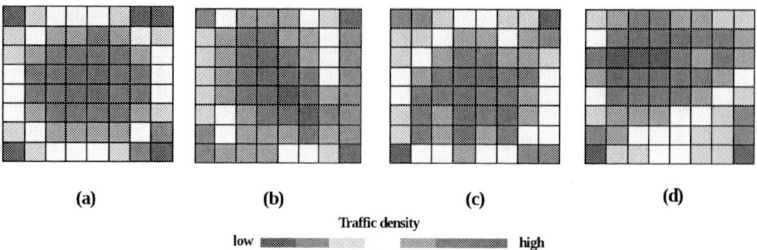

Fig. 1. Traffic Profile Graph for an 8x8 mesh NoC using (a) Uniform (b) Transpose (c) Shuffle (d) Benchmark mix traffic patterns

the paths which experience least congestion and minimum aging stress [7]. In ATDOR, a secondary network is employed for transmitting congestion messages and routing switches dynamically between XY and YX methods [8]. The problem of traffic balancing is not efficiently addressed in any of the above schemes. One major work in this direction is RCA, which uses the congestion information beyond adjacent routers to improve load balancing capability of the network [9]. GCA and GLB are also similar schemes where information on global congestion is the metric used for load balancing [10] [11]. Cool Centers follows an output port selection strategy based on prioritizing ports that route packets away from the center of the mesh [12]. This method is successful in balancing traffic flow in NoCs with input buffered routers.

Deflection routing algorithms exploit the path diversity of an NoC, hence they exhibit an inherent load balancing capability. Buffer-less deflection router BLESS performs sequential allocation of output ports to flits which are sorted in age order [3]. This routing technique reduces flit deflections to a minimum at the expense of longer critical path. CHIPPER exhibits better performance as a result of parallel port allocation and is considered to be the best buffer-less deflection router architecture [13]. A major drawback of CHIPPER is high flit deflection rate and subsequent dissipation of dynamic power. A category of minimally buffered deflection routers derived from CHIPPER reduce flit deflections without significant impact on power and area [14]. The structural limitation of the output port allocator used in all CHIPPER based router architectures causes low priority flits to be deflected randomly through vacant output ports. Hence, additional mechanisms are to be devised for achieving uniform traffic distribution for such NoCs. A recently proposed method mitigates network congestion by deflecting flits away from destination nodes that are identified as hotspots when a flit counter of the router exceeds a preset threshold [15]. A deflection routing technique which attains a uniform traffic distribution throughout the mesh NoC without compromising on the performance parameters is an open challenge.

III. MOTIVATION

In CHIPPER, productive output ports of input flits are computed using XY method and output port allocation is done using a Permutation Deflection Network (PDN). Simulations

are performed on 8x8 mesh NoC using CHIPPER with various synthetic traffic patterns and SPEC CPU 2006 benchmark mixes. The router wise traffic distribution is recorded for a simulation period of 1 million cycles with flit injection rate close to saturation point of the network. The number of flits passing through each router is used to generate a Traffic Profile Graph (TPG) [12]. Figure 1 shows the TPG for Uniform, Transpose, Shuffle and SPEC CPU 2006 benchmark mix traffic patterns. In a TPG, the 64 routers in an 8x8 mesh NoC are represented by 64 squares placed like an 8x8 matrix. The color of each square represents the amount of flit traffic moving through the corresponding router. This value is referred to as the traffic density of the router. These values are encoded by choosing an appropriate color from the color scale which shows a transition from green to red. A square with deep green color represents a router with minimum traffic passing through it. Similarly, a deep red colored square implies that the corresponding router carries heavy load.

From Figure 1, we observe that traffic density is significantly higher at the central locations of the mesh (having more red squares) than at the edges and corners in all the four cases. As seen from Figure 1(b), the uneven traffic distribution is more evident for a network intensive traffic pattern like Transpose than random patterns such as Uniform and Shuffle. Detailed flit flow analysis through the 16 central routers of the 8x8 mesh shows that 77% of it is due to flits moving in productive paths to their respective destinations and the remaining 23% is due to the deflected flits. In CHIPPER, due to unrestricted deflection schemes during port conflicts, all flits other than the highest priority flit may get deflected. For a flit whose productive port is South, assignment of East or West port is equally counter productive. Approximately 15% of these deflected flits in the central routers could be reassigned to vacant ports towards the edges/ corners of the mesh.

The 15% cases observed above show that the traffic density at the center of the mesh network could be reduced by rerouting some of the flits in each router towards the edges of the mesh without increasing the hop. In this paper, a traffic aware deflection routing mechanism that balances the load across the NoC by port reallocation of deflected flits is proposed. The additional router logic does not alter the path of flits traversing through productive directions. Hence

978-1-5386-4757-8/18 $31.00 © 2018 IEEE

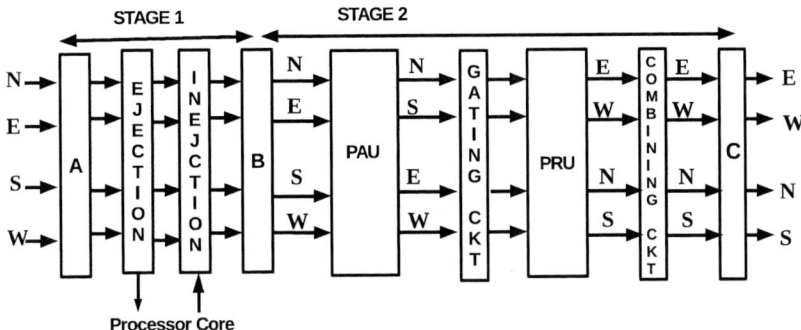

Fig. 2. Two stage pipeline diagram of the proposed router

network performance is not affected due to the proposed traffic balancing mechanism.

IV. ROUTER ARCHITECTURE

The router architecture consists of two pipeline stages as shown in Figure 2. The first stage consists of basic input functional blocks viz. ejection and injection units. The second stage consists of the output port allocation stage which is referred to as the Parallel Allocation Unit (PAU) followed by a Port Reallocation Unit (PRU) which is a newly proposed module. Each of the functional blocks in the router pipeline are explained in detail below.

A. Ejection and Injection Units

The functional blocks in the first stage of the proposed router are same as that of CHIPPER. The flits destined to the local PE are ejected out of the NoC through the Ejection Unit. This architecture supports ejection of a single flit per cycle. If there are more than one flit with the local core as destination, the flit with the highest priority is ejected while the others are deflected to neighboring routers at the end of the router pipeline. Such flits come back to the same router in subsequent cycles and compete for the ejection port. Injection of new flits from the local core is done by the Injection Unit subject to a vacancy in any of the four internal flit channels. Productive output port for the flits are computed using XY routing algorithm. After passing through the Ejection and Injection units, the flits move to register B at the end of a clock cycle.

B. Parallel Allocation Unit (PAU)

The second stage of the router pipeline begins from pipeline register B. The flits in register B are assigned output ports by the PAU on the basis of flit prioritization and port preference of each flit. The function of PAU is similar to that of PDN in CHIPPER. The PAU consists of four permuter blocks (P1, P2, P3 and P4) each having two input ports and two output ports as shown in Figure 3. Flits from the North and East internal flit channels are connected to P1 and that of South and West are connected to P2. For each input permuter block (P1 and P2), the input flit with higher priority is assigned to an

output permuter (P3 or P4) of its choice and the other input flit is assigned to the second output permuter. For example, consider two flits F1 and F2 with F1 coming through the North and F2 coming through the East internal flit channels. Let us assume that the preferred output port for F1 and F2 is South. Then, as per the flit priority (F1 > F2), F1 moves from P1 to permuter P3 to which South output port is attached. Automatically, F2 will be deflected to permuter P4 from the output of P1. Mapping of inputs to outputs in P1 and P2 is done simultaneously and the same is followed in P3 and P4. The parallel structure of permuters in PAU reduces the critical path latency of the port allocation stage.

The golden flit prioritization scheme used in CHIPPER for livelock avoidance is utilized here. The golden flit obtains desirable output ports in all the routers in its path to reach its destination without any deflection. The priority is then passed on to another flit in transit. We implement this priority mechanism so as to have a fair comparison between CHIPPER and the proposed traffic aware deflection routing method.

C. Port Reallocation Unit (PRU)

Flits from the PAU are connected to the PRU by a gating circuit as shown in Figure 2. The function of PRU is to reassign deflected flits to vacant output ports of a router. The main motive behind this is to assign a port to an already deflected flit so that the flit moves away from the center of the mesh. As shown in Figure 3, a flit, F from an output line of the PAU enters the PRU ony if all the following three conditions are satisfied.

(1) F is not assigned to a productive port

(2) The port assigned to F by the PAU will take it to a router R that is farther from the edges/corners of the mesh than the current router C (ie. C is relatively closer to the center of the mesh than R)

(3) There exists an idle output port in the current router C that will take F to a router R1 that is closer to the edges/corners of the mesh than C.

In short, the router R1 to which a flit is rerouted should be such that the minimum number of hops to reach the

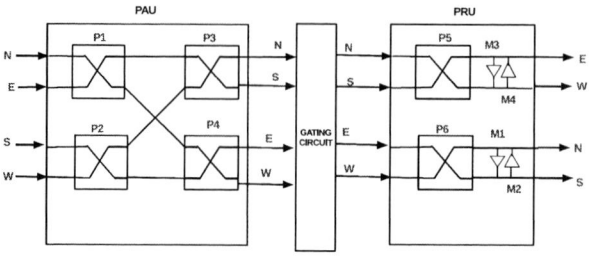

Fig. 3. Structure of Parallel Allocation Unit (PAU) and Port Reallocation Unit (PRU)

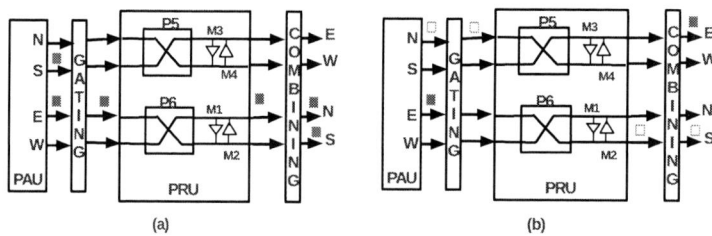

Fig. 4. Examples for port reallocation for router 50 in an 8x8 mesh network

edges/corners of the mesh from R1 is lesser than that of C and R. This conditional check and subsequent selective forwarding are done by a gating circuit.

Algorithm 1 Algorithm for PRU

Inputs : P5(Northin), P5(Southin), P6(Eastin), P6(Westin)
Outputs: P5(Eastout), P5(Westout), P6(Northout), P6(Southout)

If flit in P5(Northin) or P5(Southin)
    If flit in P5(Northin) and Enabled(P5(Eastout))
        Assign P5(Eastout) := P5(Northin)
    Else If flit in P5(Northin) and Enabled(P5(Westout))
        Assign P5(Westout) := P5(Northin)
    If flit in P5(Southin) and Enabled(P5(Westout))
        Assign P5(Westout) := P5(Southin)
    Else If flit in P5(Southin) and Enabled(P5(Eastout))
        Assign P5(Eastout):= P5(Southin)
If flit in P6(Eastin) or P6(Westin)
    If flit in P6(Eastin) and Enabled(P6(Northout))
        Assign P6(Northout) := P5(Eastin)
    Else If flit in P6(Eastin) and Enabled(P6(Southout))
        Assign P6(Southout) := P6(Eastin)
    If flit in P6(Westin) and Enabled(P6(Southout))
        Assign P6(Southout) := P6(Westin)
    Else If flit in P6(Westin) and Enabled(P6(Northout))
        Assign P6(Northout):= P6(Westin)

The PRU consists of two permuter blocks (P5, P6), each having two inputs and two outputs. P5 reallocates flits from the North and South output lines of the PAU to the East or West ports of the router if there is a flit F belonging to any of the above mentioned conditions. Similarly, P6 connects East and West outputs of PAU to North and South output ports. Algorithm 1 gives the rules for rerouting using PRU. Reallocation of flits between North and South or East and West directions are enabled using multiplexers (M1, M2, M3, M4) between output lines of P5 and P6 as shown in Figure 3. If there are no flits that satisfy the above conditions, the gating circuit bypasses the flits over the PRU by keeping the timing

constraints. The combining circuit multiplexes output lines of the PRU with corresponding output lines from the PAU.

The functionality of PRU is explained in detail with an example. Consider an 8x8 mesh topology with routers numbered from 0 (bottom left) to 63 (top right). Consider router number 50 whose South and East output ports are directed towards the center of the mesh whereas the North and West ports are towards the edges. Figure 4(a) shows PAU and PRU of this router. We assume that there are two flits at the output of PAU. The PAU assigns the East port to the green flit and South port to the red flit. Assume that East is a non-productive direction for the green flit. Hence it is ready to be deflected. At the same time, North and West ports are empty. Figure 4(a) shows that the green flit is rerouted towards the vacant North port by the permuter P6 in the PRU. Since the South port is a productive port for the red flit, it bypasses the PRU. Forwarding of green flit to the PRU and bypassing of the red flit over the PRU is done by the gating circuit by proper condition check. Figure 4(b) shows an example of a deflected flit being reassigned from North port to South port by enabling multiplexer M1. In this example, permuters P5 and P6 of the PRU are not used.

V. EXPERIMENTAL METHODOLOGY

We use an open source NoC simulator, Booksim 2.0 [16] to model the deflection router based NoC with the proposed architecture. The router pipeline is modeled with two cycle delay and the routing algorithm is implemented as described in Section IV. To compare the results, the basic CHIPPER based NoC is also modeled in Booksim. Simulations are conducted using typical synthetic traffic patterns as well as network traces generated by running multi-programmed workload mixes from

978-1-5386-4757-8/18 $31.00 © 2018 IEEE

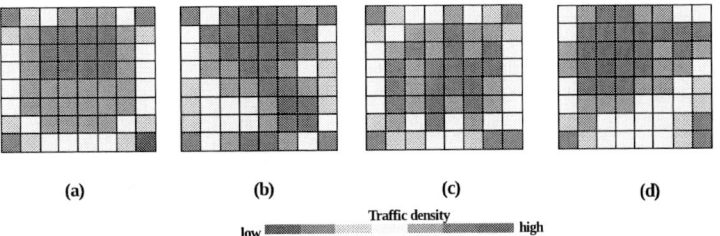

Fig. 5. Traffic Profile Graph for an 8x8 mesh NoC with traffic aware routing using (a) Uniform (b) Transpose (c) Shuffle (d) SPEC CPU 2006 benchmark mix traffic patterns.

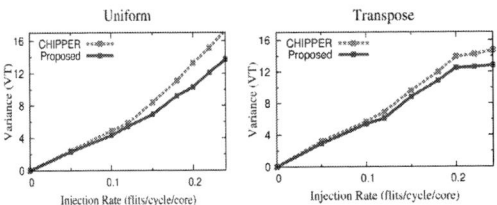

Fig. 6. Traffic Variance for Synthetic Traffic Patterns

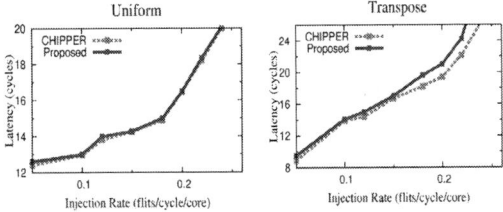

Fig. 7. Average Latency for Synthetic Traffic Patterns

SPEC CPU 2006 benchmark suite [17] on Gem5 simulator [18].

VI. EXPERIMENTAL RESULTS AND ANALYSIS

The Traffic Profile Graph for Uniform, Transpose, Shuffle and benchmark application traffic patterns for an 8x8 mesh NoC with the proposed method is depicted in Figure 5. Compared to Figure 1, we see that the port reallocation strategy used in the proposed method helps to reduce the traffic density at the center of the mesh by rerouting them to the edge/corner routers. From the Figure 5, we see that squares at the center change from red to orange and yellow colors. Similarly, the color of squares at the edges change from deep green to pale green or yellow.

A. Traffic Variance

In order to measure the traffic load and uniformity of traffic distribution across various routers in an NoC, we introduce a parameter known as traffic variance which is calculated using the formula,

$$Traffic\ Variance, V_T = \frac{\sum_{i=1}^{64} mod(A_v - T_i)}{64} \quad (1)$$

where

$$Average, A_v = \frac{\sum_{i=1}^{64}(T_i)}{64}$$

T_i is the traffic density of the ith square in the TPG.

In an NoC, lower value of traffic variance signifies higher uniformity in traffic distribution. Using Equation 1, the traffic variance for 8x8 mesh NoCs using CHIPPER and the proposed router for typical synthetic traffic patterns are calculated for various network injection rates. Figure 6 shows the graphs for uniform and transpose traffic patterns. The proposed re-routing scheme confirms lower variance compared to CHIPPER for all the cases. The lowest variance shown is 26% for uniform pattern at saturation injection rate which is approximately 0.2. Transpose pattern represents network intensive traffic with specific packet destinations. Under transpose pattern, very few flits satisfy all the three conditions for port reallocation. This accounts for the minor reduction in variance compared to uniform traffic.

Figure 7 shows the comparison of average latency for uniform and transpose traffic patterns for 8x8 mesh NoC using CHIPPER and proposed router. As flits progressing in productive paths are unaffected by the traffic re-routing mechanism, there is only negligible increase of 0.05% in average latency for the proposed mechanism. We also find that the deflections per flit reduces up to 8% since edge routers are lightly loaded and flits encounter lesser port conflicts in them.

B. Real Applications

Applications from the SPEC CPU 2006 benchmark suite are categorised as low, medium or high MPKI (misses per kilo instructions) based on the rate at which they inject packets into the network. Table I lists the applications from each category which are used in our simulations. Each simulation uses network traces generated from a 64-core TCMP running one SPEC benchmark application per core as given in the Table. Figure 8 shows the graph of normalised variance and latency of the proposed NoC with respect to CHIPPER for various benchmark mixes. Although the proposed method delivers promising results for all applications, significant reduction in

978-1-5386-4757-8/18 $31.00 © 2018 IEEE

29

	Category	Benchmark Applications
Low MPKI	C1	calculix, h264ref, gromacs, gobmk
Medium MPKI	C2	gcc, bwaves, bzip2
High MPKI	C3	leslie3d, hmmer

TABLE I
APPLICATIONS OF VARIOUS NETWORK INJECTION INTENSITY IN
SPEC CPU BENCHMARK SUITE

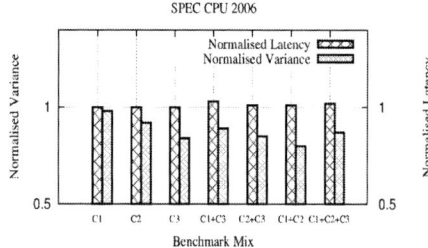

Fig. 8. Normalised variance and latency w.r.t. CHIPPER for 8x8 mesh NoC using mixes from SPEC CPU 2006 benchmark suite

traffic variance is noted for high MPKI applications Since they inject more number of flits into the network, the probability of deflections in the routers also increases. Deflected flits that satisfy the conditions for reallocation are forwarded to the edges/corner routers of the NoC, due to which traffic variance is low. It is also observed that the average latency of the proposed technique is equivalent to the normal value of CHIPPER for all application mixes.

C. Thermal Analysis

Thermal distribution across the NoC is analyzed using Hotspot 6.0 tool [19]. Dynamic power dissipation of various routers in an 8x8 mesh NoC due to varying load is extracted by modeling our router architecture in Orion 2.0, a power estimation tool [20]. Using these power traces obtained from Orion, Hotspot estimates the transient temperature variation due to the flit flow load across the 64 routers of the NoC. Simulation results confirm that there is a temperature reduction of upto 3^oK in the 16 central routers for real workloads using our proposed scheme compared to that of baseline CHIPPER.

D. Hardware Synthesis

We implement Verilog models of the proposed router and CHIPPER and synthesize using Synopsys Design Compiler with 65nm CMOS library. Router delay is the time taken by a flit to move from its input to output port through various functional units. The first stage of CHIPPER and the proposed router have the same delay because of similar functional units in both architectures. The output stage of CHIPPER consists mainly of a port allocator whereas the proposed router includes a port allocator (PAU) followed by a rerouting logic (PRU). Since both the architectures use the same port allocation logic, additional delay of 18% due to the PRU occurs in the output stage of the proposed router. Area and static power consumed by the control logic of our router are 4% and 7% higher than CHIPPER. The hardware overhead and the extended critical path of the proposed technique are justified by the significant

reduction in traffic variance that it promises. However our router pipeline frequency is same as that of CHIPPER as latency of the first stage dominates over the second stage.

VII. CONCLUSION

An adaptive deflection router which helps to achieve a uniform traffic distribution within the mesh NoC is proposed. A logic block is introduced into the router pipeline that performs output port reallocation of flits which are assigned to unproductive directions towards the center of the mesh. The merit of the proposed traffic aware routing mechanism and the corresponding thermal variation effect are quantitatively justified by the reduction in traffic variance parameter in comparison with a basic deflection router based NoC.

REFERENCES

[1] J. Balfour and W. J. Dally, "Design tradeoffs for tiled cmp on-chip networks," in *Annual International Conference on Supercomputing*, 2006, pp. 187–198.
[2] W. Dally and B. Towles, *Principles and Practices of Interconnection Networks*. USA: Morgan Kaufmann Publishers Inc., 2003.
[3] T. Moscibroda and O. Mutlu, "A Case for Bufferless Routing in On-Chip Networks," in *ISCA*, 2009, pp. 196–207.
[4] G. Chiu, "The Odd-Even Turn Model for Adaptive Routing," *IEEE Transactions on Parallel and Distributed Systems*, vol. 11, pp. 729–738, 2000.
[5] W. Dally, "Virtual-Channel Flow Control," *IEEE Transactions on Parallel and Distributed Systems.*, vol. 3, no. 2, pp. 194–205, 1992.
[6] J. Jose et. al, "BOFAR : Buffer Occupancy Factor based Adaptive Router for mesh NoC," in *International workshop on Network on Chip Architectures*, 2011, pp. 23–28.
[7] K. Bhardwaj et. al, "Towards Graceful Aging Degradation in NoCs through an Adaptive Routing Algorithm," in *Design Automation Conference*, 2012, pp. 382–391.
[8] R. Manevich et al., "A cost-effective centralized adaptive routing for networks-on-chip," in *DSD*, 2011, pp. 39–46.
[9] P. Gratz et. al, "Regional Congestion Awareness for Load Balance in Networks-on-Chip," in *International Symposium on High-Performance Computer Architecture*, 2008, pp. 203–215.
[10] M. Ramakrishna et. al, "GCA: Global Congestion Awareness for Load Balance in Networks-on-Chip," in *NoCS*, 2013, pp. 21–24.
[11] M. Ebrahimi et. al, "GLB - Efficient Global Load Balancing Method for Moderating Congestion in On-Chip Networks," in *International Symposium on Reconfigurable Communication-centric Systems-on-Chip*, 2012, pp. 1–5.
[12] J. Jose et. al, "An Energy Efficient Load Balancing Selection Strategy for Adaptive NoC Routers," in *International workshop on Network on Chip Architectures*, 2014, pp. 31–36.
[13] C. Fallin et al., "CHIPPER: A Low Complexity Bufferless Deflection Router," in *HPCA*, 2011, pp. 144–155.
[14] C. Fallin, G. Nazario et al., "MinBD: Minimally-Buffered Deflection Routing for Energy-Efficient Interconnect," in *NOCS*, 2012, pp. 1–10.
[15] R. Raj R.S. et. al, "Implementation and analysis of hotspot mitigation in mesh nocs by cost-effective deflection routing technique," in *VLSI-SoC*, 2017, pp. 1–6.
[16] N. Jiang et al., "Booksim 2.0 User's Guide." http://nocs.stanford.edu.
[17] J. Henning., "SPEC CPU 2006 Benchmark Descriptions," in *SIGARCH Computer Architecture News*, 2006.
[18] N. Binkert et al., "The gem5 simulator," *SIGARCH Computer Architecture News*, vol. 39, no. 2, pp. 1–7, 2011.
[19] S. Sharifi et al., "Hybrid dynamic energy and thermal management in heterogeneous embedded multiprocessor socs," in *ASP-DAC*, 2010, pp. 873–78.
[20] A. B. Kahng et al., "Orion 2.0: A Fast and Accurate NoC Power and Area Model for Early Stage Design Space Exploration." *IEEE Transactions on VLSI.*, vol. 20, no. 1, pp. 191–196, 2012.

978-1-5386-4757-8/18 $31.00 © 2018 IEEE

A Parallel Hardware Architecture For Quantum Annealing Algorithm Acceleration

Evelina Forno*, Andrea Acquaviva*, Yuki Kobayashi[†], Enrico Macii*, and Gianvito Urgese*
* Politecnico di Torino, Torino, Italy, 0039 011 090 7042. Email: gianvito.urgese@polito.it
[†] NEC Corporation, Kawasaki, Japan. Email: y-kobayashi@hq.jp.nec.com

Abstract—Quantum Annealing (QA) is an emerging technique, derived from Simulated Annealing, providing metaheuristics for multivariable optimisation problems. Studies have shown that it can be applied to solve NP-hard problems with faster convergence and better quality of result than other traditional heuristics, with potential applications in a variety of fields, from transport logistics to circuit synthesis and optimisation. In this paper, we present a hardware architecture implementing a QA-based solver for the Multidimensional Knapsack Problem, designed to improve the performance of the algorithm by exploiting parallelised computation. We synthesised the architecture using as a target an Altera FPGA board and simulated the execution for solving a set of benchmarks available in the literature. Simulation results show that the proposed implementation is about 100 times faster than a single-thread general-purpose CPU without impact on the accuracy of the solution.

I. INTRODUCTION

Optimization problems can be encountered in many fields of science and technology, from the synthesis of electronic circuits (Boolean satisfiability problems) to transportation and location logistics (Vehicle routing problem). In most cases, to solve such problems means to find the global optimum of a multivariable function, while fulfilling a set of constraints. Global optimization problems are NP-hard in complexity; as such, finding the exact solution is often unfeasible in terms of computation time. However, many heuristic methods have been developed to provide good quality, approximate solutions in a short time.

The list of available heuristic algorithms is long, and new variants are being developed every year. Among the most well-established [1] we find *Simulated Annealing* (SA) [2]. Inspired by the physical process of thermal annealing in materials science, SA is controlled by tuning the value of a temperature parameter T over a given period of time. The algorithm randomly generates moves in the search space that are always accepted when the cost function H is improved; moves that would result in the cost increasing are accepted with a probability $e^{\frac{\Delta H}{T}}$. SA is well-tested, efficient and robust, but requires rather long computation times.

Simulated Annealing is widely used in the VLSI field for the generation of connection paths, placement and other optimization. For example, it has been successfully employed in solving floorplanning [3] and placement [4] problems, which are classified as NP, obtaining better results than other types of heuristics such as PSO and Ant Colony. An outstanding example is the use of SA in placement algorithms within commercial software such as Quartus II [5] for FPGA design optimization. Research within the field has led to several improvements to SA, especially in parallelization efforts, which yield better results and linear speedups with respect to the classic algorithm while trying to balance the added complexity in synchronizing several solver processes.

Quantum Annealing is an emerging technique derived from Simulated Annealing. Previous papers have demonstrated that for some problems (such as the Traveling Salesman Problem (TSP) [6], the Multidimensional Knapsack Problem (MKP) [7], and the Ising spin glass [8]), QA has a faster convergence in terms of simulation steps, providing results of improved quality.

However, Quantum Annealing is even more computationally expensive than SA when executed on standard PCs. As such, there has been interest from several research teams in developing parallel hardware architectures for accelerating SA on FPGAs and GPU. Since SA is an inherently sequential algorithm, workarounds are necessary to exploit concurrency; solutions have been proposed both on CPU [4] [5] and GPU [9] which run independent SA solvers in separate threads, then choose the best solution reached among all threads. A similar approach has been attempted on FPGA [10], reporting success for relatively small problems (1024-bit, corresponding to a 32-city TSP problem). FPGA acceleration of Monte Carlo solvers, which apply concepts compatible with SA, has found great success in physical simulations of the nearest-neighbor Ising spin glass such as the *Janus II* computer [11]; however, most NP problems require higher levels of connectivity.

Regarding the Multidimensional Knapsack Problem, implementations have been made on GPU for heuristic solvers [12] [13], reporting speedup over parallel CPU solutions. Exact solvers [14] were realized on GPU for the 0-1 Knapsack Problem, but literature suggests that exact MKP solvers may be too demanding for current GPU architectures.

In this paper, we propose the design of a hardware architecture implementing a solver for the Multidimensional Knapsack Problem using *Quantum Annealing* (QA). We analyzed the behavior of the Quantum Annealing algorithm and applied modifications to improve its performance. We also parallelised computation with a multi-core architecture and applied architectural and functional optimizations to reduce calculation times. We described the QA solver in a High Level Synthesis

978-1-5386-4757-8/18 $31.00 © 2018 IEEE

language and synthesized it with an *Altera Stratix V FPGA* as target. The results in this paper are derived from RTL simulation.

The rest of the paper is organized as follows: in Section II we provide a mathematical description of the Quantum Annealing algorithm. In Section IV we discussed our proposed architecture, whereas we reported the performances achieved by our accelerator in Section IV. Finally we give some conclusions (Section V).

II. BACKGROUND

Quantum Annealing is a metaheuristic algorithm inspired by classical Simulated Annealing. Instead of applying a thermal gradient to the system, it applies a slowly diminishing *transverse field*.

The quantum annealing process can be simulated in a traditional computer using stochastic techniques like the Monte Carlo method. This algorithm involves an adaptation of the classical Metropolis-Hastings algorithm [8] to step out of local optimum solutions.

What sets apart Quantum Annealing from Simulated Annealing is the emulation of the *quantum tunneling* effect for escaping local minima. The key parameter Γ, which indicates the strength of the transverse field, represents the quantum tunneling width and determines the radius of local search. At first, the neighborhood comprises the whole search space; during the annealing this radius is gradually reduced.

A benchmark for QA [8] is the *Ising spin glass*, a mathematical model of ferromagnetism used in statistical mechanics. This model consists of a system of up/down spins organized in a graph. Each spin has a given radius of neighbor spins that it is allowed to interact with. The model is described by the Hamiltonian reported in equation 1.

$$H_c = - \sum_{\langle i,j \rangle} J_{ij} s_i s_j \tag{1}$$

Where the N spins s_i can take the values ± 1. The interaction between spins s_i and s_j on lattice sites i and j is described by the exchange coupling J_{ij}. $\langle i,j \rangle$ means that i and j are neighbor spins; the radius of neighborhood depends on the chosen model. When spins are not neighbors their interaction is $J_{ij} = 0$, therefore such pairs do not contribute to the Hamiltonian.

The Monte Carlo simulation of the Ising spin glass consists of iterating over every spin and performing an update, i.e., deciding whether or not to flip each spin based on the status of its neighbors and the strength of its interaction with them. A Monte Carlo step is concluded when all the spins in the systems have been updated.

To perform QA of Ising spin glasses, an additional term is added to the Hamiltonian by applying a transverse magnetic field Γ, as shown in equation 2.

$$H_q = - \sum_{i<j}^{N} J_{ij} \sigma_i \sigma_j - \Gamma(t) \sum_{i}^{N} \sigma_i \tag{2}$$

Γ starts out at a high value and is gradually reduced to zero during the annealing.

In computer-simulated QA [8] [7] [6], the quantum effect is simulated by mapping the partition function of the quantum Ising model to that of a classical Ising model in one higher dimension, called *imaginary time* dimension or *Trotter* dimension (as the Suzuki-Trotter expansion is used to perform this mapping). This means that the system is simulated simultaneously in R different iterations or *replicas* (Fig. 1), which start out completely independent but have a correlation factor J_t that grows in time, forcing them to converge to a single solution at the end of annealing. J_t is calculated each MCS as a function of Γ, by the formula in equation 3.

$$J_t = -\frac{1}{2} log(tanh(\Gamma)) \tag{3}$$

QA is not just SA with R copies running in parallel. Normally, SA is only able to pass to a neighboring state on the energy landscape in one step, by thermal transitions. However, by adding the J_t coupling, QA is able to tunnel through energy barriers, avoiding local maxima, and exploring the state space more effectively. This can explain the faster convergence of Quantum Annealing.

In particular, thermal transition probability is proportional to $e^{\frac{-\Delta}{k_B T}}$ (where Δ is the height of the energy barrier, k_B is the Boltzmann constant, and T the annealing temperature). This probability is dominated by the height Δ of the barrier, which means it is difficult to get out of a very deep well of local minimum by means of thermal fluctuations. However, it has been demonstrated [15] that the quantum tunneling probability through the same barrier is proportional to $e^{\frac{-\sqrt{\Delta} w}{\Gamma}}$.

The tunneling probability depends not only on the height Δ, but also on the width w of the energy barrier. This means that QA shows significant advantage on problems where the energy landscape presents a high amount of perturbation with many high and thin barriers ($w \ll \Delta$). Indeed, the search of the ground state for an Ising spin glass model is one of these problems: since many NP problems can be formulated through

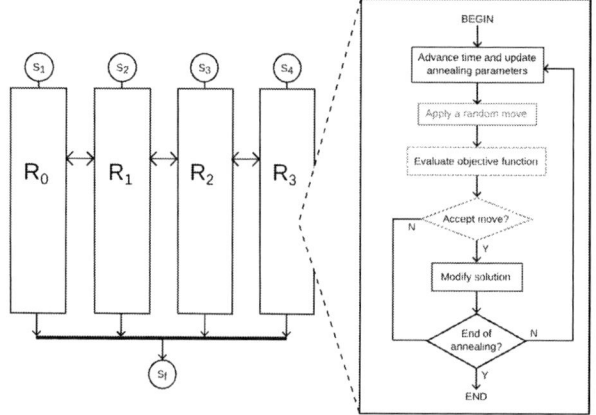

Fig. 1: The Quantum Annealing algorithm. On the left, a visual representation of replicas as a series of parallel threads exchanging information with neighbors about their local solution. On the right, the flowchart of the algorithm executed within each replica: highlighted in red are the portions of the program that can be rewritten to fit different problem types (e.g., to allow only legal moves within the constraint system). The red dashed line indicates parts that stay the same, save for the problem Hamiltonian.

the Ising model [16], we can apply QA to them and expect favorable results.

III. IMPLEMENTATION

The architecture (shown in Fig. 2) is composed of an array of R processors, each representing a Trotter replica for Quantum Annealing. Each replica shares its current *itemvector* with its neighbors.

There is also a *controller* module that ensures synchronization of the replicas during one Monte Carlo step (MCS). It fetches the J_t value for the current MCS from memory, then enables the replicas to allow calculation of the next step. It also receives the current *total profit* from each replica and detects when a new optimum has been found.

Finally, we have a *Restricted Quantum Annealing (RQA) engine* that receives the item vectors from the processors and calculates the frequency of each item across all solutions. It outputs a binary vector describing the locked items that replicas are no longer allowed to change. In the following we will examine each module in detail.

We described this circuit behaviorally in SystemC and perform high-level synthesis with the NEC *CyberWorkBench HLS compiler* (CWB) [17], exporting the components to an RTL format with the same procedure described in [18].

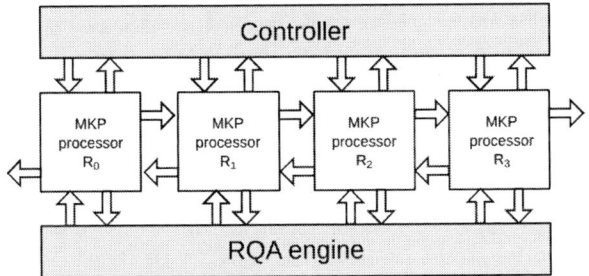

Fig. 2: Block diagram of the FPGA architecture

A. MKP Processor

The basic structure of the MKP processor is the same for all replicas. The core of this processor is a Finite State Machine that executes the operations for one Monte Carlo step of Quantum Annealing. Every MKP processor stores a copy of the problem data in local registers.

The processor is a Finite State Machine of 23 states for the largest problem (30x500); because of a few branches in the algorithm, the average latency due to pipelining is lower than 23 cycles. However, since each random number generation attempt using the LFSR introduces an extra cycle of latency, the overall latency of this module is *not deterministic*.

Indeed, a key problem of implementing a stochastic algorithm on FPGA is the quality of the Random Number Generator (RNG). We use a simple 32-bit LFSR which provides good pseudorandom performance. From this LFSR we select up to 9 bits as an item identifier (*itemRNG*) and 16 bits for the Metropolis random number (*metroRNG*). When the MKP solver needs a random number, it enables the LFSR and waits for the next clock cycle. Since the LFSR is a synchronous circuit, it necessarily introduces latency.

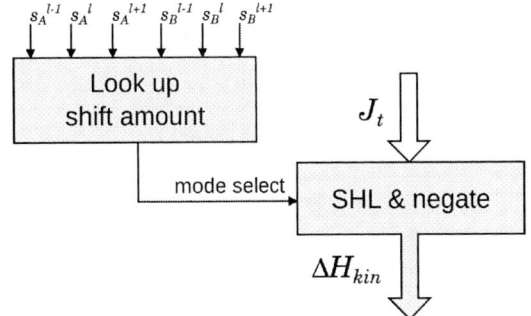

Fig. 3: Block diagram of the J_t multiplication stage

Most of the necessary instructions in the process datapath are adders, subtractors, and comparators. However, when we enter a Metropolis attempt to swap item A with item B, the calculation of the quantum portion of the Hamiltonian:

$$\Delta H_{kin} = J_t \cdot ((s_A^{l-1} + s_A^{l+1}) \cdot 2s_A^l + (s_B^{l-1} + s_B^{l+1}) \cdot 2s_B^l) \quad (4)$$

would introduce at least one 16-bit multiplier. We observe that the result of the right hand side parenthesis only has a few discrete possible values: $-8, -4, 0, +4, +8$. Then, the ΔH_{kin} calculation can be accomplished by using exclusively LUTs and shifters, as shown in Fig. 3. This saves a considerable amount of area and reduces the critical path.

The calculation of the exponential $e^{-\Delta H}$ for the Metropolis trial is also prohibitively expensive in FPGA, so we implement it with a LUT indexed by ΔH_{kin}. We experimentally determined that a high precision is not necessary for this operation and the exponential LUT only needs 24 entries of 16 bits.

When it is necessary to evaluate ΔH_{kin}, each replica needs to have a stable copy of the item vectors from its neighbors, and these vectors should be from the same annealing step that the replica is currently in. FPGA implementations of the Ising model [19] solve this problem by partitioning the spins into two groups and sharing each spin unit between two neighboring spins that are processed in separate clock cycles.

From simulation statistics we determine that calculation of ΔH_{kin} is not performed in 98-99% of annealing steps within a simulation. Then, the replicas can indeed work in parallel most of the time. In our final implementation we enable all the replicas at once; because of the low granularity of our moves, a replica can use a neighbor's result from the previous MCS, accepting a possible error of at most 2 bits. We confirm that the quality of result is equivalent to the one with the partitioned replicas. By allowing all replicas to process at the same time, we improve overall latency by about 50%.

The processor core implements the QA algorithm for MKP proposed by Bergé *et al.* [7]. We apply two modifications:

1) The mutation proposed in [7] is to try exchanging an item a outside the bag with an item b inside the bag. If the exchange is not possible, "step back" by removing item b from the bag. After making several trials, we observed that by avoiding to "step back", we are able to obtain better results.

2) The value of J_t plays a paramount role in the annealing. However, no prescription for it is given in [7]. The available benchmark problems present a very broad range

978-1-5386-4757-8/18 $31.00 © 2018 IEEE   33

$(10 - 1000)$ of possible values for the item prices, and therefore a broad range of possible values for the potential energy ΔH_{pot}. Since stepping out of the local optimum depends on the value of $\Delta H = \Delta H_{pot} + \Delta H_{kin}$, it is evident that ΔH_{kin} must be roughly of the same order of magnitude as ΔH_{pot} for the Metropolis dynamics to work. ΔH_{kin} is directly proportional to J_t, therefore we want J_t to be of the same order of magnitude as the range of item prices. One possible way to change the value of J_t is to change the value of Γ_0: however, modifying Γ_0 also influences the rate of growth of J_t. By trial and error we identified the ideal rate of growth as that corresponding to $\Gamma_0 = 3.0$, as lower values cause the J_t interaction to spike to high values too early in the annealing (preventing the system to explore the state space for most of the annealing), while high values make the growth of J_t too slow (greatly reducing the quantum tunneling effect). The method we settle on is to "amplify" J_t to fit the given problem set: before the annealing, we scan the matrix of item prices and calculate the average value. Then we use that to multiply J_t throughout the annealing. This allows us to improve results significantly, especially in problems that have high values for prices, like the ones in the Chu-Beasley benchmark.

B. Controller

The role of the controller is to keep replicas synchronized over the same Monte Carlo step. It stores the 16-bit values for J_t in a memory block of τ entries. When all the replicas are done updating, the controller disables them and fetches the next J_t value. Then the replicas are enabled and a new MCS begins. Once all J_t values have been read it raises the *QA_done* output signal and stops the annealing.

C. RQA Engine

The Restricted Quantum Annealing (RQA) engine's role is to keep track of the frequency of appearance of each item across solutions. We accomplish this by means of a SWAR algorithm for popcount (or Hamming weight counter), which is essentially a series of $\log(R) + 1$ sum, right shift and bit masking steps.

For each item i, the input to the popcount is built out of a vertical slice of all the replicas' item vectors, taking from each only the ith bit (as highlighted in red in Fig. 4). The result of the popcount is stored in a *frequency vector* at position i. If the frequency is greater than the RQA *blocking frequency*, the ith bit of the output signal *lockedItems* is set.

Using popcount lets us avoid computing long sums with the 500-position item vectors of the hardest benchmark. Every vertical slice we build is R bits long, which is generally much shorter than the item vectors.

Additionally, we can explore the design space for this component by performing loop unrolling to control the quantity of items processed at the same time and the pipeline depth of the frequency counter. We found that the smallest-area, smallest-delay, longest-latency implementation is the most efficient: we process only one item at a time, and we allow CWB's

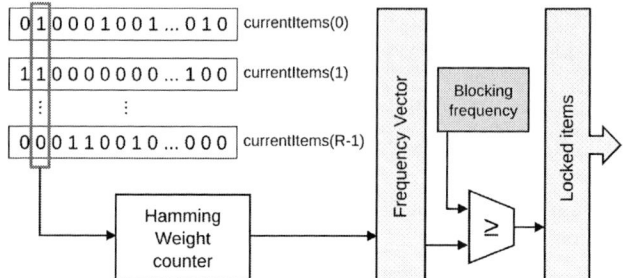

Fig. 4: Block diagram of the RQA engine

automatic scheduling to reduce the delay as much as possible, resulting in a $(\log(R) + 1)$-cycle latency.

Although the replicas don't have access to the most updated version of the *lockedItems* vector at all times, we verified in simulation that this minimized implementation has no adverse effects on the speed of convergence of RQA or the quality of result. Then, we can add RQA to the system with negligible impact on area and maximum frequency.

IV. RESULTS AND DISCUSSIONS

We performed RTL synthesis in CWB with the Altera Stratix V (5SGXEA7N2F45C2) FPGA as target. We then performed RTL simulation of the resulting Verilog files to extract the timing performance results reported in this section. The Verilog files were then input to Altera Quartus Prime for logic synthesis.

A. Area and Critical Path

In Tab. I are reported the logic synthesis results for a system with $R = 16$, $\tau = 1000000$, 500 items and 30 constraints.

The OR30x500 family of benchmarks is the largest one available to us and the one we considered for final implementation. Meanwhile, the block memory occupation also depends on τ. We synthesized up to 1 million steps, which is an appropriate simulation time for the 30x500 problem instances.

The total area of logic utilization grows linearly with the number of replicas. This is to be expected since every replica added corresponds to a new processor core in the architecture.

Examining the growth of area when varying the number of constraints and items shows that area also grows linearly with the problem size.

We compared the maximum frequency estimated by Quartus Prime over a wide range of synthesis results corresponding to various combinations of number of replicas, number of items, and number of constraints.

From Fig. 5, it is evident that the worst-case maximum frequency is decreasing as the area grows. We can identify at least two main reasons for this. i) combinational paths become

Logic utilization	147,236 / 234,720 (63 %)
Total registers	157782
Total block memory bits	18,457,600 / 52,428,800 (35 %)
Maximum frequency	164 MHz
Total Thermal Power Dissipation	8.559 mW

TABLE I: Results of QA hardware architecture logic synthesis

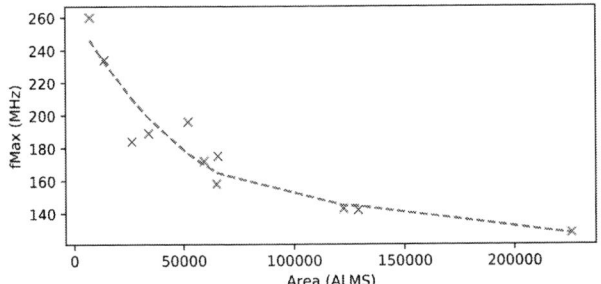

Fig. 5: Maximum frequency obtained by Quartus synthesis for varying hardware size.

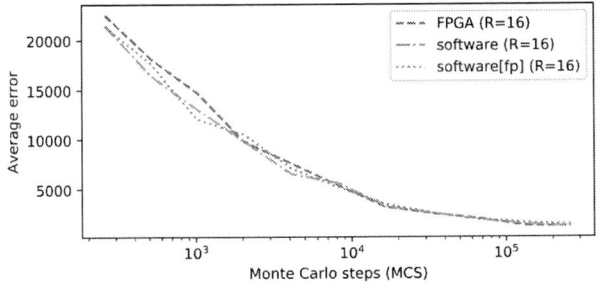

Fig. 6: Comparison of results from the FPGA and software versions of the algorithm for the benchmark OR30x500-0.25, R = 16. Quality of result is not lost in the transition from floating point (fp) to fixed point data representation.

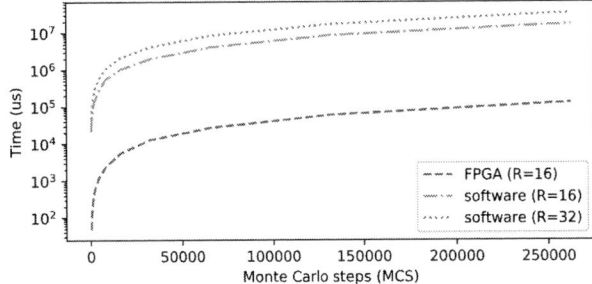

Fig. 7: Execution times for the FPGA and software versions for the benchmark OR30x500-0.25.

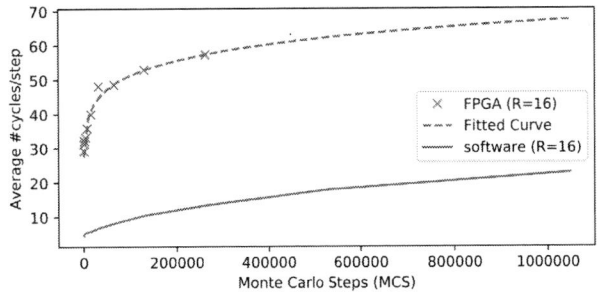

Fig. 8: Comparison of the growth in number of random trials per step as the annealing time increases, benchmark problem OR30x500-0.25. The regression function used is $f(\lambda) \approx 13.17\lambda^{0.12}$.

longer and slower as the width of parameters grows, especially of the item vector. ii) interconnections also become longer and more complex, causing increased delays. In practice, increasing the number of replicas or the size of the problem makes the system slower.

B. Quality of results compared to CPU version

We created a bit-compatible integer model in software to estimate performance of an FPGA version ahead of implementation. As shown in Fig. 6, the behavior is completely coherent with the floating point version, displaying similar convergence across a wide range of τ.

Then, it is fair to compare the behavior of the FPGA implementation (in RTL simulation) with the floating-point software version of the algorithm.

C. QA-HW vs QA-CPU

The execution time for different values of τ is charted in Fig. 7. At the maximum frequency of 164 MHz, the FPGA solution is much faster than the software at executing the same number of Monte Carlo steps. However, in the FPGA version the number of cycles per step appears to increase quite strongly as τ increases; while the speedup over software is about $350\times$ at $\tau = 2048$, at $\tau = 32768$ it is only $210\times$, and this advantage will probably continue to decrease for greater values of τ. This problem may be attributed to the LFSR's limited ability to generate an adequate number.

How much does the RNG problem affect the FPGA latency? We compared the average latency per step of the FPGA implementation with the average number of RNG trials in the software version. We suppose that the two parameters

are directly proportional as the FPGA latency per step is predominantly determined by the number of LFSR trials, with little (and generally constant) overhead from the rest of the algorithm. Fig. 8 shows how the latency due to the RNG trials, after a sharp initial growth, settles into a slowly increasing curve, compatible with that encountered in the software simulation.

Then, we expect that the FPGA implementation's speedup will maintain a reasonable advantage ($150 - 180\times$ faster) over the software version even as we increase the annealing time.

Finally, in Fig. 9 we show the speed of convergence to result for the two versions of the algorithm. It is clear that the hardware version can produce a similar quality of result as the software version in less computation time.

Our FPGA implementation appears to compare favorably with GPU results reported in literature; the parallelised QA solver reports lower computation time than the Ant Colony Optimization on GPU from [12] (408 ms for the largest instance, rather than ~ 10000 ms). We reserve comparison

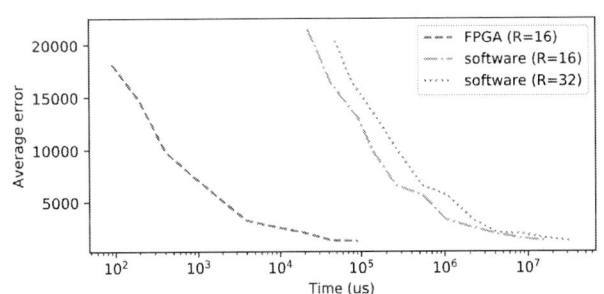

Fig. 9: Speed of convergence to result for the software and FPGA versions, benchmark problem OR30x500-0.25.

978-1-5386-4757-8/18 $31.00 © 2018 IEEE

with a GPU implementation of QA for future work.

Though synthesis provides limited information on the power consumption, it is still possible to make an optimistic estimation of the energy savings. The software version ran on a computer with an Intel i7 920 CPU; it has a maximum I_{cc} per core equal to 145 A and a typical associated V_{cc} per core of 0.131 V. Assuming our single-thread software ran on a single core at full load, that would mean an instantaneous power of 18.995 W; since the program ran for 17.319 s, the estimated energy consumption is about 329 J. Meanwhile, the FPGA energy consumption is around 8.559 W $\times$ 0.108 s = 0.924 J, leading to estimated energy savings of $\sim 350\times$.

V. CONCLUSION

In this paper we propose a parallel architecture for the solution of the Multidimensional Knapsack Problem using a Quantum Annealing solver. We implemented our hardware architecture in a High Level Synthesis language and synthesized on an FPGA target. The parametric design instantiates multiple computation cores, synchronized by a Controller module that regulates the annealing process. We also implemented an optional module for computing a Restricted Quantum Annealing solution, allowing to improve the quality of result with a negligible cost in terms of area and latency.

Simulation shows that our QA solver provides the same quality of result as a floating point software version, while outperforming a single-thread CPU. We can demonstrate analytically that the QA solver is at least $150\times$ faster than software. In addition, synthesis reports show that the implementation has a reasonable logic utilization even for the largest problems.

Future works will focus in further improving and testing the FPGA architecture. As it is, the memory containing the problem parameters is implemented in the FPGA logic elements; the block memory utilization can be improved by offloading the data to external memory. We plan to further improve the architecture and continue with the placement on an FPGA board to evaluate the real performance of the proposed accelerator.

We will also develop a CPU-GPU multithread implementation of the QA solver in order to compare the effective speedup and power savings for the FPGA version.

ACKNOWLEDGMENTS

The HLS compiler and the technical support were provided by NEC Corporation, Japan.

REFERENCES

[1] Beheshti and Shamsuddin. "A Review of Population-based Meta-Heuristic Algorithm". In: *Int. J. Advance. Soft Comput. Appl., Vol. 5, No. 1, March 2013* (2013).

[2] S. Kirkpatrick, C. D. Gelatt, and M. P. Vecchi. "Optimization by Simulated Annealing". In: *Science* 220.4598 (1983), pp. 671–680. ISSN: 00368075, 10959203. URL: http://www.jstor.org/stable/1690046.

[3] Jenifer J, Anand S, and Levingstan Y. "SIMULATED ANNEALING ALGORITHM FOR MODERN VLSI FLOORPLANNING PROBLEM". In: 2 (Apr. 2016), pp. 175–181.

[4] Atanu Roy Karthik Ganesan Pillai. "Parallel Simulated Annealing for VLSI Cell Placement Problem". In: 2009.

[5] Adrian Ludwin and Vaughn Betz. "Efficient and Deterministic Parallel Placement for FPGAs". In: *ACM Trans. Design Autom. Electr. Syst.* 16 (2011), 22:1–22:23.

[6] Roman Marto ňák, Giuseppe E. Santoro, and Erio Tosatti. "Quantum annealing of the traveling-salesman problem". In: *Phys. Rev. E* 70 (5 Nov. 2004), p. 057701. DOI: 10.1103/PhysRevE.70.057701. URL: https://link.aps.org/doi/10.1103/PhysRevE.70.057701.

[7] P. Bergé et al. "Restricting the search space to boost Quantum Annealing performance". In: *2016 IEEE Congress on Evolutionary Computation (CEC)*. July 2016, pp. 3238–3245. DOI: 10.1109/CEC.2016.7744199.

[8] Roman Marto ňák, Giuseppe E. Santoro, and Erio Tosatti. "Quantum annealing by the path-integral Monte Carlo method: The two-dimensional random Ising model". In: *Phys. Rev. B* 66 (9 Sept. 2002), p. 094203. DOI: 10.1103/PhysRevB.66.094203. URL: https://link.aps.org/doi/10.1103/PhysRevB.66.094203.

[9] Y. Han, S. Roy, and K. Chakraborty. "Optimizing simulated annealing on GPU: A case study with IC floorplanning". In: *2011 12th International Symposium on Quality Electronic Design*. Mar. 2011, pp. 1–7. DOI: 10.1109/ISQED.2011.5770735.

[10] Sanroku Tsukamoto et al. "An Accelerator Architecture for Combinatorial Optimization Problems". In: 2017.

[11] M. Baity-Jesi et al. "Janus II: A new generation application-driven computer for spin-system simulations". In: *Computer Physics Communications* 185.2 (2014), pp. 550–559. ISSN: 0010-4655. DOI: https://doi.org/10.1016/j.cpc.2013.10.019. URL: http://www.sciencedirect.com/science/article/pii/S0010465513003470.

[12] Henrique Fingler et al. "A CUDA based Solution to the Multidimensional Knapsack Problem Using the Ant Colony Optimization". In: *Procedia Computer Science* 29 (2014). 2014 International Conference on Computational Science, pp. 84–94. ISSN: 1877-0509. DOI: https://doi.org/10.1016/j.procs.2014.05.008. URL: http://www.sciencedirect.com/science/article/pii/S1877050914001859.

[13] Bianca de Almeida Dantas and Edson Norberto Cáceres. "Sequential and Parallel Implementation of GRASP for the 0-1 Multidimensional Knapsack Problem". In: *Procedia Computer Science* 51 (2015). International Conference On Computational Science, ICCS 2015, pp. 2739–2743. ISSN: 1877-0509. DOI: https://doi.org/10.1016/j.procs.2015.05.411. URL: http://www.sciencedirect.com/science/article/pii/S1877050915012193.

[14] M. E. Lalami and D. El-Baz. "GPU Implementation of the Branch and Bound Method for Knapsack Problems". In: *2012 IEEE 26th International Parallel and Distributed Processing Symposium Workshops PhD Forum*. May 2012, pp. 1769–1777. DOI: 10.1109/IPDPSW.2012.219.

[15] Arnab Das, Bikas K. Chakrabarti, and Robin B. Stinchcombe. "Quantum annealing in a kinetically constrained system". In: *Phys. Rev. E* 72 (2 Aug. 2005), p. 026701. DOI: 10.1103/PhysRevE.72.026701. URL: https://link.aps.org/doi/10.1103/PhysRevE.72.026701.

[16] Andrew Lucas. "Ising formulations of many NP problems". In: *Frontiers in Physics* 2 (2014), p. 5. ISSN: 2296-424X. DOI: 10.3389/fphy.2014.00005. URL: https://www.frontiersin.org/article/10.3389/fphy.2014.00005.

[17] Kazutoshi Wakabayashi. "CyberWorkBench: Integrated design environment based on C-based behavior synthesis and verification". In: *VLSI Design, Automation and Test, 2005.(VLSI-TSA-DAT). 2005 IEEE VLSI-TSA International Symposium on*. IEEE. 2005, pp. 173–176.

[18] Marco Bettoni et al. "A convolutional neural network fully implemented on fpga for embedded platforms". In: *CAS (NGCAS), 2017 New Generation of*. IEEE. 2017, pp. 49–52.

[19] Francisco Ortega-Zamorano et al. "FPGA Hardware Acceleration of Monte Carlo Simulations for the Ising Model". In: *CoRR* abs/1602.03016 (2016). arXiv: 1602.03016. URL: http://arxiv.org/abs/1602.03016.

An Instruction Set Architecture for Low-power, Dynamic IoT Communication

Shahzad Muzaffar and Ibrahim (Abe) M. Elfadel

Khalifa University, Abu Dhabi, UAE

{shahzad.muzaffar, ibrahim.elfadel}@ku.ac.ae

Abstract—**This paper presents an instruction set architecture (ISA) dedicated to the rapid and efficient implementation of single-channel IoT communication interfaces. The architecture is meant to provide a programming interface for the implementation of signaling protocols based on the recently introduced pulsed-index schemes. In addition to the traditional aspects of ISA design such as addressing modes, instruction types, instruction formats, registers, interrupts, and external I/O, the ISA includes special-purpose instructions that facilitate bit stream encoding and decoding based on the pulsed-index techniques. Verilog HDL is used to synthesize a fully functional processor based on this ISA and provide both an FPGA implementation and a synthesised ASIC design in GLOBALFOUNDRIES $65nm$. The ASIC design confirms the low-power features of this ISA with consumed power around $31\mu W$ and energy efficiency of less than $10pJ/bit$.**

I. Introduction

IoT nodes need to meet two conflicting requirements: high data-rate communication to support bursts of activity in sensing and communication, and low-power to improve energy autonomy. Unfortunately, existing protocols fail to meet these requirements simultaneously. Protocols providing high data rates, such as WiFi, WLAN, TCP/IP, USB, etc. [1], [2], [3], are power-hungry and involve complex controllers to handle two-way communications. On the other hand, low-power protocols such as 1-Wire [4] and UART [5] have low data rates.

To fill up the gap and address these two requirements at once, a novel family of pulsed signaling techniques for single-channel, high-data-rate, low-power dynamic communication have been recently proposed under the name of Pulsed-Index Communication (PIC) [6], [7]. The most important feature of this family of protocols is that they do not require any clock and data recovery (CDR). They are also highly tolerant of clocking differences between transmitter and receiver, and are fully adapted to the simple, low-power, area-efficient, and robust communication needs of IoT devices and sensors. These techniques are reviewed in Section II with their advantages and disadvantages clarified. The main issue that this paper addresses is to provide a flexible framework that enables the implementation of the most suitable PIC technique for a given application. The issue of selecting and implementing a communication interface in a constrained IoT node is a prevalent one, and its solution should contribute to the streamlining of communication subsystem design in IoT devices.

One candidate solution is to program all the protocols on a microprocessor and control their selection and parameters through registers. This is a standard practice that is followed for data transfer protocols such as I²C, I²S, SPI, UART, and CAN. Another possible solution is to design ASIC for the newest generation of the protocol and make it backward compatible with older versions as in the case of USB 1.0 through 3.0 [8]. Such methods increase silicon area and power consumption, and do not provide any customization features. Yet another approach is to adopt the principles of hardware-software codesign and provide a special-purpose hardware supporting a tuned or extended Instruction Set Architecture that can be used to configure and implement the various communication protocols of a given family without changing or re-designing the on-chip hardware modules. An example of such approach can be found in Cisco's routers where the main CPU (e.g., MPC860 PowerQUICC processor from Motorola/NXP) includes an on-chip Communication Processor Module (CPM) [9]. The CPM is a RISC microcontroller dedicated to several special purpose tasks such as signal processing, communication interfaces, baud-rate generation, and direct memory access (DMA). The work described in this paper is inspired with such a solution in that it proposes a flexible, fully programmable communication interface for the PIC family based on a full RISC-like ISA tailored for the efficient and seamless implementation of the PIC protocols.

Specifically, a set of special-purpose instructions and registers along with a compact assembly language is proposed to help perform the specific tasks needed for the generation of pulsed signals and to give access to all the hardware resources. The proposed ISA is called Pulsed-Index Communication Interface Architecture (PICIA) and is meant to help reduce the number of instructions required to implement a PIC family member without impacting the advantageous data rates or low-power operation of the PIC family. Verilog HDL is used to synthesize and verify a fully functional processor based on this ISA over the Spartan-6 FPGA platform. Furthermore, an ASIC design in the GLOBALFOUNDRIES $65nm$ process confirms the low-power operation with $31.4\mu W$ and energy efficiency of less than $10pJ/bit$.

II. Pulsed-Signaling Techniques

Pulsed-signaling techniques are based on the basic concept of transmitting binary word attributes rather than modulated bits. The attributes are quantified, coded as pulse counts, and transmitted as streams of pulses. The key to the success of these techniques is the encoding step whose goal is to

978-1-5386-4757-8/18 $31.00 © 2018 IEEE

minimize the pulse count. At the receiver, the decoding is based on pulse counting by detecting the rising edge of each pulse. These techniques have the distinguished feature that they don't require any clock and data recovery (CDR), which significantly contributes to their low-power and small footprint hardware implementations. Recently, three techniques based on this concept have been introduced, namely, Pulsed-Index Communication (PIC) [6], Pulsed-Decimal Communication (PDC) [7], and Pulsed-Index Communication Plus (PIC*plus*). With slight differences, these techniques apply an encoding scheme to a data word B to *minimize* the number of ON bits, and *move* them to the Least-Significant-Bit (LSB) end of the packet with the goal of lowering the number of pulses required to transmit the data bits. The encoding process includes a segmentation step where the data is broken into N independent segments of size l bits each (i.e. $N = B/l$). To maximize data rate, these use, on each segment, an encoding combination of bit inversion and/or segment reversion/flipping. For PIC and PIC*plus*, this combination is meant to reduce the number of ON bits and decrease their index values. For PDC, the same combination is meant to reduce the number of ON bits and decrease the decimal number represented by each segment. To facilitate decoding, flag pulses representing the type of encoding performed are added to each segment. Unlike PIC, the PDC segment flags of two consecutive segments and the PIC*plus* segment flags of four consecutive segments are combined in one data word flag and placed in the header. The PDC further applies a third segmentation step post-encoding whose goal is further reduce the number of pulses per segment and, therefore, further increase the data rate.

All the pieces of information including flags, the number of indices, and the indices themselves in the case of PIC and PIC*plus*, or the decimal numbers of each segment in the case of PDC, are transmitted in the form of pulse streams. Within a given packet, segment pulse streams are separated by an inter-symbol delay (α). The receiver counts the number of pulses for each pulse stream and applies the decoding according to the flags received.

III. PULSED-INDEX COMMUNICATION INTERFACE ARCHITECTURE (PICIA)

As described in Section II, the PIC family members share many ideas, some of which are used in exactly the same way and others with few changes. Their packet formats are also quite similar. There could be a number of variations that could be introduced in these techniques as per needs and choice. The proposed PICIA can be used to generate not only these standard protocols with tune-able respective communication parameters (i.e. segment size, inter-symbol delay, pulse width etc.) but it can also be used to develop other customized communications techniques that use the same underlying idea of transmitting information in the form of pulses. The PICIA is described in detail in the next subsections.

TABLE I: PICIA REGISTER SET

	Register	Type	Organization
1	R0-R7	8 bit GP[a]	8-bit Value
2	Ctrl0	8 bit SP[b]	[0,Mode,3-bit SegNum,3-bit SegSize]
3	Ctrl1	8 bit SP	8-bit Pulse Width
4	LoadReg	16 bit SP	16-bit Value

[a.] General Purpose [b.] Special Purpose

A. Register Set

The PICIA uses three types of registers. The first type includes a set of eight 8-bit registers, *R0 through R7*, which are programmer-accessible general-purpose registers. The second type is that of Control Registers *Ctrl0* and *Ctrl1* which are 8-bit registers used to store protocol configuration parameters such as mode of transaction (transmitter or receiver), segment number, segment size, and pulse width in terms of a number of clock cycles. These control registers are initially set by the programmer through specific instructions but, once set, they become accessible only to the system. The third type is the *LoadReg* register, which is a 16-bit, I/O-dedicated register used to read the I/O port, set the I/O port, and to store the updated results after an instruction is executed. Like the Control Registers, LoagReg is a privileged register accessible only to the system. These register types are summarized in Table I. In the remainder of the text, the word register will always refer to a general-purpose register.

B. Instruction Formats

The PICIA instructions are all 16-bit long and are of three different types. The first type, *I-Type 1*, handles one operand at a time and is used in operations such as to read/write the I/O port, set/clear the LoadReg, set various communication protocol parameters, and send/receive pulse streams. I-Type 1 is divided into five fragments, as shown in Fig. 1. The 5-bits *Opcode* represents the type of operation. *Type (R/C)* is used to set the type of operand (register or a constant) in an instruction. *Halt PC/WE* is used either to halt the PC during the transmission of pulse streams or to enable the store operation of received pulse-count to a specified register. The bit E sets if an extra pulse should be added to the transmitted pulse stream and/or an extra pulse should be removed from the received pulse stream. The last 8-bits long fragment of I-Type 1 is used to indicate a register number or an immediate constant value.

The second type of instruction, *I-Type 2*, handles two operands at a time and is used in operations such as updating a register with a given constant value, and jumping to a specified label in the code depending on the validity of a condition in a register. I-Type 2 is divided into three fragments, as shown in Fig. 1. The 5-bits *Opcode* represents the type of operation. The 3-bits *Register* field is used to indicate one of the general purpose registers and the 8-bits *Constant* field is used to provide a constant value or a label that is present in the code.

The third type of instruction, *I-Type 3*, handles two or three operands simultaneously. I-Type 3 is used in operations such as

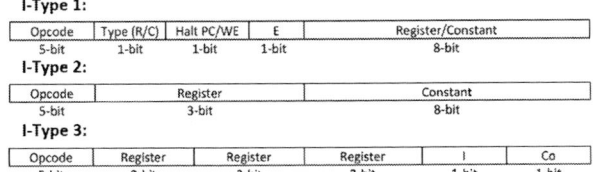

I-Type 1:

Opcode	Type (R/C)	Halt PC/WE	E	Register/Constant
5-bit	1-bit	1-bit	1-bit	8-bit

I-Type 2:

Opcode	Register	Constant
5-bit	3-bit	8-bit

I-Type 3:

Opcode	Register	Register	Register	I	Co
5-bit	3-bit	3-bit	3-bit	1-bit	1-bit

Fig. 1: PICIA Instructions Format

encoding (inversion and reversion with or without condition), combining and splitting encoding flags, and copying register contents or some other information to a specified register conditionally. I-Type 3 is divided into six fragments, as shown in Fig. 1. The 5-bits *Opcode* represents the type of operation. The 3-bits *Register* fields are used to indicate one of the general purpose registers. The combinations of 1-bit *I* and *Co* fields are used to select the source of information to be copied.

C. Addressing Modes

The PICIA employs three addressing modes: immediate, register, and auto-decrement. In the immediate mode, the source is either a constant or a label while the destination is one of the general-purpose, special-purpose, or program counter registers. In the register mode, the register contains the value of the operand. The auto-decrement mode is used only for jump operation where the branch to a label is taken and a specified register decrements by one if the register contains a non-zero number.

D. Interrupts

There are three interrupts in the PICIA supported processor. First, the I/O interrupt is generated when the data at the I/O port is available. The system remains in a halt state until the I/O interrupt is reached and the system starts the execution of instructions from the very start. Second, the transmitter interrupt is used to indicate the completion of the transmission of one pulse stream. The PICIA processor remains in a halt state, if activated, until transmitter interrupt is received and the execution continues from where it paused. Third, the receiver interrupt is generated when the reception of one pulse stream completes. The PICIA processor remains in a halt state until the receiver interrupt is received at which time, program execution is continued.

E. External I/O

Three external I/O ports are supported by the PICIA processor. One of these ports is the 16-bit data I/O port that is used to read from and write back to the external environment. To transmit and receive the packets in the form of pulse streams, a 1-bit signal I/O port is used. Another 1-bit data ready port is used to source the generation of I/O interrupts and start the execution of instructions.

IV. PICIA ASSEMBLY LANGUAGE

Before diving into the PICIA assembly language in detail, it is necessary to understand few relevant interpretations about the instructions and assembly language. These interpretations are shown in Table II. The left part of the table shows the instruction interpretations where the values of the control bits are indicated along with the corresponding effect or representation. Similarly, the right part of the table does the same but for PICIA assembly language. The PICIA supported instructions are described in the next subsections and are categorized with respect to their types. All the instructions are listed in Table III too along with their type, a brief description, and an example.

A. Type 1 Instructions (I-Type 1)

These instructions deal with only one operand at a time. The first instruction towards this is *RP*, *read the port*, that collects the data from data I/O port and stores it in the LoadReg. *WP*, *write the port*, reads data from LoadReg and updates the data I/O port. There is no operand to these instructions as the system accesses the special purpose register internally. *SSS* and *SSN* set the segment size and the segment number respectively in the Ctrl0 register. The operand for both of these instructions is an immediate constant value. The operand to SSS can be any of 0, 1, and 2 that represents the segment size of 4, 8, and 16 bits respectively. Segment size information helps the system to break the data word into smaller independent segments. The operands to SSN can be the numbers 0, 1, 2, and 3. SSN is used to select the segment that is going to be processed by all the following instructions in the program until the segment number is changed again. *SM, set the mode*, also accesses the special purpose control register Ctrl0 and sets or clears a bit representing the mode of operation. The operand to SM can either be 0 or 1 that represents the transmitter or receiver mode respectively. During transmitter mode, the signal port is used to send the pulses out and, during reception mode, the same port is used to receive the pulses from the external world. If the receiver mode is selected, the LoadReg is automatically cleared by the system to make it ready for reception. If the transmitter mode is selected, the LoadReg is updated automatically with the data present on I/O port. *SW, set pulse width*, sets the count of system-clock cycles for which the pulse remains high. The operand to SW is an 8-bit integer number.

The *SP, send the pulses*, sends a pulse stream consists of a number of consecutive pulses equal in count specified by the operand that could either be a register or an immediate constant number. The argument *h* is used to decide if the system should halt during the transmission of a pulse stream or not. If 1, halt the system unless the pulse stream transmission is complete, or continue with the next instruction if 0. The argument *E* to SP instruction informs the system if the pulse stream should include the transmission of an additional pulse at the end of stream or not. This is helpful in representing the no-pulse or zero-index condition with only one pulse as it is in the case of PIC and PIC*plus* transmission, unlike PDC where

978-1-5386-4757-8/18 $31.00 © 2018 IEEE

TABLE II: PICIA INTERPRETATIONS

Instruction Interpretation		Assembly Interpretation	
Control Bit	**Value : Effect**	**Symbol**	**Meaning**
Type (R/C)	0 : Register, 1 : Constant	R	Register Only
Halt PC	0 : No Halt, 1 : Halt	C	Constant Only
WE	0 : Register Write Disabled, 1 : Register Write Enabled	RC	Register or Constant
E	0 : Extra Pulse Disabled, 1 : Extra Pulse Enabled	R, Rx, Ry, Rs	Register Number
I	0 : No Indexing, 1 : Indexing	h	0 : No Halt, 1 : Halt
Co	0 : Copy Segment Disabled, 1 : Copy Segment Enabled		

TABLE III: PICIA ASSEMBLY LANGUAGE

	Instruction	**Type**	**Description**	**Example**
1	RP	1	Load data from Input Pins to data register.	RP
2	WP	1	Output the received data from data register to the Pins.	WP
3	SSS C	1	Set segment size (C = 0,1,2 for 4 bit,8 bit,16 bit).	SSS 1
4	SSN C	1	Select segment number (C = 0,1,2,3).	SSN 2
5	SM C	1	Set Mode (C = 0,1 for Transmitter, Receiver). Setting RX mode clears LoadReg, setting TX loads input into LoadReg.	SM 0
6	SW C	1	Set width of pulse (C = integer specifying cycle count).	SW 2
7	IV Rx,Ry	3	Inverse the selected segment. Rx=NOI & Ry=Flags (Rx/Ry= R0,R1,...R7).	IV R0,R1
8	IVC Rx,Ry	3	Inverse conditionally the selected segment if encoding condition satisfy (ON bits >Seg. Size/2). Rx=NOI & Ry=Flags (Rx/Ry= R0,R1,...R7).	IVC R0,R1
9	FL Rx,Ry	3	Flip selected segment bits. Rx=NOI & Ry=Flags (Rx/Ry= R0,R1,...R7).	FL R0,R1
10	FLC Rx,Ry	3	Flip conditionally the selected segment bits if encoding condition satisfy (Seg. >Flip(Seg.)). Rx=NOI & Ry=Flags (Rx/Ry= R0,R1,...R7).	FLC R0,R1
11	IVFL Rx,Ry	3	Invert and Flip selected segment bits. Rx=NOI & Ry=Flags (Rx/Ry= R0,R1,...R7).	IVFL R0,R1
12	SP h, E, RC	1	Send RC number of pulses (RC = register number or constant value). Halt PC if h=1 (h=0,1). (Type = 1 then it's a constant). Send one extra pulse if E=1 (E=0,1).	SP 1,1,4
13	SD h, RC	1	Inter-Symbol delay of RC number of clock cycles. Halt PC if h=1 (h=0,1).	SD 1,4
14	WR R, C	2	Write constant value to a register R (R= R0,R1,...R7).	WR R0,8
15	SRD RC	1	Set Receiver Inter-Symbol Delay equal to RC number of clock cycles.	SRD R0 or SRD 4
16	WRI WE, E, R	1	Wait for receiver pulse stream interrupt. PC halts till the interrupt arrives. Remove one extra pulse count if E=1 (E=0,1). Enable received pulse count write to register R (R= R0,R1,...R7) if WE=1 (WE=0,1).	WRI 1,1,R0
17	SDB C	1	Sets the index bits or the data bits in the LoadReg as per the received pulse stream. (C=0,1 for indexing and data respectively).	SDB 1
18	BNZD R, label	2	Branch to label and decrement R by 1 if the specified register R contains non-zero number. (R= R0,R1,...R7).	BNZD R0,loop
19	CRC R,Rs,I,Co	3	Copy register conditionally. R= Rs if I=0. R= Rs , if I=1 and LoadReg [Rs]=1 and Co=0. R=0 otherwise. R=Selected Segment, if Co=1. Rs is ignored. (R/Rs= R0,R1,...R7). Can be used to clear the register.	CRC R1,R2,1,1
20	CF R,Rx,Ry	3	Combine Flags. R={Rx[1:0], Ry[1:0]}.	CF R0,R1,R2
21	SF Rx,Ry,R	3	Split Flags. Rx=R[3:2], Ry=R[1:0].	SF R1,R2,R0
22	NOP	1	No operation.	NOP

all the pulse streams are transmitted with an additional pulse. If 1, include an extra pulse or send the exact number of pulses if 0. SD is a similar instruction but with minor differences. *SD, send the delay*, transmits an inter-symbol delay that is equal in length to the specified number of system-clock cycles. All the arguments and operands work in the same way as that of SP except that there is no choice of an extra pulse.

To set the expected number of clock cycles per inter-symbol delay during the process of reception, the instruction *SRD* is used which takes either a register number or a constant number as an operand to represent the number of clock cycles. During a reception, the system needs to wait for the incoming pulse stream so that the pulses can be counted to infer the sent information. To fulfil this task, the instruction *WRI, wait for receiver interrupt*, is used. The system goes into the halt state when this instruction is executed and returns back to the normal state at the reception of receiver interrupt that is

generated when a pulse stream is received completely. The incoming pulses are counted and the count decrements once if the argument *E* to WRI is set. The count is stored in a specified register R if the argument *WE* is set. Among different types of information chunks in a received packet, a pulse streams related to data could either represent the index number of an ON bit (as in PIC or PIC*plus*) or the decimal number for a segment (as in PDC or other custom techniques). The instruction *SDB, set data bits*, removes this confusion by informing the system if the received pulse count needs to be stored directly in the LoadReg as a segment's content (if C=1) or a bit in the LoagReg needs to be set at the index number represented by the count (if C=0). The last instruction in the category of I-Type 1 is *NOP, no operation*, that is used when there is a need to wait for some operation to complete, as in the case of instructions SP and SD, without halting the system. In this case, there should be enough number of NOPs

($PulsCount + 2$) to wait for the completion of a pulse stream transmission. All or some of these NOPs can also be replaced by other instructions in order to perform useful tasks instead of waiting for transmission.

B. Type 2 Instructions (I-Type 2)

I-Type 2 is the smallest set of instructions. As mentioned earlier, these instructions handle two operands at a time. One of the operands is a register, and the other is an immediate constant. One of these instructions is *WR*, *write register*, that is used to store an immediate constant value to a specified general purpose register. The second instruction is the jump instruction *BNZD*, *branch and decrement if not zero*. The instruction takes two arguments, a register to check the condition and a label to jump to. If the content of the specified register is a non-zero value, the program counter jumps to the label and the register value decrements once. The BNZD is helpful in writing conditional loops.

C. Type 3 Instructions (I-Type 3)

These instructions deal with either two or three operands at a time, but all of these operands must be registers. The five instructions, described next, are used in encoding the selected segment. The *IV*, *invert*, is used to complement the bits of the selected segment unconditionally and the resulting new segment replaces the corresponding segment in LoadReg. The operand register Rx stores the new number of ON bits (NOI) in the resulted segment and register Ry stores the corresponding flags to represent the encoding type, as per encoding description in PIC and PDC overview. The *IVC*, *invert conditionally*, works the same way as IV works but only if the condition of encoding is true. The condition, as mentioned earlier in the overview section, is that the number of ON bits in the selected segment should be greater than half the segment size. The Rx and Ry get updated with new NOI and Flags respectively. The *FL*, *flip*, and *FLC*, *flip conditionally*, work exactly the same way as IV and IVC, respectively, except for the base operation that is the bit wise reverse/flipping instead of inversion. The condition here for FLC is to check whether the content number of the selected segment is greater than the flipped content number of the same segment. If the condition is true, it means the ON bits are at the higher number of indices, hence, they represent a big decimal number and both of these can be reduced by relocating the ON bits to the lower index numbers. The fifth instruction that takes part in encoding is *IVFL*, *invert and flip*. The IVFL works in the same way as the other aforementioned four instructions work except it applies both the inversion and flipping together unconditionally.

The instructions *CF*, *combine the flags*, and *SF*, *split the flags*, are used for PDC, but can be used for any customized technique through PICIA. CF takes two operands, Rx and Ry, representing two flags to be combined and stores the result in the third operand register R. The first two LSBs of both Rx and Ry, in the same order, are combined to generate four LSBs in R. Similarly, SF splits the combined flags in a specified

TABLE IV: CRC INSTRUCTION FUNCTIONALITY

Co	I	LoadReg[Rs]	Description
0	0	X	R=Rs
0	1	0	R=0
		1	R=Rs
1	X	X	R=Selected Segment

register R into two separate flags and stores these in registers Rx and Ry. The Ry takes the first two LSBs of R and Rx takes the next two LSBs of R.

The last and the most complex instruction of PICIA is *CRC*, *copy register conditionally*. Based on the given settings for *I* and *Co*, the instruction performs four different copy operations, as shown in Table IV where X is the don't-care and [Rs] represents the index number of LoadReg. The CRC can be used for a simple register to register copy because the instruction copies a register Rs to R if both Co and I are cleared. If Co is cleared and I is set, the source to be copied is decided by the bit of LoadReg located at the index number represented by the contents of register Rs. If the LoadReg bit at index Rs is cleared, 0 is copied to register R, or simply Rs is copied to R otherwise. This operation is helpful in generating PIC pulse streams. Remember, PIC selects the ON bits only in data and transmits their index numbers in the form of pulse streams. Therefore, CRC with such a configuration helps in finding if the target bit is ON or not. If the bit is ON, the index number of it needs to be transmitted that is present in register Rs and that is why it is copied to R. If the bit is OFF, nothing is there to transmit and that is why 0 is copied to register R. Hence, the index numbers of the ON bits can be transmitted in a loop. If Co is set, I becomes don't care and the contents of the selected segment are copied to register R. This is helpful in generating PDC pulse streams as, unlike PIC, it transmits the contents of the sub-segments in the form of pulse streams. Hence using such a configuration for CRC, all segments of the data word can be selected and transmitted one-by-one in a loop. All the configurations of CRC instruction can be used to generate any other customized transmission techniques based on the idea of transmitting the information in the form of pulse streams.

V. EXPERIMENTAL VERIFICATION AND RESULTS

Verilog HDL is used to describe a fully functional processor based on the proposed ISA and a full experimental setup is implemented on the Xilinx Spartan-6 FPGA platform. The prototype platform is used to verify the functionality and performance of proposed PICIA. Extensive simulations and real-time hardware verification are performed to verify the results. A clock rate of 25 MHz is used for PICIA testing system. In the experimental flow, the PICIA processor's transmitter sends the 16-bit data starting at 0 with an increment of 1 at each transmission. The PICIA processor's receiver resends the same data back. The returned and original data words are compared to verify the complete round-trip chain.

In another experiment, the software aspects of two implementations are compared. In one implementation, the PIC

978-1-5386-4757-8/18 $31.00 © 2018 IEEE

TABLE V: RESULTS

	Implementation	
	PICIA	**Stand-alone**
Software Implementation Comparison		
Avg. No. Of Instructions	50-100	1300-1400
Avg. Data Rate (Mbps)	$\approx$4.1-7.1	$\approx$0.041-0.071
Hardware Synthesis Comparison		
Power (μW)	$\approx$31.14	$\approx$19-26.6
Avg. E_b (pJ/bit)	$\approx$4.2-7.6	$\approx$2.7-6.5
Area (gate count)	$\approx$4700	$\approx$2100-2400

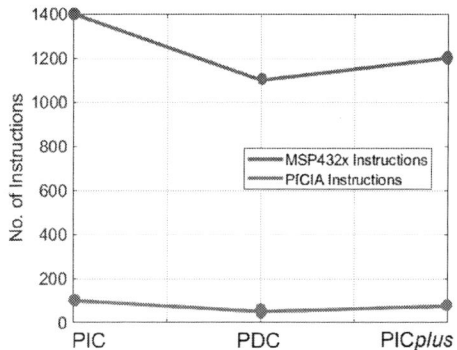

Fig. 2: PIC Family Implementation: PICIA vs. MSP432x

family member techniques are developed for MSP432X processor family. The second implementation used PICIA assembly language to develop the same techniques to run on the implemented processor. The number of instructions required to implement these techniques using MSP432X is approximately 1300 to 1400 on average whereas PICIA needs only 50 to 100 instructions. This is a notable reduction by a factor of 13 to 28, approximately. The data rates offered by the MSP432X implementation are also reduced significantly, approximately by a factor of 100. On the other hand, the data rates are preserved by the implementation of communication techniques using PICIA. The software implementation comparison is shown in Table V and Fig. 2.

We have also synthesized the PICIA processor system using GLOBALFOUNDRIES $65nm$ technology and estimated that PICIA hardware consumes around $31.14\mu W$ with a gate count of about 4700 gates. The power consumption results are promising as they remain well within the power budget of a full-hardware implementation of stand-alone pulsed-signaling techniques. Additionally, the consumption of hardware resources is comparable, data rates are preserved and the required number of instructions is reduced. Moreover, PICIA offers a customizable solution. The PICIA solution differs in that it offers a fully programmable communication interface that is specifically geared to the realization of pulsed-transmission techniques.

VI. CONCLUSIONS

The Pulsed-Index Communication Interface Architecture (PICIA) is a RISC-style special purpose ISA for single-channel, low-power, high data rate, dynamic, and robust communication based on pulsed-signaling protocols. It is

designed to facilitate the efficient generation of compact assembly code that is specific to such communication interfaces. This hardware/software co-design capability can be used to embed not only an existing PIC family member but also any custom nonstandard PIC protocol without changing the underlying hardware while greatly reducing the number of required instructions. Furthermore, such communication interface implementation will result in minimal to no impact on the data rates, power consumption, or the reliability of the protocols. The PICIA processor has been synthesized in GLOBALFUONDRIES $65nm$ technology and has been found to consume only $31.14\mu W$, which translates into an energy efficiency of less than $10pJ$ per transmitted bit. PICIA's microarchitecture and optimized hardware blocks that compactly implement the RISC-style ISA are the subject of a separate publication.

VII. ACKNOWLEDGEMENT

This work has been supported by the Semiconductor Research Corporation (SRC) under the Abu Dhabi SRC Center of Excellence on Energy-Efficient Electronic Systems (ACE[4]S), with funding from the Mubadala Development Company, Abu Dhabi, UAE.

REFERENCES

[1] S. Dayu, X. Huaiyu, S. Ruidan, and Y. Zhiqiang, "A Geo-related IoT Applications Platform based on Google Map," *7th International Conference on e-Business Engineering (ICEBE)*, pp. 380–384, Shanghai, China, November 2010.

[2] J. Byun, S. H. Kim, and D. Kim, "Lilliput: Ontology-based platform for IoT social networks," *IEEE International Conference on Services Computing*, pp. 139–146, Anchorage, AK, USA, June-July 2014.

[3] Jenq Muh Hsu and Chin Yo Chen. "A Sensor Information Gateway Based on Thing Interaction in IoT-IMS Communication Platform," *10th International Conference on Intelligent Information Hiding and Multimedia Signal Processing (IIH-MSP)*, pages 835–838, Kitakyushu, Japan, August 2014.

[4] MAXIM, *OneWireViewer User's Guide, Version 1.4*, 2009.

[5] C. dos Reis Filho, E. da Silva, E. de L. Azevedo, J. Seminario, and L. Dibb, "Monolithic data circuit-terminating unit (DCU) for a one-wire vehicle network," *Proceedings of the 24th European Solid-State Circuits Conference (ESSCIRC '98)*, pp. 228–231, Hague, Netherlands, September 1998.

[6] S. Muzaffar, A. Shabra, J. Yoo, and I. M. Elfadel, "A Pulsed-Index Technique for Single-Channel, LowPower, Dynamic Signaling," *Design, Automation and Test In Europe (DATE'15)*, pp. 1485–1490, Grenoble, France, March 2015.

[7] S. Muzaffar, and I. M. Elfadel, "A Pulsed Decimal Technique for Single-channel, Dynamic Signaling for IoT Applications," *25th IFIP/IEEE International Conference on Very Large Scale Integration (VLSI-SoC 2017)*, pp. 1–6, Abu Dhabi, UAE, October 2017.

[8] R. Teja, B. R. Jammu, M. Adimulam, and M. Ayi, "VLSI implementation of LTSSM," *International conference of Electronics, Communication and Aerospace Technology (ICECA 2017)*, pp. 129–134, Coimbatore, India, April 2017.

[9] linux-mips.org, "Cisco Systems Routers," 2012. [Online]. Available: https://www.linux-mips.org/wiki/Cisco.

978-1-5386-4757-8/18 $31.00 © 2018 IEEE

An Open-Source Verification Framework for Open-Source Cores: A RISC-V Case Study

Pasquale Davide Schiavone*, Ernesto Sanchez†, Annachiara Ruospo†*, Francesco Minervini*,
Florian Zaruba*, Germain Haugou* and Luca Benini*

*ETH Zurich

{pschiavo@iis.ee, minervif@student, zarubaf@iis.ee, haugoug@iis.ee, lbenini@iis.ee}.ethz.ch

†Politecnico di Torino

{ernesto.sanchez, annachiara.ruospo}@polito.it

Abstract—The complexity and heterogeneity of digital devices used in embedded systems is increasing everyday and delivering a bug-free design is still a very complex task. The interest for open-source hardware in real products is demanding for tools and advanced methodologies for verification to provide high reliability to open and free IPs. In this work, an open-source evolutionary optimizer has been used to create functional test programs that improve the verification test set for an open-source microprocessor, enhancing in this way, the verification level of the device. The verification programs are generated to optimize code coverage metrics and are tested against a high-level model to find device incorrectnesses during the generation time. A perturbation mechanism has been included in the verification framework to cover parts of the device under verification not reachable with only software stimuli such as interrupts or memory stalls. The proposed methodology uncovered 10 bugs still present in the RTL description of the analyzed device and demonstrated the effectiveness of open-source verification tools for the next generation of open-source RISC-V microprocessors.

1. Introduction

The increasing interest for open-source hardware is opening a new era in the silicon market. Among many, the free and open-source RISC-V *Instruction-Set Architecture* (ISA) [1] is becoming a viable option supported by industry leaders, such as Google, Micron, NXP, Microsemi, Qualcomm, Nvidia and Western Digital just to cite a fews [2]. A rich ecosystem of open-source software and hardware is growing around the RISC-V ISA. The Rocket and BOOM [3], [4] from UC Berkeley are two open-source RISC-V microprocessors already in use in companies like SiFive and Esperanto Technologies [5], [6]. The Riscy and Zero-riscy [7], [8] from ETH Zurich and University of Bologna are two more open-source RISC-V microprocessors for Internet-Of-Things (IoT) platforms used, among others, by GreenWaves Technology and Dolphin Integration. [9], [10].
Open-source cores come however with major verification challenges. Verification is already a bottleneck for the design cycle. Some surveys state that validation, verification

and testing (VV&T) require about 60% [11] of the total production costs. Companies approaching open-source IPs usually verify their functionality internally and can provide reports or fixes to the IP designers to improve the quality of the free hardware. For instance, recently the *Riscy* core has been compared and chosen to be a valid candidate in an industry project by Google [12]. For that reason, it has also been extensively verified using STING [13], a versatile design verification platform. This helped in uncovering crucial bugs in the *Riscy* multiplier and the forwarding and stall logic related to the load-and-store unit. The bugs have been reported and fixed, proving the usefulness of open-source hardware in the industry context and the needs of advanced verification strategies.

However, a complete open-source verification framework for open-hardware would allow free access to verification efforts and it would increase the reliability of free IPs. Clifford Wolf from Symbiotic EDA developed an open-source end-to-end formal verification framework for RISC-V processors [14]. However, it requires changes in the DUV interfaces to be integrated in such framework. Kami is another open-source formal verification framework that has been used to verify BlueSpec written RISC-V processors [15]. On the other side, μGP (microGP) [16] is a flexible open-source evolutionary-based tool that is able to automatically generate syntactically correct assembly programs and it has been already used to test and verify microprocessors [17]. In short, the evolutionary engine receives the description of the processor assembly syntax and the evolution parameters. As a final result, the best suited test programs, or best individuals, are included in the verification suite. In most of the cases, such programs maximize the code coverage metrics of the *Device Under Verification* (DUV) and can be used to find inconsistencies between the HDL of the DUV and the trusted golden model.

In this paper, a simulation-based verification framework based on μGP is developed to increase the verification level of the *Riscy* core with automatic test program generation. Such framework relies on an evolutionary test program generator, the DUV, a golden model to evaluate the correctness of the DUV, and an independent evaluator that promotes programs that cover most of the DUV HDL description.

978-1-5386-4757-8/18 $31.00 © 2018 IEEE

The evolutionary engine drives the test program generation targeting to maximize the high-level code coverage metrics. The verification setup takes as input an instruction library that has been split and used in several evolutionary phases to optimize the final verification test set. In addition, a perturbation module has been described to produce a noisy behavior in the processor interaction with the external world (memories and interrupts). Furthermore, the generation of the programs is combined with the verification phase to leverage all the individuals created during the evolution to increase the change to uncover bugs in the HDL description. If a bug is found during the evolution, the individual can be added to the final verification suite. The proposed verification framework is composed by free and open-source program generator and DUV[1] [2] and can be adapted for others RISC-V HDL descriptions, bringing state-of-the-art verification methods to the open-hardware ecosystem. The elements used in the proposed methodology can be easily integrated in a formal verification plan as the one defined by [18].

The main contribution of the proposed methodology and case study consists on creating stimuli for the *Riscy* core that optimize a set of code coverage metrics to thoroughly explore the verification research space. In fact, the proposed setup was able to discover bugs on arithmetic functions of the core when special operands were used. For example, infinite numbers in the core *Floating-Point Unit* (FPU) or operands that cause intermediate overflows during the computation of special function in the *Arithmetic Logic Unit* (ALU).

At the end of the proposed verification process, we obtained an automatically generated verification test-set reaching on average about 90% on a set of high-level code coverage metrics, while unveiling 10 different bugs still present in the processor description.

The rest of the paper is organized as follows: Section 2 recalls some important concepts that support the paper, and summarizes some of the most important aspect about function verification for microprocessors and introduces the proposed approach, describing the main novelties of the presented strategy. Sections 3 introduces the case study and outlines the results gathered on *Riscy*. Finally, the last sections conclude the paper and draft the future works.

2. Microprocessor Functional Verification

It is possible to define functional verification as the demonstration that the intent of a design is preserved in its implementation. Many methodologies have been proposed to reach this goal targeting digital circuit verification, and can be roughly classified as static or dynamic techniques. Static methodologies (usually called *formal*) try to demonstrate that the circuit implementation conforms the specifications.

---

1. *μGP* is open-source and freely downloadable at http://ugp3.sourceforge.net/
2. *Riscy* is open-source and freely downloadable at https://github.com/pulp-platform/riscv

Formal techniques can be classified as canonical graphical expression models, including for example BDDs [19]; and algebraic expression models, that include for example satisfiability (SAT) and integer linear programming (ILP) [20] methodologies. The main disadvantages of formal or static techniques are the huge computational resources required, even for circuits of medium complexity. Differently, dynamic methodologies (called *simulation-based*) aim at uncover design errors by exercising the actual implementation of the design. These methodologies do not suffer from the above limitations, but only cover a limited range of behaviors and will never achieve 100% confidence of correctness [21]. Interestingly, in [18], the author states that the actual success of a dynamic methodology is mainly based on a well established verification route-map. This route-map, called *verification plan*, should be composed of three main elements:

- Coverage measurement: clearly defining a verification problem, the different metrics to be used and the verification progress;
- Stimuli generation: providing the required stimulus to thoroughly exercise the devise by following the plan directives;
- Response checking: describing how to demonstrate the behavior of the device conforms the specifications.

Due to the current complexity of processor cores, a suitable solution to create a stimuli set for microprocessor verification is to resort to test or verification programs. A test program is a syntactically correct sequence of assembly instructions that is provided to the processor using the normal execution mechanisms. The test program goal is not to execute a normal task, but to try to discover any possible design or production flaw.

Since its introduction [22] in the 80's, test programs have targeted validation, verification and testing of microprocessors. One common practice for microprocessor functional verification is to resort to hand written programs made by skilled design engineers that for example exercise certain corner cases. Unfortunately, these engineers are required to have a deep knowledge of the device under verification, meaning that a considerable amount of human effort is required to create these verification programs.

A different possibility to generate functional test programs is to resort to constrained-random test generation, as proposed in [23], [24], [25]. Those techniques usually exploit program templates where some of the used values or program parameters are generated randomly, whereas others are previously defined. The main drawback of those techniques is that they are difficult to implement when targeting complex designs. A possible evolution of the previous techniques is based on the exploitation of coverage metrics able to provide feedbacks to better drive the generation process. These techniques, as described for example in [17], [26], [27], require the generation and simulation of a huge quantity of test programs, but the generation process is usually more efficient than in pure random-based approaches.

2.1. Proposed Approach

The framework developed in our case study is depicted in Figure 1. The setup includes a generation step (*stimuli generation*) combined with subsequent checking (*response checking*).

On the left part of the figure, it is possible to see the evolutionary optimizer called μGP [16], an evolutionary-based tool inspired by the Darwinian principle of reproduction and survival of the fittest. μGP receives the description of the processor assembly syntax trough the so called *Instruction Library*, in addition to configuration information that sets the parameters used during the evolution, for example, the number of test programs in the population, the number of instructions included on every test program, the maximum number of test programs to simulate, and so on. Then, the evolutionary process starts by creating a set of random programs, so called individuals, that are evaluated externally by a fitness function. At every evolutionary step or *generation*, the best individuals are improved using genetic operators such as *crossover* and *mutation*, while the worst ones are discarded.

The *Instruction Library* is devised targeting the specific DUV ISA. In this library, the different processor registers, instructions, rules of use, and for example, the available addressing modes are described to support the generations of test programs. Moreover, not only mere instructions are described, but more complex pieces of programs that define finite loops, illegal instructions, and special environmental parameters to throw exceptions. In addition, divisions by zero, infinite and not-a-number operations are also described to cover special cases in the microprocessor.

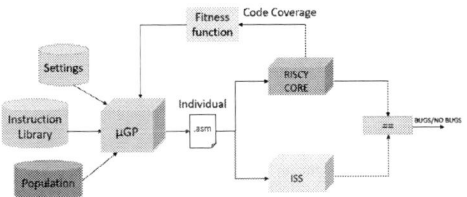

Figure 1. Verification framework

In our experiments, the verification programs are evaluated resorting to a checking scheme reported in the right part of Figure 1. In particular, the framework setup followed these steps:

1) The Instruction Library was created according to the target ISA.
2) The main μGP settings were defined.
3) A script was produced to automate the flow. Its purpose was to pick up the individuals produced by μGP generation by generation and provide it to the simulator and the ISS. Then, to create the fitness function, the code coverage percentages were collected and fed to the evolutionary algorithm again.

The fitness of every verification program is evaluated by measuring its capacity to maximize the code coverage metrics on the processor model [28] [29]. For any verification program, a fitness value was computed by resorting to an equation that includes together the different metrics; in particular, the used metrics are: *Statement Coverage, Branch Coverage, Condition Coverage, Expression Coverage, FSM State Coverage*, and *FSM Transition Coverage*.

The fitness function is computed as following: for each metric in use, the *arithmetic average*, the *variance* and the *sum* is computed accounting for all the core modules single metrics. As equation 1 highlights, this results in a vector of 18 elements which is used by μGP to drive the evolution. In particular, μGP gives more relevance to the first element of the vector fitness function and least priority to the last one. The order of elements has been decided empirically, with the first 6 elements being the average (a), the second 6 elements the variance (v) and last 6 elements the sum (s) of the *Statement Coverage* (1), *Branch Coverage* (2), *Condition Coverage* (3), *Expression Coverage* (4), *FSM State Coverage* (5), and *FSM Transition Coverage* (6).

$$fitness = \{a_1, a_2, a_3, a_4, a_5, a_6, v_1, v_2, v_3, v_4, v_5, v_6, \\ s_1, s_2, s_3, s_4, s_5, s_6\}. \tag{1}$$

Where for instance:

$$a_1 = \frac{\text{Sum of Statement Coverage of all units}}{\text{Number of units}}$$
$$a_2 = \frac{\text{Sum of Branch Coverage of all units}}{\text{Number of units}} \tag{2}$$

In order to improve the coverage results, a perturbation module has been embedded in our framework. The perturbation module is a hardware component that generates random interrupts and introduces random stalls on both data and instruction memory interfaces. It contains memory-mapped registers to configure a stall and interrupt mode (random number of stalls/interrupts or fixed). The initialization of the perturbation module is set random before executing the generated individual. However, one can consider for future works to program the perturbation registers as part of the evolutionary process.

Together with the fitness value that is used to guide the evolutionary process, any program is also used to compare the current processor model outcome (*Riscy+FPU* in Figure 1) against a high-level and reliable model (an instruction set simulator or *ISS* in Figure 1). The ISS functional model is an accurate model of the *Riscy* ISA. For every instruction, the HDL simulator pushes the instruction word to the ISS via a DPI-C wrapper. At the end of the execution, it compares the result computed by the HDL description with the one computed by the ISS. In this way, it is possible to find differences in execution between the compared models during the evolutionary generation phase rather than using only the optimized final individuals. Once a difference is found, a report is created allowing the verification engineer to evaluate the bug of the DUV.

3. Case study and experimental results

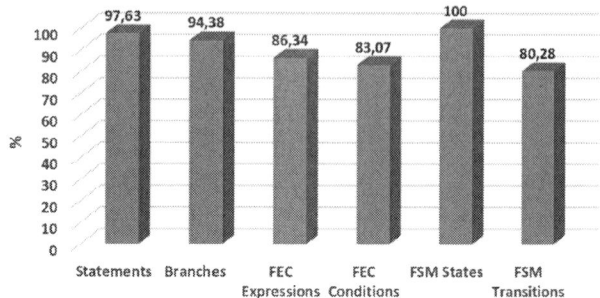

Figure 2. RI5CY Architecture

To experimentally demonstrate the effectiveness of the proposed method, the RTL description of the *Riscy* microprocessor is targeted [7]. *Riscy* is an open-source, 32bit, in-order, 4 pipeline stages microprocessor based on the opensource RV32IMFC extensions of the RISC-V ISA [1] and it is described in SystemVerilog. It has been developed under the *Parallel Ultra-Low Power Platform* (PULP) project [30] and it has been extended with HW-Loop, fixed-point, bit-manipulation and *Single-Instruction Multiple-Data* (pSIMD) instructions for energy efficient computation on signal processing algorithms [7]. It occupies only 40K nand2 gate-equivalent in the umcL65 nm technology [8] plus 30K extra gates for the FPU. The core's 4 pipeline stages are the *Instruction Fetch* (IF), *Instruction Decode* (ID), *Execution* (EX) and *Write Back* (WB) and its architecture is shown in Figure 2.

In the IF, the interaction between the core and the instruction memory bus takes place, the next program counter is calculated and the compressed instructions are decoded. The ID stage hosts the decoder, the register-file and the pipeline controller, which is also in charge to perform data-forwarding and stalling the pipeline. In the EX stage the enhanced ALU, multiplier and FPU are accommodated as well as the Control-Status (CS) register-file. The *Load-Store Unit* (LSU) sends data-memory requests during in the EX stage and receives answers in the WB stage. In addition, the WB stage takes care of two-clock-cycles FPU instructions. The instruction and data bus-interfaces implement a simple protocol via *request*, *grant* and *valid* signals. The master sends the *request* in the clock-cycle N and the arbiter can answer with the *grant* signal at clock-cycle N or later. The *valid* signal is provided by the bus-interfaces once the data from the memory is ready and it can arrive at least one clock-cycle later than the *grant* signal or later in case of multi-cycle memory accesses. *Riscy* supports up to 32 interrupts and it includes a debug-unit.

In order to verify the functionality of the DUV, it is also necessary to tests the core while handling interrupts, instruction and data bus delays (e.g., due to contention in multi-master bus or cache misses) along with the *Software* (SW) exceptions and normal instructions. Such events cannot be triggered using only instructions; therefore, external devices are involved to stimulate those conditions [31]. For this reason, a hardware implementation of the perturbation

Figure 3. Code Coverage Results

module has been designed to provide interrupt requests and bus stalls during RTL simulations. The perturbation module introduces delay-cycles both in the *grant* and *valid* signals of the core-memory interface, as well as random interrupt requests with random interrupt IDs. The memory-mapped registers allow to select the operational mode (bypass, random or not random), the number of stalls and maximum random numbers. In addition, the perturbation module can be configured to rise interrupt requests when the *Riscy* core is decoding a given instruction identified by its program counter. This mode is very useful to reproduce bugs that depend on specific conditions. The perturbation module registers are mapped in a subset of the core's debug-space so that it can be easily integrated in different platforms and its registers can be accessed by different peripherals such as JTAG or SPI or by means of load and store instructions.

The previously described framework was implemented and at the end of the experiments, the verification test set obtains a high code coverage results reported in Figure 3. It is interesting to note that the obtained results are in line with the ones obtained in [32] and described in [28]. Actually, in most of the cases, for example, in the case of the *statement coverage*, the metric saturates at the reported values. In the performed experiments, the code coverage metrics have been extracted resorting to Modelsim® HDL Simulation and used as variables of the fitness function to evaluate the program's fitness. However, any free logic simulator tool which supports code coverage estimation could be used as well (e.g. Verilator).

The *Instruction Library* contains hundreds of rules to describe the aforementioned ISA extensions and special cases, and it is split in three sub-libraries in order to allow a better exploration of the processor description and maximizing the final code coverage at the same time. The first library generates test programs that only contain RV32IMC instructions (no floats) and PULP extensions plus constraints to generate special cases as illegal instructions and functions to enter sleep mode.

To stress more complex units and corner cases, a second library has been designed to focus on FPU instructions without polluting the individuals with the generation of integer

978-1-5386-4757-8/18 $31.00 © 2018 IEEE   46

TABLE 1. CODE COVERAGE RESULTS.

Library	Code Coverage
RV32IMFC Random	52%
Single General-Purpose	64%
Proposed Optimized Splitted	85%
Proposed Optimized Splitted + Perturbation	90%

Code coverage results for different individuals.

instructions. This approach allows to further stress the FPU with high density float instruction sequences. This library also contains special cases as RV32F illegal instructions and corner cases (NaN, infinite, division by zero). Finally, the third library has been used to generate two different experiments: one to optimize the coverage of the float multiply-and-accumulate unit and the float division-and-square root unit isolated from the rest of the DUV; and one still targeting the FPU but with focus on its conditional part. This is achieved by giving more relevance to the *Condition Coverage* metrics in the aforementioned fitness function instead of the *Statement Coverage*. The final test program generation process is split in four different experiments. The union of all the best individuals results in a final code coverage of 85%. Finally, by adding random stalls on both data and instruction memory and random interrupt requests, the final code coverage increased to more than 90%. The remaining 10% was related for example to FSM transitions which never happened, special case of NaN, infinite numbers, and their combinations. It is important to note that the main drawback when dealing with evolutionary-based tools is the simulation time. In fact, to generate one of the best individuals included in the final test set, it takes about 8 hours of simulation.

Splitting the evolutionary phases is useful to specialize the verification functions to face those parts of the DUV code difficult to cover. Indeed, a single general purpose library to generate test programs based on the RV32IMFC plus extensions (but without special conditions like explicit NaN operands or illegal instructions) was able to cover only an average of 64% of the HDL code. One of the main reasons was that the generated program did not contain enough RV32F instructions keeping the FPU coverage low. A pure random program generated by the first generation of individuals from the same library had instead only 52%. Table 1 summarizes the code coverage discussed above.

The fitness function has been designed trying to boost all the coverages metrics to grow uniformly, and if two individuals have the same coverage results, the smaller code size program is preferred.

The final test set uncovered 10 design bugs during the verification process. These errors belong to different types, for example, the computation performed by the instruction *p.clip* and *p.extract* were discovered to be incorrect in a very specific case. For instance, due to an intermediated overflow, the *p.clip* instruction failed to compute the correct value with specific cases, whereas the logic shift instead of the arithmetic one was mistakenly used for the *p.extract* instruction. Another example involves bugs related to FPU square root operations with rounding modes or with NaN operations in the RISC-V *fclass* instruction.

The proposed method performs better than pure random and manual approaches as well as than using a single test program that covers the highest coverage as proposed in [17]. In fact, having the generation and verification phases split means using only the best individual to test the correctness of the core, whereas combining the two phases means testing all the programs generated during the evolutionary steps. This allows to explore a wider space of programs and increasing the probability of success. For example, the bug related to *p.clip* instruction previously mentioned was not part of the highest coverage program found by the optimizer, still it has been generated among the individuals of a prior generation to be then discarded during the evolutionary process. By combining generation and verification, every individual is not only used to calculate the code coverage, but it has the possibility of being executed and compared against the golden model (the ISS) to find design bugs. This approach shows that it is not only important to maximize the code coverage but also to leverage less important individuals during the generation phase.

4. Future Work

The success of the experiment results in a more reliable verification suite to evaluate the correctness of the RI5CY core. A similar approach can be also used to verify other microprocessors of the RISC-V open-source community but it can also be adapted for other kind of HW blocks for example *Direct Memory Access* (DMA), *HW-Accelerator* (HWA) or any peripherals. Even if this methodology was followed for a RISC-V, nothing prevents it from being performed for other ISAs. Furthermore, the presented framework did not consider the verification of the core debug unit, which can be a future extension of this work. This can be accomplished by building *Instruction Libraries* that contain pseudo-random stimuli for the DUV and by adapting the perturbation module to interact with the DUV interfaces. The increasing complexity of free HW (full microcontrollers with peripherals, DMA, cluster of multicores, event-units, debug, etc) open the opportunity for verification and testing research to spread their tools over different IPs connected together.

5. Conclusion

Today's crucial role of embedded systems requires to decrease as much as possible the current time to market times; however, it is difficult to speed up the production process due to the high amount of time and resources required during the verification step. In this paper, we proposed an evolutionary-based methodology able to automatically generate assembly programs and we applied it to enhance the verification level of the RISC-V *Riscy* processor. The proposed methodology combines the use of an evolutionary optimizer, a hardware perturbation module and a checking mechanism to rapidly create verification sets of assembly

978-1-5386-4757-8/18 $31.00 © 2018 IEEE

programs. The evolutionary optimizer, the perturbation module as well as the core are open-source and available for usage and further development. The experimental results demonstrated the effectiveness of the method by uncovering 10 bugs in the RTL description of the DUV.

Acknowledgment

This work has been funded by the Swiss National Science Foundation under grant 162524 (MicroLearn: Micropower Deep Learning).

References

[1] A. Waterman, Y. Lee, D. A. Patterson, K. Asanovic, V. I. U. level Isa, A. Waterman, Y. Lee, and D. Patterson, "The risc-v instruction set manual," 2014.

[2] "Platinum members of the RISC-V foundation," https://riscv.org/membership/wpbdp_category/platinum/.

[3] Y. Lee, "RISC-V "Rocket Chip" SoC Generator in Chisel," https://riscv.org/wp-content/uploads/2015/02/riscv-rocket-chip-generator-tutorial-hpca2015.pdf.

[4] C. Celio, P. Chiu, B. Nikolic, D. Patterson, and K. Asanovi, "BOOM v2: an open-source out-of-order RISC-V core," https://www2.eecs.berkeley.edu/Pubs/TechRpts/2017/EECS-2017-157.pdf.

[5] "Sifive," https://www.sifive.com/.

[6] "Esperanto technologies," https://www.esperanto.ai/.

[7] M. Gautschi, P. D. Schiavone, A. Traber, I. Loi, A. Pullini, D. Rossi, E. Flamand, F. K. Gurkaynak, and L. Benini, "Near-threshold RISC-v core with DSP extensions for scalable IoT endpoint devices," *IEEE Transactions on Very Large Scale Integration (VLSI) Systems*, pp. 1–14, 2017.

[8] P. D. Schiavone, F. Conti, D. Rossi, M. Gautschi, A. Pullini, E. Flamand, and L. Benini, "Slow and steady wins the race? a comparison of ultra-low-power risc-v cores for internet-of-things applications," in *Power and Timing Modeling, Optimization and Simulation (PATMOS), 2017 27th International Symposium on*. IEEE, 2017, pp. 1–8.

[9] "Green waves technologies," https://greenwaves-technologies.com/en/greenwaves-technologies-2-2/.

[10] "Dolphin integration," https://www.dolphin-integration.com/index.php/silicon_ip/ip_products/microcontrollers/overview/.

[11] "Sia," https://www.semiconductors.org/main/resources/.

[12] M. Cockrell. (2018) Use of RISC-V on Pixel Visual Core. [Online]. Available: https://tmt.knect365.com/risc-v-workshop-barcelona/speakers/matt-cockrell#workshop_use-of-risc-v-on-pixel-visual-core/

[13] "Running sting on pulpino platform," http://valtrix.in/programming/running-sting-on-pulpino.

[14] C. Wolf. End-to-end formal ISA verification of RISC-V processors with riscv-formal. [Online]. Available: http://www.clifford.at/papers/2017/riscv-formal/slides.pdf

[15] J. Choi, M. Vijayaraghavan, B. Sherman, A. Chlipala *et al.*, "Kami: a platform for high-level parametric hardware specification and its modular verification," *Proceedings of the ACM on Programming Languages*, vol. 1, no. ICFP, p. 24, 2017.

[16] E. Sanchez, M. Schillaci, and G. Squillero, *Evolutionary Optimization: the uGP toolkit*. Springer US, 2011, p. 178.

[17] F. Corno, E. Sanchez, M. S. Reorda, and G. Squillero, "Automatic test program generation: a case study," *IEEE Design Test of Computers*, vol. 21, no. 2, pp. 102–109, Mar 2004.

[18] P. Andrew, *Functional Verification Coverage Measurement and Analysis*. Springer US, 2008, pp. XVI, 216.

[19] R. E. Bryant, "Graph-based algorithms for boolean function manipulation," *IEEE Transactions on Computers*, vol. C-35, no. 8, pp. 677–691, Aug 1986.

[20] R. Brinkmann and R. Drechsler, "Rtl-datapath verification using integer linear programming," in *Proceedings of ASP-DAC/VLSI Design 2002. 7th Asia and South Pacific Design Automation Conference and 15h International Conference on VLSI Design*, 2002, pp. 741–746.

[21] V. Agrawal and M. Bushnell, *Essentials of Electronic Testing for Digital, Memory and Mixed-Signal VLSI Circuits*. Springer US, 2002, p. 690.

[22] S. M. Thatte and J. A. Abraham, "Test generation for microprocessors," *IEEE Transactions on Computers*, vol. C-29, no. 6, pp. 429–441, June 1980.

[23] A. Adir, E. Almog, L. Fournier, E. Marcus, M. Rimon, M. Vinov, and A. Ziv, "Genesys-pro: innovations in test program generation for functional processor verification," *IEEE Design Test of Computers*, vol. 21, no. 2, pp. 84–93, Mar 2004.

[24] S. K. S. Hari, V. V. R. Konda, K. V, V. M. Vedula, and K. S. Maneperambil, "Automatic constraint based test generation for behavioral hdl models," *IEEE Transactions on Very Large Scale Integration (VLSI) Systems*, vol. 16, no. 4, pp. 408–421, April 2008.

[25] Y. Zhou, T. Wang, H. Li, T. Lv, and X. Li, "Functional test generation for hard-to-reach states using path constraint solving," *IEEE Transactions on Computer-Aided Design of Integrated Circuits and Systems*, vol. 35, no. 6, pp. 999–1011, June 2016.

[26] O. Guzey and L. C. Wang, "Coverage-directed test generation through automatic constraint extraction," in *2007 IEEE International High Level Design Validation and Test Workshop*, Nov 2007, pp. 151–158.

[27] P. H. Chang, L. C. Wang, and J. Bhadra, "A kernel-based approach for functional test program generation," in *2010 IEEE International Test Conference*, Nov 2010, pp. 1–10.

[28] S. Tasiran and K. Keutzer, "Coverage metrics for functional validation of hardware designs," *IEEE Design Test of Computers*, vol. 18, no. 4, pp. 36–45, Jul 2001.

[29] E. Hung, B. Quinton, and S. J. E. Wilton, "Linking the verification and validation of complex integrated circuits through shared coverage metrics," *IEEE Design Test*, vol. 30, no. 4, pp. 8–15, Aug 2013.

[30] D. Rossi, I. Loi, F. Conti, G. Tagliavini, A. Pullini, and A. Marongiu, "Energy efficient parallel computing on the PULP platform with support for OpenMP," in *2014 IEEE 28th Convention of Electrical & Electronics Engineers in Israel (IEEEI)*. IEEE, dec 2014.

[31] F.-C. Yang, W.-K. Huang, J.-K. Zhong, and J. Huang, "Automatic verification of external interrupt behaviors for microprocessor design," *IEEE Transactions on Computer-Aided Design of Integrated Circuits and Systems*, vol. 27, no. 9, pp. 1670–1683, 2008.

[32] F. Corno, E. Sanchez, and G. Squillero, "Evolving assembly programs: how games help microprocessor validation," *IEEE Transactions on Evolutionary Computation*, vol. 9, no. 6, pp. 695–706, Dec 2005.

978-1-5386-4757-8/18 $31.00 © 2018 IEEE

On the Rectifiability of Arithmetic Circuits using Craig Interpolants in Finite Fields

Utkarsh Gupta[1] Irina Ilioaea[2] Vikas Rao[1] Arpitha Srinath[1] Priyank Kalla[1] Florian Enescu[2]
[1]Electrical and Computer Engineering, University of Utah, Salt Lake City UT, USA
[2]Mathematics and Statistics, Georgia State University, Atlanta GA, USA

Abstract—When formal verification of arithmetic circuits identifies the presence of a bug in the design, the task of rectification needs to be performed to correct the function implemented by the circuit so that it matches the given specification. This paper addresses the problem of rectification of buggy finite field arithmetic circuits. The problems are formulated by means of a set of polynomials (ideals) and solutions are proposed using concepts from computational algebraic geometry. Single-fix rectification is addressed – i.e. the case where any (set of) bugs can be rectified at a single net (gate output). We determine if single-fix rectification is possible at a particular location, formulated as the Weak Nullstellensatz test. Subsequently, we introduce the concept of Craig interpolants in polynomial algebra over finite fields and show that the rectification function can be computed using algebraic interpolants. Experimental results demonstrate the superiority of our approach against SAT-based approaches.

I. Introduction

Past few years have seen extensive investigations into formal verification of arithmetic circuits. Circuits that implement polynomial computations over large bit-vector operands are hard to verify using methods such as SAT/SMT-solvers, decision diagrams, etc. Recent techniques have investigated the use of polynomial algebra and algebraic geometry techniques for their verification. These include verification of integer arithmetic circuits [1] [2] [3] and also finite field circuits [4] [5]. While these are successful in proving correctness or detecting the presence of bugs, the problem of debugging and correction of arithmetic circuits has only just begun to be addressed [6], [7].

In this paper, we address the problem of rectification of buggy finite field arithmetic circuits. Our problem setup is as follows:

- A specification model (*Spec*) is given either as a polynomial description f_{spec} over a finite field, or as a golden model of a finite field arithmetic circuit. The finite field considered is the field of 2^k elements (denoted by $\mathbb{F}_{2^k}$), where k corresponds to the operand-width (bit-vector word length). An implementation circuit C is also given.
- Equivalence checking is performed between the *Spec* and the circuit C, and the presence of a bug is detected. No restrictions on the number, type, or locations of the bugs are assumed.

This research is funded in part by the US National Science Foundation grants CCF-1619370 and CCF-1320385.

- We assume that error-diagnosis has been performed, and a subset X of the nets of the circuit is identified as *potential rectification locations*.

Given the *Spec*, the buggy implementation circuit C, the set X of potential rectifiable locations, our objective is to determine whether or not the buggy circuit can be rectified at *one particular net (location)* $x_i \in X$. This is called **single-fix rectification** in literature [8]. If a single-fix rectification does exist at net x_i in the buggy circuit, then our subsequent objective is to derive a polynomial function $U(X_{PI})$ in terms of the set of primary input variables X_{PI}. This polynomial can be translated (synthesized) into a logic subcircuit such that $x_i = U(X_{PI})$ acts as the rectification function for the buggy circuit C so that C matches the specification.

Another important contribution of our work is that we show that the rectification function $U(X_{PI})$ can be determined based on the concept of Craig interpolants [9] in algebraic geometry. While Craig interpolation is a well-studied concept in propositional and first-order logic theories, we recently showed in [10] that polynomial algebra in finite fields also admits Craig interpolation, and described algorithms to compute interpolants. Based on our results of [10], we show how to compute a rectification function using Craig interpolation in finite fields.

Our techniques and algorithms are based on symbolic computer algebra and algebraic geometry – particularly on the concepts of the Weak Nullstellensatz and Gröbner bases [11]. We show how to apply our techniques to rectify finite field arithmetic circuits, where conventional SAT-solver based rectification approaches are infeasible.

Review of the Previous Work: Automated diagnosis and rectification of digital circuits has been addressed in [12], [13]. The paper [14] presents algorithms for synthesizing Engineering Change Order (ECO) patches. The use of interpolation for ECO has been presented in [8], [15], [16]. The single-fix rectification function approach in [15], [16] has been extended in [8] to generate multiple partial-fix functions. As these approaches are SAT based, they work well for random logic circuits but are not efficient for arithmetic circuits. In contrast to these works, our work presents a word-level formulation for single-fix rectification. Computer algebra has been utilized for circuit debugging and rectification in [17], [6], [7]. These approaches rely heavily on the structure of the circuit for coefficient calculation. If the arithmetic circuit contains redundancies, the approach may not identify the buggy gate due to ambiguity in coefficient values. On the

978-1-5386-4757-8/18 $31.00 © 2018 IEEE

other hand, our approach although more efficient for finite field arithmetic circuits, is applicable to any circuit in general.

The paper is organized as follows. The following section describes preliminary concepts in computer algebra and describe an equivalence checking framework using the Weak Nullstellensatz over finite fields. Section III presents our main theorem on identifying a rectification location and obtaining the correction function using Craig interpolants in finite fields. Section IV presents our experimental results and section V concludes the paper.

II. PRELIMINARIES

Let $\mathbb{F}_q$ denote the finite field of q elements where $q = 2^k$ and k is the operand width. Let $R = \mathbb{F}_q[x_1, \ldots, x_n]$ be the polynomial ring in n variables $x_1, \ldots, x_n$, with coefficients from $\mathbb{F}_q$. A monomial is a power product of variables $x_1^{e_1} \cdot x_2^{e_2} \cdots x_n^{e_n}$, where $e_i \in \mathbb{Z}_{\geq 0}, i \in \{1, \ldots, n\}$. A polynomial $f \in R$ is written as a finite sum of terms $f = c_1 X_1 + c_2 X_2 + \cdots + c_t X_t$, where $c_1, \ldots, c_t$ are coefficients and $X_1, \ldots, X_t$ are monomials. A monomial order $>$ (or a term order) is imposed on the ring so that the monomials of all polynomials $f = c_1 X_1 + c_2 X_2 + \cdots + c_t X_t$ are ordered w.r.t. $>$, such that $X_1 > X_2 > \cdots > X_t$, where $lm(f) = X_1$ is called the *leading monomial* of f. In this work, we employ lexicographic (lex) term orders (see Definition 1.4.3 in [11]).

We model the given circuit C by a set of multivariate polynomials $f_1, \ldots, f_s \in \mathbb{F}_{2^k}[x_1, \ldots, x_n]$; here $x_1, \ldots, x_n$ denote the nets (signals) of the circuit. Every Boolean logic gate of C is represented by a polynomial in $\mathbb{F}_2$, as $\mathbb{F}_2 \subset \mathbb{F}_{2^k}$. This is shown below. Note that in $\mathbb{F}_{2^k}$, $-1 = +1$.

$$
\begin{aligned}
z &= \neg a &\rightarrow\quad z + a + 1 &\pmod 2 \\
z &= a \wedge b &\rightarrow\quad z + a \cdot b &\pmod 2 \\
z &= a \vee b &\rightarrow\quad z + a + b + a \cdot b &\pmod 2 \\
z &= a \oplus b &\rightarrow\quad z + a + b &\pmod 2
\end{aligned}
\tag{1}
$$

Given a set of polynomials $F = \{f_1, \ldots, f_s\}$ in R, the *ideal* $J \subseteq R$ generated by them is:

$$
J = \langle f_1, \ldots, f_s \rangle = \{ \sum_{i=1}^{s} h_i \cdot f_i : h_i \in R \}.
$$

The polynomials $f_1, \ldots, f_s$ form the *generators* of J.

Let $a = (a_1, \ldots, a_n) \in \mathbb{F}_q^n$ be a point in the affine space, and f a polynomial in R. If $f(a) = 0$, we say that f *vanishes* on a. In verification, we have to analyze the *set of all common zeros* of the polynomials of F that lie within the field $\mathbb{F}_q$. In other words, we need to analyze solutions to the system of polynomial equations $f_1 = f_2 = \cdots = f_s = 0$. This zero set is called the *variety*. It depends not just on the given set of polynomials but rather on the ideal generated by them. We denote it by $V(J) = V(f_1, \ldots, f_s)$, where:

$$
V(J) = V(f_1, \ldots, f_s) = \{ a \in \mathbb{F}_q^n : \forall f \in J, f(a) = 0 \}.
$$

We denote the complement of a variety, $\mathbb{F}_q^n \setminus V(J)$, by $\overline{V(J)}$.

Algebraic Miter for Equivalence Checking: Given f_{spec} as the specification polynomial, we need to construct an *algebraic miter* between f_{spec} and C. For equivalence checking, we need

to prove that the miter is infeasible. Fig. 1 depicts how a word-level algebraic miter is setup. Suppose that $A = \{a_0, \ldots, a_{k-1}\}$ and $Z = \{z_0 \ldots, z_{k-1}\}$ denote the k-bit primary inputs and outputs of the finite field circuit. Then $A = \sum_{i=0}^{k-1} a_i \alpha^i, Z = \sum_{i=0}^{k-1} z_i \alpha^i$ correspond to the word-level polynomials for the inputs and outputs of the circuit. Here α is the primitive element of $\mathbb{F}_{2^k}$. Let Z_S be the word-level output for f_{spec}, which computes some polynomial function $\mathcal{F}(A)$ of A, so that $f_{spec} : Z_S + \mathcal{F}(A)$. The word-level outputs Z, Z_S are mitered to check if for all inputs, $Z \neq Z_S$ is infeasible.

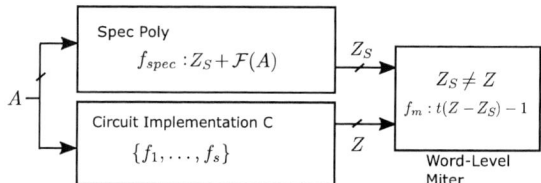

Fig. 1: Word-Level Miter

In finite fields, the disequality $Z \neq Z_S$ can be modeled as a single polynomial f_m, called the miter polynomial, where $f_m = t \cdot (Z - Z_S) - 1$, and t is introduced as a free variable. If $Z = Z_S$, $Z - Z_S = 0$. So $f_m : t \cdot 0 + 1 = 0$ has no solutions (miter is infeasible). Whereas if for some input A, $Z \neq Z_S$, then $Z - Z_S \neq 0$. Let $t^{-1} = (Z - Z_S) \neq 0$. Then $f_m : t \cdot t^{-1} - 1 = 0$ has a solution as t, t^{-1} become multiplicative inverses of each other. Thus the miter becomes feasible.

In this way, equivalence checking using the algebraic model is solved as follows: Construct an ideal $J = \langle f_{spec}, f_1, \ldots, f_s, f_m \rangle$, as described above. Then determine if the variety $V(J) = \emptyset$? If $V(J) = \emptyset$, the miter is infeasible, and C implements f_{spec}. If $V(J) \neq \emptyset$, the miter is feasible, and there exists a bug in the design.

The Weak Nullstellensatz: To ascertain whether $V(J) = \emptyset$, we employ the Weak Nullstellensatz over $\mathbb{F}_q$, for which we use the following notations. Given two ideals $J_1 = \langle f_1, \ldots, f_s \rangle, J_2 = \langle h_1, \ldots, h_r \rangle$, the sum $J_1 + J_2 = \langle f_1, \ldots, f_s, h_1 \ldots, h_r \rangle$, and $V(J_1 + J_2) = V(J_1) \cap V(J_2)$. Moreover, if $J_1 \subseteq J_2$ then $V(J_1) \supseteq V(J_2)$.

For all elements $\alpha \in \mathbb{F}_q, \alpha^q = \alpha$. Therefore, the polynomial $x^q - x$ vanishes everywhere in $\mathbb{F}_q$, and is called the vanishing polynomial of the field. Let $J_0 = \langle x_1^q - x_1, \ldots, x_n^q - x_n \rangle$ be the ideal of all vanishing polynomials in R.

Theorem II.1 (*The Weak Nullstellensatz over finite fields (from Theorem 3.3 in [18])*). *For a finite field $\mathbb{F}_q$ and the ring $R = \mathbb{F}_q[x_1, \ldots, x_n]$, let $J = \langle f_1, \ldots, f_s \rangle \subseteq R$, and let $J_0 = \langle x_1^q - x_1, \ldots, x_n^q - x_n \rangle$ be the ideal of vanishing polynomials. Then $V(J) = \emptyset \iff 1 \in J + J_0$.*

To determine whether $V(J) = \emptyset$, we need to test whether or not the unit element 1 is a member of the ideal $J + J_0$. For this *ideal membership test, we need to compute a Gröbner basis of $J + J_0$.*

Gröbner Basis of Ideals: An ideal may have many different sets of generators: $J = \langle f_1, \ldots, f_s \rangle = \cdots = \langle g_1, \ldots, g_t \rangle$. Given a non-zero ideal J, a *Gröbner basis* (GB) for J is a finite set of polynomials $G = \{g_1, \ldots, g_t\}$ satisfying $\langle \{ lm(f) \mid f \in J \} \rangle = \langle lm(g_1), \ldots, lm(g_t) \rangle$. Then $J = \langle G \rangle$ holds and so $G = GB(J)$ forms a basis for J. A GB G possesses important properties

978-1-5386-4757-8/18 $31.00 © 2018 IEEE

that allow to solve many polynomial computation and decision problems. The famous Buchberger's algorithm (see Alg. 1.7.1 in [11]) takes as input the set of polynomials $F = \{f_1, \ldots, f_s\}$ and computes the GB $G = \{g_1, \ldots, g_t\}$. A GB can be *reduced* to eliminate redundant polynomials from the basis. A reduced GB is a canonical representation of the ideal. When $1 \in J$, then $G = reduced\_GB(J) = \{1\}$.

Thus, for equivalence check, we compute a reduced GB $G = GB(J + J_0)$, and see if $G = \{1\}$. If so, $V(J) = \emptyset$ and the miter is infeasible. If $G \neq \{1\}$, then there exists a bug in the design.

Craig interpolation: The Weak Nullstellensatz is the polynomial analog of SAT/UNSAT checking. For UNSAT problems, the formal logic and verification communities have explored the notion of abstraction of functions by means of Craig interpolants, which has been applied to circuit rectification [8]. In propositional logic, the concept is defined as follows:

Definition II.1. Let (A, B) be a pair of CNF formulae (sets of clauses) such that $A \wedge B$ is unsatisfiable. Then there exists a formula I such that: (i) $A \implies I$; (ii) $I \wedge B$ is unsatisfiable; and (iii) I refers only to the common variables of A and B, i.e. $Var(I) \subseteq Var(A) \cap Var(B)$. The formula I is called the **interpolant** of (A, B).

Given the pair (A, B) and their refutation proof, a procedure called the *interpolation system* constructs the interpolant in linear time and space in the size of the proof. In our work [10], we have proposed the notion (theory and algorithms) of Craig interpolants in polynomial algebra over finite fields, based on the results of Nullstellensatz. These are presented and utilized in this paper for rectification of arithmetic circuits.

Elimination Ideals: We employ one more concept, that of elimination ideals.

Definition II.2. Given an ideal $J \subset \mathbb{F}_q[x_1, \ldots, x_n]$, the l-th elimination ideal J_l is an ideal in R defined as $J_l = J \cap \mathbb{F}_q[x_{l+1}, \ldots, x_n]$.

Theorem II.2 (*Elimination Theorem (from Theorem 2.3.4 [11])*). Given an ideal $J \subset R$ and its GB G w.r.t. the lexicographical (lex) order on the variables where $x_1 > x_2 > \cdots > x_n$, then for every $0 \leq l \leq n$ we denote by G_l the GB of l-th elimination ideal of J and compute it as:

$$G_l = G \cap \mathbb{F}_q[x_{l+1}, \ldots, x_n].$$

G_l is called the l-th elimination ideal as it eliminates the first l variables from J.

III. THEORY

This section presents our main theorem on checking whether a buggy circuit is single fix rectifiable, and a procedure for computing a correction function using the theory and algorithms on Craig interpolants in finite fields [10].

After the verification of a circuit against the specification detects the presence of a bug in the design, we are provided with a list of potential gate-output nets x_i's. The circuit may or may not be rectified at a particular x_i. First we ascertain that the circuit can indeed be rectified at some x_i and then apply a correction function $U(X_{PI})$ as $x_i = U(X_{PI})$.

A. Single Fix Rectification

In this subsection, we formally set up the problem of single fix circuit rectification. Using the Weak Nullstellensatz (Theorem II.1), we formulate the test for rectifiability at a gate output x_i in the circuit. The following proposition will be used later in this subsection.

Proposition III.1. Given two ideals J_1 and J_2 over some finite field such that $V(J_1) \cap V(J_2) = \emptyset$, there exists a polynomial U which satisfies $V(J_1) \subseteq V(U) \subseteq \overline{V(J_2)}$.

Proof. Over finite fields, $V(J_1)$ and $V(J_2)$ are finite sets of points. There exists a set of points which contains $V(J_1)$ and does not intersect with $V(J_2)$. As every set of points in finite fields is a variety, let this variety be denoted by $V(J_I)$, where J_I is the corresponding ideal. Then $V(J_1) \subseteq V(J_I) \subseteq \overline{V(J_2)}$. In addition, we can construct a polynomial U whose roots are exactly the points in $V(J_I)$ by means of the Lagrange's interpolation formula. $\square$

Now we present the theorem to check the circuit's rectifiability at some gate output. Let us assume that a potential rectifiable gate output is x_i (*i.e.* i^{th} gate) and a possible function in primary inputs that can be implemented is $x_i = U(X_{PI})$ so that the ith gate is represented by a polynomial $f_i : x_i + U(X_{PI})$. The ideal constructed from the polynomials for the gates $f_1, \ldots, f_s$ of the circuit, the specification polynomial f_{spec}, and the miter polynomial f_m, is denoted by J:

$$J = \langle f_{spec}, f_1, \ldots, f_i : x_i + U(X_{PI}), \ldots, f_s, f_m \rangle.$$

The following theorem checks whether the circuit is indeed rectifiable at gate with output net x_i.

Theorem III.1. Construct two ideals:

- $J_L = \langle f_{spec}, f_1, \ldots, f_i : x_i + 1, \ldots, f_s, f_m \rangle$ where $f_i : x_i + U(X_{PI})$ in J is replaced with $f_i : x_i + 1$.
- $J_H = \langle f_{spec}, f_1, \ldots, f_i : x_i, \ldots, f_s, f_m \rangle$ where $f_i : x_i + U(X_{PI})$ in J is replaced with $f_i : x_i$.

Compute $E_L = (J_L + J_0) \cap \mathbb{F}_{2^k}[X_{PI}]$ and $E_H = (J_H + J_0) \cap \mathbb{F}_{2^k}[X_{PI}]$ to be the respective elimination ideals, where all the non-primary input variables have been eliminated. Then the circuit can be rectified with a logic function at net x_i with the polynomial function $f_i : x_i + U(X_{PI})$ to implement the specification iff $1 \in E_L + E_H$.

Proof. We will first prove the *if* case of the theorem. Assume $1 \in E_L + E_H$, or equivalently $V_{X_{PI}}(E_L) \cap V_{X_{PI}}(E_H) = \emptyset$. Using Proposition III.1, we can find a polynomial $U(X_{PI})$ such that,

$$V_{X_{PI}}(E_L) \subseteq V_{X_{PI}}(U(X_{PI})) \subseteq \overline{V_{X_{PI}}(E_H)} \quad (2)$$

where the universal set for computing $\overline{V_{X_{PI}}(E_H)}$ is $\mathbb{F}_{2^k}^{X_{PI}}$. Let us assume that a point $\boldsymbol{p}$ exists in $V(J)$. Point $\boldsymbol{p}$ is an assignment to every variable in J such that all the generators of J are satisfied. We denote by $\boldsymbol{a}$, the projection of $\boldsymbol{p}$ on the primary inputs (the primary input assignments under $\boldsymbol{p}$). There are only two possibilities for $U(X_{PI})$,

1) $U(\boldsymbol{a}) = 1$, or in other words $\boldsymbol{a} \notin V_{X_{PI}}(U(X_{PI}))$. It also implies that the value of x_i under $\boldsymbol{p}$ must be 1 because $x_i + U(X_{PI})$ needs to be satisfied. Since the generator f_i

of J_L also forces x_i to be 1 and all its other generators are exactly the same as that of J, $\boldsymbol{p}$ is also a point in $V(J_L)$. Moreover, E_L is the elimination ideal of J_L, and therefore, $\boldsymbol{a} \in V_{X_{PI}}(E_L)$. But this a contradiction to our assumption that $V_{X_{PI}}(E_L) \subseteq V_{X_{PI}}(U(X_{PI}))$ and such a point $\boldsymbol{a}$ (and $\boldsymbol{p}$) does not exist.

2) $U(\boldsymbol{a}) = 0$, or in other words $\boldsymbol{a} \in V_{X_{PI}}(U(X_{PI}))$. Using similar argument as the previous case, we can show that $\boldsymbol{a} \in V_{X_{PI}}(E_H)$. This is again a contradiction to our assumption $V_{X_{PI}}(U(X_{PI})) \subseteq \overline{V_{X_{PI}}(E_H)}$.

In conclusion, there exists no point in $V(J)$ (or the miter is infeasible) when $U(X_{PI})$ satisfies Eqn. 2, and therefore, circuit can be rectified at x_i.

Now we will prove the *only if* direction of the proof. We show that if $1 \notin E_L + E_H$, then there exists no polynomial $U(X_{PI})$ that can rectify the circuit. If $1 \notin E_L + E_H$, then E_L and E_H have a common zero. Let $\boldsymbol{a}$ be a point in $V_{X_{PI}}(E_L)$ and $V_{X_{PI}}(E_H)$. This point can be extended to some points $\boldsymbol{p}'$ and $\boldsymbol{p}''$ in $V(J_L)$ and $V(J_H)$, respectively. Notice that in point $\boldsymbol{p}'$ the value of x_i will be 1, and in $\boldsymbol{p}''$ x_i will be 0. Any polynomial $U(X_{PI})$ will either evaluate to 0 or 1 for the assignment $\boldsymbol{a}$ to the primary inputs. If it evaluates to 1, then we can say that $\boldsymbol{p}'$ is in $V(J)$ as f_i in J forces $x_i = 1$ and all other generators of J and J_L are same. This implies that $f_m(\boldsymbol{p}') = 0$ (f_m: miter polynomial is feasible) and this choice of $U(X_{PI})$ will not rectify the circuit. If $U(X_{PI})$ evaluates to 0, then $\boldsymbol{p}''$ is a point in $V(J)$.

Therefore, no choice of $U(X_{PI})$ can rectify the circuit if $1 \notin E_L + E_H$. $\qquad \square$

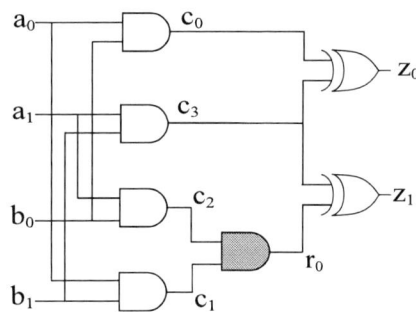

Fig. 2: A buggy 2-bit modulo multiplier circuit

Example III.1. *Consider the buggy modulo multiplier circuit in Fig. 2 where the gate output r_0 should have been the output of an XOR gate and the AND gate has been incorrectly implemented. We want to apply Thm. III.1 at r_0. The polynomials for the gates of the correct circuit implementation are,*

$$f_1 : c_0 + a_0 \cdot b_0; \quad f_2 : c_1 + a_0 \cdot b_1; \quad f_3 : c_2 + a_1 \cdot b_0;$$
$$f_4 : c_3 + a_1 \cdot b_1; \quad f_5 : r_0 + c_1 + c_2; \quad f_6 : z_0 + c_0 + c_3;$$
$$f_7 : z_1 + r_0 + c_3;$$

The problem is modeled over $\mathbb{F}_4$ and α is the primitive element of $\mathbb{F}_4$. The word-level polynomials are $f_8 : Z + z_0 + z_1\alpha$, $f_9 : A + a_0 + a_1\alpha$, and $f_{10} : B + b_0 + b_1\alpha$. The specification polynomial is $f_{spec} : Z_s + AB$. We create a miter polynomial against this specification as $f_m : t(Z - Z_s) - 1$.

The ideals J_L and J_H are as follows,

$$J_L = \langle f_{spec}, f_1, \ldots, f_4, r_0 + 1, f_6, \ldots, f_{10}, f_m \rangle$$
$$J_H = \langle f_{spec}, f_1, \ldots, f_4, r_0, f_6, \ldots, f_{10}, f_m \rangle$$

and the corresponding ideals E_L and E_H are as follows,

$$E_L = \langle a_0 b_1 + a_1 b_0, a_1 b_0 b_1 + a_1 b_0, a_0 a_1 b_0 + a_1 b_0 \rangle$$
$$E_H = \langle b_0 b_1 + b_0 + b_1 + 1, a_1 b_1 + a_1 + b_1 + 1, a_0 b_1 + a_1 b_0 + 1,$$
$$a_0 b_0 + a_0 + b_0 + 1, a_0 a_1 + a_0 + a_1 + 1 \rangle$$

If we compute a Gröbner basis of $E_L + E_H$, it results in $\{1\}$. Therefore, we can rectify this circuit at r_0.

B. Craig Interpolants in Finite Fields

If x_i is a feasible location for rectification, then the corresponding E_L and E_H satisfy $1 \in E_L + E_H$. We are now in a position to introduce the notion of Craig interpolants in finite fields which will help us in obtaining $U(X_{PI})$ from an "ideal-interpolant" J_I defined below.

Definition III.1 (*Interpolants in finite fields*). Given two ideals $J_A \subset \mathbb{F}_q[A, C]$ and $J_B \subset \mathbb{F}_q[B, C]$ where A, B, C denote the three disjoint sets of variables such that $V_{A,B,C}(J_A) \cap V_{A,B,C}(J_B) = \emptyset$. Then there exists an ideal J_I satisfying the following properties:

1) $V_{A,B,C}(J_I) \supseteq V_{A,B,C}(J_A)$
2) $V_{A,B,C}(J_I) \cap V_{A,B,C}(J_B) = \emptyset$
3) Generators of J_I contain only the C-variables; or $J_I \subseteq \mathbb{F}_q[C]$.

We call $V_{A,B,C}(J_I)$ the **interpolant** in finite fields of the pair $(V_{A,B,C}(J_A), V_{A,B,C}(J_B))$, and the corresponding ideal J_I the **ideal-interpolant**.

Example III.2. *Consider the ring $R = \mathbb{F}_2[a, b, c, d, e]$, partition the variables as $A = \{a\}, B = \{e\}, C = \{b, c, d\}$. Let ideals*

$$J_A = \langle ab, bd, bc + c, cd, bd + b + d + 1 \rangle + J_{0,A,C}$$
$$J_B = \langle b, d, ec + e + c + 1, ec \rangle + J_{0,B,C}$$

where $J_{0,A,C}$ and $J_{0,B,C}$ are the corresponding ideals of vanishing polynomials. Then, we have

$$V_{A,B,C}(J_A) = \mathbb{F}_q^B \times V_{A,C}(J_A) = (abcde):$$
$$\{01000, 00010, 01100, 10010, 01001, 00011, 01101, 10011\}$$
$$V_{A,B,C}(J_B) = \mathbb{F}_q^A \times V_{B,C}(J_B) = (abcde):$$
$$\{00001, 00100, 10001, 10100\}$$

The ideals J_A, J_B have no common zeros as $V_{A,B,C}(J_A) \cap V_{A,B,C}(J_B) = \emptyset$. The pair (J_A, J_B) admits a total of 8 interpolants:

1) $V(J_S) = (bcd) : \{001, 100, 110\}$ $J_S = \langle cd, b + d + 1 \rangle$
2) $V_C(J_1) = (bcd) : \{001, 100, 110, 101\}$
   $J_1 = \langle cd, bd + b + d + 1, bc + cd + c \rangle$
3) $V_C(J_2) = (bcd) : \{001, 100, 110, 011\}$
   $J_2 = \langle b + d + 1 \rangle$
4) $V_C(J_3) = (bcd) : \{001, 100, 110, 111\}$
   $J_3 = \langle b + cd + d + 1 \rangle$
5) $V_C(J_4) = (bcd) : \{001, 100, 110, 011, 111\}$
   $J_4 = \langle bd + b + d + 1, bc + b + cd + c + d + 1 \rangle$
6) $V_C(J_5) = (bcd) : \{001, 100, 110, 101, 111\}$
   $J_5 = \langle bc + c, bd + b + d + 1 \rangle$

978-1-5386-4757-8/18 $31.00 © 2018 IEEE

7) $V_C(J_6) = (bcd) : \{001, 100, 110, 101, 011\}$
   $J_6 = \langle bd + b + d + 1, bc + cd + c \rangle$
8) $V_C(J_L) = (bcd) : \{001, 011, 100, 101, 110, 111\}$
   $J_L = \langle bd + b + d + 1 \rangle$.

It is easy to check that all $V(J_I)$ satisfy the 3 conditions of Def. III.1. Note also that $V(J_S)$ is the smallest interpolant, contained in every other interpolant. Likewise, $V(J_L)$ contains all other interpolants and it is the largest. The other containment relationships are shown in the corresponding interpolant lattice in Fig. 3; $V_C(J_1) \subset V_C(J_5), V_C(J_1) \subset V_C(J_6)$, etc.

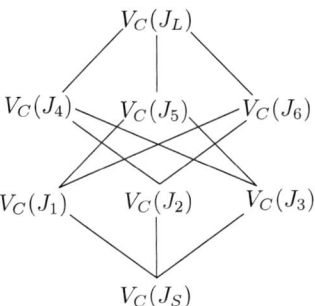

Fig. 3: Interpolant lattice for Example III.2

Theorem III.2. (from Theorem IV.1 in [10]) An ideal-interpolant J_I, and correspondingly the interpolant $V_{A,B,C}(J_I)$, as given in Def. III.1, always exists.

Another result from [10] (Theorem IV.2) that we make use of here is that the smallest interpolant can be computed as $J_I = J_A \cap \mathbb{F}_q[C]$.

Back to our formulation of single fix rectification, we have $1 \in E_L + E_H$ or $V(E_L) \cap V(E_H) = \emptyset$. E_L and E_H are elimination ideals containing only X_{PI} variables. As a result, the set of variables A, B, and C are primary inputs. Moreover, we want to compute an ideal J_I in X_{PI} such that $V_{X_{PI}}(E_L) \subseteq V_{X_{PI}}(J_I)$ and $V_{X_{PI}}(J_I) \cap V_{X_{PI}}(E_H) = \emptyset$. The *smallest ideal-interpolant* $J_I = E_L \cap \mathbb{F}_2[X_{PI}] = E_L$ itself. Therefore, we use E_L to compute the correction function $U(X_{PI})$.

C. Obtaining $U(X_{PI})$ from E_L

In finite fields, given an ideal J, it always possible to find a polynomial U such that $V(U) = V(J)$. The reason is that every ideal in a finite field has a finite variety and a polynomial with those points as its roots can always be constructed. Let the generators of J be denoted by $g_1, \dots, g_t$. We can compute U as,

$$U = (1 + g_1)(1 + g_2) \cdots (1 + g_t) + 1 \qquad (3)$$

It is easy to assert that $V(U) = V(J_I)$. Using the Eqn. 3, we can write a recursive procedure as presented in Algorithm 1 to compute U. In addition, at every recursive step we also reduce the intermediate sum by J_0 (line 6) to avoid large degree terms. In our setting, $U = U(X_{PI})$ and $J = E_L$, and therefore, we can find a correction function $x_i + U(X_{PI})$ which can be used to rectify the circuit.

Using this procedure for Example III.1, we have $U(X_{PI})$ as $a_0 b_1 + a_1 b_0$ and the correction function as $r_0 + a_0 b_1 + a_1 b_0$

Algorithm 1 Compute U from J_I such that $V(U) = V(J_I)$

1: **procedure** $compute\_U(J_I, J_0)$ //$J_I = \{f_1, \dots, f_s\}$
2:     **if** $size(J_I) = 1$ **then**
3:         **return** $(1 + J_I[1])$
4:     $subsetJ = \{J_I[1], J_I[2], \dots, J_I[size(J_I) - 1]\}$
5:     $poly\ S_1 = compute\_U(subsetJ, J_0)$
6:     Perform $S_1 \cdot J_I[size(J_I)] \xrightarrow{J_0}_{+} S_2$
7:     **return** $S_1 + S_2$

---

which can be synthesized as $r_0 = (a_0 \wedge b_1) \oplus (a_1 \wedge b_0)$ (replacing the modulo 2 product and sum with Boolean AND and XOR, respectively).

IV. EXPERIMENTAL RESULTS

We have performed experiments on finite field arithmetic circuits (used in cryptography) where the implementation is different from the specification due to exactly one gate. This is to ensure that a single fix rectification is feasible. We implement the procedure described in the previous section (Thm. III.1 and Algo. 1) using the SINGULAR symbolic algebra computation system [ver. 4-1-0][19]. The experiments were conducted on a desktop computer with a 3.5GHz Intel Core™ i7-4770K Quad-core CPU, 16 GB RAM, running 64-bit Linux OS.

We have performed experiments with three different types of finite field benchmarks. First two of them are Mastrovito and Montgomery multiplier circuits used for modular multiplication. Mastrovito multipliers compute $Z = A \times B$ (mod $P(x)$) where $P(x)$ is a given primitive polynomial for the datapath size k. Montgomery multipliers are preferred for exponentiation operations (often required in cryptosystems) over Mastrovito multipliers. The last set of benchmarks are circuits implementing point addition over elliptic curves used for encryption, decryption and authentication in elliptic curve cryptography.

TABLE I: Mastrovito multiplier rectification against Montgomery multiplier specification. Time in seconds; Time-out = 5400s; k: Operand width

k	# of Gates		SAT	Thm. III.1	Algo. 1	Mem
	Mas	Mont				
4	48	96	0.09	0.03	0.001	8.16 MB
8	292	319	158.34	0.41	0.006	20.36 MB
9	237	396	4,507	0.47	0.001	18.95 MB
10	285	480	TO	0.84	0.001	28.2 MB
16	1,836	1,152	TO	73.63	0.024	0.32 GB
32	5,482	4,352	TO	3621	0.043	2.4 GB

First we present the results for the case where the Thm. III.1 is applied at a gate location such that the circuit is completely rectifiable. Table I compares the execution time for SAT based approach [8] and our approach (Theorem III.1) for checking whether a buggy Mastrovito multiplier can be rectified at a certain location in the circuit against a Montgomery multiplier specification. We have implemented the SAT procedure using the *abc* tool [20]. We execute the command *inter* on the ON set and OFF set as described in [8]. The SAT based procedure

978-1-5386-4757-8/18 $31.00 © 2018 IEEE

is unable to perform the necessary unsatisfiability check for circuits beyond of 9 bit operand words. Using our approach, the polynomial $U(X_{PI})$ needed for rectification is computed from E_L and the time is reported in Table I in the Algo. 1 column. The last column in the table is the memory usage of our approach.

We can also perform the rectification when a polynomial specification is given instead of a specification circuit. Table II shows the result of checking whether the incorrect Mastrovito implementation can be rectified at a particular location against the word level specification polynomial $Z_S = AB$.

TABLE II: Mastrovito multiplier rectification against polynomial specification $Z_S = AB$. Time in seconds; Time-out = 5400s; k: Operand width

k	# of Gates	Thm. III.1	Algo. 1	Mem
4	48	0.01	0.001	7.24 MB
8	292	0.08	0.006	14.95 MB
16	1,836	4.83	0.038	0.2 GB
32	5,482	100.52	0.015	1.42 GB
64	21,813	4,989	0.117	12.25 GB

Point addition operation can be represented as polynomials because modern approaches represent the points in projective coordinate systems, e.g., the López-Dahab (LD) projective coordinate [21]. Each of these polynomials can be implemented as a circuit. Table III shows the result for one of these blocks. For all the experiments, the most computationally expensive part is the computation of ideals E_L and E_H.

TABLE III: Point Addition circuit rectification against polynomial specification $D = B^2 \cdot (C + aZ_1^2)$. Time in seconds; Time-out = 5400s; k: Operand width

Field Size (k)	# of Gates	Thm. III.1	Algo. 1	Mem
8	243	0.05	0.022	9.73 MB
16	1,277	3.48	0.019	88.78 MB
32	3,918	86.75	0.028	0.47 GB
64	1,5305	4,923	0.053	7.13 GB

We also performed experiments where we apply Thm. III.1 at a gate output which cannot rectify the circuit. We used Montgomery circuit as the specification and Mastrovito as the implementation as we did for the experiments in Table I. For 4 and 8 bits size cases, the execution time was comparable for Thm. III.1 and SAT based approach and was ~ 0.1 seconds. When we tried the 16 bit case, the SAT based approach was able to complete in 0.11 seconds. On the other hand, Thm. III.1 formulation resulted in a memory explosion and consumed ~ 30 GB of memory in 5-6 minutes. This is due to the fact when $1 \notin E_L + E_H$, then $GB(E_L + E_H)$ is not equal to $\{1\}$ and the Gröbner basis algorithm produces a very large generating set. To improve our approach we are working on term ordering heuristics so that our approach can perform efficiently in both cases. We also want to employ better data structures as SINGULAR's data structure is not very memory efficient and also has an upper limit on the number of variables (32,768) that can be accommodated in the system.

V. CONCLUSION

This paper considers the single-fix rectification of circuits after the verification has detected a bug in the design. A num-
ber of possible gate outputs are provided whose functionality can be changed so that circuit corresponds to the specification. We want to select one such (single) gate output so that by applying a correction function there, the circuit is rectified. We present a theorem that answers definitively whether a single fix rectification is feasible at a particular gate output. We also briefly describe the notion and definition of Craig interpolants in finite fields which is used to obtain a correction function. Experiments performed over finite field arithmetic circuits shows the efficiency of our approach and also points out the regions for improvements.

REFERENCES

[1] D. Ritirc, A. Biere, and M. Kauers, "Column-Wise Verification of Multipliers Using Computer Algebra," in *Formal Methods in Computer-Aided Design (FMCAD)*, 2017, pp. 23–30.

[2] M. Ciesielski, C. Yu, W. Brown, D. Liu, and A. Rossi, "Verification of Gate-level Arithmetic Circuits by Function Extraction," in *52nd ACM/EDAC/IEEE Design Automation Conference (DAC)*, 2015, pp. 1–6.

[3] A. Sayed-Ahmed, D. Große, U. Kühne, M. Soeken, and R. Drechsler, "Formal Verification of Integer Multipliers by Combining Gröbner Basis with Logic Reduction," in *Design, Automation Test in Europe Conference Exhibition (DATE)*, 2016, pp. 1048–1053.

[4] J. Lv, P. Kalla, and F. Enescu, "Efficient Gröbner Basis Reductions for Formal Verification of Galois Field Arithmetic Circuits," in *IEEE Trans. on CAD*, vol. 32, no. 9, 2013, pp. 1409–1420.

[5] A. Lvov, L. Lastras-Montano, B. Trager, V. Paruthi, R. Shadowen, and A. El-Zein, "Verification of Galois field based circuits by formal reasoning based on computational algebraic geometry," *Formal Methods in System Design*, vol. 45, no. 2, pp. 189–212, Oct 2014.

[6] F. Farahmandi and P. Mishra, "Automated Debugging of Arithmetic Circuits Using Incremental Gröbner Basis Reduction," in *IEEE International Conference on Computer Design (ICCD)*, 2017.

[7] ——, "Automated Test Generation for Debugging Arithmetic Circuits," in *Des. Auto. Test in Eur. Conf. (DATE)*, 2016.

[8] K. F. Tang, C. A. Wu, P. K. Huang, and C. Y. Huang, "Interpolation-Based Incremental ECO Synthesis for Multi-Error Logic Rectification," in *Proc. Design Automation Conf. (DAC)*, 2011, pp. 146–151.

[9] W. Craig, "Linear reasoning: A new form of the Herbrand-Gentzen theorem," *Journal of Symbolic Logic*, vol. 22, no. 3, pp. 250–268, 1957.

[10] U. Gupta, I. Ilioaea, P. Kalla, F. Enescu, V. Rao, and A. Srinath, "Craig Interpolants in Finite Fields using Algebraic Geometry: Theory and Applications," in *Intl. Workshop on Logic and Synthesis (IWLS)*, June 2018.

[11] W. W. Adams and P. Loustaunau, *An Introduction to Gröbner Bases*. American Mathematical Society, 1994.

[12] J. C. Madre, O. Coudert, and J. P. Billon, "Automating the Diagnosis and the Rectification of Design Errors with PRIAM," in *Proc. ICCAD*, 1989, pp. 30–33.

[13] H. T. Liaw, J. H. Tsaih, and C. S. Lin, "Efficient Automatic Diagnosis of Digital Circuits," in *Proc. ICCAD*, 1990, pp. 464–467.

[14] C. C. Lin, K. C. Chen, S. C. Chang, and M. Marek-Sadowska, "Logic Synthesis for Engineering Change," in *Proc. Design Automation Conf. (DAC)*, 1995, pp. 647–652.

[15] B. H. Wu, C. J. Yang, C. Y. Huang, and J. H. R. Jiang, "A Robust Functional ECO Engine by SAT Proof Minimization and Interpolation Techniques," in *Intl. Conf. on Comp. Aided Des.*, 2010, pp. 729–734.

[16] A. C. Ling, S. D. Brown, S. Safarpour, and J. Zhu, "Toward Automated ECOs in FPGAs," *IEEE Trans. CAD*, vol. 30, no. 1, pp. 18–30, 2011.

[17] S. Ghandali, C. Yu, D. Liu, W. Brown, and M. Ciesielski, "Logic Debugging of Arithmetic Circuits," in *IEEE Computer Society Annual Symposium on VLSI*, 2015.

[18] S. Gao, "Counting Zeros over Finite Fields with Gröbner Bases," Master's thesis, Carnegie Mellon University, 2009.

[19] W. Decker, G.-M. Greuel, G. Pfister, and H. Schönemann, "SINGULAR 4-1-0 — A computer algebra system for polynomial computations," http://www.singular.uni-kl.de, 2016.

[20] R. Brayton and A. Mishchenko, "ABC: An Academic Industrial-Strength Verification Tool," in *Comp. Aid. Verif.*, vol. 6174, 2010, pp. 24–40.

[21] J. López and R. Dahab, "Improved Algorithms for Elliptic Curve Arithmetic in GF(2^n)," in *Proceedings of the Selected Areas in Cryptography*. London, UK, UK: Springer-Verlag, 1999, pp. 201–212.

A Synthesizable Digital Low-Dropout Regulator Based on Voltage-to-Time Conversion

Naoki Ojima[1], Toru Nakura[2], Tetsuya Iizuka[1, 3] and Kunihiro Asada[1, 3]

[1]Department of Electrical Engineering and Information Systems, The University of Tokyo, Japan
[2]Department of Electrical Engineering and Computer Science, Fukuoka University, Japan
[3]VLSI Design and Education Center, The University of Tokyo, Japan
E-mail: ojima@silicon.u-tokyo.ac.jp

Abstract— This paper proposes a synthesizable digital LDO that is implemented with standard-cell-based digital design flow. With inverter chains as voltage-controlled delay lines, the difference between output and reference voltages is converted into delay difference, then compared in time-domain. Since the time-domain difference is straightforwardly captured by a phase detector that consists of a D-FF, the proposed LDO does not need an analog voltage comparator, which requires careful manual design. The prototype of the proposed LDO is fabricated in 65 nm standard CMOS technology with 0.015 mm² area occupation. The measurement results show that with 10.4 MHz internal clock the tracking response to 200 mV switching of the reference voltage is ~4.5 µs and the transient response to 5 mA change of the load current is ~6.6 µs. The quiescent current consumed by the LDO core is as low as 35.2 µA at 10 mA load current, which leads to 99.6 % current efficiency.

Index Terms—Power management, low-dropout regulator, digital LDO, circuit synthesis.

I. INTRODUCTION

Along with the exponential advancement of process technologies, performance of LSI circuits rapidly improves and many functional building blocks can be embedded in one chip, which have brought system-on-a-chip (SoC) era. Meanwhile, in order to reduce power consumption as indicated by the scaling rule, power supply voltages have been lowered. In addition, it is desirable that a power supply of each functional block is tunable according to the changing operational situation to have the optimal power efficiency. For those reasons, efficient, tunable and fast-transient power sources are in great demand for SoC, hence low-dropout (LDO) regulators are now widely used. As shown in Fig. 1(a), conventional LDOs have been designed with analog circuits and employed an error amplifier and a driver amplifier to provide voltage regulation with negative feedback. They exhibit high efficiency, fast response time and high power supply rejection (PSR) [1]–[3]. However, they have difficulty in operating at low supply voltage, which has prompted the digital implementations of LDOs [3]–[6] as shown in Fig. 1(b). Typical digital LDO has a digital controller made of logic gates that controls the number of turned-on PMOS switches at output stage, and employs an analog voltage comparator to detect the difference between reference and feedback voltages. A voltage comparator, however, often requires careful manual design so as to minimize the voltage offset between two inputs. Thus this kind of LDO requires sophisticated analog design flows, which is often time-consuming. A digital design flow,

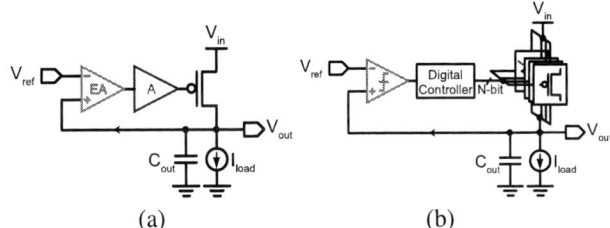

Fig. 1. LDO architectures. (a) Conventional analog LDO has a simple architecture, and includes an error amplifier and a driver amplifier. (b) Digital LDO includes a comparator and a digital controller made of logic gates.

on the other hand, requires much less design effort even when the used process technologies are updated. Thus recently many analog circuits such as analog-to-digital converters (ADC) or phase-locked loops (PLL) are widely designed through digital automated flows. We have been motivated to implement LDOs, one of the indispensable blocks for SoCs, in digital design flows. In this paper, we propose a synthesizable digital LDO whose architecture is suitable for standard-cell-based automatic place and route (P&R) in order to relax the burden of analog designs.

II. PROPOSED SYNTHESIZABLE LDO

A. Architecture

One of the design issues in constructing an LDO with standard cells is an implementation of a voltage comparison unit. The PLL-like LDO in [5] employs voltage-controlled ring oscillators in order to convert the voltage difference into the phase difference. However, this architecture is not preferable because a voltage-controlled ring oscillator has an integral characteristic that adds a pole to the system. Thus it often deteriorates the stability. Moreover, voltage-controlled ring oscillators might be a cause to increase the current consumption of the voltage comparison unit. Our proposed LDO shown in Fig. 2 employs voltage-controlled delay lines rather than voltage-controlled ring oscillators, and the difference between the reference and the output voltages are converted into the time-domain. The proposed LDO consists of two inverter chains, a bang-bang phase detector, a digital controller, a PMOS switch array, and an output capacitor. Though for this prototype a dedicated ring oscillator is used as an internal clock source and a pulse generator, these clock and pulse signals can be replaced by a clock for other blocks on the SoC. The digital controller generates 128-bit width thermometer code from 1-bit output from the bang-bang phase

978-1-5386-4757-8/18 $31.00 © 2018 IEEE   55

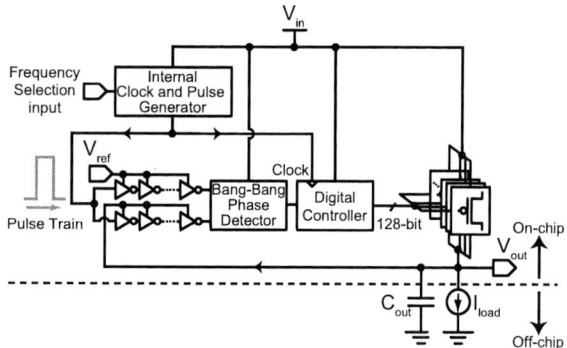

Fig. 2. Block diagram of the proposed LDO.

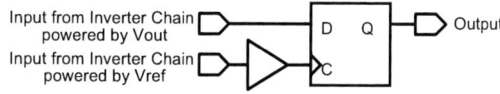

Fig. 3. Phase detector composed of standard cells.

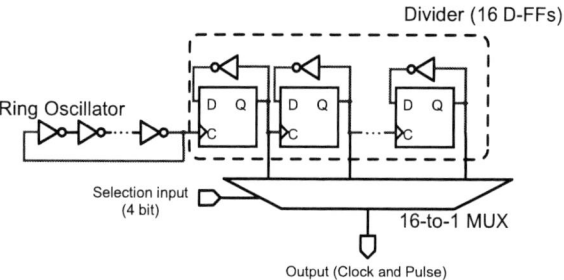

Fig. 4. Internal clock and pulse generator composed of a ring oscillator, a divider, and a multiplexer.

detector to control the switches. The PMOS switch array has parallelly-aligned 128 PMOS transistors, all of which have the same size. Each gate of the switch is connected to each bit of the thermometer code from the digital controller. The two inverter chains have the same structure, which has a series connection of 128 inverters. As shown in Fig. 3, the bang-bang phase detector is composed of a D-FF and a buffer. The buffer is connected to the clock input of the D-FF to compensate the setup time of the D-FF. As shown in Fig. 4, the internal clock and pulse generator is composed of inverters, D-FFs, and multiplexers for frequency tuning. The output capacitor is assembled off-chip. Once the switch PMOS transistor cell is added to a standard-cell library, all cells needed to generate the proposed LDO are included in the library and the LDO can be designed with Verilog gate-level netlists and synthesized with a P&R tool.

The layout of the PMOS switch follows the design rules for standard cells so that it can be placed and routed by the P&R tool. Fig. 5 shows an outline of the PMOS switch cell layout. It is designed just by removing the NMOS transistor from the inverter cell for ease of the additional cell layout. Thus, the size of the PMOS switch cell is equal to that of the inverter cell.

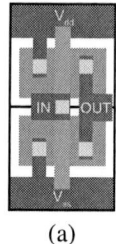 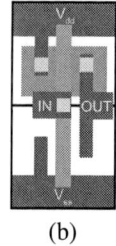

(a)   (b)

Fig. 5. (a) An inverter cell and (b) an additional PMOS transistor cell for the switch array.

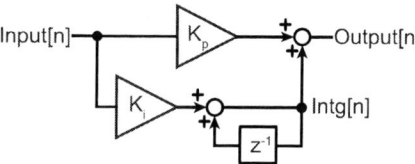

Fig. 6. Signal flow graph of the digital controller that includes proportional and integral paths.

The operation of the proposed LDO is described as follows. As shown in Fig. 2, the two inverter chains are powered by V_{ref} and V_{out}, respectively. An identical pulse train from the internal pulse generator enters into them at the same time. The inverter chains work as voltage-controlled delay lines. In other words, the inverter chains convert the voltage difference between V_{ref} and V_{out} into the delay difference that is compared by the phase detector. Based on the phase detector output, the digital controller changes the number of turned-on PMOS switches.

Fig. 6 shows the signal flow graph of the digital controller. The operation of the digital controller is expressed by the following discrete-time difference equations.

$$\text{Intg}[n] = \text{Intg}[n-1] + K_i \times \text{Input}[n] \quad (1)$$

$$\text{Output}[n] = \text{Intg}[n] + K_p \times \text{Input}[n] \quad (2)$$

The digital controller includes proportional and integral paths. Output[n] is 7-bit binary. Then the binary output is decoded into thermometer code so that the number of turned-on PMOS switches can be controlled one by one. When V_{out} is higher than V_{ref}, the phase detector output becomes HIGH and the digital controller decreases the number of turned-on switches. On the contrary, when V_{out} is lower than V_{ref}, the phase detector output becomes LOW and the digital controller increases the number of turned-on switches. In this way V_{out} approaches to V_{ref}. As the divider and the multiplexer is attached to the internal clock and pulse generator, in this prototype the clock frequency can be easily tuned by the multiplexer for test purpose.

The components other than the inverter chains are powered by V_{dd}. The HIGH-level voltage of the pulse trains which travel through the inverter chains are equal to their power source voltage, V_{ref} or V_{out}. Therefore, if V_{ref} is lower than the logic threshold voltage of the phase detector powered by V_{dd}, the phase detector cannot be driven by the pulse from the inverter chain powered by V_{ref}. Thus, the lower limit of V_{ref} is determined as the logic threshold voltage of the standard cells powered by V_{dd}.

978-1-5386-4757-8/18 $31.00 © 2018 IEEE

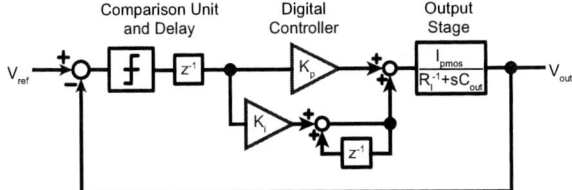

Fig. 7. Signal flow graph of the proposed LDO.

B. Transfer Function of the Control Loop

Fig. 7 shows the signal flow graph of the proposed LDO. The comparison unit composed of inverter chains and a D-FF generates an error sample. Since the comparison is done based on the pulse train which is the same signal as the clock, one clock delay occurs there at every sample. As previously described, the digital controller has proportional and integral paths. In order to investigate the loop stability in continuous-time domain, the approximation below is applied:

$$z \approx 1 + sT_s \quad (3)$$

where T_s represents the sampling period. The gain of the digital controller is thus approximated as follows.

$$H_{ctrl} = K_p + \frac{1}{1 - z^{-1}} K_i \quad (4)$$

$$\approx K_p + \frac{1 + sT_s}{sT_s} K_i \quad (5)$$

The output stage is composed of the switch array, the output capacitor, and the effective resistance. I_{pmos} is the current through a single PMOS switch. According to [3], the effective resistance R_l can be approximated as V_{out}/I_{load}. Using (3) and (5), the continuous-time open-loop transfer function $G(s)$ is given by

$$G(s) = \left(\frac{K_p}{1 + sT_s} + \frac{K_i}{sT_s} \right) \cdot \frac{I_{pmos}}{R_l^{-1} + sC_{out}}. \quad (6)$$

When I_{load} is small, R_l becomes big so that $R_l^{-1} \ll sC_{out}$. Then, $G(s)$ approximates

$$G(s) \approx \left(\frac{K_p}{1 + sT_s} + \frac{K_i}{sT_s} \right) \cdot \frac{I_{pmos}}{sC_{out}}. \quad (7)$$

If $K_p = 0$, the poles of the closed-loop transfer function are close to the imaginary axis and the system tends to be unstable. To avoid oscillation, we add the proportional gain K_p to the digital controller. Fig. 8 shows the bode plots of the open loop transfer function with small I_{load} of 100 μA for $K_P = 0$ and $K_p = 1$, respectively. When $K_p = 0$, the phase margin is 17°, whereas it is 27° when $K_p = 1$, which suppresses the abrupt phase change around 1 MHz.

C. Design Procedure of the Proposed LDO

The design procedure of the proposed LDO is explained in this section. First, the PMOS switch cell is designed and added to the standard-cell library. Second, the RTL Verilog code of the digital controller is prepared, then synthesized to have the gate-level Verilog netlist. The gate-level Verilog netlists of other components, such as the inverter chain,

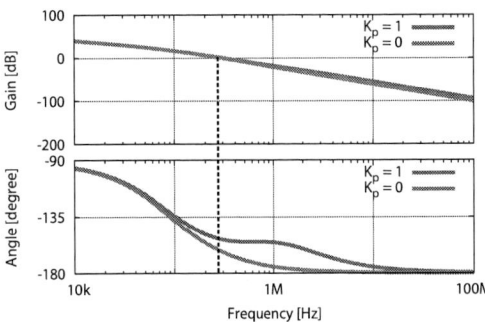

Fig. 8. Bode plots of the open loop system with the small I_{load} of 100 μA for $K_P = 0$ and $K_p = 1$, respectively.

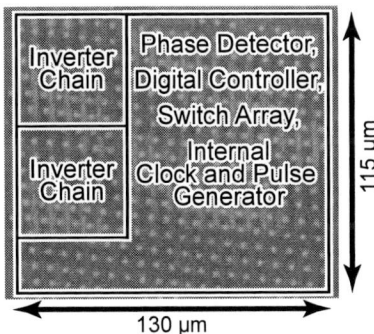

Fig. 9. Chip photo of the proposed LDO that occupies 115 μm × 130 μm in 65 nm standard CMOS technology.

the phase detector, the switch array, and the internal clock and pulse generator are generated by a dedicated script as the third step. Since all the building blocks have simple standard-cell-based structure, the gate-level netlist generation is simply implemented. Fourth, the layout of each building block is individually placed and routed by a P&R tool because they have different power supplies. Finally, all the layouts are connected together. It takes few hours to generate the whole layout of the proposed LDO from scratch, which is much less time than that in the case for the conventional LDO with analog design flows.

III. PROTOTYPE IMPLEMENTATION AND MEASUREMENT RESULTS

Based on the architecture described above, the prototype of the proposed LDO is fabricated in 65 nm standard CMOS technology. Fig. 9 shows the chip photo. The active area of the proposed LDO is 0.015 μm².

Fig. 10 shows the measured tracking response of V_{out} with 10.4 MHz clock when V_{ref}, which is externally supplied in this measurement, switches between 600 mV and 800 mV. Here, I_{load} and C_{out} is 10 mA and 220 pF, respectively. When V_{ref} changes from 600 mV to 800 mV, the settling time is 4.5 μs, whereas it is 4.4 μs when V_{ref} changes from 800 mV to 600 mV. Fig. 11 shows the measured transient response of V_{out} with 10.4 MHz clock when I_{load} changes between 5 mA and 10 mA. V_{ref} of 800 mV and C_{out} of 220 pF is used in this experiment. When I_{load} changes from 5 mA to 10 mA, the settling time is 6.6 μs and the undershoot is 303 mV.

978-1-5386-4757-8/18 $31.00 © 2018 IEEE   57

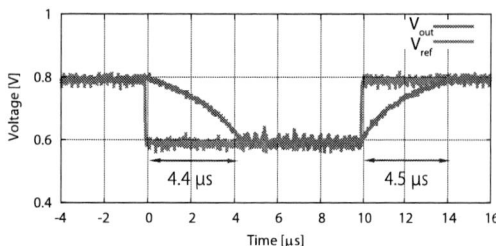

Fig. 10. Measured tracking response of V_{out} when V_{ref} switches between 600 mV and 800 mV with 10.4 MHz-clock, I_{load} of 10 mA and C_{out} of 220 pF.

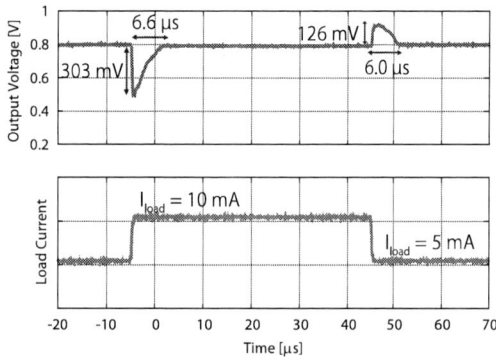

Fig. 11. Measured transient response of V_{out} when I_{load} changes between 5 mA and 10 mA with 10.4 MHz-clock, V_{ref} of 800 mV and C_{out} of 220 pF.

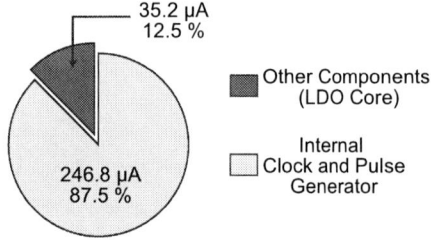

Fig. 12. Breakdown of the current consumption with I_{load} of 10 mA. Ratio of the current is calculated by circuit simulation.

When I_{load} changes from 10 mA to 5 mA, the settling time is 6.0 μs and the overshoot is 126 mV. Since the operation region of PMOS temporarily enters into saturation region when the undershoot occurs while it does not for the case of overshoot, the waveforms of V_{out} transient become different in these two cases.

The overall current consumption with 10.4 MHz clock and I_{load} of 10 mA is 282 μA including the current consumed at the internal clock and pulse generator, which is not essential in the actual use because it can be substituted by an internal clock for other functional blocks on the SoC. Based on the circuit simulation result, the LDO core consumes 12.5 % of the total current as shown in Fig. 12. Thus, the quiescent current of the LDO core is assumed to be 35.2 μA, which leads to 99.6 % current efficiency.

In comparison to prior digital LDOs in Table I, this work realizes a competitive current efficiency and the best FOM$_T$ with synthesizable architecture.

TABLE I
COMPARISON OF PERFORMANCE

	This work	[4]	[5]	[6]
Process	65 nm	65 nm	32 nm	65 nm
Active area [mm^2]	0.015	0.042	0.008	0.0023
V_{in} [V]	1.0	0.5	0.7-1.0	0.5-1.0
V_{out} [V]	0.8	0.45	0.5-0.9	0.3-0.45
$I_{load,max}$ [mA]	10	0.2	5	2
C_{out} [pF]	220	100000	100	400
Quiescent I_q [μA]	35.2	2.7	92	14
Peak current efficiency [%]	99.6	98.7	98.2	99.8
Transient ΔV_{out} @ load step ΔI_{load}	300 mV @ 5 mA	40 mV @ 0.2 mA	150 mV @ 0.8 mA	40 mV @ 1.06 mA
FOM$_T$ [ps]*	93	270000	1150	199

*FOM$_T$ = $(C_{out} \times \Delta V_{out} \times I_q)/\Delta I_{load}^2$ [6]

IV. CONCLUSION

This paper proposes a synthesizable digital LDO which is designed by a P&R tool. In the proposed LDO, by using inverter chains as voltage-controlled delay lines, the difference between the output and the reference voltages is converted into the delay difference that can be compared by a phase detector. The voltage control loop is all composed of standard cells and synthesizable, which drastically relaxes the design burden. The prototype is fabricated in 65 nm standard CMOS technology with 0.015 mm^2 area occupation. According to the measurement results of the prototype, with 10.4 MHz clock and C_{out} of 220 pF the tracking response time when V_{ref} switches between 600 mV and 800 mV is ~4.5 μs with I_{load} of 10 mA, and the transient response time when I_{load} changes between 5 mA and 10 mA is ~6.6 μs with V_{ref} of 800 mV. The quiescent current consumed by the LDO core is as low as 35.2 μA at 10 mA load current, which leads to 99.6 % current efficiency.

V. ACKNOWLEDGMENT

This work is partly supported by JSPS KAKENHI Grant Number 17H03244, and is supported by VLSI Design and Education Center (VDEC), the University of Tokyo in collaboration with Synopsys, Inc., Cadence Design Systems, Inc., and Mentor Graphics, Inc.

REFERENCES

[1] Y. H. Lam and W. H. Ki, "A 0.9 V 0.35 μm Adaptively Biased CMOS LDO Regulator with Fast Transient Response," *IEEE Int. Solid-State Circuits Conf. Dig. Tech. Papers,* pp. 442-626, Feb. 2008.

[2] R. J. Milliken, J. Silva-Martinez, and E. Sanchez-Sinencio, "Full On-Chip CMOS Low-Dropout Voltage Regulator," *IEEE Trans. Circuits Syst. I: Regular Papers,* vol. 54, no. 9, pp. 1879-1890, Sept. 2007.

[3] S. B. Nasir, S. Gangopadhyay, and A. Raychowdhury, "All-Digital Low-Dropout Regulator With Adaptive Control and Reduced Dynamic Stability for Digital Load Circuits," *IEEE Trans. Power Electronics,* vol 31, no. 12, pp. 8293-8302, Dec. 2016.

[4] Y. Okuma *et al.,* "0.5-V input digital LDO with 98.7% current efficiency and 2.7-μA quiescent current in 65 nm CMOS," *Proc. of IEEE Custom Integrated Circuits Conf.,* pp. 1-4, Sept. 2010.

[5] S. Gangopadhyay, D. Somasekhar, J. W. Tschanz, and A. Raychowdhury, "A 32 nm Embedded, Fully-Digital, Phase-Locked Low Dropout Regulator for Fine Grained Power Management in Digital Circuits," *IEEE J. Solid-State Circuits,* vol. 49, no. 11, pp. 2684-2693, Nov. 2014.

[6] L. G. Salem, J. Warchall, and P. P. Mercier, "A Successive Approximation Recursive Digital Low-Dropout Voltage Regulator With PD Compensation and Sub-LSB Duty Control," *IEEE J. Solid-State Circuits,* vol 53, no. 1, pp. 35-49, Jan. 2018.

An analysis of test solutions for COTS-based systems in space applications

R. Cantoro, S. Carbonara, A. Floridia, E. Sanchez, M. Sonza Reorda
Politecnico di Torino, Torino, Italy

Jan-Gerd Mess
DLR, Bremen, Germany

Abstract—One of the current trends in space electronics is towards considering the adoption of COTS components, mainly to widen the spectrum of available products. When substituting space-qualified components with COTS ones a major challenge lies in guaranteeing the same level of reliability. To achieve this goal, a mix of different solutions can be considered, including effective test techniques, able to guarantee a high level of permanent fault coverage while matching several constraints in terms of system accessibility and hardware complexity. In this paper, we describe an approach based on Software-based Self-test, which is currently being adopted within the MaMMoTH-Up project, targeting the development of an innovative COTS-based system to be used on the Ariane5 launcher. The approach aims at testing the OR1200 processor adopted in the system, combined with new and effective techniques for identifying the safe faults. Results also include a comparison between functional and structural test approaches.

I. INTRODUCTION

Space applications are known to be extremely challenging from a reliability point of view, since they are supposed to work in a harsh environment (not only in terms of radiation but also from the point of view of stresses coming from extreme temperature, pression, vibration, etc.) with strong requirements in terms of reliability. In order to reduce cost and especially to increase device availability, there is a trend towards the adoption of Commercial Off-The-Shelf (COTS) components instead of the space qualified ones. Obviously, this trend requires evaluating the costs and efforts for guaranteeing that the resulting reliability still reaches the target threshold [2]. A special niche within the general domain of space applications relates to launchers. In this case, the mission time is more reduced, while the radiation environment corresponds to all the layers from ground up to the geostationary orbit (GEO). The MaMMoTH-Up project [3], funded by the European Commission within the frame of the Horizon 2020 research and innovation programme, aims at developing and evaluating a COTS-based system to be used in the telemetry unit of the Ariane5 (A5) launcher. More in details, the MaMMoTH-Up system is composed of several boards targeting data acquisition and processing, power management, and data transmission. All these boards use COTS components, including a flash-based FPGA where several IPs are mapped, including an OpenRISC1200 (OR1200) processor [4] whose design has been properly modified to harden it with respect to radiation effects. The adoption of such processor allows the MaMMoTH-Up system to perform significantly more powerful functions than the system it is going to substitute, e.g., in terms of data analysis and compression. In order to match the strict reliability targets of A5, the MaMMoTH-Up system must be protected not only from the radiation effects, which are mainly responsible for Latch-up and transient fault effects, but also from

possible permanent faults arising during both the manufacturing process and the following system life. To target permanent faults several test steps have been identified, which are performed during and at the end of the manufacturing process, at the end of the assembly step, and after the system is mounted in the final position. Some test is also performed during the mission. The fault coverage which can be achieved by these test steps is important, since it directly impacts the achieved reliability level. To estimate the Fault Rate of the different components, we followed the FIDES guidelines [5], taking into account the stress conditions which are applied to the system before and during the mission. The Failure Rate is then derived by applying an FMECA (Failure Mode, Effects, and Criticality Analysis) procedure [11] which identifies the fault effects (and their criticality) and takes into account the timing and effectiveness (i.e., the fault coverage) of the different test steps. Remarkably, some of them have to be performed while the system is already mounted in its final position. Hence, they must basically correspond to a self-test, during which some command is sent to the system, the system performs a test of the hardware, and then results are sent outside. In the previous versions of the target system, which was based on much simpler space qualified hardware, a *functional test* was used for this purpose, where the system was asked to perform some basic operations, and a check on the computed results was sufficient to identify possible faults. Due to the much higher complexity of the MaMMoTH-Up system this approach can hardly guarantee the achievement of the required fault coverage, especially on the OR1200 core. Hence, a *structural approach* has been devised, based on a set of self-test procedures in charge of checking the possible presence of permanent faults affecting the processor core. The key difference between the two approaches lies in the fact that the functional one checks whether the system is able to deliver the expected functions, while the structural one identifies first some fault model related to the implementation of the underlying circuit, and then tries to detect the resulting faults. A major advantage of the structural approach clearly lies in the fact that the adopted fault coverage metric can be more deterministically and quantitatively evaluated than for the functional approach. Moreover, while for simple systems the functional approach (if suitably implemented) can achieve a sufficient testing quality, for more complex modules (such as the OR1200 one) the same is not true, as we will experimentally show in the paper. When dealing with the OR1200 processor, the self-test procedures implementing the structural approach follow the Software-based Self-test (SBST) paradigm [6]. Their code is integrated in the application software and, when activated, forces the processor to execute a proper sequence of instructions. The produced results are compacted into a signature which is returned to the calling

978-1-5386-4757-8/18 $31.00 © 2018 IEEE

program, which can thus check the possible presence of a fault by comparing it with the expected one.

The contribution of this paper lies first in describing a case of study (corresponding to the OR1200 core) where the characteristics of a functional and structural approach can be compared (not only in terms of achieved fault coverage, but also of memory footprint and duration). Secondly, it describes a scenario, where SBST can be effectively adopted, matching the several requirements of the qualification, acceptance and in-fly test of a space application. Finally, since the target system is expected to perform a well-defined set of functions and in a very specific configuration (e.g., in terms of memory address space), the FMECA is in charge of identifying which faults within the OR1200 core can produce any failure, and which faults will never be able to do so, e.g., because they relate to some hardware part which is not used by the application. While several techniques are available to automatically identify some categories of untestable faults, we focus here on those faults called *Safe faults*. These faults cannot produce any failure due to the specific (hardware or software) constraints the system matches during its normal operation. The paper shows that the number of safe faults is far from being negligible and uses an improved version of the method proposed in [7] to partly automate their identification. Due to the impact of the considered scenario, the fault coverage results reported in this paper are not directly comparable with those in [8], which focus on end-of-manufacturing test, although they refer to the same processor. The paper is organized as follows. Section II summarizes the main characteristics of the MaMMoTH-Up system, both in terms of underlying hardware and performed functions. Section III compares the functional and structural approaches, while Section IV focuses on the identification of safe faults. Section V finally draws some conclusions.

II. THE MaMMoTH-Up SYSTEM

A. General architecture and functions

The MaMMoTH-Up system shall provide an experiment and data acquisition opportunity on board the Ariane5 upper stage [3]. It is designed to offer the following functionalities:

1. Acquire measurement data
2. Configure and control the experiment
3. Provide a power supply
4. Perform self-testing and fault management.

To meet these functional requirements, a COTS-based system including one experiment controller (TCM-S), two computing nodes (OBC-S), two data acquisition boards (AQB) and a power supply unit (PSU) was developed. The system is housed in a foam-cushioned container to protect it from the harsh environment on board the launch vehicle. In order to collect sensor data and communicate with the Ariane5 upper stage, the system offers analogue acquisition channels for temperature, acceleration, vibration, shock and pressure sensors and a RS422 interface for data downlink. Synchronization with the launcher timeline and direct status reporting is done using three closed-current loops as inputs and eight discrete output pins. During the mission, the system steps through a number of different acquisition schemes according to the specific mission profile. An acquisition scheme determines which sensors are activated at which sampling rates up to 10 kHz. The data is collected and preprocessed by the computing nodes and then sent to the experiment controller using the internal SpaceWire bus. On the experiment controller, the data is analyzed, compressed and

stored on a flash-based mass memory before it is sent to the Ariane5 and downlinked using the launcher's telemetry chain. The complete data flow including its allocation to the different boards is depicted in Figure 1.

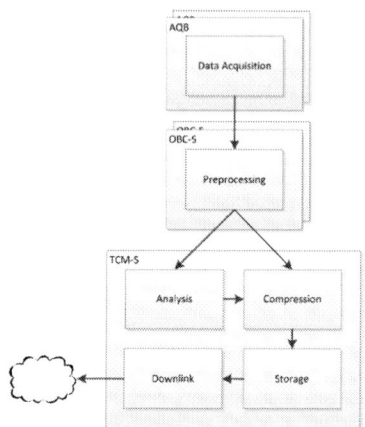

Figure 1 Data Flow

Each OBC-S board as well as the TCM-S board include a flash-based IGLOO2 FPGA. Each of these is holding an OR1200 soft core as well as accompanying IP cores, i.e. for SpaceWire communication amongst the boards. The required software images are kept in a two gigabyte NAND-flash memory that is implemented on each board. For data storage, the TCM-S is equipped with an additional sixteen gigabyte NAND-flash. The data acquisition is performed by a custom IP core that samples ADC channels and returns a block of samples to the software. Preprocessing, analysis and compression are then performed by software run by the OR1200 processors. The data compression algorithm consists of two steps whose load is divided between the OBC-S and the TCM-S. The OBC-S boards will perform a wavelet transform. The transformed data is then sent to the TCM-S. From the received wavelet transform, certain characteristics of the underlying data (e.g., value range and maximum gradient) are deduced. The transformed coefficients are then encoded into an embedded bitstream. According to the deduced characteristics of a given block, a certain number of bytes in the downstream are allocated for this bitstream. All other bits are cut to save downlink budget. The complete compression scheme is described in [9]. From a reliability point of view, although the OR1200 processors on the FPGAs and especially their memories and registers are hardened by duplication or triplication of some of the underlying flip-flops, there is no redundancy at the unit or system level. If the failure is not permanent, the system is able to recover by performing a software reset or power-cycle on the affected board. Should this not be successful because the failure proves to be permanent, the board has to be deactivated, inevitably resulting in a loss of the connected sensor channels. In this case, the MaMMoTH-Up system follows the concept of graceful degradation: although parts of the sensors cannot be acquired anymore, the remaining transfer budget can be reallocated to use it as efficiently as possible.

B. The OR1200 processor

The OR1200 is the only major RTL implementation of the OR1K architecture spec. The OR1200 is a 32-bit scalar RISC with Harvard micro-architecture and 5 stage integer pipeline. The OR1200 core is mainly intended for embedded, portable and networking applications. Fig. 2 shows its architecture.

978-1-5386-4757-8/18 $31.00 © 2018 IEEE

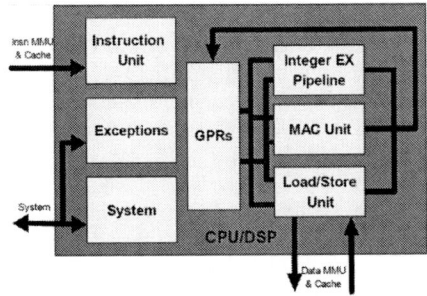

Figure 2. OR1200 CPU Architecture

III. COMPARING THE FUNCTIONAL AND STRUCTURAL APPROACHES

A. Background

In the frame of the actions to evaluate the reliability of the MaMMoTH-Up system and to guarantee that the target figures are matched, a key role is played by the test solutions adopted to identify possible permanent faults. These solutions are activated in different phases of the product life time, since the qualification step until the operational phase (i.e., during the launch). We underline that these test solutions should be usable and effective when applied in a scenario, where the target modules (e.g., the FPGA implementing the processor) have been already mounted on their boards, and each board has been included in the final box corresponding to the telemetry unit, which has been installed in its final location within the launcher. Hence, the whole test should be performed with very limited support from the outside, and should be minimally invasive with respect to the target system. In order to evaluate the effectiveness of these test solutions and to use meaningful fault coverage figures during the reliability evaluation process, a metric must be first identified. Traditionally, a functional metric is adopted. Since the early specification phases, the list of functions that the system must support is defined. For each of them, a *functional test* is then developed, aimed at verifying that the target function is correctly performed by the system. In this scenario a qualitative metric is adopted, which guarantees that the system is not affected by any fault if all the functional tests for all the functions are successful. When moving to more complex systems including COTS components, a different metric can be considered, which first identifies a structural fault model which is supposed to represent the possible permanent faults in the target device, and then computes the percentage of structural faults which are detected by the considered test solution. One of the goals of the MaMMoTH-Up project is to define new procedures for reliability evaluation, able to match the characteristics of COTS-based systems. Given their complexity, the project partners decided to assess the effectiveness of the functional approach when a structural fault model was adopted, at least for the most complex module, corresponding to the OR1200 CPU. Since the detailed information about the structure of the adopted FPGA were missed, we decided to perform such an assessment resorting to the popular stuck-at fault model, computing first the fault coverage achieved by the functional test when the CPU circuitry mapped on the FPGA was synthesized with a generic gate-level library. This approach is partly supported by the results of [20], showing that the stuck-at fault model, when applied to FPGAs, provides Fault Coverage results which are not far from those which can be obtained resorting to more accurate fault models, based on the knowledge of the internal implementation of the device (which is not available in our case).

Moreover, we developed a set of SBST test procedures targeting the stuck-at faults inside the OR1200 core. These procedures (that we cumulatively call *structural test* in the following) are integrated within the application software of the system and can be easily launched from the outside or by the system itself when required. Each of them returns a signature compacting the results produced by the test code, which can be compared with the expected one. A mismatch means that a permanent fault exists in the CPU core. In the following, we first report some information and figures (Table I) about the functional test and the structural one (based on SBST procedures) we developed. We will then report the experimental results aimed at comparing the effectiveness of the two test approaches (Table II).

B. The functional test

The functional test is composed of a compression algorithm that imposes a high workload on the arithmetic units of the processors. It is essential that the processor is fault-free, because even small changes in single bits of the output stream can result in a completely different set of data after decompression. Since it is impossible to predict the exact sensor readings, the processor cannot be checked using live data. Instead, precompiled blocks of sensor data together with expected values for the resulting transformation coefficients and bitstream are used. By comparing the output of the compression algorithm with the expected values, it is checked whether the calculations can be executed as planned. However, in case of an error, no diagnostic conclusions about the affected units within the processor can be drawn. The second line of Table I reports the size and duration of the functional test in terms of amount of memory to store the code and test time execution.

C. The structural test

The structural test is based on a suite of test procedures that target the different modules of the processor: program counter generator (*genpc*), instruction fetch (*if*), control unit (*ctrl*), register file (*rf*), operand muxes (*opmux*), arithmetic logic unit (*alu*), multiply and accumulate unit (*multmac*), load and store unit (*lsu*) and write back multiplexer (*wbmux*). Each test program executes a sequence of instructions aimed at stimulating as much as possible the target unit. At the end of the test, a signature is stored in memory: if the produced result is different than the expected one, it means that the CPU module is affected by a fault. All the test procedures have been written manually following the guidelines provided in [8]. In the following, we provide the most important characteristics of every one of the developed test programs.

The *genpc* and *if* modules are tested together using a single program. Any type of instruction from the Instruction Set must be tested. The program is written in such a way, that each type of instruction is followed by an unconditional jump to a procedure to the bottom of the code that updates the value of the signature and then it jumps back again to the top. In this way, the program counter adder inside the *genpc* is well tested, since it continuously jumps backward and forward, so performing additions and subtractions. In order to test the *ctrl* module, it is necessary to give it as inputs all the possible instructions from the Instruction Set: arithmetic, logic, branch, jump, compare, multiply, load and store, immediate or register-to-register. Since the *ctrl* module also generates signals to freeze some selected stages of the pipeline or to activate the forwarding when data hazards occur, it is important to include some instruction sequences with suitable data dependency in order to stimulate those signals. The values of the operands are not so important in this case, so random values are chosen for the operations. The *rf* module is tested using register to register operations. Basically, the test consists in writing a value into a register and then reading it. The test is divided into four parts. In the first part of the test, the stack pointer and the link registers are tested. In the second part, the first half of the registers (r2-r15) is tested, assuming the other part is not faulty and using one among these registers to hold the signature. In the third part of the test, the second half is considered in turn, assuming the first part is not faulty. The values written in the registers are 0x55555555 and 0xAAAAAAAA. To protect the CPU core against temporary faults the register file has been duplicated and the first operand is read from one register, while the second operand is read from the second register; the write back operation updates both registers. Hence, it is necessary to perform each instruction twice, swapping the two operands, in order to read the values from both registers. The *opmux* module selects the operands for the execution units, choosing between values coming from the register file or from the various pipeline registers when forwarding is needed. The idea to test this module is to choose arithmetic, logic, load/store and multiply instructions in such a sequence that causes data dependencies in different stages of the pipeline. The *alu* module test addresses all the possible arithmetic/logic instructions of the instruction set. Some special values generated resorting to an Automatic Test Pattern Generation (ATPG) tool launched on the combinational part are chosen as operands to better test its functionalities and all the operations are performed choosing as operands all the possible combinations between the values above. The test of the *mult_mac* module depends significantly on the values chosen for the operands. Therefore, an ATPG tool has been used again to generate proper input values. The test program consists in a series of multiplications (also with immediate, signed and unsigned operands), multiply-accumulate and multiply-subtract instructions of the computed random operands. Division instructions also involve the *mult_mac* module to operate and it has also been tested. Since the *mac* instruction uses special purpose registers to accumulate, it is necessary to read the values written in these registers after each multiply-accumulate instruction. For testing the *lsu* module, all kinds of load and store instructions are considered: load/store byte or word, extended to zero or signed. The program is constituted of a sequence of instructions to write and read contiguous locations in memory; each block is composed of instructions performing the following three steps: a) Storing a value in a memory location, b) Reading the value from the same location, c) Updating the signature. The values chosen to be written in memory are random and the offset to be added to the base address is a large value (from 16,380 to 17,380). The *wbmux* module chooses the value to be written back into the register file, whether it comes from the memory system (for a load instruction) or from the execution units. Since this module basically corresponds to a mux, the program is very similar to that developed for the *opmux* module. Table I summarizes the characteristics of the Functional and Structural tests in terms of size and duration. For the Structural test, we detailed these figures for each test procedure.

D. Results

For the purpose of our experiments, we created a simulation setup where the OR1200 processor core lies in a system composed also of a 64 MB RAM, as in the MaMMoTH-Up OBC-S boards. The processor core has been synthesized targeting the NanGate 45nm Open Cell Library. The obtained netlist is used to perform the fault simulation experiments with a

commercial tool. Using this setup we evaluated the stuck-at Fault Coverage obtained by both the functional and the structural test described above. Results are reported in Table II for the whole CPU and for each component module. As the reader can notice, the Fault Coverage achieved by the Functional test is far lower than the one of the Structural test. This supports the claim that Functional test cannot be effectively used when complex COTS-based systems are used. It is also worth underlining that the comparison between the two tests provides very different results depending on the considered module. For some of them (e.g., genpc) the fault coverage achieved by the Functional test is slightly higher than by the Structural test. This is basically due to the fact that some modules can be tested in a good way by executing long programs, and the Functional test is much longer than the Structural one. However, for modules that include large combinational parts (e.g., alu and mult-mac) or require a specific sequence of operations to be tested (e.g., rf) the Structural approach is far more effective.

Table I. Characteristics of the test programs

	Size [Byte]	Duration [#clock cycles]
Functional test	17,360	379,815
Structural test	25,676	74,761
genpc-if	2,896	41,635
ctrl	980	980
rf	10,076	7,281
opmux	544	508
alu	3,184	10,497
multmac	2,996	9,962
lsu	4,244	3,224
wbmux	756	674

IV. SAFE FAULTS

A. Safe faults

We denote as *Safe Faults* those faults that can never produce a failure in the considered system[1]. One of the goals of the FMECA is their identification, since they do not contribute to the Failure Rate, and should thus be removed from the Fault List to be considered when evaluating the Fault Coverage achieved by the test procedures. The ISO 26262 standard for automotive applications define them as "application dependent safe faults". Clearly, Safe Faults include untestable faults. Hence, it can be useful to review the different categories included in the set of Safe Faults for a given system:

- *Structurally (or combinationally) untestable faults*, i.e., faults for which a test does not exist even if the combinational block where the fault is located is fully controllable and observable (e.g., via scan). Examples of faults belonging to this category include faults that cannot be tested due to some redundancy in the combinational logic. If a gate-level description of the device is available, an ATPG tool can identify these faults.

- *Sequentially untestable faults*, i.e., faults that do not belong to the previous group, but cannot be tested due to the sequential behavior of the circuit, for example, because the circuit cannot

---

[1] When performing FMECA, it is common to also distinguish between failures depending on their criticality, i.e., on how serious the effects of these failures are. Reliability figures typically depend only on *critical* safe faults. For the purpose of this paper we ignore any distinction within the set of safe faults.

reach any of the states required for their test. Several works proposed techniques to automatically identify these faults, either in a generic circuit [12][13][14][17][18] or specifically in a CPU [15].

- *On-line functionally untestable faults* [10], i.e., faults that do not belong to the previous groups, but cannot be tested in a functional manner (i.e., without resorting to Design for Testability) in the operational conditions the target device works in. On-line functionally untestable faults can be related for example to the specific memory configuration adopted by the system [16]. Several bits in the processor Program Counter or in the registers storing the addresses in the Load-Store Units become untestable if the memory area storing the code and the data is less than the maximum one.

Table II. Stuck-at (SA) fault coverage of functional and structural test on OR1200.

Module	Total SA faults	Functional test	Structural test
CPU	124,612	32.09 %	81.89 %
genpc	4,906	60.80 %	57.97 %
if	2,268	50.57 %	71.12 %
ctrl	4,320	71.53 %	80.25 %
rf	39,056	33.97 %	90.93 %
opmux	2,530	90.51 %	96.05 %
alu	14,532	46.04 %	78.50 %
mult_mac	39,398	13.91 %	95.77 %
sprs	5,522	8.31 %	37.61 %
lsu	2,708	67.61 %	65.99 %
wbmux	2,286	69.29 %	78.83 %
freeze	126	75.40 %	76.98 %
except	6,716	15.86 %	18.92 %
cfgr	232	0.00 %	0.00 %

Safe faults include and extend the previous categories. In the following we report some examples of safe faults:

- The debug circuitry possibly existing in a processor generates safe faults, since debug facilities are not used during the normal behavior, and several faults within it cannot impact the system behavior and produce any failure.

- Several faults in the Design for Testability hardware (e.g., the scan chains) used for end-of-manufacturing test also correspond to safe faults: for example, faults on the scan-in input of the scan flip-flops are safe faults.

In [7], we reported some results concerning the identification of safe faults in the openMSP430 processor. In that paper, we also considered those safe faults that cannot produce any failure, due to the specific application code executed by the CPU. As a simple example, if the system application only uses integer arithmetic, faults in the Floating Point Unit become untestable.

B. Safe faults identification

The typical approach for safe faults identification is based on manual analysis. In many project teams, the designers, test engineers, and reliability/functional safety experts systematically meet to categorize faults and (based on their effects) identify safe faults. Clearly, this process is extremely time consuming (and hence expensive), as well as prone to possible errors. For this reason, in [7] we recently proposed an approach, which aims at partly automating the safe faults identification process taking into account all the constraints coming from the application scenario, including the application software to be run by the CPU. Some preliminary results coming from the application of the same method to the OR1200 CPU have been reported in [19]. In this paper, we improve the procedure used in [7] and [19],

which is now able to identify a larger number of safe faults, thanks to a mechanism allowing to exploit the power of a commercial ATPG tool. Our method for safe fault identification is based on the following steps:

1. We identify the set of all inputs to the CPU module which will remain at a fixed value during the system operation (e.g., the Normal/Test signal). Let us call PI_{fixed} this set.

2. We perform several simulation experiments on the CPU module running the actual application and with different but realistic data input sequences and using toggle activity to identify the set $FF_{possibly-fixed}$ of flip flops which never toggle.

3. We focus on $FF_{possibly-fixed}$, and manually check whether any of the flip flops in this set may possibly toggle if a different sequence of input data and events is considered. The remaining set of flip flops, called FF_{fixed} is composed of those flip flops that will never toggle in the operating conditions.

4. We resort to an ATPG tool to identify the faults in the combinational logic of the processor that become untestable once the constraints coming from the fixed values of the PI_{fixed} and FF_{fixed} signals are applied. In other words, we specify the inputs of the combinational logic whose value always remains fixed during the operational phase, and ask the ATPG tool to identify untestable faults under these conditions. These faults correspond to safe faults for the system.

The reader should note that in [7] and [19] the last step was performed resorting to a simple topological analysis of the effects of the fixed values in the combinational logic: the analysis identified for each gate the possible untestable faults caused by any fixed value on the inputs of the fault. Hence, the analysis was only able to identify redundant signals or gates in the circuit leading to untestable faults. To perform the same step, in this paper we resort to an ATPG tool, so that a larger number of safe faults can be identified, taking into account the constraints on the input signals of the combinational logic. This allowed us to increase by about 3% the number of safe faults identified by this step with respect to the results presented in [19]. The usage of a commercial tool also makes the applicability of the proposed method easier. It is important to underline that our method cannot identify all safe faults in the system. However, we claim that it can identify a significant number of them and represents a first step towards the automation of the whole safe fault identification procedure, thus contributing to significantly reducing its cost.

C. Results

We implemented a tool based on a set of TCL scripts interacting with a logic simulator to implement the procedure described in the previous sub-section. The required time to run the simulation campaign to gather the data for the Toggle analysis and to process them to extract the list of Safe faults (including the ATPG step) is in the order of a few hours. By using the same commercial ATPG tool we also identified the number of structurally untestable faults in the OR1200, which amounts to 80. Following the proposed procedure and referring to the environment and application code of the OBC-S board, we identified a set of safe faults in the OR1200 processor, as reported in the second column of Table III. We also computed the Safe Fault Coverage (*SFC*) for the Functional and Structural tests (columns 4 and 5), defined as:

978-1-5386-4757-8/18 $31.00 © 2018 IEEE

$$SFC = \frac{\#detected\ faults}{\#faults - \#safe\ faults}$$

The reported results show that:

- The number of safe faults is relevant, accounting for about 13% of the whole stuck-at fault list.
- The percentage of safe faults varies widely from one module to another. It is about 20% for modules such as *mult_mac* and *sprs* (dealing with special purpose registers, which are not significantly used by the application). It is also significant for modules such as *if*, *genpc* and *rf*, which are not fully used by the application code.
- The SFC figure achieved by the structural test procedures is quite high (taking also into account the observability constraints of the test environment) and allows (combined with some further test techniques implemented at a higher level) to fully match the reliability requirements for the MaMMoTH-Up system.
- We are still working to improve the achieved SFC by developing some test procedures targeting the few small modules which are not well covered by the current test procedures, resorting to more sophisticated solutions (e.g., triggering some exceptions or moving to supervisor mode to test the special-purpose registers) which can only be used before the launch.

Table III. Safe stuck-at fault coverage (SFC) of functional and structural tests after untestable analysis on OR1200.

Module	Safe faults	Safe Faults w.r.t. Total SA faults	SFC	
			Functional Test	Structural Test
CPU	16,183	12.98 %	36.88 %	84.41 %
genpc	425	8.66 %	66.57 %	63.24 %
if	204	9.00 %	55.57 %	76.74 %
ctrl	13	0.30 %	71.74 %	80.42 %
rf	5,550	14.21 %	39.60 %	92.89 %
opmux	41	1.62 %	92.00 %	96.47 %
alu	75	0.51 %	46.28 %	78.54 %
mult_mac	7,861	19.95 %	17.38 %	97.04 %
sprs	1,070	19.38 %	10.31 %	44.41 %
lsu	7	0.26 %	67.79 %	66.16 %
wbmux	0	0.00 %	69.29 %	78.83 %
freeze	7	5.55 %	79.83 %	79.51 %
except	912	13.58 %	18.35 %	21.85 %
cfgr	18	7.76 %	0.00 %	0.00 %

V. CONCLUSIONS

This paper deals with the adoption of COTS components in the design and manufacturing of systems to be used on a launcher. We focused on the MaMMoTH-Up system to be used on board the Ariane5 launcher, which represents a testbench for developing a suitable design and manufacturing flow compatible with the adoption of COTS components. In particular, we focused on the test of the CPU core used within such a system, showing first that the functional test is not able to achieve a sufficient test quality, while structural SBST test procedures can be much more effective. We also focused on the identification of safe faults, i.e., those faults that cannot produce any failure due the hardware and software constraints provided by the application environment. We proposed a semi-automated method able to significantly reduce the cost and effort for safe faults identification, showing that the method can identify a significant number of safe faults. We reported experimental results on the OR1200 processor core used within the MaMMoTH-Up system. Although the proposed method has been experimentally evaluated referring to stuck-at faults, only, the same approach can be adopted to deal with other fault models (e.g., transition delay faults, or bridges), if required. We are currently working towards the development of further improved techniques for safe faults identification and towards a new and more effective release of our SBST procedures.

ACKNOWLEDGEMENTS

This work has been supported by the European Commission through the Horizon 2020 Project No. 637616 (MaMMoTH-Up).

REFERENCES

[1] S. Avramenko; M. Sonza Reorda; M. Violante; G. Fey; J. -G. Mess; R. Schmidt, "On the robustness of DCT-based compression algorithms for space applications", 2016 IEEE 22nd International Symposium on On-Line Testing and Robust System Design (IOLTS)

[2] Michel Pignol, "COTS-based applications in space avionics", 2010 Design, Automation & Test in Europe Conference & Exhibition (DATE 2010)

[3] http://www.mammoth-up.eu/

[4] https://openrisc.io/

[5] UTE FIDES guide 2009, Edition A, September 2010

[6] M. Psarakis et al., "Microprocessor Software-Based Self-Testing", IEEE Design & Test of Computers, vol. 27, no. 3. May-June 2010, pp. 4-19

[7] R. Cantoro, A. Firrincieli, D. Piumatti, E. Sanchez, M. Sonza Reorda, M. Restifo, "About functionally untestable fault identification in microprocessor cores for safety-critical applications", IEEE Latin-American Test Symposium (LATS), 2018

[8] N. Kranitis; A. Merentitis; G. Theodorou; A. Paschalis; D. Gizopoulos, "Hybrid-SBST Methodology for Efficient Testing of Processor Cores", IEEE Design & Test of Computers, 2008, Volume: 25, Issue: 1, pp. 64 – 75

[9] J.-G. Mess, R. Schmidt, G. Fey, "Adaptive Compression Schemes for Housekeeping Data", 2017 IEEE Aerospace Conference

[10] P. Bernardi, M. Bonazza, E. Sanchez, M. Sonza Reorda, and O. Ballan, "On-line functionally untestable fault identification in embedded processor cores", Proc. Design, Autom. Test Eur. Conf. Exhibit. (DATE), Mar. 2013, pp. 1462–1467

[11] W. M. Globe, "Control Systems Safety Evaluation and Reliability", third edition, ISA, ISBN 978-1-934394-80-9

[12] J. Raik, H. Fujiwara, R. Ubar, A. Krivenko, "Untestable Fault Identification in Sequential Circuits Using Model-Checking", Proc. IEEE Asian Test Symposium, 2008, pp. 21-26

[13] Syal, M.; Hsiao, M.S., "New techniques for untestable fault identification in sequential circuits", IEEE Transactions on Computer-Aided Design of Integrated Circuits and Systems, vol. 5, no. 6, 2006, pp. 1117 – 1131

[14] H.-C. Liang; C. L. Lee; Chen, J.E., "Identifying Untestable Faults in Sequential Circuits", IEEE Design & Test of Computers, Vol. 12 , No. 3, 1995, pp. 14-23

[15] W.-C. Lai; Krstic, A.; Kwang-Ting Cheng, "Functionally testable path delay faults on a microprocessor", IEEE Design & Test of Computers, vol. 17, no. 4, 2000, pp. 6-14

[16] A. Riefert; R. Cantoro; M. Sauer; M. Sonza Reorda; B. Becker, "A Flexible Framework for the Automatic Generation of SBST Programs", IEEE Transactions on Very Large Scale Integration (VLSI) Systems, 2016, Volume: 24, Issue: 10, pp. 3055 – 3066

[17] David E. Long, Mahesh A. Iyer, Miron Abramovici, "FILL and FUNI: algorithms to identify illegal states and sequentially untestable faults", ACM Transactions on Design Automation of Electronic Systems (TODAES), v. 5, n. 3, pp. 631-657, July 2000

[18] Daniel Tille, Rolf Drechsler, "A fast untestability proof for SAT-based ATPG", 12th International Symposium on Design and Diagnostics of Electronic Circuits&Systems, pp. 38-43, April 15-17, 2009

[19] S. Carbonara, A. Firrincieli, M. Sonza Reorda, J.-G. Mess, "On the test of a COTS-based system for space applications", 24th IEEE International Symposium on On-Line Testing and Robust System Design, 2018, poster session

[20] Jaroslav Borecky; Martin Kohlik; Pavel Kubalik; Hana Kubatova, "Fault Models Usability Study for On-line Tested FPGA", 14th Euromicro Conference on Digital System Design, 2011, pp. 287-290

Robust Detection of Bridge Defects in STT-MRAM Cells Under Process Variations

Andres F. Gomez[1,2], Freddy Forero[1], Kaushik Roy[3], Victor Champac[1]

[1]National Institute for Astrophysics, Optics and Electronics - Mexico.
[2]Manuela Beltrán University - Colombia
[3]Purdue University - USA

Abstract—Spin-Transfer-Torque Magnetic RAM (STT-MRAM) is a promising memory technology due to its ultra-integration density capability; nanosecond read and write operation speeds and CMOS/FinFET fabrication process compatibility. As every silicon technology, STT-MRAMs may be affected by fabrication defects, which may be difficult to detect under process variability in deeply scaled transistor technology. This paper proposes a Design-For-Test (DFT) circuit to detect short defects in the STT-MRAM cells. The proposed methodology is based on the observation that a short defect makes different the amplitude of the current entering and leaving the memory cell. The proposed DFT circuitry is robust to process-induced parameters variations in the memory cell. In such way, defects detection probabilities are increased, and a high-quality product can be guaranteed.

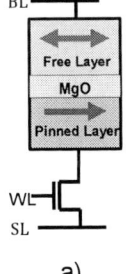

Node	Write		Read
	$P \rightarrow AP$	$AP \rightarrow P$	
WL	V_{DD}	V_{DD}	V_{DD}
BL	0	V_{DD}	V_{READ}
SL	V_{DD}	0	0

Figure 1. STT-MRAM cell (a) Circuit Schematic (b) Voltages for write and read operation

I. INTRODUCTION

Spin-Transfer-Torque Magnetic RAM (STT-MRAM) is an emerging non-volatile memory technology with high endurance and CMOS/FinFET compatibility [1]. For these reasons, the International Technology Roadmap for Semiconductor (ITRS) highlighted STT-MRAM as a promising candidate for future on-chip memory applications [2].

The basic structure of an STT-MRAM cell is shown in Figure 1(a). It uses a Magnetic Tunneling Junction (MTJ) as storing element and a transistor as access device to the MTJ. Due to its superior gate controllability, larger "ON" current, and lower variability, the use of FinFET transistor as access device of STT-MRAM cells is preferred over CMOS planar transistor [3]. The MTJ consist of two ferromagnetic layers separated by a thin oxide barrier. The magnetic orientation of one layer, called Pinned Layer (PL), is fixed, whereas the magnetic orientation of the other layer, called Free Layer (FL), is allowed to switch between parallel (P) and antiparallel (AP) states with respect to the PL. The resistance of the MTJ is low (high) when the MTJ is in the parallel (antiparallel) state. These resistance states are used as logic values. The write and read operation are performed by applying appropriate voltages to the bit-line (BL), source-line (SL) and Word-Line (WL) terminals of the cell (See Figure 1(b)). For a write operation, the current that flows through the MTJ has to be large enough to switch the magnetization orientation of the free-layer successfully [4]. The read operation is performed by sensing the resistance of the MTJ.

Correct STT-MRAM behavior may be affected by manufacturing defects. A comprehensive analysis of all the possible resistive short/open defects in an STT-MRAM array was presented in [5]. It was shown that some fault models and test techniques used for SRAM technology are not extendable to STT-MRAM technology.

In this paper, we present a Design-For-Test (DFT) circuit to detect resistive-short defects between an internal node of an STT-MRAM and an external node. The proposed test circuit performs defect-oriented test, which it is adequate for detection of weak defects that may escape classic logic test and degrade the memory block reliability [6]. The proposed test circuit is based on the observation that a short defect makes different the amplitude of the current entering and leaving a cell in the memory array. It is different from classic I_{DDQ} test in that the proposed circuit can measure the current flowing through a single memory cell, which provides higher defect detection and diagnosis capabilities. Furthermore, weak short defects may escape classic I_{DDQ} test due to process variations. The capability of the proposed test circuit to detect sizes of resistive-short defects that can escape conventional logic test is validated under process variations effects, and it is shown that the defect detectability is improved.

The rest of this paper is organized as follows: Section II describes the read and write operation of the STT-MRAM cell and the required circuitry to perform such operations. Section III presents an analysis of the electrical behavior of the cell under the presence of short defects. Section IV presents

978-1-5386-4757-8/18 $31.00 © 2018 IEEE

the proposed test technique. Finally, section V presents the conclusion of this work.

II. WRITE AND READ OPERATION OF STT-MRAM CELLS

The basic the read and write circuitry of an STT-MRAM array is shown in Figure 2. For illustration purposes, only one cell is shown, but it should be noted that an entire column is connected to the BL and SL nodes of the read-write circuitry, while the acceded cell from a given row is selected with the WL signal. The resistive-short defect model is also illustrated in Figure 2. Let us not consider this model by now. The memory cell was designed in a standard-connected fashion. The access transistor of the cell, which is driven by the word-line (WL) signal, controls whether current can flow through the MTJ to perform write and read operations.

A. Write Operation

The write circuit is activated with the Write Enable (WE) signal. It applies a bi-directional current to the memory cell depending on the Data Input (DI) to be written. The AP state is written if $DI = 1$ while the parallel state is written if $DI = 0$. The transistors of the write circuit were made large enough to provide sufficient current capability for correct write operation.

B. Read Operation

The read circuit is activated with the Read Enable (RE) signal. A current I_{REF} is generated using a reference cell whose resistance is designed to be the mean of the parallel and anti-parallel resistance of a cell. This current is copied and applied to the accessed cell. The voltages generated at the current mirror terminals (V_{REF} and V_{cell}) are compared to determine whether the accessed cell is in the P or AP state.

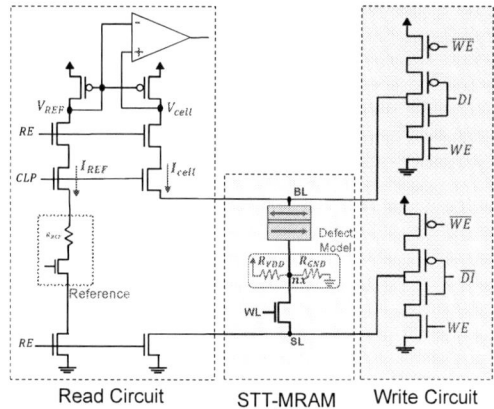

Figure 2. Read and write circuits to simulate the behavior of the STT-MRAM cell under short defects.

III. ANALYSIS OF STT-MRAM BEHAVIOR UNDER SHORT DEFECTS

The defect model shown in Figure 2 considers resistive-short defects between the internal node of an STT-MRAM cell and an external node. These type of defects are modeled by a resistance connected from the internal node of the cell

to either V_{DD} (R_{VDD}) or GND (R_{GND}). However, the proposed methodology is also valid to detect other types of short defects as explained later in section IV-D.

To simulate the STT-MRAM cell in SPICE, a Predictive 20nm FinFET technology [7] is used along with the SPICE-compatible MTJ model proposed in [8]. The MTJ has dimensions of $60nm \times 40nm \times 1.4nm$. A stability factor of $\Delta = 70$ is considered. Other MTJ parameters are set as in [8]. A FinFET access transistor with 6 Fins was used to provide enough current capability to perform a successful write operation in $3.5ns$. Variations of the MTJ were assumed of 15% for the cross-sectional area (A_t) and 5% for the oxide barrier thickness (t_{ox}). For the access transistor, 30% of threshold voltage (V_{th}) variation due to the work function variation was assumed. For simplicity purposes, process parameters variations were only considered for the memory cell.

A. Impact of Short Defects on Write Operation

The nominal current that flows through the MTJ (I_{MTJ}) during writing a P state and an AP state is shown in Figures 3(a) and 3(b), respectively. The writing time (t_{wr}) to switch the magnetization orientation of the MTJ depends on the amplitude of the current that initially flows through the MTJ. If a parallel state is to be written, the current that flows through a *"good"* cell is close to $145\mu A$. When a resistive defect R_{VDD} exists, the write time increases because this defect limits the current that can flow through the MTJ since the access transistor has to drive both the MTJ current and the defect current. On the other hand, defect R_{GND} slightly reduces the write time because it adds a discharging current path. If an anti-parallel state is to be written (See Figure 3(b)), the current that flows through a *"good"* cell is close to $139\mu A$. The defect R_{VDD} increases the current and reduces the write time, while R_{GND} reduces the current flowing through the MTJ and increases the delay.

Impact of Process Variations

The write pulse duration constraint for the designed cell was of $3.5ns$, which is the largest write time that a defect-free cell can take under process variations, according to our results from 1000 Monte-Carlo simulations. Note that this write time corresponds to the $P \rightarrow AP$ operation, which is slower than the $AP \rightarrow P$ operation due to source degeneration of the access transistor [9]. If the cell's write time becomes larger than $3.5ns$ due to a short defect, an incorrect write operation is performed, and the defect presence can be detected using conventional logic test.

Some delay histograms of the $AP \rightarrow P$ write operation for good and defective cells with strong short defects of $2k\Omega$ are shown in Figure 4(a). R_{GND} slightly reduces the delay and therefore does not cause any logic fault, while R_{VDD} increases the delay and may cause a fault if the defect resistance is small enough. However, it can be seen that a large number of the defective cells may perform within desired delay margins and escape test. For the $P \rightarrow AP$ write operation (Figure 4(b)), defect R_{VDD} reduces the write time, and consequently, this defect does not trigger any logic fault. On the other hand, defect R_{GND} increases the delay

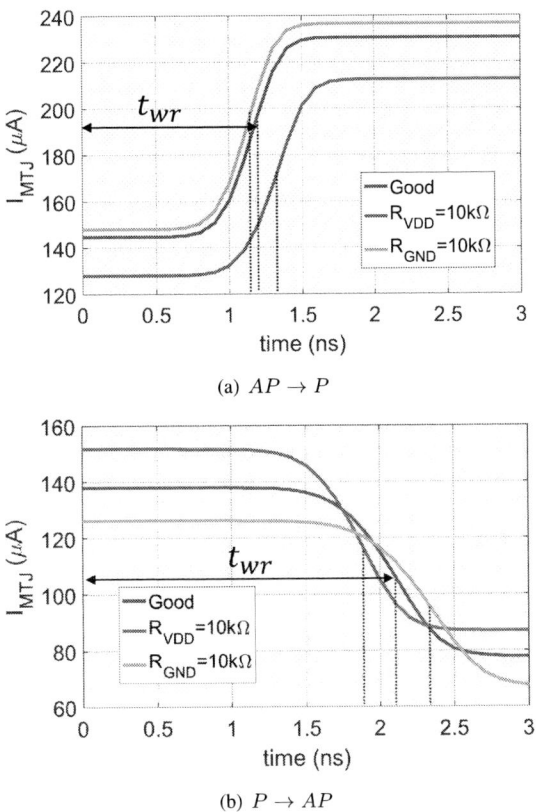

(a) $AP \rightarrow P$

(b) $P \rightarrow AP$

Figure 3. Current waveform for both $P \rightarrow AP$ and $AP \rightarrow P$ write transitions.

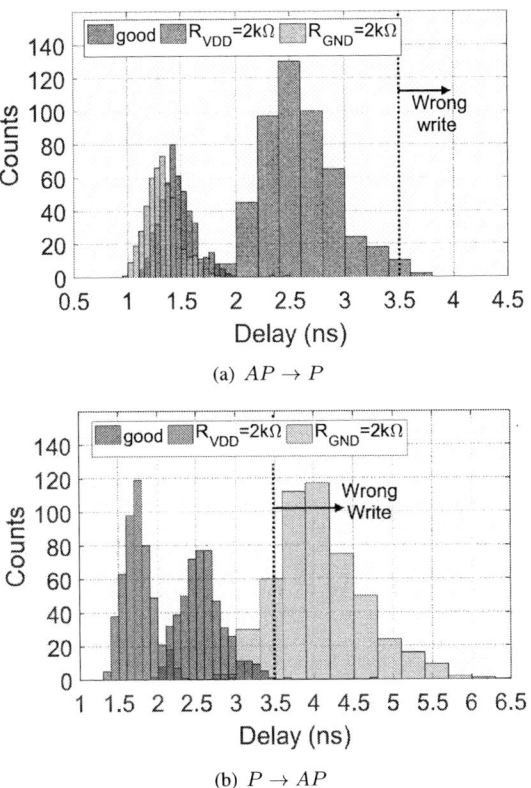

(a) $AP \rightarrow P$

(b) $P \rightarrow AP$

Figure 4. Histograms for write time for good and defective cells.

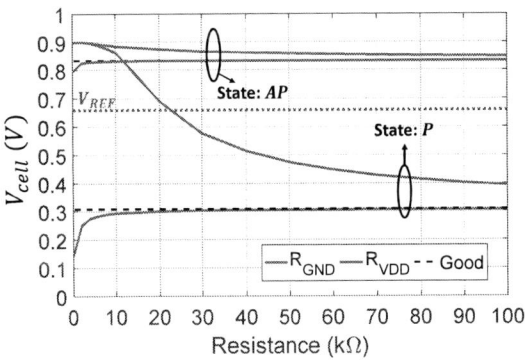

Figure 5. Cell voltage (See Figure 2) generated during the read operation as function of short defect resistance.

and can be partially detectable. However, even though this short defect of small size is strong, it may escape conventional logic test due to process variations. Only short defects of very small resistance can be completely distinguished from good cells, even under process variations. Figure 4 suggest that conventional logic test with write operation may not be effective enough as the defect detectability is low.

B. Impact of Short Defects on Read Operation

Figure 5 shows the voltage generated at the node V_{cell} of the read circuit (See Figure 2) as function of the short defect resistance for both R_{GND} and R_{VDD} defects. The black dashed lines correspond to the case of a "good" cell without defects, where the generated voltage V_{cell} is $0.30V$ and $0.85V$ when the MTJ is at the P and AP state, respectively. The voltage generated by the reference cell (V_{REF}) is $0.66V$. Defect R_{VDD} injects extra current to the internal node of the cell, which has to be driven by the access transistor. As a consequence, the resistance seen from the BL node increases and the generated V_{cell} increases too. Defect R_{GND} is placed in parallel with the access transistor. When the resistance of the short defect is large, its effect on the overall cell resistance is negligible. When the resistance of the short defect is small, the equivalent resistance reduces until it becomes closer to the pure resistance of the MTJ. It can be seen in Figure 5 that the generated V_{cell} reduces as R_{GND} becomes smaller.

An incorrect read operation occurs when the voltage generated for reading an AP state becomes lower than the reference voltage (V_{REF}) or when the voltage generated for reading a P state becomes larger than the reference voltage (V_{REF}). Figure 5 shows that an incorrect read occurs for a read P operation only and for a small enough R_{VDD}. The R_{GND} defect does not trigger incorrect read operations. Therefore, this defect can not be detected by a read operation using conventional logic fault test.

Impact of Process Variations

Figure 6 shows histograms for the generated V_{cell} during a read operation for good and defective cells with shorts defects

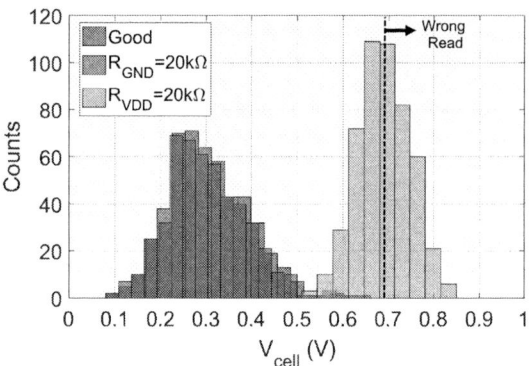

Figure 6. Histogram of generated cell voltage during reading P state operation.

of $20k\Omega$ at R_{GND} and R_{VDD}. As expected, when the defect R_{GND} is presented, it does not trigger an incorrect read. Instead, it makes V_{cell} even lower than the reference voltage. It can be seen that some of the defective cells with R_{VDD} may not cause an incorrect read (logic fault) and may escape logic fault test.

IV. PROPOSED TEST TECHNIQUE

A. Fundamental of the proposed test technique

The proposed test technique is based on the observation that a resistive-short defect between an external node and the internal node of the memory cell modifies the current flowing into the cell (through BL terminal) and out of the cell (through SL terminal).

As shown in Figure 7(a), a good cell behaves as a single current path, where the current provided by the read circuitry at BL terminal flows through the MTJ and the access transistor till the SL terminal. When a R_{VDD} short exist (Figure 7(b)), it injects current to the cell, so that I_{SL} becomes greater than I_{BL}. Similarly, when R_{GND} short exist, it removes current from the cell and I_{BL} becomes greater than I_{SL}.

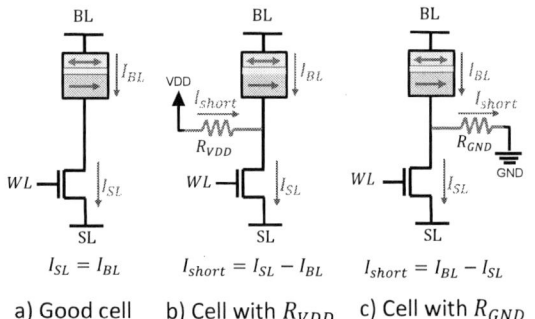

Figure 7. STT-MRAM cells with different short defects.

Figures 8(a) and 8(b) show the amplitude of the current flowing through BL and SL terminals during a read operation as a function of the short defect resistance R_{VDD} and R_{GND}, respectively. As R_{VDD} becomes smaller, the current injected by the defect pass through the access transistor and the SL

terminal. This has two effects: 1) The current flowing through the SL increases, and 2) the voltage at the internal node of the cell increases, which tend to turn-off the transistors of the current mirror in the read circuit, and as a consequence, the current flowing through the BL reduces. Similarly, as R_{GND} becomes smaller, two effects occur: 1) the access transistor of the cell drives less current, which reduces the current flowing through the SL, and 2) The voltage at the internal cell node slightly reduces, which causes a small increase in the current flowing through the BL terminal. These Figures show that as the short defect becomes smaller (stronger) the difference between the current that flows through BL and SL increases.

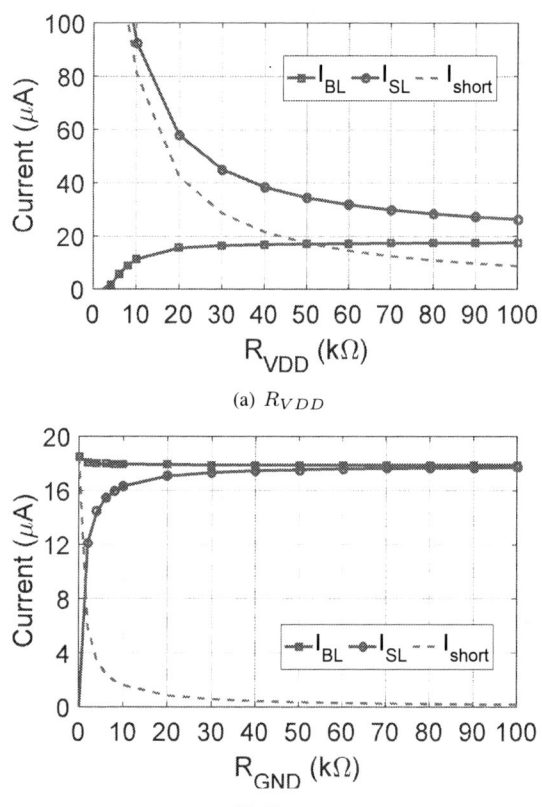

(a) R_{VDD}

(b) R_{GND}

Figure 8. BL and SL current behavior as function of short defect resistance.

B. Proposed Test Circuitry

Based on the above observations, we propose to modify the basic readout circuit of a column in the memory block so that it can measure the difference between the current flowing into and out of any cell in the column. A large current difference would indicate the presence a defect.

Figure 9 shows the modified readout circuit with test capability. Differential current amplifiers are introduced to both the reference branch and the memory cell branch of the read circuit. The modified circuit has three possible modes of operation: 1) Normal mode, 2) Test mode 1, and 3) Test mode 2. Normal mode is activated by setting "Normal" voltage to V_{DD}, and "Test 1" and "Test 2" voltages to 0. Under these

978-1-5386-4757-8/18 $31.00 © 2018 IEEE

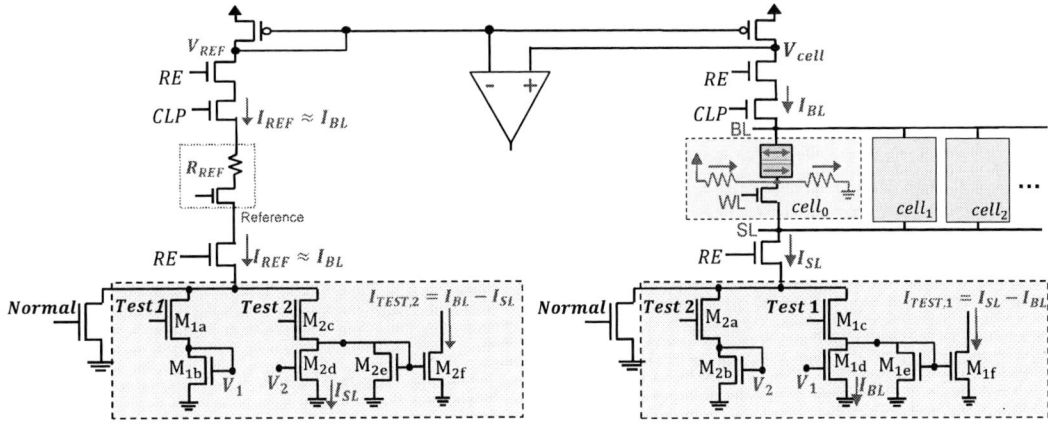

Figure 9. Modified read circuit with test capability.

conditions, normal functionality of the read circuit is obtained. Test Mode 1 is activated by setting *"Test 1"* voltage to V_{DD}, and *"Normal"* and *"Test 2"* voltages to 0. In this mode, the reference current (I_{REF}) passes through transistors M_{1a} and M_{1b} generating a voltage $V1$, which is used to copy I_{REF} to transistor M_{1d}. Note that $I_{REF} \approx I_{BL}$, thus the copied current represents I_{BL}. Since I_{SL} flows through M_{1c}, the current difference $I_{TEST,1} = I_{SL} - I_{BL}$ flows through transistor M_{1e}. This current is copied to M_{1f} which is the output transistor of the differential current amplifier. In summary, in Test Mode 1 the operation $I_{TEST,1} = I_{SL} - I_{BL}$ is performed to generate an output current when I_{SL} is bigger than I_{BL} (i.e. due to short R_{GND}). Test Mode 2 is activated by setting *"Test 2"* voltage to V_{DD}, and *"Normal"* and *"Test 1"* voltages to 0. Test Mode 2 operation is very similar to Test 1 Mode, but in this case the operation $I_{TEST,2} = I_{BL} - I_{SL}$ is performed to generate an output current at M_{2f} when I_{BL} is greater than I_{SL} (i.e. due to short R_{VDD}).

The detection capability of the proposed circuit was validated under the effect of process variations. Figures 10(a) and 10(b) show histograms for the differential current between BL and SL at the output of the proposed test circuit for both Test Mode 1 and Test Mode 2, respectively, for 1000 Montecarlo simulations. Figure 10(a) shows that defective cells with weak short R_{VDD} as large as $100k\Omega$ can be fully distinguished from good cells as there is no overlap between the distribution of the defective cell and the good one. Furthermore, when the defect becomes stronger (smaller resistance), the test output current increases making easier the detection of the defect. Defect R_{GND} is more difficult to detect. For defect $R_{GND} = 100k\Omega$, there is still a large overlap between the current distribution of a good cell and the defective cell. However, short defects of $20k\Omega$ can be fully detectable, which is still a significant improvement of the detection capability compared to logic test.

The obtained results suggest that the proposed test circuit is effective to detect short defects of sizes that are not detectable using conventional logic test. Moreover, the proposed test circuit could be used to diagnose the severity of a short defect

by defining thresholds over the differential output current.

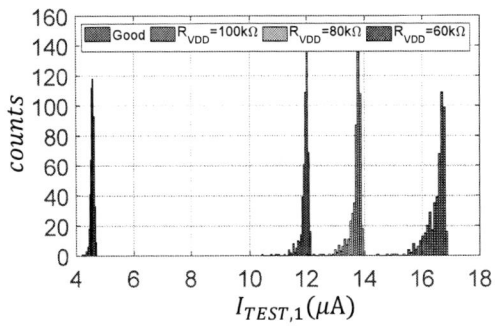

(a) Short R_{VDD} (Test Mode 1)

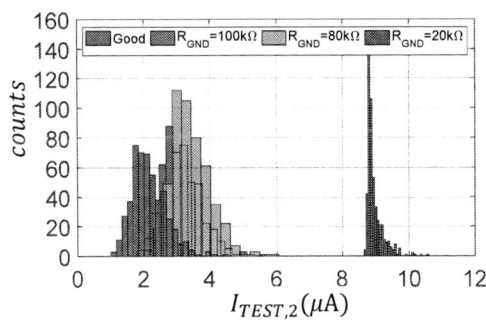

(b) Short R_{GND} (Test Mode 2)

Figure 10. Histograms of the measured differential current for various defects sizes.

C. Comparison with Logic Test

Table I shows a comparison of the proposed test approach with the conventional logic test. The proposed method offers a higher defect detectability of both weak and strong short detects whereas fault-based test only detects strong defects causing logic faults. Those test-escapes from classic logic faults may damage circuits lifetime reliability. Regarding the area overhead, the logic test does not require any additional transistors. According to Figure 9, the proposed DFT approach requires 6 additional transistors in the read circuit for the cell and other 6 transistors in the read circuit of the reference cell. However, it must be noted that only one read/test circuit need to be used for several memory cells, i.e., those cells in the same column of the array. Thus, the area overhead is very low, compared to the size of the entire memory array. Regarding test time, the logic test requires to perform multiple writes and reads operations, which makes the test procedure slow. On the other hand, the proposed test approach only requires performing a single read operation with each of the test modes activated at a time. Thus few clock cycles would be required. Regarding robustness against process variations, even some strong defects may be masked by process variations in the conventional logic test. On the other hand, the proposed approach is robust to process variations because the differential current amplifier cancels most of the variations in the memory, as they impact in a similar way the current flowing into and out of the cell.

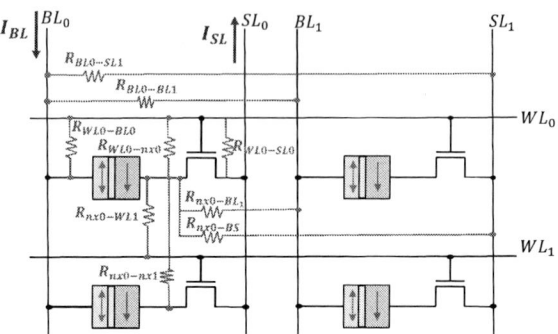

Figure 11. Other resistive-shorts that can be detected using the proposed approach.

Table I
COMPARISON WITH FAULT LOGIC TEST

Feature	Logic Test	Prop. Technique
Detectability	Strong Defects	Strong and Weak Defects
Area Overhead	No	Low: 12 transistors per column
Test Time	Long	Short
PV Robustness	No	Yes

D. Detection of Other Short Defects

The analysis above has been focused on short defects between supply rails (VDD and GND) and the internal node (nx) of the STT-MRAM cell. However, the proposed technique is also valid to detect all those short defects that create an unbalanced current between BL and SL nodes. Note that resistive-open defects can not be detected with the proposed approach, as these defects do not make different the current flowing through BL and SL terminals. Figure 11 shows a memory array, where some short defects that can be detected using the proposed approach are highlighted. The probability of ocurrence of the short defects depends on the memory array architecture and how its layout is made. Typically, wider lines placed closer to other lines are more likely to present bridge defects [10]. The defects shown in Figure 11 exhibit similar behavior to R_{VDD} and the R_{GND} defects analyzed in this paper. Gate short defects between the WL0 and either BL0, nx0, or SL0 behave as R_{VDD} because WL signal is settled to V_{DD} during a read operation to activate the access to the cell. Similarly, inter-cell shorts could be detected by previously setting the voltages of the adjacent cell at adequate values. For example, $R_{BL0-SL1}$ behaves as R_{VDD} is SL1 terminal is set to V_{DD}. Therefore, the proposed test technique can cover a wide variety of manufacturing short defects.

V. CONCLUSIONS

This paper has proposed a circuit for a defect-oriented test of resistive-shorts in an STT-MRAM cell. The proposed

circuit senses the difference of the current that flows through BL and SL terminals of the cell. This current difference is caused by the current that the short defect introduces/removes from internal nodes of the cell. The proposed approach is robust to process variations and significantly improves the detectability of resistive short defects that otherwise could escape conventional logic test.

ACKNOWLEDGMENT – This work was supported by CONACYT (Mexico) through the Ph.D. scholarship number 420129/264560.

REFERENCES

[1] Arundhati Bhattacharya, Soumitra Pal, and Aminul Islam. Implementation of finfet based stt-mram bitcell. *2014 IEEE International Conference on Advanced Communications, Control and Computing Technologies*, pages 435–439, 2014.
[2] ITRS International Technology Roadmap for Semiconductor. Available on: http://www.itrs2.net/.
[3] A. Shafaei, Y. Wang, and M. Pedram. Low write-energy stt-mrams using finfet-based access transistors. In *2014 IEEE 32nd International Conference on Computer Design (ICCD)*, pages 374–379, Oct 2014.
[4] X. Fong, S. H. Choday, and K. Roy. Bit-cell level optimization for non-volatile memories using magnetic tunnel junctions and spin-transfer torque switching. *IEEE Transactions on Nanotechnology*, 11(1):172–181, Jan 2012.
[5] A. Chintaluri, H. Naeimi, S. Natarajan, and A. Raychowdhury. Analysis of defects and variations in embedded spin transfer torque (stt) mram arrays. *IEEE Journal on Emerging and Selected Topics in Circuits and Systems*, 6(3):319–329, Sept 2016.
[6] A.F. Gomez, F. Lavratti, G. Medeiros, M. Sartori, L. Bolzani Poehls, V. Champac, and F. Vargas. Effectiveness of a hardware-based approach to detect resistive-open defects in sram cells under process variations. *Microelectronics Reliability*, 67:150 – 158, 2016.
[7] Predictive technology models, available online: http://ptm.asu.edu/.
[8] Xuanyao Fong, Sri Harsha Choday, Panagopoulos Georgios, Charles Augustine, and Kaushik Roy. Spice models for magnetic tunnel junctions based on monodomain approximation, Aug 2013.
[9] Y. Zhang, X. Wang, Y. Li, A. K. Jones, and Y. Chen. Asymmetry of mtj switching and its implication to stt-ram designs. In *2012 Design, Automation Test in Europe Conference Exhibition (DATE)*, pages 1313–1318, March 2012.
[10] Freddy Forero, Jean-Marc Galliere, Michel Renovell, and Victor Champac. Detectability challenges of bridge defects in finfet based logic cells. *Journal of Electronic Testing*, 34(2):123–134, Apr 2018.

978-1-5386-4757-8/18 $31.00 © 2018 IEEE

Evaluating the Impact of Process Variability and Radiation Effects on Different Transistor Arrangements

Leonardo H. Brendler[1], Alexandra L. Zimpeck[1], Cristina Meinhardt[2] and Ricardo Reis[1]

[1]*Instituto de Informática*, PPGC/PGMicro - Universidade Federal do Rio Grande do Sul (UFRGS)

[2]*Departamento de Informática e Estatística* - Universidade Federal de Santa Catarina (UFSC)

{lhbrendler, alzimpeck, reis}@inf.ufrgs.br, cristina.meinhardt@ufsc.br

Abstract—The high integration capacity of digital circuits, which occurs due to technological scaling, presents new challenges for nanotechnology designs. The evolution of integrated circuits has made them more susceptible to faults, besides increasing the process variability, which can lead to circuits operating outside their specification ranges. This work evaluates the effects of process variability and radiation faults on complex gates. These effects are compared to alternative circuits that implement the same functions but exploring a multi-level of basic cells as NAND2, NOR2 and Inverters. The technology adopted is 7nm FinFET ASAP. Results show that although complex cells present better timing and power results, multi-level circuits are up to 28% less sensible to radiation faults and about 40% more stable under process variability.

Index Terms—process variability, radiation effects, complex gates, FinFET technology

I. INTRODUCTION

The evolution of the transistors manufacturing process has been happening at an impressive speed. The technology scaling increases the integration capacity of integrated circuits and also allows operating frequencies to become increasingly high. Nowadays, integrated circuits have an essential role in practically every one of our daily tasks, at the cost of these circuits have become increasingly dense and complex.

By contrast, new challenges were introduced in the design of integrated circuits due to scale down, such as increased manufacturing process variability [1], Short-Channel Effects (SCE), leakage current [2] and increased susceptibility to radiation effects. Moreover, Bulk CMOS technology has reached its geometrical and physical limit. Bulk MOSFET devices have been used in the integrated circuits design for several decades. However, at each new technology node, Bulk MOSFET devices suffer from the undesirable leakage currents, soft errors and SCE [3] [4]. The use of multi-gate devices is an option to overcome these obstacles and continue the technology scaling because these devices provide better control of SCE, lower leakage and better yield [5]. FinFET (Fin-Shaped Field Effect Transistor) technology is used as the main multi-gate device replacing MOSFET devices in sub-22nm technology nodes [6].

The focus of this work is the variability and the faults arising from the radiation effects. Few works address these two

effects together on the circuit design. Few solutions are found in the literature to mitigate the impact of process variability. Transistor arrangement is a technique explored to design faster circuits [7], but also to deal with Bias Temperature Instability (BTI) effects [8] or improve design robustness against permanent and transient faults. Adoption of complex gates reduce the transistor number that is correlated to the area and also reduce the delay and power consumption, but complex gates could introduce challenges related to regularity and reliability that might be avoided with more regular and basic cells.

The main objective of this work is to evaluate the impact of process variability and Single Event Transient (SET) on a set of complex logic gates considering different transistor topologies. A comparison is made between complex logic gates in their traditional versions and a multi-level of basic logic gates that implement the same function using NAND2, NOR2 and Inverter cells in FinFET technology.

Section II introduces some FinFET properties. The process variability and radiation effects concepts are shown in Section III. The methodology utilized in experiments is presented in Section IV. Section V explains the impact of process variability and radiation effects on different transistor arrangements. Finally, conclusions are displayed in Section VI.

II. FINFET TECHNOLOGY

FinFET devices have vertical silicon structures to form the channel region and to connect the source and drain regions at each end [6]. The gate region is wrapped around this vertical structure, named as the fin. MOS channels are formed at the two sidewalls. The ON current (I_{ON}) of these devices is a function of the sum of the drive currents contributed by the two side-gate transistors. This fin-like geometry, where the depletion regions reach from the gates entirely into the body region implies that no free charge carriers are available, making the suppression of SCE possible in FinFETs [9]. Fig. 1, presents FinFET key geometric parameters [10]: Gate length (L_G/L_{FIN}), fin height (H_{FIN}) and fin width/thickness (W_{FIN}/T_{SI}).

FinFET devices variability hardly affects the currents behavior, especially the static current [11]. On multi-gate devices,

978-1-5386-4757-8/18 $31.00 © 2018 IEEE

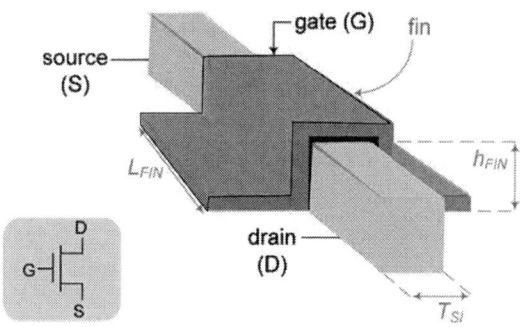

Fig. 1. Structure and geometric parameters of FinFETs [10].

variability effects are mainly due to the work function fluctuation (WFF) of the metal gate [6] [12].

III. PROCESS VARIABILITY AND RADIATION EFFECTS

The reduction of device dimensions has increased the importance of studying the reliability issues of integrated circuits. Thus, this section presents the main characteristics of two of the most significant adverse effects on the reliability of these circuits: Process Variability and Radiation Effects.

Circuits exposed to radiation, may suffer alterations and disturbances, impairing their correct functioning. The integrated circuits that experience the interaction of ionizing particles suffer two types of degradation: those of singular character, that occur due to the incidence of a single particle, and the ones of cumulative nature, which occur due to the accumulation of doses of ionizing radiation over the life of the circuit. This work addresses the effects of the incidence of a single particle, the Single-Event Effects (SEE).

An SEE occurs when a single particle hit the device, ionizing and releasing energy that can damage the device permanently or induce transient effects. The most common transient effects on combinational circuits are the SET, where the incidence of an ionized particle produces a transient pulse that can propagate through a logic path and be latched by memory elements. At a particle collision with silicon, the effects are: 1) the deposition of an additional charge on the affected device; and 2) the charge collection step proceeds through mainly Drift and Diffusion process.

Drift occurs when the high-intensity electric field rapidly collects the additional carriers deposited by the ion in this region if the resulting ionization path crosses the depletion region formed at the P-N junctions [13]. However, diffusion dominates the collection process by collecting all the remaining carriers that were generated besides the depletion layer [14]. After this collection process, a transient pulse is observed in the affected node. At nanometer nodes, it is verified an increase in the susceptibility of the circuit relative to the noise from the environment and particularly the incidence of particles of radiation [15]. Even particles with low energy that reach the Earth surface, previously overlooked, they are now able to interfere with the operation of a circuit.

Variability can be divided into three factors: environmental factors, reliability factors and physical factors [16]. The environmental factors appear during the operation of a circuit. Power supply and temperature variations are examples of environmental factors. The reliability factors are related to the aging of the transistor, due to the high electrical fields presented in modern circuits. Finally, the physical factors are associated with variations in the electrical parameters, which can occur due to the manufacturing process.

The primary sources of process variability at nanometer nodes are due to the sub-wavelength lithography [17] [16]. Variability on geometric parameters due to lithography impact directly the transistor threshold voltage (V_{th}). These variations can compromise entire blocks of cells or reduce the performance and energy efficiency of the chip.

From the beginning of FinFET devices adoption on digital circuits, many works strengthened the relevance of considering the impact of variability on FinFET devices to estimate how it will impact the design and guaranteed functional devices [3] [18] [19] [20].

The most significant parameter for the process variability effects is WFF. WFF is caused by the dependency of metal work-function on the orientation of its grains, as depicted in Fig. 2. A high correlation between the variability in I_{ON} and I_{OFF} currents and threshold voltage fluctuation in the presence of granularity of the metal gate (MGG) [19] is presented by the parameter. In the ideal fabrication process, metal gates devices have the gates produced with metal uniformly aligned and very lower WFF deviation. Nevertheless, in the real manufacturing process, metal gate devices are generally built with metals having different work-functions (Φm) randomly aligned which causes higher WF variation.

IV. METHODOLOGY

This work evaluates the impact of process variability and radiation effects on different transistor arrangements. Three logic functions were chosen (AOI22, OAI211 and XOR) and four different transistor topologies are explored (complex gate, only NAND2, only NOR2 and NAND2/NOR2/INV) to perform this analysis. For the complex gate transistor topology, the functions are optimized and designed as a complex logic gate

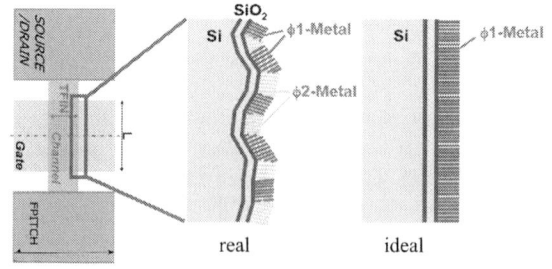

Fig. 2. Metal gate fabrication in the ideal and real aspects [11].

978-1-5386-4757-8/18 $31.00 © 2018 IEEE

CMOS topology. Then, the functions are converted using De Morgan's theorem into the three other transistor arrangements, in a way that only basic cells are employed. Tables I, II and III show the complex functions of each complex gate and its respective functions converted using only NAND2 gates, only NOR2 gates and NAND2/NOR2/INV gates.

TABLE I
COMPLEX AND CONVERTED FUNCTIONS OF XOR GATE

Transistor Arrangement	XOR
Complex Gate	Y = A.B' + A'.B
NAND2	Y = ((A . (B.B)')' . ((A.A)' . B)')'
NOR2	Y = ((((A+A)' + B)' + (A + (B+B)')')')' + (((A+A)' + B)' + (A + (B+B)')')')'
NAND2/NOR2/INV	Y = ((((A.B')')' + ((A'.B)')')')'

TABLE II
COMPLEX AND CONVERTED FUNCTIONS OF OAI211 GATE

Transistor Arrangement	OAI211
Complex Gate	Y = (A+B . C.D)'
NAND2	Y = (((A.A)' . (B.B)')' . ((C.D)' . (C.D)')')'
NOR2	Y = (((A+B)' + ((C+D)' + (C+D)')')' + ((A+B)' + ((C+D)' + (C+D)')')')'
NAND2/NOR2/INV	Y = ((((A'.B')')' + (C.D)')')'

TABLE III
COMPLEX AND CONVERTED FUNCTIONS OF AOI22 GATE

Transistor Arrangement	AOI22
Complex Gate	Y = (A.B + C.D)'
NAND2	Y = (((A.B) . (C.D)')' . ((A.B) . (C.D)')')'
NOR2	Y = (((A+A)' + (B+B)')' + ((C+C)' + (D+D)')')'
NAND2/NOR2/INV	Y = ((((A'+B')')' . (C+D)')')'

Electrical simulations with the HSPICE tool were made using the 7nm FinFET technology from ASAP7 [21] to perform all steps of this work. Table IV resumes the main parameters of the 7nm FinFET ASAP7 technology. In the experiments, the transistor sizing considers all transistors with three fins for all circuits [22]. The nominal supply voltage of the model adopted is 0.7V. The minimum switching frequency of the input signals was 500MHz, and four inverters (Fanout 4) were used as the load at the output of the circuit.

The analysis of this work can be divided into three stages: (1) nominal, (2) process variability and (3) radiation effects. In addition to comparing the results obtained in each stage, a general comparison of the results is also performed. The objective is to highlight which transistor arrangement presents the best results aiming only process variability and only transient faults, but also to give the topology that has an ideal average behavior in the three stages studied.

A. Nominal Conditions

This stage evaluates the circuits at nominal conditions, i.e., process variability and radiation effects are not considered. In this step, the propagation times and the total power consumption of the three complex gates in their different

TABLE IV
7NM FINFET ASAP7 MAIN PARAMETERS [21]

Parameter		7nm
Supply Voltage		0.7V
Gate Length (L_G)		21nm
Fin Width (W_{FIN})		6.5nm
Fin Height (H_{FIN})		32nm
Oxide Thickness (Tox)		2.1nm
Channel Doping		$1 \times 10^{22} m^{-3}$
Source/Drain Doping		$2 \times 10^{26} m^{-3}$
Work Function	NFET	4.3720eV
	PFET	4.8108eV

arrangements are compared. The objective is to analyze the typical characteristics of each gate and each arrangement. Nominal values are used as a form of reference values to evaluate the variability and radiation effects.

B. Process Variability

The analysis considering the effects of process variability is performed at the second stage. Metal gate work-function exhibits a multi-nominal distribution, which can be approximated by a Gaussian distribution if the number of grains on the surface of metal-gate is high enough (>10) which corresponds to the FinFET ASAP7 model characteristics. Thus, the WFF of each device is varied according to a Gaussian distribution with a 3-sigma deviation of 3% [11] the WFF. Two thousand simulations were run for each logic gate [23]. Timing and power consumption measurements were taken for each Monte Carlo simulation. Robustness analysis was performed using the sigma/mean ratios (called deviation) of each delay arc, always aiming at the worst case.

C. Radiation Effects

The third stage considers the effects of radiation through the insertion of transient faults. The SET fault injection is modeled as the Messenger's equation shown in Eq. 1, where $Qcoll$ is the collected charge, τ_α ($1.64 \times 10^{-10}s$) is the collect charge timing constant, τ_β ($5 \times 10^{-11}s$) is the timing constant to establish the ion track and L ($2\mu m$) is the charge collection profundity [24]. This effect is reproduced on the SPICE simulation as a current source, simulating the SET effects on the transistors.

$$I(t) = \frac{Qcoll}{\tau_\alpha - \tau_\beta}(e^{-\frac{t}{\tau_\alpha}} - e^{-\frac{t}{\tau_\beta}})$$

(1)

$$Qcoll = 10.8 \times L \times LET$$

This work follows the parameter and methodology presented in [25], investigating the effects of SET 010 and 101 in all devices of the two inverter circuits. Thus, a current source is inserted into each internal node and the output of the circuit. Also, this evaluation considers all the input vectors for the selected logic gates.

The simulation adopts a linear energy transfer (LET) of $1 MeV - cm^2/mg$. The fault is detected if the output of

978-1-5386-4757-8/18 $31.00 © 2018 IEEE

the circuit is bigger than $V_{DD}/2$ for logic level '0' and smaller than $V_{DD}/2$ for logic level '1'; otherwise, the fault is considered masked. The fault masking is determined by Eq. 2, considering the ratio between the number of faults detected and the total inserted faults, i.e., a fault entered in each internal node and the output, for each test vector of the circuit. For example, a logic gate with four inputs has 16 test vectors, if this same gate has five internal nodes plus the output, there would be 96 inserted faults in the circuit.

$$Fault\ Masking = \frac{Faults\ Detected}{Total\ Inserted\ Faults} \qquad (2)$$

V. RESULTS

To compare the different transistor topology and to identify common characteristics to mitigate process variability and radiation effects, the results of the set of experiments performed in this work is organized into three main parts, as the methodology described: nominal behavior, process variability impact and radiation effects on the circuits.

A. Nominal Behavior

Complex gates reduce the number of literals in the equations and, consequently, this reflects on the fact that with this transistor arrangement, all the three functions evaluated presented better timing and power results. Fig. 3, compares the maximum transition time for all arches of each evaluated function. NAND2 version is about 18% slower than Complex gate versions, but NOR2 circuits could insert more than twice times of delay degradation on the OAI211 gate.

The impact of the use of different arrangements is even worse, when the total power consumption considering all timing arches, is analyzed. Fig. 4, shows that multi-level versions (NAND2, NOR2 and NAND2/NOR2/INV) consumes at least 49% more than complex gate circuits. The XOR gate has the smallest increase in power consumption (49.4%) in NAND2 version, while the AOI22 gate in NOR2 version has the most significant growth, around 129%.

Thus, to optimized circuits addressing power and timing reduction, the complex gate is the best alternative to transistor arrangement. In sequence, the evaluation of the process variability and radiation effects will throw light on these effects on complex gates and the alternative circuits evaluated.

B. Process Variability

Process variability inserts oscillations on the delay and power consumption compared to the results presented in the nominal behavior evaluation. In general, complex gate circuits under process variability show the most significant variation in the delay compared to nominal behavior, reaching up to 5% on the worst case delay. Multi-level circuits presented a difference on the mean delay of 4%, 3% and 2% for NAND2, NOR2 and NAND2/NOR2/INV alternative circuits. However, complex gate and NAND2/NOR2/INV versions present mean power values statistically identical to the nominal results,

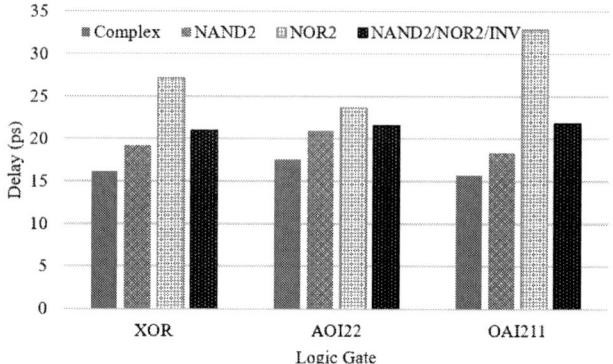

Fig. 3. Delay at nominal conditions.

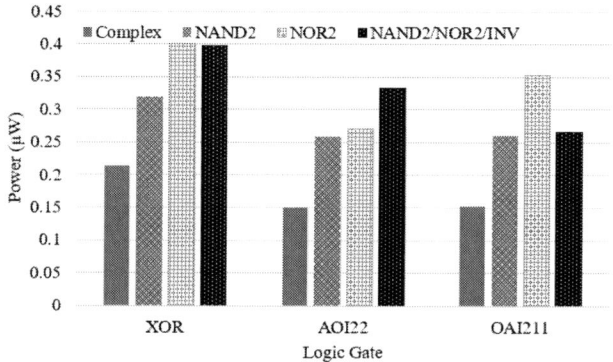

Fig. 4. Power at nominal conditions.

despite the large standard deviations. Table V shows the mean and standard deviation values obtained from the Monte Carlo simulation.

To compare the effects of process variability, considering both mean and standard deviation, are presented in Fig. 5 and Fig. 6 respectively, the deviation results for delay and power. All the circuits demonstrate more sensibility on delay due to process variability than on power results, as we are analyzing the normalized standard deviation and not the absolute values

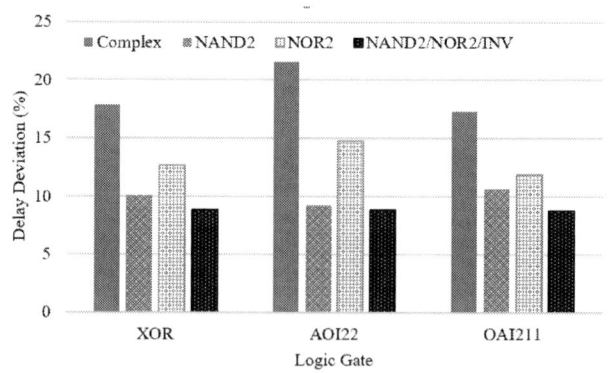

Fig. 5. Delay deviation due to process variability.

TABLE V
MEAN AND STANDARD DEVIATION RESULTS FOR POWER AND DELAY

Complex Gate	Measures	Complex		NAND2		NOR2		NAND2/NOR2/INV	
		Mean	Sigma	Mean	Sigma	Mean	Sigma	Mean	Sigma
XOR	Worst delay (ps)	17.1	3.1	19.7	1.9	28.2	3.6	21.5	1.9
	Power (nW)	213.8	9.8	311.8	15.6	401.3	17.8	394.6	23.2
AOI22	Worst delay (ps)	17.3	3.7	21.6	1.9	24.7	3.6	22.0	1.9
	Power (nW)	150.4	6.3	259.9	10.7	278.5	12.8	332.2	14.1
OAI211	Worst delay (ps)	16.5	2.8	19.1	2.1	34.2	4.1	22.4	1.9
	Power (nW)	153.3	6.5	281.1	13.2	363.8	17.5	264.9	13.9

this behavior is possible. Complex gates significantly suffer the influence of process variability on delay. For all the three functions evaluated, this alternative shows the worst results with more than 15% of delay deviation. Multi-gate with basic cells alternatives demonstrates better delay robustness to process variability. NAND2/NOR2/INV shows a slight advantage compared to NAND2 versions, with a difference of 1.5%. However, NAND2 versions present less power sensibility. One of the factors that contribute to the greater robustness of the multi-level topologies is the existence of identical elements in the vicinity, this guarantees an easy impression and final verification.

In the comparison of the normalized deviation of the multi-level alternatives with the complex gates, for all functions, multi-level circuits improved the process variability robustness over 30%. AOI22 is the cell most beneficiated with the multi-level arrangements, presenting over 50% of the reduction in delay deviation.

C. Radiation Effects

From the analysis of fault masking between the complex gate and the different arrangements with basic cells, the advantage of using basic cells is evident when the objective is to mitigate transient faults. In Fig. 7, this advantage can be analyzed mainly in the topologies that use only NAND2 and only NOR2 gates. For the XOR gate, the NAND2 and NOR2 arrangements present decreases of 7% and 14% in the sensitivity to transient faults about the use of complex gates, respectively. For the AOI22 gate, this reduction is even higher,

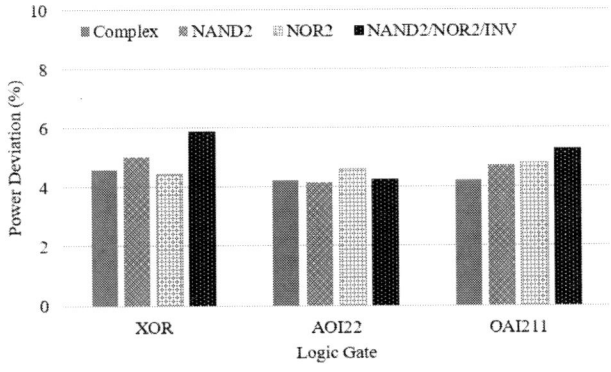

Fig. 6. Power deviation due to process variability.

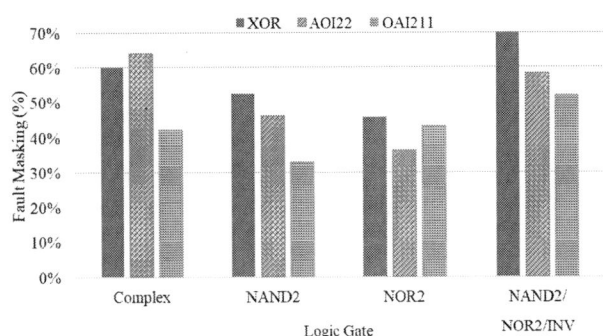

Fig. 7. Fault masking comparison.

being 18% for the topology with only NAND2 and 28% for the topology with the only NOR2.

The arrangement with NAND2 continues with the reduction of the sensitivity for the OAI211 gate, around 9%. However, the NOR2 arrangement has a small increase (1%) in the sensitivity compared with the complex gates. In general, the arrangement using NAND2, NOR2 and Inverters together does not perform well about fault masking. Only for the AOI22 gate, this topology has a sensitivity reduction (6%), whereas, for the XOR and OAI211 gates, there is an increase of 10% in the fault sensitivity. Although the NOR2 arrangement presents the highest percentages of sensitivity decrease to transient faults, the NAND2 topology is the most stable, having a significant reduction for all three gates analyzed in this study. Table VI shows the total injected faults in each logic function examined. Moreover, the number of faults masked and detected are also displayed.

VI. CONCLUSIONS

This work presented an evaluation of the influence of process variability and the radiation effects on 7nm FinFET ASAP technology. At nominal conditions, it is evident that the complex gate arrangement is the best choice to obtain less power consumption and a decrease in the propagation delay. It occurs because the complex gate topology reduces the number of transistors and the connections between them significantly. On the other hand, when the behavior of the logic gates is investigated with radiation or process variability effects, the multi-level arrangements are the best option. In comparison to complex gate topology, the XOR and AOI22 logic gates composed only by NOR2 gates can be up to 14%

978-1-5386-4757-8/18 $31.00 © 2018 IEEE

TABLE VI
FAULT ANALYSIS OF ALL FUNCTIONS

Fault Analysis/ Transistor Arrangement	Logic Function											
	XOR				AOI22				OAI211			
	Complex	NAND2	NOR2	NAND2/ NOR2/INV	Complex	NAND2	NOR2	NAND2/ NOR2/INV	Complex	NAND2	NOR2	NAND2/ NOR2/INV
Inserted	20	40	48	44	64	128	224	208	64	192	192	160
Detected	12	21	22	31	41	59	81	121	27	63	83	83
Masked	8	19	26	13	23	69	143	87	37	129	109	77
Fault Masking	60.0%	52.5%	45.8%	70.5%	64.1%	46.1%	36.2%	58.2%	42.2%	32.8%	43.2%	51.9%

and 28% more fault tolerant, respectively. Similar behavior can be seen with the OAI211 gate where the arrangement with only NAND2 reaches around of 9% more fault tolerance about to the topology of complex gates.

Complex gates suffer more influence of process variations on propagation delays than the power consumption. For all logic gates analyzed, this alternative presented more than 15% of delay deviation. Compared to multi-level topologies, complex gates delay deviation impact over 30% of the stability of the functions. Finally, under the impact of radiation or process variations, better results can be found using the topologies based on multi-level arrangements. Aiming at future work more in-depth analyzes will be carried out regarding the effects of radiation. Also, the impact of the transistors sizing on these effects will also be analyzed.

ACKNOWLEDGMENT

This work is partially supported by the Brazilian National Council of Scientific and Technologic Development (CNPq), by the Brazilian Coordination of Improvement of Higher Level Personnel (CAPES) and by the Research Support Foundation of the State of Rio Grande do Sul (FAPERGS).

REFERENCES

[1] M. Orshansky, S. R. Nassif, and D. Boning, "Introduction," *Design for Manufacturability and Statistical Design: A Constructive Approach*, pp. 1–8, 2008.

[2] Y. Taur, D. A. Buchanan, W. Chen, D. J. Frank, K. E. Ismail, S.-H. Lo, G. A. Sai-Halasz, R. G. Viswanathan, H.-J. Wann, S. J. Wind *et al.*, "Cmos scaling into the nanometer regime," *Proceedings of the IEEE*, vol. 85, no. 4, pp. 486–504, 1997.

[3] T.-J. King, "Finfets for nanoscale cmos digital integrated circuits," in *Proceedings of the 2005 IEEE/ACM International conference on Computer-aided design.* IEEE Computer Society, 2005, pp. 207–210.

[4] D. J. Frank, R. H. Dennard, E. Nowak, P. M. Solomon, Y. Taur, and H.-S. P. Wong, "Device scaling limits of si mosfets and their application dependencies," *Proceedings of the IEEE*, vol. 89, no. 3, pp. 259–288, 2001.

[5] ITRS. (2011) International technology roadmap for semiconductors. [Online]. Available: http://www.itrs2.net/2011-itrs.html File: 011Exec-Sum.pdf

[6] A. R. Brown, J. R. Watling, and A. Asenov, "Intrinsic parameter fluctuations due to random grain orientations in high-κ gate stacks," *Journal of Computational Electronics*, vol. 5, no. 4, pp. 333–336, 2006.

[7] P. F. Butzen, V. Dal Bem, A. I. Reis, and R. P. Ribas, "Transistor network restructuring against nbti degradation," *Microelectronics Reliability*, vol. 50, no. 9-11, pp. 1298–1303, 2010.

[8] D. N. da Silva, A. I. Reis, and R. P. Ribas, "Cmos logic gate performance variability related to transistor network arrangements," *Microelectronics Reliability*, vol. 49, no. 9-11, pp. 977–981, 2009.

[9] X. Huang, W.-C. Lee, C. Kuo, D. Hisamoto, L. Chang, J. Kedzierski, E. Anderson, H. Takeuchi, Y.-K. Choi, K. Asano *et al.*, "Sub 50-nm finfet: Pmos," in *Electron Devices Meeting, 1999. IEDM'99. Technical Digest. International.* IEEE, 1999, pp. 67–70.

[10] M. Alioto, "Comparative evaluation of layout density in 3t, 4t, and mt finfet standard cells," *IEEE Transactions on Very Large Scale Integration (VLSI) Systems*, vol. 19, no. 5, pp. 751–762, 2011.

[11] C. Meinhardt, A. L. Zimpeck, and R. A. Reis, "Predictive evaluation of electrical characteristics of sub-22 nm finfet technologies under device geometry variations," *Microelectronics Reliability*, vol. 54, no. 9-10, pp. 2319–2324, 2014.

[12] V. B. Kleeberger, H. Graeb, and U. Schlichtmann, "Predicting future product performance: Modeling and evaluation of standard cells in finfet technologies," in *Proceedings of the 50th Annual Design Automation Conference.* ACM, 2013, p. 33.

[13] M. Agarwal, B. C. Paul, M. Zhang, and S. Mitra, "Circuit failure prediction and its application to transistor aging," in *VLSI Test Symposium, 2007. 25th IEEE.* IEEE, 2007, pp. 277–286.

[14] D. Munteanu and J.-L. Autran, "Modeling and simulation of single-event effects in digital devices and ics," *IEEE Transactions on Nuclear science*, vol. 55, no. 4, pp. 1854–1878, 2008.

[15] R. C. Baumann, "Radiation-induced soft errors in advanced semiconductor technologies," *IEEE Transactions on Device and materials reliability*, vol. 5, no. 3, pp. 305–316, 2005.

[16] S. R. Nassif, "Process variability at the 65nm node and beyond," in *Custom Integrated Circuits Conference, 2008. CICC 2008. IEEE.* IEEE, 2008, pp. 1–8.

[17] H. F. Dadgour, K. Endo, V. K. De, and K. Banerjee, "Grain-orientation induced work function variation in nanoscale metal-gate transistors — part ii: Implications for process, device, and circuit design," *IEEE Transactions on Electron Devices*, vol. 57, no. 10, pp. 2515–2525, 2010.

[18] K. Endo, T. Matsukawa, Y. Ishikawa, Y. Liu, S. O'uchi, K. Sakamoto, J. Tsukada, H. Yamauchi, and M. Masahara, "Variation analysis of tin finfets," in *Semiconductor Device Research Symposium, 2009. IS-DRS'09. International.* IEEE, 2009, pp. 1–2.

[19] X. Wang, A. R. Brown, B. Cheng, and A. Asenov, "Statistical variability and reliability in nanoscale finfets," in *Electron Devices Meeting (IEDM), 2011 IEEE International.* IEEE, 2011, pp. 5–4.

[20] H. Dadgour, V. De, and K. Banerjee, "Statistical modeling of metal-gate work-function variability in emerging device technologies and implications for circuit design," in *Computer-Aided Design, 2008. ICCAD 2008. IEEE/ACM International Conference on.* IEEE, 2008, pp. 270–277.

[21] L. T. Clark, V. Vashishtha, L. Shifren, A. Gujja, S. Sinha, B. Cline, C. Ramamurthy, and G. Yeric, "Asap7: A 7-nm finfet predictive process design kit," *Microelectronics Journal*, vol. 53, pp. 105–115, 2016. [Online]. Available: http://asap.asu.edu/asap/

[22] B. Chava, D. Rio, Y. Sherazi, D. Trivkovic, W. Gillijns, P. Debacker, P. Raghavan, A. Elsaid, M. Dusa, A. Mercha *et al.*, "Standard cell design in n7: Euv vs. immersion," in *Design-Process-Technology Co-optimization for Manufacturability IX*, vol. 9427. International Society for Optics and Photonics, 2015, p. 94270E.

[23] M. Alioto, E. Consoli, and G. Palumbo, "Variations in nanometer cmos flip-flops: Part i—impact of process variations on timing," *IEEE Transactions on Circuits and Systems I: Regular Papers*, vol. 62, no. 8, pp. 2035–2043, 2015.

[24] G. Messenger, "Collection of charge on junction nodes from ion tracks," *IEEE Transactions on nuclear science*, vol. 29, no. 6, pp. 2024–2031, 1982.

[25] S. Uznanski, G. Gasiot, P. Roche, C. Tavernier, and J.-L. Autran, "Single event upset and multiple cell upset modeling in commercial bulk 65-nm cmos srams and flip-flops," *IEEE Transactions on Nuclear Science*, vol. 57, no. 4, pp. 1876–1883, 2010.

978-1-5386-4757-8/18 $31.00 © 2018 IEEE

An accurate novel gate-sizing metric to optimize circuit performance under local intra-die process variations

Zahira Perez
Dept. Electronics
INAOE
Luis Enrique Erro 1, Tonanzintla
Puebla, Mexico C.P. 72840
arihaz.zerep24@gmail.com

Hector Villacorta
Dept.Posgraduate and Research
Polytechnic University of Aguascalientes
Paseo San Gerardo 207, Fracc.San Gerardo
Aguascalientes, Mexico C.P.20342
hvillamina@gmail.com

Victor Champac
Dept. Electronics
INAOE
Luis Enrique Erro 1, Tonanzintla
Puebla, Mexico C.P. 72840
champac@inaoep.mx

Abstract—Due to aggressive technology scaling in nanometer regime, the impact of process variations on IC design has become a challenge. Process variations are usually classified into two types: inter-die and intra-die variations. The focus of this paper is on intra-die variations, which were ignored in the past, and are now significantly impacting the performance of modern circuits. A novel statistical gate-sizing metric is proposed to enhance the performance of digital integrated circuits in the presence of local intra-die process variations. The metric selects the most beneficial gates to be resized for improving circuit performance at a lower area cost. The proposed gate-sizing metric provides a means to increase yield leading to better chip revenue.

Index Terms—process variations, metric, gate-sizing

I. INTRODUCTION

In newer technologies, intra-die process variations have become an important contribution to the variations of the devices and are growing with technology scaling [1] [2]. Intra-die variations can be spatially correlated and non-correlated (local intra-die) variations [2]. The latter type of variations, which are the subject of this work, cause that even neighbour devices present different electrical characteristics. Random dopant fluctuations (RDF), line edge roughness, and work function variations are among the main causes of intra-die local variations in advanced technology nodes [3]. Process variations impact the speed of the circuits limiting their performance and reducing chip revenue. Time margin is usually used to assure correct circuit functionality due to process variations, but its use increases the penalisation in the chip speed for advanced nodes.

Gate sizing technique is a widely used circuit optimisation approach to cope with process variations. Optimization techniques based on Lagrange multipliers [6] [8] consumes a considerable amount of computing time. On the other hand, the use of heuristics with metrics [4] [5] [7] [9] [10] [11] [12] for selective gate sizing allows a significant reduction of the computing time. The importance of the metric lies in the efficient selection of gates for re-size to improve the circuit

performance. In [4] is proposed a Heuristic with a metric to optimize the path delay, the metric is the rate between the variation in the p-percentile and the transistor channel width. The process to calculate the metric for each gate in the circuit could take a long time. In [5] a heuristic with a metric is proposed to improve the path delay, in this case, the metric is the rate between the area and the delay for each gate in each stage in pipeline circuits. In recent works [10], a Heuristic with a metric is used to improve the soft error. The metric statistical vulnerability window (SVW) which is used to find the SET (Single-event transient) in a given circuit node to be latched as a soft error. A sensitivity measurement determines the relative merits of each sizing action. In [12] the criticality-sensitivity (metric) provides a ranking of gates by their impact on the improvement of the circuit reliability after ageing. A group of n highest-ranking gates is chosen for optimisation, after optimisation the ageing-aware statistical timing analysis is carried out to update the ageing effect, timing, and reliability of the circuit. The previous works show that due to process variations the circuits have issues like delay, soft errors and ageing among others. An efficient way to overcome these issues is sizing-up gates, using selection metric. This paper proposes a novel gate-sizing metric selecting those gates providing more benefit in the reduction of the path delay standard deviation at a lower area cost. We focus on local intra-die process variations due to RDF affecting the transistor threshold voltage, but the extension of the results to consider other types of local intra-die variations is straightforward.

The rest of the paper is organized as follows. Section II presents the fundamentals of the basic metric. Section III presents key issues of the proposed metric and its formulation. Section IV shows the metric application to two ad-hoc logic paths. Section V shows the metric application to ISCAS benchmark circuits. Finally, Section VI presents the conclusions of this work.

978-1-5386-4757-8/18 $31.00 © 2018 IEEE

II. BASIC METRIC

Fig. 1 plots the delay standard deviation of an inverter with a fixed load capacitance as a function of the gate size. K=1 states for a minimum symmetrically-sized inverter, K=2 states for an inverter with transistor sizes twice of the minimum symmetrically-sized inverter, and so on. Slopes of tangent lines at three values of K are illustrated in Fig. 1. The slope indicates the rate of change of the delay standard deviation to a change in the size of the gate. The slope can be obtained with the derivative of the function of the delay standard deviation with respect to a change in the gate size evaluated at a given gate size. A larger value of the derivative indicates that the delay standard deviation changes more significantly with a small change in the inverter size. Thus, sizing-up an inverter with a larger value of the derivative is an area-cost effective way of reducing the delay standard deviation of the inverter as shown for the inverter with K = 2 in Fig. 1. On the other hand, inverter gates with sizes K = 4 and K = 6 present smaller derivative values. This means that sizing-up these inverters, by the same amount, yields a lower reduction of the gate delay standard deviation in comparison to the inverter with K = 2. Then, the derivative of the delay standard deviation of a gate i with respect to its size can be used to select the best gates to size-up to reduce the standard delay deviation at a lower area cost. We called *basic metric* to [6],

$$m_i = \sigma_{D_i} \frac{d\sigma_{D_i}}{dK_i} \qquad (1)$$

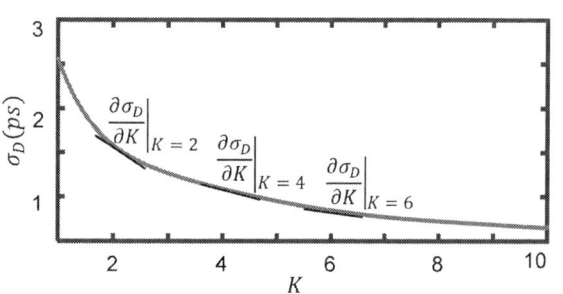

Fig. 1. Delay standard deviation of an inverter gate as a function of its gate size.

III. PROPOSED METRIC

A. Key issues for the proposed metric

The *basic metric* indicates the change of the derivative of the standard delay deviation of an isolated gate with respect to its size. However, this metric may fail to represent the benefits of sizing-up a gate that is part of a logic path. Two critical issues will be addressed next: a) path segment analysis, and b) input transition time is a normal distribution.

1) Path segment analysis: Let us consider the inverter chain in Fig. 2. Fig. 3 shows the delay standard deviation of each gate as gate G_i is sized-up by a factor K_i. The following takes place when G_i is sized-up.

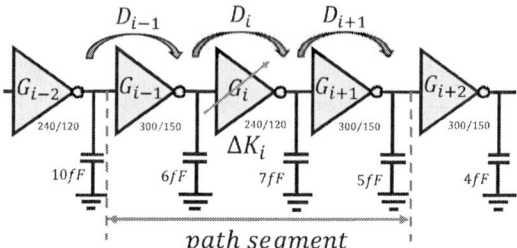

Fig. 2. Logic path 1: a 5-inverters chain.

- The delay standard deviation of the sized-up inverter G_i reduces due to the increase of its gate size.
- Sizing-up inverter G_i increases its driving current and load capacitance. The dominant effect is a faster input transition time at the input of the gate G_{i+1}. As a consequence, the standard delay deviation of inverter gate G_{i+1} reduces.
- Sizing-up inverter G_i increases the load capacitance driven by gate G_{i-1}. As a consequence, the delay standard deviation of inverter gate G_{i-1} increases.

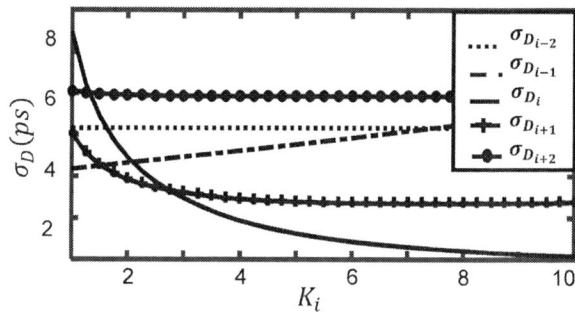

Fig. 3. Delay standard deviation of each gate of the logic path 1 as gate G_i is sized-up by a factor K_i.

On the other hand, the delay standard deviation of gates G_{i-2} and G_{i+2} do not suffer significant changes when gate G_i is sized-up as illustrated in Fig. 3. Let us define that a *path segment* is composed by the sized-up gate, the driven gate, and the preceding gate. The previous observations suggest that the delay variations of a *path segment* represents closely the variations of the entire logic path as a single logic gate is sized-up.

2) Input transition time : The second important issue is that the input transition time of a logic gate is a normal distribution [13]. For instance, the propagation delay distribution of gate G_i (See Fig. 2) depends on the normal distributions due to both variations of the transistor threshold voltages in gate G_i and variations of its input transition time. Besides, there is a covariance between the delays of two consecutive gates (e.g., $Cov(D_i, D_{i+1})$) because the output transition time of a gate is the input transition time of the next gate.

B. New Metric Formulation

Let us denote the delay distribution of a logic gate as,

978-1-5386-4757-8/18 $31.00 © 2018 IEEE

$$D = D_{Vth} + D_{Sin} \qquad (2)$$

where D_{Vth} is the component of the normal delay distribution of the gate due to local intra-die variations in the transistor threshold voltage, and D_{Sin} is the component of the normal delay distribution of the gate due to its input transition time. Thus, the delay variance of the *path segment* (σ_{Ds}^2) shown in Fig. 2 can be expressed as,

$$\sigma_{Ds}^2 = \sigma_{D_{i-1}}^2 + \sigma_{D_i}^2 + \sigma_{D_{i+1}}^2 + 2Cov(D_{i-1}, D_i) \\ + 2Cov(D_i, D_{i+1}) + 2Cov(D_{i-1}, D_{i+1}) \qquad (3)$$

where the first three terms are the delay variances of each logic gate, and the other terms are the delay covariances between gates.

The covariance between delay distributions of two adjacent gates (e.g., between D_{i-1} and D_i) is given by,

$$Cov(D_{i-1}, D_i) = Cov((D_{Vth_{i-1}} + D_{Sin_{i-1}}), \\ (D_{Vth_i} + D_{Sin_i})) \\ = 0 + Cov(D_{Vth_{i-1}} + D_{Sin_{i-1}}, D_{Sin_i}) \\ = \sigma_{D_{i-1}} \sigma_{D_i, Sin} \qquad (4)$$

Two facts were considered to obtain the final equation in (4). First, the delay covariances involving variations due to V_{th} are zero as this term states for local intra-die variations that are non-correlated. Second, for two adjacent gates, the component of the delay distribution of a gate due to its input transition time can be assumed fully correlated with the delay distribution of the preceding gate [13]. Considering these observations the result of (4) is substituted in (3), and the delay variance of the path segment reduces,

$$\sigma_{Ds}^2 = \sigma_{D_{i-1}}^2 + \sigma_{D_i}^2 + \sigma_{D_{i+1}}^2 \\ + 2\sigma_{D_{i-1}}\sigma_{D_i, Sin} + 2\sigma_{D_i}\sigma_{D_{i+1}, Sin} \\ + 2\rho_{(D_{i-1}, D_{i+1})}\sigma_{D_{i-1}}\sigma_{D_{i+1}, Sin} \qquad (5)$$

where $\sigma_{D_i, Sin}$ and $\sigma_{D_{i+1}, Sin}$ are the delay standard deviations due to their input transition times of gates G_i and G_{i+1}, respectively. Deriving (5) with respect to K_i, an expression to compute the rate of improvement in the path delay variance when the gate G_i sized-up by a certain amount is obtained.

The term associated with the derivative of the last term in (5) is neglected because of its small contribution to the total variation of the path segment. Developing the rest of the derivatives in (5) and factoring we have.

$$M_i = (\sigma_{D_{i-1}} + \sigma_{D_i, Sin})\frac{d\sigma_{D_{i-1}}}{dK_i} + (\sigma_{Di} + \sigma_{D_{(i+1),Sin}})\frac{d\sigma_{D_i}}{dK_i} \\ + \sigma_{D_{i+1}}\frac{d\sigma_{D_{i+1}}}{dK_i} + \sigma_{D_{i-1}}\frac{d\sigma_{D_i, Sin}}{dK_i} + \sigma_{D_i}\frac{d\sigma_{D_{(i+1),Sin}}}{dK_i} \qquad (6)$$

Metric M_i improves metric m_i which only considers independent process variations of the transistor threshold voltage in a single logic gate. Thus, M_i represents closely the total variation of the entire logic path. Applying the previous equation to each one of the *path segments* of a logic path, the gate with highest metric score gives more benefits in the delay variance reduction at the same cost of the extra area.

C. Metric considering area

The relative area cost of the gates has to be considered in the new metric. For instance, increasing the inverter size by a certain amount ΔK_i has a different area cost than increasing a 3-Nand gate by the same amount ΔK_i. This different cost is because more transistors increase their area in a 3-Nand gate than in an inverter gate. Hence, the score of the metric is divided by the area of the sized-up gate.

$$M_{if} = \frac{M_i}{A_i} \qquad (7)$$

the gate area is computed by,

$$A_i = A_{min}K \qquad (8)$$

where A_{min} is the gate area with minimum symmetrical dimensions allowed by the technology.

D. Optimization Methodology

The linear model [2] is used to compute (6). In this model, the standard deviation of the gate delay is expressed in terms of delay sensitivities and variations of a parameter process variations. As the delay distribution is due to local intra-die variations (V_{th}) and input transition time variations (S_{in}), the gate delay variation expressed in terms of delay sensitivities is given by,

$$\Delta D_i = S_{\Delta V_{thi}}^{D_i}\sigma_{\Delta V_{thi}} + S_{\Delta S_{ini}}^{D_i}\sigma_{\Delta S_{ini}} \qquad (9)$$

where $S_{\Delta V_{thi}}^{D_i}$ and $S_{\Delta S_{ini}}^{D_i}$ are the delay sensitivities to variations in Vth and Sin, respectively. $\sigma_{\Delta V_{thi}}$ and $\sigma_{\Delta S_{ini}}$ are the standard deviation of Vth and Sin, respectively, due to process variations given by process fabrication. In a similar way, the output transition time variation of a gate is:

$$\Delta Srout_i = S_{\Delta V_{thi}}^{Srout_i}\sigma_{\Delta V_{thi}} + S_{\Delta S_{ini}}^{Srout_i}\sigma_{\Delta S_{ini}} \qquad (10)$$

where $S_{\Delta V_{thi}}^{Srout_i}$ and $S_{\Delta S_{ini}}^{Srout_i}$ are the sensitivities of output transition time to variations in Vth and Sin, respectively. The delay variance of the gate can be expressed as,

$$Cov(\Delta D_i, \Delta D_i) = \sigma_{D_i}^2 \qquad (11)$$

Using (9) in (11) we have

$$\sigma_{D_i}^2 = (S_{\Delta V_{thi}}^{D_i})^2\sigma_{V_{thi}}^2 + (S_{\Delta S_{ini}}^{D_i})^2\sigma_{S_{ini}}^2 \qquad (12)$$

In the same way, the variance of output transition time can be obtained,

978-1-5386-4757-8/18 $31.00 © 2018 IEEE

$$\sigma^2_{Srout_i} = (S^{Srout_i}_{\Delta V_{thi}})^2 \sigma^2_{V_{thi}} + (S^{Srout_i}_{\Delta S_{ini}})^2 \sigma^2_{S_{ini}} \quad (13)$$

It is known that,

$$Srout_{i-1} = Sin_i \quad (14)$$

Thus,

$$\sigma_{Srout_{i-1}} = \sigma_{Sin_i} \quad (15)$$

Then, the expression for the variance of delay for any gate inside the path is:

$$\sigma^2_{D_i} = (S^{D_i}_{\Delta V_{thi}})^2 \sigma^2_{V_{thi}}$$
$$+ (S^{D_i}_{\Delta S_{ini}})^2 ((S^{Srout_{i-1}}_{\Delta V_{thi-1}})^2 \sigma^2_{V_{thi-1}} + (S^{Srout_{i-1}}_{\Delta S_{ini-1}})^2 \sigma^2_{S_{ini-1}})$$
$$(16)$$

To compute each term in (16) is required to know the delay sensitivities of each gate as a function of gate size, load capacitance and input transition time. Doing SPICE simulations, polynomial expressions of the gate delay, output transition time and delay sensitivities to changes in V_{th} and S_{in} are obtained. The polynomial expressions are a function of the gate size, load capacitance, and input transition time. It must be noted that this process is just made once for the entire digital library of a given technology.

IV. METRIC APPLICATION TO AD-HOC LOGIC PATHS

The proposed gate selection metric has been applied to two ad-hoc logic paths implemented in a CMOS 65nm technology. The proposed metric is applied to each one of the *path segments* of the logic path. Both metrics M_i and M_{if} are applied to the two ad-hoc logic paths. A variation of 20% of the threshold voltage due to RDF has been assumed for a minimum-sized transistor.

A. Logic path 1

The results for *Logic path 1* composed only by inverter gates (See Fig. 2) are shown in Table I. It can be observed that the gate with the highest metric score (M_i) provides the highest optimisation of the delay standard deviation of the logic path, and the degree of optimisation reduces as the metric score of the gate decreases. The results for the metric considering area (M_{if}) are shown in the right side of Table I. It can be observed that degree of optimisation of the delay standard deviation is equal in comparison to the metric M_i, and also, the ranking of the gates using M_{if} is the same than using M_i. It is because all gates in the logic path are of the same type, and hence, the area cost to resize any gate is the same.

TABLE I
METRIC SCORES AND PERCENTAGE OPTIMIZATION IN $\sigma_{D_{path}}$ FOR LOGIC PATH 1

Gate	Score (M_i)	%Opt	Gate	Score (M_{if})	%Opt
G_{i-2}	124.2	9.8	G_{i-2}	758.2	9.8
G_i	76.1	6.3	G_i	371.4	6.3
G_{i-1}	29.4	2.5	G_{i-1}	179.3	2.5
G_{i+1}	16.1	1.2	G_{i+1}	78.6	1.2
G_{i+2}	8.3	0.6	G_{i+2}	40.5	0.6

TABLE II
METRIC SCORES AND PERCENTAGE OPTIMIZATION IN $\sigma_{D_{path}}$ FOR LOGIC PATH 2

Gate	Score (M_i)	%Opt	Gate	Score (M_{if})	%Opt
G_{i-2}	31.2	8.5	G_{i-2}	190.5	8.5
G_{i+1}	13.8	3.9	G_{i-1}	33.6	3.7
G_{i-1}	13.2	3.7	G_{i+2}	30.9	1.9
G_i	8.5	2.1	G_{i+1}	20.3	3.9
G_{i+2}	6.3	1.9	G_i	9.8	2.1

B. Logic path 2

Figure 4 shows the second analysed logic path, which is composed of a different types of gates such as inverter, Nand and Nor gates. The transistor channel widths are also indicated. Table II shows the results for both the metric M_i and the metric M_{if}. It can be observed that the gate with the highest metric score M_i provides the highest optimisation of the delay standard deviation of the logic path, and the degree of optimisation reduces as the metric score of the gate decreases. However, it must be noted that the metric M_i does not consider area cost. The results for the metric considering area (M_{if}) are shown in the right side of Table II. It can be observed that the gate with the higher metric score does not always provide the highest optimisation of the delay standard deviation of the logic path. In this case, the metric M_{if} allows selecting those gates to be re-sized for improving circuit performance at a lower area cost.

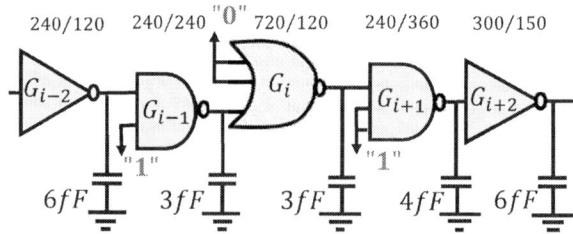

Fig. 4. Logic path 2: path with different types of gates and different fan-in

Fig. 5 compares the optimization results of the *logic path 2* (See Fig. 4) using the proposed gate metric M_{if} against SPICE Monte Carlo simulations. The size of the gates in the logic path is changed at each SPICE simulation. The delay standard deviation and area of the path are measured at each simulation and plotted with a small circle. The dotted line is obtained applying repeatedly the metric and optimising the logic path until no further improvement in the path delay standard deviation is obtained. It can be observed that the lower part of the small circles (SPICE), which corresponds to the optimized path with the lowest cost in the area, agrees with the results obtained with the proposed metric.

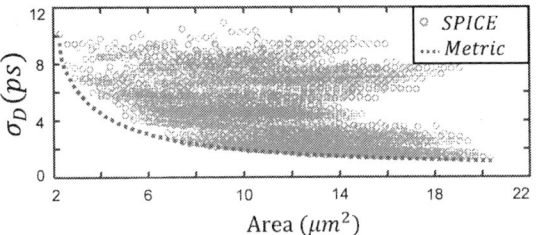

Fig. 5. Optimization results comparison between the proposed metric and SPICE

V. METRIC APPLICATION ISCAS CIRCUITS AND COMPARISON

The metric has also been applied to the longest path of ISCAS benchmark circuits. The metric is applied to optimize the standard deviation of the path delay subject to 5% of area restriction. This experiment was made using Lagrange multipliers methodology, Derv-Var-Path (the complete derivative of the delay variance of the path) and the proposed Metric (M_{if}). Fig. 6 shows the percentage of optimisation of $\sigma_{D_{path}}$ of each methodology for each one of the circuits. It can be observed that the percentage of optimisation of Der-Var-Path and Metric are similar. It means that the proposed metric M_{if} tracks very well the variations of the entire logic path. Also, the optimisation results using the proposed metric are close in comparison to those using the Lagrange methodology. The error is up to 2-3%. The computing time using the three methodologies are shown in Fig. 7. As expected, the computing time using Lagrange methodology is the largest one. The computing time using the proposed metric is the lowest one.

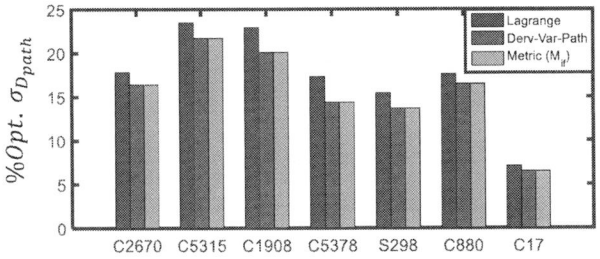

Fig. 6. Percentage optimization in $\sigma_{D_{path}}$ in the longest path after process optimization with 5% restriction in area.

Is important to mention that in this work we are working only with basic logic gates (NOT, NAND and NOR) the gates XOR, XNOR and AOI were not studied yet. The RC of the

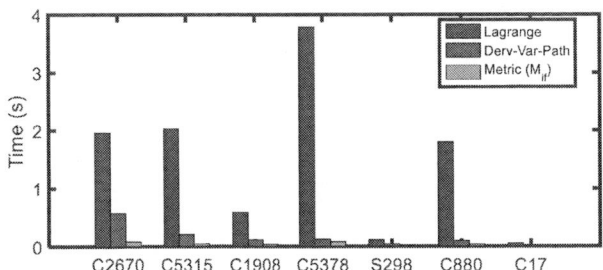

Fig. 7. Computer Time after optimization $\sigma_{D_{path}}$ in the longest path after process optimization with 5% restriction in area.

wire in the path segment were not considered yet to simplify the development of the metric. It could be characterised to consider in the metric in the future.

VI. CONCLUSIONS

A novel statistical gate-sizing metric to optimize circuit performance under local intra-die variations has been proposed. The metric allows the selection of the best gates providing more benefit in the reduction of the standard deviation of the path delay at a lower area cost. Two critical issues are addressed in the metric: a) the interactions between adjacent gates (path segment), and b) the input transition time of a gate is assumed a normal distribution. The metric has been extensively validated in two ad-hoc logic paths. The metric can rank the most favourable gates of the logic paths correctly. Then, the proposed metric is applied to the longest path of ISCAS benchmark circuits. Lagrange and the complete derivative of the delay variance of the path is also applied for comparison purposes. The proposed metric provides optimisation results close to using Lagrange multipliers method, and it provides the lowest computing time of the three methodologies. The proposed gate-sizing metric provides a means to increase yield leading to better chip revenue.

ACKNOWLEDGMENT

This work has been supported by The CONACYT (Mexico) through the PhD scholarship number 291025/68145

REFERENCES

[1] Nassif, S.: "Modeling and Analysis of Manufacturing Variations", *IEEE Custom Integrated Circuits Conference*, 2001, pp. 223-228

[2] Blaauw,D., Chopra, K., Sirvastava, A., Scheffer, L.: "Statistical Timing Analysis: From Basic Principles to State of the Art", *IEEE Transactions on CAD of CI and Syst.*, 2008,27,(4), pp. 589-607

[3] Hamed F.,Dadgour, Kazuhiko Endo, Vivek K.,De. and Kaustav Banerjee: "Grain-Orientation Induced Work Function Variation in Nanoscale Metal-Gate Transistors - Part II: Implications for Process, Devices and Circuit Design", *IEEE Transactions on Electron Devices*, Oct 2010,on vol.10.

[4] Agarwal, A., Chopra, K., Blaauw, D.and Zolotov, V.: "Circuits Optimization Using Statistical Static Timing Analysis", *Proceedings of the 42nd DAC'05*, 2005, pp. 321-324

[5] Datta, A., Bhunia, S., Mukhopahyay, S. and Roy, K.: "A Statistical approach to area-constraint yield enhancement for pipeline circuits under parameter variations", *14th IEEE Asia Test Symposium(ATS'05)*, 2005, pp.170-175

[6] Champac, V., Reyes, A. and Gomez, A.: "Circuit Performance Optimization for local Intra-die Process Variations using Gate Selection Metric", *(VLSI-SoC) IFIP/IEEE International Conference*, 2015, pp.165-170

[7] Davoodi,A. and Siravastava, A.: "Variability Driven Gate Sizing for Bininig Yield Optimization", *Very Large Scale Integration (VLSI) Systems,IEEE Transactions IEEE International Conference*, Jun 2008.on vol.16.no.6, pp.683,692.

[8] Seung Hoon Choi, Paul,B.C and Roy, K.: "Novel sizing algorithm for yield improvement under process variation in nanometer technology", *Design Automation Conference, 2004. Precedings*, 41st, Jul 7-11, 2004, pp.454-459.

[9] Visweswariah,C.,Ravindran, K., Kalafala, K., Walker, S.G.,Narayan, S.,Beece, D.K., Piaget, J., Venkateswaran, N. and Hemmett,J.G.: "First Order Incremental Block-Based Statistical Timing Analysis", *Computer-Aided Design of Integrated Circuits and Systems, IEEE Transactions*, Oct 2006, on vol.25, no.10, pp.2170,2180.

[10] Mohsen Raji and Behnam Ghavami: "Soft Error Rate Reduction of Combinational Circuits Using Gate Sizing in the Presence of process variations", *IEEE Transactions on very Large Scale Integration (VLSI) system*, Jan 2017.

[11] Chang-Lin Tsai, Chao-Wei Cheng, Ning-Chi Huang and Kai-Chiang Wu: "Analysis and Optimization of Variable-Latency Design in the Presence of Timing Variability", *Design, Automation, Test in Europe Conference Exhibition (DATE)*, 2017, pag.1219-1224.

[12] Yinghai Lu, Li Shang, Hai Zhou, Hengliang Zhu, Fan Yang ang Xuan Zeng: "Statistical Reliability Analysis under process variation and aging effects", *46th ACM/IEEE Design Automation Conference*, 2009, pag.514-519.

[13] Kouno,T. and Onodera, H.: "Consideration of Transition-Time variability in statistical Timing Analysis", *IEEE International SOC Conference*, 2006.

On the Effectiveness of the Satisfiability Attack on Split Manufactured Circuits

Suyuan Chen and Ranga Vemuri

Department of Electrical Engineering and Computing Systems
University of Cincinnati, Cincinnati, Ohio
Email: chens6@mail.uc.edu, vemurir@ucmail.uc.edu

Abstract—Split manufacturing of integrated circuits was proposed as a strong defense technique against reverse engineering at untrusted foundries. Split manufacturing has been shown to be vulnerable to proximity-based attacks and suitable defenses against these attacks have been proposed. In this paper, we apply an attack against split manufactured circuits based on satisfiability (SAT) solving, without requiring any proximity information. Our method formulates the problem of recovering the hidden signals as a Boolean decryption problem of determining the control signals (keys) of a multiplexer network, determines logical constraints to avoid solutions which could enable cyclic paths through the circuit and reduces the number of these constraints by eliminating redundant constraints. The resulting constrained decryption problem is solved by a satisfiability attack. We demonstrate the effectiveness of the attack on a set of benchmarks. In addition, we discuss a split manufacturing approach to potentially deter the SAT attack.

I. INTRODUCTION

Integrated circuit (IC) design, fabrication and test process has become a global enterprise due to the rising number of intellectual property (IP) vendors and fab-less design companies [1]. The major driver for this global design flow is the prohibitive cost of owning a state-of-the-art foundry [2]. However, this global flow involves numerous vulnerabilities in security and IP protection since any uncontrollable node in this process is potentially untrusted [3]. In addition, motivated attackers try to implant malicious circuits or reverse engineer the circuit for valuable intellectual property [4].

Split manufacturing (SM) was introduced to defend against reverse engineering an IC design at an untrusted foundry while keeping the fabrication cost still affordable [5]. In SM, an IC design is split into two sets of layers: Front-End-Of-Line (FEOL) and Back-End-Of-Line (BEOL). FEOL layers contain all the transistors and most of the connections and are fabricated in an advanced but potentially untrusted foundry. BEOL layers, usually the upper metal layers, include the rest of the signal connections and are fabricated in a secure foundry. Since BEOL layers are thicker and have larger pitch, BEOL fabrication and their bonding with the FEOL part in a secure facility is possible with a relatively low budget [6]. SM aims to thwart potential malicious reverse engineering at untrusted foundries by omitting some key signals of the circuit from the attacker. Even if the attacker has the opportunity to fully access the FEOL layers the circuit design is still protected as the BEOL connections are unknown to the attacker.

Attackers can attempt to use any available information to recover the BEOL signals. Since the IC design is usually produced using widely available commercial EDA tools, cell libraries and design methodologies, attackers can use any

available insights to enable effective BEOL signal recovery. In addition, fully packaged ICs available in open markets can be used to validate the correctness of the recovered design against test vectors.

Rajendran et al. [7] proposed an attack named *proximity attack*. This attack utilizes insights into widely used physical CAD techniques such as min-cut bi-partitioning and minimum wire length placement and logical constraints such as the absence of cycles in combinational circuits to recover the missing BEOL signals. Wang et al. [8] developed an enhanced attack called the *network flow based attack* by adding hints pertaining to gate mapping and detailed routing. This method makes full use of design rules and restrictions through the physical design flow. However, these hints usually help recover most but not all of the 'hidden' signals. In these attacks, the packaged IC from the open market is used to calculate the Hamming Distance (HD) between the output vectors generated by the recovered circuit and the IC itself as a measure of correctness of the recovered circuit. Any none-zero HD indicates incorrect connections in the recovered circuit. Defense methods against proximity-based attacks based on distributing related pins apart from each other by introducing appropriate layout constraints were proposed in [7], [8].

In this paper, we develop an effective methodology to apply a Boolean satisfiability based strategy that successfully recovers all the BEOL signals. Further, this attack methodology does not need or depend on any proximity information or other insights into the nature of the physical EDA tools used in the design process. Accordingly, this attack can thwart any defense method introduced solely against proximity attacks such as those in [7], [9].

We summarize our contributions as follows:

- We introduce an interconnect network based on key-controlled multiplexers which allows all possible signal connection combinations between the FEOL partitions. In order to recover the correct BEOL signal connections, we need to determine the correct key values to activate the correct connections in the key-controlled network.
- We propose an efficient method to locate all possible combinational cycles introduced by the redundancy of the interconnect network and generate a reduced set of key constraints for the satisfiability solver.
- We evaluate the attack method using benchmarks from the ISCAS-85 and ITC-99 sets which have been used in other attacks on SM and hardware security [7], [8]. Results show that we are able to recover 100% of the BEOL connections for all benchmarks.
- We discuss a SM method that could defend against both proximity and satisfiability attacks and show its

978-1-5386-4757-8/18 $31.00 © 2018 IEEE

effectiveness against the SAT attack.

II. RELATED WORK

Proximity attack and its derivative network flow based attack algorithm use hints based on the physical design flow [7], [8]. An attacker who already knows the FEOL circuits recovered at the untrusted foundry generates constraints based on insights into commonly used optimal CAD strategies to direct the BEOL recovery attack to eliminate spurious connections and recover the correct ones. However, these attacks can be thwarted by introducing certain amount of physical design overhead as these sub-optimal designs no longer reveal the uniqueness features which are conducive to constraint generation facilitating the attack [7], [8].

The SAT attack, proposed by Subramanyan et al. [10], has been proven to be a powerful strategy against logic encryption/obfuscation/locking and circuit camouflaging. In this method, a logic circuit encrypted by inserting key gates can be decrypted by using satisfiability checking along with strategically selected combinations of input-output vectors generated using an accessible IC or an equivalent oracle.

Shamsi et al. introduced cyclic obfuscations to thwart the SAT attack [11]. In this method, key gates are introduced such that incorrect key values introduce cyclic signal dependencies in combinational logic thereby confusing the SAT solver which would produce incorrect results and hang.

More recently, Zhou et al. [12] proposed CycSAT which enables the SAT attack on combinational circuits with cycles generated by encryption algorithms. By carefully examining all logical loops, the attacker can generate a set of logic constraints which can guide the SAT solver in the right track. Cyclic encryption algorithms in logic locking tend to generate relatively small number of loops since large number of loops require longer wires and more components, thus increasing the performance and area overhead [11].

To mount the SAT attack against SM, we need to transform the BEOL connection recovery problem into a logic decryption problem on an encrypted circuit with key gates. As we will discuss, such transformation introduces cyclic signal dependencies into the logic circuit. Elimination of these dependencies using the CycSAT approach is hard due to the fact that the number of cycles introduced is orders of magnitude larger than those introduced by cyclic obfuscation methods. This necessitates an efficient method to locate the cycles and generate appropriate Boolean constraints to direct the SAT attack. We will discuss this further in Section IV.

III. PROBLEM FORMULATION

A. Problem Description

Let $G = (V, E)$ be a directed acyclic hypergraph representation of a combinational logic circuit. Vertices $v_i, 1 \leq i \leq |V|$ represent logic gates and hyperedges $e_j, 1 \leq j \leq |E|$ represent signal nets. Nodes with no incoming (outgoing) edges represent the primary input (output) terminals. Let $H \subseteq E$ be the subset of nets selected to be assigned to the 'hidden' BEOL layers. Remaining circuit $(V, E - H)$ is assigned to the FEOL layers. We assume that the attacker already knows $(V, E - H)$. For a certain input pattern I_x, the correct output from a packaged IC is O_x. The problem is to recover H correctly by using several (strategically selected) (I_x, O_x) pairs that the attacker is assumed to have access to a packaged IC.

Removal of BEOL nets H induces non-primary input and output nodes (with no incoming and outgoing edges respectively) in FEOL circuit $(V, E - H)$. These non-primary input (output) nodes of the FEOL circuit are also referred to as the output (input) nodes or terminals of BEOL nets.

B. Transformation to Equivalent Logic Decryption Problem

The missing H signals, when reintroduced, should connect certain output terminals to certain input terminals in the FEOL circuit. These terminals are internal to the original circuit but available as external terminals of the FEOL circuit. We introduce a key-controlled network to connect these output terminals to the input terminals. Key inputs represent the uncertainty about output-input mapping. Correct key will program the network correctly to recover the missing BEOL connections and hence the correct circuit.

For convenience, we assume that the FEOL circuit is identified by partitioning the original logic circuit into two or more partitions and the cut-set signals are assigned to BEOL as discussed in [7]. To minimize the size of the cut-set, the BEOL signals are routed with fan-in/fan-out of one. Hence, the missing connections among the output-input terminals are mapped one to one (one output connected to a unique input).

A key controlled crossbar network, such as the Benes network [13] would be a good candidate interconnect network to model this connectivity. This network can establish all possible connections between inputs and outputs based on the key values. However, the number of the gates to represent this network is too large leading to too many clauses for the SAT solver. So an interconnect network which results in fewer SAT clauses is preferred.

We will use a multiplexer network as an approximation of the crossbar network. The N-to-1 multiplexer (MUX) allows its output to be connected to any of its inputs by controlling the select signals. A N-to-1 multiplexer M has inputs $i_1, i_2, \cdots, i_N$ and $n = \lceil log_2(N) \rceil$ select signals which are also called the *key bits*. We will denote these key bits by a vector of Boolean variables $K_M = <k_1, k_2, \cdots k_n>$. Each input signal i of the MUX is associated with a unique binary-valued assignment $B_M^i = <b_1, b_2, \cdots, b_n>$ to the key bits such that when $K_M = B_M^i$, the output of the MUX is connected to i.

The structure of a MUX based interconnect network is shown in Fig. 1 assuming two FEOL partitions. For the missing BEOL nets $e \in H$, all the source nodes of e constitute the input terminals of the MUX network and the target nodes of e constitute the output terminals of the MUX network as illustrated in Figure 1. This network structure contains more connection combinations than the crossbar network since it allows for a BEOL input terminal to be mapped to multiple BEOL output terminals. However this is not an issue since the SAT solver will automatically preclude the key combinations that lead to functionally incorrect circuits. The key bits of all the multiplexers need to be determined by the SAT attack subject to the constraint that each BEOL

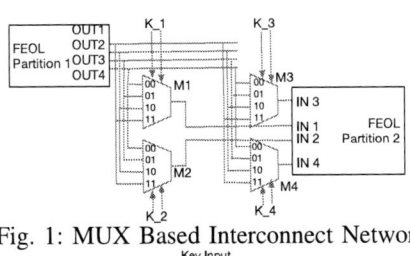

Fig. 1: MUX Based Interconnect Network

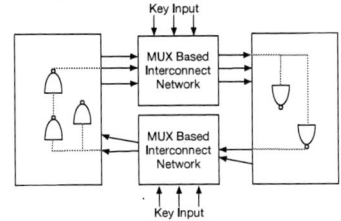

Fig. 2: MUX Network for a Bipartitioned FEOL Circuit

output terminal must be driven by a unique BEOL input terminal. For example, referring to Figure 1, for connecting OUT1→IN1, OUT2→IN2, OUT3→IN3 and OUT4→IN4, the key bits recovered by the SAT attack should be $11, 10, 01$ and 00 respectively for multiplexers M1, M2, M3 and M4.

The structure of MUX network is much simpler than the crossbar switch resulting in a smaller SAT problem with fewer clauses resulting in higher attack efficiency. If a circuit was bipartitioned to separate the FEOL and BEOL layers before manufacturing and the attacker is able to recover and separate the FEOL circuit into the two partitions correctly as discussed in [7] then a two-way MUX network is generated between the two FEOL partitions as shown in Fig.2. If the attacker recovered the FEOL circuit but is unable to separate it correctly into the two partitions then the MUX network shown in Fig. 3 is used. In any case, all of the control signals (keys) of the multiplexers need to be 'decrypted' by the SAT attack to recover the BEOL signals, hence the original circuit, correctly.

IV. ATTACK METHODOLOGY

Introduction of the MUX network leads to potential cyclic paths in the circuit. While numerous cyclic paths are physically present, none of the paths will of course be activated with the correct MUX key bits. However, there is a possibility to generate many combinational cycles during the attack process corresponding to incorrect key guesses and these cycles can confuse the SAT solver and thwart the SAT attack. We generate constraints on the key values that would avoid activating the cyclic paths. We need to identify all the cycles in order to find the complete set of key constraints for all combinational cycles. However, if we analyze each cycle through each gate similar to CycSAT [12], the NC (no cycle) conditions will not necessarily cover all possible combinational cycles in polynomial time as discussed in detail by Roshanisefat et al. [14]. CycSAT application context dealt with cyclic obfuscations which tend to introduce far fewer and less complex cycles than our MUX networks. In this section, we introduce an efficient method to generate the cycle avoidance constraints.

A. Identification of All Cyclic Paths

The extracted FEOL circuit is cycle free if the original circuit is cycle free. So all cycles will be generated by the

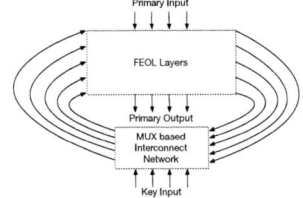

Fig. 3: MUX Network for a General FEOL Circuit

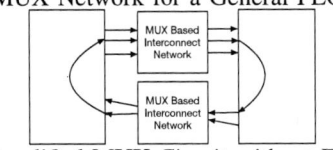

Fig. 4: Simplified MUX Circuit without FEOL Gates

MUX networks we introduced earlier. In order to locate all the cycles in the complete circuit including the MUX networks, the complete circuit needs to be analyzed using depth first search [15]. To increase the efficiency of this search, a simplified MUX graph G' is produced. The idea is illustrated with this example: In Fig. 2, a complete circuit with MUX based interconnect networks is shown. In order to identify all the cycles, we need to traverse all the gates in the circuit. In Fig. 4, the internal gates are removed and direct edges in place of directed paths are introduced between the multiplexers. The attacker only needs to check in this simplified graph.

We construct a simplified graph called the MUX graph G' as follows: We represent each MUX by a node in G'. We introduce directed edges from node $m1$ to node $m2$ (represeing MUXes $M1$ and $M2$ respectively) in G' as follows: For each $i - o$ pair, where i is a FEOL input port and o is a FEOL output port, if $M1$ drives i, o drives one of the inputs of $M2$ and i is in the input cone of o, we introduce a *unique* directed edge from $m1$ to $m2$.

The MUX graph G' represents only the multiplexers and the combinational paths among the multiplexers through the FEOL circuit. None of the circuit gates are directly represented. However, we make the following observations: (1) Each node in G' has exactly as many incoming edges as the number of inputs to the corresponding MUX. (2) Every MUX to MUX cyclic path in the circuit has a corresponding cycle in G'.

Clearly G' is sufficient to determine all the cycles introduced by the MUX network in the FEOL circuit. We use the depth first search algorithm described in [15] to generate all the cycles in the MUX graph.

B. Generation of Cycle Constraints

Once we locate all possible combinational cycles, constraints are generated for all the MUXes in each combinational cycle. Suppose a cyclic path contains multiplexers $M_1, M_2, \cdots M_m$ controlled by keys $K_1, K_2, \cdots K_m$ respectively where each K is a vector of Boolean variables. Intuitively, at least one of these MUXes should be "disabled" in order to disable the cyclic path. That is, at least one of $K_i, 1 \le i \le m$, must be set to a binary vector such that the cyclic path is *disabled* through M_i. Suppose key K_i is the Boolean variable vector $K_i = <k_{i_1}, k_{i_2}, \cdots k_{i_{|K_i|}}>$ and that the variables must assume the binary values $B_i = <b_{i_1}, b_{i_2}, \cdots b_{i_{|K_i|}}>, b_{i_j} \in \{0, 1\}$ in order to enable the correct path through the multiplexer M_i. This cyclic path would be

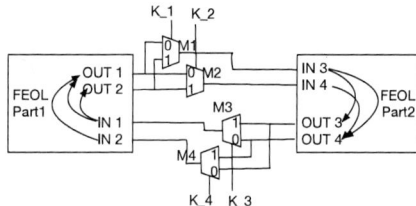

Fig. 5: Example for Cycle Constraints

enabled if every MUX M_i in the path is enabled correctly. This condition can be expressed as,

$$\bigwedge_{i \in \{1,m\}} (K_i = B_i) \tag{1}$$

Referring to Fig. 5, the cyclic path $OUT1 \rightarrow IN3 \rightarrow OUT3 \rightarrow IN1 \rightarrow OUT1$ would be enabled if $(K_1 = 0) \wedge (K_3 = 1)$.

For M_i to be disabled, we need ensure that at least one of these keys is incorrectly set. This can be expressed as a logical constraint of the form,

$$\bigvee_{i \in \{1,m\}} (K_i \neq B_i) \tag{2}$$

In the above example, the cyclic path would be disabled provided $(K_1 \neq 0) \vee (K_3 \neq 1)$.

Since each key K_i consists of several bits $K_i =< k_{i_1}, k_{i_2}, \cdots k_{i_{|K_i|}} >$, for $K_i \neq B_i$, at least one of these bits must be incorrectly set. Hence, the logical condition for disabling the cyclic path can be expressed as,

$$\bigvee_{i \in \{1,m\}, j \in \{1,|K_i|\}} (k_{i_j} \oplus b_{i_j}) \tag{3}$$

For the example above, assuming that $K_1 =< k_{11} >$ and $K_3 =< k_{31} >$, the following SAT constraint clause form can be generated to disable the cycle: $k_{11} \vee \bar{k}_{31}$. For each cycle in the MUX graph G', a SAT constraint clause of the form shown in Eqn. 3 should be generated to disable that cycle.

C. Cycle Constraint Optimization

The cycle constraint generator can potentially locate all cyclic paths for large circuits and generate the corresponding CNF clauses. For large circuits, these clauses can exceed the capability of state-of-the-art SAT solvers. Hence, we further reduce these constraints based on the following observation: If P_1 and P_2 are two cyclic paths in the graph G' such that every edge in $P1$ is also an edge in $P2$, then disabling the $P1$ path in the circuit implies that the $P2$ is also disabled. Hence, it is sufficient to include the constraints for $P1$ and omit those for $P2$. That is, if a constraint for a cycle involving a specific sequence of multiplexers $M_x, \cdots M_y$ with a specific set of key values $(K_x = B_x, .., K_y = B_y)$ is identified and corresponding constraint is generated, then there is no need to introduce a constraint for another cyclic path which contains multiplexers with the same key values for each multiplexer.

In the example shown in Fig. 5, consider a cycle $P1$: $OUT1 \rightarrow IN4 \rightarrow OUT4 \rightarrow IN1 \rightarrow OUT1$. This cycle is enabled if $(K_2 = 0) \wedge (K_3 = 0)$. Disabling constraint $C1$ for this cycle is $(K_2 \neq 0) \vee (K_3 \neq 0)$. Consider another cycle $P2$: $OUT1 \rightarrow IN4 \rightarrow OUT4 \rightarrow IN1 \rightarrow OUT2 \rightarrow IN3 \rightarrow OUT3 \rightarrow IN2 \rightarrow OUT1$. This cycle is enabled if $(K_2 = 0) \wedge (K_3 = 0) \wedge (K_1 = 0) \wedge (K_4 = 0)$. Hence the

deactivation constraint $C2$ for this cycle is $(K_2 \neq 0) \vee (K_3 \neq 0) \vee (K_1 \neq 0) \vee (K_4 \neq 0)$. Comparing the disabling constraints $C1$ and $C2$, clearly if the former constraint is enforced then the later constraint is satisfied and is redundant.

In order for a clause $C2$ to be covered by a clause $C1$ (both of the form Eqn. 2), the disjuncts in $C1$ can appear in any order in $C2$. Changing the order of MUX disabling conditions doesn't change the overall path disabling condition. Checking if $C1$ implies $C2$ itself can be formulated as a SAT problem. Hence determining all redundant cycle constraint by comparing every cycle constraint with every other cycle constraint is prohibitively expensive in time.

We use a simple heuristic to identify and eliminate redundant constraints. First, we formulate constraints of the form Eqn. 2 for all the cycles in G'. Then, we sort the constraints in increasing order of the number of disjuncts in the clauses, ie. the number of multiplexers in the cycle. (In fact, this sorting occurs naturally due to the depth-first search algorithm [16] used to identify all the cycles in G'. Depth-first search identifies cyclic paths in increasing order of path lengths.) Next, we compare every clause $C1$ with every other clause $C2$, where $C2$ is longer than $C1$, to determine if $C1$ occurs as a sub-clause of $C2$ at the beginning of $C2$. That is, we check if $C2$ is of the form $(C1 \vee \cdots)$. (This is same as checking if the MUXes in the first cyclic path appear at the beginning of the second cyclic path in the same order and the same enabling conditions as in the first one.) If so, we delete $C2$ from the clause list. This method is fast and, as the experimental results show, eliminates a significant number of redundant clauses. But, it cannot eliminate all redundancy since only the leading subclauses of $C2$ are examined.

In the previous example, referring to Fig. 5, $C1$ occurs at the beginning of $C2$ as a subclause. Hence, our method will eliminate $C2$.

Following redundant clause elimination, the remaining clauses are refined into the form of Eqn. 3, combined with the SAT model of the FEOL circuit and the MUX network, and submitted to the SAT attack.

The procedure of constraint generation and optimization is summarized in Algorithm 1.

Algorithm 1: Generate Cycle Constraints

Input: FEOL Circuit with MUX Network
Output: Cycle Constraints

1 G' = Generate Graph from FEOL & MUX network;
2 C = Identify the set of all cycles in G';
3 NC = Generate cycle constraints for all cycles in C in the form of Eqn. 2;
4 NC = Arrange NC in increasing order of the number of disjuncts in each clause;
5 NC_{opt} = Eliminating redundant clauses in NC;
6 NC_{opt} = Express each clause in the form of Eqn. 3;
7 **return** NC_{opt}

D. SAT Attack Algorithm

We use a SAT solver based attack method based on CycSAT proposed by Zhou [12] which is in turn based on the SAT

Algorithm 2: SAT Attack on Split Manufactured Circuit

Input: FEOL Circuit with MUX Network, Packaged IC.
Output: MUX Keys to yield Correct BEOL Nets.

1 NC_{opt} = Generate_Cycle_Constraints();
2 $g(x, k)$ = circuit with inputs x and keys k;
3 $g(x, k1)$ = copy of $g(x, k)$;
4 $g(x, k) = g(x, k) \wedge NC_{opt}$;
5 $g(x, k1) = g(x, k1) \wedge NC_{opt}$;
6 **while** $\hat{x} = SAT(g(x, k) \neq g(x, k1))$ **do**
7    $\hat{y} = IC\_eval(\hat{x})$;
8    $g(x, k) = g(x, k) \wedge (g(\hat{x}, k) = \hat{y})$;
9    $g(x, k1) = g(x, k1) \wedge (g(\hat{x}, k1) = \hat{y})$;
10 **end**
11 $k^\star = SAT(g(x, k))$;
12 **return** $k^\star$;

TABLE I: Benchmark Details

ISCAS-85				ITC-99			
Circuit	#In	#Out	#Gates	Circuit	#In	#Out	#Gates
c499	41	32	202	b14	277	299	9,767
c880	60	26	383	b15	485	519	8,367
c1355	41	32	546	b20	522	512	19,682
c1908	33	25	880	b21	522	512	20,027
c2670	157	64	1,193	b17	1,452	1,512	30,777
c3540	50	22	1,669	b18	3,357	3,343	111,241
c5315	178	123	2,307				
c7552	207	108	3,512				

attack for encrypted circuits [10]. The pseudo-code of the detailed algorithm is shown in Algorithm 2. Step 1 generates the cycle constraints NC_{opt} as discussed before. $g(x, k)$ is the circuit model with inputs x and MUX control signals (keys) k. IC_eval is the process of applying inputs $\hat{x}$ to the packaged IC and reading outputs $\hat{y}$. SAT is a call on a satisfiability solver. Steps 3-12 constitute the standard SAT attack as discussed in [10]. In each iteration of the while loop a distinguishing input pattern is determined to eliminate a set of keys. At the end the algorithm returns a set of MUX control signals which when enabled recover the original circuit or its logical equivalent. The connections made by the MUXes yield the BEOL nets.

V. EXPERIMENTAL RESULTS

A. Experimental Setup

We evaluate the proposed attack methodology using benchmarks from the ISCAS-85 and ITC-99 suites which are widely used in related research on SM [7], [8]. Details of these benchmarks are shown in Table I. We used a DELL Precision M4800 with CPU@2.7GHz for all experiments.

For each benchmark, the BEOL signals are selected using a min-cut bipartitioning process similar to [7]. This keeps the number of BEOL signals to a minimum, hence, the cost of expensive BEOL fabrication is minimized. We used a multilevel hypergraph partitioning tool named PaToH [16] which is based on the Fiduccia-Mattheyses(FM) partitioning algorithm. After bipartitioning the nets in the cut-set (ie. BEOL signals) are removed and, for each removed net, a pair of unconnected terminals are left in the FEOL circuit.

Then the attack methodology discussed in this paper is applied to recover the missing BEOL nets. After inserting the MUX network shown in Fig. 2 and generating the SAT

TABLE II: Cut Set, Key Size, Cycles before/after Optimization and Gate count for FEOL circuit and MUX Network

Ckt	Cutset	Key Size	Cycles	$Cycles_{opt}$	FEOL Gates	MUX NW Gates
c499	17	60	17,792	3,360	202	222
c880	20	91	0	0	383	409
c1355	17	60	17,792	6,880	546	222
c1908	31	144	240	240	880	696
c2670	19	76	0	0	1,193	276
c3540	57	320	1,836	1,836	1,669	2,614
c5315	33	153	12	12	2,307	1,043
c7552	28	129	0	0	3,512	567
b14	128	842	19,560	8,300	9,767	11,900
b15	139	1,077	0	0	8,367	18,689
b20	88	528	3,178	3,178	19,682	4,688
b21	89	508	1,040	1,040	20,027	5,242
b17	197	1,480	168	168	30,777	29,926
b18	114	778	38	38	111,241	15,735

constraints corresponding to the cyclic paths in the network, our implementation of Algorithm 2 uses the 'sld' program which was developed and successfully used for SAT-based logic decryption [10]. As each cycle generates a CNF clause for SAT, we assume 1,000,000 cycles as the limit for staying within the capacity of the SAT solver. For problems with cycles exceed the limit, the SAT solver either cannot handle the size or takes prohibitively long execution time ($> 48\ hrs$).

B. Experimental Results

1) Attack Correctness: Our attack could recover the correct logic circuit for all the benchmarks in normal min-cut partition mode. For 7 of the benchmarks (c1908, c2670, c5315, c7552, b14, b15, b17), the connections recovered are identical to the BEOL connections in cut-set identified during SM. For the remaining benchmarks, the recovered connections are not identical but logically equivalent to the removed cut-set. The reason for this is that some pairs of signals assigned to the BEOL layers are functionally equivalent to each other. That is, there are functionally equivalent sub-circuits driving the internal terminals in the FEOL circuit. Our SAT attack is free to swap these sub-circuit outputs or use the same sub-circuit for the equivalent connections (by appropriate setting of the MUX control signals during the attack). All recovered circuits are compared with the original circuits for correctness using a SAT based equivalence checker called 'lcmp' from from [10]. Attack data and results are summarized in Table II and Fig. 6. Since each cycle constraint is expressed as one clause, the 4th and 5th columns in the table denote cycle constraint clauses before/after reduction.

2) Combinational Cycles and Cutset Sizes: The number of combinational cycles varies by the internal logic of a circuit and its partition result. Results in Table II show that c499, c1355 and b14 have a larger number of combinational cycles. Our constraint optimization method successfully eliminated more than 50% of the cycle constraints and this greatly increases the attack efficiency. Also, cutset size greatly affects the attack time. Though b14 and b15 have fewer gates compared with b20 and b21, results in Fig. 6 show that b14 and b15 needed more time to attack. This can be explained by the large cutset size for b14 and 15. Larger MUX networks and more loop constraints lead to more clauses in CNF problem for SAT. This further leads to larger iteration count during the SAT attack (while loop in Algorithm 2) and more attack time.

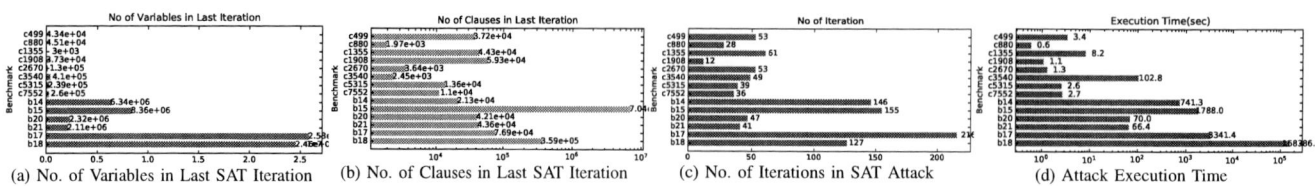

(a) No. of Variables in Last SAT Iteration   (b) No. of Clauses in Last SAT Iteration   (c) No. of Iterations in SAT Attack   (d) Attack Execution Time

Fig. 6: Variables, Clauses and Iterations in SAT Attack and Attack Execution Time

VI. DEFENSE AGAINST SAT AND PROXIMITY ATTACKS

From the experimental results, we observe that our attack efficiency and feasibility mainly depend on the number of the clauses we send to the SAT solver. To thwart the SAT attack against SM, BEOL nets have to be identified such that insertion of the MUX network leads to an exceedingly large number of cycles leading to extremely large SAT problems far beyond the capacity of state-of-the-art SAT solvers. To model a N input MUX, we need N AND gates with N inputs and one OR gate with N inputs. This single MUX generates $(N + 1)^2$ clauses. If we double the number of BEOL signals, the number of the clauses for interconnect network will be 8 times that of the original one. In addition, a large number of cycles will be introduced. Based on these observations, number of clauses can be increased in at least two ways:

a) Camouflage Partitioning Information: In our experimentation above we have assumed that the attacker is able to recover the FEOL partitioning information (as in [7]) and insert the MUX network following the scheme in Fig. 2. However, if the BEOL nets could be identified and layout could be done such that the partitioning information is not clear to the attacker, then he is forced to use the MUX network in the scheme in Fig. 3. This MUX network scheme generates much larger number of cycles, exceeding 1 million and often reaching 10 million even for moderately sized benchmarks.

Table III shows experimental results comparing the MUX schemes of Figs. 2 and 3 (known partition vs. unknown partition) for the same cutset size (BEOL nets). We are able to attack only 5 small benchmarks of the 14 benchmarks. For the remaining 9 benchmarks the number of cycles exceeded the 1 Million limit. (N/A in the table means Not Attackable.) Note the significant increase in the number of cycles even for the small examples we were able to attack.

Although this method thwarts the SAT attack, it has two drawbacks: (1) It is unclear as to how the attacker can be effectively prevented from recovering the partition information. (2) The method may still be vulnerable to proximity attacks.

b) Combined Defense Against SAT and Proximity Attacks: Recently Chen et al. [9] proposed a novel defense against proximity attacks by carefully selecting the BEOL nets based on a signal priory factor which takes the signal's effect on the output into account. This method is not based on minimum cutset partitioning, hence, the attacker cannot use the scheme in Fig. 2. We have used Chen et al.'s signal selection method to identify and hide the BEOL nets and use the general MUX network scheme of Fig. 3 followed by our SAT attack method to recover these BEOL nets. The results are shown in the last three columns of Table III. In this case we are able to attack only 6 small benchmarks out of 14. However, note the significantly larger numbers of cycles and longer attack

TABLE III: Key Bits, Number of Cycles and Execution Time (s) for Three Defense Methods

Circuit	Fig.2			Fig.3			[9]		
	Keys	Cycles	Time	Keys	Cycles	Time	Keys	Cycles	Time
c499	60	3,360	3.3	85	5,986	4.1	172	167,244	12
c880	91	0	0.6	100	3,862	4.2	220	90	44.4
c1355	60	6,880	8.3	N/A	N/A	N/A	173	99,946	623.6
c1908	147	612	1	N/A	N/A	N/A	290	42,470	32.8
c2670	68	0	0.9	95	56	1.5	310	432	3,942.8
c5315	170	16	25.5	198	1,185	9.2	729	31	629.9
c7552	129	0	2.7	140	9,454	12.7	N/A	N/A	N/A

times when compared with the previous defense method. For larger benchmarks, the number of cycles simply prevents the SAT attack due to the problem size. These preliminary results indicate the signal ranking functions such as the one used in [9] hold promise to thwart both attacks against SM.

VII. CONCLUSION

Unlike the proximity guess based attack algorithms against split manufactured circuits, the proposed SAT attack methodology is very effective without requiring any proximity information. Although preliminary results show promising approaches, effective defenses against combined proximity and SAT attacks require further research.

REFERENCES

[1] R. Kumar, "Simply Fabless!" *IEEE Solid-state Circuits Magazine*, vol. 3, no. 4, pp. 8–14, 2011.

[2] DIGITIMES_Research, "Trends in the global IC design service market," 2012. [Online]. Available: http://www.digitimes.com/news/a20120313RS400.html?chid=2,%202012

[3] W. Chen, S. Ray, J. Bhadra, M. Abadir, and L.-C. Wang, "Challenges and trends in modern SoC design verification," *IEEE Design & Test*, vol. 34, no. 5, pp. 7–22, 2017.

[4] E. Love, Y. Jin, and Y. Makris, "Enhancing security via provably trustworthy hardware intellectual property," in *HOST, 2011*, pp. 12–17.

[5] R. Jarvis and M. McIntyre, "Split manufacturing method for advanced semiconductor circuits," U.S. Patent US7 195 931, Mar. 27, 2007.

[6] K. Vaidyanathan, B. P. Das, E. Sumbul, R. Liu, and L. Pileggi, "Building trusted ICs using split fabrication," in *2014 IEEE International Symposium on HOST*, Arlington, VA, May 2014, pp. 1–6.

[7] J. Rajendran, O. Sinanoglu, and R. Karri, "Is split manufacturing secure?" in *2013 DATE*, Grenoble, France, Mar. 2013, pp. 1259–1264.

[8] Y. Wang, P. Chen, J. Hu, and J. J. Rajendran, "The cat and mouse in split manufacturing," in *DAC, 2016*. IEEE, 2016, pp. 1–6.

[9] S. Chen and R. Vemuri, "Improving the security of split manufacturing using a novel beol signal selection method," in *Proceedings of the 2018 on Great Lakes Symposium on VLSI*. ACM, 2018, pp. 135–140.

[10] P. Subramanyan, S. Ray, and S. Malik, "Evaluating the security of logic encryption algorithms," in *HOST, 2015*. IEEE, 2015, pp. 137–143.

[11] K. Shamsi, M. Li, T. Meade, Z. Zhao, D. Z. Pan, and Y. Jin, "Cyclic obfuscation for creating sat-unresolvable circuits," in *GLSVLSI 2017*. ACM, 2017, pp. 173–178.

[12] H. Zhou, R. Jiang, and S. Kong, "CycSAT: SAT-based attack on cyclic logic encryptions," in *ICCAD, 2017*. IEEE, 2017, pp. 49–56.

[13] V. E. Beneš, *Mathematical theory of connecting networks and telephone traffic*. Academic press, 1965, vol. 17.

[14] S. Roshanisefat, H. M. Kamali, and A. Sasan, "SRCLock: SAT-resistant cyclic logic locking for protecting the hardware," *arXiv preprint arXiv:1804.09162*, 2018.

[15] D. B. Johnson, "Finding all the elementary circuits of a directed graph," *SIAM Journal on Computing*, vol. 4, no. 1, pp. 77–84, 1975.

[16] J.-S. Cherng, S.-J. Chen, C.-C. Tsai, and J.-M. Ho, "An efficient two-level partitioning algorithm for VLSI circuits," in *ASP-DAC, 1999*. IEEE, 1999, pp. 69–72.

Secure and Compact Full NTRU Hardware Implementation

Konstantin Braun*, Tim Fritzmann*, Georg Maringer*, Thomas Schamberger*, Johanna Sepúlveda*

*Technical University of Munich, Munich, Germany

email: {konstantin.braun, tim.fritzmann, georg.maringer, t.schamberger, johanna.sepulveda}@tum.de

Abstract—The foreseeable breakthrough of quantum computers represents a risk for secure communications. In order to prepare for such an event, electronic systems must integrate secure quantum-computer-resistant (post-quantum) cryptography protected against implementation attacks. The NTRU cryptosystem is one of the main alternatives for practical implementations of post-quantum public-key cryptography. The standardized version of NTRU (IEEE 1363.1) provides security against chosen ciphertext attacks (CCA) through a padding scheme that limits ciphertext malleability, thus restricting a large range of attacks. So far, previous NTRU hardware implementations do not include the NTRU padding scheme. Moreover, a previously proposed NTRU optimization of the polynomial multiplication leads to a degradation of the security level. Therefore, previous works provide a wrong impression regarding the real implementation cost of NTRU. In this work, we present two contributions: i) the first complete and compact NTRU hardware implementation; and ii) the analysis of the security degradation due to the NTRU multiplication optimization proposed in previous works.

I. INTRODUCTION

Public-key cryptography (PKC) provides the basis for establishing secured communication channels between multiple parties. It allows confidentiality, authenticity and non-repudiation of electronic communications and data storage. Internet-of-Things (IoT) and Cloud computing are some of the technologies that use PKC to secure channels. The fast development of quantum computers represents a risk for many public-key cryptosystems. Almost all established approaches rely on the hardness of factoring large integers (RSA) or computing discrete logarithms (ECC). It is known that cryptographic systems based on these problems will be threatened when a large enough quantum computer is built. Shor's quantum algorithm will solve the problems on which PKC currently relies in polynomial time.

To ensure long term communication security quantum-resistant (also called post-quantum) cryptography must be adopted. Post-quantum cryptography relies on mathematical problems that remain hard to solve even with a quantum computer. In order to provide resistance against quantum attacks, the National Institute of Standards and Technology (NIST) announced the beginning of the transition towards post-quantum cryptography [1]. The NTRU lattice-based cryptosystem is one of the main alternatives for practical implementations of post-quantum PKC. NTRU is characterized by small key sizes (low memory footprint) and computational efficiency when compared to other post-quantum approaches [2], [3]. NTRU has been standardized in the IEEE Standard Specification for Public Key Cryptographic Techniques Based on Hard Problems over Lattices (IEEE-1363.1) [4].

Empowering electronic devices with strong security poses a challenging problem due to limited resources and tight performance requirements. Moreover, embedded implementations must be resistant to implementation attacks, such as Chosen-Ciphertext Attack (CCA) or Side-Channel Attacks (SCA). Adversaries can recover the secret key by gathering information obtained through the decryption of fabricated ciphertexts or by the physical information leakage during the cryptographic operation (power consumption, timing and electromagnetic radiation). While CCA can be avoided by adopting a padding scheme, inhibiting SCA requires a careful implementation of the cryptographic algorithm. The NTRU standard (IEEE-1363.1) defines the Short Vector Encryption Scheme (SVES) as the padding scheme to avoid CCA.

NTRU hardware implementations have been previously demonstrated in [5]–[10]. While current works about NTRU focus on efficient convolution techniques, a complete implementation of the standardized NTRU is still missing. Moreover, security aspects of the implementation are still largely unexplored. The works presented in [5]–[10] do not implement the SVES padding scheme. Furthermore, the NTRU optimization presented in [10] reduces the security of NTRU by leaking information regarding the secret key, as the execution time of this implementation depends on the value of the secret key. This makes an implementation impractical for real applications.

In this work, we present the first complete, compact and secure NTRU hardware implementation. In addition, we demonstrate the security reduction of a previous NTRU implementation. In summary, the contributions of the paper are:

- First complete NTRU hardware implementation which includes the SVES padding scheme;
- A compact NTRU implementation able to execute encryption and decryption operations;
- Performance and cost evaluation of our NTRU implementation;
- A security analysis of the previous NTRU implementation presented in [10] and demonstration of the security reduction.

The remainder of this article is organized as follows: Section II presents the previous work on NTRU hardware implementations. Section III describes the instantiation of NTRU with the SVES padding scheme. Section IV and Section V present our complete NTRU implementation and the security analysis of the optimized NTRU presented in [10]. The experimental results are presented in section VI. A conclusion is given in section VII.

II. RELATED WORKS

The probably first NTRU encryption hardware implementation was proposed by Bailey *et al.* in 2001 [5]. To speed up the polynomial multiplication, which is usually the performance bottleneck of NTRU, the authors propose to scan the

978-1-5386-4757-8/18 $31.00 © 2018 IEEE

coefficients of the blinding polynomial r. For each non-zero coefficient, the public key h is added to a temporary result. Atici proposed the first encryption and decryption NTRU hardware implementation. The architecture includes power saving methods, such as clock gating and partially rotating registers [6]. The implementation of Kamal *et al.* uses the special structure of the public key, which has a large number of zero coefficients to optimize performance [7]. In [9], Liu *et al.* use the fact that the polynomial multiplication in the truncated polynomial ring of NTRU can be modeled with a linear feedback shift register (LFSR) to implement the polynomial multiplication. In [10], they speed up their implementation by skipping the multiplication operation when two consecutive zero coefficients in the ternary polynomial are detected. Thus, the multiplication time depends on the number of double zeros contained in the polynomial. This information decreases the NTRU security level as discussed in Section V.

Moreover, so far none of the existing works proposed a full hardware implementation of NTRU with the SVES padding scheme as defined in the IEEE-1363.1 standard. As the integration of SVES is mandatory to inhibit a CCA, these implementations show a misleading picture of the implementation cost. A commonly used tool for transforming cryptographic algorithms into CCA secured schemes is the NAEP [11] transformation. SVES is a concrete instantiation of the NAEP transform, which was specially designed for NTRU. The first iterations, SVES-1 and SVES-2, are vulnerable to attacks exploiting decryption errors [12]. The latest iteration, SVES-3, which is sometimes only referred to as SVES, does not show this vulnerability. It is standardized in IEEE-1363.1 [4]. In contrast to previous works, we present a CCA secure NTRU hardware implementation compliant with the standard.

III. NTRU

A. Notation

The main elements of NTRU are the polynomials in the following integer rings:

$$R_{N,p} = \frac{(\mathbb{Z}/p\mathbb{Z})[x]}{(x^N - 1)}, \quad R_{N,q} = \frac{(\mathbb{Z}/q\mathbb{Z})[x]}{(x^N - 1)} \ . \quad (1)$$

These rings define each polynomial to be at most of degree $N - 1$ and to have integer coefficients. For $R_{N,p}$ and $R_{N,q}$ these coefficients are reduced modulo p and modulo q, respectively. Unless otherwise noted, all polynomials are elements of the ring $R_{N,q}$.

For the standardized parameter sets of NTRU the modulus p is fixed to a small prime $p = 3$. In this case, the elements of the ring $R_{N,p}$ are called ternary polynomials. A ternary polynomial $\mathcal{T}_N(d, e)$ has d coefficients equal to one and e coefficients equal to minus one, while the remaining coefficients are set to zero. NTRU ternary polynomials are sparse, that is, the majority of coefficients are set to zero. The values d and e are part of the parameter set and can be changed to achieve different security levels.

Additionally, the parameter q is fixed to $q = 2048$, which simplifies the implementation of the algorithm. By choosing the modulus to be a power of two, the modulo operation can be performed without additional cost by reducing the result register size.

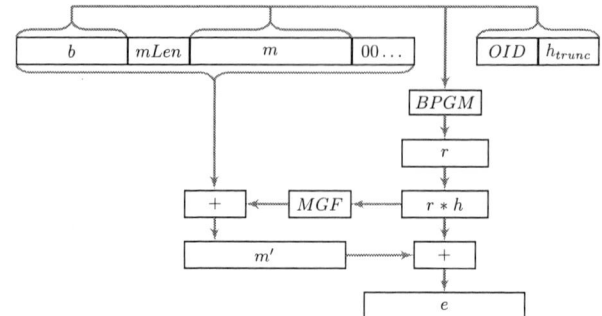

Fig. 1. NTRU encryption with SVES

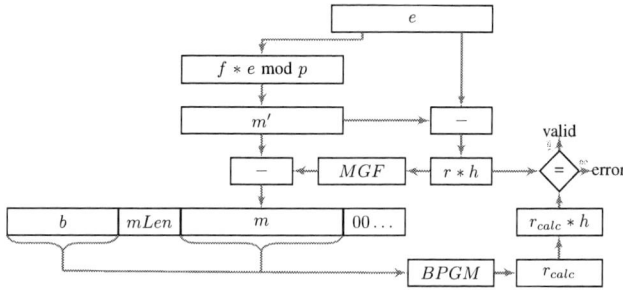

Fig. 2. NTRU decryption with SVES

B. Short Vector Encryption Scheme (SVES)

Padding schemes prevent cryptanalysis by hiding the characteristics of the ciphertext. For instance, the length of the encrypted message may leak information regarding the original message m. The SVES pads the message by using a variable number of zeros and uses two auxiliary methods: i) Blinding Polynomial Generation Method (BPGM) and ii) Mask Generation Function (MGF).

1) Blinding Polynomial Generation Method (BPGM): This method generates a ephemeral blinding polynomial r in a deterministic way with the use of a pseudo-random number generator (PRNG). This PRNG is based on a hash function **G** and is initialized by a seed consisting of four values:

$$BPGM(OID, b, m, h_{trunc}) \ . \quad (2)$$

The identifier OID is an unique three-byte value for each parameter set. The parameter b is a random number and m the message to be encrypted. The last part of the seed h_{trunc} consists of a defined number of bits of the public key h.

2) Mask Generation Function (MGF): Similar to BPGM, the MGF uses a hash function **G** to generate a mask. The input of **G** is the result of the polynomial multiplication of the ephemeral blinding polynomial and the public key ($r * h$). The resulting mask is added to m.

C. NTRU with SVES

The NTRU scheme instantiated with SVES consists of the three cryptographic operations: key generation, encryption and decryption.

The key generation step creates a key pair, consisting of the public key h with its corresponding secret key f, through three steps. The first step generates two random ternary polynomials,

$F \in R_{N,p}$ and $g \in R_{N,p}$. The positions of the polynomial coefficients with value one and minus one are selected based on an uniform distribution. The second step calculates the private key f as $f = 1 + pF$ together with its inverse f^{-1} modulo q. Not all the polynomials have an inverse in the corresponding ring. Therefore, it is possible that the inverse f^{-1} can not be found. In this case, the key generation is restarted until a key with a valid inverse is found. The third step computes the public key h as $h = f^{-1} * g$.

The NTRU encryption is shown in Fig. 1. It transforms a message m into a ciphertext e through five steps. The first step formats m into a ternary polynomial representation and concatenates this polynomial together with the random b, the identifier OID, h_{trunc}, $mLen$ and zeros for padding as $m_{pad} = (OID \| b \| m \| h_{trunc} \| mLen \| 00 \dots)$. The second step uses m_{pad} as the input of the BPGM to create the ephemeral polynomial r as $r = BPGM(m_{pad})$. The third step multiplies r with the public key polynomial h. The result is masked through the MGF in order to obtain $m_{Mask} = MFG(r * h)$. The fourth step adds the mask to the concatenated message to produce $m' = m + m_{Mask}$. The final step computes the ciphertext as $e = m' + r * h$.

The NTRU decryption is shown in Fig. 2. It retrieves the original message m from the ciphertext e through four steps. The first step retrieves m' by multiplying the ciphertext e with the private key f as $m' = f * e \pmod{p}$. In the second step $r * h$ can be retrieved, by subtracting m' from the ciphertext e, as the equation $r * h = e - m'$ holds. The third step uses the resulting product as an input to the MGF to retrieve the concatenated message as $m = m' - MGF(r * h)$. The fourth step checks the validity of the ciphertext by applying the BPGM to the corresponding elements of m to produce the value r_{calc}. The multiplication of $r_{calc} * h$ is now compared with the polynomial $r * h$ from the second decryption step. If both polynomials are equal, the algorithm outputs the concatenated message m. Otherwise an invalid ciphertext is detected and the algorithm outputs an error message.

IV. NTRU ARCHITECTURE

Our proposed hardware NTRU architecture is illustrated in Fig. 3. The encryption and decryption flows are highlighted in green and red, respectively. To keep the area costs low, the encryption and decryption operation share common hardware modules. The resource sharing is managed by a small controller that sets the data selector values of all multiplexers. The NTRU architecture is composed of four main hardware modules: CONV, BPGM, MGF and MOD p. Their implementation is described in the following subsections.

A. Convolution (CONV)

In this work, we adopted the convolution architecture of [9] because of its efficiency and simplicity. This architecture is able to multiply a ternary polynomial with a regular polynomial in $R_{N,q}$. However, in order to support the encryption and decryption operations, the following modifications are required: i) integration of the *control unit* to manage the convolution during encryption and decryption operations; and ii) support for the regular multiplication ($f * e$). The enhanced convolution circuit (CONV) is shown in Fig. 4.

CONV multiplies a ternary polynomial $A \in \mathcal{T}_N$ with coefficients $\{-1, 0, 1\}$ and a regular polynomial $B \in R_{N,q}$.

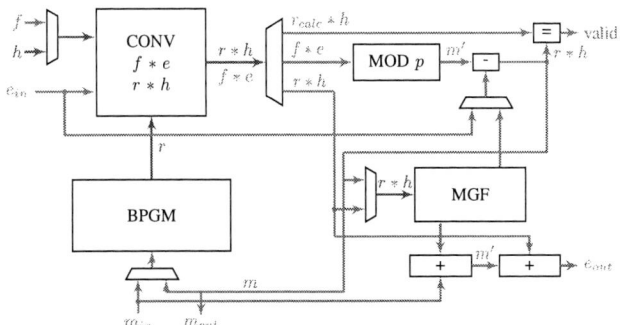

Fig. 3. NTRU architecture, green encryption, red decryption, blue shared

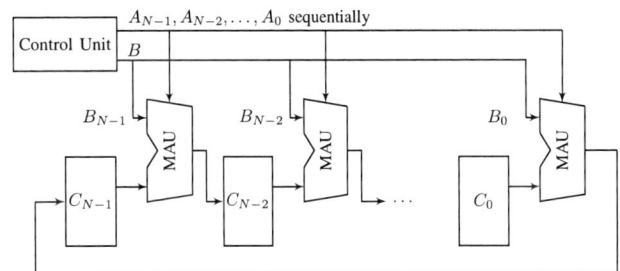

Fig. 4. Circular convolution model

The circularity of the convolution is realized by shifting the result values C in an linear-feedback shift register (LSFR). Depending on the sequentially inputted coefficient of A, the modular arithmetic unit (*MAU*) either adds B_i to C_i, subtracts B_i from C_i or keeps C_i unchanged, where $i = 0, 1, \dots N - 1$. The result of each *MAU* is forwarded to the next register.

During encryption, the ternary polynomial r and a regular polynomial h are multiplied through *CONV*, thus generating $r * h$. However, the decryption requires a multiplication of two regular polynomials f and e. In order to use *CONV* for this multiplication, we use the definition of f given in the IEEE-1363.1 standard, such that $f = 1 + pF$, where F is a ternary polynomial. As a result, $f * e = (1 + pF) * e = e + pF * e$. To obtain $pF * e$, we repeat the convolution of $F * e$ two more times ($p = 3$) without resetting the registers after each round. At the end of this operation, the *control unit* inputs a ternary polynomial such that the value of the first coefficient is '1' and '0' otherwise. The second input will remain with the polynomial e. This procedure for calculating the addition of $pF * e$ with e takes one round. To avoid this round, the registers can also be preloaded with e at the beginning of the decryption process. In addition to the calculation of $f * e$, the decryption has to calculate $r_{calc} * h$, which requires one additional round. The proposed process increases the convolution processing time during the decryption operation by a factor of four. However, it avoids the integration of additional multipliers, thus decreasing the required area for the decryption.

B. Blinding Polynomial Generation Method (BPGM)

Hash functions are the core of the *BPGM* and *MGF* modules. The IEEE-1363.1 standard suggests the use of SHA-1 or SHA-256, depending of the desired security level. SHA-1 and SHA-256 have a 512 bit input and 160 and 256 bit output,

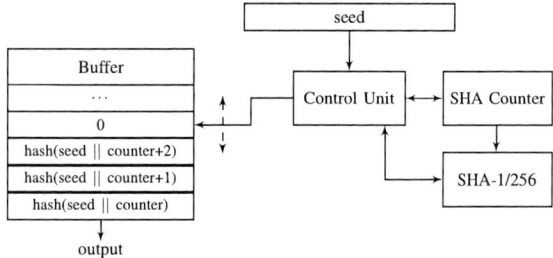

Fig. 5. Buffer generation

respectively. The NTRU parameter set defines the number of calls of the SHA function in order to generate the ephemeral blinding polynomial r. More specifically, the SHA function is executed $minCallsR$ times, whose value is defined in the standard. The seed varies in each hash call, by using the four values described in Subsection III-B (OID, b, m, h_{trunc}) concatenated with a *counter* value, which is increased after each hash call. Our NTRU architecture uses the *SHA-1/256* open core modules from [13]. The *control unit* manages the generation of r. As shown in Fig. 5, the generated hash values are stored in a buffer. These values are used for determining the indexes of ones and minus ones of r.

C. Mask Generation Function (MGF)

The MGF shares the *buffer generation module* presented in Fig. 5 with the $BPGM$. However, instead of using the buffer output for finding the value of the indexes of r, the MGF transforms the output from a binary into a ternary representation. A look-up table is employed to perform this transformation.

D. Modulo reduction (MOD p)

The coefficients of the polynomials can be reduced modulo p by subtracting p from each coefficient of the polynomial until the coefficient is smaller than p. However, as $p = 3$ is a Mersenne prime number, a faster method to calculate the modulo reduction can be employed, as shown in [14] and optimized in [3]. Algorithm 1 presents the modulo reduction employed in our NTRU architecture. To improve the NTRU performance, the $MOD\ p$ block can be instantiated multiple times.

Algorithm 1: Mersenne prime modulo division ($p = 3$)

Input: Integer a
Result: Integer a mod 3
additional_reduction $= \{0, 1, 2, 0, 1, 2\}$
`// reduce a`
$a = (a \gg 8) + (a\ \&\ 0xFF)$
$a = (a \gg 4) + (a\ \&\ 0xF)$
$a = (a \gg 2) + (a\ \&\ 0x3)$
$a = (a \gg 2) + (a\ \&\ 0x3)$
`// at this point a < 6`
a = additional_reduction[a]

V. SECURITY ANALYSIS

In [10], Liu *et al.* show an optimization of their implementation presented in [9]. It is based on scanning the coefficients of the ternary polynomial. When two consecutive zeros are detected, the multiplication can be skipped. In the following subsections, we describe the optimized architecture and present the security vulnerability caused by this optimization.

A. Optimized Architecture

The optimized architecture is able to detect two consecutive zeros in the ternary input polynomial A (Figure 4). The processing of a zero coefficient during the convolution can be seen as a single circular shift of the coefficients C_i. Therefore, two zeros can be substituted by a single shift of two places within one clock cycle. This results in a reduction of one cycle in the total multiplication time for each pair of consecutive zeros. The implementation of Liu *et al.* requires an additional multiplexer for the MAU, which is connected to the preceding register output of the result coefficient C_i. In comparison to their original and non-optimized implementation in [9], the authors report a reduction of 36.7 % of the execution time for the convolution with the parameter set *ees541ep1*.

B. Vulnerability

The optimized implementation of [10] leaks information regarding the secret key through a timing side-channel because the convolution time depends on the structure of the secret key. More specifically, it depends on the amount of consecutive double zeros in the secret key polynomial F. This vulnerability cannot be fixed because the optimization is solely based on the amount of double-zeros. A time protected version will remove the performance gains and will behave equally to the unoptimized implementation. Exploiting the side-channel by observing the execution time of the algorithm reveals the amount of double-zeros in the private key polynomial F. This additional information limits the search space and therefore reduces the effective length of the private key polynomial and thus the security level.

Theorem 1. *The number of valid ternary polynomials for a given number of double-zeros d_z is upper bounded by*

$$u_r(n, d_f, d_z) = \frac{(2d_f + d_z)!}{(d_f!)^2 d_z!} \binom{2d_f + d_z + 1}{n - 2d_f - 2d_z}, \quad (3)$$

where n is the number of coefficients of the polynomial and d_f the number of ones and minus ones in the private key F.

Proof. Valid keys consist of zeros, double zeros (two consecutive zeros), ones and minus ones. All possible patterns are constructed using these elements following the constraints imposed by the public parameters (n, d_f) as well as the number of double zeros d_z, gained from the timing side-channel. For simplicity, we discuss the two factors of Equation 3 independently.

The first factor represents the quantity of possible key patterns under the assumption that the locations of single zeros are fixed. It counts the number of all pattern changes that can be made by swapping double zeros, ones and minus ones. As a visualization, we denote the interchangeable elements with the symbol $*$ and the fixed single zero coefficients with 0.

$$*0**0***\cdots*0**$$

978-1-5386-4757-8/18 $31.00 © 2018 IEEE

The number of $*$ symbols in the pattern is $2d_f + d_z$, which gives

$$\frac{(2d_f + d_z)!}{(d_f!)^2 d_z!} \qquad (4)$$

possible combinations (using the multiset permutation formula).

In the next step, we take into account that the locations of single zeros are not fixed anymore and introduce the second part of Equation 3. This factor increases the amount of possible patterns as it takes the different possibilities of single zero locations into account. It is upper bounded to

$$\binom{n - d_z - l + 1}{l}, \qquad (5)$$

where l denotes the number of single zeros. The parameter l is set to $n - 2d_f - 2d_z$, which can be included in Equation 5 to get the second factor of Equation 3.

The proof for Equation 5 is done by induction. First, we consider only two single zeros ($l = 2$) in the pattern and fix the rightmost zero at the last position.

$$* * * * * * * * * * * * \cdots * 0 * 0$$

In this case, the first zero has $(n - d_z - 2)$ possible positions in the pattern. Shifting the rightmost zero by one position to the left reduces the number of possibilities for the first zero by one, resulting in $(n - d_z - 3)$ possibilities. The summation of these quantities for all valid positions of the rightmost zero results in

$$\frac{(n - d_z - 1)(n - d_z - 2)}{2} = \binom{n - d_z - 1}{2}. \qquad (6)$$

Next, the inductive step ($l \rightarrow l+1$) is shown. In the following, the last two elements, containing one of the single zeros, are separated from the other elements by the symbol $\|$.

$$* * * * * * * * \ldots 0 * 0 * \cdots * 0 * 0 \| * 0$$

All single zeros l are located left of this border. Therefore, we can use the induction hypothesis to compute the amount of possible configurations for the single zeros given by

$$\binom{(n - d_z - 2) - l + 1}{l}. \qquad (7)$$

Similar to the proof for $l = 2$, the rightmost zero is shifted step by step to the left. By adding up the resulting possibilities, it follows

$$\sum_{k=0}^{n - d_z - 2l - 1} \binom{l + k}{l} = \binom{n - d_z - l}{l + 1}, \qquad (8)$$

by using the equation

$$\sum_{k=0}^{m} \binom{n + k}{n} = \binom{n + m + 1}{n + 1}. \qquad (9)$$

$\square$

Figure 6 illustrates the complexity reduction factor α for different parameter sets. This factor can be computed by

$$\alpha = \frac{u_r(n, d_f, d_z)}{K_c}, \quad \text{with} \quad K_c = \frac{n!}{(d_f!)^2 (n - 2d_f)!}, \qquad (10)$$

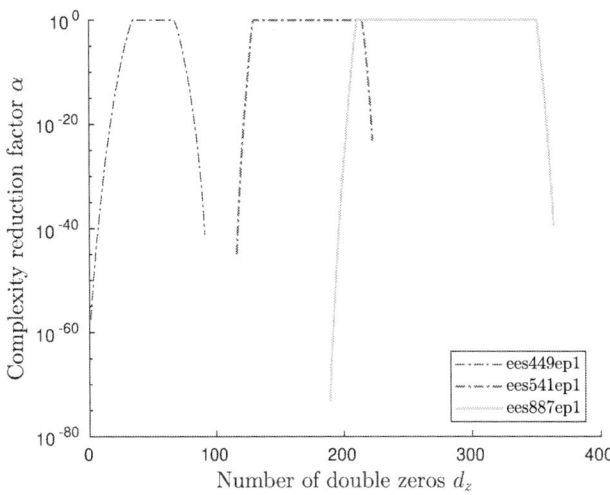

Fig. 6. Complexity reduction of the exhaustive search space of the private key F for a known amount of double-zeros. For the parameter set *ess887ep1*, the secret key F consists of $d_f = 81$ ones and minus ones, which results in a maximum of 362 double zeros. The complexity reduction factor is bounded by a minimal amount of double zeros as it is possible that the number of non-zero elements is not sufficient to separate the remaining zeros.

where K_c denotes the cardinality of the key space. It shows the complexity reduction of an exhaustive search for F given a known amount of double zeros.

VI. RESULTS

Our proposed NTRU hardware architecture was implemented on the Zedboard, which is equipped with a Xilinx Zynq-7000. The IEEE-1363.1 standard provides different parameter sets for different security levels and optimization goals. Table I summarizes the results of our proposed system. It contains the total number of LUTs and the required number of clock cycles for encryption and decryption. The results show that the number of LUTs scales with the parameter n, which determines the size of the polynomials.

Figure 7 provides a more detailed view of the required clock cycles for the encryption. The time required for the convolution depends directly on the value of n, as n clock cycles are required for the circular shift within the LSFR. Results show that the impact of the padding scheme on the cost and performance of NTRU is not negligible. For some NTRU configurations, the convolution has a minor influence on the computation cost when compared to the padding scheme. For the parameter set *ees401ep1*, the padding scheme takes nearly 90 % of the encryption time.

The main bottleneck of the padding scheme is the hash function. The number of clock cycles spent by the BPGM depends on the parameter d_r, which determines the number of ones and minus ones of r and thus the number of hash calls.

Figure 8 illustrates the computation costs for the decryption. Compared to the encryption, the number of clock cycles of the convolution during the decryption is four times higher. Both, the convolution and the modulo p operation directly scale with n. For the modulo p operation only one module was implemented. To decrease the computation cost, several instances of MOD p modules can be used.

TABLE I
PARAMETER SET DEFINED IN IEEE-1363.1

Security Level	Parameter Set	n	LUTs	#CC Enc.	#CC Dec.
Low	ees401ep1	401	29545	3423	5430
	ees541ep1	541	38240	2409	5116
	ees659ep1	659	46124	2413	5711
Middle	ees449ep1	449	32907	3642	5890
	ees613ep1	613	44797	2675	5743
	ees761ep1	761	53338	2799	6606
High	ees677ep1	677	49786	4020	7407
	ees887ep1	887	63861	3113	7551
	ees1087ep1	1087	74280	3760	9197
Highest	ees1087ep2	1087	74620	4723	10159
	ees1171ep1	1171	83896	4345	10202
	ees1499ep1	1499	99717	4715	12212

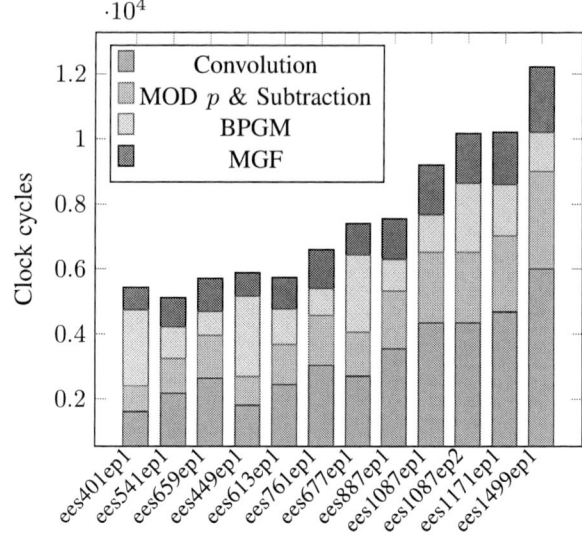

Fig. 8. Clock cycles for decryption

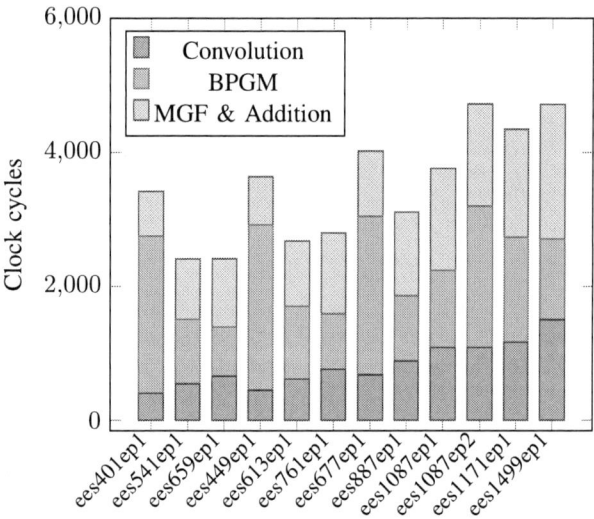

Fig. 7. Clock cycles for encryption

VII. CONCLUSION

Efficient and secure post-quantum cryptography is mandatory to ensure long-term security. In this work, we proposed the first compact, complete and secure NTRU architecture. It is compact due to its shared resources for the encryption and decryption process. In contrast to previous works, our NTRU architecture is complete because we included the SVES padding scheme, defined in IEEE-1363.1. Our architecture is secure because it avoids CCA attacks by including the padding scheme, and SCA attacks by implementing a constant time convolution. We show that previous NTRU works may decrease the security level of NTRU. Moreover, we show that the integration of the SVES scheme requires, for some parameters, nearly 90 % of the encryption time and thus its implementation cost cannot be neglected.

Acknowledgments. This work was partly funded by the Fraunhofer High Performance Center for Secure Connected Systems of Munich.

REFERENCES

[1] National Institute of Standards and Technology, "Announcing request for nominations for public-key post-quantum cryptographic algorithms," 2016, https://csrc.nist.gov/news/2016/public-key-post-quantum-cryptographic-algorithms.

[2] J. Hoffstein, J. Pipher, J. M. Schanck, J. H. Silverman, W. Whyte, and Z. Zhang, "Choosing parameters for NTRUEncrypt," *IACR ePrint*, vol. 2015, p. 708, 2015. [Online]. Available: http://eprint.iacr.org/2015/708

[3] O. M. Guillen, T. Pöppelmann, J. M. B. Mera, E. F. Bongenaar, G. Sigl, and J. Sepulveda, "Towards post-quantum security for IoT endpoints with NTRU," in *Design, Automation Test in Europe Conference Exhibition (DATE), 2017*, March 2017, pp. 698–703.

[4] IEEE, "IEEE Standard Specification for Public Key Cryptographic Techniques Based on Hard Problems over Lattices," *IEEE Std 1363.1-2008*, pp. C1–69, March 2009.

[5] D. V. Bailey, D. Coffin, A. J. Elbirt, J. H. Silverman, and A. D. Woodbury, "NTRU in constrained devices," in *Cryptographic Hardware and Embedded Systems - CHES 2001, Third International Workshop, Paris, France, May 14-16, 2001, Proceedings*, no. Generators, 2001, pp. 262–272.

[6] A. C. Atici, L. Batina, J. Fan, I. Verbauwhede, and S. B. Örs, "Low-cost implementations of NTRU for pervasive security," in *19th IEEE International Conference on Application-Specific Systems, Architectures and Processors, ASAP 2008, July 2-4, 2008, Leuven, Belgium*, 2008, pp. 79–84.

[7] A. A. Kamal and A. M. Youssef, "An FPGA implementation of the NTRUEncrypt cryptosystem," in *Microelectronics (ICM), 2009 International Conference on*. IEEE, 2009, pp. 209–212.

[8] X. Zhan, R. Zhang, Z. Xiong, Z. Zheng, and Z. Liu, "Efficient Implementations of NTRU in Wireless Network," *Communications and Network*, vol. 5, no. 03, p. 485, 2013.

[9] B. Liu and H. Wu, "Efficient architecture and implementation for NTRUEncrypt system," in *Circuits and Systems (MWSCAS), 2015 IEEE 58th International Midwest Symposium on*. IEEE, 2015, pp. 1–4.

[10] ——, "Efficient multiplication architecture over truncated polynomial ring for NTRUEncrypt system," in *IEEE International Symposium on Circuits and Systems, ISCAS 2016, Montréal, QC, Canada, May 22-25, 2016*, 2016, pp. 1174–1177.

[11] N. Howgrave-Graham, J. H. Silverman, A. Singer, and W. Whyte, "NAEP: Provable security in the presence of decryption failures." *IACR Cryptology ePrint Archive*, vol. 2003, p. 172, 2003.

[12] N. Howgrave-Graham, P. Q. Nguyen, D. Pointcheval, J. Proos, J. H. Silverman, A. Singer, and W. Whyte, "The impact of decryption failures on the security of NTRU encryption," in *Annual International Cryptology Conference*. Springer, 2003, pp. 226–246.

[13] A. de la Piedra, "SHA-256 Core," 2013, https://opencores.org/project/sha256core/.

[14] D. W. Jones, "Modulus without division, a tutorial," 2001. [Online]. Available: http://homepage.cs.uiowa.edu/ jones/bcd/mod.shtml

Lightweight and High Performance SHA-256 using Architectural Folding and 4-2 Adder Compressor

Ming Ming Wong, Vikramkumar Pudi and Anupam Chattopadhyay

Hardware and Embedded Systems Lab (HESL)
School of Computer Science and Engineering (SCSE)
Nanyang Technological University (NTU) Singapore
{mmwong, pudi, anupam}@ntu.edu.sg

Abstract—**The modern era of Internet-of-Things (IoT) is naturally imposing a tight area/runtime constraint on the computing kernels. Security kernels, as part of the standardized protocols as well as custom defense techniques, are among the most common tasks executed on every digital device. Therefore, low area cost and high performance implementation of security kernels is an important goal of current system designers. In this paper, we revisit the state-of-the-art implementations of SHA-256, a standardized security primitive for authentication and propose novel optimizations. Our optimizations, based on architectural folding and 4-2 adder compressor, are geared toward both lightweight and high performance implementations. Detailed experiments of our optimized architecture on different FPGA fabrics clearly demonstrate their benefits. Our presented design point successfully attained the highest hardware efficiency (throughput/area) figures among the published literature so far.**

Index Terms—SHA-256, folding, 4-2 adder compressor, lightweight, high performance.

I. INTRODUCTION

The ratification of Secure Hash Algorithm (SHA) in 1995 and the increasing demand to enforce a higher security level in electronic systems have given rise to various research work in designing optimized hardware implementation of SHA algorithm. The SHA-2 family, which was published in 2002 by the National Institute of Standards and Technology (NIST) [1] is a more robust version than its predecessor (i.e. SHA-0 and SHA-1). It is a one-way function used to obtain compressed representation of data and the algorithm is widely deployed in various authentication protocols such as digital signal generation, SSH, PGP, and IPSec.

Despite the existence of new hash algorithms such as SHA-3 in recent years, SHA-2 family, particularly the SHA-256 algorithm, still find its popularity in a wide variety of area constrained applications such as the IoT (Internet-of-Things) and mobile devices. These applications are operated on less powerful processors with limited RAM and memory storage capacities. Therefore, the need for compact SHA-256 hardware implementation with reasonable throughput performance has become increasingly important ever since.

The challenge in designing efficient hardware implementation of SHA-256 are mainly due to its **high data dependency, long critical path** and **large number of adders**. In this paper, we propose new design optimization techniques that employ **architectural folding** and **4-2 adder compressor**. Architectural folding exploits regularity and repetitive computational patterns in the algorithm to achieve maximum resource sharing in the hardware circuit. Resource sharing will efficiently lead to area cost reduction as the same computational unit will be shared (and reused) to perform two or more operations in the system. In addition to that, the adder compressor which has started to gain its popularity over the conventional binary adder, has the capability to reduce the number of adders in partial product accumulation stages. Hence, it will contribute towards a significant improvement in terms of the computation time.

As far as we know, this is the first work to employ the above-mentioned approaches for SHA-256 hardware implementation. As a result of this study, we produced two new and optimized SHA-256 designs which we have denoted as Case I and Case II respectively. The former is a rescheduled and transformed architecture with folding factor of 5 and the latter architecture utilized folding factor of 2 along with a f4-2 adder compressor. Both architectures are designed, verified and synthesized on Xilinx Virtex-2,4,5,6 FPGA.

The experimental results showed that the smallest design is found in Case I (with 150 slices) and the most efficient design is reflected in Case II (with 197 slices). Both cases have the efficiency (throughput/area) of 4.44 and 7.13 respectively.

The rest of this document is organized as follows. First, the overview of the SHA-256 algorithm is briefed in Section II while the considerations of its hardware implementation is summarized in Section III. The new and optimized SHA-256 designs are presented next in Section IV. Experimental results and the comparisons against the related works are discussed in detailed in Section V. Last, the concluding remarks of this work is drawn in Section VI.

II. SHA-256 ALGORITHM

The SHA-256 [1] hash algorithm takes an input message with arbitrary length between $0 - 2^{64}$ and generates a 256-bit digest message. The output, which is the digest message, also serves as the concise representation of the original message. This is due to the property of the hash algorithm that the slightest change from the original message will cause significant differences in the corresponding digest. Generally, SHA-256 offers a security level of 128-bits, which implies that collision (through a birthday attack) will occur in $O(2^{128})$ time.

978-1-5386-4757-8/18 $31.00 © 2018 IEEE

The complete hashing process essentially consists of four stages as illustrated in Fig. 1. It begins with the **Pre-processing** stage where the input message is processed into data block of 512 bits as described below.

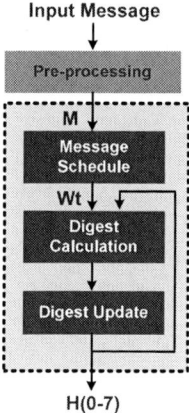

Fig. 1. SHA-256 Algorithm Block Diagram

- The input message which is composed of n bits, is appended with bit '1', and followed by k zero bits. The value of k has to be the smallest solution to the equation $n + 1 + k \equiv 448 \bmod 512$.
- The message is later appended with a 64-bits value n, which is the binary representation of the size of the original input message.
- Lastly, the padded message is parsed into N 512-bit blocks notated as M^i with $i = 0, 1, 2, \ldots, N - 1$.

The 512-bit block (M^i), is sequentially streamed to the second stage, namely the **Message Scheduler**, or also known as Message Expansion, where it is further expanded into sixty-four 32-bit words (W_t).

Technically, every computation round in SHA-256 requires a 32-bit word (W_t) deduced from the data block M^i. The data block from the first stage comprises of sixteen 32-bit words and thus, expansion is performed to produce the remaining words required for computation rounds 17 to 64 individually.

The formulation of these W_t words is based on the following equations.

$$\sigma_0(x) = ROTR^7(x) \oplus ROTR^{18}(x) \oplus SHR^3(x)$$
$$\sigma_1(x) = ROTR^{17}(x) \oplus ROTR^{19}(x) \oplus SHR^{10}(x)$$

$$W(t) = \begin{cases} M_t, & 0 \leq t \leq 15 \\ \sigma_1(W_{t-2}) + W_{t-7} + \sigma_0(W_{t-15}) + W_{t-16} & 16 \leq t \leq 63 \end{cases} \quad (1)$$

where ROTR denotes a circular rotation to the right, when SHR is shifting to the right functions. All additions in SHA-256 algorithm are modulo 2^{32}.

Following next, each of the W_t is put into the third stage, which is the **Digest Calculation** or also known as Compres-

sion Function. This is the core of hash algorithm and also appears to be the most computationally expensive stage where data will be processed for 64 iterative loops.

There are eight 32-bit working variables labeled A to H with their initialization values being from H_0 to H_7 respectively (given in [1]), and two 32-bit intermediate values, T_1 and T_2. In every round, these values will be calculated and updated based on a series of dedicated functions as follows.

$$\begin{aligned} T_1 &= H + \Sigma_1(E) + Ch(E, F, G) + K_t + W_t \\ T_2 &= \Sigma_0(A) + Maj(A, B, C) \end{aligned}$$

$$\begin{aligned} A &= T_1 + T_2 & E &= D + T_1 \\ B &= A & F &= E \\ C &= B & G &= F \\ D &= C & H &= G \end{aligned}$$

where

$$\begin{aligned} Ch\{x, y, z\} &= (x \wedge y) \oplus (\bar{x} \wedge z) \\ Maj\{x, y, z\} &= (x \wedge y) \oplus (x \wedge z) \oplus (y \wedge z) \\ \Sigma_0(x) &= ROTR^2(x) \oplus ROTR^{13}(x) \oplus ROTR^{22}(x) \\ \Sigma_1(x) &= ROTR^6(x) \oplus ROTR^{11}(x) \oplus ROTR^{25}(x) \end{aligned}$$

where K_t are sixty-four 32-bit constants as specified in [1]. A single computation round in SHA-256 is depicted in Fig. 2.

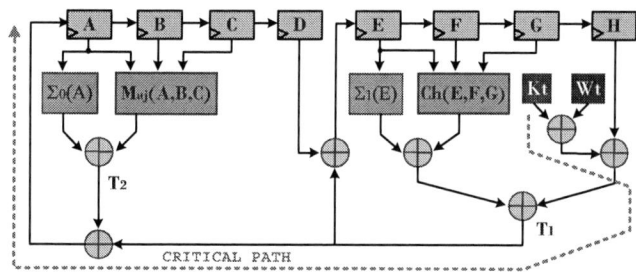

Fig. 2. SHA-256 Transformation Round

Upon the completion of 64 iterations, the intermediate hash value H^i will be calculated in the last stage **Digest Update**. This process, listed below, will produce the partial digest message with respect to the current M^i 512-bit data block.

$$H_0^i = A + H_0^{(i-1)}; \quad H_1^i = B + H_1^{(i-1)}; \quad H_2^i = C + H_2^{(i-1)};$$
$$H_3^i = D + H_3^{(i-1)}; \quad H_4^i = E + H_4^{(i-1)}; \quad H_5^i = F + H_5^{(i-1)};$$
$$H_6^i = G + H_6^{(i-1)}; \quad H_7^i = H + H_7^{(i-1)}$$

The SHA-256 hashing computation (from Stage 2 to 4) shall repeat with another 512-bit block (in Stage 1) until all the N data blocks have been processed. The final 256-bit digest message output, H^N for a given input message is found as a result of concatenating the final hash values.

$$H^N = H_0^N \| H_1^N \| H_2^N \| H_3^N \| H_4^N \| H_5^N \| H_6^N \| H_7^N$$

III. Hardware Architectures for Hash Function

Research studies on SHA-256 that were reported in the literature have highlighted the main challenges of the hash core implementation on hardware, which are summarized as follows.

- **High data dependency** in SHA-256 algorithm complicates the design effort to realize hardware implementation with high performance. In each computation round, the working variables A through H are calculated and updated using the respective values obtained from previous cycles. Subsequently, several architectural optimization approaches could not be deployed effectively to achieve significant speed enhancement.
- **Long critical path,** which involved chains of logical and arithmetic operations is found in computing variable A. This longest data path, shown in Fig. 2, covers seven binary adders (modulo 2^{32}) becomes bottleneck in achieving fast throughput in SHA-256 implementations.
- **Large number of 32-bit adders** that are needed in SHA-256 core are rather computational intensive in terms of the total execution time leading to high power consumption as well.

Due to the above-mentioned constraints, specialized hardware cores implemented on re-configurable hardware (FPGA) in recent years generally focus on either one of the two main directions, i.e. **compact** design or **high throughput** implementation. For instance, the work reported in [2], [3], [4], [5] aimed in achieving a compact hardware implementation via resource sharing approaches, where throughput performance is often traded off for hardware cost saving.

On the other hand, other works which are reported in [6], [7], [8] focused on high throughput hardware implementation through various optimization techniques, of which the computation cycles are effectively reduced at the cost of excessive hardware resources. Some of these optimization techniques are summarized below.

- **Carry-Save Adder (CSA)** is utilized in order to improve the critical path in the SHA-256 core. The CSA diligently separates the sum and the carry paths which efficiently reduces the delay caused by the carry propagation time. Furthermore, as the CSA operates on 3-inputs, this allows variable A to be computed using five CSAs in the SHA-256 circuit.
- **Pipelining** effectively contributes to circuit throughput improvement because it allows multiple independent data streams to be processed simultaneously. Due to data dependency, by itself pipelining is unable to offer significant improvement in SHA execution time. Therefore, rescheduling needs to be performed prior to pipelining in order to maximize the attainable computation throughput. This approach consumes large areas due to pipeline registers, as well as the duplication of hardware for every round computation.
- **Unrolling (partially and fully)** is another architectural technique that can potentially improve the data dependency in SHA-256 algorithm and lead to throughput enhancement. Basically, in a k-unrolled architecture, execution of k steps can be performed simultaneously in the same clock cycle. Hence the total number of clock cycles required to compute a single iteration of message digest is effectively reduced by k factor. However, such an advantage results in a large hardware area cost.

Taking into account both design challenges, our work aims at achieving a breakthrough in a SHA-256 design that has minimal hardware area size and maximal throughput performance.

IV. The New and Optimized SHA-256 Implementation

Two optimization techniques are employed in this work. First, **architectural folding** is applied to enable the reusing and sharing of a binary adder in the hash computation (refer Section IV-A). Second, a **4-2 adder compressor** is utilized in order to minimize the propagation delay in the adders (refer Section IV-B).

This study produced two SHA-256 hardware architectures (labeled as Case I and Case II), which are presented in Section IV-C. To the best of our knowledge, our work is the first to employ architectural folding and 4-2 adder compressor in SHA-256 implementation.

A similar approach was reported by S. Dominikus [9] where a folding technique was deployed in an integrated hash circuit. However, the technique was used to enable resource sharing among four different hash algorithms of the MD4 family implemented in their system. In another work, fast adder compressor was also utilized by Ling and Li in [6]. The authors proposed a new 7-3-2 compressor which is a combination of 7-3 and 3-2 adder compressors.

A. Architectural Folding and Resource Sharing

The concept of folding [10], [11], which is different from unrolling, is an architectural transformation technique used for minimizing the number of functional units in the digital system. In short, with K being the folding factor, a functional unit in the transformed system could be reused to perform K operations in the original system.

In this work, we exploited the regularity in SHA-256, where the chain of additions (modulo 2^{32}) in variable A computation is scheduled into mutually exclusive clock cycles in order to minimize the number of adders required in the circuit. In addition to this, it is essential that the scheduling is planned in such way that it does not impose excessive registers and multiplexers in the folded architecture.

B. 4-2 Adder Compressor

The 4-2 adder compressor is part of the fast adder evolution that computes four additions simultaneously. This higher order parallelism reduces the critical path as well as the internal glitching of the sum tree, which in turn reduces the dynamic power dissipation as well [12].

In general, a 4-2 adder compressor takes 5 inputs (A, B, C, D and Cin) and generates 3 outputs (Sum, $Carry$ and $Cout$) as shown in Fig. 3. The 4 inputs A, B, C and D and the output Sum are of the same weightage. Meanwhile, Cin is the output from a previous lower significant compressor and $Cout$ is the input for the compressor in the next significant stage. Note that the computation $Cout$ is independent of Cin and this improves the speed of the carry save adder.

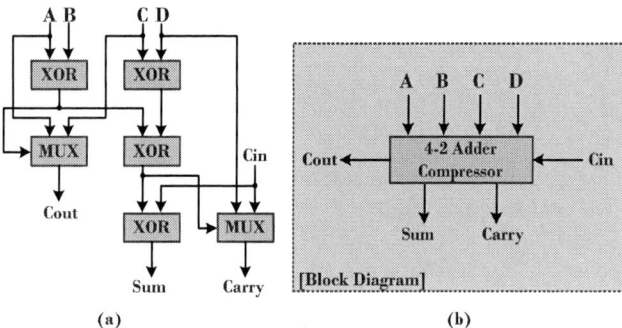

Fig. 3. (a) Architecture of 4-2 Adder Compressor (b) Block Diagram

Different structures of 4-2 compressors were presented in the literature such as [12], [13], [14] but all of them comply with the fundamental equation in (2).

$$Sum + 2*(Carry + Cout) = A + B + C + D + Cin \quad (2)$$

The most common 4-2 adder compressor architecture comprised of XOR and MUX modules which are architecturally similar to the carry save adder (3-2 compressor). Its calculation is governed by the equations in (3).

$$
\begin{aligned}
Sum &= A \oplus B \oplus C \oplus D \oplus Cin \\
Cout &= (A \oplus B)*C + \overline{(A \oplus B)}*A \\
Carry &= (A \oplus B \oplus C \oplus D)*Cin + \overline{(A \oplus B \oplus C \oplus D)}*D
\end{aligned}
\quad (3)
$$

C. Proposed SHA-256 Architecture

Based on (1), the first 16 W_t that are needed in the hashing core are equal to the corresponding 512-bit data block but the subsequent W_t for t from 16 to 63 would require additional computations using the previous values. In our proposed architecture, only the recent 16 W_t will be stored during each hashing computation round for efficient memory usage.

D. Case I: Folded-5 SHA-256 Circuit

As discussed earlier, the critical path in SHA-256 is the computation of variable A (also refer to Fig. 2). Full calculation of A is expressed in (4) where it is the sum of seven numerical values. In order to reduce the number of adders, the computation in SHA-256 core can be rescheduled such that

the architecture can be effectively folded to enable resource sharing.

The aim is to break the chains of seven additions into several clock cycles diligently. This appears to be similar to the attempt to reduce the critical path, but in a way that could also maximize the chances of functional unit sharing within the hash core. Besides, this rescheduling has to take into account the data path E as its computation also requires additions (as expressed in (5)).

$$
\begin{aligned}
A &= H + \Sigma_1(E) + \Sigma_0(A) + Maj(A,B,C) \\
&\quad + Ch(E,F,G) + K_t + W_t \quad (4) \\
E &= D + H + \Sigma_1(E) + Ch(E,F,G) + K_t + W_t \quad (5)
\end{aligned}
$$

The proposed architecture in Case I is transformed and folded by factor 5 where the additions of A and E are scheduled into 5 clock cycles (as depicted in Fig. 4). As an end result, only two 32-bit adders are required and they will be shared (and reused) to perform all the arithmetic additions in SHA-256.

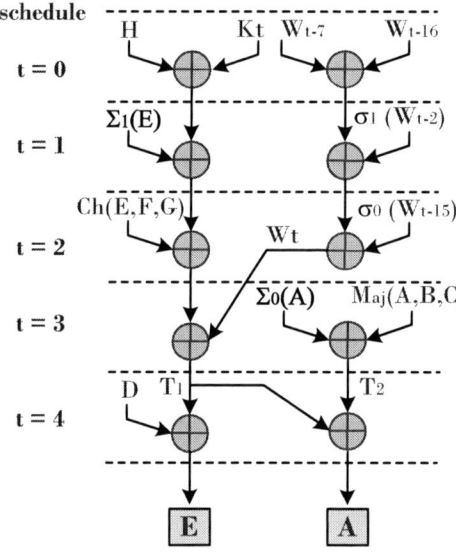

Fig. 4. Scheduling for SHA-256 with folding=5)

Architectural folding inevitably imposes substantial timing overhead in the hash computation. The total clock cycles needed for a complete hashing process will increase linearly with the corresponding folding factor. In our proposed Case I, the SHA-256 core computation will consume a total of $64*5 = 320$ clock cycles. However, as the rescheduling was performed on the longest data path, the new critical path in the folded architecture is diligently reduced from 7 additions to 2 additions. With that, the new design can be clocked at a faster rate leading to a constructive improvement in the overall throughput performance.

978-1-5386-4757-8/18 $31.00 © 2018 IEEE

E. Case II: Folded-2 SHA-256 Circuit with 4-2 Adder Compressor

In an attempt to achieve optimal efficiency that strikes a balance between achieving small hardware area size and high throughput performance, we had incorporated adder compressor in our folded architecture. Similar to Case I, we proposed another folded SHA-256 with two adders but using 4-2 adder compressors instead of the conventional binary adders. Considering the fact that the adder compressor could compute 4 operands simultaneously, the additions in variable A can be effectively scheduled with less clock cycles.

To be exact, the new architecture Case II is folded by factor 2, which is illustrated in Fig. 5. With that, the hashing computation in Case II will consume a total of $64 * 2 = 128$ clock cycles, which is less amount of overhead compared to Case I. Not only that, the 4-2 adder compressor is computationally more efficient than the conventional binary adder. Hence, this further increases the highest achievable speed rate in the circuit. Subsequently, the throughput performance of the SHA-256 will be improved significantly without much trade-off in terms of the hardware area cost.

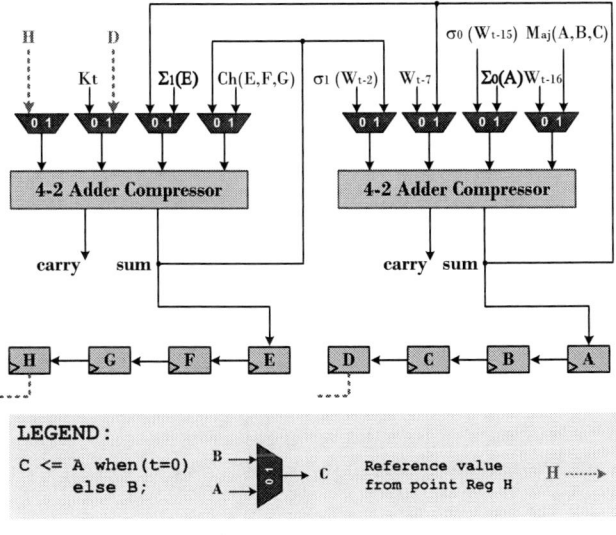

Fig. 5. Proposed SHA-256 Architecture (Case II)

V. EXPERIMENTAL RESULTS AND PERFORMANCE ANALYSIS

The SHA-256 hardware architectures (for Cases I and II) are described using VHDL and synthesized with Xilinx ISE 14.7 on Virtex-6. For fair comparison with existing works, both architectures are also implemented on Xilinx's older platforms, Virtex-2,4,5 using Xilinx ISE 9.2. The FPGA implementation results are investigated in terms of the area, frequency (maximum), throughput and hardware efficiency. These metrics are as tabulated in Table I. The throughput performance and hardware efficiency for multi-message hashing are computed as follows (refer (6) and (7)),

$$\text{Throughput} = \frac{\#blocksize \times F_{max}}{\#clockcycle} \times \#N_{msg} \quad (6)$$

$$\text{Efficiency} = \frac{\text{Throughput}}{\#Slice} \quad (7)$$

where $\#blocksize$ refers to the number of processed bits, $\#clockcycle$ corresponds to the total clock cycles between successive messages which generate each digest message and F_{max} is the highest attainable frequency in the implementation. $\#N_{msg}$ is the number of messages that can be simultaneously hashed at a given time. In both our designs, the $\#blocksize$ is 512-bits and $\#N_{msg}$ is 1 as our architectures are not pipelined. The $\#clockcycle$ for Case I and Case II are 321 cycles and 129 cycles respectively.

Among the **compact (C) implementations** presented in Table I, both of our proposed architectures exhibit the highest throughput (Mbps) with minimal hardware cost (slices). In other words, our SHA-256 designs portray better hardware efficiency (throughput/area) i.e. a better compromise between hardware area and performance, in comparison to the related studies.

The reason being is that during rescheduling in the folding transformation registers were inserted for temporary data storage. In our cases, the SHA-256 critical path is rescheduled and hence, the insertion of registers in every shared adder brings the same effect as that of pipelining. This in turn, effectively reduced the critical path in the folded architectures. As a result, the proposed new technique not only minimized the number of adders (hardware size) needed but also improved the speed performance (throughput).

Using Xilinx Virtex-5 implementation, the smallest design is presented in [5] with 139 slices whereas our Cases I and II that utilized 205 and 273 slices respectively. However, both our designs exhibit a more desirable efficiency ratio (2.55 and 3.95) than their work (0.85).

In Table I, the benchmark with the **high throughput (H) implementations** [6], [7], [8] served as a reference to analyze the amount of hardware cost saving obtained in relation to the speed and throughput performance. The most relevant work was reported in [6] where the authors had also utilized adder compressor (7-3-2 adder compressor) to attain performance enhancement. However, our Case II architecture surpassed their results, both in throughput gain and area cost reduction even though our design's total clock cycles for the hash computation doubled theirs.

Not only that, the efficiency attained in Case II (on Virtex-6 platform) has also outperformed the high throughput implementation which is reported in [7]. It is also worth noting that our architecture required only 197 slices as opposed to their loop-unrolled design that utilized a total of 1,831 slices.

VI. CONCLUSION

Overall, we propose lightweight and high performance SHA-256 architectures which are suitable for resource constrained devices. From an architectural perspective, this study

TABLE I

EXPERIMENTAL RESULTS FOR CASES I AND II ARCHITECTURES, AS WELL AS THE EXISTING COMPACT (C) AND HIGH THROUGHPUT (H) SHA-256 HASH IMPLEMENTATIONS.

Work	FPGA Devices	Area (Slices)	Fmax (MHz)	Clock Cycles per block	Design Aim (C / H)	Throughput (Mbps)	Efficiency (TP/Slice)
Our work (Case I) **Folded-5**	Virtex-2 (-6)	392	147	321	C	234.5	0.60
	Virtex-4 (-12)	382	238	321	C	379.6	0.99
	Virtex-5 (-3)	205	328	321	C	523.2	2.55
	Virtex-6 (-3)	150	418	321	C	666.7	4.44
Our work (Case II) **Folded-2 with** **4-2 Adder Compressor**	Virtex-2 (-6)	502	134	129	C	531.8	1.06
	Virtex-4 (-12)	485	222	129	C	881.1	1.82
	Virtex-5 (-3)	273	272	129	C	1,080	3.95
	Virtex-6 (-3)	**197**	**354**	**129**	**C**	**1,405**	**7.13**
Kim et al. [2] (2009)	Virtex-2	779	72	490	C	74.7	0.1
Cao et al. [3] (2011)	Virtex-2	639	85	1,120	C	38.9	0.06
	Virtex-4	615	102	1,120	C	46.6	0.08
Kim et al. [4] (2012)	Virtex-2	1,210	85	355	C	122.6	0.1
Garcia et al. [5] (2014)	Virtex-4	422	50	280	C	91.5	0.22
	Virtex-5	139	64	280	C	117.8	0.85
Ling and Li [6](2009)	Altera Cyclone III	2,633(LE)	83	65	H	654	0.25(TP/LE)
Michail et al. [7](2012)	Virtex-5	1,885	169	65	H	10,816	5.74
	Virtex-6	1,831	172	65	H	11,008	6.01
Jeong and Kim [8](2014)	Virtex-5	2,796	179	65	H	1,410	0.50

presents a new optimization measure which is different from the conventional approach in the sense that we aim to strike an optimal balance between the hardware area cost and its throughput performance. The hardware synthesis results deduced from Cases I and II precisely reflected that our proposed approach which employed architectural folding with resource sharing, together with the usage of 4-2 adder compressor had successfully produced efficient hardware implementations of SHA-256. With implementation on Xilinx Virtex-6 FPGA, we have presented the highest efficiency (throughput/area) SHA-256 core in Case II with only 197 slices and efficiency of 7.13.

REFERENCES

[1] NIST, "FIPS 180-2, Secure Hash Standard, Federal Information Processing Standard (FIPS), Publication 180-2," DEPARTMENT OF COMMERCE, Tech. Rep., Aug. 2002.

[2] M. Kim, J. Ryou, and S. Jun, "Efficient hardware architecture of SHA-256 algorithm for trusted mobile computing," *Information Security and Cryptology*, pp. 240–252, 2009. [Online]. Available: http://dx.doi.org/10.1007/978-3-642-01440-6_19

[3] X. Cao, L. Lu, and M. ONeill, "A compact SHA-256 architecture for RFID tags," in *Proceedings of the 22nd IET Irish signals and systems conference (ISSC) Trinity College Dublin*, 2011.

[4] M. Kim, D. G. Lee, and J. Ryou, "Compact and unified hardware architecture for SHA-1 and SHA-256 of trusted mobile computing," *Personal and Ubiquitous Computing*, vol. 17, no. 5, pp. 921–932, Jun 2013. [Online]. Available: https://doi.org/10.1007/s00779-012-0543-0

[5] R. García, I. Algredo-Badillo, M. Morales-Sandoval, C. Feregrino-Uribe, and R. Cumplido, "A compact FPGA-based processor for the secure hash algorithm SHA-256," *Comput. Electr. Eng.*, vol. 40, no. 1, pp. 194–202, Jan. 2014. [Online]. Available: http://dx.doi.org/10.1016/j.compeleceng.2013.11.014

[6] L. Bai and S. Li, "VLSI implementation of high-speed SHA-256," in *2009 IEEE 8th International Conference on ASIC*, Oct 2009, pp. 131–134.

[7] H. E. Michail, G. S. Athanasiou, V. Kelefouras, G. Theodoridis, and C. E. Goutis, "On the exploitation of a high-throughput SHA-256 FPGA design for HMAC," *ACM Trans. Reconfigurable Technol. Syst.*, vol. 5, no. 1, pp. 2:1–2:28, Mar. 2012.

[8] C. Jeong and Y. Kim, "Implementation of efficient SHA-256 hash algorithm for secure vehicle communication using FPGA," in *2014 International SoC Design Conference (ISOCC)*, Nov 2014, pp. 224–225.

[9] S. Dominikus, "A hardware implementation of MD4-family hash algorithms," in *9th International Conference on Electronics, Circuits and Systems*, vol. 3, 2002, pp. 1143–1146 vol.3.

[10] R. Mehra and J. Rabaey, "Exploiting regularity for low-power design," in *Proceedings of International Conference on Computer Aided Design*, Nov 1996, pp. 166–172.

[11] A. Canis, J. H. Anderson, and S. D. Brown, "Multi-pumping for resource reduction in FPGA high-level synthesis," in *2013 Design, Automation Test in Europe Conference Exhibition (DATE)*, March 2013, pp. 194–197.

[12] R. Dornelles, G. Paim, B. Silveira, M. Fonseca, E. Costa, and S. Bampi, "A power-efficient 4−2 adder compressor topology," in *2017 15th IEEE International New Circuits and Systems Conference (NEWCAS)*, June 2017, pp. 281–284.

[13] S. Kumar and M. Kumar, "4−2 compressor design with new XOR-XNOR module," in *2014 Fourth International Conference on Advanced Computing Communication Technologies*, Feb 2014, pp. 106–111.

[14] J. Tonfat and R. Reis, "Low power 3−2 and 4−2 adder compressors implemented using ASTRAN," in *2012 IEEE 3rd Latin American Symposium on Circuits and Systems (LASCAS)*, Feb 2012, pp. 1–4.

Low-budget Energy Sector Cyberattacks
via Open Source Exploitation

Anastasis Keliris
Tandon School of Engineering
New York University
Email: anastasis.keliris@nyu.edu

Charalambos Konstantinou
Center for Advanced Power Systems
Florida State University
Email: konstantinou@caps.fsu.edu

Marios Sazos, Michail Maniatakos
Center for Cyber Security
New York University Abu Dhabi
Email: {marios.sazos, michail.maniatakos}@nyu.edu

Abstract—Modern cyber warfare involves penetration of a nation's computers and networks, aiming to cause extensive damage and/or disruption. Such actions are generally deemed feasible only by resource-wealthy nation state actors. In this work, we challenge this perception and introduce a methodology dubbed Open Source Exploitation (OSEXP), which leverages public infrastructure to execute an advanced cyber attack on critical infrastructure. In particular, we characterize and verify an effective and reusable OSEXP attack vector based on time spoofing of Global Positioning System (GPS) signals. Our GPS attack employs commercial devices and open source software, and manipulates the time synchronization of carefully selected power grid equipment in a manner that can lead to large scale blackouts. We experimentally verify the feasibility of our GPS OSEXP methodology, and demonstrate that an actor with limited budget has the ability to cause significant disruption to a nation.

I. INTRODUCTION

The first public electric power systems were established in the 1880's for providing street lighting [1]. Since then, power systems have significantly evolved and grew to become essential in our everyday life. Today, electric power systems can be considered the "backbone" of critical infrastructure as several sectors, such as water treatment and desalination, heavy industry, and military defense systems, rely on electric power. The far-reaching effects of power outages (also known as *blackouts*), further demonstrate our strong dependence on uninterrupted supply of electricity. Typical causes of blackouts include extreme weather and natural phenomena, misoperation, human errors, equipment failures, and animals [2].

Over the past years, *cyberattacks* have been added to the list of potential threats against the stable and continuous operation of power systems. The Aurora Generator Test was one of the first demonstrations of how cyberattacks can damage physical power grid components [3]. This test was performed in 2007 by the Idaho National Laboratory, and used a software program to exploit a vulnerability in diesel generators causing them to explode. Moving from tests in controlled environments to cyberattacks against real world systems, two sophisticated attacks targeting the Ukrainian power grid in 2015 and 2016 led to partial blackouts in the Ivano-Frankivsk region and Kiev, respectively [4], [5].

Despite these tangible incidents, there remains a lot of uncertainty among power industry stakeholders regarding the real threat cyberattacks pose to power systems. Such attacks are usually considered to only be within the reach of resource-wealthy nation state actors. However, given the plethora of publicly available information regarding power systems and the dependency of these systems on public infrastructure, we argue that it is feasible for actors with lower budgets to instantiate disrupting attacks. To that end, we introduce a low-budget methodology capable of causing wide area blackouts, solely relying on public information and public infrastructure.

Similar to Open Source Reconnaissance,[1] where meaningful information is extracted from public sources [6], we introduce *Open Source Exploitation* (OSEXP). OSEXP leverages public resources and infrastructure to construct low budget attack vectors against judiciously selected power grid locations. We present an in-depth study of a specific OSEXP vector, namely Global Positioning System (GPS) time spoofing, that exploits the reliance of power systems on GPS for time synchronization. Although the potential of GPS spoofing attacks to affect power system measurements has been demonstrated in literature, it required specialized, expensive equipment and extended technical knowledge [7], [8].

Our main contribution in this work is the characterization and experimental verification of GPS time spoofing attacks against carefully selected power grid devices *using low-cost Commercial-Off-The-Shelf (COTS) equipment and open source software*. Our approach significantly reduces the cost and complexity of GPS time spoofing, "open sourcing" the exploitation phase of campaigns targeting power systems. Furthermore, it enables a one-time design of an exploitation vector and reuse of the same vector worldwide, as GPS is employed for time synchronization purposes in systems across the globe.

The rest of the paper is structured as follows: Section II provides preliminaries on power systems and GPS. We present an end-to-end open source approach for constructing attack vectors against power systems and introduce OSEXP in Section III. We elaborate on a specific OSEXP attack, GPS time spoofing, in Section IV, and experimentally verify the feasibility of this attack using low-cost equipment and open source software in Section V. Related work is discussed in Section VI, and we conclude the paper in Section VII.

---

[1]The term open source is not related to open source software throughout this work, unless explicitly stated.

II. PRELIMINARIES

A. Power systems

Power systems are collections of networked components that generate, transfer, and utilize electric power. Large scale power systems are also known as power grids, because of the interconnected network topology of their components. In general, power systems are comprised of four stages, namely generation, transmission, distribution, and consumption.

The first stage is *generation*, where electricity is produced in power plants through the conversion of other forms of primary energy to electrical energy. Electricity is then transferred in the *transmission* stage, utilizing an elaborate network of interconnected high voltage lines and substations spanning large geographical distances. In general, overhead transmission lines are preferred to underground lines due to reduced costs, with 97% of U.S. transmission lines being overhead [9]. *Distribution* is the stage at which electricity is distributed to end consumers, using lines that span smaller distances and operate at lower voltages compared to transmission lines and step down transformers to reduce the voltage levels to ranges that match the operational voltages of end consumers. Finally, electricity is utilized in the *consumption* stage.

B. Protection and control equipment

Protection and control devices, such as Circuit Breakers (CBs) and relays, are employed for ensuring the stable and secure power system operation. Their operation guarantees normal grid equipment operation by separating the system into protective zones, and isolating faulty zones as necessary. This separation can limit or prevent damages to equipment and personnel in the case of overloads or faults.

According to the North American Electric Reliability Council (NERC), 70% of the major disturbances in the U.S. are associated with faulty operation of relay controllers [10]. Optimal attack strategies may require changing the breaker status signal at only one transmission line [11], highlighting the necessity of constant and reliable operation of CBs and relay controllers for avoiding severe consequences.

C. Grid modernization

Cyberattacks against power systems are mainly enabled by an ongoing modernization, materialized through the convergence of Operational Technology (OT) and Information Technology (IT). Components in power grid are being upgraded with "smart" counterparts that enable fine-grain control and faster incident response times. To achieve these goals while keeping development costs low, vendors of power equipment typically leverage COTS hardware and software, use common general-purpose microprocessor architectures (e.g., ARM, Intel x86) and real-time versions of commercial operating systems (e.g., Windows and Linux) [12].

In general, Intelligent Electronic Devices (IEDs) deployed in power systems observe the variables and state of the system, store necessary data, make decisions, and take protection and control actions towards preserving performance and stability. For example, Wide Area Monitoring Systems (WAMS) highly rely on IEDs to gather system information from multiple sources. WAMS are mainly enabled by Phasor Measurement Units (PMUs) that take synchronized snapshots of electrical quantities across the system, and use the comparative measurements to estimate the health and power quality of the grid. PMUs are deployed primarily in the transmission stage and provide synchronized phasor (synchrophasor) measurements of voltage and current at several locations to provide time-stamped information of the system's state. Given the dispersed topology of the power grid, accurate time synchronization between such devices is essential for their operation. To that end the majority of PMUs rely on timing provided by GPS modules for capturing synchronized snapshots of the system across geographically dispersed locations.

D. Global Positioning System

Global Navigation Satellite Systems (GNSS), an example of which is GPS, use a collection of earth-orbiting satellites equipped with transmitters. The transmitters include an atomic clock synchronized to the Coordinated Universal Time (UTC) and their location is assumed known at all times, as the satellites follow predetermined trajectories. Each satellite broadcasts a navigation signal including time stamp data and any deviation from its predetermined trajectory [13]. Receivers obtain such signals from satellites within their field of view, and use the signal propagation delays to calculate their three-dimensional location data and time [14]. GPS signals are categorized in those available for civilian use (L1 C/A transmitted at 1575.42 MHz), and encrypted restricted signals (L2 transmitted at 1227.60 MHz), typically used by military applications. In this work we focus on L1 signals, as PMUs utilize these signals for time synchronization.

III. OPEN SOURCING MALICIOUS CAMPAIGNS AGAINST POWER SYSTEMS

In this section we present an end-to-end open source approach for campaigns aiming to disrupt national power grids. Retracing the steps of a malicious actor whose objective is causing a wide area blackout, we identify three main requirements for achieving this goal. These are:

1) Construct a model of the target system. This model is necessary for understanding the system, its dependencies and interconnections, and identifying its weak spots.
2) Analyze the model to identify and select critical targets. By carrying out analytical, data-driven studies of the system model, adversaries can identify critical locations that when attacked could lead to cascading failures.
3) Construct attack vectors targeting the identified critical locations, disrupting power system operations.

A. Threat model

The threat actors we consider in this work are adversaries that aim to cause power system disruption and large scale power outages. We assume that the adversaries have technical expertise of power system operations and that they can, if required, be in physical proximity to power grid assets.

However, we do not consider them to possess confidential information, or have network access to the equipment and control center of the target power system. In our threat model, we assume the adversaries can leverage publicly available information and infrastructure to achieve their objective, and are thus not limited to resource-wealthy nation states.

B. Modeling a system using public information

Most impactful blackouts to date have been the result of faults occurring at the transmission stage. Thus, adversaries are most likely to focus on this stage to cause power outages. To that end, a model including the network topology of transmission lines, transmission substations and their interconnections, as well as general load and capacity estimations can enable further studies of the target system.

Given the plethora of publicly available sources of information concerning power systems, adversaries may reconstruct a system model from such sources. Evidently, this form of "open source reconnaissance" is employed in an ongoing campaign against U.S. systems [15]. Examples of sources include: a) public reports, such as blackout reports and expansion planning reports, b) power system databases, such as Enipedia [16] and Open Energy Information [17], and c) press releases and success stories from power utilities and power grid equipment vendors. By combining and fusing information from such sources, a model of a target power system can be constructed for carrying out subsequent analyses. Construction of a power system model using public information is out of the scope of this work, and a detailed example can be found in [18].

C. Identifying critical locations with contingency analysis

Power studies of the constructed model can enable judicious selection of target locations for materializing an attack. Considering the objective of power system disruption, contingency analysis studies in particular are useful to malicious actors. Contingency studies aim to analyze unscheduled events (e.g., generator, transformer and/or transmission line failures) in a power system, assessing and ranking their impact [19]. Note that in general power systems must be able to sustain $N-1$ contingencies (e.g., to enable maintenance). However, for all systems there exists a number of contingencies that leads to non-sustainable scenarios and can also lead to cascading failures and blackouts. By applying such contingency analysis techniques on the model constructed in the previous step, adversaries can identify these critical transmission lines and interconnections and direct their attacks against them. Contingency analysis is described in detail in [19].

D. Open Source Exploitation

With knowledge of the critical points of a power system, adversaries need to construct attack vectors against the system. More specifically, they need to devise means towards disconnecting critical transmission lines capable of a nonsustained contingency scenario. In this work, we focus on the exploitation of *public infrastructure*, in a process we name Open Source Exploitation (OSEXP). OSEXP techniques can be

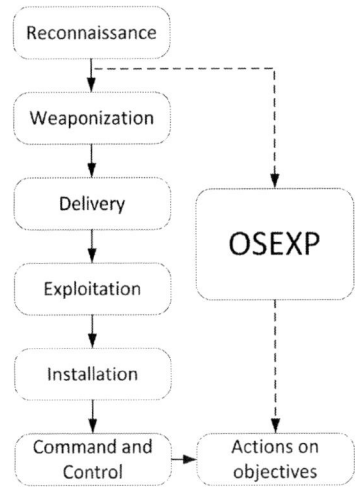

Fig. 1. Conventional Cyber Kill Chain with our proposed OSEXP step.

used in conjunction with conventional cyberattack techniques (e.g., phishing, credential harvesting, lateral movement, etc.), depending on the campaign objectives. For a campaign whose target is to cause large scale power outages rather than just get information and leverage on a target system, we argue that OSEXP techniques can be advantageous.

In general, the Cyber Kill Chain (CKC) is used to describe the structure of a cyberattack [20]. The steps of CKC are: 1) reconnaissance, where information is gathered, 2) weaponization, where a payload is designed, 3) delivery of the payload, 4) exploitation, where a vulnerability of the target system is exploited, 5) installation, where the payload is installed and executed on the target system, 6) command and control, where adversaries remotely tweak and instruct the payload and finally 7) actions on objectives, where adversaries fulfill the objectives of their campaign. By exploiting public infrastructure using OSEXP, steps 2 to 6 are replaced with an OSEXP step leading to an alternative path in the CKC, depicted in Fig. 1. The resulting CKC using OSEXP attacks has fewer steps, is reusable and leaves less evidence behind, making forensic studies and attribution harder.

IV. OSEXP: GPS TIME SPOOFING AGAINST PMUS

As outlined in Section II-C, PMUs can take protective control actions in addition to their monitoring role. Taking advantage of this, judicious manipulation of PMU measurements can destabilize a system, making PMUs attractive targets for malicious actors. The OSEXP attack against PMUs we describe in this section exploits the reliance of PMUs on GPS (which is a public resource) for time synchronization. Our OSEXP attack introduces erroneous PMU measurements by manipulating the timing source of PMUs, effectively disconnecting selected PMU-controlled transmission links.

Corroborating the feasibility of OSEXP GPS attacks, information, implementation details, and software regarding GPS are part of the public domain. Open source implementations of GPS receivers and transmitters for Software Defined Radios (SDRs), software GPS simulators and available literature

978-1-5386-4757-8/18 $31.00 © 2018 IEEE

lower the technical requirements for successful GPS spoofing attacks [21]. Furthermore, the global nature of GPS ensures that a GPS spoofing attack can be reused in diverse systems employing different hardware across the globe. In contrast, techniques that require identifying and exploiting deployed devices, software, and network channels are system-specific and require undertaking laborious research for each system.

GPS receivers inherently trust the signals they receive, assuming the signals have not been tampered with. In most countries in the world, any transmission in the frequency band of GPS is illegal, addressing the risk with policy safeguards. However, from a technical standpoint L1 GPS signals do not have any built-in integrity protection mechanisms. With OSEXP GPS spoofing attacks, we challenge the inherent trust in the integrity of these signals, arguing that adversaries with far-reaching agendas, such as causing blackouts, will not be bound by ethical and legal concerns.

Given the reliance of PMUs on GPS for capturing the state of a power system in a synchronized manner, we describe the process of introducing errors in PMU measurements by manipulating GPS signals in their vicinity. This can cause desynchronized snapshots of the system state from PMUs in different geographical locations, leading to system destabilization and even cascading failures. In particular, GPS time spoofing attacks can introduce errors in the absolute time perceived by the affected PMUs. For an f-Hz signal the relationship between the clock offset error $\tilde{t}_\delta - t_\delta$ and the phase angle measurement error ϵ are described by the following equation [22]:

$$\epsilon = [f \times (\tilde{t}_\delta - t_\delta) \times 360^\circ] \ (\mathrm{mod}\ 360^\circ) \qquad (1)$$

PMUs with control capabilities have a preconfigured threshold for allowed phase angle difference, that is dependent on the specifics of the system they are deployed in. Phase differences larger than this threshold cause connected CBs to open for avoiding fault propagation and protecting the equipment. However, introducing timing errors with GPS spoofing to instantly change the perceived state of the system for a PMU to exceed this threshold is not possible, because of the standards that govern PMUs. In particular, the IEEE standard for Synchrophasor Measurements for Power Systems (C37.118) dictates that clock synchronization errors between any two measurements from different PMUs should not exceed $31(26)\ \mu s$ for 50(60) Hz systems [23]. For a successful attack, it is thus necessary to slowly drift angle measurements, without exceeding these limits.

Another requirement for a successful GPS spoofing attack is knowledge of the legitimate GPS signal as it is perceived by the target receiver, including location information. This requirement can be fulfilled by co-locating the spoofing equipment in the physical vicinity of the target. The location information of a receiver is static, as the antenna is mounted on a building. Towards measuring the receiver location, attackers can measure their relative distance from the receiving antennas and calculate the offset, for example by employing drones equipped with cameras and GPS receivers. By flying directly

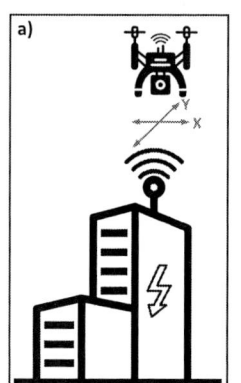
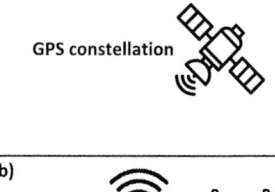
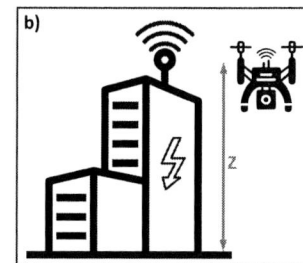

Fig. 2. Estimation of 3D location of a static GPS receiver using a drone. a) x,y coordinates. b) z coordinate.

over the target antenna, adversaries can capture the x,y location using the drone mounted GPS receiver. Subsequently, the z coordinate can be measured independently. Fig. 2 illustrates this scenario.

In addition to identifying receiver location, generation of appropriate synthetic GPS signals requires that the spoofed and legitimate GPS signals are time-synchronized [24]. This enables attackers to concurrently transmit a spoofed signal that is synchronized with the legitimate signal, gradually increase the transmitting power overtaking the GPS receivers in the affected vicinity, and then introduce time delays that will cause erroneous PMU measurements. The naive approach of recording legitimate GPS signals and replaying them after introducing the necessary time delays is not possible due to non-deterministic delays introduced by the retransmitting equipment's hardware components and the strict timing requirements of the IEEE C37.118 synchrophasor standard. To overcome this challenge, attackers can generate a *leading* GPS signal, and then gradually introduce appropriate delays to achieve synchronization between their spoofed and the legitimate GPS signals. The equipment required for this are two GPS receivers (one for the legitimate and one for the spoofed signal) and means to measure the time difference between the two signals.

An observation regarding the GPS OSEXP attack is that it requires simultaneous physical proximity to all target locations is required, meaning that adversaries need to coordinate an attack at k locations in the case of attacking a system to trigger an $N - k$ contingency. For most power systems, opening CBs at two judiciously selected locations is sufficient to destabilize the system. We argue here that requiring two to three field agents for launching an attack of this scale and impact is realistic and by no means prohibitive.

V. EXPERIMENTAL EVALUATION OF GPS OSEXP

In this section we evaluate the feasibility of a low cost GPS time spoofing OSEXP attack. In particular, we verify that open source software and Software Defined Radio (SDR) platforms are capable of launching GPS time spoofing attacks with the necessary granularity as this is defined by IEEE C37.118.

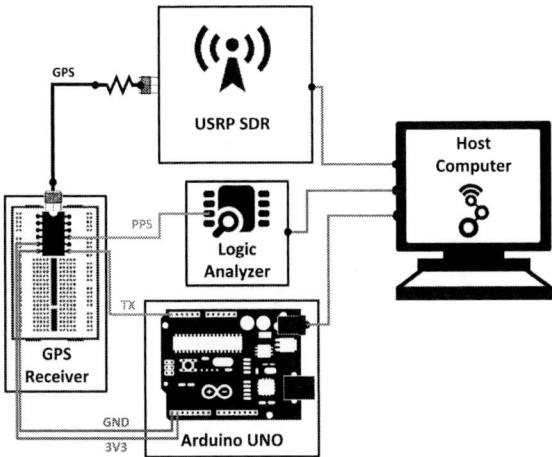

Fig. 3. Experimental setup for GPS spoofing attack.

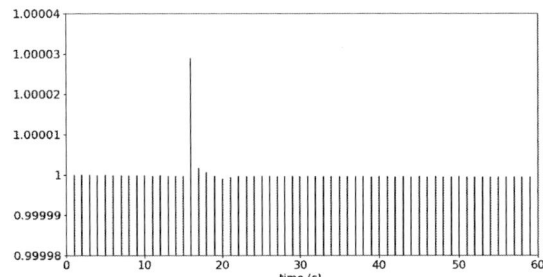

Fig. 4. Experimental results showing GPS receiver output PPS duration. The GPS spoofing attack is launched at $t = 16$ seconds.

This can desynchronize PMUs measurements, causing CBs at critical locations to open, and leading to wide area blackouts.

In our experiments we assume that attackers have synchronized their synthetic signals to legitimate GPS signals and have taken over control of the GPS receiver. These are realistic assumptions if an attacker can introduce *arbitrary delays* to a GPS signal, as arbitrary delays can be leveraged to achieve synchronization of leading signals with the legitimate ones. After the two signals are synchronized, attackers can gradually increase the spoofed signal power, overtaking control of receivers within their vicinity [7].

A. Experimental setup

The hardware in our experimental setup consist of a GPS receiver, an Arduino board, a SDR, a logic analyzer and a host computer. The GPS receiver employs the Venus638FLPx chip, which is a commercial, high performance receiver with 29 seconds cold start time-to-first-fix, up to 20 Hz update rate, and built-in jamming detection and mitigation. The GPS receiver is powered by an Arduino UNO board, which is also connected to the host computer for receiving and outputting the decoded NMEA messages. We utilize a Saleae Logic Pro 8 logic analyzer for sampling the Pulse-Per-Second (PPS) output pin of the receiver at a sampling rate of 10 MHz, which is satisfactory given the GPS receiver's PPS measured accuracy of 2 μs. For transmitting GPS signals we use an Ettus USRP N210 SDR, equipped with a GPSDO kit and a 40 MHz SBX 400-4400 MHz Rx/Tx. Respecting the legal framework concerning GPS signal transmission over-the-air, we conduct all of our experiments using cable connections and never transmit signals over-the-air, without loss of generality. To further ensure no side-effects we attenuate the USRP output to -140dBm, which is close to the minimum required signal by our GPS receiver for a fix (-148 dBm) and enclose the experimental setup in RF shielding fabric. Our experimental setup is depicted in Fig. 3.

In terms of software, we rely solely on open source software. For generating synthetic GPS data we use the Software-Defined GPS Signal Simulator (`gps-sdr-sim`) [25]. We download the required ephemerides data that indicate the current state of the satellite constellation from the Crustal Dynamics Data Information System [26]. Using `gps-sdr-sim` and the current ephemerides we create a raw synthetic static L1 GPS signal with a 2.5 MHz sampling rate, that is leading the current wall time by a few seconds. We input this signal to GNU Radio to perform the necessary type conversions, and add a delay block of user-specified duration between the file source and the USRP sink. This entire process is automated.

B. Experimental results

Assuming a 50 Hz frequency for the target power system, our goal is to introduce a 30 μs delay, which lies within the allowed range according to IEEE C37.118. To that end, we select 30 μs as the user-specified delay duration in our GNU Radio flowchart and launch our automated script. We present the experimental results regarding time as it is perceived by the receiver in Fig. 4. In particular, the figure presents the absolute duration of PPS signals as it is perceived by the GPS receiver and measured by the logic analyzer. We observe that up to $t = 16s$ (which is when the attack is launched), each PPS signal is received exactly every one second, as expected. After the attack is launched, the particular pulse duration at $t = 16s$ becomes 1.0000289 s, indicating a shift in the perceived time by the GPS receiver as a result of our GPS signal manipulation. The introduced delay of 28.9 μs is below the 31 μs threshold, verifying the feasibility of using COTS equipment and open source software to launch fine-grain GPS time spoofing attacks.

Regarding the impact of our GPS time spoofing on a power system, the delay we introduce on PMU measurements results in a shift of 0.54° in the measured angle, calculated using (1). By repeatedly applying the same time-shifting technique we can introduce *delays of arbitrary duration*. Accumulation of such delays can gradually increase the phase difference between actual and measured angles, reaching the preprogrammed threshold at which the respective CBs are tripped, leading to sectionalization and cascading failures. Note that in addition to introducing erroneous measurements to PMUs, the same time-shifting technique can be employed to synchronize leading synthetic signals and legitimate GPS signals.

978-1-5386-4757-8/18 $31.00 © 2018 IEEE

C. Budget

The cost of the equipment we utilize for GPS time spoofing mainly consists of the Ettus USRP SDR (and its respective add-on modules) and the Saleae Pro 8 logic analyzer. Their costs are $3529 USD and $699 USD respectively, for a total of $4228 USD. Launching a concurrent attack against k locations to materialize an $N - k$ contingency would require $k\times$ $4228 USD (typically $k = 2$ or $k = 3$ locations are sufficient), which is low given the attack's far-reaching impact.

Our equipment costs are dominated by the Ettus USRP SDR and they can be further reduced by replacing it with cheaper hardware, such as the bladeRF ($420 USD), or HackRF ($295 USD). An inherent limitation of these lower-cost devices is the reduced accuracy of their built-in oscillator, which is not adequate for transmitting GPS signals. However, this problem can be alleviated with OSEXP by leveraging another public infrastructure; *GSM base stations* [27]. As cell towers must be accurate within 0.5 parts-per-million (which is sufficient for GPS transmission), we can initially configure SDRs as GSM receivers. Using GSM signals, we can calculate the internal clock drift of our SDR with reference to the GSM base station clock, and then reconfigure the SDR as a GPS spoofer to carry out the spoofing technique as described above.

VI. Related work

The general requirements for successful GPS spoofing attacks are discussed in [24], and an investigation of whether a GPS spoofing attack against PMUs is theoretically feasible is presented in [22]. To the best of our knowledge, the only unclassified device capable of practical GPS spoofing is custom [28]. Among other applications, it was used to spoof GPS signals and desynchronize PMUs in a controlled demonstration [7]. In comparison to previous works, in this paper we demonstrate that COTS SDR hardware and open source software are sufficient for launching practical attacks against power systems. To thwart GPS spoofing attacks several countermeasures have been proposed in literature, including techniques based on signal processing, encryption, drift monitoring, signal direction of arrival, etc. [29]. However, to the best of our knowledge, such countermeasures are not implemented in commercial GPS receivers, including the COTS components used in power grid IEDs.

VII. Conclusion

In this work we introduce OSEXP, a technique that leverages public infrastructure to materialize attacks against power systems. We experimentally verify a specific OSEXP vector, GPS time spoofing, that can desynchronize phase angle measurements of judiciously selected PMUs to cause wide area blackouts. Our GPS spoofing technique relies on COTS hardware and open source software, enabling reusable low-budget high-impact attacks against power systems. With this study we aim to challenge the perception these attacks are feasible only by resource-wealthy nation state actors, and assist stakeholders and regulators take informed decisions to secure power grids around the world.

References

[1] R. Lobenstein and C. Sulzberger, "Eyewitness to DC history," *IEEE Power and Energy Magazine*, vol. 6, no. 3, pp. 84–90, 2008.

[2] Eaton, "Blackout tracker: United States annual report 2017," 2017.

[3] J. Weiss, "Aurora generator test," in *Handbook of SCADA/Control Systems Security, Second Edition*. CRC Press, 2016, pp. 107–114.

[4] R. M. Lee, M. J. Assante, and T. Conway, "Analysis of the cyber attack on the Ukrainian power grid," SANS ICS, Tech. Rep., 2016.

[5] P. Polityuk, O. Vukmanovic, and S. Jewkes, "Ukraine's power outage was a cyber attack: Ukrenergo," http://www.reuters.com, [Accessed 9-May-2018].

[6] R. D. Steele, "Open source intelligence," *Handbook of intelligence studies*, pp. 129–147, 2007.

[7] D. P. Shepard, T. E. Humphreys, and A. A. Fansler, "Evaluation of the vulnerability of phasor measurement units to GPS spoofing attacks," *International Journal of Critical Infrastructure Protection*, vol. 5, no. 3, pp. 146–153, 2012.

[8] C. Konstantinou, M. Sazos, A. S. Musleh, A. Keliris, A. Al-Durra, and M. Maniatakos, "GPS spoofing effect on phase angle monitoring and control in a real-time digital simulator-based hardware-in-the-loop environment," *IET CPS: Theory & Applications*, 2017.

[9] F. Alonso and C. Greenwell, "Underground vs. Overhead: Power line installation-cost comparison and mitigation," *Electric Light and Power*, vol. 22, 2016.

[10] North American Electric Reliability Council, New Jersey, "NERC Disturbance Reports 1992-2009."

[11] D. Deka, R. Baldick, and S. Vishwanath, "One breaker is enough: Hidden topology attacks on power grids," in *Power & Energy Society General Meeting, 2015 IEEE*. IEEE, 2015, pp. 1–5.

[12] K. Stouffer, J. Falco, and K. Scarfone, "Guide to industrial control systems security," *NIST special publication SP 800-82*, 2011.

[13] U.S. Government, "Official U.S. government information about the Global Positioning System (GPS) and related topics," http://www.gps.gov, [Accessed 9-May-2018].

[14] E. Kaplan and C. Hegarty, *Understanding GPS: Principles and applications*. Artech house, 2005.

[15] U.S. DHS and FBI, "Russian government cyber activity targeting energy and other critical infrastructure sectors," https://www.us-cert.gov/ncas/alerts/TA18-074A, [Accessed 9-May-2018].

[16] C. Davis, A. Chmieliauskas, and I. Nikolic, "Enipedia," *Energy & Industry group, TU Delft*, 2015.

[17] "Open energy information," http://openei.org, [Accessed 9-May-2018].

[18] C. Konstantinou, M. Sazos, and M. Maniatakos, "Attacking the smart grid using public information," in *IEEE Latin-American Test Symposium*, 2016, pp. 105–110.

[19] S. Pajic, "Power system state estimation and contingency constrained optimal power flow: A numerically robust implementation," 2007.

[20] Lockheed Martin, "Cyber Kill Chain," https://www.lockheedmartin.com, 2014, [Accessed 9-May-2018].

[21] E. Blossom, "GNU radio: tools for exploring the radio frequency spectrum," *Linux journal*, vol. 2004, no. 122, p. 4, 2004.

[22] X. Jiang, "Spoofing GPS receiver clock offset of phasor measurement units," Master's thesis, UIUC, 2012.

[23] P. S. R. Committee, "IEEE Standards for synchrophasor measurements for power systems C37.118," *New York, USA*, 2011.

[24] N. O. Tippenhauer, C. Pöpper, K. B. Rasmussen, and S. Capkun, "On the requirements for successful GPS spoofing attacks," in *Proceedings of the 18th ACM conference on Computer and communications security*. ACM, 2011, pp. 75–86.

[25] T. Ebinuma, "Software-Defined GPS signal simulator," https://github.com/osqzss/gps-sdr-sim, [Accessed 9-May-2018].

[26] C. E. Noll, "The Crustal Dynamics data information system: A resource to support scientific analysis using space geodesy," *Advances in Space Research*, vol. 45, no. 12, pp. 1421–1440, 2010.

[27] N. G. Varma, U. Sahu, and G. P. G. Charan, "Robust frequency burst detection algorithm for GSM/GPRS," in *60th IEEE Conference on Vehicular Technology*, vol. 6. IEEE, 2004, pp. 3843–3846.

[28] T. E. Humphreys, B. M. Ledvina, M. L. Psiaki, B. W. O'Hanlon, and P. M. Kintner Jr., "Assessing the spoofing threat: Development of a portable GPS civilian spoofer," in *Proceedings of the ION GNSS International technical meeting of the satellite division*, vol. 55, 2008.

[29] M. L. Psiaki and T. E. Humphreys, "GNSS spoofing and detection," *Proceedings of the IEEE*, vol. 104, no. 6, pp. 1258–1270, 2016.

978-1-5386-4757-8/18 $31.00 © 2018 IEEE

Differential Power Analysis Mitigation Technique Using Three-Independent-Gate Field Effect Transistors

Edouard Giacomin and Pierre-Emmanuel Gaillardon
Electrical and Computer Engineering Department
University of Utah
Salt Lake City, Utah, USA
pierre-emmanuel.gaillardon@utah.edu

Abstract—**Hardware security vulnerabilities are a major concern for embedded computing devices which are now used in many application such as credit cards, SIM cards, or financial systems, putting sensible data at risk. Such systems are often targeted by differential power attacks, where the power trace can be monitored in order to get access to the sensible data. To alleviate this issue, a possible technique proposed in literature is to use a complementary gate (*e.g.*, computing both XOR and XNOR operations in parallel) in order to have a symmetrical power trace for all possible input combinations. However, this technique results in a large area and power overhead since it approximatively requires twice the number of transistors. Recently, novel technologies such as *Three-Independent-Gate Field Effect Transistors* (TIGFETs) have been shown to be able to realize compact logic gates using less transistors when compared to *Complementary Metal Oxide Semiconductor* (CMOS) technology. In this paper, we investigate the benefits of using TIGFETs in terms of hardware security. First, we show that using the complementary gate technique with TIGFETs can reduce the transistor count, the power trace variation, the switching power and leakage by $2\times$, 57%, 36% and $8\times$ respectively, when compared to CMOS. In addition, we show that for the same transistor count and similar switching power, using TIGFETs can reduce the power trace variation and the leakage by 81% and $6.7\times$ respectively when compared to CMOS.**

I. INTRODUCTION

With the fast development of the semiconductor industry and the Internet of Things in the past few years, many critical applications now rely on the use of *Integrated Circuits* (ICs): credit cards, SIM cards, electronic banking, e-commerce, *etc.* Such systems are generally used to deal with important data such as credit card or social security numbers, insider markets information, making them a target for attackers. As a protection, the data is usually encrypted through a circuit which can be subject to hardware security vulnerabilities, threatening the successful deployment of such computing devices. Indeed, circuits may leak some important information, such as timing delay [1], power [2] or electromagnetic radiation [3], making them vulnerable to external attacks.

Attacks using information leaking from the cryptographic circuits are known as *Side Channel Attacks* (SCAs) and were first introduced by Kocher in 1996 [1]. In particular, *Differential Power Analysis* (DPA) [2] exploits the correlation between

the power trace of the circuit and the cipher encryption technique. This is based on the fact that logic gates such as XOR, which are widely used in cryptographic algorithms [4], have a power trace that strongly depends on the input data [2]. As a result, such attacks have since been shown to be successful at extracting the cryptographic key of several algorithms running on different kind of digital circuits [5], [6], making hardware security a major concern for modern ICs. To this end, some work proposed to use complementary output for logic gates (*e.g.*, computing both XOR/XNOR or OR/NOR operations in parallel) [7]–[9] in order to make the power consumption more independent from the input values. However, it requires many more transistors, considerably increasing the area footprint and the overall power consumption of the circuit.

Recently, novel technologies such as *Three-Independent-Gate Field Effect Transistors* (TIGFETs) [10] have been proposed as an alternative to FinFETs. By having three independent gate terminals, TIGFETs present richer switching capabilities for a given transistor. They have been shown to be able to realize compact logic gates, such as XOR or XNOR that only require 4 transistors instead of 8 for conventional CMOS gates [10]. As a result, when using the complementary gate technique to reduce the power trace variation against DPAs, the area and power consumption overhead can be greatly reduced with TIGFET devices. In this paper, we investigate the benefits of using TIGFET technology in term of hardware security with the complementary gate technique using XOR and XNOR gates. We evaluate the benefits of our circuits at the 22nm technology node by comparing *Silicon NanoWire TIGFETs* (TIG SiNWFETs) with CMOS FinFET *Low-STandby Power* (LSTP) structures [11]. The contributions of this paper are:

- We propose a new organization for the gate inputs of the TIGFET XOR cell, leading to a reduction of the input current gate variation by $1.87\times$ when compared to the conventional design.
- When compared to CMOS, using the proposed TIGFET XOR and XNOR gates can reduce the transistor count, the power trace variation, the switching power and the leakage by $2\times$, 57%, 26% and $8\times$ respectively.

978-1-5386-4757-8/18 $31.00 © 2018 IEEE

- We show that for the same number of transistors and a similar switching power, using TIGFETs decreases the power trace variation and the leakage by 81% and 6.7× respectively, when compared to CMOS.

The rest of this paper is organized as follows: In Section II, we provide the technical background about TIGFET technology and review some side channel attack protection techniques. Section III presents the proposed TIGFET-based design. Section IV shows experimental results. Section V concludes this paper.

II. TECHNICAL BACKGROUND

In this section, we first present the necessary background on TIGFET technology and review some generalities about side channel attacks and possible mitigation techniques.

A. Operations of TIGFET

Fig. 1 shows a scanning electron microscopy view of a fabricated TIGFET device [12] with its three independent gate contacts: the *Control Gate* (CG) which controls the potential barrier in the channel and the *Polarity Gate at Source* (PGS) and the *Polarity Gate at Drain* (PGD) which modulate the Schottky barriers at source and drain. Depending on the three gate voltage biasings, a TIGFET can be configured in 4 modes: low-V_T n-type, high-V_T n-type, low-V_T p-type and high-V_T p-type. In addition to polarity control, a unique feature of TIGFET technology is their ability to act as two series n-type or p-type transistors by biasing one input and controlling the two others. Therefore, by using TIGFET, more compact logic gates can be built. More details on this feature of TIGFETs can be found in [10].

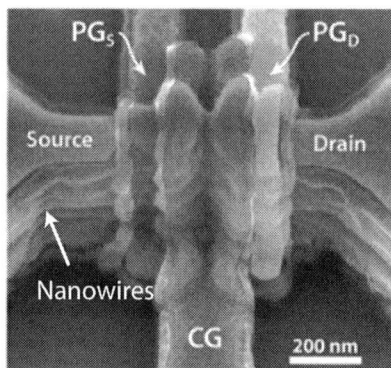

Fig. 1: Scanning Electron Microscopy image of a fabricated TIGFET device using four vertically stacked silicon nanowires [12]. The three gates are surrounding the silicon nanowires in a gate-all-around fashion.

TIGFET devices have been successfully fabricated with several channel technologies such as FinFET [13], 2D [14] or SiNWFET [10]. In this paper, we will consider the latter since it provides better electrostatic control over the channel and better scalability properties than FinFETs [15] while being fully compatible with CMOS. Being based on Schottky barriers, TIGFET devices exhibit very low leakage compared to their

CMOS counterpart [10], as it will be demonstrated in Section IV. In the considered TIGFET structure, the nanowires have a diameter d of 15nm while the length of the gates are both 24nm long. The dielectric layer is HfO_2 with a thickness of 5.1nm and an equivalent oxide thickness of 0.8nm. These materials were selected to ensure full compatibility with standard CMOS processes. The dual-V_T I-V curves of a single nanowire TIGFET, simulated with TCAD Sentaurus [16], are shown in Fig. 2. The solid lines are the low-V_T configurations and the dashed lines are the high-V_T configuration. The extracted on-current is about $33.5 \mu A$. While the current drive is lower than CMOS ($51.5 \mu A$), TIGFET devices unlock new circuits opportunities and bring very competitive advantages in terms of standby energy or energy-delay product when compared to CMOS [17]. In addition, even though the required V_{DD} for the considered TIGFET technology ($1.2V$) is higher than the nominal voltage of the 22nm FinFET technology node ($0.9V$), TIGFET can still achieve a significant power reduction, as it will be demonstrated in Section IV.

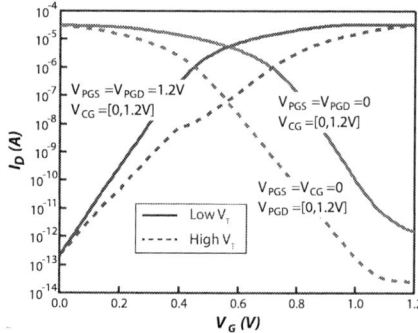

Fig. 2: Simulated I-V curve of a TIG SiNWFET for $V_{DS} = 1.2V$ in logarithmic scale [10].

B. TIGFET Based Compact Logic Gates

Due to their richer switching capabilities, TIGFETs can be used to design compact logic gates. For instance, a TIGFET-based XOR can be realized with only 4 transistors [10], as shown in Fig. 3 (a), while a CMOS XOR requires 8 transistors. In the same manner, other compact logic gates such as tristate inverter [18], used to build multiplexers or majority gates, proven to be very promising operators [19], can be realized using TIGFETs, as depicted in Fig. 3 (b) and (c) respectively.

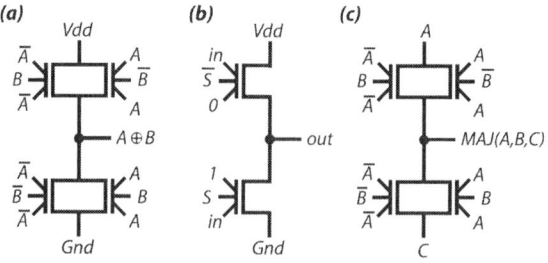

Fig. 3: TIGFET compact logic gates: (a) 2-input XOR; (b) tristate inverter; (c) 3-input majority gate.

978-1-5386-4757-8/18 $31.00 © 2018 IEEE   108

C. Side Channel Attacks and Mitigation Techniques

To protect the data at the hardware level against DPAs, several works proposed [20], [21] to randomize the correlation between the power trace and the input combination. While such kind of technique was proven to be efficient against DPAs, later works showed that it was still at threat against template attacks [22], [23]. Another possible technique is to reduce the dependency between the power trace and the input combination, considerably increasing the number of trace measurements required to get the private key and thus the cost required for an external attack to be successful [24]. The first technique of this kind was proposed by Tiri *et al.* [7] where they showed that by having a constant output load capacitance, the power trace was more independent of the input values. To do so, they proposed a sense amplifier based logic where each gate was redesigned to produce the regular output (such as AND or OR) as well as the complementary output (such as NAND or NOR). While the power trace variation was greatly reduced compared to conventional CMOS logic gates, this techniques came at the price of twice the area of the power, making it difficult to expand for large circuits. On the other hand, using TIGFETs with this kind of technique can reduce the overall area overhead as well as the power consumption since they require less transistors to build logic gates when compared to CMOS, as explained in Section II-B.

III. TIGFET-BASED POWER TRACE VARIATION MITIGATION TECHNIQUE

The origin of the power trace variation in conventional logic gates is due to two factors: (i) the input and output loads of the gate seen for each transition; (ii) the current characteristic of *n*-type or *p*-type transistors. Symmetrizing both current and capacitances ensure that the same energy is consumed/dissipated for each transition and hence leads to the same power trace. In this section, we first show how to rearrange the gate inputs for the TIGFET XOR in order to symmetrize the input loads. Without loss of generality, we focus in this paper on the XOR operator since it is widely used in cryptography for its mathematical properties (such as the ability to realize an addition modulo 2) [4]. Then, we present how to reduce the output power trace with the complementary gate technique by using TIGFET XOR and XNOR gates.

A. Symmetrical TIGFET XOR Design

As for the CMOS structure, the conventional TIGFET XOR [10], depicted in Fig. 4 (a), presents different power traces coming from the different input loads on the polarity and control gates. To mitigate this effect, the inputs can be rearranged in a certain way in order to symmetrize the input gate load, without changing the XOR functionality, as depicted in Fig. 4 (b). With this configuration, each input (A, B, $\overline{A}$ and $\overline{B}$) sees one control gate and two polarity gates so the same equivalent input gate capacitance. We will see in the experimental section that this technique leads to a lower input gate power variation when compared to the conventional design.

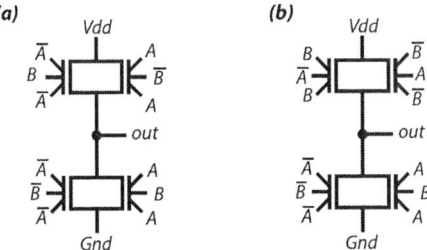

Fig. 4: TIGFET 2-input XOR gate: (a) conventional [10]; (b) symmetrical.

B. Complementary TIGFET XOR/XNOR Circuit

In Section II-C, we reported that using complementary gates (AND and NAND or XOR and XNOR) was a possible mitigation scheme against side channel attack by balancing the output load between the different possible input combinations. Fig. 5 (a) shows CMOS XOR and XNOR gates. By having complementary outputs, if A is switching from 0 to 1, one output will switch from 0 to 1 and the other one from 1 to 0. As a result, one gate is loading while the other is discharging, so the power output trace is less dependent on the input values, making it harder for external attackers to find a correlation with the private key. While it helps reducing

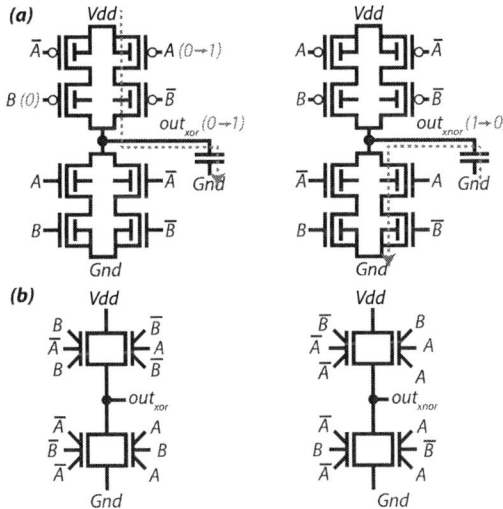

Fig. 5: (a) CMOS XOR and XNOR cells; (b) TIGFET symmetrized XOR and XNOR gates.

the power trace variation, it comes roughly at twice the cost in term of area and power consumption [7]. Thanks to their ability to realize compact logic gate as reviewed in Section II-B, TIGFET-based logic gates can take advantage of this technique without leading to a large area and power overhead. As shown in Fig 5 (a) and (b), CMOS XOR and XNOR cells require a total of 16 transistors while TIGFET XOR and XNOR gates only requires 8 transistors. Therefore, by having twice less transistors, TIGFET XOR and XNOR gates are expected to reduce the power consumption when compared to CMOS XOR and XNOR gates since the datapath goes

978-1-5386-4757-8/18 $31.00 © 2018 IEEE   109

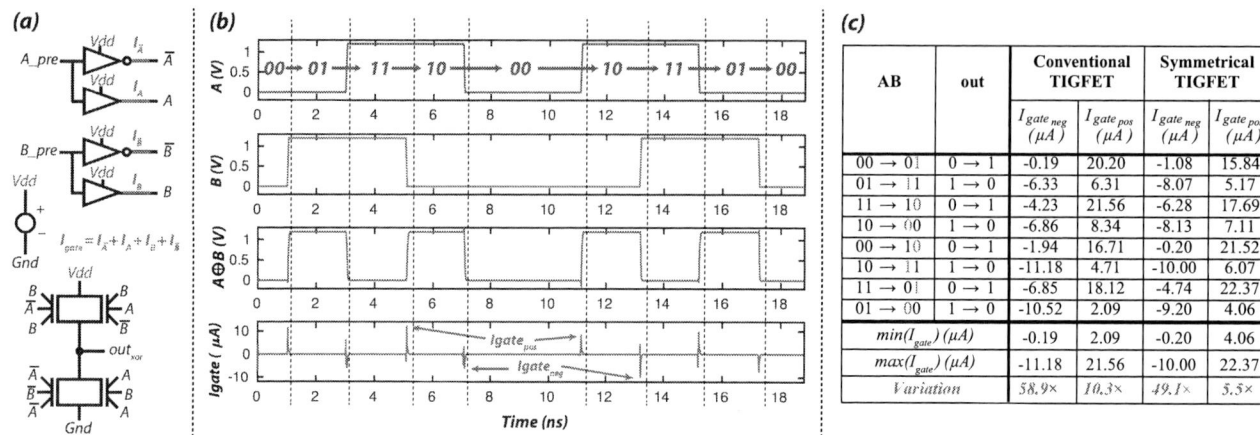

Fig. 7: (a) I_{gate} current evaluation methodology; (b) Transient curves of a TIGFET-based XOR for different input combinations and I_{gate} evaluation methodology; (c) I_{gate} current values comparison between conventional and symmetrical TIGFET XORs, for different input combinations.

AB	out	Conventional TIGFET		Symmetrical TIGFET	
		$I_{gate_{neg}}$ (μA)	$I_{gate_{pos}}$ (μA)	$I_{gate_{neg}}$ (μA)	$I_{gate_{pos}}$ (μA)
$00 \rightarrow 01$	$0 \rightarrow 1$	-0.19	20.20	-1.08	15.84
$01 \rightarrow 11$	$1 \rightarrow 0$	-6.33	6.31	-8.07	5.17
$11 \rightarrow 10$	$0 \rightarrow 1$	-4.23	21.56	-6.28	17.69
$10 \rightarrow 00$	$1 \rightarrow 0$	-6.86	8.34	-8.13	7.11
$00 \rightarrow 10$	$0 \rightarrow 1$	-1.94	16.71	-0.20	21.52
$10 \rightarrow 11$	$1 \rightarrow 0$	-11.18	4.71	-10.00	6.07
$11 \rightarrow 01$	$0 \rightarrow 1$	-6.85	18.12	-4.74	22.37
$01 \rightarrow 00$	$1 \rightarrow 0$	-10.52	2.09	-9.20	4.06
$min(I_{gate})$ (μA)		-0.19	2.09	-0.20	4.06
$max(I_{gate})$ (μA)		-11.18	21.56	-10.00	22.37
Variation		*58.9×*	*10.3×*	*49.1×*	*5.5×*

through less diffusion capacitances. In addition, a standalone CMOS XOR gate requires the same number of transistors (8) than a TIGFET using the complementary gate technique (XOR and XNOR gates combined). As a result, for the same number of transistor, TIGFET can reduce the output power trace variation, leading to more robust designs against side channel attacks. Both of those predictions will be verified in Section IV.

IV. EXPERIMENTAL RESULTS

In this section, we demonstrate the benefits of using TIGFETs to realize robust logic gates against DPAs. We first introduce our experimental methodology and then show the benefits of the proposed symmetrical TIGFET design. Then, we compare the power trace as well as the switching and leakage power of TIGFET and CMOS XOR and XNOR gates.

A. Experimental Methodology

For circuit-level simulations, the VerilogA table model from [10] has been used for the TIGFET model. Its equivalent RC circuit of is depicted in Fig. 6. The current values were extracted from TCAD Sentaurus [16] simulations under the assumptions discussed in Section II-A. Capacitance values have been extracted from TCAD simulations as an average value under all possible bias configurations to model intrinsic capacitances of the device. The supply voltage used for the TIG SiNWFETs is 1.2V. The *Predictive Technology Model* (PTM) 20nm-FinFET *Low-STandby-Power* (LSTP) [11] is used in the circuits of the CMOS logic gates. They have a nominal supply voltage $V_{DD} = 0.9V$. Current and power values are extracted from HSPICE simulations [25].

B. Conventional and Symmetrical TIGFET XOR Cells Comparison

As explained in Section III-A, TIGFET XOR inputs can be rearranged in order to symmetrize the gate loads without changing the XOR functionality. In that way, each input sees

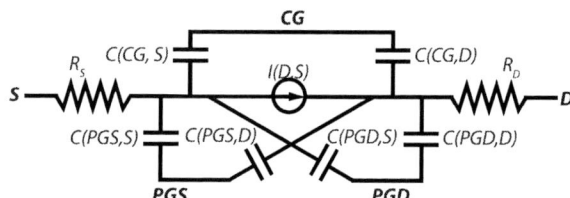

Fig. 6: Equivalent RC circuit of a SiNWFET.

the same equivalent input capacitance, reducing the overall power trace variation. To evaluate the benefits of this re-arrangement, we evaluated the gate current values for each possible input combinations. To do so, we monitored the total output current coming from the input inverters, which is also the input gate currents, as shown in Fig. 7 (a). A_{pre} and B_{pre} denotes signals A and B before conditioning. As illustrated in Fig. 7 (b), there are some "positive" (denoted as $I_{gate_{pos}}$) and "negative" (denoted as $I_{gate_{neg}}$) current peaks depending on the input transition. In order to study the current gate variation for the conventional and symmetrical TIGFET XOR gates, we report the minimum and maximum input currents for both $I_{gate_{pos}}$ and $I_{gate_{neg}}$ for each transition in Fig. 7 (c). As predicted, the symmetrical design can reduce the input gate currents, up to $1.87\times$ compared to the conventional implementation. In the rest of this paper, we will only consider symmetrical TIGFET XOR and XNOR cells.

C. CMOS and TIGFET Complementary Gates (XOR and XNOR) Comparison

To evaluate the power trace of both CMOS and TIGFET circuits, we evaluate the minimum peak current I_{supply} for every possible transitions, as shown in Fig. 8 (a) on the orange curve. From this, we compute the minimum and maximum I_{supply} peak values ($I_{supply_{min}}$ and $I_{supply_{max}}$ respectively) and derive the I_{supply} variation for both architectures. The supply voltage being constant, monitoring I_{supply} peak values is the same as monitoring the total power trace values. For

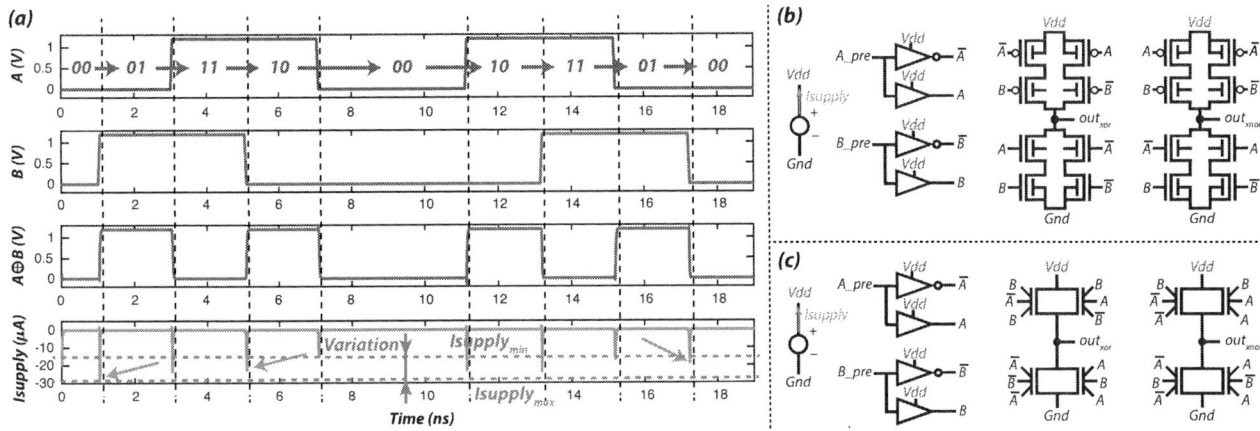

Fig. 8: (a) Transient curves of CMOS XOR/XNOR for different input combinations and I_{supply} evaluation methodology; Simulation setup schematics of: (b) CMOS XOR/XNOR and (c) TIGFET XOR/XNOR.

the rest of this paper, we refer to current trace variation by power trace variation. Here, we first compare I_{supply} peak values for both CMOS and TIGFET circuits using the complementary gate technique with XOR and XNOR cells, whose simulation schematics are shown in Fig. 8 (b) and (c) respectively. For the rest of this paper, we denote XOR and XNOR gates by XOR/XNOR. As shown in Table I, the maximum power variation for TIGFET is 12.5% lower than CMOS, making TIGFET-based design more robust against DPAs. Note that the power trace variation could be further improved against CMOS, by using an engineered TIGFET device with a high level of symmetry between *n*-type and *p*-type configurations, as presented in [26]. In addition, the TIGFET circuit only requires 8 transistors while the CMOS circuit requires 16, decreasing the overall area and reducing the power consumption, as it will be showed in Section IV-E.

D. TIGFET Complementary Gates (XOR and XNOR) Comparison with Standalone CMOS XOR Gate

As anticipated in Section II-A, due to their richer switching capabilities, TIGFET can realize smaller logic gates in term

TABLE I: I_{supply} current values comparison between CMOS XOR/XNOR and symmetrical TIGFET XOR/XNOR, for different input combinations.

AB	out	*Isupply* CMOS (μA)	*Isupply* TIGFET (μA)
$00 \rightarrow 01^{(1)}$	$0 \rightarrow 1$	-76.361	-26.571
$01 \rightarrow 11$	$1 \rightarrow 0$	-76.392	-29.047
$11 \rightarrow 10$	$0 \rightarrow 1$	-92.779	-28.400
$10 \rightarrow 00$	$1 \rightarrow 0$	-92.957	-29.703
$00 \rightarrow 10$	$0 \rightarrow 1$	-72.760	-28.881
$10 \rightarrow 11$	$1 \rightarrow 0$	-82.231	-26.605
$11 \rightarrow 01$	$0 \rightarrow 1$	-92.670	-29.861
$01 \rightarrow 00$	$1 \rightarrow 0$	-92.698	-28.275
$min(I_{supply})(\mu A)$		-92.957	-29.861
$max(I_{supply})(\mu A)$		-72.760	-26.871
$Variation$ (%)		28 %	12 %

(1) The red logical value denotes which input is switching.

of transistor count when compared to CMOS. As we saw, the technique of using a complementary gate to reduce the power trace variation can be applied with TIGFETs, without compromising the overall area. More particularly, the TIGFET XOR and XNOR cells require the same number of transistors (8) as a standard CMOS XOR. Fig. 9 (a) shows the power trace values for different input combinations for the TIGFET XOR/XNOR gates and the CMOS XOR gate. While the CMOS XOR has a maximum power trace variation of 64%, the symmetrized TIGFET XOR/XNOR circuit only presents a 12% variation, leading to a power trace variation reduction of 81%, as shown in 9 (b). Note that this reduction is realized with the same number of transistors between both structures.

E. Summary and Power Consumption Evaluation

In this part, we summarize the power trace variation for 4 architectures: CMOS XOR, CMOS XOR/XNOR, TIGFET XOR and TIGFET XOR/XNOR. As explained before, the power trace variation denotes the maximum variation on the power trace for every possible input combinations. In addition, for each architecture, we evaluated the switching and leakage power (as the average value of all switching and leakage power values for each input combinations respectively), as well as the transistor count, as shown in Table II. By using the complementary gate technique with both XOR/XNOR, TIGFET circuits can reduce the power trace variation by 57%, the transistor count by 2×, the switching power by 26% and the leakage by 8× against CMOS XOR/XNOR. In addition, when compared to a standalone CMOS XOR, using the TIGFET XOR/XNOR circuit can reduce the power trace variation by 81% and the leakage by 6.7× for the same transistor count and switching power.

V. CONCLUSION

In this paper, we investigated the benefits of using TIGFET technology in term of hardware security. We showed that TIGFET can reduce the power trace variation when compared to CMOS, making TIGFET-based designs more robust against

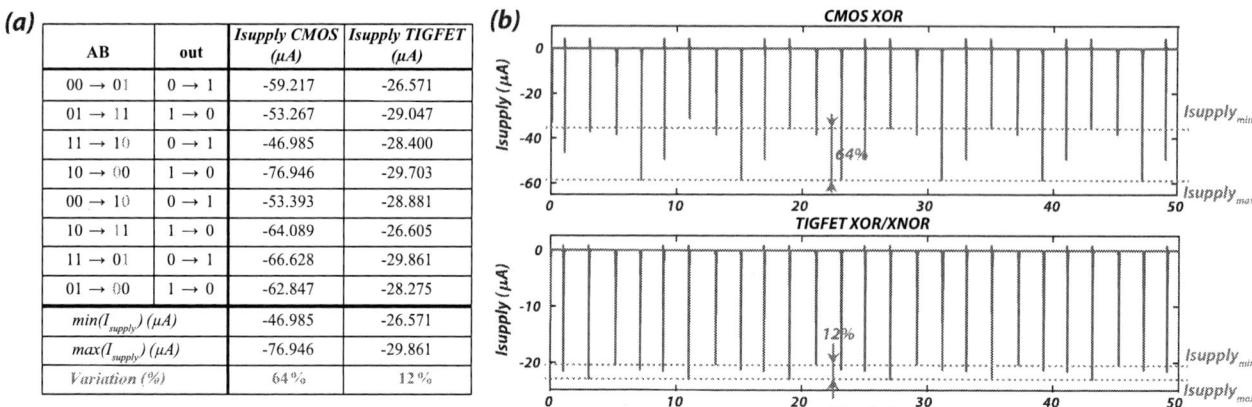

Fig. 9: (a) I_{supply} current values and (b) I_{supply} transient curves comparison between CMOS XOR and symmetrical TIGFET XOR/XNOR, for different input combinations.

TABLE II : Switching power, power trace variation, transistor count and leakage for the different circuits.

Design	Switching Power (μW)	Power Trace Variation (%)	Transistor Count	Leakage (nW)
CMOS XOR	9.192	64	8	5.738
CMOS XOR/XNOR	12.611	28	16	6.918
TIGFET XOR	6.604	75	4	0.594
TIGFET XOR/XNOR	9.299	12	8	0.854

DPAs attacks. In particular, we showed that when using the complementary gate technique with XOR and XNOR gates, TIGFET circuits can reduce the power trace variation by 57%, the transistor count by $2\times$, the switching power by 26% and the leakage by $8\times$ against CMOS. When compared to a standalone CMOS XOR, using the TIGFET XOR/XNOR circuit can reduce the power trace variation by up to 81% and the leakage power by $6.7\times$ for the same transistor count and a similar switching power. This work shows that emerging technology such as TIGFET can bring novel advantages beyond traditional power and area performances when compared to regular CMOS, here as improved hardware security features.

ACKNOWLEDGMENTS

This work was supported by the NSF Career Award number 1751064, the IMEC core partners' CMOS program and the SRC Contract 2018-IN-2834. The authors would like to acknowledge Michael Niemer and Sharon Hu from The University of Notre Dame for the fruitful discussion.

REFERENCES

[1] P-C. Kocher, *Timing Attacks on Implementations of Diffie-Hellman, RSA, DSS, and Other Systems*, CRYPTO, 1996.
[2] P-C. Kocher *et al.*, *Differential power analysis*, CRYPTO, 1999.
[3] J. Quisquater *et al.*, *ElectroMagnetic analysis (EMA): Measures and counter-measures for smart cards*, E-smart, 2001.
[4] G. Oded, *Foundations of Cryptography: Volume 2, Basic Applications*, Cambridge University Press New York, 2009.

[5] F.-X. Standaert *et al.*, *Differential Power Analysis of FPGAs : How Practical is the Attack?*, FPL, 2003.
[6] S. B. Ors *et al.*, *Power- Analysis Attack on an ASIC AES implementation*, ITCC, 2004.
[7] K. Tiri *et al.*, *A Dynamic and Differential CMOS Logic with Signal Independent Power Consumption to Withstand Differential Power Analysis on Smart Cards*, ESSCIRC 2002.
[8] M. Bucci *et al.*, *Three-phase dual- rail pre-charge logic*, CHES, 2006.
[9] J. Quan *et al.*, *A New Method to Reduce the Side-Channel Leakage Caused by Unbalanced Capacitances of Differential Interconnections in Dual-Rail Logic Styles*, ITNG, 2009.
[10] J. Zhang *et al.*, *Configurable Circuits Featuring Dual-Threshold-Voltage Design With Three-Independent-Gate Silicon Nanowire FETs*, IEEE TCAS I, 61(10): 2851-2861, 2014.
[11] Predictive technology model. [Online]. Available at: http://ptm.asu.edu/
[12] J. Zhang *et al.*, *Polarity-Controllable Silicon Nanowire Transistors with Dual Threshold Voltages*, IEEE TED, 61(11): 3654-3660, 2014.
[13] J. Zhang *et al.*, *A Schottky-Barrier Silicon FinFET with 6.0mV/dec Subthreshold Slope over 5 Decades of Current*, IEDM, 2014.
[14] G. V. Resta *et al.*, *Polarity Control in WSe_2 Double-Gate Transistors*, Scientific Reports, 6, 2016.
[15] W. Lu *et al.*, *Nanowire Transistor Performance Limits and Applications*, IEEE TED, 55(11): 2859-2876, 2008.
[16] *Sentaurus Device User Guide, Version N-2017.09*, Synopsys Inc.
[17] J. Romero-Gonzalez *et al.*, *BCB Evaluation of High-Performance and Low-Leakage Three-Independent-Gate Field Effect Transistors*, IEEE JESSCDC, 4(1): 1-9, 2018.
[18] E. Giacomin *et al.*, *Low-Power Multiplexer Designs Using Three-Independent-Gate Field Effect Transistors* NanoArch, 2017.
[19] L. Amarú *et al.*, *IMajority-Inverter Graph: A New Paradigm for Logic Optimization*, IEEE TCAD ICS, 35(5): 806-819, 2016.
[20] L. Goubin *et al.*, *DES and Differential Power Analysis – The "Duplication" Method*, CHES, 1999.
[21] D. Suzuki *et al.*, *Random Switching Logic: A Countermeasure against DPA based on Transition Probability*, IACR Cryptology, 2004.
[22] P. Schaumont *et al.*, *Masking and Dual-Rail Logic don't add up*, LNCS, vol. 4727, pp. 95-106, 2007.
[23] E. Oswald *et al.*, *Template Attacks on Masking - Resistance Is Futile*, CT-RSA, 2007.
[24] P. Kocher *et al.*, *Introduction to differential power analysis*, J. Cryptogr. Eng., 1(1): 5–27, 2011.
[25] *HSPICE User Guide: Basic Simulation and Analysis, Version M-2017.3*, Synopsys Inc.
[26] D.-Y. Jeon *et al.*, *Operation regimes and electrical transport of steep slope Schottky Si-FinFETs*, Journal of Applied Physics, 121(6): 064504-1-064504-7, 2017.

Energy-Driven Precision Scaling for Fixed-Point ConvNets

Valentino Peluso, Andrea Calimera

Politecnico di Torino, 10129 Torino, Italy

{valentino.peluso, andrea.calimera}@polito.it

Abstract—**Data precision scaling is a well-known technique for power/energy minimization in error-resilient applications. It has proven particularly suited for embedded Convolutional Neural Networks (ConvNets) made run on fixed-point arithmetic co-processors. The key observation is that methods that only account for accuracy during the precision assignment process may lead to sub-optimal energy minimization. This work introduces an energy-driven optimization that delivers per-layer quantization under a user-defined accuracy constraint. The tool is conceived for accelerators that dynamically adapt their energy and accuracy through software-programmable multiprecision Multiply&Accumulate (MAC) units. Simulation results collected on different ConvNets trained with public data-set show substantial energy savings and improved energy-accuracy tradeoffs w.r.t. conventional fixed-point methods.**

I. INTRODUCTION

Embedded Convolutional Neural Networks (ConvNets hereafter) represent an attractive solution for near-sensor data-analytics in the IoT. The big challenge is to reduce their complexity thus to reach an energy/throughput ratio that meets the constraints of low-power end-nodes. The research community is attacking the problem from opposite directions. From the bottom, with the design of custom ICs [1], [2] which show an energy consumption of few pico-Joules/operation. From the top, with new learning strategies [3]. Lots of room is available in the middle, during software-to-silicon mapping. That's the role of design automation, which contributes with cross-layer optimizations [4]. That's the target of this work.

The iterative error compensation strategy used during the training of ConvNets (e.g. Gradient Descent) may lead to highly redundant models; such a redundancy reflects into ConvNets with high resilience to errors, and hence, large margins for simplification [5]. ConvNets compression through weights pruning, weight sharing, and data precision scaling are common strategies [6] indeed. This work deals with precision scaling.

The use of fixed-point arithmetic (INT) instead of floating-point (FP) is a promising approach. Integer arithmetic units take less silicon area, consume less power, are faster and they also give the flexibility of a wide precision range. ConvNets implemented with 32-, 16-, 8-bit [7], and even below, 6, 4-bit [2], till the limit of a single bit [8], achieve remarkable results. As it will be discussed later in the text, 16-bit INT gives accuracies that are comparable to that of FP models.

Previous works on optimal precision scaling are mainly accuracy-driven, while how to achieve the minimum energy fixed-point representation still is an open issue. A key point is that maximizing the number of layers at low-precision does not necessarily minimize the total energy. That's intuitive as layers show different size and complexity. A recent paper dealing with neuron pruning [6] confirms this trend. Moreover, precision scaling has been originally introduced as a static optimization, that is, precision is defined at design time and then kept constant at run-time. However, the availability of adaptive strategies that reduce the Quality-of-Results to save energy consumption is paramount in many portable applications [9]. Dynamic precision scaling might represent a practical knob to control the energy-accuracy tradeoff of ConvNets at run-time [10]. This work elaborates this concept and it introduces a cross-layer precision scaling strategy where energy consumption is brought in the optimization loop as a direct variable. The proposed solution is conceived for software-programmable ConvNet accelerators where precision scaling is implemented by means of multiprecision/variable-latency Multiply&Accumulate (MAC) instructions; this is compliant with emerging industrial platforms (e.g. Google TPU [11]) which provide users with FP, 16-bit INT and 8-bit INT ISA. We describe both the hardware implementation (Section III) and the precision mapping tool (Section IV). The latter consists of an optimization engine that takes as input the layout of the ConvNet and it searches for the most suited data precision of each layer (both weights and activations) s.t. (i) the overall energy consumption is minimized and (ii) the accuracy constraint is met. Different precision settings can be used at run-time, depending on the actual context, thus to achieve the best accuracy-energy tradeoff. As a key feature, the proposed strategy makes use of a unique set of weights whose arithmetic precision can be programmed through software flags. Experimental results (Section V) conducted on three ConvNets trained over well-recognized data-set, i.e. SVHN, CIFAR-10 and CIFAR-100, validate the proposed strategy and quantify the energy savings (52.1% best-case).

II. TAXONOMY AND CLASSIFICATION

A survey on the optimization techniques for ConvNets can be found in [6]. The objective of this section is to provide a taxonomy of precision scaling techniques for fixed-point accelerators.

Complexity reduction through precision scaling exploits the characteristics of the weight distributions in order to find the best data representation with the minimum loss of accuracy [12], [13]. Precision scaling encompasses two main stages: *bitwidth selection* and *radix-point shifting* [14]. The former aims at defining the minimum number of bits (for both input maps and weights), the latter finds the position of the radix point that minimizes the impact of quantization.

978-1-5386-4757-8/18 $31.00 © 2018 IEEE

Precision scaling can be applied at different levels of *spatial* and *temporal* granularity. In terms of spatial granularity two options are available: *per-net* or *per-layer*. In the former case all the layers share the same representation; in the latter case, each layer has its own representation. Since the weights distribution may substantially differ from layer to layer, a finer, i.e. per-layer, strategy achieves lower accuracy loss [12]. Also in terms of temporal granularity, the options are two: *design-time* or *run-time*. While most of previous works deal with design-time optimizations [14], [12], recent trends for embedded systems propose a precision scaling mechanism that leverages some knob, e.g. Dynamic Voltage Frequency Scaling (DVFS) [15], to reduce energy at run-time. However, DVFS needs layout customization not always applicable (e.g. in FPGAs).

Whatever the level of granularity applied, existing works, mainly from the machine-learning community, e.g. [12], [14], focused on accuracy-driven optimal precision scaling. Only a few papers from the embedded system community take hardware resources into account, e.g. [13] and [7]. The authors of [13] briefly describe a greedy approach where low precision is assigned starting from the first layer of the net (topological order) without considering the complexity of the layer. In [7] authors describe the design of embedded ConvNets for FPGAs and propose a per-layer precision scaling that is aware of the number of memory accesses.

A common practice when dealing with precision scaling in ConvNets is to apply a retraining stage. Even thought retraining allows a partial recovery of accuracy loss, the resulting network and so its weights are shaped for that specific accuracy. This imposes the storage of multiple sets of weights, potentially one for each level of accuracy. By contrast, a solution where precision scaling is obtained through multiprecision instructions run over the *same* set of weights (as proposed in this work) gives higher flexibility.

Following the above taxonomy, our strategy can be classified as a retraining-free, per-layer precision scaling (both bitwidth selection and radix point shifting) for energy-accuracy tradeoff at run-time; the optimization loop is aware of the underlying hardware, hence energy-driven.

III. PRECISION SCALABLE ARITHMETIC

A. Software: dot-Product Algorithm

Given two $N \times N$ matrices (e.g., the input map $\underline{\mathbf{I}}$ and the weight $\underline{\mathbf{W}}$), their convolution is the dot-product r of the two unrolled vectors I and W (weight) of length $M = N \times N$:

$$r = \sum_{i=0}^{M-1} I_i \times W_i \qquad (1)$$

Assuming each item in $\underline{\mathbf{I}}$ and $\underline{\mathbf{W}}$ has a bitwidth $N = 2^i$ ($i = 4$ in this work), and given $K = N/2$ the bitwidth of the most significant part, the scalar product $I_i \times W_i$ can be decomposed as iteration of multiply, shift and accumulate operations between the most significant parts ($I_i^{\mathrm{H}}, W_i^{\mathrm{H}}$) and the least significant parts ($I_i^{\mathrm{L}}, W_i^{\mathrm{L}}$) [16]. Algorithm 1 describes the iterative dot-product multiplication; to notice that I_i^{L} and W_i^{L} are taken as unsigned integers, whereas I_i^{H} and W_i^{H} are

---

Algorithm 1: Iterative multiply-accumulate algorithm

Input: $I^{\mathrm{H}}, I^{\mathrm{L}}, W^{\mathrm{H}}, W^{\mathrm{L}}$, *precision*
Output: Dot-Product r

1 **for** $i = 0;\ i < M;\ i = i + 1$ **do**
2 $\quad$ $r = r + I_i^{\mathrm{H}} \times W_i^{\mathrm{H}}$
3 $r = r \ll K$
4 **if** (*precision==half*) **then return** r ;   `// half:KxK`
5 **for** $i = 0;\ i < M;\ i = i + 1$ **do**
6 $\quad$ $r = r + I_i^{\mathrm{H}} \times W_i^{\mathrm{L}}$
7 **if** (*precision==mid*) **then return** r ;   `// mid:NxK`
8 **for** $i = 0;\ i < M;\ i = i + 1$ **do**
9 $\quad$ $r = r + I_i^{\mathrm{L}} \times W_i^{\mathrm{H}}$
10 $r = r \ll K$
11 **for** $i = 0;\ i < M;\ i = i + 1$ **do**
12 $\quad$ $r = r + I_i^{\mathrm{L}} \times W_i^{\mathrm{L}}$
13 **return** r ;   `// full:NxN`

---

signed. This straightforward algorithm offers a simple way to adjust quality of results and computational resources.

The precision options are: *half* ($K \times K$, i.e. 8×8 in this work), *mid* ($N \times K$, i.e. 16×8), *full* ($N \times N$, i.e. 16×16). All of them can be implemented on the same $K \times K$ MAC unit (more details in the next subsection). We assume the availability of memories that support both word (N-bit) and half-word (K-bit) accesses [17].

The advantage of this approach is twofold: (i) enable the implementation of precision scaling through a lightweight $K \times K$ variable-latency MAC unit; (ii) reduce the number of memory access according to the desired precision. As a result, both computational (due to (i)) and memory (due to (ii)) energy consumptions get smaller as precision reduces.

B. Hardware: 8×8 MAC Unit

The RTL structural view of our 8×8 ($K \times K$) MAC unit is reported in Figure 1. The kernel is a 9×9 multiplier, where the 9th bit is used for the sign extension of the operands. Indeed, as described in the previous subsection, the most significant parts ($I_i^{\mathrm{H}}, W_i^{\mathrm{H}}$) are signed, while the least significant parts ($I_i^{\mathrm{L}}, W_i^{\mathrm{L}}$) are unsigned. Therefore, the MSB of ($I_i^{\mathrm{L}}, W_i^{\mathrm{L}}$) belongs to the module, while that of ($I_i^{\mathrm{H}}, W_i^{\mathrm{H}}$) is the sign. In order to account for this issue we implemented the following mechanism: when ($I_i^{\mathrm{H}}, W_i^{\mathrm{H}}$) are processed, the sign is extended to the 9th by concatenating the MSB (i.e. the sign) of I and W; when ($I_i^{\mathrm{L}}, W_i^{\mathrm{L}}$) are processed a 0 is concatenated. The selection is done through the control signals *signed-I* and *signed-W* driven by the local control unit (omitted in the picture for the sake of space). The same control unit is in charge of feeding the MAC with the right sequence of data (H or L) fetched from a local memory.

The accumulator has 16 guard bits and an embedded saturation logic to handle underflow and overflow. The role of the programmable shifter is two-fold. First, to shift the partial results when needed (see Algorithm 1). Second, to implement the dynamic fixed point arithmetic by moving the radix point of the final accumulation result depending on the desired fractional length [18]. A *range check* logic drives bit saturation if the result does not fit the word-length.

In order to minimize the dynamic power consumption, the *zero-skipping* strategy [15] is implemented by means of

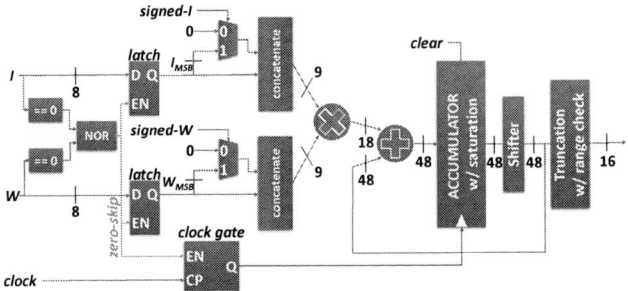

Figure 1: 8×8 HW unit for multi-precision MAC.

Table I: Energy/MAC vs Precision

# bits ($I \times W$)	N_{cycles}	E_{MAC} (pJ)
16×16	4	3.80
16×8	2	1.90
8×16	2	1.90
8×8	1	0.95

latch-based operand isolation and clock-gating. If one of the operands is zero, then (i) the latches prevent the propagation of inputs minimizing the switching activity, (ii) the clock-gating cell disables the clock signal thus reducing the equivalent load capacitance of the clock signal.

C. Precision-Energy Tradeoff for Single MAC Operation

The proposed SW-HW precision scaling strategy can be implemented using both FPGA and ASIC technologies. In this work we designed and characterized the 8×8 MAC unit using a commercial $28\,\text{nm}$ UTBB FDSOI technology and the Synopsys Galaxy Platform, versions L-2016.03. The frequency constraint is set to $1\,\text{GHz}$ at $0.90\,\text{V}$ in a typical process corner (compliant with recent works that used the same technology [19]). Power consumption is extracted using Synopsys PrimeTime L-2016.06 with SAIF back-annotation. Collected results show a standard cell area of $1443\,\mu\text{m}^2$ and total average power consumption of $0.95\,\text{mW}$. Compared to a traditional $8{\times}8$ MAC unit, the proposed architecture shows 3.7% area penalty. The overhead is due to the sign extension logic and the additional bit of the multiplier.

Table I reports latency (N_{cycles}) and energy consumption per MAC (E_{MAC}) under different precision constraints. Each row in the table corresponds to a different implementation point in the precision-energy-latency space. If one of the two operands is zero, energy E_{zero} reduces substantially due to the zero-skipping logic: $E_{zero}= 0.103 E_{MAC}$.

IV. ENERGY-DRIVEN OPTIMIZATION

A. Problem Formulation

For a ConvNet of L layers, the precision scaling encompasses the choice of the bitwidth for the *input map* ($\underline{\mathbf{I}}$) and the *weight* ($\underline{\mathbf{W}}$) matrices of each layer i, and that of the *output map* matrix ($\underline{\mathbf{O}}$) of the last layer; the precision of $\underline{\mathbf{O}}$ does not impact computation as it only affects the number of memory accesses. Assuming the availability of the four accuracy options described in Section III, i.e. *full* (16×16), *mid* (16×8 or 8×16), *half* (8×8), the precision for $\underline{\mathbf{I}}$ and $\underline{\mathbf{W}}$ of each layer, and that of $\underline{\mathbf{O}}$ for last layer, can be assigned to 8-bit or 16-bit. We encode the unknown of the problem as a vector X of ($2{\times}L{+}1$)

Algorithm 2: Simulated Annealing

Input: T_0, T_f, X_0, cooling, K_b, iter, λ_{max}, valset
Output: X
1   T = T_0
2   E = energy(X_0)
3   E_{max} = *energy* (*ones* $(2L + 1)$); E_{min} = *energy* (*zeros* $(2L + 1)$)
4   **while** ($T \geq T_f$) **do**
5   **for** *i = 0; i < iter; i = i+1* **do**
6   *next_state = move (current_state)*
7   **if** *accuracy_drop(next_state, valset, tested) $< \lambda_{max}$*
       then
8   E_next = *energy* (next_state)
9   ΔE = (E_next-E_current) / (E_{max} - E_{min})
10   **if** *(dE < 0) or (exp[-ΔE/K_b·T] > random(0, 1))*
         then
11   current_state = next_state
12   E_current = E_new
13   **if** *E_current < E_best* **then**
14   E_best = E_current
15   best_state = current_state
16   *update(tested)*
17   T = T · *cooling*
18   **return** *best_state*

Boolean variables x_i, where the variable $x_{2 \times L+1}$ refers to $\underline{\mathbf{O}}$. The encoding map is: $x{=}0{\rightarrow}$8-bit, $x{=}1{\rightarrow}$16-bit.

The optimal assignment is the x that minimizes the total energy consumption $E(X)$ while ensuring an accuracy loss $\lambda(X)$ lower than a user-defined constraint λ_{max}.

B. Optimization Engine

The optimization kernel is built on a modified version of Simulated Annealing (SA). Even though SA is known to be time-consuming, the lack of information on the best move in the optimization space has driven us towards this meta-heuristic. SA allows a wide exploration of possible solutions, while collected results will provide a guideline to implement further (and faster) heuristics. Algorithm 2 shows the pseudo-code of the SA. It gets as inputs the following parameters:

- T_0: initial temperature;
- T_f: final temperature;
- X_0: starting solution, all 1s in our case, i.e., full-precision (16-bit) to all the L layers (both $\underline{\mathbf{W}}$ and $\underline{\mathbf{W}}$, and $\underline{\mathbf{O}}$);
- *cooling*: temperature derating factor (*geometric* in our case);
- K_b: normalization factor of the acceptance probability.
- *iter*: number of iterations for each temperature T;
- λ_{max}: the user-defined accuracy drop (percentage);
- *valset*: set of images used for the estimation of accuracy drop through the function *accuracy_drop*;

At each iteration, the next state is generated as a random perturbation of the current state (line-6). For those states that satisfy the accuracy constraint (line-7), the energy cost function E is evaluated (line-8) through the function *energy*. If ΔE (lines-9) reduces (lines-10), the new state is accepted (line-11:12). If not, the new state is accepted following a Boltzmann probability function (line-10:12); the acceptance ratio gets smaller as T reduces. States that show minimum energy are iteratively saved as best solution (line-13:15). Once the total number of iterations is reached (line-5), the temperature T is cooled down (line-19). The process iterates till the minimum temperature T_f is reached (line-4).

The bottleneck of the algorithm is the call to the function *accuracy_drop*. For this reason, the algorithm takes trace of already processed states; this information is fed to the *accuracy_drop* function which can eventually by-pass accuracy estimation. For instance, if a state does match with the accuracy constraint, all the states with a superior precision will match the constraint as well, hence they do not need any accuracy test.

C. Energy Model

Most of the ConvNet accelerators described in the literature, e.g. [2], show a similar system architecture made up of: (i) a planar array of processing elements (PE), in our case the MAC units described in Section III; (ii) a set of SRAM buffers for temporal data (Input Buffer, Weight Buffer, and Output Buffer); (iii) an off-chip memory and its DMA; (iv) a control unit (RISC) that schedules the operations.

With no lack of generality, the total energy consumption E is the sum of two main contributions: $E = E^{comp} + E^{mem}$. E^{comp} is the energy consumed by the PE array, E^{mem} is the energy consumed due to data movement through the memory hierarchy. The first term is defined as:

$$E^{comp} = \sum_{i=1}^{L} E^{MAC} \cdot N_{cycles}(x_i) \cdot N_i^{MAC} + E^{zero} \cdot N_i^{zero} \quad (2)$$

L is the number of layers of the ConvNet. E^{MAC} is the energy consumption of the *half* precision MAC (row 8×8 in table I). N_{cycles} is the latency of a single MAC operation of the i-th layer; it is given as multiple of the latency of the *half* precision MAC (row 8×8 in Table I) and it is function of the precision x_i. N_i^{MAC} is the number of non-zero MAC operations of the i-th layer. E_i^{zero} is the energy consumed under zero-skipping (mostly due to leakage). N_i^{zero} is the number of zero MAC. The second term is defined as:

$$E^{mem} = \sum_{i=1}^{2 \cdot L + 1} E^{MAC} \cdot [\alpha_i(x_i) + \beta_i(x_i) + \gamma_i(x_i] \quad (3)$$

E^{MAC} is the same as in Equation 2, while α_i, β_i and γ_i are three parameters that describe the energy consumed by the i-th layer due to reading/writing the *input map* (α_i), the *weights* (β_i), the *output map* (γ_i). More specifically they represent the ratio between the energy consumption of the memory and the energy consumption of the PE array; here again, the energy unit is the *half* precision MAC (row 8×8 in table I) [2]. Obviously, α and β do not contribute for the final output layer: $\alpha_{L+1}=0$ and $\beta_{L+1}=0$.

All the three parameters are function of the layer precision x_i: both fetch and write-back operations depend on (i) the accuracy of the MAC algorithm, and (ii) the number of zero-multiplications (switching activity to/from memory may change substantially). Moreover α_i, β_i, γ_i change depending on the ConvNet model: number and size of weights/channels per layer, stride and padding. Finally, they also differ depending on the size of the hardware components (PE array, and global buffers). Since the target of this work is not the energy model per se, not even the evaluation of different architectural solutions, α_i, β_i, γ_i are extracted for the architecture proposed in [2] and then scaled to our precision reduction strategy. The same E^{mem} model applies to different architectures by a proper tuning of the three parameters.

D. Accuracy Drop Assessment

The loss function λ quantifies the difference between the output of the full-precision ConvNet and that inferred with the precision-scaled ConvNet. It is computed as the ratio between the number of miss-classified images and the total number of images in the *val_set* (see Algorithm 2).

The estimation of λ encompasses the execution of S feed-forward inferences, with S as the cardinality of *val_set*. Apart from the intrinsic time complexity of the inference process, the actual issue is that current framework for ConvNets, e.g. TensorFlow, make use of GPUs or CPU with BLAS instruction set. Neither of them supports fixed-point operations (at least those of common available products). We therefore resort to *fake* quantization [14], a SW strategy that emulates the loss of information due to fixed-point arithmetic still using floating-point data-type. More in details, we add *quantization* wrappers at the input/output of each ConvNet layer; the wrapper converts input data (32-bit floating-point) into a *fake* integer, namely, still a 32-bit floating-point number subtracted of an amount equal to the error that the fixed-point representation would have brought [14]. The advantage is that all the fixed-point operations are physically run by the high-performance FP-units. Fake quantization is quite accurate and very efficient.

V. RESULTS

A. ConvNet Benchmarks and Set-up

The proposed adaptive scaling strategy has been validated on three ConvNet benchmarks with different depth and size as shown in Table II. Each convolutional and dense layer is followed by a ReLU activation function, except for the last one which is followed by a softmax classifier.

The networks are trained and tested using three datasets: SVHN [20] for CNN-1, CIFAR-10 for CNN-2 and CIFAR-100 [21] for CNN-3. The training is driven by a gradient descent minimization based on the Adam algorithm. The learning rate is 1e-4 and the batch size 32; other hyper-parameters are set as indicated in [22]. Overfitting is managed through the insertion of dropout layers placed after pooling layers (with rate 0.25) and after the ReLU of dense layers (with rate 0.50). The simulation results reported in this section refer to the learned model with the highest accuracy over 200 training epochs. Both training and inference are performed using the Tensorflow software library [23] (version 1.2.1-rc1) run with single-precision floating point on a Intel(R) Core(TM) i7-2600 CPU @ 3.40GHz (4 cores, 8 threads). The fake quantization method is used to emulate fixed-point MAC.

The per-layer precision strategy proposed in this work integrates both bitwidth selection and radix-point shifting (Section II). Bitwidth optimization is run through the SA engine (Section IV) embedded into the Tensorflow framework. Main parameters of the SA algorithm are: $T_0 = 512$; $T_f = 2.5$; K_b = 1E-2; *cooling* = 0.5; *iter* is set to 10 for CNN-1, 10^2 for CNN-3, and 10^4 for CNN-3 $\lambda_{max} = [1\%, 5\%, 10\%]$ w.r.t. the full-precision model (16); *valset*=5000 images randomly

Table II: Datasets and ConvNet benchmarks

	CNN-1	CNN-2	CNN-3
Network Architecture	input 32×32×3 conv 3×3×32 conv 3×3×64 maxpool dense 128 dense 10 softmax	input 32×32×3 conv 3×3×32 conv 3×3×32 maxpool conv 3×3×64 conv 3×3×64 maxpool dense 512 dense 10 softmax	input 32×32×3 conv 3×3×32 conv 3×3×32 maxpool conv 3×3×64 conv 3×3×64 maxpool conv 3×3×128 conv 3×3×128 maxpool dense 1024 dense 512 dense 10 softmax
Dataset	SVHN	CIFAR-10	CIFAR-100
#MACs	16 914 314	20 828 426	25 718 538

Table III: Per-Net Precision Scaling: Accuracy

	32-bit FP	16×16 Fix	8×16 Fix	16×8 Fix	8×8 Fix
CNN-1	90.02%	90.01%	89.66%	76.64%	74.54%
CNN-2	81.77%	81.75%	80.99%	77.80%	72.17%
CNN-3	55.80%	55.82%	52.79%	22.27%	14.69%

picked from the original data-set. Concerning optimal radix-point shifting, the integer length IL is set on the base of the maximum absolute value. Since weights show different dynamic ranges across the layers [12], we implemented a per-layer fixed point shifting as suggested in [24].

B. Results

We first provide the results of a state-of-art *per-net* precision scaling scheme implemented with the proposed MAC arithmetic unit. Table III shows the top-1 prediction accuracy. To notice that we do not run any retraining after quantization. The full-precision fixed-point ConvNets (column 16×16 Fix) keeps almost the same accuracy of the original floating-point model (maximum relative drop 0.02% for CNN-2). This result is in line with previous works and motivates the choice of using 16-bit as the reference to compare with. Regarding the mid-precision option: 8×16 assigns 8-bit to input maps (I) and 16-bit to the weights (W); 16×8 does the opposite. The 8×16 option is by far more accurate than 16×8; it shows a max drop below 1% for CNN-1 and CNN-2, and 5.4% for CNN-3. The half-precision (column 8×8) shows substantial accuracy drop, ranging from 11.7% (CNN-2) to 73.7% (CNN-3).

Those numbers suggest a preliminary conclusion. While 8×16 is a good candidate for energy-accuracy scaling at run-time, the other two options 16×8 and 8×8 are impractical. The per-net granularity strictly limits the margins to apply an accuracy-energy trade-off (16×16 and 8×16 as the only viable options). A finer accuracy assignment, i.e. per-layer, will open up new Pareto points in the optimization space.

Concerning energy minimization, the key assumption is: maximizing the number of layers at low precision is not enough, to reduce the precision at the proper layer is important. The results collected in Figure 2 provide an empirical proof. The plot shows the energy consumption (normalized w.r.t. full-precision) for different precision assignments explored during the evolution of the SA algorithm. Numbers refer to CNN-2 with 10% accuracy constraint. The x-axis reports the total number of "operands" (i.e. input maps $\underline{I}$, weight $\underline{W}$, output $\underline{O}$ of different layers) assigned to 16-bit. The energy cost is

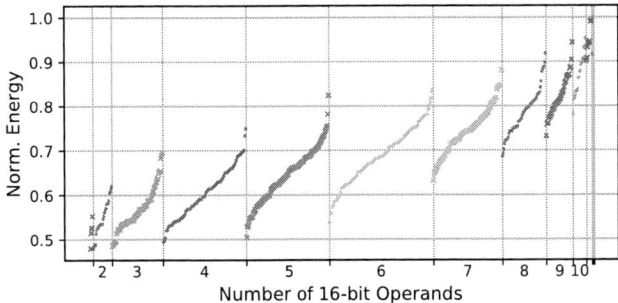

Figure 2: Energy during SA evolution: CNN-2/drop-10%

not monotonic and configurations with more operands at full-precision may get more energy efficient. For instance, some configurations with 6 operands at full-precision consumes less energy that configurations with 5 operands at full-precision. The minimum energy is achieved by assigning one single operand to full-precision; however, other configurations with one operand at full-precision get worse than some of those with two full-precision operands.

Table IV shows the accuracy of the optimal per-layer precision assignment returned by our SA algorithm under the three accuracy constraints under analysis. Percentages in the brackets refer to accuracy loss w.r.t. 16×16. Table V collects the precision mapping returned by SA.

Table IV: Per-Layer Precision Scaling*: Accuracy (Drop)

Max. Drop λ_{max}	1%	5%	10%
CNN-1	89.69% (0.35%)	86.82% (3.55%)	86.82% (3.55%)
CNN-2	80.76% (1.21%)	77.77% (4.87%)	75.57% (7.56%)
CNN-3	55.20% (1.11%)	54.01% (3.24%)	51.54% (7.67%)

*Precision assignment obtained with the proposed SA Optimizer

Table V: Per-Layer Precision Scaling

	CNN-1			CNN-2			CNN-3	
1%	5%	10%	1%	5%	10%	1%	5%	10%
8×16	8×8	8×8	8×16	8×8	8×8	16×16	8×16	8×16
8×16	8×8	8×8	8×16	8×16	8×8	16×16	8×16	8×16
8×16	8×16	8×16	8×8	8×8	8×8	16×16	8×16	8×16
8×8	8×8	8×8	8×8	8×8	8×8	8×16	8×16	8×16
8	8	8	8×16	8×8	8×8	8×16	8×16	8×16
			16×8	8×16	8×8	16×16	8×16	8×8
			16	16	16	8×16	8×16	8×8
						8×8	8×8	8×8
						16×8	8×8	8×8
						16	16	16

Our per-layer precision assignment allows a fine-tuning of the ConvNets by enabling configurations that match different accuracy drops. This is an important achievement that gives a proof of feasibility for the proposed adaptive strategy. The availability of a multiprecision instruction set (as the one proposed in Section III or those currently available in commercial products [11]) is a practical option for adaptive ConvNets. The optimal precision configurations (those reported in Table V) are stored at the SW level (LookUp-Table) and then accessed at run-time (during set-up stages) to bring ConvNets to the desired accuracy. It is worth empathizing that different accuracy constraints come with different precision assignments. As shown in Table V, the precision of each layer may change with accuracy. Hence, solutions that make use of different weight models (one for each accuracy constraint) might result

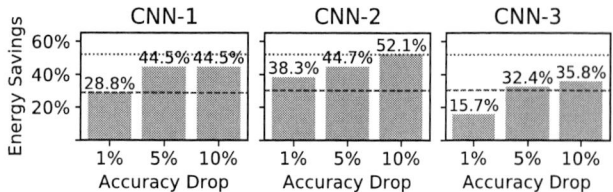

Dotted Upper-line: 8×8 - Dotted Bottom-line: 8×16

Figure 3: Energy-Accuracy Scaling

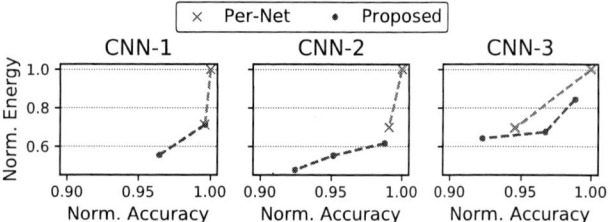

Figure 4: Energy-Accuracy Tradeoff Extension.

impractical. With the proposed strategy what changes are the precision flags and not the weights.

The barcharts in Fig. 3 plots the energy savings achieved with our strategy; numbers are normalized w.r.t. the full-precision model (16×16). Horizontal dotted lines highlight the energy savings obtained with a per-net assignment at half-precision 8×8 (upper line) and mid-precision 8×16 (bottom line). To notice that 8×8 achieves the maximum energy savings; solutions returned by our SA engine get very close to those values. For instance, for both CNN-1 and CNN-2 at 10% drop, energy savings are just 7.5% and 0.01% below the upper line. It is a remarkable result if one considers that using 8×8 for the whole net the accuracy loss is huge (see Table III). For CNN-3, the distance from the maximum energy savings is larger (16.0%), but then again, the accuracy of 8×8 is a mere 14.69%. To be also noted that savings improve w.r.t. 8×16, which is the per-net precision closest to 16×16 (Table III); the only exception is CNN-3 with 1%-drop constraint.

A more interesting analysis concerns the energy-accuracy tradeoff. The plots in Figure 4 show a comparison between the Pareto curves returned by our SA algorithm and those obtained with a per-net precision assignment. Both accuracy and energy are normalized w.r.t. the full-precision 16×16. While a per-net strategy just offers two optimal operating points ($\times$ marker), i.e. full-precision 16×16 and mid-precision 8×16, the proposed strategy extend and improve the Pareto points ($\bullet$ marker). This gives IoT applications more degrees of freedom, and hence the ability to adapt to the context. For instance, reduce the accuracy depending on the complexity of the running task and/or the battery lifetime.

VI. CONCLUSIONS

This work introduces an energy-driven per-layer precision scaling strategy that enables software-scalable ConvNets. Experiments on different ConvNets models show significant energy reduction (52.1% best-case). The proposed framework improves the energy/accuracy tradeoff, thus leading to the flexibility of trading accuracy vs. energy at run-time with low design overhead.

REFERENCES

[1] R. Andri *et al.*, "Yodann: An architecture for ultra-low power binary-weight cnn acceleration," *IEEE Transactions on Computer-Aided Design of Integrated Circuits and Systems*, 2017.

[2] Y.-H. Chen *et al.*, "Eyeriss: An energy-efficient reconfigurable accelerator for deep convolutional neural networks," *IEEE Journal of Solid-State Circuits*, vol. 52, no. 1, pp. 127–138, 2017.

[3] J. Gu *et al.*, "Recent advances in convolutional neural networks," *Pattern Recognition*, 2017. [Online]. Available: http://www.sciencedirect.com/science/article/pii/S0031320317304120

[4] V. Sze *et al.*, "Efficient processing of deep neural networks: A tutorial and survey," *arXiv preprint arXiv:1703.09039*, 2017.

[5] Z. Du *et al.*, "Leveraging the error resilience of neural networks for designing highly energy efficient accelerators," *IEEE Transactions on Computer-Aided Design of Integrated Circuits and Systems*, vol. 34, no. 8, pp. 1223–1235, 2015.

[6] T. J. Yang *et al.*, "Designing energy-efficient convolutional neural networks using energy-aware pruning," in *2017 IEEE Conference on Computer Vision and Pattern Recognition (CVPR)*, July 2017, pp. 6071–6079.

[7] C. Szegedy *et al.*, "Going deeper with convolutions," in *Proceedings of the IEEE conference on computer vision and pattern recognition*, 2015, pp. 1–9.

[8] M. Courbariaux *et al.*, "Binaryconnect: Training deep neural networks with binary weights during propagations," in *Advances in Neural Information Processing Systems*, 2015, pp. 3123–3131.

[9] M. Shafique *et al.*, "Adaptive and energy-efficient architectures for machine learning: Challenges, opportunities, and research roadmap," in *VLSI (ISVLSI), 2017 IEEE Computer Society Annual Symposium on.* IEEE, 2017, pp. 627–632.

[10] V. Peluso *et al.*, "Weak-mac: Arithmetic relaxation for dynamic energy-accuracy scaling in convnets," in *Circuits and Systems (ISCAS), 2018 IEEE International Symposium on.* IEEE, 2018, pp. 1–5.

[11] N. P. Jouppi *et al.*, "In-datacenter performance analysis of a tensor processing unit," in *Proceedings of the 44th Annual International Symposium on Computer Architecture*, ser. ISCA '17. New York, NY, USA: ACM, 2017, pp. 1–12. [Online]. Available: http://doi.acm.org/10.1145/3079856.3080246

[12] D. Lin *et al.*, "Fixed point quantization of deep convolutional networks," in *International Conference on Machine Learning*, 2016, pp. 2849–2858.

[13] B. Moons *et al.*, "Energy-efficient convnets through approximate computing," in *Applications of Computer Vision (WACV), 2016 IEEE Winter Conference on.* IEEE, 2016, pp. 1–8.

[14] L. Shan *et al.*, *A Dynamic Multi-precision Fixed-Point Data Quantization Strategy for Convolutional Neural Network.* Singapore: Springer Singapore, 2016, pp. 102–111.

[15] B. Moons *et al.*, "14.5 envision: A 0.26-to-10tops/w subword-parallel dynamic-voltage-accuracy-frequency-scalable convolutional neural network processor in 28nm fdsoi," in *Solid-State Circuits Conference (ISSCC), 2017 IEEE International.* IEEE, 2017, pp. 246–247.

[16] I. Koren, *Computer arithmetic algorithms.* Universities Press, 2002.

[17] S. R. Jahnke *et al.*, "Micro-controller direct memory access (dma) operation with adjustable word size transfers and address alignment/incrementing," Nov. 9 2004, uS Patent 6,816,921.

[18] M. Courbariaux *et al.*, "Training deep neural networks with low precision multiplications," *arXiv preprint arXiv:1412.7024*, 2014.

[19] G. Desoli *et al.*, "14.1 a 2.9 tops/w deep convolutional neural network soc in fd-soi 28nm for intelligent embedded systems," in *Solid-State Circuits Conference (ISSCC), 2017 IEEE International.* IEEE, 2017, pp. 238–239.

[20] Y. Netzer *et al.*, "Reading digits in natural images with unsupervised feature learning," in *NIPS workshop on deep learning and unsupervised feature learning*, 2011, pp. 1–5.

[21] A. Krizhevsky *et al.*, "Learning multiple layers of features from tiny images," 2009.

[22] D. Kingma *et al.*, "Adam: A method for stochastic optimization," *arXiv preprint arXiv:1412.6980*, 2014.

[23] M. Abadi *et al.*, "Tensorflow: A system for large-scale machine learning." in *OSDI*, vol. 16, 2016, pp. 265–283.

[24] J. Qiu *et al.*, "Going deeper with embedded fpga platform for convolutional neural network," in *Proceedings of the 2016 ACM/SIGDA International Symposium on Field-Programmable Gate Arrays.* ACM, 2016, pp. 26–35.

978-1-5386-4757-8/18 $31.00 © 2018 IEEE

Enhancing Performance of Computer Vision Applications on Low-Power Embedded Systems Through Heterogeneous Parallel Programming

Stefano Aldegheri, Silvia Manzato, Nicola Bombieri
Department of Computer Science
University of Verona
Email: name.surname@univr.it

Abstract—Enabling computer vision applications on low-power embedded systems gives rise to new challenges for embedded SW developers. Such applications implement different functionalities, like image recognition based on deep learning, simultaneous localization and mapping tasks. They are characterized by stringent performance constraints to guarantee real-time behaviors and, at the same time, energy constraints to save battery on the mobile platform. Even though heterogeneous embedded boards are getting pervasive for their high computational power at low power costs, they need a time consuming customization of the whole application (i.e., mapping of application blocks to CPU-GPU processing elements and their synchronization) to efficiently exploit their potentiality. Different languages and environments have been proposed for such an embedded SW customization. Nevertheless, they often find limitations on complex real cases, as their application is mutual exclusive. This paper presents a comprehensive framework that relies on a heterogeneous parallel programming model, which combines OpenMP, PThreads, OpenVX, OpenCV, and CUDA to best exploit different levels of parallelism while guaranteeing a semi-automatic customization. The paper shows how such languages and API platforms have been interfaced, synchronized, and applied to customize an ORB-SLAM application for an NVIDIA Jetson TX2 board.

I. Introduction

Computer vision is becoming pervasive in modern cyber-physical systems [1], [2]. Its main goal is the use of digital processing and intelligent algorithms to interpret meaning from images or video streams. Computer vision applications, in the context of cyber-physical systems, generally consists of several computational-intensive kernels that implement different functionalities, ranging from image recognition to simultaneously mapping and localization (SLAM).

Developing and optimizing such applications for an embedded system is not an immediate and simple task. Beside functional correctness, developers have to deal with non-functional aspects like performance, power consumption, energy efficiency and real time constraints. The optimization is architecture dependent and spans across two main dimensions: block-level and system-level. The first is more intuitive and involves the re-implementation/parallelization of single kernels for the target board accelerators (e.g., GPU, DSP,

or multi-cores) through specific languages or programming environments like CUDA, OpenCL, Pthreads or OpenMP. The system-level optimization targets the overall system power consumption, memory bandwidth, and inter-process communication overhead. Mapping space exploration means exploring the different strategies to map each of such kernels to the right processing elements of the board and analyzing the corresponding impact on the design constraints.

OpenVX [3] is increasingly gaining consensus in the embedded vision community as programming environment and API library for system-level optimizations [4]. Such a platform is designed to maximize functional and performance portability across different hardware platforms, providing a computer vision framework that efficiently addresses different hardware architectures with minimal impact on software applications. Starting from a graph model of the embedded application, it allows for automatic system-level optimizations and synthesis on the target architecture by optimizing performance, power consumption and energy efficiency [5], [6], [7], [8].

Nevertheless, due to the limitation of OpenVX to model complex applications through data-flow graphs and to the incompleteness of the OpenVX primitive library, any real embedded vision application requires the integration of OpenVX with user-defined C/C++ code. On the one hand, the user-defined code can benefit from parallelization techniques for multi-cores, thus providing heterogeneous parallel environments (i.e., multi-core + GPU parallelism). On the other hand, due to the private and not user-controlled memory stack of OpenVX, such an integration leads to the sequentialization of the different execution environments, with a consequent strong impact on the system-level optimization.

This paper presents a framework for heterogeneous parallel programming of embedded vision applications. It allows combining different programming environments, i.e., OpenMP, PThreads, OpenVX, OpenCV, and CUDA to best exploit different levels of parallelism while guaranteeing the semi-automatic customization.

The paper presents an analysis of the limitations found by applying the state-of-the-art parallel programming environments to customize a modern SLAM application for the

978-1-5386-4757-8/18 $31.00 © 2018 IEEE

widespread NVIDIA Jetson TX2 board. Finally, it presents the results of the mapping space exploration we performed by considering performance, power consumption, energy efficiency, and result quality design constraints. The paper is organized as follows. Section II presents an introduction on OpenVX and the related work. Section III presents an analysis of parallel programming environments for embedded vision applications through a case study. Section IV presents the proposed framework. Section V presents the experimental results, while Section VI is devoted to the concluding remarks.

II. BACKGROUND AND RELATED WORK

OpenVX is a framework to develop and optimize embedded vision applications by considering different design constraints (i.e., performance, power consumption). It relies on a graph-based model to define a high-level and architecture independent representation of the application. Such a representation is modularly built by the user through the use of a set of primitives, which are provided by the framework and that represent the most commonly used functionalities and data objects in computer vision algorithms, such as scalars, arrays, matrices and images, as well as high-level data objects like histograms, image pyramids, and look-up tables.

The high-level representation (i.e., the *graph*) is then automatically optimized and embedded thanks to the libraries of architecture-oriented implementation of the primitives and data-structures provided by the board vendor.

The developer defines a computer vision algorithm by instantiating kernels as nodes and data objects as parameters (see the example in Fig. 1). Each node of the graph is identified as a function kernel that can run on any processing unit of target heterogeneous board. Indeed, the application graph represents the partitioning of the whole application into blocks, which can be be executed across different hardware accelerators (e.g., CPU cores, GPUs, DSPs).

The programming flow starts by creating an OpenVX *context* to manage references to all used objects. Based on this context, the code builds the graph and generates all required data objects. Then, it instantiates the kernel as *graph nodes* and generates their connections. The framework first checks the graph integrity and correctness (e.g., checking of data type coherence between nodes and absence of cycles) and, finally, it processes the graph. At the end of the code execution, it releases all created data objects, the graph, and the context.

In the example of the Fig. 1, the application computes the gradient magnitude and gradient phase from a blurred input image. The *Magnitude* and *Phase* nodes are independently computed, in that each does not depend on the output of the other. OpenVX does not mandate that they are run simultaneously or in parallel, but it allows the runtime manager of the board vendor to decide on the mapping and execution strategy.

By adopting any vendor library that implements the graph nodes as Computer Vision primitives, OpenVX allows applying different mapping strategies between nodes and processing

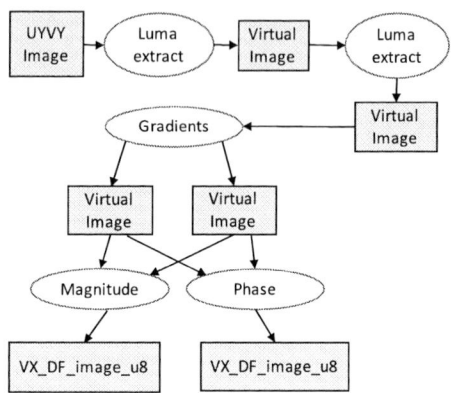

Fig. 1. OpenVX sample application (graph diagram)

elements of the heterogeneous board, by targeting different design constraints (e.g., performance, power, energy efficiency).

Different works have been presented to analyze the use of OpenVX for embedded vision [6], [7], [5], [8]. In [6], the authors present a new implementation of OpenVX targeting CPUs and GPU-based devices by leveraging different analytical optimzation techniques. In [7], the authors examine how OpenVX responds to different data access patterns, by testing three different OpenVX optimizations: kernels merge, data tiling and parallelization via OpenMP. In [5], the authors introduce ADRENALINE, a novel framework for fast prototyping and optimization of OpenVX applications for heterogeneous SoCs with many-core accelerators. In [8], we proposed a methodology to integrate a model-based design environment to OpenVX. The methodology allows applying Matlab/Simulink for the model-based design, parametrization, and validation of computer vision applications. Then, it allows for the automatic synthesis of the application model into an OpenVX description for the hardware and constraints-aware application tuning.

III. ANALYSIS OF PARALLEL PROGRAMMING ENVIRONMENTS FOR EMBEDDED VISION APPLICATIONS THROUGH THE ORB-SLAM CASE STUDY

In order to understand the limitations of the state-of-the-art environments for parallel programming embedded vision applications and the contribution of the proposed framework, we first present the case study, which will be used as a model in the subsequent sections. The case study, ORB-SLAM [9], represents a typical real embedded application, which is applied in different contexts, ranging from automotive to robotic systems. NVIDIA Jetson TX2, which is a widespread and low-cost embedded board, is the target platform.

ORB-SLAM solves the simultaneous localization and mapping problem when RGB camera sensors are adopted. It computes, in real-time, the camera trajectory and a sparse 3D reconstruction of the scene in a wide variety of environments, ranging from small hand-held sequences of a desk to a car driven around several city blocks. It builds a 3D map starting from an input stream and/or it performs localization

978-1-5386-4757-8/18 $31.00 © 2018 IEEE

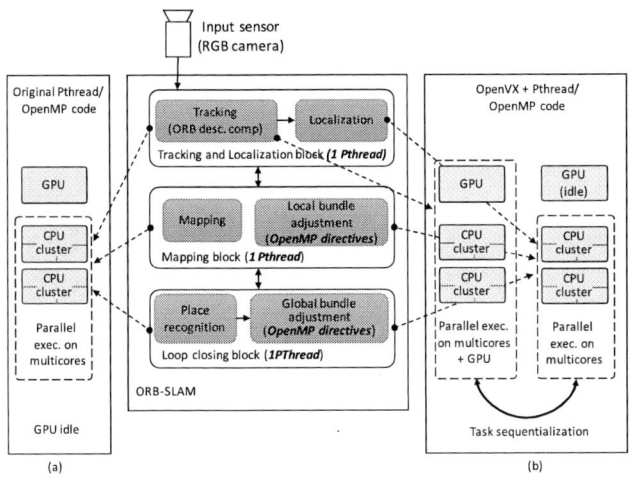

Fig. 2. Overview of ORB-SLAM application and execution models: (a) the original code (parallelized for multicore), (b) the state-of-the-art OpenVX implementation.

by considering the current map. The application consists of three main blocks (see Figure 2):

- The *tracking and localization* block computes visual features, it localizes the agent in the environment, and, in case of significant discrepancies between an already saved map and the input stream, it communicates updating information of the map to the mapping block. The processing rate (i.e., the supported frame rate per second) and the main power consumption of the whole application strongly depend on this block performance.

- The *mapping* block updates the environment map by using information (detected map changes) sent by the localization block. In case of a well consolidated map, this module can be shut down to save system resources.

- The *loop closing* block aims at adjusting the scale drift error accumulated during the input analysis, which is unavoidable when adopting a monocular vision system (i.e., RGB camera). When a loop in the agent pathway is detected, this block updates the mapped information through a high latency heavy computation, during which the first two blocks must be suspended. This can lead the agent to loose tracking and localization information and, as a consequence, the agent to get temporary lost. As a consequence, the computation efficiency of this block (run on-demand) is crucial for the quality of the whole application results.

In the best ORB-SLAM implementation at the state of the art [9], due to their concurrent execution model, the three blocks are implemented to be run in parallel through PThreads on shared-memory multiprocessors. In addition, since the bundle adjustment task, both local in the mapping block and global in the loop closing block, can have long latencies, it is a primary target for parallelization. Its nested and data-independent loops well apply for directive-based *automatic* parallelization. Thus, the state of the art code is available with

OpenMP directives for parallel execution on multi-cores. No block is originally considered for parallel execution on GPU (see Figure 2(a)).

The manual implementation of any sub-block for GPU is out of the scope of this work. Rather, due to the complexity of such a parallelization task for this application class yet considering different design constraints (power consumption and energy efficiency beside performance), we consider the semi-automatic *embedding* of the application through OpenVX.

We rely on standard libraries of computer vision functions, which are provided by the target board vendors (i.e., Vision-Works [10] for NVIDIA boards). The library can be extended through user-defined or third-party CUDA kernels, which are integrated in the OpenVX implementation as *custom nodes*.

On the other hand, due to the limitation of OpenVX to model complex applications through data-flow graphs and to the incompleteness of the vendor library, the OpenVX application has to be often integrated to standard C/C++ code. In the ORB-SLAM case study, only the tracking sub-block can be modelled through a data-flow graph and is worth to be optimized for CPU/GPU execution. Even though the rest of the code can still run on multicores, the two environment execution (OpenVX+CUDA and the rest) is sequentialized to allow for communication and synchronization, as explained in Section IV. The proposed method aims at integrating the two environments. Such an integration involves several advantages, such as, multi-level parallel execution of the application and better mapping space between tasks and processing elements to be explored.

IV. METHOD

Figure 3 depicts the overview of the proposed framework. We consider six different languages and parallel programming environments (*environments* in the following): C/C++, Pthreads, OpenMP, OpenCV, OpenVX, and CUDA. The environment heterogeneity allows implementing different application blocks with the most appropriate style, such as C/C++ for control parts, Pthreads for concurrent execution functions on the CPUs, OpenMP for directive-based automatic parallelization of code chunks, CUDA for any kernel (if available) acceleration on GPU, and OpenVX for primitive-based parallelization of data-flow routines. OpenCV has been chosen to implement standard I/O communication protocols of computer vision applications through standard data-structures and APIs. This allows the embedded vision applications to be portable and efficiently integrated to any other application compliant to the standard.

For the sake of clarity and without loss of generality, we consider, as a running example, the widespread and most popular NVIDIA Jetson TX2 as the target platform. Such an embedded board relies on a shared-memory architecture, in which two different clusters of CPUs (four cores Cortex-A57 CPUs and two cores Denver CPUs) and a GPU with two symmetric multiprocessors share an unified memory space.

978-1-5386-4757-8/18 $31.00 © 2018 IEEE

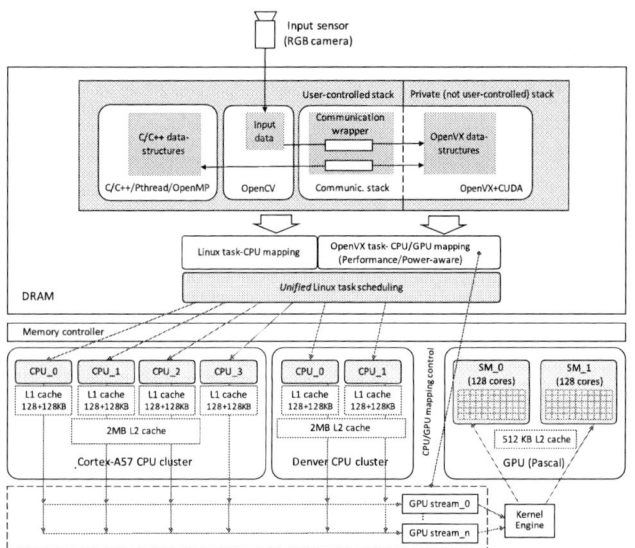

Fig. 3. Framework overview: memory stack, task mapping, and task scheduling layers of an embedded vision application developed with the proposed method on the NVIDIA Jetson TX2 board.

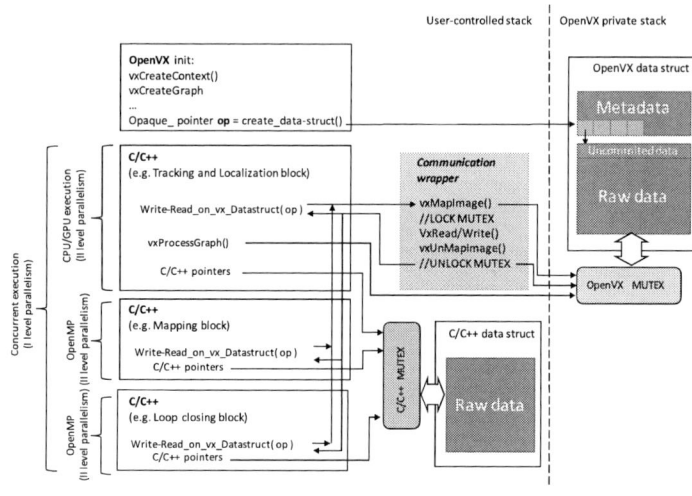

Fig. 4. Overview of the communication wrapper and its integration in the system.

The top of Figure 3 depicts the stack layer involved by the concurrent execution of each environment. It relies on two main parts:

- *The user-controlled stack*, which allows for shared memory-based communication among processes running on different CPUs. They include C/C++ processes, OpenCV APIs, Pthreads, and processes generated by OpenMP.

- *The private (not user-controlled) stack*, which is created and handled by OpenVX and allows for communication between OpenVX graph nodes running on different CPUs or on the GPU.

The tasks related to the user-controlled stack are mapped to the CPU cores by the operating system (i.e., Linux Ubuntu for the NVIDIA Jetson). The OpenVX tasks are mapped to the CPU cores or GPU multiprocessors by the OpenVX runtime system.

To enable the full concurrency of the two parts, to avoid sequentialization of the two sets of tasks, and to avoid the consequent synchronization overhead, we associate the two parts to a single *unified* scheduling engine. This allows all the tasks mapped to the CPU cores (of both stack parts) to be scheduled by the operating system, while the GPU task scheduling, the CPU-to-GPU communication and synchronization (i.e., GPU stream and kernel engine) to be controlled by the OpenVX runtime system. To do that, we propose a C/C++-OpenVX template-based communication wrapper, which allows for memory accesses to the OpenVX data structures on the private stack and for full control of the OpenVX context execution by the C/C++ environment.

Figure 4 gives an overview of the wrapper and its integration in the system. The OpenVX initialization phase generates the graph context and allocates the private data structures. Such allocation returns *opaque pointers* to the allocated memory segments, i.e., pointers to private memory areas which layout is unknown to the programmer.

OpenVX read and write primitives (`Write-Read_on_vx_Datastructure()` in the Figure) have been defined to access the private data structures through the opaque pointers. The primitives are invoked from the C/C++ context and, through the communication wrapper APIs, they set a mutex mechanism to safety access the OpenVX data structures. The same mutex is shared with the OpenVX runtime system for the overall graph processing (`vxProcessGraph()` in the Figure). As a consequence, the mechanism guarantees synchronization during the accesses to the shared data structures between the OpenVX and C/C++ contexts when run concurrently on multicores. It is important to note that the invocation of the overall graph processing, which is performed in the C/C++ environment, starts the execution of the data-flow oriented OpenVX code. As shown in Figure 4, such an invocation can be performed concurrently by different C/C++ threads, and each invocation involves a mapping and scheduling of the corresponding graph instance. The proposed communication wrapper and mutex system allow for synchronization among the different concurrent OpenVX graph executions and the C/C++ calling environments.

Standard mutex mechanisms are adopted to synchronize all the other C/C++ based contexts belonging to the user-controlled stack, when accessing shared data structures.

The mutex-based communication wrapper allows for multi-level parallel execution of the application. Considering for example the ORB-SLAM case study, the first level of parallelism is implemented by the Pthreads, which run the three main modules of the application on different CPU cores.

Then, the tracking block of the first module is implemented

978-1-5386-4757-8/18 $31.00 © 2018 IEEE 122

in OpenVX and run on a CPU core and on the GPU. The parallel implementation of the graph nodes offloaded on the GPU is provided by the OpenVX library vendor (i.e., NVIDIA VisionWorks for our case study) and are optimized for the specific GPU architecture. In case two nodes of the OpenVX graph are independent (see the the example of Fig. 1), they are executed concurrently.

Finally, OpenMP provides another level of parallelism when a block is enriched with parallel directives (e.g., Mapping and Loop closing blocks in the example). Each of these blocks is executed in parallel by the threads generated automatically by the compiler, which run on the available CPU cores.

V. Experimental Results

The framework has been applied to embed the ORB-SLAM application on the Jetson TX2 incrementally. We started from the most efficient parallel implementation at the state of the art [11]. We then integrated modularly the different parallel environments supported by the framework as follows:

- *Version 1 (Pthreads)*: It is the starting version [11], in which the three main blocks (Tracking and localization, Mapping, and Loop closing blocks) are run concurrently by Pthreads on the CPU cores.
- *Version 2 (Phtreds+OpenMP)*: It extends version 1, by enabling OpenMP parallelism. In particular, it parallelizes the bundle adjustment task, both local in the mapping block and global in the loop closing block.
- *Version 3 (Pthreads+OpenVX)*: It extends version 1 (i.e., with Pthreads, without OpenMP parallelism) by implementing the tracking sub-block in OpenVX.
- *Version 4 (Pthreads+OpenMP+OpenVX)*: It extends version 3 by enabling also OpenMP.
- *Version 5 (Pthreads+OpenVX+CUDA)*: starting from version 3, we reused a CUDA kernel that implements the *ORB* primitive in the tracking sub-block. We modularly replaced the corresponding OpenVX VisionWork primitive with such a more optimized kernel.
- *Version 6 (Pthreads+OpenMP+OpenVX+CUDA)*: It extends version 5 by enabling also OpenMP.

We validated and evaluated all the versions by using the *KITTI* dataset [12], which is a standard and widespread benchmark for vision applications. The dataset consists of video streams captured by driving around a car equipped with RGB camera in the mid-size city of Karlsruhe, in rural areas and on highways. For the sake of space, we present the results obtained on the sequence number 13, since it is the most meaningful to show the variance of workload in all the three blocks of ORB-SLAM and the corresponding effects on the design constraints.

For the evaluation, we set the Jetson TX2 board with two different configurations: medium frequency (75%) and maximum frequency (100%). They represent the frequency setting of 4 board components, i.e., the four cores Cortex-A57 cluster (1.42 GHz and 2.035 GHz as medium and maximum

frequency, respectively), the two cores Denver cluster (1.42 GHz and 2.035 GHz), the GPU (1.032 GHz and 1.3 GHz), and the memory (1.062 GHz and 1.866 GHz).

Tables I and II show the results for the medium and maximum frequency setting, respectively. The best results are reported in bold. The mapping columns report the number of processing elements used by the different versions during computation. The Pthreads guarantee the minimum level of parallelism, by enabling one core per block. OpenMP has been set to use the maximum number of available CPU cores (6). The GPU is enabled only by OpenVX/CUDA.

Columns *FPS* and *Time per frame* report information about the application performance, and, in particular, *FPS* represents the maximum number of frames per second supported by the embedded system. The columns underline how each level of parallelism influences the overall performance.

To understand the effect of the different versions on power and energy efficiency, the tables report the total energy spent for the computation of the whole stream, the average and peak power, and the average energy per frame.

Finally, the tables report information about the quality of service (QoS) results. It includes the number of frames correctly processed against those skipped for the overloading of the processing elements. Frame skipping is caused by the mapping and loop closing blocks that run the bundle adjustment computation and, due to the work overload, their latency prevent the tracking block in analysing new frames. The maximum number of frames skipped tolerated is a design constraints, since it involves the QoS of the application like the number of times the system gets lost (see Section III).

The tables underline that, as expected, the performance (FPS) provided by the different versions are strictly correlated with the power consumption. Enabling all the processing elements through the different levels of parallelism leads to the best performance at the cost of the higher peak power. However, we found that OpenMP allows improving the performance in the overall heterogeneous context not in all cases. Version 6 is an example, in which switching on the OpenMP parallelism does not provide better performance than the Pthread+OpenVX+CUDA version while it increases the peak power consumption. On the other hand, version 6 provides better QoS in the maximum frequency configuration. This is due to the fact that OpenMP is strictly involved by the bundle adjustment phases, which affect the frame skipped while they do not affect the supported FPS. This does not happen in the medium frequency setting as the CPU frequency in which such a kernel is run does not allow the tracking block to respect the real time constraints.

We found that, for the medium frequency configuration, version 3 is the most energy efficient and provides the best QoS results. Version 5 provides the best performance and does not involve the worst power consumption. Version 6 provides the best performance, and it pays the almost best QoS (99.8%) with the highest power consumption.

978-1-5386-4757-8/18 $31.00 © 2018 IEEE

TABLE I
AVERAGE FPS AND TIME PER FRAME VALUES ON KITTI, SEQUENCE 13, 75% OF THE FREQUENCIES

Version	Mapping			FPS	Time per frame (ms)	Energy (J)	Avg Power (W)	Peak power (W)	% frame processed	Energy per frame (J)
	A57	Denver	GPU SM							
Version 1	3	-	-	13.1	76.4	1,205	**3.37**	**4.05**	3,097/3,281 (94.4%)	**0.357**
Version 2	4	2	-	12.9	77.2	1,125	3.43	5.24	3,021/3,281 (92.1%)	0.372
Version 3	3	-	2	18.3	54.7	**1,039**	3.78	5.62	**3,279/3,281 (99.9%)**	0.378
Version 4	4	2	2	19.8	50.6	1,242	5.79	7.85	3,260/3,281 (99.4%)	0.381
Version 5	3	-	2	**23.2**	**43.2**	1,184	3.61	5.46	3,269/3,281 (99.0%)	0.362
Version 6	4	2	2	**23.2**	**43.2**	1,197	5.65	7.92	**3,271/3,281 (99.8%)**	0.366

TABLE II
AVERAGE FPS AND TIME PER FRAME VALUES ON KITTI, SEQUENCE 13, 100% OF THE FREQUENCIES

Version	Mapping			FPS	Time per frame (ms)	Energy (J)	Avg Power (W)	Peak power (W)	% frame processed	Energy per frame (J)
	A57	Denver	GPU SM							
Version 1	3	-	-	16.9	59.1	1,917	**5.84**	**8.54**	3,264/3,281 (99.5%)	0.587
Version 2	4	2	-	17.2	58.2	1,967	6.00	9.61	3,269/3,281 (99.6%)	0.602
Version 3	3	-	2	21.2	47.2	1,954	5.96	9.54	3,276/3,281 (99.8%)	0.597
Version 4	4	2	2	21.7	46.0	1,945	7.93	11.01	**3,280/3,281 (100%)**	0.593
Version 5	3	-	2	**29.3**	**34.1**	1,895	5.98	9.65	3,274/3,281 (99.0%)	0.579
Version 6	4	2	2	**29.2**	**34.3**	**1,843**	7.62	11.70	**3,279/3,281 (99.9%)**	**0.562**

For the maximum frequency configuration, version 5 provides the best tradeoff in terms of performance and power consumption, while version 6 provides the best tradeoff in terms of performance, energy efficiency, and QoS.

In conclusion, the experimental results show how the different versions and, for each of them a frequency configuration of the single processing elements, provide a very large mapping space to be explored (which is out of the scope of this work). Such a space can provide the best solution for each of the considered design constraints like performance, power consumption, energy efficiency, and quality of service.

VI. CONCLUSION

This paper presented a framework for heterogeneous parallel programming of embedded vision applications. The paper first presented an analysis of the actual limitations of the most meaningful and used environments for parallel programming embedded vision applications at state of the art when applied singularly. The paper then presented how the framework allows combining such environments, i.e., OpenMP, PThreads, OpenVX, OpenCV, and CUDA to best exploit different levels of parallelism while guaranteeing the semi-automatic customization. The paper presented the analysis and framework through a real case of study, i.e., an ORB-SLAM application, which has been customized for an NVIDIA Jetson TX2 embedded board. The paper finally showed that, thanks to the larger mapping space generated and the multi-level parallelism provided by such a heterogeneous parallel programming, we obtained sensibly better results over different design constraints i.e., performance, power consumption, energy efficiency, and result quality.

REFERENCES

[1] B. Meus, T. Kryjak, and M. Gorgon, "Embedded vision system for pedestrian detection based on hog+svm and use of motion information implemented in zynq heterogeneous device," vol. 2017-September, 2017, pp. 406–411.

[2] B. Z. C. Z. K. M. jizhong zhao; Nanning Zheng, "Hierarchical and parallel pipelined heterogeneous soc for embedded vision processing," vol. 99, no. 99, 2017.

[3] Khronos Group, "OpenVX: Portable, Power-efficient Vision Processing," https://www.khronos.org/openvx.

[4] S. Aldegheri, D. D. Bloisi, J. J. Blum, N. Bombieri, and A. Farinelli, "Fast and power-efficient embedded software implementation of digital image stabilization for low-cost autonomous boats," in *Proc. of 11th International Conference Field and Service Robotics (FSR)*, 2017, pp. 129–144.

[5] G. Tagliavini, G. Haugou, A. Marongiu, and L. Benini, "Adrenaline: An openvx environment to optimize embedded vision applications on many-core accelerators," in *International Symposium on Embedded Multicore/Many-core Systems-on-Chip*, 2015, pp. 289–296.

[6] K. Yang, G. A. Elliott, and J. H. Anderson, "Analysis for supporting real-time computer vision workloads using openvx on multicore+gpu platforms," in *Proceedings of the 23rd International Conference on Real Time and Networks Systems*, ser. RTNS '15, 2015, pp. 77–86.

[7] D. Dekkiche, B. Vincke, and A. Merigot, "Investigation and performance analysis of openvx optimizations on computer vision applications," in *14th International Conference on Control, Automation, Robotics and Vision*, 2016, pp. 1–6.

[8] S. Aldegheri and N. Bombieri, "Extending OpenVX for model-based design of embedded vision applications," in *Proceedings of the 2010 18th IEEE/IFIP International Conference on VLSI and System-on-Chip, VLSI-SoC*, 2017, pp. 1–6.

[9] R. Mur-Artal, J. M. M. Montiel, and J. D. Tards, "Orb-slam: A versatile and accurate monocular slam system," *IEEE Transactions on Robotics*, vol. 31, no. 5, pp. 1147–1163, Oct 2015.

[10] NVIDIA Inc., "VisionWorks," https://developer.nvidia.com/embedded/visionworks.

[11] G. Klein and D. Murray, "Parallel tracking and mapping for small ar workspaces," in *2007 6th IEEE and ACM International Symposium on Mixed and Augmented Reality*, Nov 2007, pp. 225–234.

[12] A. Geiger, P. Lenz, C. Stiller, and R. Urtasun, "Vision meets robotics: The kitti dataset," *International Journal of Robotics Research (IJRR)*, 2013.

978-1-5386-4757-8/18 $31.00 © 2018 IEEE

An FPGA-based Hardware Accelerator for Scene Text Character Recognition

Luiz Antonio de Oliveira Junior
Centro de Informática
Universidade Federal de Pernambuco
Recife, Brazil
laoj2@cin.ufpe.br

Edna Barros
Centro de Informática
Universidade Federal de Pernambuco
Recife, Brazil
ensb@cin.ufpe.br

Abstract—Scene text character recognition is a challenging task in Computer Vision since natural scene images usually have cluttered background and the character's size, font, orientation, texture, brightness, and alignment in the picture are variable and non-predictable. Furthermore, most systems including scene text character recognition are usually embedded in a system on a chip (SoC), which has critical requirements, such as low latency, low area, mobility, and flexibility, at the same time that they require high accuracy. In this context, in this work we propose a heterogeneous system for embedded applications with time, area and power constraints, that combines hardware and software to accelerate a technique for scene text character recognition, based on Histogram of Oriented Gradients (HOG) for feature extraction and a neural network Extreme Learning Machine (ELM) as a classifier. The system was prototyped and experimented in the Terasic embedded platform DE2i-150 and the results showed that the system has accuracy of 65.5% in the Chars74k-15 dataset and is able to process up to 11 frames per second, having a good trade-off between processing time and accuracy in embedded environments. Moreover, it occupies only 11% logic elements of the Altera Cyclone IV FPGA, enabling its use in embedded systems.

Index Terms—**Character Recognition, FPGA, HOG, ELM, Computer Vision.**

I. Introduction

Scene text character recognition (STCR) approaches in Computer Vision aim to develop systems that can automatically recognize characters information in natural scene images. Although this kind of systems have been widely used in industrial and commercial applications, such as mobile navigation and scene understanding [1] and license plate recognition, STCR still is a challenging computational problem, since natural images usually have cluttered background and the character's size, font, orientation, texture, brightness, and alignment in the image are variable and non-predictable. Figure 1 shows how characters from different classes look similar due to the natural scene image characteristics aforementioned.

Three common steps of an STCR algorithm are (1) pre-processing, (2) feature extraction and (3) character classification.

During the pre-processing phase, algorithms are used to improve the quality of visual information by keeping only the information related to the character itself and removing noise-

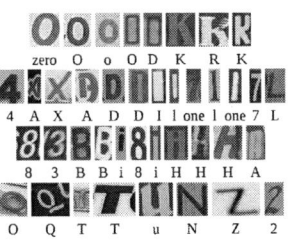

Fig. 1. Examples of natural scenes character images from the Chars74k dataset.

related information. The most commonly used preprocessing techniques are color space conversion and threshold, morphological operations such as erosion and dilation for noise removal and reconstruction of objects; analysis of connected components used to remove noise or to isolate the character of other objects in the image. Also, most feature extraction algorithms, such as Histogram of Oriented Gradients (HOG), [2] and Convolutional Neural Networks (CNN) [3] require that all input images are of a specific size. However, character images in actual scenes can range from tens to thousands of pixels, depending on the distance between the character and the camera and the type of the source, for example. Therefore, it is necessary to apply some image resizing method, such as the spline cubic interpolation algorithms and bicubic interpolation [4].

Given an input preprocessed image, the step of extracting features obtains image characteristics that are intrinsic to the character class to which it belongs. In this context, a good extractor must produce similar attributes for all characters of the same type, even if there are: (1) variations in the size, (2) rotation and (3) translation. Three main types of feature extractors have been used [5]: Histogram of Oriented Gradients (HOG); Convolutional Neural Networks (CNN); and Structural character characteristics [6].

In the step of character recognition, a classifier receives the image characteristics vector and classifies the image in its respective class. The most used classification techniques are *Support Vector Machine* (SVM), Convolutional Neural

Networks (CNN), Random Forests, and Neural Networks.

In this work, we propose a heterogeneous architecture (processor and FPGA) to accelerate an STCR technique, to address the limitations of the existing software-based and hardware-based techniques. The main contributions of this work include: (1) a novel method for SCTR, (2) an FPGA-based architecture for bicubic interpolation algorithm and (3) an STCR architecture that is suitable for embedded applications.

This paper is organized as follows: in section 2, we describe related works. In section 3, we introduce our proposed technique for STCR. In section 4 we present the proposed CPU/FPGA-based architecture implementing the STCR technique. Finally, we conclude the in section 5.

II. RELATED WORKS

To tackle the challenges of the STCR problem, many efforts have been devoted to developing accurate techniques. Campos et al. [7] introduced the scene text characters benchmark Chars74K and showed that commercial Optical Character Recognition (OCR) tools do not have a good performance when used to classify natural scenes characters. In this work, the authors addressed the STCR problem on the Chars74k-15 dataset by testing combinations of feature extractors and classifiers, such as Multiple Kernel Learning (MKL). The techniques were implemented only in software and the processing times were not mentioned. Yi et al. [1] proposed a scene text recognition method for Android applications that employs stroke configuration map as feature representation and a cascade AdaBoost learning scheme to train the classifier. The system was evaluated on the Chars74k dataset and was implemented on a Samsung Galaxy II, and the authors reported an approximated time of 1 second to process each frame.

The authors in [8] proposed a new method for scene character recognition that uses a CNN to extract features from the input image and uses these features to train stroke detectors. Although the authors reported an accuracy of 76.1% on the Chars74k-15 database, this method may not be suitable for embedded systems, since CNNs are both memory and computationally intensive [9], such that the execution time can be very high if there are not enough resources.

The authors in [10] employ Fisher Vectors derived from Gaussian Mixture Models (GMM) and a linear classifier on character recognition. The results showed that this method is able to achieve up to 74.8% of accuracy on Chars74k-15 and has a inference time of 12.5 ms. However, the time analysis was performed using a PC with a Intel i7 CPU and 12GB of RAM, therefore this method can not be fairly compared with the ones implemented in embedded environments.

Most of the STCR techniques proposes a combination of a feature extractor and a classifier aiming to achieve high accuracy, and they do not regard to reduce the processing time, a critical metric for real-time applications. Moreover, these techniques are usually executing on a high-performance CPU, which can be not appropriated for embedded applications. On the other hand, a growing number of systems that need scene

text character recognition are usually embedded and have critical requirements. Such systems must have low latency, low area, mobility, and flexibility, at the same time that they require high accuracies, such as autonomous vehicles [11] and walking assistants for blind people [12].

Some recent works have used FPGA to accelerate character recognition techniques [13], [14] and the processing times are indeed lower than the software approaches. In [13], the proposed architecture can classify about 12 characters per second, whereas in [14] the proposed architecture processes one image per 6ms. These works, however, have been evaluated using MINIST dataset, a very basic dataset of handwritten digits or some proprietary datasets with a uniform background, size, and fonts.

III. THE STCR PROPOSED METHOD

The biggest challenge in STCR is to choose the algorithms for each STCR step maintaining a compromise between accuracy and processing time. In this context, we propose a novel technique for STCR, based on resize, HOG and ELM neural network. Figure 2 illustrates the execution flow of the proposed approach. The proposed system receives as input

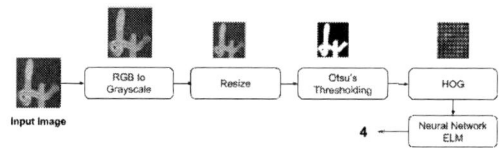

Fig. 2. Stream of execution of the proposed STCR system

an RGB image. In the pre-processing step, the input image is first converted to the gray-scale color space. The gray-scale image is then resized to a default size of 128×128 pixels, to standardize the size of the input images. Finally, the image size of 128×128 pixels is thresholded, using the Otsu [15] technique, to segment the *background* character. In the next step, a Characteristic Extractor based HOG [2] is applied to the threshold image with the purpose of generating a 1296 dimensionality vector containing the characteristics of the image character. Finally, this feature vector is the input of a neural network Extreme Learning Machine (ELM) [16], which classifies the character.

In order to verify the accuracy of the proposed approach we used the Chars74K-15 benchmark. In this sense, we implemented the proposed approach in C ++, using a fixed point in Q1.8.8 format to represent the real numbers, under an Ubuntu 16.04 LTS OS, running on an Intel i7-4500U dual-core 3GHz, with 8GB of RAM and 1TB hard drive.

The parameters in the ELM neural networks are the number of neurons in the hidden layer, L and the activation function g. Therefore, we performed two experiments for finding the number of neurons in the hidden layer and the activation function that resulted in the best performance of the proposed technique: (1) ELM with linear activation function, ie $g(x) = x$, and (2) ELM with hyperbolic tangent activation

978-1-5386-4757-8/18 $31.00 © 2018 IEEE

function that is $g(x) = tanh(x)$. The results indicate that the best accuracy occurs when we use an ELM neural network with 18000 neurons in the hidden layer, $L = 18000$, and the hyperbolic tangent activation function, $g(x) = tanh(x)$. In this configuration, the system has the accuracy of 67.2%, similar or higher accuracy in comparison with related works.

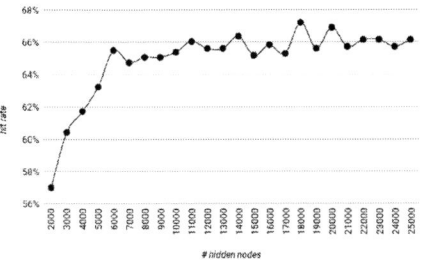

Fig. 3. System accuracy as a function of the number of neurons of the ELM hidden layer

IV. THE HETEROGENEOUS ARCHITECTURE FOR THE STCR TECHNIQUE

The first step was to define which tasks to perform in software (in the CPU) and which ones to execute in hardware (in the FPGA). For this purpose, we implemented the STCR method in C ++ in the target architecture and measured the average time each step took to execute on the Intel Atom N2600 processor, to identify the more time-consuming functions to be hardware accelerated. For this purpose we used *profiling* tool Gprof [17] and the functions of the *time.h* Library. The result of the analysis shows that the *Resize* function is the bottleneck, consuming 71.75% of the total execution time.

In this way, we developed a hardware module, the Bicubic Interpolation Accelerator (BIA), to accelerate the *Resize* step and integrated it into the target platform for accelerating the STCR algorithm.

Figure 4 also illustrates the execution flow of the STCR algorithm and the division of tasks between hardware and software. The tasks running on the ATOM were implemented in C++, while the tasks running on the FPGA were specified in RTL-level SystemVerilog.

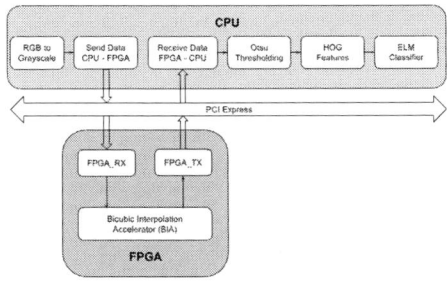

Fig. 4. Proposed Architecture

According to figure 4, the proposed system receives an input RGB image, which is converted in the Grayscale color system.

In the second step, *Send Data CPU - FPGA*, the *grayscale* image is converted into a stream of untyped bytes, which are sent to the FPGA via PCI-E bus. The FPGA_RX module controls reading transactions on the PCI-E bus to receive the stream of bytes. Then, the BIA module resizes the gray-scale image to an image size of 128×128 pixels, through the bicubic interpolation algorithm. The resized image is sent back to the software in through the FPGA_TX module. The module *Receive Data FPGA - CPU* converts the stream of bytes sent by the FPGA back to a typed image. Finally, the HOG-based feature extractor is applied to the image, and the classifier based on an EML neural network classifies the input image into one of the 62 possible characters.

A. Resize Technique

As mentioned, we use the bicubic interpolation technique to resize input images of any size to a standard size of 128×128 pixels. An interpolation function in one dimension can be defined as a convolution between a discrete function g and an interpolation kernel r (equation 1). The general form of a resized image through an interpolation function is given by equation 1:

$$f(x') = [r * g](x') = \sum_{m=-\infty}^{+\infty} r(dx - m).g(x + m) \quad (1)$$

where x' is the point where we want to estimate the value of f, where $x = \lfloor x' \rfloor$ and $dx = x' - x$.

The main difference between the various interpolation functions is the convolution *kernel*. After analyzing different kernels we chose the bicubic interpolation method, which is composed of third degree polynomials defined in the intervals: $(-2, -1)$, $(-1, 0)$, $(0, 1)$ and $(1, 2)$. Outside this interval, it assumes the value zero (equation 2).

$$r_{cub}(x, a) = \begin{cases} (-a+2)|x|^3 + (a-3)|x|^2 - 1, & \text{if } 0 \le |x| < 1, \\ -a|x|^3 + 5a|x|^2 - 8a|x| + 4a, & \text{if } 1 \le |x| < 2, \\ 0, & \text{if } |x| \ge 2 \end{cases} \quad (2)$$

such as $r_cub = 0$ when $|x|2$. Only four discrete values of $g(u)$ are required for the convolution operation. Therefore, the equation 1 can be rewritten as described below.

$$f(x') = [r * g](x') = \sum_{m=-1}^{+2} r_{cub}(dx - m).g(x + m) \quad (3)$$

Equation 3 defines the one-dimensional case. As digital images are two-dimensional, we need to redefine the interpolation function for the 2D case. So let I be an input image of $a \times b$ pixels and $R_{cub2D}(x, y) = r_{cub}(x)r_{cub}(y)$ or *kernel* of the 2-D bicubic interpolation, we can define the resized image I_{red} (of $p \times q$ pixels) through the following equation:

$$I_{red}(x', y') = \sum_{m=-1}^{+2} \sum_{n=-1}^{+2} I(x+m, y+n)r_{cub}(dx-m)r_{cub}(dy-n)$$

$$(4)$$

978-1-5386-4757-8/18 $31.00 © 2018 IEEE

in which (x', y') are the coordinates of any one pixel in the output image, where $(x, y) = (\lfloor x' * \frac{a}{p} \rfloor, \lfloor y' * \frac{b}{q} \rfloor)$, $dx = x' * \frac{a}{p} - x$ e $dy = y' * \frac{b}{q} - y$.

This resizing method is described in algorithm 1. In this work, the number of rows and columns in the output images, O_x and O_y, are 128. Algorithm 1 describes a convolution between the input image and the kernel using a neighborhood of 4×4 pixels (window) to produce each pixel of the image output.

Algorithm 1 Image resizing using the bicubic interpolation method

---

Input: I, the original image; I_x e I_y, the number of rows and columns of the original image, respectively; O_x e O_y, the number of rows and columns in the output image

Output: O, the resized image;

  for $i \leftarrow 0 : O_x$ **do**
    for $j \leftarrow 0 : O_y$ **do** $x \leftarrow \lfloor i * \frac{I_x}{O_x} \rfloor$ $y \leftarrow \lfloor j * \frac{I_y}{O_y} \rfloor$ $O(i,j) \leftarrow 0$
      for $m \leftarrow -1 : 2$ **do**
        for $n \leftarrow -1 : 2$ **do** $dx \leftarrow i * \frac{I_x}{O_x} - y$ $dy \leftarrow j * \frac{I_y}{O_y} - x$
          if $x + m < I_x$ **and** $y + n < I_y$ **then** $O(i,j) \leftarrow O(i,j) + I(x+m, y+n) * r_{cub}(dx - m)|_{a=0.5} * r_{cub}(dy - n)|_{a=0.5}$
          end if
        end for
      end for
    end for
  end for

Figure 5 illustrate the process described by algorithm 1. For each pixel of the output image, a region of 4×4 pixels adjacent to the input image (window) is selected, and each pixel in that region is multiplied by a corresponding weight, as defined by the interpolation kernel.

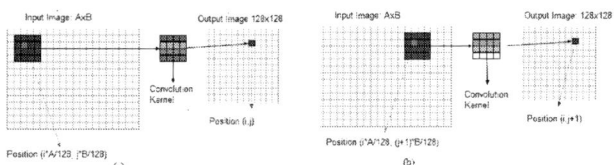

Fig. 5. Bicubic Interpolation Method

Although interpolation is quite similar to filtering, there are two main differences to be considered. In bicubic interpolation for large input images, distant points in the input image may result in points near the output image (see figures 5-a and 5-b). In the case of filtering, near points in the input image always generate near points in the output image. Additionally, in bicubic interpolation the convolution kernel values, r_{cub}, vary at each iteration depending on the values of dx and dy, which can be observed in line 15 of the algorithm 1. In the case of filtering, the filter values are static.

The shaded region over the input image represents the neighborhood of 4×4 pixels (window) used to produce a pixel in the output image. It can be seen in Figure 5 that near points in the output image ((i, j) and $(i, j+1)$) are generated by distant points in the input image. Additionally, for two different output pixels, the kernel values are not the same.

B. The BIA Module

The Bicubic Interpolation Accelerator is the proposed hardware module for resizing an input image of any size to the size of 128 x 128 pixels, through the bicubic interpolation algorithm. This module was designed in RTL-level and prototyped in FPGA.

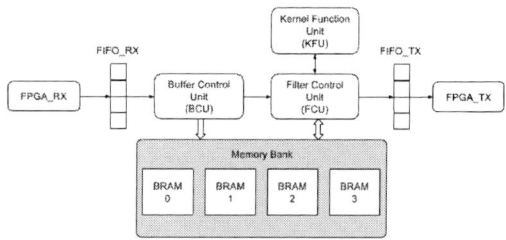

Fig. 6. Bicubic Interpolation Accelerator architecture.

The proposed architecture for the BIA module consists of three processing units: (1) Buffer Control Unit (BCU), (2) Filter Control Unit (FCU) and (3) Kernel Function Unit (KFU) and four on-chip memory banks of the BRAM type. This architecture was designed to be integrated with the processor as a standalone unit, flexible and able to adapt to various sizes of input images. The module was implemented as a pipeline and can provide one output pixel every six cycles of a 50 MHz clock. The BIA exploits the parallelism of the bicubic interpolation algorithm by processing 4 pixels (of the 16 pixels required to compute an output pixel) at each clock cycle. According to algorithm 1, given an input image I of any size, the method constructs the output image iteratively, calculating each output pixel, through the equation 4.

Two fundamental features increase the complexity when designing the bicubic interpolation hardware accelerator. In bicubic interpolation, the filter weights, i.e., the values of r_{cub}, are calculated dynamically for each input pixel, so it is not possible to use optimizations based on static filters. To cope with this, we have developed the *Kernel Function Unit*. For computing a given pixel of the output image, the bicubic interpolation algorithm accesses 16 pixels of the input image, which is the neighborhood of 4×4 pixels described before and indicated by the loops of algorithm 1. From this point, we will refer to this region as a window.

Unlike a filtering process, the coordinates of the pixels contained in the window are calculated dynamically for each output pixel, since they depend on the dimensions of the input image and the coordinates of the output pixel. Therefore, the use of simple input image storage mechanisms (i.e *shift registers*), widely used in filtering operations, is not suitable for an interpolation algorithm.

The Buffer Control Unit controls the storage of the input image. Input pixels are stored in 4 BRAMs since this storage structure are on-chip memories, and their read latency is only one clock cycle. At each clock cycle, this unit receives 4 consecutive pixels from the input image, and stores each of them in a different BRAM. That is, let p_i be the i-th pixel of the input image, p_i is stored in the $(i \mod 4)$th BRAM, in the address $i/4$. This pixel storage arrangement allows the Filter Control Unit module to access four consecutive pixels of the input image in parallel since two or more consecutive pixels will never be stored in the same BRAM. Since each BRAM stores, at maximum, one fourth of the input image, we set their sizes to $\lceil \frac{M*N}{4} \rceil$ bytes, where M and N are the height and width of the largest image in the training and test set. Therefore, in our experiments each BRAM has 81KB.

The Kernel Function Unit (KFU) is the unit responsible for dynamically computing weights for each input pixel, according to equation 2, where $a = 0.5$. That is, given an input value (x, y), this module returns the value of $r_{cub}(x) * r_{cub}(y)$. The input values of the function r_{cub}, $dx - m$ and $dy - n$ are real numbers. Therefore, in the hardware module, we represent the numbers in the fixed-point format $Q1.8.8$. That is, 17-bit registers are used to store numerical values, in which the first most significant bit is used as the signal bit, the next eight most significant bits represent the integer part of the number, and the least significant 8 bits represent the fractional part. Since we are using a fixed-point representation for $dx - m \in [-2, 2]$ and $dy - n \in [-2, 2]$, all possible input values of the KFU belong to the range $-512..512$ ($512 = 2 << 8$), so there are only 1025 possible outputs of r_{cub} (1025 is the number of integers in the range $-512..512$). For this reason, we are using a look-up table (LUT) to implement the kernel function, $r_{cub}(x)$. The main advantages of using a LUT in this unit are: (1) to avoid performing a large number of multiplications and additions for each input value and (2) the output of the module is available with the latency of one clock cycle.

As the BIA module exploits the 4-pixel parallelism in a window row, the KFU unit inputs are: (1) 4 ordered pairs, which correspond to the values of (dx, dy) for each pixel, and (2) the value of $m \in [-1, 2]$, which depends on which line is currently being processed. The value of n is known and fixed for all entries since we are handling all the pixels of a line in parallel. The Filter Control Unit (FCU) calculates for each output pixel the coordinates of the pixels belonging to the window; reads the BRAM memory pixels, based on the coordinates calculated; calculates the input values of the Kernel Function Unit, dx, dy and m; and calculates the input values of the Kernel Function Unit, dx, dy and m. In the design of this unit, we explore hardware parallelism by computing a line (i.e., 4 out of 16 pixels) of the 4×4 pixel window in a clock cycle. The operation of the FCU is implemented as a finite state machine. It controls the reading of the pixels of the input image, i.e., how the FCU controls access to BRAM memories and how to control the multiplication between the pixels read and the corresponding kernel weights.

V. RESULTS

The proposed approach was evaluated considering the accuracy and execution time. We performed experiments in two different scenarios. In the first one, we implement the proposed STCR only in software (C++) under Ubuntu 14.04, running on an Intel ATOM dual-core 1.6GHz, with 2GB of RAM at the DE2i-150 platform. For a fair comparison, the software implementation was also based on fixed point of format Q1.8.8 and the values of $r_{cub}(x)$ were computed through a *hash table*, to simulate a *look-up table*. In the second scenario, the part in software (C ++) runs on the ATOM processor. The hardware part was prototyped in a Cyclone IV. Communication between the processor and the FPGA is done through a PCI-E bus. Besides, we use the RIFFA framework [18] to interface the FPGA with the CPU. Both scenarios were evaluated using the Chars74K-15 benchmark [7]. It contains 930 training images and 930 test images, divided into 62 classes. Therefore, in all subsequent tests, it is assumed that Chars74K-15 was used.

The ELM parameters must be adjusted for the DE2I platform. To chose the number of neurons that results in a better performance and accuracy trade-off, we evaluate the impact of the number of neurons in the hidden layer on the average time for classifying an image in the target platform. Figure 7 summarizes the result. According to figure 7, the classification time of the ELM neural network grows approximately linearly with the number of neurons in the hidden layer. A good performance trade-off was obtained using 6000 neurons in the hidden layer. With this configuration, the system accuracy was 65.5%, a better result in a comparison of recent works, but also only 1.7 percentage points lower than the rate considering 18,000 neurons (for I7 based platform). Moreover, in this configuration, the average time for classifying an image was 50.7 ms, which is three times smaller than the average classification time of an ELM neural network with $L = 18000$ (151.9 ms).

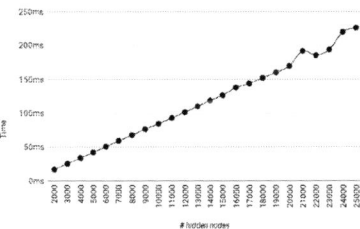

Fig. 7. Classification timing as a function of number of neurons.

For evaluating the average execution time of the proposed hybrid architecture for the STCR method, the individual times of each task were measured using the profiling tool Gprof [17] and the C library time.h. The execution time for the module *Resize* shown in table I is the CPU-FPGA communication time plus the BIA execution time. The results indicate that the BIA module is able to resize images up to 137 times faster than the C++ implementation.

Since we use a fixed point for representing real numbers in the *Resize* step, we evaluated its impact on the system's accuracy. Figure 8 indicates that there is no significant impact on the accuracy with the use of fixed point.

STCR Task	Atom N2600 (ms)	Atom + FPGA (ms)
RGB to Grayscale	1.1	1.1
Resize	220.34	**1.6**
Otsu	1.2	1.2
HOG	30.9	30.9
ELM	50.7	50.7

TABLE I

AVERAGE TIME EXECUTION FOR SOFTWARE ONLY AND HYBRID ARCHITECTURE

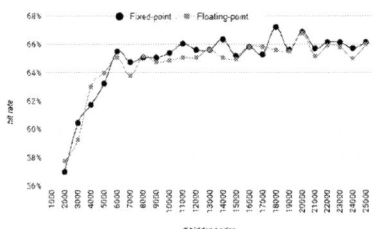

Fig. 8. System Accuracy for fixed-point and floating-point implementations

The proposed system was also compared to related works for the Chars74K-15 database. The table II indicates that the proposed architecture has a good trade-off between processing time and accuracy in embedded environments, being able to process up to 11 frames per second. The proposed CPU-FPGA hybrid architecture is 3.6 times faster than software-only implementation and up to 11.7 times faster than the another method implemented in an embedded system [1].

Meth	time (s)	Arch	Accuracy(%)
MKL [7]	-	-	55.8
Tensor+NN [19]	-	-	57.1
Fisher Vectors [10]	0.012	Intel i7	74.8
CNN + Stroke [8]	-	-	76.1
HOG+Adaboost [1]	1	ARM-A9 +GPU	60.0
Proposed in SW	0.308	Intel Atom	65.5
Proposed in HW-SW	0.085	Atom+FPGA	65.5

TABLE II

COMPARISON STCR APPROACHES USING DATASET CHARS74K-15

VI. CONCLUSION

In this work, we present a novel technique based on resize, HOG and ELM for character recognition in natural scenes (STCR). This technique achieves an accuracy of 65.5% in the dataset Chars74K-15, a good result compared with related works in an embedded environment. For accelerating the frame rate, we developed the BIA module, an accelerator for image resizing based on Bicubic Interpolation. The whole system was prototyped on the DE2i-150 embedded platform, which includes an ATOM processor and FPGA. The heterogeneous architecture is up to 11.7 times faster than the related works and can process up to 11 frames per second. The BIA module occupies only 11% of the FPGA area so that it can be used it in embedded systems with time and area constraints.

As future work, more FPGA accelerators can be developed for the other time-consuming parts of the algorithm, such as the ELM Neural Network and the HOG method. The proposed

STCR system can be integrated to a Scene Text Localization system to accelerate text recognition in natural end-to-end scenes. Finally, because the BIA is a very flexible unit, it can be used to accelerate any computer vision system that needs image resizing.

REFERENCES

[1] C. Yi and Y. Tian, "Scene text recognition in mobile applications by character descriptor and structure configuration," *IEEE Transactions on Image Processing*, vol. 23, no. 7, pp. 2972–2982, 2014.

[2] N. Dalal and B. Triggs, "Histograms of oriented gradients for human detection," in *Computer Vision and Pattern Recognition, 2005. CVPR 2005. IEEE Computer Society Conference on*, vol. 1. IEEE, 2005, pp. 886–893.

[3] A. Krizhevsky, I. Sutskever, and G. E. Hinton, "Imagenet classification with deep convolutional neural networks," in *Advances in neural information processing systems*, 2012, pp. 1097–1105.

[4] R. Keys, "Cubic convolution interpolation for digital image processing," *IEEE transactions on acoustics, speech, and signal processing*, vol. 29, no. 6, pp. 1153–1160, 1981.

[5] C. Chen, D.-H. Wang, and H. Wang, "Scene character and text recognition: The state-of-the-art," in *International Conference on Image and Graphics*. Springer, 2015, pp. 310–320.

[6] C. Shi, C. Wang, B. Xiao, S. Gao, and J. Hu, "End-to-end scene text recognition using tree-structured models," *Pattern Recognition*, vol. 47, no. 9, pp. 2853–2866, 2014.

[7] T. E. de Campos, B. R. Babu, and M. Varma, "Character Recognition in Natural Images," *Visapp (2)*, pp. 273–280, 2009.

[8] Z. Zhang, H. Wang, S. Liu, and B. Xiao, "Deep contextual stroke pooling for scene character recognition," *IEEE Access*, vol. 6, pp. 16 454–16 463, 2018.

[9] C. Zhang and V. Prasanna, "Frequency domain acceleration of convolutional neural networks on cpu-fpga shared memory system," in *Proceedings of the 2017 ACM/SIGDA International Symposium on Field-Programmable Gate Arrays*. ACM, 2017, pp. 35–44.

[10] C. Shi, Y. Wang, F. Jia, K. He, C. Wang, and B. Xiao, "Fisher vector for scene character recognition," *Pattern Recogn.*, vol. 72, no. C, pp. 1–14, Dec. 2017. [Online]. Available: https://doi.org/10.1016/j.patcog.2017.06.022

[11] X. Rong, C. Yi, and Y. Tian, "Recognizing Text-based Traffic Guide Panels with Cascaded Localization Network," *European Conference on Computer Vision*, pp. 1–14, 2016.

[12] M. Aggravi, A. Colombo, D. Fontanelli, A. Giannitrapani, D. Macii, F. Moro, P. Nazemzadeh, L. Palopoli, R. Passerone, D. Prattichizzo, and Others, "A Smart Walking Assistant for Safe Navigation in Complex Indoor Environments," in *Ambient Assisted Living*. Springer, 2015, pp. 487–497.

[13] K. Sanni, G. Garreau, J. L. Molin, and A. G. Andreou, "FPGA implementation of a Deep Belief Network architecture for character recognition using stochastic computation," in *2015 49th Annual Conference on Information Sciences and Systems (CISS)*, 2015, pp. 1–5.

[14] H. Zho, G. Zhu, and Y. Peng, "A RMB optical character recognition system using FPGA," in *2016 IEEE International Conference on Signal and Image Processing (ICSIP)*, 2016, pp. 539–542.

[15] N. Otsu, "A Threshold Selection Method from Gray-Level Histograms," *IEEE Transactions on Systems, Man, and Cybernetics*, vol. 9, no. 1, pp. 62–66, jan 1979.

[16] G.-B. Huang, Q.-Y. Zhu, and C.-K. Siew, "Extreme learning machine: theory and applications," *Neurocomputing*, vol. 70, no. 1, pp. 489–501, 2006.

[17] S. L. Graham, P. B. Kessler, and M. K. Mckusick, "Gprof: A call graph execution profiler," in *ACM Sigplan Notices*, vol. 17, no. 6. ACM, 1982, pp. 120–126.

[18] M. Jacobsen, D. Richmond, M. Hogains, and R. Kastner, "RIFFA 2.1: A reusable integration framework for FPGA accelerators," *ACM Transactions on Reconfigurable Technology and Systems (TRETS)*, vol. 8, no. 4, p. 22, 2015.

[19] M. Ali and H. Foroosh, "Natural Scene Character Recognition Without Dependency on Specific Features." in *VISAPP (2)*, 2015, pp. 368–376.

VLSI-Architecture of Radix-2/4/8 SISO Decoder for Turbo Decoding at Multiple Data-rates

Rahul Shrestha, *Member, IEEE*
School of Computing and Electrical Engineering
Indian Institute of Technology Mandi
Mandi-175005, India
rahul_shrestha@iitmandi.ac.in

Ashutosh Sharma
Center for VLSI and Embedded System Technologies
International Institute of Information Technology Hyderabad
Hyderbad-500032, India
ashutosh.sharma@research.iiit.ac.in

Abstract—We propose flexible-architecture for soft-input soft-output (SISO) decoder with radix-2/4/8 modes to support multiple data-rates. This work presents designs of major internal blocks of SISO decoder using extensive steering logic to support multiple radix operating-modes. These architectures enable efficient clock-gating of our decoder for low-power consumption in different operating modes. Subsequently, we have aggregated eight SISO decoders with quadratic-permutation-polynomial (QPP) interleavers/de-interleavers to design parallel-turbo decoder architecture which can operate in multi-radix mode. Suggested SISO decoder is ASIC synthesized and post-layout simulated in UMC 65 nm-CMOS process. Performance analyses in AWGN channel environment showed that the bit-error-rate (BER) of 10^{-4} could be achieved at 5 dB and 0.8 dB for SISO and turbo decoders respectively. Implementation result shows that the suggested SISO decoder could achieve throughput in the range $270-810$ Mbps with the corresponding power consumption range of $12.24-37.67$ mW. In comparison to the state-of-the-art, our design achieved 38% higher throughput and 61% lower power consumption. Similarly, our multi-radix parallel-turbo decoder is hardware prototyped in 28 nm-CMOS Zynq-FPGA board. It delivers a range of data-rates from $80-320$ Mbps operating at 160 MHz of clock frequency for 8 iterations.

Index Terms—SISO decoding, turbo code, parallel-turbo decoder, communication, VLSI design, digital architectures, FGPA.

I. INTRODUCTION

In the recent time, third-generation partnership-project (3GPP) has standardize the specifications for phase-I (Release 15) of 5G new-radio physical layer and the prospective 5G network will be commercially deployed in the year 2020 [1], [2]. This release has specified low-density parity-check (LDPC) and polar codes for data and control information encoding respectively [3]–[5]. Thereby, the dominant turbo encoding/decoding [6] from 4G technology has been excluded in the phase-1 of 5G technology. The primary reason being LDPC and polar decoders are lower in complexity and achieves adequate performance with area and power efficient hardware. On the other hand, new radio for 5G must support additional features like IoTs, vehicular communications and cloud computing [2]. These are short-range communications that takes place in dynamic network where the device need not operate at fixed data-rate consuming power at constant rate.

As we know that the channel decoder is the most power and area consuming blocks in the physical layer, it is high-time to conceive a flexible decoder architecture for supporting various data-rates. In comparison to LDPC and polar decoders, turbo decoder is more flexible and it is one of its key feature for the adaptation in 4G-LTE-Advanced standard where it supports 188 different block sizes. Hence, it is the most suitable channel coding scheme for the aforementioned features of 5G new radio and there is high probability of its inclusion in future 5G standardization (phase-2).

Soft-input soft-output (SISO) decoder based on Bahl-Cocke-Jelinek-Raviv (BCJR) algorithm is the prime block responsible for the design of turbo decoder [9] and thereby, it is essential to first design flexible SISO-architecture which can be aggregated to build turbo decoder for multiple data-rates. Literature is flooded with SISO decoder implementations and the selected ones are as follows. Chen-Hun Lin et al. presented scalable SISO-decoder (which performs single & double binary turbo-decoding) to support various 3G wireless standards and delivered dual-mode throughputs up to 500 Mbps [10]. However, it is necessary to enhance this throughput range with improved flexibility in the decoder architecture and maximize its achievable throughput. In this work, we are proposing reconfigurable very-large scale-integration (VLSI)-architecture for the SISO decoder in order to support different data-rates for various features of new radio. This design can be configured to perform decoding in either of radix-2/4/8 operating modes delivering various data rates. Internal architectures of SISO decoder like state-metric and logarithmic-likelihood-ratio (LLR) computation units are redesigned for supporting these modes. We have additionally used clock-gating technique to lower the power consumption. Eventually, we have combined eight such SISO-decoders with quadratic-permutation-polynomial (QPP) interleaver and de-interleaver blocks to design a new parallel-turbo decoder-architecture to support multiple data-rates. Rest of this paper is organized as follows: section II presents specification and theoretical background of SISO decoding. The proposed VLSI architectures along with qualitative analysis for SISO & turbo decoding are included in section III. Section IV incorporates bit-error-rate (BER) performance analysis, ASIC synthesis & post-layout simulation results and comparison with reported implementations. This section also

978-1-5386-4757-8/18 $31.00 © 2018 IEEE

presents field-programmable gate-array (FPGA) prototyping of parallel-turbo decoder and its implementation results. Finally, this paper concludes in section V.

II. DESIGN SPECIFICATIONS AND THEORETICAL PRELIMINARIES

We incorporated channel-coding specifications of 4G-LTE-Advanced standard [11] for our decoder architecture based on 8-states radix-2 trellis-graph which is generated using the convolutional-encoder (CE) transfer function G(D) = [1, $(1+D+D^3)/(1+D^2+D^3)$] with the constraint length (K_r) of 4, as shown in Fig. 1. Each CE used in the construction of turbo code for 4G-LTE-A standard must have a code-rate of r = 1/2 and thereby, we considered the same in this work [12]. The proposed decoder-architecture is design for the sliding-window based SISO decoding [13] for which the block-size (N) and sliding-window-size (M) are 6144 and 24 respectively. Since r = 1/2 for the systematic CE, there is a transmission of information bits u_j and encoded parity bits x_j $\forall$ j = {1, 2, 3, ... N}. At the receiver end, the soft demodulator provides $2 \times N$ a-priori LLR values to the SISO decoder input-side and are represented as λu_j and λx_j $\forall$ j = {1, 2, 3, ... N}. For every state transition in i^{th} stage of the trellis graph with N stages, there are 16 state transitions and each of them is associated with a branch metric which is denoted by $\gamma_i(s_m, s_n)$ $\forall$ {m, n} $\in$ {0, 1, 2, ... 7} where s_m and s_n are previous and present states respectively. Thereby, the branch metric for each state transition in j^{th} trellis stage can be computed as $\gamma_j(s_m, s_n) = \{u_j \times L(u_j) / 2\} + l_c/2 \times \{u_j \times \lambda u_j + x_j \times \lambda x_j\}$ where l_c and $L(u_j)$ are channel reliability and logarithmic a-priori probability respectively [13]. Rather than

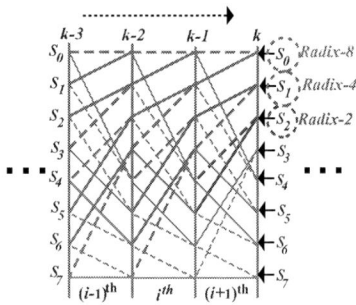

Fig. 1. Trellis graph for radix-2/4/8 SISO decoder.

computing all $2 \times S_N$ branch metrics where S_N is total number of states in the trellis graph (S_N = 8 in this work), we need to compute only $\log_2(2 \times S_N)$ parent branch metrics and use them in repetitive fashion to cover all state transitions [14]. In this work, state metric computation is performed in three different ways: radix-2, 4 & 8 which process single, double and triple trellis stages, respectively, in each clock cycle. Three stages of radix-2 trellis graph where s_0, s_1 and s_2 represent different states for radix-2, 4 & 8 forward state-metric computations

respectively. Thereby for radix-8, one of the forward state metric of state s_0 is computed as [15]

$$\alpha_k(0) = max\{\alpha_{k-1}(0) + \gamma_{i+1}(0,0), \alpha_{k-1}(1) + \gamma_{i+1}(1,0)\} \quad (1)$$

where

$$\alpha_{k-1}(0) = max\{\alpha_{k-2}(0) + \gamma_i(0,0), \alpha_{k-2}(1) + \gamma_i(1,0)\} \quad \& \quad (2)$$

$$\alpha_{k-1}(1) = max\{\alpha_{k-2}(2) + \gamma_i(2,1), \alpha_{k-2}(3) + \gamma_i(3,1)\} \quad (3)$$

where

$$\alpha_{k-2}(0) = max\{\alpha_{k-3}(0) + \gamma_{i-1}(0,0), \alpha_{k-3}(1) + \gamma_{i-1}(1,0)\}, \quad (4)$$

$$\alpha_{k-2}(1) = max\{\alpha_{k-3}(2) + \gamma_{i-1}(2,1), \alpha_{k-3}(3) + \gamma_{i-1}(3,1)\}, \quad (5)$$

$$\alpha_{k-2}(2) = max\{\alpha_{k-3}(4) + \gamma_{i-1}(4,2), \alpha_{k-3}(5) + \gamma_{i-1}(5,2)\} \quad \& \quad (6)$$

$$\alpha_{k-2}(3) = max\{\alpha_{k-3}(6) + \gamma_{i-1}(6,3), \alpha_{k-3}(7) + \gamma_{i-1}(7,3)\}. \quad (7)$$

Similarly, radix-4 computation of forward state-metric for state s_1 is performed as

$$\alpha_k(1) = max\{\alpha_{k-1}(2) + \gamma_{i+1}(2,1), \alpha_{k-1}(3) + \gamma_{i+1}(3,1)\} \quad (8)$$

where

$$\alpha_{k-1}(2) = max\{\alpha_{k-2}(4) + \gamma_i(4,2), \alpha_{k-2}(5) + \gamma_i(5,2)\} \quad \& \quad (9)$$

$$\alpha_{k-1}(3) = max\{\alpha_{k-2}(6) + \gamma_i(6,3), \alpha_{k-2}(7) + \gamma_i(7,3)\}. \quad (10)$$

Finally, the radix-2 forward-state metric for s_2 is computed as [15]

$$\alpha_k(2) = max\{\alpha_{k-1}(4) + \gamma_{i+1}(4,2), \alpha_{k-1}(5) + \gamma_{i+1}(5,2)\}. \quad (11)$$

On completing the computation of all branch, forward and backward state metrics, a-posteriori LLRs Γ_i $\forall$ j = {1, 2, 3, ... N} are determined for radix-2, 4 & 8 depending on the mode of operation. Detail discussion on the qualitative analysis of this process is presented in section III along with the proposed VLSI architectures.

III. PROPOSED VLSI ARCHITECTURES

A. State-Metric Computation-Unit Architecture

State-metric computation-unit (SMCU) of SISO decoder computes a forward/backward state metric for each state and thereby, S_N such units are stacked to compute all the state metrics of each trellis stage. We propose a flexible architecture for SMCU that computes state metrics for radix-2/4/8 SISO decoding. This architecture has been presented based on equations (1-11), as discussed earlier. The major challenge is to reliably route the branch and state metrics for different operating modes. As the radix value scales up (from radix-2 to radix-8), traversing of trellis becomes deeper for the state-metric computation of present state and the number of branch as well as state metrics of previous states, required for state metric calculation, gradually increases. The proposed SMCU architecture is presented in Fig. 2. On analyzing this design, Rad2-SMCU computes the state metric for radix-2 operating mode from (11). Subsequently, three such units are aggregated to construct Rad4-SMCU block for the calculation of radix-4 state metric using (8-10). These units are integrated with the multiplexer/de-multiplexer network to build reconfigurable SMCU which computes radix-8/4/2 state metrics for a given

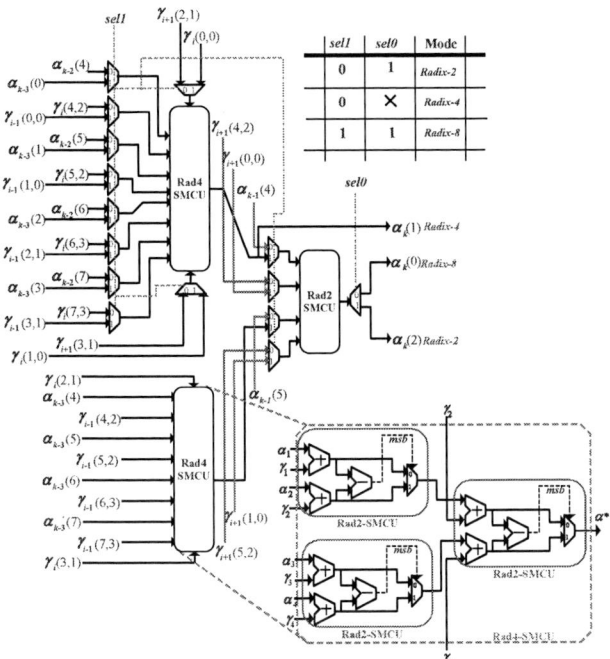

Fig. 2. Proposed VLSI architecture of reconfigurable SMCU for radix-2/4/8 operating modes.

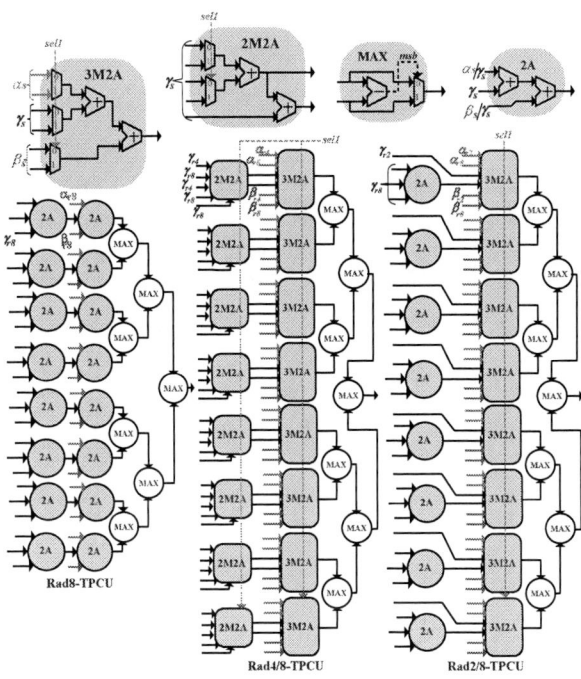

Fig. 3. Proposed reconfigurable VLSI-architectures of submodules used for computing transition probabilities in LCU.

state depending on the values of select lines *sel0* and *sel1*, as shown in Fig. 2. For example, when the value of *sel1* is high then all the state metrics of $(k-3)^{th}$ stage and branch metrics of $(i-1)^{th}$ stage, for radix-8 computation, are passed to Rad4-SMCU blocks through first stage of multiplexers. Subsequently, next multiplexer-stage releases two computed radix-4 state-metrics which are then fed to Rad2-SMCU along with respective branch metrics. Eventually, if *sel0* is high then de-multiplexer releases radix-8 state metric $\alpha_k(0)$, as shown in Fig. 2. In the same manner, this architecture can be used for radix-2/4 state-metric computation by changing the values of select lines.

B. Flexible Architectures of LLR Computation Unit & SISO Decoder

Computation of LLR for the multi-radix SISO decoder is a crucial and complex operation where radix-2, 4 and 8 operating modes must produce 1, 2 and 3 LLRs, respectively, in every computation cycle. Each stage of trellis graph for radix-4 and 8 is shown in Fig. 1. Considering radix-2 LLR computation, the transition probability for any arbitrary state transition $s_p \rightarrow s_q$ is computed as [13]

$$\lambda_{p:q} = \alpha_{k-1}(p) + \gamma_{i+1}(s_p, s_q) + \beta_k(q) \quad \forall \ p \in S_N \ \& \ q \in S_N. \quad (12)$$

Referring Fig. 1, LLR of $(i+1)^{th}$ stage is computed as

$$\Gamma_{i+1} = max_{\delta:u_j=1}\{\lambda_{p:q}\} - max_{\delta:u_j=0}\{\lambda_{p:q}\} = \lambda_{p:q}^1 - \lambda_{p:q}^0 \quad (13)$$

where $\delta: u_j = 1/0$ represents set of all state transitions for u_j is 0 or 1 [6]. Similarly, LLRs are computed for all the trellis stages to complete radix-2 decoding. In case of radix-4 LLR

computation, two of the radix-2 trellis stages are combined as one stage, as shown in Fig. 1. If $s_{q'}$ represents an intermediate state between $s_p \rightarrow s_q$ transition then its probability is computed as

$$\lambda_{p:q':q} = \alpha_{k-3}(p) + \gamma_{i-1}(s_p, s_{q'}) + \gamma_i(s_{q'}, s_q) + \beta_{k-1}(q). \quad (14)$$

Fig. 1 shows that there are four state transitions from previous stage towards each of the sink states in radix-4 trellis. Thereby, four transition probabilities for each of such states are represented as $\lambda_{p:q':q}^{00}$, $\lambda_{p:q':q}^{10}$, $\lambda_{p:q':q}^{11}$ and $\lambda_{p:q':q}^{01}$. Thereby, LLRs for $(i-1)^{th}$ and i^{th} trellis stages those are computed in one computation cycle are [16]

$$\Gamma_{i-1} = max_{\delta:u_j=1}\{\lambda_{p:q':q}^{10}, \lambda_{p:q':q}^{11}\} - max_{\delta:u_j=0}\{\lambda_{p:q':q}^{00}, \lambda_{p:q':q}^{01}\}, \quad (15)$$
$$\Gamma_i = max_{\delta:u_j=1}\{\lambda_{p:q':q}^{11}, \lambda_{p:q':q}^{01}\} - max_{\delta:u_j=0}\{\lambda_{p:q':q}^{00}, \lambda_{p:q':q}^{10}\}. \quad (16)$$

Similarly, there are eight state-transitions towards each sink state of radix-8 trellis stage, as shown in Fig. 1, where each transition probability is computed as

$$\lambda_{p:r:s:q} = \alpha_{k-3}(p) + \gamma_{i-1}(s_p, s_r) + \gamma_i(s_r, s_s) + \gamma_{i+1}(s_s, s_q) + \beta_k(q) \quad (17)$$

where s_r and s_s are intermediate states at $k-2$ and $k-1$ stages respectively. Therefore, three LLRs in each computation cycle of radix-8 decoding are calculated as [16]

$$\Gamma_{i-1} = max_{\delta:u_j=1}\{\lambda_{p:r:s:q}^{100}, \lambda_{p:r:s:q}^{110}, \lambda_{p:r:s:q}^{111}, \lambda_{p:r:s:q}^{101}\} - max_{\delta:u_j=0}\{\lambda_{p:r:s:q}^{000}, \lambda_{p:r:s:q}^{010}, \lambda_{p:r:s:q}^{011}, \lambda_{p:r:s:q}^{001}\}, \quad (18)$$

$$\Gamma_i = max_{\delta:u_j=1}\{\lambda_{p:r:s:q}^{110}, \lambda_{p:r:s:q}^{010}, \lambda_{p:r:s:q}^{011}, \lambda_{p:r:s:q}^{111}\} - max_{\delta:u_j=0}\{\lambda_{p:r:s:q}^{000}, \lambda_{p:r:s:q}^{100}, \lambda_{p:r:s:q}^{101}, \lambda_{p:r:s:q}^{001}\}, \quad (19)$$

$$\Gamma_{i+1} = \max_{\delta:u_j=1}\{\lambda_{p:r:s:q}^{011}, \lambda_{p:r:s:q}^{111}, \lambda_{p:r:s:q}^{101}, \lambda_{p:r:s:q}^{001}\} -$$
$$\max_{\delta:u_j=0}\{\lambda_{p:r:s:q}^{000}, \lambda_{p:r:s:q}^{100}, \lambda_{p:r:s:q}^{110}, \lambda_{p:r:s:q}^{010}\}. \qquad (20)$$

The transition probability computations from (12), (14)

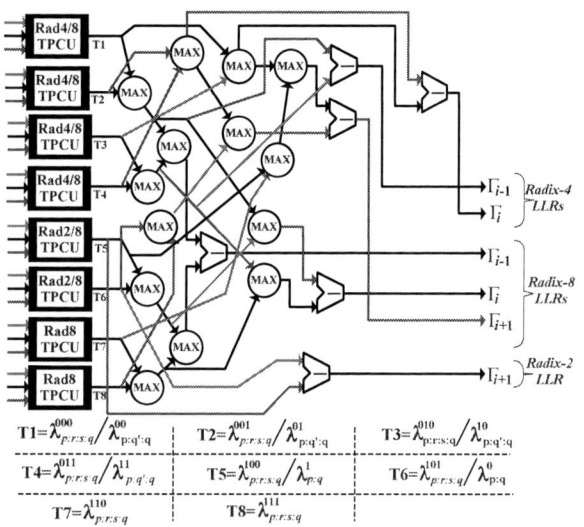

Fig. 4. Proposed VLSI-architecture of LLR computation unit for multi-radix SISO decoding.

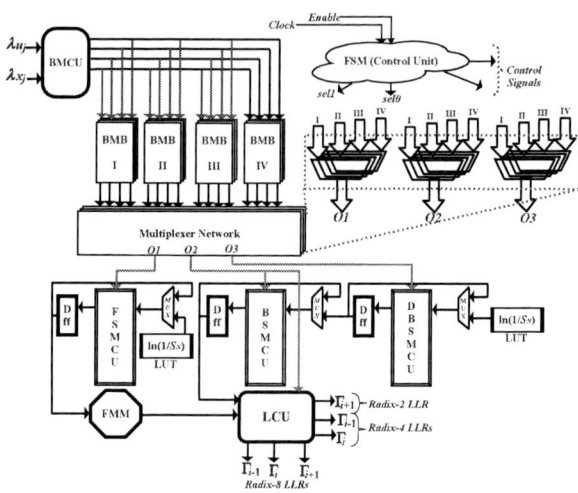

Fig. 5. Overall reconfigurable VLSI-architecture of multi-radix SISO decoder.

and (17) indicate that the number of branch metrics involved in such computation varies for different operating modes of multi-radix SISO decoding. Prior to the design of LLR computation unit (LCU), we propose architectures for its three integral modules: radix-8 transition-probability computation-unit (Rad8-TPCU), flexible-radix-4/8 transition-probability computation-unit (Rad4/8-TPCU) and flexible-radix-2/8 transition-probability computation-unit (Rad2/8-TPCU). Fig. 3 shows these new flexible architectures where γ_{r8}, γ_{r4} and γ_{r2} represent branch metrics for radix-8, 4 and 2 operating modes respectively. Similarly, α_{r8}/β_{r8}, α_{r4}/β_{r4} and α_{r2}/β_{r2} represent forward/backward state metrics for the same. Internal architectures of their sub-modules like 3M2A, 2M2A, 2A and MAX have been presented in Fig. 3. Suggested architectures of Rad4/8-TPCU and Rad2/8-TPCU can be reconfigured using *sel1* signal to compute transition probabilities of radix-4/8 and radix-2/8 operating modes respectively. Unlike, Rad8-TPCU computes transition probability for only radix-8 mode. On integrating these computation units along with the MAX comparator network, we finally present the proposed VLSI architecture of LCU in Fig. 4. It shows that the transition probabilities computed using eight TPCUs are fed to the network of MAX comparators which calculates the LLRs for radix-2, 4 and 8 operating modes based on (13), (15-16) and (18-20) respectively.

We designed sliding-window based SISO decoding where the entire trellis length is segregated into N/M windows and each of which comprises of M trellis stages [13]. Fig. 5 shows the suggested decoder architecture which is an aggregation of various sub-modules like: branch metric computation unit (BMCU), branch-metric memory banks (BMBs), multiplexer

network (MuN), forward-SMCU (FSMCU), backward-SMCU (BSMCU), dummy-BSMCU (DBSMCU), forward-state metric memory (FMM) and LCU. BMCU computes parent branch metrics based on the equations discussed in section II and BMBs are used for storing these values for different sliding windows. FSMCU as well as BSMCU comprises of S_N SMCUs for computing all the forward/backward state metrics of each trellis stage. DBSMCU has been included to perform dummy backward recursion to estimate the values of backward state-metric while starting the actual backward recursion [14]. Subsequently, this unit is initialized using $\ln(1/S_N)$ equiprobable value (stored in LUT) to start the dummy backward recursion, as shown in Fig. 5. MuN has been incorporated to route the branch metrics into FSMCU, BSMCU and DBSMCU for different sliding windows. Additionally, a finite state machine (FSM) based control unit has been designed to generate various control signals (like *sel1* and *sel0*) to operate this data-path of SISO decoder in radix-2/4/8 operating mode. Finally, computed values of forward & backward state-metrics and branch metrics are fed to LCU for computing the LLRs, as discussed earlier.

C. Overall parallel-turbo decoder architecture

We have integrated 8 proposed-SISO decoders in parallel with two inter-connecting networks (ICNWs) which is a Batchers network based QPP interleaver/de-interleaver [6]. In order to complete one iteration of turbo decoding, we have two stages and each of them is explained in this section. In stage-I, step1: received LLRs stored in memory banks *LLRx* and *LLRxp1* are read sequentially and their outputs Wi and $Yi \ \forall \ i=\{1, 2, 3, \dots 8\}$ are fed to 8 SISOs, as shown in Fig. 6. Step2: Radix-2/4/8 LLRs from SISOs are buffered in the register bank (REG). Now, both ICNW1 & ICNW2 are activated and the read address ($Oa92$) is used for fetching the LLRs from REG those are fed to ICNW2. Step3: ICNW1 starts functioning and its outputs $Xi \ \forall \ i=\{1, 2, 3, \dots 8\}$ are supplied

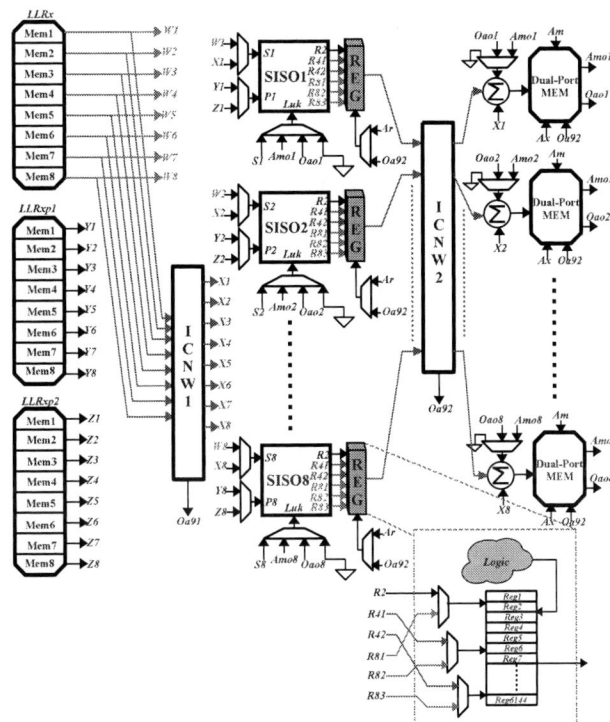

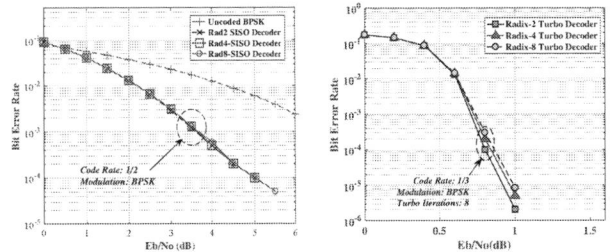

Fig. 7. Performance analysis of the proposed SISO and turbo decoders, compliant to LTE-Advanced wireless standard, under AWGN channel environment.

Fig. 6. Overall flexible architecture of parallel-turbo decoder with eight proposed multi-radix SISO decoders.

in Fig. 5, critical path lies in FSMCU/ BSMCU/ DBSMCU and all other units are feed-forward architectures which are coarse-grain pipelined in this work. Therefore, maximum clock frequency of SISO decoder operating in radix-8 mode is 276 MHz. Suggested architectures from Fig. 2 & 3 indicates that minimum and maximum number of units are activated while operating in radix-2 and 8 modes respectively. Thus, our SISO architecture can deliver throughputs from 270 to 810 Mbps with variable power consumptions, using clock-gating technique, as plotted in Fig. 8. Implementation results

to SISO decoders and subtractors which computes the extrinsic information that is stored in dual-port MEM in sequential fashion (using address Am). Step4: SISO decoders again computes multiple radix LLRs using the extrinsic information and stores them in REG. In stage2, step1: LLRs from REG are fetched sequentially and are fed to ICNW2. At the same time, ICNW1 is activated and $LLRx$ data are fetched using $Oa91$ address line and are fed to subtractor for computing extrinsic information. Step2: extrinsic information stored in dual-port MEM are fetched sequentially and stored in dual-port MEM using the write address $Oa92$ generated from ICNW2, as shown in Fig. 6. These two stages completes one iteration for turbo decoding and our decoder has been designed for 8 such iterations.

IV. EXPERIMENTAL RESULTS & COMPARISON

We performed Monte-Carlo simulation of the reconfigurable SISO and turbo decoding algorithms, in the additive-white Gaussian-noise (AWGN) channel environment, based on the specifications of LTE-A wireless standard discussed in section II. Fig. 7 shows the BER plots at different E_b/N_0 values (in decibel) for radix-2, 4 and 8 SISO as well as turbo decoding. It can be observed that the suggested decoders for different operating modes delivers identical performance. SISO decoder has been synthesized and post-layout simulated using UMC 65 nm-CMOS process, and its final chip-layout and implementation results are presented in Fig. 8 and Table I respectively. In the proposed SISO-decoder architecture shown

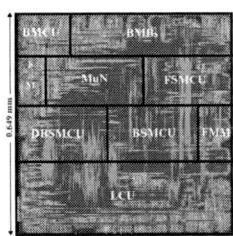 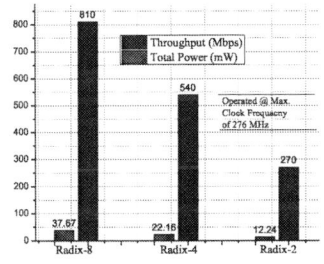

Fig. 8. ASIC chip layout incorporating 397k cell count and power-&-throughput analysis of the proposed reconfigurable SISO decoder when operated at maximum clock frequency.

of the proposed SISO decoder are compared with the reported works from literature, as listed in Table I. On the other side, the suggested parallel-turbo decoder has been hardware implemented in FPGA platform and it is functionally verified using valid test vectors which are obtained through high-level simulations. Fig. 9 shows the output waveform of the hard decoded bits along with the inputs like clock signal, input LLRs, reset for control counter, and select lines for choosing the mode of operation. Table II shows the implementation results of FPGA prototyped parallel-turbo decoder.

V. CONCLUSION

This paper presented new VLSI-architectures for various internal modules (like FSMCU, BSMCU, DBSMCU and LCU) and overall SISO decoder which supported multi-radix operating modes. Such SISO decoders has been suitably

978-1-5386-4757-8/18 $31.00 © 2018 IEEE

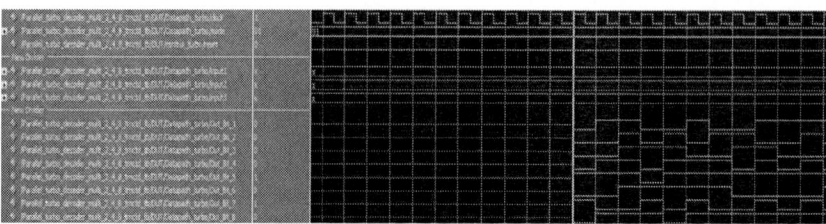

Fig. 9. Input/output waveform of our parallel-turbo decoder (operating in radix-4 mode) obtained after the post-route simulation in FPGA platform for functional verification.

TABLE I
COMPARISON OF THE PROPOSED SISO DECODER WITH THE REPORTED WORKS

	[17]	[18]	[10]	[6]	This work
Technology (nm)	180	130	130	130	**65**
Supply (V)	1.8	1.2	1.2	1.2	**1**
Area (mm²)	8.7	1.96	1.28	2.12	**0.42**
Scaled Area @ 65 nm (mm²)	1.13	0.49	0.32	0.53	**0.42**
Max. Clk. Freq. (MHz)	285	238	125	526	**276**
Total Power (mW)	330	528	97.83	NA	**12.24 — 37.67**
Radix-r	$r = 2$	$r = 16$	$r = 4$	$r = 2$	**$r = 2$ /4 /8**
Throughput (Mbps)	285	952	500	526	**270 — 810**

TABLE II
SUMMARY OF FPGA IMPLEMENTATION RESULTS OF 8-SISOS PARALLEL-TURBO DECODER OF OUR WORK

Target Device	xc7z100-2ffg900
FPGA Family	Zynq
No. of Slice Registers	40813 out of 554800 (7% utilization)
No. of Slice LUTs	177599 out of 277400 (64% utilization)
No. of LUT-FF pairs	33005 out of 185407 (17% utilization)
No. of Bonded IOBs	36 out of 362 (9% utilization)
No. of Block RAMs/FIFOs	24 out of 755 (3% utilization)
No. of BUFG/BUFGCTRLs	1 out of 32 (3% utilization)
Max. Clock Frequency	160.135 MHz (minimum delay 6.245 ns)

integrated to design parallel-turbo decoder to support various contemporary features like IoTs, cloud computing and vehicular communications. Our design consumed moderately lower power, delivered higher throughput and supported largest dynamic range in comparison to the reported works. This design can be prospective channel decoder for the physical layer specification of future 5G communication standard.

VI. ACKNOWLEDGMENT

The authors would like to thank Science and Engineering Research Board (SERB), Department of Science and Technology (DST), Govt. of India, for supporting this research work.

REFERENCES

[1] Y. Yifei and W. Xinhui, "5G new radio: Physical layer overview," *ZTE Communications*, vol. 15, no. S1, pp. 3-10, Jun. 2017.

[2] F. Boccardi, R. W. Heath, A. Lozano, T. L. Marzetta, P. Popovski, "Five disruptive technology directions for 5G," *IEEE Communications Magazine*, vol. 52, no. 2, pp. 74−80, Feb. 2014.

[3] E. Arikan, "Channel polarization: A method for constructing capacity-achieving codes for symmetric binary-input memoryless channels," *IEEE Trans. Inf. Theory*, vol. 55, no. 7, pp. 3051-3073, Jul. 2009.

[4] S. Kumawat, R. Shrestha, N. Daga, and R. P. Paily, "High-throughput LDPC decoder architecture using efficient comparison techniques & dynamic multi-frame processing schedule," *IEEE Trans. Circuits Syst. I, Reg. Papers*, vol. 62, no. 5, pp. 1421−1430, May 2015.

[5] T. Richardson and S. Kudekar, "Design of low-density parity check codes for 5G new radio," *IEEE Commun. Mag.*, vol. 56, no. 3, pp. 28-34, Mar. 2018.

[6] R. Shrestha and R. P. Paily, "High-throughput turbo decoder with parallel architecture for LTE wireless communication standards," *IEEE Trans. Circuits Syst. I, Reg. Papers*, vol. 61, no. 9, pp. 2699−2710, Sep. 2014.

[7] C. Berrou, A. Glavieux, and P. Thitimajshima. "Near Shannon limit error correcting coding and decoding: Turbo codes," *in Proc. IEEE Int. Conf. Commun.*, vol. 40, no. 8, pp. 1064−1070, 1993.

[8] T. Ilnseher, F. Kienle, C. Weis, and N. Wehn, "A 2.15 GBit/s turbo code decoder for LTE advanced base station applications," *in Proc. Int. Symp. Turbo Codes and Iterative Information Processing (ISTC)*, pp. 21−25, 2012.

[9] L. R. Bahl, J. Cocke, F. Jelinek, and J. Raviv, "Optimal decoding of linear codes for minimizing symbol error rate," *IEEE Trans. Inf. Theory*, vol. 20, no. 2, pp. 284−287, Mar. 1974.

[10] C.-H. Lin, C.-Y. Chen, and A.-Y. Wu, "Area-efficient scalable MAP processor design for high-throughput multistandard convolutional turbo decoding," *IEEE Trans. Very Large Scale Integr. (VLSI)*, vol. 19, no. 2, pp. 305−318, Feb. 2017.

[11] 3GPP; Technical Specification Group Radio Access Network, "E-UTRA; Multiplexing and Channel Coding (Release 10) 3GPP," *3GPP, TS 36.212, Rev. 10.0.0*, 2011, Std.

[12] P. Bhat, S. Nagata, L. Campoy, I. Berberana, T. Derham, G. Liu, X. Shen, P. Zong and J. Yang, "LTE-Advanced: An Operator Perspective," *IEEE Commun. Mag.*, vol. 50, no. 2, pp. 104−114, 2012.

[13] J. P. Woodard and L. Hanzo, "Comparative study of turbo decoding techniques: An overview," *IEEE Trans. Veh. Technol.*, vol. 49, no. 6, pp. 2208−2233, Nov. 2000.

[14] C. Benkeser, A. Burg, T. Cupaiuolo, and Q. Huang, "Design and optimization of an HSDPA turbo decoder ASIC," *IEEE J. Solid-State Circuits*, vol. 44, no. 1, pp. 98−106, Jan. 2009.

[15] C. Studer, C. Benkeser, S. Belfanti, and Q. Huang, "Design and implementation of a parallel turbo-decoder ASIC for 3GPP-LTE," *IEEE J. Solid-State Circuits*, vol. 46, no. 1, pp. 8−17, Jan. 2011.

[16] Y. Sun, Y. Zhu, M. Goel, and J. R. Cavallaro, "Configurable and scalable high throughput turbo decoder architecture for multiple 4G wireless standards," *in Proc. Int. Conf. Application-Specific Syst., Arch. and Processors*, pp. 209−214, 2008.

[17] S.-J. Lee, N. R. Shanbhag, and A. C. Singer, "A 285-MHz pipelined MAP decoder in 0.18 um CMOS," *IEEE J. Solid-State Circuits*, vol. 40, no. 8, pp. 1718−1725, Aug. 2005.

[18] C.-H. Tang, C.-C. Wong, C.-L. Chen, C.-C. Lin, and H.-C. Chang, "A 952MS/s max-log MAP decoder chip using radix-4..4 ACS architecture," *in Proc. IEEE Asian Solid-State Circuits Conf. (A-SSCC)*, pp. 79−82, 2006.

Two Combinatorial Problems on the Layout of Switching Lattices

Anna Bernasconi   Antonio Boffa   Fabrizio Luccio   Linda Pagli

Dipartimento di Informatica, Università di Pisa, Italy

{anna.bernasconi, antonio.boffa, fabrizio.luccio, linda.pagli}@unipi.it

Abstract—A non classical approach to the logic synthesis of Boolean functions based on switching lattices is considered, for which deriving a feasible layout has not been previously studied. The problem presents new interesting combinatorial and algorithmic aspects. Our basic assumptions are that the positions of the switches in the lattice are fixed in the synthesis stage, and the layout for connecting the subsets of switches with the same input literal must be realized in superimposed planes through vias that take the same switch area. The overall goal is to minimize the number of layers needed. Since multiple choices of input literals are possible for each switch, we first study how to assign a single literal to each switch, to minimize the number of lattice portions of adjacent cells associated to the same literal (Problem 1). Then we study how to derive a feasible layout by building connections onto different layers, to minimize the number of layers (Problem 2). Problem 1 is NP-hard. Problem 2 seems to be also intractable, and exhibits limit instances that require an exceedingly number of layers or are even unsolvable. Heuristic algorithms are then developed for both problems and their encouraging performances are proved on a set of known benchmarks.

Index Terms—Circuit Layout, Switching Lattices, Logic Synthesis, NP-Complete Problems

I. INTRODUCTION

The logic synthesis of a Boolean function is the procedure that implements the function into an electronic circuit. The literature on this subject is extremely vast and large part of it is devoted to *two-level* logic synthesis, where the function is implemented in a circuit of maximal depth 2 [9]. In this paper, we focus on a different synthesis method, based on *switching lattices*. A switching lattice is a two-dimensional array of four-terminal switches implemented in its cells. Each switch is linked to the four neighbors and is connected with them when the switch is ON, or is disconnected when the switch is OFF. The idea of using regular two-dimensional arrays of switches to implement Boolean functions dates back to a seminal paper by Akers in 1972 [1]. Recently, with the advent of a variety of emerging nanoscale technologies based on regular arrays of switches, synthesis methods targeting lattices of multi-terminal switches have found a renewed interest [2], [3], [6].

A Boolean function can be implemented in a lattice with the following rules:
- each switch is controlled by a Boolean literal;
- if a literal takes the value 1 all corresponding switches are connected to their four neighbors, else they are not connected;

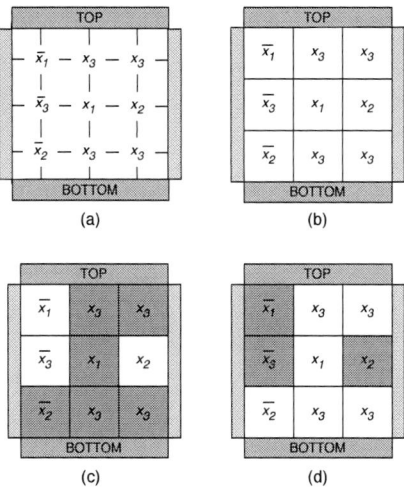

Fig. 1: A four terminal switching network implementing the function $f = \overline{x}_1\overline{x}_2\overline{x}_3 + x_1x_3 + x_2x_3$ (a); the corresponding lattice (b); the lattice evaluated on the assignments 1,0,1 (c) and 0, 1, 0 (d), with grey and white squares representing ON and OFF switches, respectively.

- the function evaluates to 1 for any input assignment that produces a connected path between two opposing edges of the lattice, e.g., the top and the bottom edges; the function evaluates to 0 for any input assignment that does not produce such a path.

For instance, the 3×3 network of switches in Fig. 1 (a) corresponds to the lattice form depicted in Fig. 1 (b), which implements the function $f = \overline{x}_1\overline{x}_2\overline{x}_3 + x_1x_3 + x_2x_3$. If we assign the values 1, 0, 1 to the variables x_1, x_2, x_3, respectively, we obtain paths of gray square connecting the top and the bottom edges of the lattices (Fig. 1 (c)), and f evaluates to 1. On the contrary, the assignment $x_1 = 0, x_2 = 1, x_3 = 0$, on which f evaluates to 0, does not produce any path from the top to the bottom edge (Fig. 1 (d)).

The synthesis problem on a lattice consists of finding an assignment of literals to switches implementing a given target function with a lattice of minimal size, measured as the number of switches in the lattice. In [2], [3], Altun and Riedel developed a synthesis method for switching lattices that assigns *at least* one literal to each lattice position, with the literal controlling the corresponding switch. If several literals are assigned to a switch the choice of the controlling literal is arbitrary.

978-1-5386-4757-8/18 $31.00 © 2018 IEEE   137

Starting from the lattice obtained by the Altun-Riedel method we consider two problems related to the physical implementation of the circuit, both motivated by the following considerations and assumptions.

1) Equal literals must be connected together, and to an external terminal on one side (e.g. the top edge) of the lattice. This may require using different layers, and vias to connect cells of adjacent layers.
2) Connections can be laid out horizontally or vertically (but not diagonally) between adjacent cells.
3) Each cell can be occupied by a switch, or by a portion of a connecting wire, or by a via. No two such elements can share a cell on the same layer.
4) The overall target is designing a layout with the minimum number of layers. Since the problem is hard, it will be relaxed to finding a reasonable layout by heuristic techniques.

As a consequence the circuit will be generally built starting from the original $N \times M$ lattice and superimposing to it a certain number H of layers, to give rise to a multidimensional grid of size $N \times M \times H$. Note that the switches associated with the same literal cannot be generally connected all together on the same layer, so several subsets of these switches will be connected on different layers and then connected through vias. The degree of freedom arising from the multiple choices of literals is exploited to enlarge these subsets (Problem 1). Then the connection of the different subsets with the same literal by themselves and by the external lead is addressed (Problem 2).

As said, we first consider the problem of assigning one literal to each switch in case of different choices at the switch. Consider the $N \times M$ lattice as a non-directed graph $G = (V, E)$ whose vertices correspond to the switches (then $|V| = NM$) and whose edges correspond to the horizontal and vertical connections between adjacent switches (then $E = 2NM - N - M$). We shall refer indifferently to the lattice or to the graph; to switches or to vertices; and to connections or to edges. Occasionally a vertex will be indicated with v_i, with $1 \leq i \leq N \cdot M$, or with a pair of integers (h, k) denoting the row and the column of the lattice where the vertex lays, with $1 \leq h \leq N$ and $1 \leq k \leq M$. Obviously the vertices have degree 2 or 3 if they lay on the corners or on the borders of the of the lattice, and have degree 4 if they are internal to the lattice. Finally let L be the set of literals occurring in the Boolean function. Each vertex v_i is associated with a non-void subset L_i of L, from which one literal has to be eventually assigned to v_i. Once a single literal assignment has been done for each vertex, an *area* denotes a *maximal connected subgraph* of G (or connected portion of the lattice) where all vertices have the same literal assigned. Note that if two areas A_1, A_2 have the same literal they must be disjoint and no two vertices $a_1 \in A_1, a_2 \in A_2$ may be adjacent in G. We pose:

Problem 1. *Find a literal assignment that minimizes the number of areas.*

Any literal assignment solving Problem 1, and the corresponding family of areas, is called an *MPA* for *minimal partition*

assignment. The problem has been studied in [8] for general graphs and for some of its variations, showing that is NP-hard on a lattice. Hence we will study how to solve it heuristically.

After Problem 1 is solved, we have to choose how to connect the different areas associated with the same literal and then connect them to the external input leads. To this end different layers are needed to attain all non-crossing connections. Formally we pose the following problem, whose complexity is discussed in Section III-B:

Problem 2: *Find a minimum number of layers allowing to connect together all areas with the same literal, and to connect them to the input leads, using non-crossing connections.*

The solution of Problem 1 gives the input for Problem 2. Since both problems are hard we solve them heuristically, showing experimentally that, for reasonable sizes of the lattice and of the number of variables, our heuristics allow to find efficient solutions.

II. SOLVING PROBLEM 1

The solution of Problem 1 may be simplified if a preliminary examen of the lattice is performed with the attempt of reducing the number of literals contained in the subsets associated to the vertices. For this purpose the following Rule 1 may be tested and applied if possible.

Rule 1. Let v_j be a vertex; v_1, v_2, v_3, v_4 be the four vertices adjacent to v_i (if any); L_j, L_1, L_2, L_3, L_4 be the relative subsets of literals. Apply in sequence the following steps:

Step 1. Let $|L_j| > 1$. If a literal $x \in L_j$ does not appear in any of the sets L_i, for $1 \leq i \leq 4$, cancel x from L_j and repeat the step until at least one element remains in L_j.

Step 2. Let $|L_j| > 1$, and let $L_k \subset L_j$ with $k \in \{1, 2, 3, 4\}$. If a literal $x \in L_j$ appears in exactly one set L_h with $h \in \{1, 2, 3, 4\}$ and $h \neq k$, then cancel x from L_j and repeat the step until at least the literals of L_k remain in L_j.

The following proposition be easily proved:

Proposition 1. *The application of Rule 1 does not prevent finding an MPA.*

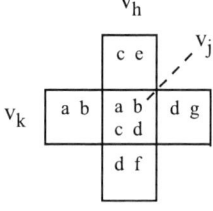

Fig. 2: Canceling a literal from a multiple choice using step 2 of Rule 1. Literals are denoted by a, b, c, d, e, f, g. Literal c in cells v_j, v_h is canceled from v_j.

An example of application of step 2 of Rule 1 is shown in Fig. 2. A literal cancelation from L_j may induce a further cancelation in an adjacent cell. In the example of Fig. 2, if all the cells adjacent to v_h except for v_j do not contain the literal c, the cancelation of c from L_j induces the cancelation of c from L_h if step 1 of Rule 1 is subsequently applied to v_h. Before running an algorithm for solving Problem 1, the sets L_i may be reduced using Rule 1 through a scanning of the lattice.

978-1-5386-4757-8/18 $31.00 © 2018 IEEE

1	1	5	5	2	2
1	4	2	1	5	5
3	3	2	1	5	7
3	3	2	4	6	6
5	6	6	3	3	7

Fig. 3: Example of starting lattice.

Moreover, as we have seen several successive scans may be applied for further reduction until no change occurs in a whole scan, although this is likely to produce much less cancelations than the first scan. These operations constitutes the first phase of any algorithm. Then, as the problem is computationally intractable, an heuristic must be applied. The one proposed here is the simplest possible, namely:

- scan the lattice row-wise: for any vertex v_i reduce the associated set of literals L_i to just one of its elements chosen at random;
- for any vertex v_i not yet included in an area, build a tree T_i spanning the maximal connected subgraph whose vertices hold the same label of T_i and include the vertices of T_i in a new area.

The experimental results discussed in the last section are derived with this simplified approach. Better results would be possibly obtained with more skilled heuristics at the cost of a greater running time.

III. Solving Problem 2

In order to better understand the nature of the problem let us explain it with an example. The starting point for Problem 2 is a lattice of $N \times M$, positions, each associated to one of the $2n$ literals, corresponding to the n input variables and their complements; in practical applications we have $2n < N \times M$, hence there are positions assigned to the same literal. As we have mentioned before, all these positions must be connected together in order to be reached in parallel from outside.

Fig. 3 shows an example of starting lattice of size 5×6 and 7 literals, (in this case directed variables only) indicated by numbers and not by x_i for simplicity. Let us suppose that the side devoted to connections to outside is the top side of the lattice. In the first layer the only connections we can lay out are those of the areas on the top row with outside and those connecting positions inside the same area. The connections of the first layer are shown in Fig. 4 (a). The first layer can be implemented only in this way for this example. In the first layer there is no room for other connections, hence a new layer must be added. For the already connected areas, it is enough that a single position of the area goes to the next layer. Obviously the final solution can be affected also by the selection of the position we choose, however we do not consider this possibility and we arbitrary choose this position: in our example of Fig. 4 (a), the underlined literals indicate that the corresponding position is selected.

A possible implementation of the second layer is shown in Fig. 4 (b). Recall that connections cannot cross, hence not all areas can be connected. Note that areas already connected

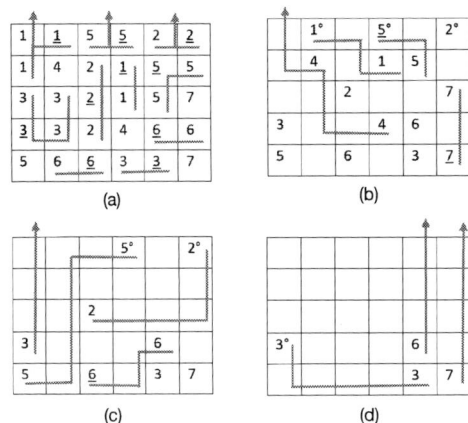

Fig. 4: (a) The first layer, (b) the second layer, (c) the third layer, and (d) the last layer.

	0	1	2
0	1	2	3
1	2	3	1
2	3	1	2

Fig. 5: A non solvable instance.

to the top side, even if not completely connected, don't need to be connected to the top side again. For example, the area associated to literal 1, already connected to outside in layer 1, is not connected to the top side in layer 2; areas associated to literals 3, 6, 7 instead have still to be connected. Note also that all literals with label 4 have been connected in a single area and outside, therefore there is no need to connect this area to the next layer. The next layer, layer 3 is depicted in Fig. 4 (c), and the fourth and last layer is shown in Fig. 4 (d). The way of selecting the connections is also arbitrary for layer 3, while all the rest is connected in the last layer.

In principle we do not know if the given problem can be solved in less than 4 layers. In fact, as we discuss below, the problem is computationally quite difficult, if not at all impossible to be solved. For a formal approach we define a *legal wiring* an assignment of the connections to a layer such that all the rules $(1-4)$ introduced in Section I are respected. Then Problem 2 is formalized as follows:

Instance: An array of size $N \times M$ positions, containing integers in the range $1, \dots k$, $k < N \times M$ or empty.

Goal: Connect together all integers with the same value and lead the connections to row 1, in a legal wiring, to obtain the layout with the minimal number H of layers.

A. Impossible Instances

It is not difficult to verify that not all possible configurations of the input lattice are solvable. For instance consider an array where each row contains a cyclic shift of the literals in the previous row, as in the example of Fig. 5. Since no cell can be connected with others with the same literal, no connections are possible in any layer. We now show that the vast majority of problem instances are theoretically solvable, although some may require an exceedingly high number of layers to be practically solved. We have:

Proposition 2. *A problem instance cannot be solved if and only if in the initial literal assignment no two adjacent cells share a same literal and, no matter how the multiple assignments are resolved in Problem 1, each cell in row zero contains a literal that occurs also in another cell.*

Proof. *If part.* If no two adjacent cells share a same literal initially, in the assignment of Problem 1 all areas contain exactly one literal. Although the cells of row zero can be connected to the output, there is no way to connect them to the other cells with the same literal since all cells will be occupied by a connection in all layers.

Only-if part. If at least one of the conditions stated in the proposition does not hold, at least one cell in layer 2 is made available (or "free") for routing due to an area built in layer 1, or to a connection to the output in that layer. Once a free cell arises, it can be "moved" to any cell of the array by consecutive movements of adjacent literals as in the well known *15-slide game*, and any literal adjacent to the free cell can similarly be moved around to be brought adjacent to a cell with the same literal. Proceeding with this strategy all cells with a same literal can be linked together and brought to the output.   □

Note that the strategy indicated in the only-if part of the above proof may require a very large number of layers if only a small number ν of free cells exist in a layer, as only ν movements can be done in that layer. In particular, if only a few cells are made free by the solution of Problem 1, i.e. if layer 1 contains a large number of small areas, the routing mechanism may not apply in practice. As we shall see, an answer must be left to simulations on significant examples.

B. Hardness of Problem 2

Solving Problem 2, the cells containing the same literal in any layer will be connected as trees (not as general subgraphs) to avoid useless occupation of free cells. The problem of minimizing the number of layers is related to the one of building the maximum number of such trees in any layer whose edges do not intersect. If a 15-slide movement of free cells is required the problem is NP-hard [10]. If such movements are not required the problem has strong similarities with other known NP-hard problems dealing with grid embedding of graphs, as for example determining the Steiner tree among k vertices on a grid [4], or determining the rectilinear crossing number of a graph [5], etc. We have not been able to prove that Problem 2 is NP-hard also in this simpler case, however for its solution we rely on a heuristic algorithm that produces satisfying results on a large class of benchmark instances. If no tree can be directly built in a layer, as discussed in the previous subsection, the heuristic stops declaring that routing is impossible. Otherwise we have:

Proposition 3. *Let α be the number of areas generated by Problem 1 and k be the number of literals. An upper and a lower bound to the number of layers are given by α and $\lceil k/M \rceil$, respectively.*

Proposition 3 is immediately proved by noting that at least one pair of cells holding the same literal are connected in each

layer (upper bound), and k external leads must be reached from the M cells of the upper row of the grid (lower bound). In the example of Fig. 4 we have $\alpha = 15$, $k = 7$, and $M = 6$. The proposed layout with $H = 4$ layers is far from approaching the upper bound 15, while is reasonably close to the lower bound $\lceil 7/6 \rceil = 2$.

C. Heuristics for Problem 2

We propose a heuristic algorithm for solving Problem 2. In this first approach, we never "move" free cells of the array, as in a 15-slide game, as this strategy might lead to layouts with a very high number of layers; thus we consider impossible all instances whose solvability requires such moves. This limitation only slightly affects our experimental results on benchmark circuits, as our algorithm failed to find layout for theoretically solvable lattices only for about 4% of the lattices (see Section IV).

A general greedy heuristic for Problem 2, could consist in the following two main steps:

(i) Connect all adjacent cells with the same literal on the first layer and then connect them to the external input leads, whenever possible;

(ii) while there are literals still to be connected between them and/or to the outside:
- add a new layer
- try to connect each pair of cells assigned to the same literal
- try to connect each literal to outside, if not already connected to the external leads.

Step (i) can be implemented in a standard way, visiting the lattice to search and connect all positions inside an area of adjacent cells assigned to the same literal. Moreover, all areas with a cell on the top row can be connected to the external leads. Note that this initial step is optimal, i.e., an optimal minimization algorithm for the number of layers cannot do better on the first layer.

To implement the second and main step of the heuristic, we introduce the concept of *free area* and *boundary cells*, that will be exploited to check whether two cells with the same literal can be connected, and to search a path between them.

Definition 1. *A* free area *in a lattice is a subset of free adjacent cells. The* boundary cells *of a free area are the cells surrounding it.*

Free areas are computed through a scanning of the lattice, in time linear in its dimension. An example of lattice with three different free areas is shown in Fig. 6. Observe that all boundary cells assigned to a single literal facing the same free area can be connected, since we can use the free cells inside the area to lay out the connections. Of course, this holds only for the first literal of the boundary cell that will be processed, while the others boundary cells, assigned to different literals, can be connected only if the required connections do not cross those already laid out on that area, since different connections cannot share a position.

Our idea is to use free areas to avoid the search for connections that are impossible to implement: we limit the

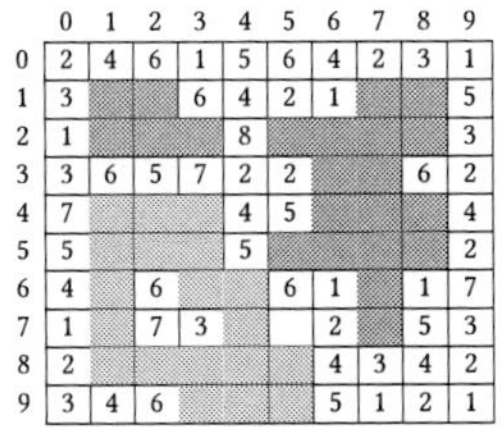

Fig. 6: A lattice with three free areas.

Algorithm for layout computation

Function *Thread (Free Area A)*
   for each *pair of boundary cells c_1, c_2 with the same literal*
      SearchHeuristic(c_1, c_2)

Function *main ()*
   Connect areas of adjacent cells with the same literal
   Connect cells on the top row to the external leads
   while *there are cells still to be connected*
      Select a single position in each area for the next layer
      Add a new layer
      Compute the free areas
      for each *(free area A)*   StartNewThread *(Thread (A))*
      JoinThread()

Fig. 7: Layout computation.

---

search to the connections traversing a free area, leaving the search for connections between cells facing different areas to the next iterations. In this way we can save computational time, as cells around different areas cannot be directly connected (see for example the two cells $(1,0)$ and $(2,9)$ assigned to the literal 3 in the lattice in Fig. 6). Therefore, we structure step *(ii)* of the proposed heuristic as follows: *first compute all free areas, then try to connect all boundary cells assigned to the same literal facing the same free area*. Since free areas are mutually disjoint, the searches for connections can be performed in parallel creating a thread for each free area. The only portion of the lattice shared by multiple threads are the boundary cells facing different free areas. These situations are managed with appropriate lock variables, which force the threads to access those cells in mutual exclusion.

As already mentioned, a drawback of this strategy is that it does not completely eliminate the search for connections that are impossible to implement on the current layer. Indeed, as soon as two cells have been connected through a path on a free area, some other cells facing this same area become unreachable from one another, since the first connection divided the free area in separate subareas. We could solve this issue recomputing the free areas after each connection, but this approach is computationally very heavy. Therefore, we compute free areas only once, at the beginning of each iteration, and then apply *non-exhaustive search algorithms* within each area, in order to limit the search for non-existing connections, still guaranteeing that mutually reachable cells will be connected with high probability. The overall heuristic is described in Fig. 7.

Let us now briefly discuss the possible implementations of

the search heuristics within each area. We are given a boundary cell c_1 that must be connected to a target cell c_2, through a path of distinct free cells in the free area A. This can be formalized as a state space search, where the state space is of size $O(4^{N \times M})$ as the number of cells in the area is $O(N \times M)$ and there are at most four possible moves from each cell. As a search in this space would be prohibitively expensive, we can use heuristics to find solutions of high quality as quickly as possible. We have considered *best-first, beam searches, greedy beam search, and hill-climbing* heuristics [7], [11]. These heuristics select the next cell to visit according to an evaluation function h that provides an estimate of the distance from the target cell, both under Euclidean and Manhattan distance. The first two heuristics provide better results, but are computationally very expensive (their time complexity is $O(4^{N \times M})$ and $O(2^{N \times M})$, respectively) and can be applied only to small size lattices. The last two have time complexity linear in the lattice size, but produce worse quality results, as they fail in connecting some mutually reachable cells on a given layer and the final layout may contain a very high number of layers. Depending on the lattice size and on the specific application, we can therefore select one of the four heuristics, trading-off quality of results vs. scalability.

IV. EXPERIMENTAL RESULTS

In this section we report the experimental results related to the physical implementation of switching lattices, according to the rules (1-4) described in Section I. The aim of our experimentation is to determine whether the physical implementation of the lattices, shaped as a multidimensional grid of size $N \times M \times H$, where N and M are the number of rows and columns of the lattice, and H is the number of layers, could be considered technologically feasible. In our work we have considered the lattices obtained applying the Altun-Riedel method to the benchmarks taken from LGSynth93 [12], where each output has been treated as a separate Boolean function. Due to the limited space available, we report in the following only a significant subset of the functions as representative indicators of our experiments.

The experiments have been run on a IntelCore i7-4710HQ 2.50GHz CPU with 8 GB of main memory, running Linux Ubuntu 17.10. The algorithm for computing the number of layers has been implemented in C.

In this first experimental evaluation, we have analyzed lattices where the literals assigned to the switches have been chosen arbitrarily, in all cases of different choices at the switch. In Table I we report (a subset of) the results of this experimentations. The first column reports the name and the number of the separate output function of the benchmark circuit. The following two columns report the number of different literals occurring in the lattice and its dimension ($N \times M$). Finally, the last four columns report the number H of layers computed by the Algorithm in Fig. 7 with the *best-first* and with the *greedy beam search* heuristic, together with the corresponding running time (in seconds). The last row reports the sum of the values of the corresponding column. The cases

978-1-5386-4757-8/18 $31.00 © 2018 IEEE

TABLE I: Number of layers for the lattice layout of a subset of standard benchmark circuits, for lattices with arbitrary selection of literals in all cases of different choices at the switch.

Bench	lit	N×M	Best-first H	Best-first Time(s)	Greedy Beam H	Greedy Beam Time(s)
add6(5)	24	156×156	8	757.59	9	0.01
adr4(1)	16	36×36	21	0.85	24	0.08
alu2(2)	16	10×11	4	0.01	5	0.01
alu2(5)	20	13×14	5	0.01	5	0.01
alu3(0)	8	4×5	4	0.01	4	0.01
alu3(1)	12	7×8	5	0.01	5	0.01
b12(0)	7	6×4	4	0.01	4	0.01
b12(1)	9	5×7	5	0.01	5	0.01
b12(2)	10	6×7	7	0.01	7	0.01
bcc(5)	28	27×9	11	0.04	12	0.01
bcc(7)	29	31×11	19	0.08	26	0.02
bcc(8)	29	31×12	17	0.07	20	0.02
bcc(27)	28	39×19	18	0.21	20	0.06
bcc(43)	28	20×10	9	0.02	10	0.01
bench1(2)	18	45×24	22	0.43	29	0.15
bench1(3)	18	31×16	29	0.19	45	0.04
bench1(5)	18	50×27	20	0.51	25	0.20
bench1(6)	18	35×21	25	0.33	32	0.08
bench1(7)	18	43×21	19	0.24	22	0.11
bench1(8)	18	44×24	22	0.45	28	0.14
bench(6)	10	8×4	6	0.01	6	0.01
br2(4)	18	18×8	9	0.01	10	0.01
br2(5)	19	14×4	16	0.01	16	0.01
br2(6)	19	16×5	14	0.01	14	0.01
clpl(3)	11	6×6	2	0.01	2	0.01
clpl(4)	9	5×5	2	0.01	2	0.01
co14(0)	28	92×14	12	0.38	14	0.05
dc1(4)	7	5×4	6	0.01	6	0.01
dc2(4)	11	10×9	8	0.01	8	0.01
dc2(5)	9	6×6	7	0.01	7	0.01
dk17(1)	10	8×2	6	0.01	6	0.01
dk17(3)	11	11×3	9	0.01	9	0.01
dk17(4)	12	9×3	−	0.01	−	0.01
ex1010(0)	20	91×46	34	29.22	40	1.35
ex4(4)	13	17×6	7	0.01	7	0.01
ex4(5)	27	35×45	14	1.12	15	0.08
ex5(32)	14	4×10	8	0.01	8	0.01
ex5(36)	11	2×8	4	0.01	4	0.01
ex5(38)	13	4×9	5	0.01	5	0.01
ex5(40)	15	6×12	10	0.01	14	0.01
ex5(43)	15	8×14	11	0.01	15	0.01
exam(5)	13	11×6	8	0.01	10	0.01
exam(9)	20	59×30	27	2.19	33	0.43
max128(5)	14	14×17	11	0.01	13	0.01
max128(8)	13	5×10	10	0.01	12	0.01
max128(17)	14	26×25	20	0.18	19	0.04
max1024(5)	20	117×122	31	1087.12	33	13.83
mp2d(6)	14	10×6	10	0.01	10	0.01
mp2d(9)	14	6×8	4	0.01	4	0.01
mp2d(10)	10	6×3	−	0.01	−	0.01
sym10(0)	20	130 × 210	11	2938.34	13	11.57
tial(5)	28	181×181	21	5622.57	30	40.99
z4(0)	7	15×15	7	0.01	9	0.01
z4(1)	14	28×28	11	0.12	13	0.01
Z5xp1(2)	14	12×11	10	0.01	10	0.01
Z5xp1(3)	14	18×18	13	0.03	16	0.01
			658	10442,61	770	72,32

where the algorithm failed in finding a layout for theoretically unsolvable lattices are marked with −. Considering the whole set of benchmarks analyzed, the algorithm did not find a layout for about 4% of the lattices.

By comparing the results, the values show that, as expected, we obtain a better layout using the best-search heuristic, at the expense of the computational time. Moreover, we note that the increase in the number of layers computed with the faster greedy beam search heuristic appears quite limited on average, while it can be relevant on single lattices (see for example benchmarks $bench1(3)$ and $tial(5)$).

V. CONCLUDING REMARKS

We have presented the first study on connection layout for two-dimensional switching lattices referring to the network implementation proposed by Altun and Riedel [3]. We have shown how to build a stack of consecutive layers where the connections between switches driven by the same variable can be laid without crossings, with the aim of minimizing the number of layers. Since the problem is computationally intractable we have designed a family of heuristics for finding satisfactory solutions, reporting only the results of the fastest and the slowest of the two on a standard subset of Boolean functions, for space reasons.

Countless improvements are open. For theoretical completeness, the NP-hardness of step 4 of our approach has to be proved to fully justify the use of heuristics. Better heuristics could be studied, and tested on larger data samples. The layout for other switching lattices should be considered. The layout rules should be possibly changed, in particular allowing more than one wire traversing a switch area in the higher layers. We are presently working on all these issues.

REFERENCES

[1] S. B. Akers, "A Rectangular Logic Array," *IEEE Trans. Comput.*, vol. 21, no. 8, pp. 848–857, Aug. 1972.

[2] M. Altun and M. D. Riedel, "Lattice-Based Computation of Boolean Functions," in *Proceedings of the 47th Design Automation Conference, DAC 2010, Anaheim, California, USA, July 13-18, 2010*, 2010, pp. 609–612.

[3] ——, "Logic Synthesis for Switching Lattices," *IEEE Trans. Computers*, vol. 61, no. 11, pp. 1588–1600, 2012.

[4] C. C. N. Chu and Y. Wong, "FLUTE: fast lookup table based rectilinear steiner minimal tree algorithm for VLSI design," *IEEE Trans. on CAD of Integrated Circuits and Systems*, vol. 27, no. 1, pp. 70–83, 2008. [Online]. Available: https://doi.org/10.1109/TCAD.2007.907068

[5] J. Fox, J. Pach, and A. Suk, "Approximating the rectilinear crossing number," *CoRR*, vol. abs/1606.03753, 2016. [Online]. Available: http://arxiv.org/abs/1606.03753

[6] G. Gange, H. Søndergaard, and P. J. Stuckey, "Synthesizing Optimal Switching Lattices," *ACM Trans. Design Autom. Electr. Syst.*, vol. 20, no. 1, pp. 6:1–6:14, 2014.

[7] P. E. Hart, N. J. Nilsson, and B. Raphael, "A formal basis for the heuristic determination of minimum cost paths," *IEEE Trans. Systems Science and Cybernetics*, vol. 4, no. 2, pp. 100–107, 1968. [Online]. Available: https://doi.org/10.1109/TSSC.1968.300136

[8] F. Luccio and M. Xia, "The MPA graph problem: definition and basic properties." *Department of Informatics, University of Pisa. Technical Report.*, 2018.

[9] G. D. Micheli, *Synthesis and Optimization of Switching Theory.* Mc-Grow Hill, 1994.

[10] D. Ratner and M. K. Warmuth, "Finding a shortest solution for the N × N extension of the 15-puzzle is intractable," in *Proceedings of the 5th National Conference on Artificial Intelligence. Philadelphia, PA, August 11-15, 1986. Volume 1: Science.*, 1986, pp. 168–172.

[11] S. J. Russell and P. Norvig, *Artificial intelligence - a modern approach, 2nd Edition*, ser. Prentice Hall series in artificial intelligence. Prentice Hall, 2003. [Online]. Available: http://www.worldcat.org/oclc/314283679

[12] S. Yang, "Logic Synthesis and Optimization Benchmarks User Guide Version 3.0," Microelectronic Center, User Guide, 1991.

HLS Support for Polymorphic Parallel Memories

L. Stornaiuolo*, M. Rabozzi*, D. Sciuto*, M. D. Santambrogio*, G. Stramondo[+], C. Ciobanu[+], A. L. Varbanescu[+]

*Politecnico di Milano, Milan, Italy; [+]Universiteit van Amsterdam, Amsterdam, Netherlands

{luca.stornaiuolo, marco.rabozzi, donatella.sciuto, marco.santambrogio}@polimi.it

{g.stramondo, c.b.ciobanu, a.l.varbanescu}@uva.nl

Abstract—The importance of High-Level Languages in abstracting machine language to enhance productivity has been proved in many sectors, and has recently encouraged the spread of reconfigurable hardware for general purpose computing. At the same time, Field Programmable Gate Arrays (FPGAs) become popular for data-intensive applications, because they promise customized hardware accelerators and achieve high-performance with low power consumption. However, taking advantage of parallel accesses to the local memories of FPGAs remains difficult, as it currently requires application re-engineering. A solution to this challenge is PolyMem, an easy-to-use parallel memory. In this work, we investigate the implementation, integration, and performance of PolyMem for HLS applications. To this end, we present a novel open-source implementation of PolyMem, optimized for the Xilinx Design Suite. We further demonstrate the use of PolyMem for three different case studies, implemented using both the Vivado workflow with a Virtex-7 VC707, and the SDx workflow with a Kintex Ultrascale 3 ADM-PCIE. Finally, we provide a thorough empirical analysis of these three cases studies in terms of latency, hardware resources, and productivity. Our results demonstrate that PolyMem delivers the expected performance, while enhancing productivity at the cost of a small increase in resources.

I. INTRODUCTION

The success of High-Level Languages (HLLs) for non-traditional computing systems, like Graphics Processing Units (GPUs) and Field Programmable Gate Arrays (FPGAs), have accelerated the adoption of these platforms for general purpose computing. In particular, the main hardware vendors released tools and frameworks to support their products by allowing the design of optimized kernels using HLLs. This is the case, for example, for Xilinx, which allows using C++ or OpenCL within the Vivado Design Suite [1] to target FPGAs.

Moreover, FPGAs are increasingly used for data-intensive applications, because they enable users to create custom hardware accelerators, and achieve high-performance implementations with low power consumption. However, one aspect still lagging behind is the efficient use of BRAMs, the FPGA distributed, high-bandwidth, on-chip memories [2]. BRAMs can provide memory-system parallelism, but their use remains challenging due to the many different ways in which data can be partitioned in order to achieve efficient parallel data accesses. Changing data access patterns on the application side is the current state-of-the-art approach, which does parallelize operations and reduces the kernel execution time, but also requires extensive modification of the application code.

To address the challenges related to the design and practical use of parallel memory systems for FPGA-based applica-

tions, PolyMem, a Polymorphic Parallel Memory, was proposed [3]. PolyMem is envisioned as a high-bandwidth, two-dimensional (2D) memory *used to cache performance-critical data on the FPGA chip*, making use of the existing distributed memory banks (the BRAMs). PolyMem is inspired by the Polymorphic Register File (PRF) [4], a runtime customizable register file for Single Instruction, Multiple Data (SIMD) co-processors. PolyMem is tailored for FPGA accelerators which require high bandwidth, even if they do not implement full-blown SIMD co-processors on the reconfigurable fabric.

The first hardware implementation of the Polymorphic Register File was designed in System Verilog [5]. MAX-PolyMem is the first prototype of PolyMem written entirely in MaxJ, and targeted at Maxeler DFEs [6], [7]. Our new HLS PolyMem is an alternative HLL solution, proven to be easily integrated with the Xilinx toolchains.

Figure 1 depicts the envisioned system architecture. The FPGA board (with a high-capacity DRAM memory), is connected to the host CPU through a PCI Express link. Poly-Mem acts as a high-bandwidth, 2D parallel software cache, able to feed an on-chip application kernel with multiple data elements every clock cycle. The focus of this work is to provide an efficient implementation of PolyMem in Vivado HLS, and employ it to maximize memory-accesses parallelism by exploiting BRAMs; we empirically demonstrate the gains we get from PolyMem by comparison against the partitioning of BRAMs, as provided by Xilinx tools, for three case-studies.

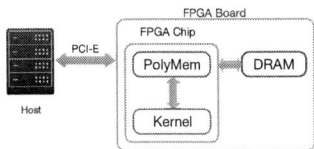

Fig. 1. System organization using PolyMem as a parallel cache.

Specifically, the main contributions of this paper are:

1) A novel, open-source implementation[1] of PolyMem for Vivado HLS, that allows its integration within the Xilinx Hardware-Software Co-Design Workflow;
2) Optimizations of the previously proposed PolyMem interface by adding masked methods to avoid overwrites and reduce latency;
3) Comparisons in terms of performance, resource-utilization, and productivity, of HLS PolyMem against standard memory partitioning techniques for three case-studies.

[1]https://github.com/storna/hls_prf

978-1-5386-4757-8/18 $31.00 © 2018 IEEE

II. BACKGROUND

A. The PRF and PolyMem

A PRF is a parameterizable register file, which can be logically reorganized by the programmer or a runtime system to support multiple register dimensions and sizes simultaneously [4]. The simultaneous support for multiple conflict-free access patterns, called *multiview*, is crucial, providing flexibility and improved performance for target applications. The *polymorphism* aspect refers to the support for adjusting the sizes and shapes of the registers at runtime. Table I presents the PRF *multiview* schemes (ReRo, ReCo, RoCo and ReTr), each supporting a combination of at least two conflict-free access patterns. A scheme is used to store data within the memory banks of the PRF, such that it allows different parallel *access types*. The different *access types* refer to the actual data elements that can be accessed in parallel.

TABLE I
THE PRF MEMORY ACCESS SCHEMES

PRF Schemes	Available Access Types
ReO	Rectangle
ReRo	Rectangle, Row, Main/Secondary Diagonals
ReCo	Rectangle, Column, Main/Secondary Diagonals
RoCo	Row, Column, Rectangle
ReTr	Rectangle, Transposed Rectangle

PolyMem reuses the PRF conflict-free parallel storage techniques and patterns, as well as the polymorphism idea. Figure 2 illustrates the set of access patterns supported by the PRF and, ultimately, by PolyMem. In this example, a 2D logical address space of 8×9 elements contains 10 memory Regions (R), each with different size and location: matrix, transposed matrix, row, column, main and secondary diagonals. Assuming a hardware implementation with eight memory banks, each of these regions can be read using one (R1-R9) or several (R0) parallel accesses.

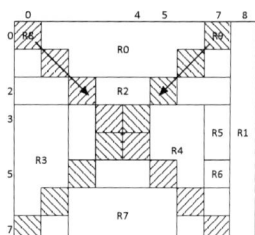

Fig. 2. PolyMem supported access patterns

By design, the PRF optimizes the memory throughput for a set of predefined memory access patterns. For PolyMem, we consider $p \times q$ memory modules and the five parallel access schemes presented in Table I. Each scheme supports dense, conflict-free access to $p \cdot q$ elements. When implemented in reconfigurable technology, PolyMem allows application-driven customization: its capacity, number of read/write ports, and the number of lanes to best support the application needs.

The block diagram in Figure 3 shows, at high level, the PEF architecture. The multi-bank memory is composed of a bi-dimensional matrix containing $p \times q$ memory modules. This enables parallel access to $p \cdot q$ elements in one memory

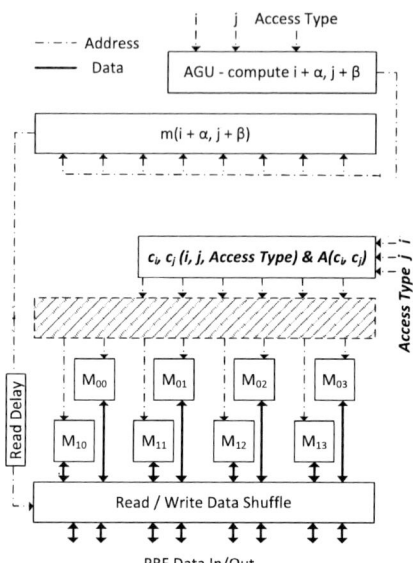

Fig. 3. Block diagram of the PRF [4]. The inputs are the matrix indexes (i, j) pointing to the first cell of the block of data the user wants to read/write in parallel, and the AccessType to compute the other addresses to point the right PRF banks of memory.

operation. The inputs of the PRF are shown at the top of the diagram. AccessType represents the parallel access pattern. The indexes (i, j) are the top-left coordinate of the parallel accesses. The list of elements to access is generated by the AGU module and is sent to the A module and to the m module. The A module generates one in-memory address for each memory bank in the PRF; the m module, applies the mapping function relative to the implemented scheme and computes for each accessed element the respective memory bank where it is stored. The Data Shuffle block reorders the addresses, generated by the m module, to the respective memory banks, then reorder the Data In/Out ensuring that the user of the PRF obtains the accessed data in their original order.

B. Matrix storage in a parallel memory

Figure 4 compares two ways for a 6×6 matrix to be mapped in BRAMs to enable parallel accesses. Thus, the default Vivado HLS partitioning techniques with a factor of 3 is compared against a PolyMem with 3 memory banks, organized exploiting the PolyMem RoCo scheme. The memory banks, in this case, are organized in a 1×3 structure, allowing parallel access to rows and columns of three, eventually unaligned, elements. The left side of the Figure shows an example of a matrix that the user wishes to store on partitioned BRAMs to achieve hardware parallelism in data reads/writes. The right side illustrates the techniques used to partition the matrix. Taking two random, unaligned, parallel accesses of 3 elements and using a RoCo scheme, starting respectively from the cells contain elements 8 and 23, it is possible to see that each element of each access type is mapped within PolyMem on a different memory bank. Hence, with one memory operation performed in parallel on each different memory bank, it is possible to read/write 3 elements in parallel. This small-scale

978-1-5386-4757-8/18 $31.00 © 2018 IEEE   144

example is included for visualization purposes only. Real-applications are like to use more memory banks, allowing parallel accesses to larger data blocks.

III. IMPLEMENTATION DETAILS

This section describes the main components of our Poly-Mem implementation for Vivado HLS. The goal of integrating PolyMemin the Xilinx workflow is to provide users with an easy-to-use solution to exploit parallelism when accessing data stored on the on-chip memory with different patterns.

Our Vivado HLS PolyMem implementation exploits one of the five schemes, the RoCo, to store on the BRAMs of the FPGA the data required to perform the application operations. Compared to the default Vivado memory partitioning techniques, which allow hardware parallelism in one dimension without consuming too many hardware resources, a Poly-Mem configured with the RoCo scheme can manage two types of access patterns simultaneously.

We implemented a template-based class prf that exploits loop unrolling to parallelize memory accesses. When the HLS PolyMem is instantiated within the user application code, it is possible to specify PRF_DATA_T, i.e., the type of data to be stored, the $(p \times q)$ number of internal banks of memory (which also represents the level of parallelism), the $(N \times M)$ dimension of the matrix to be stored (also used to compute the depth of each bank of data), and the $scheme$ to organize data within the different banks of memory.

In Listing 1 the interfaces of methods that allow accesses to data stored within PolyMem are presented. Simple **read** and **write** methods use the m and A functions to compute, respectively, the address and the depth of the bank of memory in which the required data is stored or needs to be saved. On the other hand, the **read_block** and the **write_block** exploit optimized versions of m and A to read/write $(q \cdot p)$ elements in parallel, while limiting the hardware resources used to reorder data. Finally, we optimized the memory access operations by implementing a **write_block_masked** method to specify which data in the block has to be overwritten within PolyMem. As an example, this method is useful when PolyMem supports a wide parallel access (e.g., 8 elements), but the user has less data to be stored (e.g., 5 elements), and wants avoid overwriting existing data (e.g., the remaining 3 elements).

Listing 1. List of the methods interfaces to allow user read/write data by used sequential or parallel accesses

```
PRF_DATA_T read(int i, int j);
void write(PRF_DATA_T data, int i, int j);
void read_block(int i, int j, PRF_DATA_T out[p * q],
                int PRF_ACCESS_TYPE);
void write_block(PRF_DATA_T in[p * q], int i, int j,
                int PRF_ACCESS_TYPE);
void write_block_masked(PRF_DATA_T in[p * q],
                ap_uint<p * q> mask,
                int i, int j,
                int PRF_ACCESS_TYPE);
```

IV. EXPERIMENTAL EVALUATION

In this section, we propose three applications (i.e., matrix multiplication, Markov chain, and LU decomposition) that

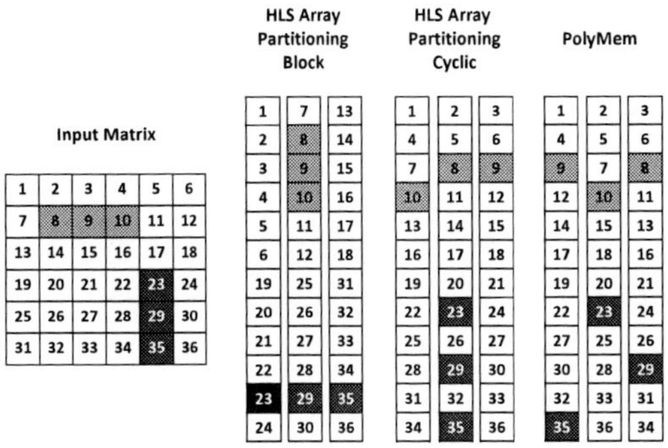

Fig. 4. Comparison between different partitioning techniques offered by Vivado HLS (factor = 3) and the RoCo scheme of PolyMem, with 3 memory banks, for data stored in a 6x6 matrix. PolyMem allows 3 parallel data reads/writes, from the rows and the columns of the original matrix. Unaligned blocks are also supported.

exploit our HLS PolyMem to parallelize accesses to matrix data by using the RoCo scheme. Each application demonstrates different features of our HLS PolyMem. In the matrix multiplication case-study, we show how our approach outperforms implementations that use the default partitioning of Vivado HLS. For the Markov Chain application, we show how HLS PolyMem enables performance gain with minimal changes to the original software code. Finally, we present the use of the masked methods in the LU decomposition implementation.

A. Matrix multiplication

In this case study, we analyze the multiplication between two square matrices, B and C, of size DIM, that are stored by using either the default HLS array partitioning techniques or the HLS PolyMem implementation. Since the multiplication $B \times C$ is performed by accessing the rows of B and multiply-accumulating the data with the columns of C, it is convenient, when using HLS default partitioning, to partition B on the second dimension and C on the first one. Indeed, this allows to achieve parallel accesses to the rows of B and columns of C in the innermost loop of the computation. On the other hand, for the HLS PolyMem implementation, we store both B and C in the HLS PolyMem , configured with a RoCo scheme, because it allows parallel accesses to both rows and columns.

Listings 2 and 3 show the declaration of the matrices and their partitioning using the HLS default partitioning and the HLS PolyMem, respectively. In both listings, a parallel factor of 16 has been used. The B and C HLS PolyMem instances are initialized with $p = 4$ and $q = 4$, which results in partitioning the data onto 16 memory banks.

Listing 2. Declaration and partitioning of matrices to parallelize accesses to rows (dim=2) of B and to columns (dim=1) of C with a parallel factor of 16.

```
float B[DIM][DIM];
#pragma HLS array_partition variable=B
                block factor=16 dim=2
float C[DIM][DIM];
#pragma HLS array_partition variable=C
                block factor=16 dim=1
```

Listing 3. Declaration of the matrices stored by using the HLS PolyMem with the RoCo scheme with a parallel factor of $4 \cdot 4 = 16$.

```
#include "hls_prf.h"
hls::prf<float, 4, 4, DIM, DIM, SCHEME_RoCo> B;
hls::prf<float, 4, 4, DIM, DIM, SCHEME_RoCo> C;
```

Listings 4 and 5 show the matrix multiplication code when using the HLS default partitioning and the HLS PolyMem, respectively.

Listing 4. Matrix multiplication code that leverages default HLS partitioning to perform parallel accesses.

```
// B*C matrix multiplication
for (int i = 0; i < DIM; ++i)
  for (int j = 0; j < DIM; ++j) {
#pragma HLS PIPELINE II=1
    float sum = 0;
    for (int k = 0; k < DIM; ++k)
      sum += B[i][k] * C[k][j];
    OUT[i][j] = sum;
  }
```

Listing 5. Matrix multiplication code that exploits the HLS PolyMem with RoCo scheme to perform parallel accesses.

```
// B*C matrix multiplication
for (int i = 0; i < DIM; ++i)
  for (int j = 0; j < DIM; ++j) {
#pragma HLS PIPELINE II=1
    float sum = 0;
    for (int k = 0; k < DIM; k += 16) {
      B.read_block(i, k, temp_row, ACCESS_Ro);
      C.read_block(k, j, temp_col, ACCESS_Co);
      for (int t = 0; t < 16; t++)
        sum += temp_row[t] * temp_col[t];
    }
    OUT[i][j] = sum;
  }
```

Even though both approaches achieve the goal of computing the matrix multiplication by accessing 16 matrix elements in parallel, the HLS PolyMem solution provides more flexibility when additional data access patterns are required, which is often the case for larger kernels. In order to highlight this aspect, we also consider a kernel function in which both the $B \times C$ and the $C \times B$ products need to be computed.

Table II reports the latency and resources utilization estimated by Vivado HLS when computing the matrix multiplication $B \times C$ (rows 1,2), and when computing $B \times C$ followed by $C \times B$ (rows 3,4 and 5,6) for the two approaches. By using the default Vivado HLS partitioning techniques, the second multiplication $B \times C$ cannot be computed efficiently due to the way in which the matrix data is partitioned into the memory banks, as described in Section II. Indeed, C can only be accessed in parallel by rows and B by columns. On the other hand, the implementation based on HLS PolyMem is also capable of performing the matrix product $C \times B$ efficiently. This is also reflected in the estimated latency reported in Table II, which is the same for both products.

It is also worth noting that for matrix size of 32, the two approaches have similar resource consumption, while for

TABLE II
LATENCY AND HARDWARE RESOURCES FOR MATRIX MULTIPLICATION WITH DIFFERENT MEMORY CONFIGURATIONS AND MATRIX DIMENSIONS

Memory	Matrix size	Parallel factor	Latency		Hardware resources			
			$B \times C$	$C \times B$	BRAM	DSP	FF	LUT
HLS	32	4	4227	n.a.	18	40	6162	6485
PolyMem	32	$2 \cdot 2$	4227	n.a.	18	40	6153	6018
HLS	32	4	4227	16503	18	40	7444	9197
PolyMem	32	$2 \cdot 2$	4227	4227	18	40	7367	7364
HLS	96	16	28033	442722	96	164	28554	40474
PolyMem	96	$4 \cdot 4$	28033	28033	96	160	30969	43636

matrices with larger dimensions and a parallel factor of 16, the HLS PolyMem has a resource consumption overhead in terms of FF and LUT of at most 8.5% compared to the HLS default partitioning schemes. Finally, in order to empirically validate the designs, we implemented the kernel module performing both $B \times C$ and $C \times B$ with matrix size of 96 and a parallel factor of 16 on a Xilinx Virtex-7 VC707 with a target frequency of 100MHz. The HLS PolyMem achieved a read and write throughput of 0.4 GB/s and a speedup of 5x compared to the implementation based on default HLS memory partitioning.

B. Markov Chain and the Matrix power operation

A Markov Chain is a stochastic model used to describe real-world processes. Some of the most relevant applications are found in queuing theory and study of population growths [8], while they are also used in stochastic simulation methods such as Gibbs sampling [9] and Markov Chain Monte Carlo [10]. Moreover, Page Rank [11], an algorithm used to rank websites by search engines, leverages a time-continuous variant of this model. A Markov Chain can also describe a system composed of multiple discrete states, where the probability of being in a state depends only on the previous state of the system. A Markov Transition Matrix A, which is a variant of an adjacency matrix, can be used to represent a Markov Chain. In this matrix, each row contains the probability to move from the current state to any other state of the system. More specifically, given two states i and j, the probability to transition from i to j is $a_{i,j}$, where $a_{i,j}$ is the element at row i and column j of the transition matrix A.

Computing the h-th power of the Markov Transition Matrix is a way to determine what is the probability to transition from an initial state to a final state in h steps. Furthermore, when the number of steps h tends to infinity, the result of A^h can be used to recover the stationary distribution of the Markov Chain, if it exists. From a computational perspective, an approximate value for the result of $lim_{x \to \infty} A^x$ is obtained for large enough values of x. In our implementation, matrix A is stored in a HLS PolyMem, so that both rows and columns can be accessed in parallel, then, we compute A^2 and save the result into a support matrix A_temp, partitioned on the second dimension. After A^2 is computed, we can easily compute A^{2^h} by copying back results to the HLS PolyMem and iterating the overall computation h times. Implementing the same algorithm by using the HLS partitioning techniques, as presented in the previous case study, results in poor exploitation of the available

978-1-5386-4757-8/18 $31.00 © 2018 IEEE

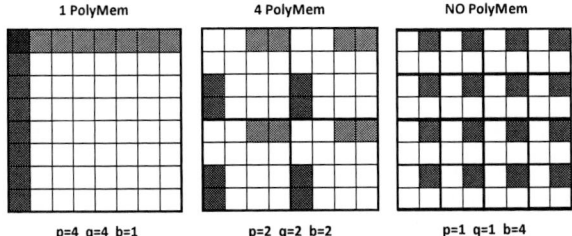

1 PolyMem	4 PolyMem	NO PolyMem
p=4 q=4 b=1	p=2 q=2 b=2	p=1 q=1 b=4

Fig. 5. Comparison between different partitioning of the input matrix in a grid of b^2 components implemented by PolyMem with a level of parallelism of $p \times q$. When both p and q are set to 1, it is possible to remove the HLS PolyMem logic.

parallelism, or in duplicated data, since A needs to be accessed both by rows and columns.

The HLS PolyMem enables paralell accesses to matrix A for both rows and columns, but adds to the design an overhead in terms of hardware resources and complexity of the logic to shuffle data within the right memory banks. The resources overhead has a quadratic growth with respect to the number $p \cdot q$ of parallel memories used to store data [4].

A possible solution to this problem, is to reduce the dimension of PolyMem by dividing the A input matrix and store the values in a grid of multiple PolyMems. If A has dimension $DIM \times DIM$, it is possible to organize the on-chip memory to store data in a grid of $b \times b$ squared blocks each having size $\frac{DIM}{b} \times \frac{DIM}{b}$. In order to preserve the same level of parallelism, we can re-engineer the original computation to work in parallel on the data stored in each memory within the grid. Instead of computing a single vectorized row-column product, it is possible to perform the computation on multiple row-column products in parallel and reduce the final results.

Figure 5 shows how the input matrix is divided in multiple memories according to the choice of the parameters p, q and b. Moreover, the figure also shows which is the data accessed concurrently at each step of the computation. As an example, for the case $p = q = b = 2$ there are 4 row-column products performed in parallel (b^2) and for each of them 4 values are processed in parallel ($p \cdot q$).

It is important to notice that when $p = q = 1$ the PolyMems reduce to memories in which a single element is accessed in parallel. In this case, each PolyMem can be removed and substituted by a single memory bank.

In Table III we report the latency and the resource utilization estimated by Vivado HLS together with the number of lines of code (LOC) for different configurations of the parameters p, q and b on 8 iterations of the power operation for a 384x384 matrix. As can be seen, by re-engineering the code and the access patterns ($b > 1$), it is possible to achieve a smaller overall latency. However, this comes at the cost of a more convoluted code which is approximately twice the lines of code of the original version. On the contrary, by using a single PolyMem ($b = 1$) we can still obtain higher performance than using the default HLS array partitioning techniques, with a much smaller and simpler code base. Indeed, PolyMem allows to reduce the time to develop an optimized FPGA-based implementation of the algorithm with minor modifications to

TABLE III
LATENCY, HARDWARE RESOURCES AND LINES OF CODE, FOR 8 ITERATIONS OF THE MATRIX POWER OPERATION WITH DIFFERENT MEMORY CONFIGURATIONS AND A MATRIX SIZE OF 384

Memory	p	q	b	Latency	Hardware resources				LOC
					BRAM	DSP	FF	LUT	
PolyMem	2	2	1	1,557,835,871	1,036	14	9,936	11,071	98
PolyMem	2	4	1	840,333,407	1,044	17	19,678	28,855	98
PolyMem	4	4	1	488,632,423	1,060	31	36,138	53,621	98
multi PolyMem	1	1	2	758,085,955	1,036	14	6,967	5,572	188
multi PolyMem	1	2	2	394,149,976	1,044	28	14,709	12,934	188
multi PolyMem	2	2	2	214,032,480	1,060	45	24,845	22,418	188
NO PolyMem	1	1	4	101,848,419	1,124	76	32,852	13,706	188

the original software code. Thanks to HLS PolyMem we raise the level of abstraction of parallel memory accesses, thus enhancing the overall design experience and productivity.

Finally, to validate the flexibility the HLS PolyMem library, we implemented and tested the application by using Xilinx SDx tool, that enables OpenCL integration and automatically generates the PCIe drivers for communication. We synthesized a design for a matrix size of 256 and parameters $p = q = b = 2$ at 200MHz, and we benchmarked its performance on the Xilinx Kintex Ultrascale ADM-PCIE-KU3 platform, obtaining a read and write throughput of 1.6 GB/s.

C. LU decomposition

The last case study we present is the LU decomposition algorithm. This algorithm allows to decompose an input matrix A into a product of a lower triangular matrix L and an upper triangular matrix U:

$$A = LU$$

$$\begin{bmatrix} a_{00} & a_{01} & a_{02} \\ a_{10} & a_{11} & a_{12} \\ a_{20} & a_{21} & a_{22} \end{bmatrix} = \begin{bmatrix} 1 & 0 & 0 \\ l_{10} & 1 & 0 \\ l_{20} & l_{21} & 1 \end{bmatrix} \begin{bmatrix} u_{00} & u_{01} & u_{02} \\ 0 & u_{11} & u_{12} \\ 0 & 0 & u_{22} \end{bmatrix}$$

This factorization is used in many applications, such as solving linear equations or compute the inverse of a matrix. Furthermore, such application has been proved to be suitable for hardware acceleration [12]–[14].

Listing 6 shows the LU decomposition algorithm. Matrix A is given as input, and matrices L and U are computed while matrix A is zeroed.

Listing 6. LU decomposition algorithm

```
for(k=0; k<DIM; k++){
  for(i=k; i<DIM; i++){
    L[i][k] = A[i][k] / A[k][k];
    U[k][i] = A[k][i];
  }
  for(i=k; i<DIM; i++)
    for(j=k; j<DIM; j++)
      A[i][j] = A[i][j] - L[i][k] * U[k][j];
}
```

Analyzing the loops, it is possible to see that the algorithm works on successive sub-matrices, identified by the iterator of the outermost loop k. In the first nested loop, matrix A is accessed both column-wise and row-wise. The results of those statements are respectively stored column-wise in matrix L and row-wise in matrix U. Finally, the second nested loop updates matrix A before starting the new iteration.

978-1-5386-4757-8/18 $31.00 © 2018 IEEE

This brief analysis shows that this implementation of the LU decomposition algorithm uses interleaved row-wise and column-wise accesses to the same matrix; moreover, due to the offset introduced by iterator k, those accesses could be unaligned. Even in this case, HLS PolyMem represents a valid solution to parallelize the read and write operations.

Matrix A stored using a PolyMem with p=2, q=2

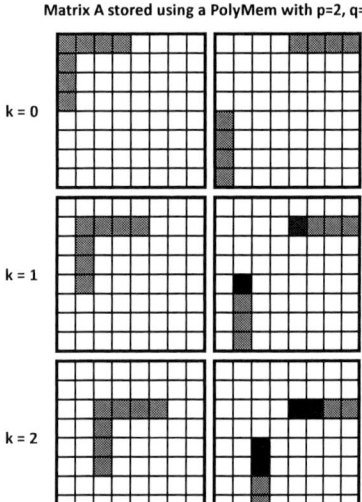

Fig. 6. Three iterations of loop k (rows of the figure) and the iterations of the nested loop (columns of the figure) to update in parallel blocks of values of the HLS PolyMem where input and output are stored. The writes are performed by using the **write_block_masked** method: the black cells of the matrices correspond to the 0 values of the passed mask and they are not written in the HLS PolyMem memory.

Unaligned accesses to the matrix represent an interesting case study and they are supported by the HLS PolyMem implementation. Furthermore, by exploiting the fact that the values of the matrix A are iteratively zeroed and the structure of the L and U matrices, we can store the entire computation on a single HLS PolyMem memory. The final L and U matrices are stored as follows:

$$\begin{bmatrix} u_{00} & u_{01} & u_{02} \\ l_{10} & u_{11} & u_{12} \\ l_{20} & l_{21} & u_{22} \end{bmatrix}$$

Depending on the value of the iterator k, the amount of data to be processed might not be a multiple of the parallel factor being used. For this reason, special care must be taken when dealing with the last block being processed as shown in figure 6. To avoid to overflow the matrix dimensions, the last block is computed out of the loop, and always starts at an offset of $DIM - (p \cdot q)$. Then, an appropriate write mask is applied to ensure that only the needed data is written to the HLS PolyMem memory. In order to enable this solution, we implemented the method **write_block_masked** that allows to pass a mask of bits that represent the positions of values within the block that need to be written. Since this method adds some logic to solve the mask, we use it only while updating the last block, out of the nested loop. Thanks to the introduction of this class of methods, we simplify the adoption of the HLS PolyMem for a broader set of applications.

V. CONCLUSIONS

In this paper, we presented a C++ implementation of PolyMem optimized for Vivado HLS, ready-to-use as a library for applications requiring parallel memories. Our implementation exposes an easy-to-use interface to enhance design productivity for FGPA-based applications. Furthermore, we extended the original PolyMem implementation to support masked on-chip parallel accesses.

We proved the flexibility of the library among the Xilinx Design Tools, by implementing the kernels for *both* the Vivado workflow with a Virtex-7 VC707 and the SDx workflow with a Kintex Ultrascale 3 ADM-PCIE. Our empirical analysis of our library on three case studies (Matrix multiplication, Markov Chains, and LU decomposition) demonstrated competitive results in terms of latency, low code complexity, but also a small overhead in terms of hardware resource utilization.

Our future work focuses on (1) including support for additional PolyMem schemes optimized for Vivado HLS, (2) designing an automatic framework to analyze the user application code and suggest how to improve its performance with HLS PolyMem, and (3) improving the HLS PolyMem shuffle module by exploiting a Butterfly Network [15] for the memory banks connections.

REFERENCES

[1] Xilinx, "Vivado high-level synthesis." [Online]. Available: https://www.xilinx.com/products/design-tools/vivado/integration/esl-design.html

[2] M. Weinhardt and W. Luk, "Memory access optimisation for reconfigurable systems," *IEE Proceedings-Computers and Digital Techniques*, vol. 148, no. 3, pp. 105–112, 2001.

[3] C. B. Ciobanu *et al.*, "MAX-PolyMem: High-Bandwidth Polymorphic Parallel Memories for DFEs," in *RAW2018 (to appear)*, pp. 1–8.

[4] C. Ciobanu, "Customizable Register Files for Multidimensional SIMD Architectures," Ph.D. dissertation, TUDelft, The Netherlands, 2013.

[5] C. Ciobanu *et al.*, "Scalability Study of Polymorphic Register Files," in *Proc. of DSD*, 2012, pp. 803–808.

[6] C. B. Ciobanu *et al.*, "Max-polymem: High-bandwidth polymorphic parallel memories for dfes," in *IEEE IPDPSW - RAW'18*, May 2018, pp. 107–114.

[7] ——, "EXTRA: An Open Platform for Reconfigurable Architectures," in *Proceedings of SAMOS XVIII (to appear)*, July 2018, pp. 1–10.

[8] J. J. Arsanjani *et al.*, "Integration of logistic regression, markov chain and cellular automata models to simulate urban expansion," *International Journal of Applied Earth Observation and Geoinformation*, vol. 21, pp. 265–275, 2013.

[9] A. F. Smith and G. O. Roberts, "Bayesian computation via the gibbs sampler and related markov chain monte carlo methods," *Journal of the Royal Statistical Society. Series B (Methodological)*, pp. 3–23, 1993.

[10] W. R. Gilks *et al.*, *Markov chain Monte Carlo in practice*. CRC press, 1995.

[11] S. D. Kamvar *et al.*, "Extrapolation methods for accelerating pagerank computations," in *Proceedings of the 12th international conference on World Wide Web*. ACM, 2003, pp. 261–270.

[12] G. Govindu *et al.*, "A high-performance and energy-efficient architecture for floating-point based lu decomposition on fpgas," in *Parallel and Distributed Processing Symposium, 2004. Proceedings. 18th International*. IEEE, 2004, p. 149.

[13] V. Daga *et al.*, "Efficient floating-point based block lu decomposition on fpgas," in *International Conference on Engineering of Reconfigurable Systems and Algorithms, Las Vegas*, 2004, pp. 21–24.

[14] M. K. Jaiswal and N. Chandrachoodan, "Fpga-based high-performance and scalable block lu decomposition architecture," *IEEE Transactions on Computers*, vol. 61, no. 1, pp. 60–72, 2012.

[15] A. Avior *et al.*, "A tight layout of the butterfly network," *Theory of Computing Systems*, vol. 31, no. 4, pp. 475–488, 1998.

978-1-5386-4757-8/18 $31.00 © 2018 IEEE

Inferential Logic: a Machine Learning Inspired Paradigm for Combinational Circuits

Valerio Tenace, Andrea Calimera

Dipartimento di Automatica e Informatica, Politecnico di Torino, Italy

Abstract—**Machine learning (ML) theories and tools suggest alternative forms to conceive and represent relationships among data. The same theories find their application in the Boolean domain, where logic functions can be described as inference rules. This paper introduces *Inferential Logic*, a novel paradigm that leverages the ML concept of statistical inference for the design of combinational logic circuits, the *Inferential Logic Circuits* (ILCs). This new design concept is conceived for low-power circuits that run *quasi-exact* computation in error-resilient applications, but it also provides an *exact* run-mode that can be dynamically enabled when accuracy scaling is not an option.**

I. INTRODUCTION

Machine learning (ML) is a paradigm in which hardware or software systems replicate a few simple learning/reasoning mechanisms proper of the human brain to resolve complex relationships among data [1]. The strength of ML tools lies under their ubiquity; they orthogonally apply to problems of different nature using statistical models on the collected data, just as the brain does through inductive reasoning over previous acquired knowledge.

Although the first evidence of such techniques dates back to the mid-20th century, recent advancements in computer architectures and massively parallel computing have prompted renewed interest on the matter. ML is an inspiring new research trend which might have huge impact on many commercial and scientific fields. The Electronic Design Automation (EDA) is not an exception. The EDA community is mainly focusing on efficient hardware mapping of brain-inspired computing models, e.g., deep learning, while little effort is being spent on investigating how to take advantage of biological mechanisms to solve EDA problems. In this context, most noticeable contributions include the works by Li Wang et al. [2], [3], where ML techniques are shaped to address testing and verification of digital circuits by means of several supervised and unsupervised techniques. Inspired by a previous work [4], Li Wang et al. also describe the concept of a supervised learning mechanism for Boolean functions where a statistical model evaluates the output of the function in order to reconstruct the original logic. Apart from these examples, very few has been done in terms of design strategies; for instance, a proper flow to build logic circuits using ML is currently not available.

In our view, the potential of ML techniques should be exploited more efficiently in order to create logic circuits that mimic how the human brain works, namely, to implement inaccurate, yet fast logic reasoning [5], [6], [7]. This paradigm

shift encompasses the replacement of exact logic rules and Boolean operators in favor of statistical models and inference tools.

Moving towards this objective, in this paper we describe a new design methodology where the representation of a generic Boolean function is obtained by means of a learning problem. The goal is to represent the behavior of a logic function through a more compact abstract model that works as a statistical description of the function itself. Such description is then used to *infer* the outcomes of the Boolean function up to a certain level of accuracy. Once mapped on a piece of hardware, the resulting circuit runs the *quasi-exact* computation of the logic function; we called this block *Inference Unit*. However, since for some applications accuracy degradation is not an option, we describe how the inferential model can be reinforced in order to reach the full description of the logic function, hence, the *exact* computation. In such a case the Inference Unit is supported by an auxiliary unit called *Supervisor*. We refer to this new class of circuits as Inferential Logic Circuits (ILCs) and we provide a complete framework for their logic synthesis and optimization.

Experimental results collected from an ILC embedded into an error-resilient application demonstrate that quasi-exact computation achieves an average accuracy of 94%, yet using $22\times$ less devices w.r.t. a circuit counterpart obtained with a state-of-the-art standard design flow. Moreover, we quantify the figures of merit of other ILC benchmarks thus to give a fair comparison against a multi-level logic optimizer that obeys Boolean rules.

II. BACKGROUND

In the panorama of machine learning, Classification Trees (CTs) represent a class of methods for the construction of classification models. Solving a classification problem encompasses two main stages: the *training* stage, which aims at generating the statistical abstract representation of the model, i.e., the CT; the *validation* stage, which quantifies the level of accuracy of the trained CT.

The training stage makes use of a labeled dataset with n observations; each observation is described by p predictor variables $X = \{x_1, \dots x_p\}$ and is labeled by one of the m available classes $y_i \in Y = \{y_1, \dots y_m\}$. During the validation stage, the CT is used to classify never occurred samples described by X over the label-set Y; the accuracy is given by the number of new samples that are correctly classified.

Building a CT implies the search for a "good" partitioning of the input space [8]. A CT implements a recursive partitioning using a proper sequence of comparisons between predictor

978-1-5386-4757-8/18 $31.00 © 2018 IEEE

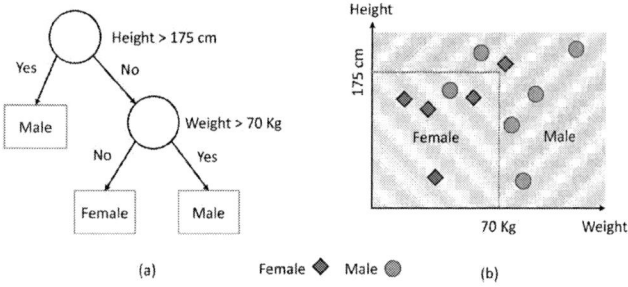

Figure 1. Visual description of a Classification Tree. Logical structure (left), domain partitioning (right).

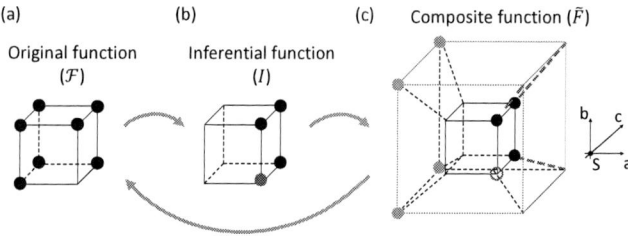

Figure 2. Positional cube notation of the example functions.

variables and numeric thresholds. Let us assume to have a training dataset of people living in a city described through several predictors, or features, such as X={*Age, Weight, Salary, Height, Nationality*}. If the objective is to classify people by gender, i.e., $Y = \{Male, Female\}$, an abstract model with reasonably good accuracy would select which predictors are the most significant among the available predictors X in order to separate the training population into two different, and ideally disjoint clusters. This concept is graphically depicted in Figure 1, where the selected predictors are *Height* and *Weight*. The most important components learned during training are: the best split predictors, i.e., for each iteration which variable $x_k \in X$ is able to separate the training dataset into different groups with the lowest level of impurity [9], and the evaluation of the optimal threshold for each split. These two concepts are combined together into a statistical dispersion index called Gini index [9]. It is therefore possible to assume that a CT can be used to select which predictors better describe the training population.

Both predictor variables and labels can assume any type of value, e.g., *Weight* and *Height* are real numbers, *Male* or *Female* are categorical values; entirely-binary classification problems come with predictors and labels in the form of *binary* variables, i.e., *True* or *False*, 1 or 0.

III. COMPUTING THROUGH INFERENCE

A. Exact and Quasi-Exact Representation

A generic Boolean function $\mathcal{F}$ can be defined as an entirely-binary classification problem with a training set described by its truth table. The evaluation of an input pattern encompasses the classification on two possible labels: logic-0 or logic-1. As per the description given in Section II, a CT selects the most significant predictors, namely, the primary inputs that best describe the function $\mathcal{F}$. Since the selection is done through a statistical method, it is unlikely that a CT gives a complete cover of $\mathcal{F}$. In that sense, a CT is a *quasi-exact* description of $\mathcal{F}$. To notice that this is slightly different from the classical concept of approximate logic function. Indeed, the CT may cover both the ON-set and the OFF-set of $\mathcal{F}$.

In order to better understand this concept and describe how to achieve an *exact* representation, let us resort to a simple function $\mathcal{F} : \mathbb{B}^3 \to \mathbb{B}$ defined as in (1). It represents the function to be learned, and its ON-set is graphically depicted in Figure 2-a by means of a positional cube representation.

$$\mathcal{F}(a,b,c) = (a \wedge b) \vee (a \wedge c) \vee \neg a \quad (1)$$

Let us then consider a function $\mathcal{I} : \mathbb{B}^3 \to \mathbb{B}$ defined as in (2), being its cube representation reported in Figure 2-b.

$$\mathcal{I}(a,b,c) = a \quad (2)$$

Assume that $\mathcal{I}$ is used to mimic $\mathcal{F}$: three over eight minterms are covered, i.e., $\mathcal{I} \equiv \mathcal{F}$ for three input patterns over eight. $\mathcal{I}$ covers a sub-set of the ON-set of $\mathcal{F}$ and part of the OFF-set of $\mathcal{F}$ (red bullet in Figure 2-a). As mentioned above, this is what we refer as the *quasi-exact* representation. In order to achieve an exact representation, $\mathcal{I}$ has to be flanked by an extra function $\mathcal{S} : \mathbb{B}^3 \to \mathbb{B}$ which fires *iff* function $\mathcal{I}$ infers a wrong output value. A possible expression for $\mathcal{S}$ is given by (3).

$$\mathcal{S}(a,b,c) = \neg a \vee (\neg b \wedge \neg c) \quad (3)$$

It is therefore clear that $\mathcal{I}$ and $\mathcal{S}$ should collaborate to achieve the exact representation of $\mathcal{F}$. Indeed, $\mathcal{S}$ plays the role of a *Supervisor* that can be used as a flag to decide whether $\mathcal{I}$ has to be corrected.

The composite function $\widetilde{\mathcal{F}} = \mathcal{S} \circ \mathcal{I}$ can be graphically represented as the projection of the cube of $\mathcal{I}$ (Figure 2-b) into the 4th dimension $\mathcal{S}$ (Figure 2-c). The resulting hypercube is described by the following algebraic expression:

$$\widetilde{\mathcal{F}}(a,b,c,\mathcal{S}) = \mathcal{S} \wedge \neg a \vee \neg \mathcal{S} \wedge a \quad (4)$$

under the following satisfiability don't care (SDC) condition:

$$SDC(\mathcal{S}) = \neg \mathcal{S} \wedge \mathcal{F} \oplus \mathcal{I} \quad (5)$$

e.g., $\{\neg \mathcal{S} \wedge \neg a\}$ in our example. Such SDC is imposed by construction: $\mathcal{S}$ fires *iff* $\mathcal{I}$ is wrong.

$\widetilde{\mathcal{F}}$ can be graphically simplified by: (i) collapsing minterms on the versor $\widehat{\mathcal{S}}$ on the inner cube, (ii) dropping unreachable vertices (bold dotted lines in the picture). Indeed, both $\mathcal{I}$ and $\mathcal{S}$ have the same support-set. The result is the 3-D cube of $\mathcal{F}$. The equivalent Boolean operation is the substitution $\mathcal{S} \to \widetilde{\mathcal{F}}$, which leads to the final relationship in (6). This transformation brings back to the original Boolean function $\mathcal{F}$.

$$\begin{aligned} \widetilde{\mathcal{F}}(a,b,c) &= ((\neg a \vee (\neg b \wedge \neg c)) \wedge \neg a) \vee (\neg(\neg a \vee (\neg b \wedge \neg c)) \wedge a) \\ &= (a \wedge b) \vee (a \wedge c) \vee \neg a \\ &= \mathcal{F}(a,b,c) \end{aligned}$$

$$(6)$$

978-1-5386-4757-8/18 $31.00 © 2018 IEEE

To summarize: *Any n-ary Boolean function $\mathcal{F}$ can be described as:*

$$\mathcal{F} = (\mathcal{S} \wedge \neg \mathcal{I}) \vee (\neg \mathcal{S} \wedge \mathcal{I}), \qquad (7)$$

*where $\mathcal{I}$ is a k-ary inferential function with $k < n$, and $\mathcal{S}$ is a n-ary supervisor function that fires **iff** $\mathcal{I} \neq \mathcal{F}$.*

It is important to notice that this transformation does not impose, nor require, any specific characteristic on $\mathcal{F}$.

B. Inferential Logic

The architecture of an *Inferential Logic Circuit* (ILC), Figure 3, is a straightforward implementation of the Boolean formulation described in Section III-A. It consists of two main logical blocks: (i) the *Inferential Unit* (IU), which implements the CT function $\mathcal{I}$; (ii) the *Supervisor Unit* (SU), which restores the output of $\mathcal{F}$ when $\mathcal{I}$ yields to an incorrect prediction.

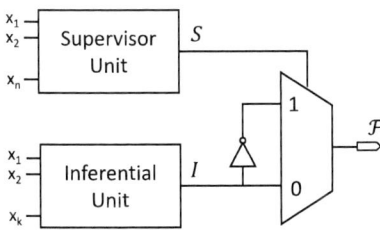

Figure 3. Inferential Logic Circuit architecture.

An ILC can run under two operating modes.

Quasi-exact computing: inputs $X_k = (x_1, \dots, x_n)$ are evaluated by the IU; its output is *similar* to the Boolean function $\mathcal{F}$. The accuracy of the IU is defined as the ratio between the number of correctly-classified inputs and the total number of inputs, i.e., 2^n.

Exact computing: inputs $X_k = (x_1, \dots, x_n)$ are evaluated by both the IU and the SU. Two are the possible outcomes: (i) the input pattern belongs to the set of misclassified patterns, hence the output inferred by the IU is wrong and the SU drives the complement of IU towards the primary output; (ii) the input pattern belongs to the set of correctly classified samples, hence the value inferred by IU is propagated to the output.

C. Design Flow

Figure 4 gives an overview of the proposed design flow. It starts with an exhaustive description of the digital circuit defined through a truth table (2^n permutations). Such table represents the training set. Building the CT is demanded to a Matlab script that leverages the `fitctree` function. Once the CT is generated, the validation phase takes place; each entry of the truth table is classified by means of the CT and results are collected into two separate sets: the classified set $\mathcal{C}$, and the misclassified set $\mathcal{M}$. The former is a Boolean description of the CT, that is, the function $\mathcal{I} : \mathbb{B}^k \rightarrow \mathbb{B}$. By means of an in-house software package, described later in Section IV-B, the CT structure is synthesized, optimized and mapped onto a MUX-INV netlist. The resulting logic circuit is the *Inferential Unit* of Figure 3. $\mathcal{M}$ is a Boolean description of the *Supervisor Unit* (SU), i.e., the function $\mathcal{S} : \mathbb{B}^n \rightarrow \mathbb{B}$, and it

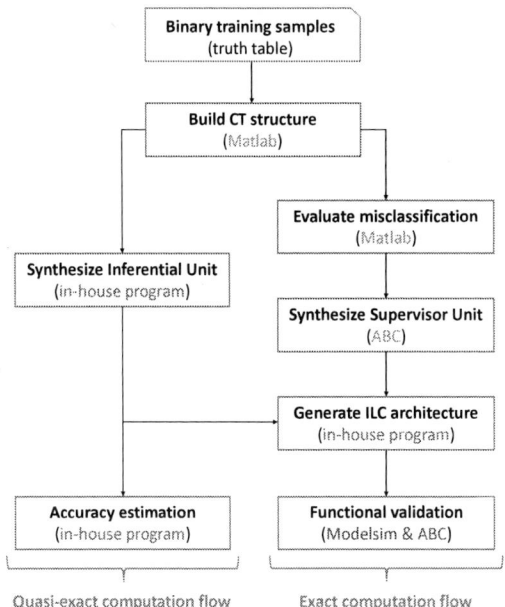

Figure 4. Design flow generating exact and quasi-exact computing flows.

is synthesized with the ABC synthesis tool [10]. According to (7), output signals of the Inferential and Supervisor units are then plugged into a MUX. The obtained ILU architecture is then validated thought exhaustive functional simulations and Boolean equivalence checking.

IV. Logic Synthesis of the Inferential Unit

As previously discussed, the Supervisor is synthesized with a standard multi-level flow. Here we focus our discussion on an efficient synthesis flow tailored on quasi-exact Boolean function representations.

A. From CTs to Mux-Inverter Trees

A CT is a rooted and directed acyclic graph defined as $\Gamma = (\Phi \cup D \cup \Theta \cup E)$. The set of decision vertices $d \in D$ with two outgoing edges $e_0, e_1 \in E$ are associated with a threshold operation on a primary input variable $x_k \in \mathcal{X}$. Such nodes drive the connection between nodes at the lower levels with those at the upper levels of the tree structure. For instance, if $x_k > T_h$ (being T_h the threshold selected at training time by the Gini index), then the left branch is connected to the current node's output, right branch otherwise. Terminal nodes $\theta \in \Theta$ with out-degree 0 represent the class to which a given input pattern belongs to. Finally, root nodes $\phi \in \Phi$ with in-degree 0 represent the output value assumed by the Boolean logic function $\mathcal{F}(\mathcal{X})$. From the definition above, the representation of decision nodes using Boolean gates may look prohibitive due to the complexity involved in designing a threshold comparator. However, since training samples are always described by means of Boolean variables, the CT algorithm will always select a threshold subject to (8), where $max(k)$ and $min(k)$ represent the maximum and minimum values assumed by the input feature k, namely 1 and 0.

$$T_h = \frac{max(k) + min(k)}{2} = 0.5 \qquad (8)$$

978-1-5386-4757-8/18 $31.00 © 2018 IEEE

It is thereby possible to reduce the threshold comparator through the equivalences reported in (9).

$$(x_k \geq 0.5) \equiv (x_k = 1)$$
$$(x_k < 0.5) \equiv (x_k = 0) \tag{9}$$

Such equations suggest that decision vertices can be reduced to a simple identity comparator, not dissimilar to decision nodes in Binary Decision Diagrams (BDDs) [11]. The built-in functionality can then be reduced to an If-Then-Else (ITE) primitive, where the selection variable is represented by the primary input $x_k \in \mathcal{X}$, and the *true* and *false* branches are represented by the left and right child nodes, respectively. As a consequence, the most simple transposition of a decision node is a single Multiplexer (MUX). Without loss of generality, we can say that any CT grown on a fully-Boolean training sample can be represented by means of a tree of MUXes. Obviously, other circuit implementations do exist.

B. Synthesis Flow

Figure 5 describes the adopted synthesis flow for the Inferential Unit. First, the CT structure described as a list of

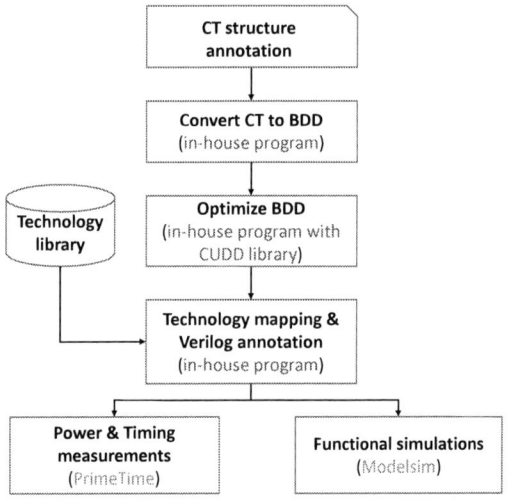

Figure 5. Synthesis flow for the Inferential Unit.

ITE statements is converted into a BDD structure and then optimized according to the following three rules:

1) Node elimination: a node having both left and right branches connected to the same child node represents a don't care. Therefore, it is eliminated from the structure.
2) Node sharing: if two or more nodes share the same child nodes, then they are said to be equivalent. It is then possible to merge them into a single node.
3) Variable ordering: a good variable order is selected as to guarantee the lowest cardinality for the decision diagram [12].

Those transformations are performed with an in-house C program that leverages the CUDD package [13] for decision diagram manipulations.

As a final stage, the reduced and ordered BDD is mapped into a MUX-INV tree, where MUXes are used to map decision

nodes, and INVs are used to eliminate constant terminal nodes and to invert signals along negated edges [5]. The obtained annotated Verilog netlist is then validated through functional simulations and post-synthesis analyses.

V. EXPERIMENTAL RESULTS

In this section we demonstrate the effectiveness of the proposed ILC paradigm. First, we show that Inferential Computing is particularly effective in error-resilient applications; we provide a case-study analysis for a widely-used edge detection technique, i.e., the Sobel operator [14]. We describe in detail the adopted implementation and provide an accurate analysis of the obtained results in terms of output accuracy, area, and power efficiency. Second, we quantify area and power of ILC circuits against an open-source multi-level synthesis flow: the ABC synthesis tool [10]. Quality-of-result for quasi-exact and exact computing is also assessed.

A. Edge Detection Through The Sobel Operator

In order to appreciate the potential of the proposed technique, we resort to a real-life error-resilient application. In the field of machine vision, edge detection is the operation of detecting significant local changes in an image. Given a matrix of pixels I, a *step edge* is associated to a peak in the first derivative of I. Edges can be detected by means of a gradient operator. In our analysis, we employee the Sobel operator, defined as in (10).

$$G = \begin{pmatrix} 1 & 0 & -1 \\ 2 & 0 & -2 \\ 1 & 0 & -1 \end{pmatrix} \tag{10}$$

Let us assume to have a 3x3 grayscale image I, where each element $i_{x,y}$ of the matrix represents a pixel in the image. The Sobel operator G applied to I returns the matrix convolution: an element-wise multiply and accumulate function, as in (11); the result of such convolution represents the intensity of the pixel stored in the output matrix C at position (x,y).

$$C_{x,y} = I \otimes G = \begin{pmatrix} i_{x-1,y-1} & i_{x-1,y} & i_{x-1,y+1} \\ i_{x,y-1} & i_{x,y} & i_{x,y+1} \\ i_{x+1,y-1} & i_{x+1,y} & i_{x+1,y+1} \end{pmatrix} \otimes G$$
$$= i_{x-1,y-1} \cdot 1 + i_{x-1,y} \cdot 0 + \cdots + i_{x+1,y+1} \cdot (-1) \tag{11}$$

Such operation is then iterated over all the pixels contained in I, thus to generate a new image containing information on horizontal edges.

The circuit we implemented performs the convolution reported in (11) between G and a grayscale source image where each pixel is in the range $[0 - 255]$. The input of the circuit is composed of the value of the 6 pixels involved in the convolution. Input space cardinality was reduced by considering only the three most significant bits for each pixels, with a total of 18 primary inputs. A threshold operator on the result of (11) was also applied as to eliminate noise on obtained images. As a consequence, the output of the circuit is represented by a single bit subject to the following rule: if C is below the intensity threshold, then the output pixel does not represent an edge (output 0); otherwise the function evaluates to 1.

978-1-5386-4757-8/18 $31.00 © 2018 IEEE

	E-MS	Q-IU
Number of gates:	308	14
Power savings:	1X	7.3X

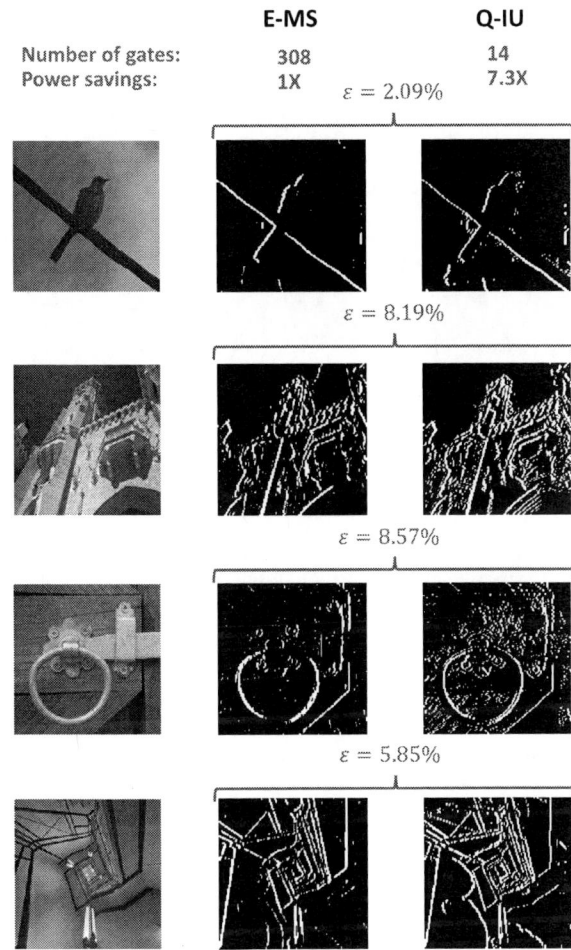

$\varepsilon = 2.09\%$

$\varepsilon = 8.19\%$

$\varepsilon = 8.57\%$

$\varepsilon = 5.85\%$

Figure 6. Result comparison with circuit implementing the Sobel operator. Original image (left), E-MS (center), Q-IU (right).

The exhaustive truth table was then generated by means of a Matlab script. The experimental setup is composed of two design flows, summarized as follows.

Quasi-Exact Computing via Inferential Unit (Q-IU): only the Inferential Unit composing the architecture illustrated in Section III is employed. Mapping and optimization as described in Section IV are also applied.

Exact Computing via Multi-level Synthesis (E-MS): the benchmark is processed with the ABC synthesis tool and mapped on a technology library at the 45nm node consisting of 32 logic primitives. The design is optimized by using the built-in `collapse` command. We first analyze the quality-of-result of the Q-IU implementation. Figure 6 shows a few sample images[1] of size 150×150 pixels used for the evaluation process. Error rate (ε) was computed as the pixel-wise difference between the exact solution provided by the E-MS circuit, and the quasi-exact counterpart obtained through the Q-IU. Starting from the topmost picture in Figure 6, error rates for each image are: 2.09%, 8.19%, 8.57%, and 5.85%.

[1]Considered images were obtained through an open-source repository. All images are released as public domain.

These values lead to a fundamental observation: Inferential logic is actually capable to extrapolate the most significant operations of a Boolean function, i.e., the ones with a higher computational and expressive power. This observation is supported by comparing the performance of the two implementations: the Q-IU is composed of 14 gates only, thus ensuring a $22\times$ smaller area w.r.t. the E-MS counterpart. Such a huge area saving translates to $7.3\times$ less dynamic power consumption.

B. ILC vs. Multi-level

It is important to understand that the objective of this section is not to demonstrate how ILC circuits could replace standard synthesis flows, but rather to provide a clearer picture of pro's and con's of adopting the inferential paradigm in accuracy-critical applications. The adopted experimental setup can be summarized as follows.

ILC performing exact computing (E-ILC): benchmarks are represented with the architecture illustrated in Section III; the Inference Unit is designed by means of the mapping and optimization steps described in Section IV; the Supervisor Unit is obtained with a standard multi-level ABC synthesis flow. Notice that the design of the Supervisor Unit is not optimized, as to provide a worst-case analysis. Multi-output logic functions are elaborated by isolating and processing each output cone separately.

ILC performing quasi-exact computing (Q-ILC): in this case each benchmark is composed of the Inference Unit alone; the design encompasses the stages described in Section IV.

Multi-level Synthesis (MS): benchmarks are processed with the ABC synthesis tool. Each benchmark is optimized by means of the built-in `fx` command.

Considered benchmarks are general-purpose open-source circuits belonging to the LGSynth91 suite. Designs are elaborated by means of their exhaustive truth table descriptions with digital files compliant to the Espresso format definition. In all considered synthesis flows we adopt the same CMOS technology library at the 45 nm node consisting of 32 logic primitives. Power estimations are conducted by means of Synopsys PrimeTime simulations. It is important to underline that truth tables represent a readily available circuit description for our purposes; obviously, they represent a bottleneck, especially with big circuits. However, the objective of this work is not to provide an ultimate solution, but rather to illustrate a proof-of-concept for future ML-driven design flows. Alternative design methodologies to overcome this issue, e.g., derive CTs from RTL descriptions, are part of future works.

Figure 7 shows the accuracy of both E-ILC and Q-ILC implementations; numbers have been obtained by means of exhaustive functional simulations for each considered benchmark. The plot demonstrates that the E-ILC architecture yields designs which are fully compliant with the original Boolean network specifications. On the other hand, Q-ILC designs achieve a remarkable accuracy: 90% on average.

Table I reports the total number of devices for each benchmark processed with the three considered implementations. Notice that the area of the E-ILC is due to the contributions of both the

978-1-5386-4757-8/18 $31.00 © 2018 IEEE

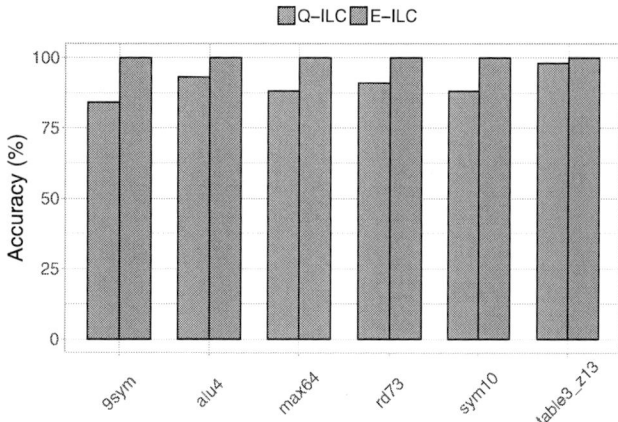

Figure 7. Quality of result analysis: E-ILC vs. Q-ILC.

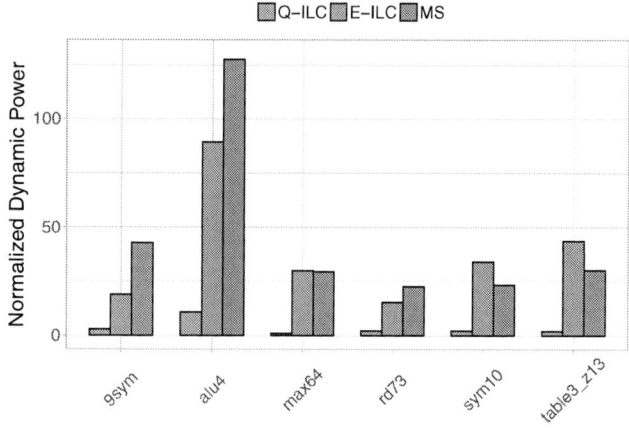

Figure 8. Normalized dynamic power for the three considered design flows.

Table I
OBTAINED NUMBER OF DEVICES WITH Q-ILC, E-ILC, AND MS DESIGN
FLOWS. REPORTED SAVINGS ARE W.R.T. MS.

	ILC					MS
	Q-ILC	Savings (%)	Supervisor	E-ILC	Savings (%)	
9sym	13	87.61	52	65	38.09	105
max64	8	91.11	100	108	-20	90
rd73	18	73.13	42	60	10.44	67
sym10	10	93.71	107	117	26.41	159
table3_z13	9	93.83	137	146	0	146
alu4	288	84.72	1409	1697	9.97	1885
Average	**57.66**	**87.35**	**307.83**	**365.5**	**10.56**	**408.66**

Supervisor and the Inferential units. Numbers suggest that E-ILC implementations achieve excellent results. Indeed, E-ILC shows an average gate count similar to the MS counterparts (10.56% smaller). As one might observe, there are cases where area gets larger, e.g., the max64. Nonetheless, these results suggest that the proposed ICL architecture can allow to achieve acceptable results in terms of area occupation. On the other hand, the portion of area taken by the Q-ILC is substantially smaller, with an average gate saving of 87.35%. Another important analysis concerns power consumption. Figure 8 depicts the normalized dynamic power recorded for each implementation. On average, the Q-ILC implementation achieves 13× (10×) lower dynamic power consumption w.r.t. MS (E-ILC). The E-ILC almost matches the power profile of the MS implementation, still with a 16% saving on average. Also in this case, there might be benchmarks for which E-ILC consumes more power, e.g., max64.

It is worth to emphasize that, even without aggressive logic optimizations, ILC circuits show similar performance of multi-level counterparts, yet with an additional feature: the possibility to readily adopt an on-line, adaptive, trade-off strategy between accuracy and power consumption. Indeed, through a dedicated *power-knob*, ILCs could switch between (*i*) a quasi-exact computation for devices with scarce power budgets, e.g., a mobile device running out of power, that achieves minimal power consumptions still having a good quality of result, or (*ii*) an exact computation, which can be used whenever accuracy scaling is not an option.

VI. CONCLUSIONS

In this paper we introduced a design paradigm for combinational circuits, the *Inferential Logic Circuits*. The proposed ML-inspired synthesis flow yields digital circuits that leverage statistical inference to process the outcome. Experimental results demonstrate that the proposed quasi-exact computation is a viable solution for error-resilient applications; it achieves 90% accuracy using 87.35% fewer devices w.r.t. a multi-level logic design; the same circuits can also deliver exact computations with 100% accuracy.

REFERENCES

[1] H. Brink, J. Richards, and M. Fetherolf, *Real-world machine learning*. Manning Publications Co., 2016.

[2] L.-C. Wang and M. S. Abadir, "Data mining in eda - basic principles, promises, and constraints," in *Proceedings of the 51st Annual Design Automation Conference*, ser. DAC '14. ACM, 2014, pp. 159:1–159:6.

[3] L.-C. Wang, "Experience of data analytics in eda and testprinciples, promises, and challenges," *IEEE Transactions on Computer-Aided Design of Integrated Circuits and Systems*, vol. 36, no. 6, pp. 885–898, 2017.

[4] O. Guzey *et al.*, "Extracting a simplified view of design functionality based on vector simulation," in *Haifa Verification Conference*. Springer, 2006, pp. 34–49.

[5] V. Tenace and A. Calimera, "Quasi-exact logic functions through classification trees," *Integration*, 2018.

[6] V. Tenace and A. Calimera, "Activation-kernel extraction through machine learning," in *CAS (NGCAS), 2017 New Generation of*. IEEE, 2017, pp. 5–8.

[7] R. G. Rizzo, V. Tenace, and A. Calimera, "Multiplication by inference using classification trees: A case-study analysis," in *Circuits and Systems (ISCAS), 2018 IEEE International Symposium on*. IEEE, 2018, pp. 1–5.

[8] A. J. Izenman, "Modern multivariate statistical techniques," *Regression, classification and manifold learning*, 2008.

[9] J. L. Gastwirth, "The estimation of the lorenz curve and gini index," *The review of economics and statistics*, pp. 306–316, 1972.

[10] B. L. Synthesis and V. Group, "ABC: A System for Sequential Synthesis and Verification," http://www.eecs.berkeley.edu/ alanmi/abc/, 2016.

[11] S. B. Akers, "Binary decision diagrams," *IEEE Trans. Computers*, vol. 27, no. 6, pp. 509–516, 1978.

[12] R. E. Bryant, "Symbolic boolean manipulation with ordered binary-decision diagrams," *ACM Computing Surveys (CSUR)*, vol. 24, no. 3, pp. 293–318, 1992.

[13] F. Somenzi, "CUDD: CU decision diagram package-release 2.4. 0," *University of Colorado at Boulder*, 2009.

[14] R. Jain, R. Kasturi, and B. G. Schunck, *Machine vision*. McGraw-Hill New York, 1995, vol. 5.

Implantable IoT System for Closed-Loop Epilepsy Control based on Electrical Neuromodulation

Reza Ranjandish*, and Alexandre Schmid*

*Microelectronic Systems Laboratory, Swiss Federal Institute of Technology (EPFL), Switzerland
reza.ranjandish@epfl.ch

Abstract—A closed-loop system aiming at epilepsy control is proposed in this paper, in which electrical stimulation is triggered upon a decision aggragating different biological signatures of the seizure such as changes in the heart rate, blood flow in the brain, along side the changes in the iEEG signals. iEEG signals are recorded and processed in the implantable part of the system. Electrical stimulation of deep-brain or vagus nerve targets which are known to be effective in seizure abortion is started upon detection of a seizre onset. This paper focuses on the implantable part of the system including a high dynamic-range amplifier, a sub-ranging/amplification stage, a line-length feature extractor and a stimulator. The implantable part is integrated using a 0.18 μm technology.

Index Terms—Closed-loop electrical stimulation, IoT solutions for health-care, High dynamic-range amplifier.

I. INTRODUCTION

Epilepsy is a major neurological disorder in terms of prevalence, cumulative incident count, as well as in terms of induced trauma and induced social disability. According to the World Health Organization (WHO) report on epilepsy, 80% of patients with epilepsy live in low- and mid-income countries. Patient responding to prevalent cure like medications and surgery are approximately 70%, while approximately 30% of the patients are untreated or poorly treated because of the following reasons, i) seizures diffuse over an excessively large area, or ii) seizures occur in sensitive area of eloquent cortex that may not be surgically treated, iii) seizures have multiple focus (multifocal seizures) which are thus difficult to individual localization and in practice impossible to surgically treat, and finally iv) surgery may not be tolerable due to specific medical conditions.

Electrical stimulation has been proposed as a therapy to reducing the number of seizures of patients suffering from refractory epilepsy. Open and closed-loop stimulation approaches have been successfully demonstrated. In open-loop systems, stimulation is delivered obeying a time pattern. The Vagus Nerve Stimulator (VNS) is an open-loop and FDA approved device, which has been shown to be effective to reduce the rate of seizures as much as 90%. Nevertheless, open-loop systems are deemed power hungry since stimulation is carried out on a time-pattern basis, *i.e.*, continuously, and it is not based on the detection of the seizure onset.

Closed-loop stimulation has been proposed as an approach that allows reducing the power consumption of stimulators. The seizure onset must be detected and only upon this event,

a burst of electrical stimulation signals is triggered to abort the seizure. Hence, closed-loop stimulation also eludes some issues related to the adaptation to a permanent stimuli that may become less effective. Closed-loop based system stimulators require a feature extractor to detect the onset of a seizure. Energy, line-length, moving average, autocorrelation, etc. are the common types of features that are routinely extracted from recorded cortical signals which are EEG or iEEG, and analyzed to detect the onset of a seizure. The quality of detection in a closed-loop system is extremely important, and can be defined in terms of sensitivity and specificity of the seizure detection. An ideal feature extractor presents a sensitivity of 100% and a specificity of 100%. The quality of detection should be high in order to reduce power consumption in the entire system. In addition to cortical signals, other biological markers or parameters can be used to increase the quality of detection. For example, the pattern of heart rate is shown to change during a seizure [1]. This topic is the subject of clinical research aiming at discovering markers of early signs of seizure, that, combined to cortical signals may enable a safe predictions of seizure. A reliable prediction is prerequisite to any seizure-abortion therapy that are under study nowadays, including electrical, and chemichal stimulation, optogenetic-based, or electro-magnetic and electro-accoustic approaches, and thus to restoring a decent quality of life to patients suffering from intractable epilepsy.

For the first time, in this paper we propose an implantable IoT system for closed-loop epilepsy control based on electrical neuromodulation that consists of an implantable, invasive, as well as external, non-invasive parts.

II. THE IoT APPROACH TO AN EPILEPSY CONTROL SYSTEM

Emerging applications of the Internet of Things (IoT) offer lifesaving applications within the health-care industry. IoT health-related solutions generally consists of collecting data from wearable or bedside devices, identifying patient information, processing the acquired data and performing real-time diagnosis. Using IoTs, the entire patient care could be significantly improved by reducing the potential data loss and/or potentially mistakes in diagnosis. Effective healthcare depends on speed and accuracy which can be improved taking benefit of the information transfer between different healthcare devices as IoT offers by its nature. Providing a common language and single platform for different devices, the numerous advantages

978-1-5386-4757-8/18 $31.00 © 2018 IEEE

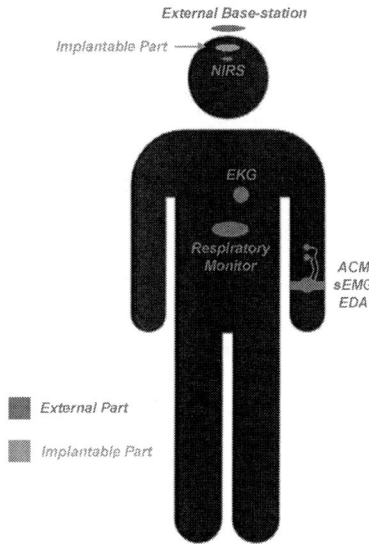

Fig. 1. Illustration of an IoT healthcare system constructed from implantable part and external nodes.

of IoTs will be deployed. The IoT-based closed-loop system healthcare approach and application is introduced in this paper for the first time.

A. Overview of the System

The overall sketch of the proposed system is presented in Fig. 1. The system comprises an implantable module as well as external units. The implantable part is employed to record ECoG signals from the cortex and performs electrical stimulation, for example when a seizure is detected. The implantable part comprises a high dynamic-range amplifier, a sub-ranging/amplification stage, ADC, feature extractor and an electrical stimulator. The external parts collect additional data and information related to essential and relevant biological parameters of the patient that are pertinent to the detection of the seizure onset. External sensors systems include surface electromyography (sEMG), Electrodermal activity (EDA), Electrocardiogram (EKG), Accelerometry (ACM), Near-infrared spectroscopy (NIRS), skin temperature, and respiratory monitoring. Each of these sensors techniques are briefly explained in the following subsections. The information originating from the external parts are delivered to the implantable part using an external base-station. The communication between external parts and the base-station operates on the principle of far-field communication, while the communication between external base-station with the implantable part is based on near-field communication. The external base-station also receives the ECoG signals, when this is required for data screening purposes. In addition, the external-base station delivers energy to the implant using wireless power transfer (WPT). Depending on the type of seizure, one or several external parts can be employed to increase the detection quality.

1) Surface Electromyography (sEMG): Using sEMG, the electrical activity of muscles is monitored using surface electrodes. Since most of the seizures comprise a motor compo-

nent, measurements obtained from sEMG provide information about the seizure onset [2].

2) Electrodermal Activity (EDA): Electrodermal activity (EDA) is the modulation of the skin impedance which is measured by applying a small current. In [3] epileptic patients are studied by monitoring EDA. This study shows that EDA significantly increases, exactly after the electrical seizure onset (EEG seizure).

3) Electrocardiogram (EKG): Cardiovascular failure is the main reason of death related to the sudden unexplained death in epilepsy (SUDEP). In addition, vascular changes have been studied before, during and after seizures and shown to be correlated with seizure onset [1].

4) Accelerometry (ACM): Measuring changes in the velocity and direction of the movements of the body or some specific limb is helpful for motor seizure detection [4]. Accelerometry (ACM) is a method aiming at detecting these movements using an accelerometer attached to a limb.

5) Near-Infrared Spectroscopy (NIRS): Near-infrared spectroscopy (NIRS) employs specific wavelengths to measure the cerebral oxygen saturation. This method is based on the specific absorption properties of the tissues in the near infrared range. A spectrophotometer is placed on the forehead that emits infrared light into the tissue from the surface of the scalp. Subsequently, the light is collected using a detector located close to the emitter and the oxygen saturation is processed. NIRS is helpful in the detection of the focal seizures, including focal dyscognitive, focal without impairment of consciousness, and focal with secondary generalization as well as absence seizures [5].

6) Skin temperature: Skin temperature is another variable that has been shown to be a predictive parameter of seizure. A study reported in [6] demonstrates a relation between the probability of seizure onset and the ratio of the mean temperature during sleep and awake periods.

7) Respiratory monitor: It is suggested that sudden unexplained death in epilepsy (SUDEP) is caused by cardiac and/or respiratory failure during a seizure [7]. Hence, monitoring respiration during seizures may be a crucial measurement that the external part performs. Using an elastic belt or adhesive patches that maintain electrodes used for impedance measurements, respiration is monitored and any detected abnormality is reported to the implantable part to increase the quality of the seizure detection.

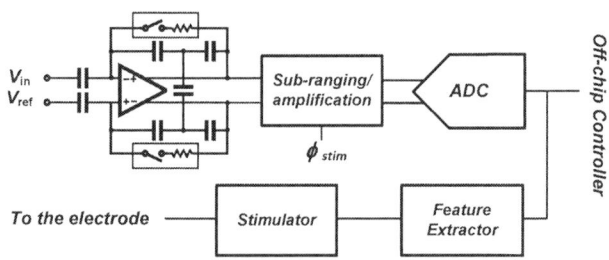

Fig. 2. Structure of the implantable part of the system.

978-1-5386-4757-8/18 $31.00 © 2018 IEEE

III. A High Dynamic-Range AFE-Based Implantable Part

One of the major part of the proposed system is the implantable part. As shown in Fig. 2, the implantable part consists of a high-dynamic range AFE that records the neural signals even during the electrical stimulation, a sub-ranging/amplification stage used to increase the resolution of the ADC during stimulation (ϕ_{stim}) and increase the amplification during the non-stimulation period. A 10-bit SAR ADC is used to digitize the samples acquired by the AFE. A coastline feature extractor monitors the acquired signals and detects a seizure onset. Electrical stimulation pulses are triggered if a seizure onset is detected by the feature extractor. In stimulation mode, an artifact rejection algorithm is employed using an off-chip controller to remove the artifacts from the recorded signals and to reconstruct the main neural signal. Hence, the activity of the brain is monitored during the stimulation and the stimulation is stopped when the activity of the brain becomes normal. More details of each section of the implantable unit are provided in the following. As discussed above, the additional data is collected by the external base station using far-field wireless communication and is transferred to the internal part using near-field communication. These parts are not in the scope of this paper and they are not integrated on the silicon. For artifacts rejection, an off-chip controller is used to calculate the artifact coefficients and to subtract it from the received signal. During normal condition, the sub-ranging/amplification stage is in amplification mode and consequently digital data is fed into the feature extractor to detect the seizure onset. If a seizure is detected, the stimulator is triggered. A burst of electrical pulses is then started to abort the seizure. During the stimulation, the sub-ranging/amplification stage mode is changed to the sub-ranging. In this mode, this stage performs as a sub-range ADC. A two-stages sub-ranging ADC is chosen to extract 3 bits prior the main SAR ADC. Therefore, during the stimulation, the sub-ranging/amplification stage and the main SAR ADC perform as a single unit SAR-assisted pipe-line ADC. Thanks to the sub-ranging stages, the neural signals are distinguishable

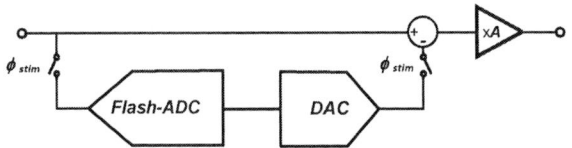

Fig. 4. General structure of single stage of the subranging/amplification stage.

from the artifacts since the resolution is increased.

A. High Dynamic-range Amplifier

The structure of the high dynamic-range amplifier is depicted in Fig. 3. The high dynamic-range amplifier uses switched-resistor technique to substitute a non-linear pseudo-resistor with a linear resistor. The amplifier architecture is a folded cascode amplifier. The bandwidth of the amplifier is limited to 1 KHz using capacitor at the output node of the amplifier. The capacitors are much larger than the input capacitors of the next stage. Failing to do so, the bandwidth of the high-dynamic range amplifier may change. The amplification of the first stage is chosen at 20 dB. The input capacitance of this stage is important for the noise performance and the area of the amplifier. A large input capacitance increases the noise performance of the circuit; however, it increases the area of the amplifier which is not desirablee in multichannel neural recording systems.

B. Sub-ranging/amplification Stage

The structure of the sub-ranging/amplification stage is shown in Fig. 4. As discussed above, during normal operation of the system (no electrical stimulation), this stage works as 2-stage amplifier to increase the total amplification of the input neural signals. Hence, alongside with the 20 dB amplification at the high-dynamic range amplifier, the total amplification of signals before the SAR ADC is 38 dB. Since in this mode there is no artifact introduced in the recorded signals, the non-linearity introduced by the sub-ranging/amplification stage is negligible. During the stimulation mode, the sub-ranging/amplification stage performs as a coarse ADC to extract 3 bits from the recorded signals. In this mode, sub-ranging/amplification stage and the SAR ADC perform as a SAR-assisted pipeline ADC.

C. Electrical Stimulation and Charge Balancing

Regarding biological safety issues, the injected net charge in an electrical stimulation must be equal to zero. Any non-conformities leads to a pH shift or tissue damage. in addition, a pH shift causes electrode dissolution due to the electrolysis and consequently leads to releasing toxic substances into the biological environment. Therefore, stimulators must ensure that the remaining net charge on the electrode after each stimulation period is equal to zero or is kept within a safe window (e.g. ± 100 mV), making use of a charge balancer. Charge balancing can be performed as an active or passive technique. In active charge balancing, active circuits are used to insure that the voltage on the tissue interface is controlled

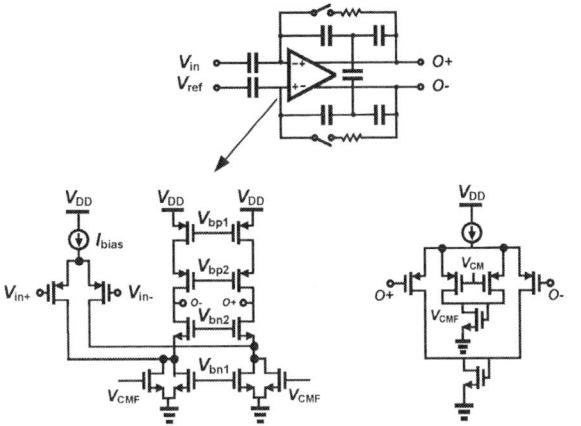

Fig. 3. Structure of the high dynamic-range amplifier employed in the system.

978-1-5386-4757-8/18 $31.00 © 2018 IEEE   157

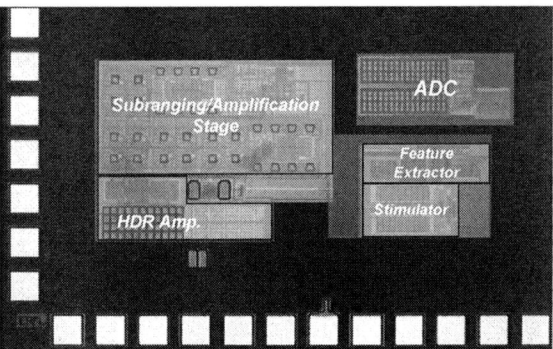

Fig. 5. Die photograph of the implantable part.

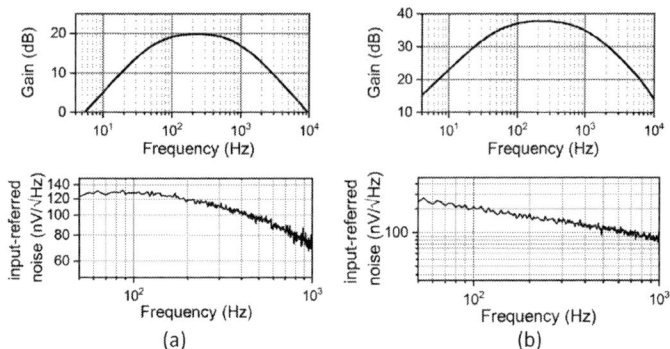

Fig. 6. Performance of the (a) high dynamic range amplifier and (b) the chain of high dynamic-range amplifier connected to the subranging/amplification stage.

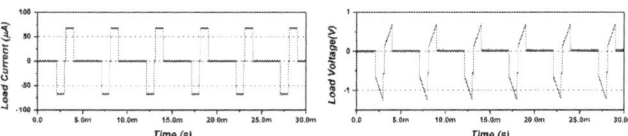

Fig. 7. Measured results obtained from the electrical stimulator using an electrical model of the tissue.

to its initial value after cathodic phase. In a passive charge balancer no active components is used and charge cancellation is carried out by shorting the electrode.

IV. EXPERIMENTAL RESULTS

Using a 0.18 μm technology, the implantable part of the proposed system is implemented to evaluate its performance. The chip photograph of the implemented system is depicted in Fig. 5. Fig. 6a presents the performance of the high dynamic-range amplifier in terms of the frequency response and input-referred-noise. The high dynamic-range amplifier exhibits a high-pass frequency corner of 50 Hz and a low-pass frequency of 1 kHz. The integrated input referred noise is equal to 3 μV_{rms}. By activating the normal operation of the system (no stimulation) the total frequency response and noise level (including the amplification of the sub-ranging/amplification stage) is depicted in Fig. 6b. The high dynamic range amplifier draw 0.5 μA from 1.5 V supply voltage which leads to a power consumption of 750 nW. Furthermore, the subranging/amplification stage draws 0.1 μA from 1.5 V supply which leads to a 150 nW power consumption. The performance of the electrical stimulator is depicted in Fig. 7. In order to validate the performance of the electrical stimulator and its safety, the current pulses are applied to the electrical model of the tissue. An electrical model including a C_{dl} of 100 nF and an R_s of 10 kΩ is employed. Fig. 7 presents the applied current pulses and recorded voltage on the tissue model. The remaining voltage on the tissue model after each stimulation is in a safe window which ensures the safety of the implemented electrical stimulator.

V. CONCLUSION

The recent emergence of the IoT concept and technology offers new perspective to the development of therapeutic systems, which take benefit of aggragating various information delivered from a network of heteregeneous sensors to improving the quality of a an automated decision of treatment delivery. The intricacy of refractory epilepsy control in closed loop evidences the need of such advances IoT systems that include a network of external nodes and also an implantable system. Power consumption is a major issue of implantable electornic systems that take benefit of an accurate decision of electrical stimulation delivery. An implantable closed-loop IoT node that performs signal recording, seizure detection and subsequent electrical stimulation is presented, that uses a high dynamic range front-end that supports simultaneous cortical stimulation and recording.

ACKNOWLEDGMENT

This work has been supported by the Swiss NSF under grant number 200020-175790.

REFERENCES

[1] S. Behbahani, N.J. Dabanloo, A.M. Nasrabadi, C.A. Teixeira, A. Dourado, "Pre-ictal heart rate variability assessment of epileptic seizures by means of linear and non-linear analyses," Anadolu Kardiyol Derg, 13 (8) (2013), pp. 797-

[2] S.N. Larsen, I. Conradsen, S. Beniczky, H.B.Sorensen, "Detection of tonic epileptic seizures based on surface electromyography," Conf Proc IEEE Eng Med Biol Soc, 2014 (2014), pp. 942-945

[3] M.Z. Poh, T. Loddenkemper, N.C. Swenson, S. Goyal, J.R. Madsen, R.W. Picard Continuous monitoring of electrodermal activity during epileptic seizures using a wearable sensor Conf Proc IEEE Eng Med Biol Soc, 2010 (2010), pp. 4415-4418

[4] T.M. Nijsen, J.B. Arends, P.A. Griep, P.J.Cluitmans, "The potential value of three-dimensional accelerometry for detection of motor seizures in severe epilepsy," Epilepsy Behav, 7 (1) (2005), pp. 74-84

[5] M. Seyal, "Frontal hemodynamic changes precede EEG onset of temporal lobe seizures," Clin Neurophysiol, 125 (3) (2014), pp. 442-448

[6] B. Kim, A.B. Nogueira, S. Thome-Souza, K.Kapur, J. Klehm, M. Jackson, et al., "Diurnal and nocturnal patterns of autonomic neurophysiological measurements are related to timing of seizures." Poster presented at: CNS 2015. Child Neurology Society annual meeting, 2015, October 7–10, Washington, DC (2015)

[7] K. Singh, E.S. Katz, M. Zarowski, T.Loddenkemper, N. Llewellyn, S. Manganaro, et al., "Cardiopulmonary complications during pediatric seizures: a prelude to understanding SUDEP," Epilepsia, 54 (6) (2013), pp. 1083-1091

Mobile Phones *Hematophagous Diptera* Surveillance in the field using Deep Learning and Wing Interference Patterns

M. Souchaud[1], P. Jacob[1], C. Simon-Chane[1], A. Histace[1],
O. Romain[1], M. Tchuenté[2], D. Sereno[3]

[1]*ETIS UMR 8051, Université Paris Seine, UCP, ENSEA, CNRS, F-95000, Cergy, France*
[2]*UMMISCO, IRD, Université de Yaoundé 1, Cameroun*
[3]*UMR IRD 177 IRD-CIRAD, Montpellier Cedex*

Abstract—Real-time monitoring of *hematophagous diptera* (such as mosquitoes) populations in the field is a crucial challenge to foresee vaccination campaigns and to restrain potential diseases spreading. However, current methods heavily rely on costly DNA extraction which is destructive, costly, time consuming and requires experts. The contributions of this work are: 1) the usage of a new type of imaging, named Wing Interference Patterns (WIPs), which is non-destructive and easier to produce during in the field experiments; 2) a deep learning architecture which is optimized for very low computation cost, memory usage and a short inference time; 3) the use of a dataset of more than 50 medically important species of *hematophagous diptera* with more than 3000 images of WIPs. With these contributions, we demonstrate that WIPs are an excellent medium to automatically recognize a large amount of *hematophagous diptera* species with very high accuracy and low computational cost convolutional neural network.

Index Terms—Deep Learning, WIP, Image Classification, Hematophageous Diptera.

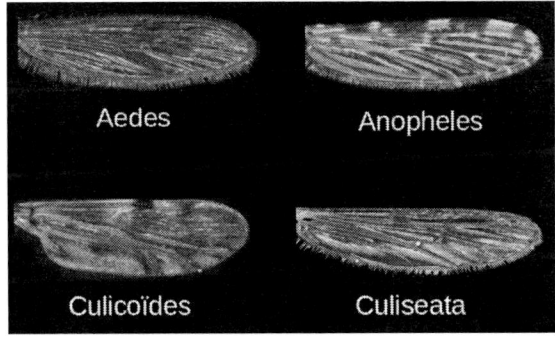

Fig. 1: Example of WIPs for four different genera. While *Culicoïdes* is different in term of shape, the three others are very different thanks to the interference pattern.

I. INTRODUCTION

Blood-sucking insects are known vectors of a wide variety of diseases and viruses, such as malaria, yellow fever or dengue. Consequently, it is necessary to study both their habitat preferences and their implication in the diseases transmission. To that aim, lots of research works are based on building a taxonomy [1] to classify *hematophagous diptera* and thus to identify which classes are the most common vector transmissions for a given disease or virus. Such knowledge allowed the development of control systems and sampling methods for the most important families of *hematophagous diptera* . In order to build this taxonomy, standard practices combine both morphological and molecular techniques [2]. Thus, DNA barcode characterization has been recognized as the most accurate identification protocol and it is the most used method for classifying such genus. However for in the field experiments, DNA extraction is not available as it needs very specific and non-transportable equipment. To cope with these difficulties, recent works proposed to classify these blood-sucking insects with other innovative means, *e.g.* with the use

of matrix-assisted laser desorption/ionization to measure time of flight by mass spectrometry or with the record of wingbeat sounds. In imaging, Wing Interference Patterns (WIPs) [3] have been demonstrated as an efficient tool to determine in which genus the given *hematophagous diptera* image belongs to [3]. This type of imaging has the advantage to be between laboratory imaging and on the field imaging. In fact, the necessary tools are much more cheaper and it is simpler to generate such interference patterns during in the field experiments.

At the same time, major improvements in image analysis have been achieved with local features extraction, global representations and most recently with the use of deep learning [4], [5]. In the case of insects recognition, the majority of works use hand-crafted features with machine learning algorithm such as SVM to classify either a set of insects or a set of genus that belong to the same insect, but they neither test their method with a more fine-grained taxonomy such as species and sub-species nor take advantages of deep learning architecture.

In this work, we gather WIP imaging and deep architecture for in the field experiments and automatic classification by designing a deep learning architecture for embedded systems

978-1-5386-4757-8/18 $31.00 © 2018 IEEE

(*i.e.* with few parameters and small inference time). This architecture is able to classify more than 50 medically important genus and species of *hematophagous diptera* with high accuracy while using WIP images. Finally, as a proof-of-concept, we develop a real-time mobile application for automatic genus/species recognition based on these WIPs and our architecture.

II. RELATED WORK

A. Insect Classification

Martineau *et al.* [6] present a recent survey of insect classification. We invite the reader to refer to this article for more references. The key elements from this study are the image acquisition protocols, the features extraction and the automatic classification algorithms. Image acquisition protocols are divided into two parts: the laboratory-based images and the field-based images. In the case of laboratory-based images, there is a fix protocol for image acquisition from the insect trapping to its placement and the material. The insect is manually positioned following this setup and then the image is acquired to be processed in a future step. Such protocols are usually heavily constrained to ensure low variability between samples and make the classification easier. In the case of field-based, image are taken without constraints directly in cultivated fields. However, such approach is usually much more harder than the previous one mainly due to the very high variability in samples. In this paper, we use the recent WIPs that have been demonstrated to contain precise taxonomy information [3] and that have the advantage to be easier to produce than the aforementioned laboratory-based images but without the variability of field-based images. Furthermore, Sereno *et al.* [7] patent a portable system and a dataset of WIPs for in the field experiments and propose an automatic recognition system to classify *hematophagous diptera* genus. In this work, we propose to go further in the taxonomy by automatically classifying species using the aforementioned dataset.

When images are acquired, features have to be extracted from these images to be able to classify them. One more time, features extraction are either based on manually extracted features such as wing's venation or wing's geometry or extracted with hand-crafted features such as SIFT or texture. Sereno *et al.* [7] extract SIFT to build global representations with VLAT [8] and use linear SVM to classify these genus. However, manually extracted features are time consuming and do not fully take advantages of effective methods such as SIFT. Here we propose to replace SIFT extractor by a convolutional neural network (CNN) which has been successfully used as a features extractor in recent works [4], [5], [9]. However, the idea is to build not only an accuracy system but a fast for an easy in-the-field use. Consequently, we study the recent architectures that have been proposed for mobile phone applications or real-time use.

B. Mobile phone CNN architectures

With the popularization of deep learning with AlexNet [9] and then VGG-16/19 [4], many researchers have studied more efficient architecture in terms of computational cost and number of parameters. Convolutions are the most time-consuming part as the standard process computes a convolution with supports $D_h \times D_w$ for each input channel and repeats this step as many times as needed to have the expected number of output channels. To cope with such computation overload, MobileNet introduces the depth-wise convolution. Unlike the standard process, depth-wise convolution compute in a first time one $D_h \times D_w$ convolution per input channel and in a second time a 1×1 full convolution is applied to generate the expected number of output channels. This has the advantage to greatly reduce the computation cost but also the number of parameters. Indeed, with N, M the number of input and output filters respectively, the computational factor of depth-wise convolution over standard convolution is (from [10]) : $\frac{DepthConv}{FullConv} = \frac{1}{N} + \frac{1}{D_w D_h}$. Another way to decrease the computation time for standard convolution is to use the principle of network in network as it is used in YOLOv2 [11]. This principle is used to reduce the computational cost of 3×3 convolution (but also the number of parameters) and allows their best architecture to process 544x544 pixels images at 40fps on GPU. In this work, we build light architectures inspired by MobileNet, ResNet and YOLOv2 and compare them in accuracy, number of parameters and inference time.

III. OUR METHOD

In this section, we first present the dataset, and the training protocol. Then, we detail our small architectures inspired by MobileNet [10], ResNet [5] and YOLOv2 [11]. Finally, we give some implementation details to train our networks.

A. WIPs dataset

The Wing Interference Patterns dataset is the one patented by Sereno *et al.* [7]. In its most recent version, it is composed by 3311 images of *hematophagous diptera* in high resolution (around 1500x500 pixels) and contains 22 different genus for more than 50 species. Moreover, due to the limited size of this dataset, we do cross-validation by splitting the dataset in 5 parts: we train 5 different networks on 4 parts and we test on the last one and average the results. Note that we divided the dataset while keeping the population distribution. As metric, we report accuracy and per class accuracy, *i.e.* the number of images correctly predicted (for a given class) over the total number of images (for the given class) respectively.

We train our networks for three classification tasks. The first one is the classification of genus (Table II), similarly to [7]. For this task, we train our architectures on only the most represented genus, *i.e.* 11 genus over 22 which represent 3254 images over 3312. The second one is the first test to go further in the taxonomy: for a given genus, we classify its most represented species (Figure 2), *i.e.* 25 species from 5 genus for 2405 images over 3311. The last one is species classification without the knowledge of the genus (Table III and Figure 3). For this test, we keep 58 species from 11 genus which represent 2918 over 3311 images.

type	filters	strides	input shape
DepthConv	3x3, 32	2	256x116
DepthConv	3x3, 64	2	128x58
DepthConv	3x3, 128	2	64x29
Dropout	25%	1	32x15
DepthConv	3x3, 256	2	16x8
Dropout	25%	1	16x8
DepthConv	3x3, 256	2	8x4
Dropout	25%	1	8x4
DepthConv	3x3, 512	2	4x2
Dropout	25%	1	4x2
Fully Connected	256	-	8
Dropout	50%	1	256
Fully Connected	K	-	256
Softmax	-	-	K

TABLE I: Our mobilenet-like architecture.

B. Our architectures

As this task is much more constrained than natural images recognition tasks, we use smaller architecture than the original MobileNet [10], ResNet [5] and YOLOv2 [11]. The first architecture is greatly inspired by MobileNet and takes advantage of depthwise convolution [10]. Our smaller architecture use one scale less (we do not use the scale with 1024 filters) and only one depthwise convolution per scale compared to MobileNet which has 2 depthwise convolution. We also use batch normalization to speedup and stabilize training. Contrary to MobileNet, we use two fully connected layers similarly to VGG [4]. Table I details this architecture. In the case of YOLOv2, we reproduce architecture of DarNet-19 [11]. As the entire architecture tends to over-fit the training set (see Figure 2), we test two reduced architectures, *i.e.* using 1 or 2 scales less than the original network. We call them DarkNet-9 (8 convolution layers and 1 classification layer) and DarkNet-14 (13 convolution layers and 1 classification layer). Results are reported in Figure 2. Finally, we reproduce the ResNet18 architecture from [5] and train it from a random initialization. Even if this architecture seems too deep for this task compared to our others architectures, residual connections allow convergence of the training procedure and excellent results (see Figure 2).

We also test a more standard approach based of the extraction of SURF descriptors, a Bag of Features (BoF) representation using a 4000 codewords dictionary and a SVM with polynomial kernel. This is a similar approach to Sereno *et al.* [7]: they used SIFT with VLAT representations and a linear SVM as classifier. VLAT which is a second-order statistics based method has much more information than the zero-order BoF one but has the advantage to be smaller and faster.

C. Implementation details

All the aforementioned networks are trained for classification tasks. For each task, we only use 1 fully connected layer with the softmax activation to predict the probability that an image belongs to the correct class. We train our networks using Stochastic Gradient Descent (SGD) with a learning rate of 10^{-2} and a momentum of 0.9 for 30 epochs. We also do data augmentation due to the limited size of this dataset. We resize images from their full resolution to 270x120 pixels randomly

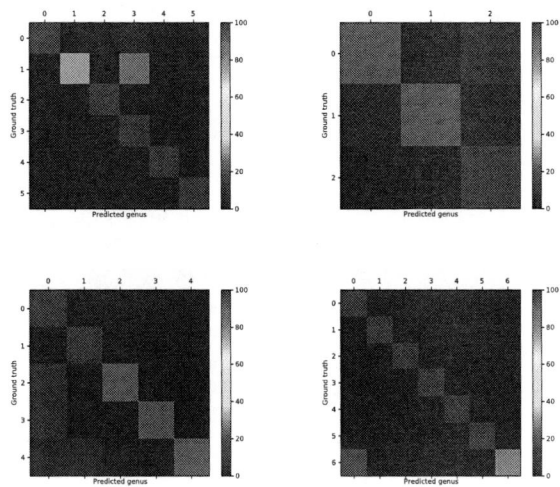

Fig. 2: Confusion matrices for Anopheles (top-left), Culex (top-right), Glossina (bottom-left) and Phlebotomus species (bottom-right). *Best viewed in color.*

cropped to the fixed resolution 256x116, we randomly rotate the images by $\pm$ 5 degrees and randomly horizontally or vertically flip the images. At the testing time, we resize the images to 256x116.

IV. RESULTS

In this section, we report results for the three classification tasks, *i.e.* genus classification (Table II), species classification for given genus (Figure 2), and global species classification (Figure 2).

For this first task, we only test our MobileNet-like architecture presented in the previous section and the standard model which uses SURF+VLAT+SVM. The second one approach reaches 84.8% accuracy on the whole dataset (see Table II). While confirming that WIPs are a good medium for *hematophagous diptera* recognition, this also demonstrates that automatic recognition is makes automatic recognition easier. At the same time, our MobileNet-like architecture achieved the astonishing results of 98.3% accuracy also on the whole dataset (see Table II) and outperforms the more standard method by nearly 14% and the 93.0 reported by Sereno *et al.* [7] is outperformed by 7%. This demonstrates the efficiency of our architecture to classify genus. For the second task, we only test our MobileNet-like architecture and report confusion matrices (see Figure 2). In term of accuracy, Anopheles species are classified with 97.0% accuracy, Glossina with 97.5%, Culex with 97.0%, Aedes with 97.1% and Phlebotomus with 96.2%. These results are close to the one in genus classification, that demonstrates the efficiency of our CNN to also classify accurately species. For the last classification task, we test our 3 architectures and report all global classification accuracy in Table III. Moreover, we report confusion matrix in Figure 3 for the MobileNet-like architecture. Table III demonstrates that our CNN architectures

	Culiseata	Aedes	Tryciclea	Lutzia	Anopheles	Phlebotomus	Culicoides	Chrysops	Culex	Orthopodomyia	Glossina	Global
# Test images	2	97	6	4	173	54	6	1	109	3	195	650
Sereno *et al.* [7]	-	-	-	-	-	-	-	-	-	-	93.0	-
Ours (BoF)	-	-	-	-	-	-	-	-	-	-	-	84.8
Ours (CNN)	50.0	97.9	50.0	75.0	98.8	100.0	100.0	100.0	98.2	100.0	100.0	**98.3**

TABLE II: Classification of genus. Results are reported in accuracy. #Image is the average number of images in the test base. Note that the three lowest accuracy are for three underrepresented genus.

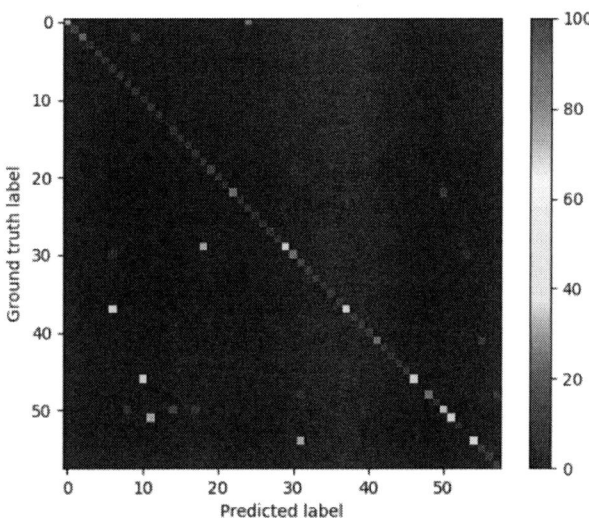

Fig. 3: Confusion matrix for the 58 species computed from the MobileNet-like architecture. *Best viewed in color.*

Method	Acc(%)	Params(M)	s/batch	Speed-up
DarkNet-19	81.0	19.9	1.86	1
DarkNet-14	87.7	4.6	1.60	1.2
DarkNet-9	92.5	**0.8**	1.26	1.5
ResNet-18	88.4	9.5	1.68	1.1
MobileNet-like	**95.9**	4.5	**0.094**	**19.8**

TABLE III: Accuracy for the species classification task.

are able to classify with high accuracy each specie from a large set of genus. Sereno *et al.* [7] report an average of 65% accuracy for a one-vs-all classification task that is similar to this third task. Consequently, we are able to reach the astonishing 95.9% accuracy that is more than 30% over [7]. Moreover, the confusion matrix in Figure 3 demonstrated that only few classes are really miss-classified, confirming the efficiency and the semantic power of CNN but also the great interest of WIP imaging. Moreover, even if the given timing are computed on a workstation with a quad-core CPU at 3.0GHz with 16Go RAM, the MobileNet-like architecture is able to compute more than 10 batches of 32 images per second. For mobile phone application, the system needs to process only 1 image at time, allowing real-time processing.

V. CONCLUSION

In this paper, we proposed to use the recent Wing Interference Patterns (WIPs) imaging coupled with the efficiency of low computing convolutional neural networks. Three state-of-the-art architectures for mobile phone applications have been tested for three different classification tasks, *i.e.* genus classification, species classification with the knowledge of the genus and all species classification directly. Our three architectures have shown astonishing results in term of accuracy, reaching 98.3% accuracy for the first task, 97% in average for the second one and 95.9% accuracy for the last one. The third task outperformed the previous results on this dataset by more than 30% accuracy, while having few classes with very bad precision scores. Moreover, the proposed architectures are able to compute in real-time images even on mobile phone application by the use of an optimized architecture for such application with a small inference time. Future works include the improvement of the proposed architecture or new ones, the sub-species classification and a new framework to process underrepresented species and sub-species.

REFERENCES

[1] N. G. Gratz, *Vector- and Rodent-Borne Diseases in Europe and North America Distribution, Public Health Burden, and Control.* Cambridge University Press, November 2006.

[2] M. Hajibabaei, G. A. Singer, P. D. Hebert, and D. A. Hickey, "Dna barcoding: how it complements taxonomy, molecular phylogenetics and population genetics," *TRENDS in Genetics*, vol. 23, no. 4, pp. 167–172, 2007.

[3] E. Shevtsova, C. Hansson, D. H. Janzen, and J. Kjærandsen, "Stable structural color patterns displayed on transparent insect wings," *Proceedings of the National Academy of Sciences*, vol. 108, no. 2, pp. 668–673, 2011.

[4] K. Simonyan and A. Zisserman, "Very deep convolutional networks for large-scale image recognition," *International Conference on Learning Representations (ICLR)*, 2015.

[5] K. He, X. Zhang, S. Ren, and J. Sun, "Deep residual learning for image recognition," in *Proceedings of the IEEE Conference on Computer Vision and Pattern Recognition (CVPR)*, 2016, pp. 770–778.

[6] M. Martineau, D. Conte, R. Raveaux, I. Arnault, D. Munier, and G. Venturini, "A survey on image-based insect classification," *Pattern Recognition*, vol. 65, pp. 273–284, 2017.

[7] D. Sereno, A. Cannet, M. Akhoundi, O. Romain, and A. Histace, "Système et procédé d'identification automatisée de diptères hématophages," French Patent WO2 016 097 499A1, Dec., 2015.

[8] D. Picard and P.-H. Gosselin, "Improving Image Similarity With Vectors of Locally Aggregated Tensors," in *2011 IEEE International Conference on Image Processing (IEEE ICIP2011)*, Brussels, Belgium, Sep. 2011, pp. 669 – 672.

[9] A. Krizhevsky, I. Sutskever, and G. E. Hinton, "Imagenet classification with deep convolutional neural networks," in *Advances in Neural Information Processing Systems 25*, F. Pereira, C. J. C. Burges, L. Bottou, and K. Q. Weinberger, Eds. Curran Associates, Inc., 2012, pp. 1097–1105. [Online]. Available: http://papers.nips.cc/paper/4824-imagenet-classification-with-deep-convolutional-neural-networks.pdf

[10] A. G. Howard, M. Zhu, B. Chen, D. Kalenichenko, W. Wang, T. Weyand, M. Andreetto, and H. Adam, "Mobilenets: Efficient convolutional neural networks for mobile vision applications," *arXiv preprint arXiv:1704.04861*, 2017.

[11] J. Redmon and A. Farhadi, "Yolo9000: Better, faster, stronger," in *The IEEE Conference on Computer Vision and Pattern Recognition (CVPR)*, July 2017.

978-1-5386-4757-8/18 $31.00 © 2018 IEEE

Radar for assisted living in the context of Internet of Things for Health and beyond

Julien Le Kernec[1,2*], *Senior Member, IEEE*, Francesco Fioranelli[1], Member, IEEE,
Shufan Yang[1], *Member, IEEE*, Jordane Lorandel[3], *Member, IEEE*, Olivier Romain[3], *Member, IEEE*

[1]*Communication, Sensing and Imaging group, School of Engineering, University of Glasgow, Glasgow, UK*
[2]*School of Information and Electronics, Univ. of Electronic Science and Technology of China, Chengdu, China*
[3]*ETIS-ASTRE, Université Cergy-Pontoise, Cergy-Pontoise, France*

Abstract—**This paper discusses the place of radar for assisted living in the context of IoT for Health and beyond. First, the context of assisted living and the urgency to address the problem is described. The second part gives a literature review of existing sensing modalities for assisted living and explains why radar is an upcoming preferred modality to address this issue. The third section presents developments in machine learning that helps improve performances in classification especially with deep learning with a reflection on lessons learned from it. The fourth section introduces recent published work from our research group in the area that shows promise with multimodal sensor fusion for classification and long short-term memory applied to early stages in the radar signal processing chain. Finally, we conclude with open challenges still to be addressed in the area and open to future research directions in animal welfare.**

Index Terms—**Human activity classification, fall detection, ambient assisted living, inertial sensors, magnetic sensors, radar sensors, multisensory data fusion, feature selection, machine learning, micro-Doppler signatures, feature extraction.**

I. INTRODUCTION

Internet of Things (IoT) in healthcare was evaluated at $60 billion and will reach $136 billion by 2021 [1]. IoT comprises intermediary components, such as devices, network connectivity, electronic systems, and software. It is networked smart electronic devices sharing information autonomously leveraging machine learning. In healthcare, this technology will facilitate managing and mining patient data and resources.

Life expectancy is increasing and poses challenges for health services as it comes with medical issues (chronic illnesses, multi-morbidity) and an alarming rise in the population over 60 predicted to reach 30% by 2050 worldwide [1-2]. This trend is not new but accelerating especially in developed countries.

In 2016, signal processing magazine had a special issue on assisted living [3–8]. It covered a range of technologies such as inertial measurement units, wearables, ambient sensors (pyroelectric infrared (PIR), vibration sensors, accelerometers, cameras, depth sensors and microphones) and radio waves with existing infrastructure (Wifi) present on site or active devices such as radar. For all sensing modalities, enhancing accuracy, lowering computational complexity, reducing power

consumption, exploiting multiple domains and modalities for complementarity and robustness, are crucial in developing technology enabled self-dependent living in-home care.

II. EXISTING SENSING MODALITIES

Many systems have been proposed to tackle this problem [5,8-9] including radar sensors or a combination of these systems, whereby their information is used concurrently and fused at different levels to optimize the overall performance.

Monitoring people in their daily life poses a privacy issue; there is a correlation between the perceived privacy and richness of information collected by sensors [9]. Video provides very rich information but is perceived as intrusive; PIR sensors are not perceived as invasive but provide little information.

A review of healthcare using mobile wireless technologies shows major challenges (data acquisition, processing data locally, wireless data, quality of service over cellular network, cloud storage, security, user interface and platforms) before being feasible [11]. It also suffers from integration problems where a lot has to come together before it is practical to use and requires the lifting of technological barriers as well.

Wearable sensors despite giving good classification results [12] greater than 98%, suffer from several major problems [13]:

- require user compliance as they need to be worn or to think about it if you wake during the night to go to the washroom.

- easily broken if dropped, crushed while sitting or falling.

In [14], [15], entire apartments have been fitted with sensors PIR motion sensors, stove sensors, floor sensors,… and provide good density maps for activities of daily living at the macro level. However, they cannot provide a finer granularity for gait analysis change detection as well as requiring transformations in a persons living environment.

An extensive review [16] of RGB cameras, depth sensors and radar technologies for assisted living highlighting open challenges for deployment in residences or specialized homes:

- For cameras, the main challenges are occlusions, working at night, dead zones in 3D, accuracy, precision, resolution and respecting privacy.

- For radar systems, the presence of strong scatterers and clutter in indoor environments may generate multipath and ghost targets which is comparable to occlusion in cameras. The compliance of radar system with emission regulations limits.

The technological challenges are greater for radar technology, but the fact there are no judicial issues regarding rights to image and plain images are not recorded, thus respecting privacy, facilitating acceptance of end users and investors. For these reasons, the radar sensing modality is an interesting research trend however still underutilized in specialized homes.

Radar is attractive due to reliability, low power emissions for indoor use (similar to WiFi), safety, which brings it at the frontier of indoor monitoring modalities rivaling video cameras and wearable devices for health. Radar can be used for fall detection, gait analysis and activities of daily living (ADL) to provide supplemental information to detect early signs of deteriorating physical/cognitive health. It would allow greater healthcare coverage, better quality of provision through 24/7 monitoring of the elderly well-being while respecting privacy.

Furthermore, the elderly may suffer from reduced cognitive capabilities and memory loss. To enable assistive technologies to help them to deal with ADL and monitoring their condition, a system requiring no intervention from their part is more suited.

Existing radar systems can be used to monitor activities [12,16–20], but it could create a paradigm shift in health monitoring moving from reactive technologies to preventive. If they are made smart enough to learn the daily activity pattern of an end user, and identify deviations/anomalies linked to declining health, they could foresee the occurrence of possible critical events (e.g. falls, strokes).

Radar will enable prompt emergency responses following critical events (reactive), continuous in-home health monitoring for medical professionals to improve diagnostics and develop precision medicine for individuals (predictive). It would also enable persuasive feedback to individuals to advise/influence behaviours for safer and better practice, when variations in their routine are identified (prevention & assistance).

III. MACHINE LEARNING PERSPECTIVE

Machine learning is becoming an integral part of technology development given the advantages it provides, radar system applications are also leveraging machine learning for enhanced performances and accuracy in activity classification.

Generally, to classify activities, radar micro-Doppler (mD) signatures are used as a base. The relative motion of limbs and head with respect to the torso generates unique signatures in the time-frequency domain of the radar returns. Different activities create uniquely identifiable features in mD signatures used for classification. A comprehensive coverage of the subject can be found in [21-22]. Spectrograms are then processed to extract features [23] followed by different classifiers [24-25].

Here is a non-exhaustive list of machine learning techniques for classification: Fisher Discriminant Analysis [26-27], K-nearest neighbors [28-29], Naïve Bayes [30], Ensembles (e.g. Bagging [31]) and Support Vector Machine (SVM) [32].

A review classifiers for activity classification [16] advise to use multiple sensors to enhance classification accuracy by covering multiple aspect angles and combat occlusions. Another way to improve accuracy is to fuse data and select the most salient features [12,18,33-34]. Many classifiers are used in activity classification of which SVM is the most common [35]. The choice of classifier is important, but choosing the most salient features has a greater impact on accuracy than the classifier [36]. There is a wealth of contributions trying to extract features and classify activities from mD signatures [16].

Beyond machine learning lies deep learning thanks to advances in computational power (GPUs). Feature extraction is an expert-knowledge based task. Deep learning techniques however can figure out relevant features for classification, sparse representations and time-dependencies through several layers of neurons with activation functions e.g. recognize faces with convolutional neural networks [37]. Another class of deep learning algorithms used for speech recognition are Recurrent Neural Networks (RNN) with Gated Recurrent Units [38], [39] and Long Short-Term Memory (LSTM) [40].

A general belief is that deep learning requires "Big Data" to be effective; but small datasets also produce good results [41-42] via data augmentation and transfer learning.

Figure 1 summarizes the research on activity classification using deep learning for enhanced accuracy [43–57] yielding precisions from 80 to almost 100%.

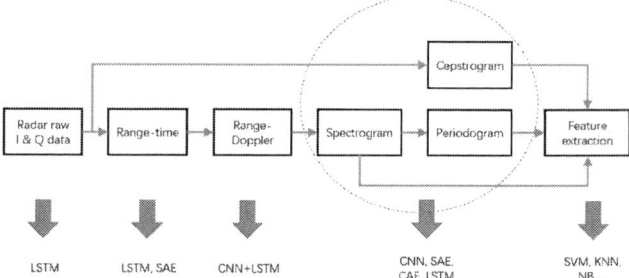

Figure 1: typical radar signal processing chain and associated machine/deep learning method from the state of the art (SAE: stacked Autoencoders, CAE: convolutional autoencoders, LSTM: Long Short-Term Memory, CNN: Convolutional Neural Network)

It is hard to assess the different performances since all the deep learning algorithms are ad hoc and the size and the nature of datasets vary. Because the intra class variance for similar activities is smaller than for different ones, therefore advertised accuracy varies in meaning. Deep learning is already showing better performance than expertly pre-trained models [58].

The problematic of multiple people in the field of view [47, 54, 56, 59] is rarely studied as they mostly consider only one person. In [16, 19, 24, 25, 60–63], the multi-static radar approach is utilized for classification from spectrograms using feature extraction. The difficulty in research with multi-static radar is the synchronization requirement between radar units and they are not commercially available. Aspect angle dependence in classification is rarely discussed although it has a large effect on accuracy [36, 53, 60]; most studies adopt actions happening in the radial direction of the radar.

Generally, the activities are looked at different activity snapshots and not in a continuum like in [51] for wearables.

The lessons from the literature are that CNN can recognize elaborate features from signals/images for particular snapshots at a given time where RNN of which LSTM is the leading technique takes into consideration time dependencies between snapshots. [51] shows a combination of CNN and LSTM for wearables and [49] presents a multimodal CNN multi-stream in parallel with LSTM with fusion; showing new ways to think about classifying data as a continuum.

Great efforts should go on preparing datasets, neural network architectures, training/optimizing to avoid overfitting, bias and ensure the model generalizes the activities to recognize unseen data or people accurately even with small datasets.

IV. RECENT RESULTS FROM THE COMMUNICATIONS, SENSING AND IMAGING GROUP AT UNIVERSITY OF GLASGOW

Now we have gone around the state of the art and the context, it is time to present some results from our recent studies.

A. Multisensor approach for remote health monitoring of older people [12], [18]

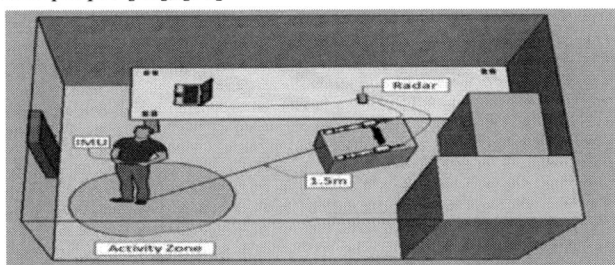

Figure 2: Experimental setup with radar and inertial motion unit (accelerometer, gyroscope, magnetometer, inertial) [12]

The experimental setup (Figure 2) shows the placement of the radar and the wearable sensor. The activities (walk, walk while carrying an object, sit down, stand up, pick up an object, crouch to tie shoe laces, drink, answer the phone, frontal fall, check under a piece of furniture) were measured with 9 volunteers giving 270 samples in total. 177 features were extracted using the inertial sensor in time and frequency domains and 28 features from spectrograms using radar.

The classifiers were quadratic kernel SVM and 10-NN trained using 10-folds randomly (9 for training and 1 for testing). The results in Table I are the result of the average of 10 folds. Notice, radar underperforms compared to wearables.

Feature selection (Fscore [64], ReliefF [65], SFS [64]) on single modality significantly increases classification accuracy.

Table I: Classification accuracy using a single sensor [12]

Classification Accuracy (%)	SVM	KNN
Accelerometer	85.2	79.6
Gyroscope	84.1	79.6
Magnetometer	80.4	69.6
Inertial	89.3	85.2
Radar	77.9	70.7

Table II: Improvements wit feature selection methods for IMU and radar in terms of accuracy and number of features

	IMU		Radar	
Method	Accuracy(%)	Features #	Accuracy(%)	Features #
Fscore SVM	90.7	73	78.8	17
Fscore KNN	88.2	76	74.1	17
ReliefF SVM	91.1	164	74	20
ReliefF KNN	89.3	58	67	18
SFS SVM	95.6	35	85.6	20
SFS KNN	88.25	69	79.8	19

Table II shows 5-9% improvement for both sensors. So "less is more": having more features does not improve accuracy but using/identifying the salient features does.

To increase accuracy, it is interesting to explore the benefit of using multimodal fusion at various levels (signal, feature and decision) [66–68] (Table III).

Table III: classification accuracy improvement with fusion

Fusion method	Accuracy (%)
Feature level	97.4
Decision level logP [69]	96.7
Decision level fuzzy logic [66]	94.8
Decision level voting [12]	97.8

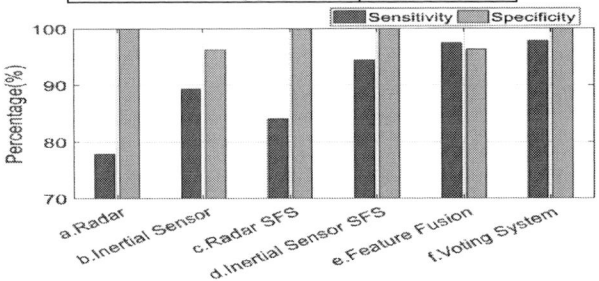

Figure 3: sensitivity and specificity for the fall action using various classifications methods [12]

Figure 3 shows an improvement overall without affecting fall specificity (reaching 100% with voting) by applying suitable feature selection and fusion.

B. Activity Classification Using Raw Range and I & Q Radar Data With Long Short Term Memory Layers [70]

LSTM are used to classify directly from raw data and range maps for binary classification every 0.5 s of action recorded. 5 subjects contributed, actions were recorded continuously for 60s giving 19 recordings (10 'walk' and 9 'sitting & standing').

For both, 2,280 samples were obtained by dividing the recording in 0.5s snapshots and the data is presented to the LSTM as described in Table IV and Figure 4.

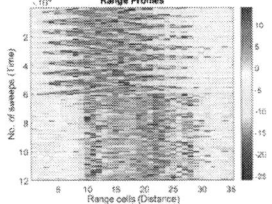 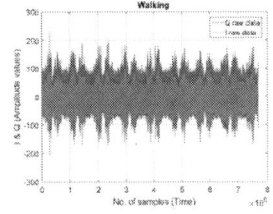

Figure IV: left) Range profiles for walking (60s) and sitting&standing movements (60s). right) illustrate I & Q data for walking movement. The time patterns that exist in the signals are exploited through the LSTM layers. [70]

Table IV: LSTM data parameters for classification

LSTM data	Number of time samples	Number of features
I&Q	64000	2
Range profiles	500	35

Table V: results of preliminary binary classification and metrics

Metrics	I&Q data	Range Profiles
LSTM units	4	35

978-1-5386-4757-8/18 $31.00 © 2018 IEEE

Mean test Accuracy	97.67%	94.16%
Standard deviation	1.14%	2.02%
optimizer	RMSprop	RMSprop
Learning rate	0.001	0.001
Batch size	1	1
epochs	10	50
Layers	2	2
Prediction time	2s	2ms

The samples were shuffled in a stratified manner (80% for training and 20% for testing) under the 5-fold scheme. Table V shows that I&Q data yields better accuracy than range profiles and LSTM are able to process backscattered data as time series. The drawback is the time for an estimation is 2s for raw data and 2ms for range profiles for 0.5s of a data continuum.

V. Conclusion

From the literature and our recent work, the radar community is very active in the development of robust classification algorithms for elderly care using a range of algorithms and modalities. The advent of deep learning will certainly help improve algorithms gradually. Still some very important open challenges remain in this area such as how much data is enough data? how to teach a network to learn fast? what about community data sharing regulations? how to get relevant data and moving from detection to prediction?. The linchpin challenge is the real-time implementation of those algorithms on hardware while maintaining the accuracy obtained with offline processing.

Furthermore, IoT can be extended to animal welfare applications where the dairy industry, farm animals (sheep, cattle, pigs) and horses (Thoroughbreds and leisure) can benefit for lameness assessment [33], [34] and connected farms with IoT will improve significantly productivity and animal monitoring for better yield for our growing needs.

References

[1] Mohd. Adnan Malik, "Internet of Things (IoT) Healthcare Market by Component (Implantable Sensor Devices, Wearable Sensor Devices, System and Software), Application (Patient Monitoring, Clinical Operation and Workflow Optimization, Clinical Imaging, Fitness and Wellness Measu," 2016.

[2] WHO, "Ageing," *WHO*, 2016. .

[3] P. D. United Nations, Department of Economic and Social Affairs, "World Population Prospects The 2017 Revision Key Findings and Advance Tables," *World Popul. Prospect. 2017*, pp. 1–46, 2017.

[4] M. G. Amin, Y. D. Zhang, F. Ahmad, and K. C. D. Ho, "Radar signal processing for elderly fall detection: The future for in-home monitoring," *IEEE Signal Process. Mag.*, vol. 33, no. 2, pp. 71–80, 2016.

[5] T. R. Bennett, J. Wu, N. Kehtarnavaz, and R. Jafari, "Inertial Measurement Unit-Based Wearable Computers for Assisted Living Applications: A signal processing perspective," *IEEE Signal Process. Mag.*, vol. 33, no. 2, pp. 28–35, 2016.

[6] F. Erden, S. Velipasalar, A. Z. Alkar, and A. E. Cetin, "Sensors in Assisted Living," *IEEE Signal Process. Mag.*, no. March, pp. 36–44, 2016.

[7] S. Savazzi, S. Sigg, and M. Nicoli, "Device-Free Radio Vision for Assisted Living," *IEEE Signal Process. Mag.*, vol. 33, no. 2, pp. 45–58, 2016.

[8] K. Witrisal et al., "High - Accuracy Localization for Assisted Living," *IEEE Signal Process. Mag.*, vol. 33, no. 2, pp. 59–70, 2016.

[9] C. Debes, A. Merentitis, S. Sukhanov, M. Niessen, N. Frangiadakis, and A. Bauer, "Monitoring Activities of Daily Living in Smart Homes: Understanding human behavior," *IEEE Signal Process. Mag.*, vol. 33, no. 2, pp. 81–94, 2016.

[10] R. Igual, C. Medrano, and I. Plaza, "Challenges, issues and trends in fall detection systems," *BioMedical Engineering Online*, vol. 12, no. 1. 2013.

[11] G. Chiarini, P. Ray, S. Akter, C. Masella, and A. Ganz, "MHealth technologies for chronic diseases and elders: A systematic review," *IEEE J. Sel. Areas Commun.*, vol. 31, no. 9, pp. 6–18, 2013.

[12] H. Li, A. Shrestha, H. Heidari, J. L. Kernec, and F. Fioranelli, "A Multisensory Approach for Remote Health Monitoring of Older People," *IEEE J. Electromagn. RF Microwaves Med. Biol.*, vol. 2, no. 2, pp. 102–108, 2018.

[13] H. Aykan, "Report to Congress: Aging Services Technology Study," Washington, 2012.

[14] P. Calyam, I. Jahnke, A. Mishra, R. B. Antequera, D. Chemodanov, and M. Skubic, "Toward an ElderCare Living Lab for Sensor-Based Health Assessment and Physical Therapy," *IEEE Cloud Comput.*, vol. 4, no. 3, pp. 30–39, 2017.

[15] S. Wang, M. Skubic, and Y. Zhu, "Activity density map visualization and dissimilarity comparison for eldercare monitoring," *IEEE Trans. Inf. Technol. Biomed.*, vol. 16, no. 4, pp. 607–614, 2012.

[16] E. Cippitelli, F. Fioranelli, E. Gambi, and S. Spinsante, "Radar and RGB-depth sensors for fall detection: A review," *IEEE Sensors Journal*, vol. 17, no. 12. pp. 3585–3604, 2017.

[17] A. Shrestha, J. Le Kernec, F. Fioranelli, E. Cippitellii, E. Gambi, and S. Spinsante, "Feature diversity for fall detection and human indoor activities classification using radar systems," in *International Conference on Radar Systems (Radar 2017)*, 2017, pp. 1–6.

[18] H. Li et al., "Multisensor data fusion for human activities classification and fall detection," in *2017 IEEE SENSORS*, 2017, pp. 1–3.

[19] F. Fioranelli, M. Ritchie, S. Z. Gürbüz, and H. Griffiths, "Feature Diversity for Optimized Human Micro-Doppler Classification Using Multistatic Radar," *IEEE Trans. Aerosp. Electron. Syst.*, vol. 53, no. 2, pp. 640–654, 2017.

[20] L. Yang, G. Li, M. Ritchie, F. Fioranelli, and H. Griffiths, "Gait classification based on micro-Doppler features," in *2016 CIE International Conference on Radar, RADAR 2016*, 2017.

[21] D. Tahmoush, "Review of micro-Doppler signatures," *IET Radar, Sonar Navig.*, vol. 9, no. 9, pp. 1140–1146, 2015.

[22] V. C. Chen, D. Tahmoush, and W. J. Miceli, "Radar Micro-Doppler signatures: processing and applications," in *Micro-Doppler signatures - Review, challenges, and Perspectives*, The Institute of Engineering and Technology, 2014.

[23] S. Björklund, H. Petersson, and G. Hendeby, "Features for micro-Doppler based activity classification," *IET Radar, Sonar Navig.*, vol. 9, no. 9, pp. 1181–1187, 2015.

[24] F. Fioranelli, M. Ritchie, and H. Griffiths, "Performance analysis of centroid and SVD features for personnel recognition using multistatic micro-Doppler," *IEEE Geosci. Remote Sens. Lett.*, vol. 13, no. 5, pp. 725–729, 2016.

[25] F. Fioranelli, M. Ritchie, and H. Griffiths, "Multistatic human micro-Doppler classification of armed/unarmed personnel," *IET Radar, Sonar Navig.*, vol. 9, no. 7, pp. 857–865, 2015.

[26] J. Ye, R. Janardan, and Q. Li, "Two-Dimensional Linear Discriminant Analysis," in *Advances in Neural Information Processing Systems*, 2005, pp. 1569–1576.

[27] S. Mika, G. Ratsch, J. Weston, B. Schölkopf, and K.-R. Muller, "Fisher discriminant analysis with kernels," *Ieee*, pp. 41–48, 1999.

[28] P. Cunningham and S. J. Delany, "K -Nearest Neighbour Classifiers," *Mult. Classif. Syst.*, pp. 1–17, 2007.

[29] J. M. Kraus, L. Lausser, and H. A. Kestler, "Exhaustive k-nearest-neighbour subspace clustering," *J. Stat. Comput. Simul.*, vol. 85, no. 1, pp. 30–46, 2015.

[30] S.-B. Kim, K.-S. Han, H.-C. Rim, and S. H. Myaeng, "Some Effective Techniques for Naive Bayes Text Classification," *IEEE Trans. Knowl. Data Eng.*, vol. 18, no. 11, pp. 1457–1466, 2006.

[31] T. K. Ho, "A data complexity analysis of comparative advantages of decision forest constructors," *Pattern Anal. Appl.*, vol. 5, no. 2, pp. 102–112, 2002.

[32] S. Tong and D. Koller, "Support vector machine active learning with applications to text classification," *J. Mach. Learn. Res.*, vol. 2, pp. 45–66, 2002.

[33] A. Shrestha, J. Le Kernec, F. Fioranelli, J. F. Marshall, and L. Voute, "Gait analysis of horses for lameness detection with radar sensors," in *International Conference on Radar Systems (Radar 2017)*, 2017, pp. 1–6.

[34] A. Shrestha *et al.*, "Animal Lameness Detection With Radar Sensing," *IEEE Geosci. Remote Sens. Lett.*, pp. 1–5, 2018.

[35] Y. Kim and H. Ling, "Human activity classification based on micro-doppler signatures using a support vector machine," *IEEE Trans. Geosci. Remote Sens.*, vol. 47, no. 5, pp. 1328–1337, 2009.

[36] B. Çağliyan and S. Z. Gürbüz, "Micro-Doppler-Based Human Activity Classification Using the Mote-Scale BumbleBee Radar," *IEEE Geosci. Remote Sens. Lett.*, vol. 12, no. 10, pp. 2135–2139, 2015.

[37] K. He, X. Zhang, S. Ren, and J. Sun, "Deep Residual Learning for Image Recognition," in *2016 IEEE Conference on Computer Vision and Pattern Recognition (CVPR)*, 2016, pp. 770–778.

[38] K. Cho, B. Van Merrienboer, D. Bahdanau, and Y. Bengio, "On the Properties of Neural Machine Translation : Encoder – Decoder Approaches," *Ssst-2014*, pp. 103–111, 2014.

[39] J. Chung, C. Gulcehre, K. Cho, and Y. Bengio, "Empirical evaluation of gated recurrent neural networks on sequence modeling," in *NIPS 2014 Workshop on Deep Learning*, 2014.

[40] S. Hochreiter and J. J. Schmidhuber, "Long short-term memory," *Neural Comput.*, vol. 9, no. 8, pp. 1–32, 1997.

[41] Q. Wu, Y. D. Zhang, W. Tao, and M. G. Amin, "Radar-based fall detection based on Doppler time–frequency signatures for assisted living," *IET Radar, Sonar Navig.*, vol. 9, no. 2, pp. 164–172, 2015.

[42] Y. Sawada and K. Kozuka, "Transfer learning method using multi-prediction deep Boltzmann machines for a small scale dataset," in *2015 14th IAPR International Conference on Machine Vision Applications (MVA)*, 2015, pp. 110–113.

[43] B. Jokanovic and M. Amin, "Fall Detection Using Deep Learning in Range-Doppler Radars," *IEEE Trans. Aerosp. Electron. Syst.*, vol. 54, no. 1, pp. 180–189, 2018.

[44] Y. Shao, S. Guo, L. Sun, and W. Chen, "Human Motion Classification Based on Range Information with Deep Convolutional Neural Network," in *2017 4th International Conference on Information Science and Control Engineering (ICISCE)*, 2017, pp. 1519–1523.

[45] B. Chikhaoui and F. Gouineau, "Towards Automatic Feature Extraction for Activity Recognition from Wearable Sensors: A Deep Learning Approach," in *2017 IEEE International Conference on Data Mining Workshops (ICDMW)*, 2017, pp. 693–702.

[46] M. Z. Uddin, W. Khaksar, and J. Torresen, "Human activity recognition using robust spatiotemporal features and convolutional neural network," in *2017 IEEE International Conference on Multisensor Fusion and Integration for Intelligent Systems (MFI)*, 2017, pp. 144–149.

[47] Y. Zhang *et al.*, "Poster Abstract: CAR - A Deep Learning Structure for Concurrent Activity Recognition," in *2017 16th ACM/IEEE International Conference on Information Processing in Sensor Networks (IPSN)*, 2017, pp. 299–300.

[48] T. Zebin, P. J. Scully, and K. B. Ozanyan, "Human activity recognition with inertial sensors using a deep learning approach," in *2016 IEEE SENSORS*, 2016, pp. 1–3.

[49] S. Song *et al.*, "Multimodal Multi-Stream Deep Learning for Egocentric Activity Recognition," in *2016 IEEE Conference on Computer Vision and Pattern Recognition Workshops (CVPRW)*, 2016, pp. 378–385.

[50] R. Saeedi, S. Norgaard, and A. H. Gebremedhin, "A closed-loop deep learning architecture for robust activity recognition using wearable sensors," in *2017 IEEE International Conference on Big Data (Big Data)*, 2017, pp. 473–479.

[51] J. F. Ordóñez and D. Roggen, "Deep Convolutional and LSTM Recurrent Neural Networks for Multimodal Wearable Activity Recognition," *Sensors*, vol. 16, no. 1. 2016.

[52] M. S. Seyfioglu, A. M. Ozbayoglu, and S. Z. Gurbuz, "Deep Convolutional Autoencoder for Radar-Based Classification of Similar Aided and Unaided Human Activities," *IEEE Transactions on Aerospace and Electronic Systems*, 2018.

[53] Y. Kim and B. Toomajian, "Hand Gesture Recognition Using Micro-Doppler Signatures With Convolutional Neural Network," *IEEE Access*, vol. 4, pp. 7125–7130, 2016.

[54] R. P. Trommel, R. I. A. Harmanny, L. Cifola, and J. N. Driessen, "Multi-target human gait classification using deep convolutional neural networks on micro-doppler spectrograms," in *2016 European Radar Conference (EuRAD)*, 2016, pp. 81–84.

[55] Y. Kim and T. Moon, "Human Detection and Activity Classification Based on Micro-Doppler Signatures Using Deep Convolutional Neural Networks," *IEEE Geosci. Remote Sens. Lett.*, vol. 13, no. 1, pp. 8–12, 2016.

[56] G. Klarenbeek, R. I. A. Harmanny, and L. Cifola, "Multi-target human gait classification using LSTM recurrent neural networks applied to micro-Doppler," in *2017 European Radar Conference (EURAD)*, 2017, pp. 167–170.

[57] M. Gochoo, T. H. Tan, S. C. Huang, S. H. Liu, and F. S. Alnajjar, "DCNN-based elderly activity recognition using binary sensors," in *2017 International Conference on Electrical and Computing Technologies and Applications (ICECTA)*, 2017, pp. 1–5.

[58] S. Cao and R. Nevatia, "Exploring deep learning based solutions in fine grained activity recognition in the wild," in *2016 23rd International Conference on Pattern Recognition (ICPR)*, 2016, pp. 384–389.

[59] F. Adib, Z. Kabelac, and D. Katabi, "Multi-Person Localization via {RF} Body Reflections," in *12th {USENIX} Symposium on Networked Systems Design and Implementation ({NSDI} 15)*, 2015, pp. 279–292.

[60] F. Fioranelli, M. Ritchie, and H. Griffiths, "Aspect angle dependence and multistatic data fusion for micro-Doppler classification of armed/unarmed personnel," *IET Radar, Sonar Navig.*, vol. 9, no. 9, pp. 1231–1239, 2015.

[61] F. Fioranelli, M. Ritchie, and H. Griffiths, "Classification of Unarmed/Armed Personnel Using the NetRAD Multistatic Radar for Micro-Doppler and Singular Value Decomposition Features," *IEEE Geosci. Remote Sens. Lett.*, vol. 12, no. 9, pp. 1933–1937, 2015.

[62] F. Fioranelli, M. Ritchie, and H. Griffiths, "Bistatic human micro-Doppler signatures for classification of indoor activities," in *2017 IEEE Radar Conference (RadarConf)*, 2017, pp. 610–615.

[63] J. S. Patel, F. Fioranelli, M. Ritchie, and H. Griffiths, "Multistatic radar classification of armed vs unarmed personnel using neural networks," *Evol. Syst.*, vol. 9, no. 2, pp. 135–144, 2018.

[64] S. Z. Gürbüz, B. Erol, B. Çağlıyan, and B. Tekeli, "Operational assessment and adaptive selection of micro-Doppler features," *IET Radar, Sonar Navig.*, vol. 9, no. 9, pp. 1196–1204, 2015.

[65] R. Durgabai and Y. Ravi Bhushan, "Feature selection using reliefF algorithm," *Int. J. Adv. Res. Comput. Commun. Eng.*, vol. 3, no. 10, pp. 8215–8218, 2014.

[66] R. C. King, E. Villeneuve, R. J. White, R. S. Sherratt, W. Holderbaum, and W. S. Harwin, "Application of data fusion techniques and technologies for wearable health monitoring," *Med. Eng. Phys.*, vol. 42, pp. 1–12, 2017.

[67] D. L. Hall and J. Llinas, "An introduction to multisensor data fusion," *Proc. IEEE*, vol. 85, no. 1, pp. 6–23, 1997.

[68] F. CasA Review of Data Fusion Techniquestanedo, "A Review of Data Fusion Techniques," *Sci. World J.*, vol. 2013, 2013.

[69] C. Chen, R. Jafari, and N. Kehtarnavaz, "A Real-Time Human Action Recognition System Using Depth and Inertial Sensor Fusion," *IEEE Sens. J.*, vol. 16, no. 3, pp. 773–781, 2016.

[70] C. Loukas, F. Fioranelli, J. Le Kernec, and S. Yang, "Activity Classification Using Raw Range and I & Q Radar Data With Long Short Term Memory Layers," in *Cyber Science and technology congress*, 2018, pp. 1–5.

A Hybrid Bioimpedance Spectroscopy Architecture for a Wide Frequency Exploration of Tissue Electrical Properties

Achraf Lamlih[1][2], Philippe Freitas[1], Mohamed-Moez Belhaj[1], Jérémie Salles[1],
Vincent Kerzérho[1], Fabien Soulier[1], Serge Bernard[1], Tristan Rouyer[2] and Sylvain Bonhommeau[3]

[1]University of Montpellier, LIRMM, CNRS, 161 rue Ada 34095 Montpellier Cedex 5, FRANCE
[2]Ifremer/MARBEC, Av. J. Monnet, 34203 Sète, France
[3]Ifremer/DOI, rue J. Bertho, 97822 Le Port, France
contact: firstname.lastname@lirmm.fr / firstname.lastname@ifremer.fr

Abstract—**Bioimpedance spectroscopy (BIS) is a technique increasingly used for measuring the electrical properties of biological tissues. Choosing an integrated system architecture for bioimpedance spectroscopy is very dependent on the application and ruled by several constraints such as precision, bandwidth and measurement time. This paper presents a hybrid architecture providing fast measurement time while maximizing precision. This new architecture has been defined for a wide exploration of electrical properties of biological tissues. It combines the frequency sweep and multitone measurement techniques. Using the multitone measurement over the α dispersion and a frequency sweep over the β dispersion, enable the system architect to overcome the design challenges faced when using each technique separately. Its critical blocks are optimized for a bandwidth up to 10 MHz, thus covering the α and β frequency ranges, an example of the design optimization is detailed for the current driver.**

I. INTRODUCTION

Bioimpedance spectroscopy (BIS) is a technique increasingly used for measuring the electrical properties of biological tissues [1]. Since bioimpedance is a function of the physiological processes, variations of the composition of biological tissues can be seen as variations of the electrical impedance of the tissue.

In order to measure bioimpedance, current stimulation is generally used. Thus, a current containing a single or multiple frequencies is injected in the tissue under test via electrodes, the bio-modulated response voltage is then measured to determine the impedance transfer function.

Due to its non-invasive nature, its low cost and its ability to achieve low-power consumption, BIS is a suitable technique for long periods of monitoring. In fact, its use can improve the quality of healthcare delivered through earlier diagnosis and less invasive procedures.

Choosing an integrated system architecture for BIS is a difficult task as it is ruled by several constraints such as precision, bandwidth and measurement time. In the literature, two types of architectures could be found, frequency sweep based architectures and multitone-based architectures. These architectures provide different trade-offs between measurement time and precision. This paper presents a hybrid architecture providing fast measurements while maximizing precision. It has been defined for biological tissue electrical properties exploration over a wide frequency range. Its critical blocks are optimized for a bandwidth up to 10 MHz, an example of the design optimization is detailed for the current driver.

II. BIS ARCHITECTURES

A. Constraints

Building an integrated system for BIS is very dependent on the application and ruled by several constraints. The main constraints facing a BIS integrated system designer are bandwidth, precision and measurement time.

In the literature, many applications using single frequency bioimpedance measurements can be found. However, for a rigorous characterization of the biological tissue composition, the complex impedance needs to be analyzed over a large range of frequencies. In fact, each tissue molecule has a different response to a given frequency. Indeed the electrical permittivity of biological tissues decreases in three main steps corresponding to three dispersions [2] : the α dispersion (10 Hz to 10 KHz), the β dispersion (10 KHz to 10 MHz) and the γ dispersion ($\geq$ 10 MHz). Since α and β dispersions are associated with the most relevant biological aspects [3], BIS integrated systems should be able to cover at least over these frequency ranges (10 Hz to 10 MHz). The Signal to Noise Ratio (SNR) is used as a precision metric. SNR cannot be improved by increasing the amplitude of the excitation signal, since it is limited by the safety requirements. Thus, care must be taken in order to maximize SNR over the frequency range of interest. Finally, in order to respect the system invariance criterion, measurement time should be as short as possible.

In general, two types of architectures could be found in the literature. Frequency sweep based architectures provide maximum precision while having long measurement times. On the other hand, multitone-based architectures provide short measurement times but their precision decreases with the number of frequency components in the excitation signal.

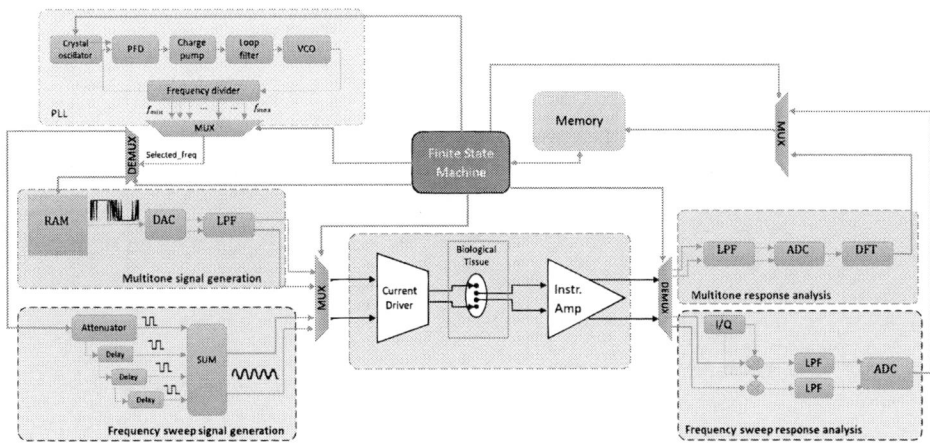

Fig. 1. Hybrid bioimpedance spectroscopy architecture

B. Multitone architectures

In BIS applications where the properties of the tissue under test are rapidly changing over time, fast BIS should be performed in order to respect the system invariance criterion. A solution is to use multitone signals in order to cover the chosen frequency range in a short time frame. However the energy of the signal is spread among its frequency points. Thus, multitone-based architectures provide short measurement times but their precision decreases with the number of frequency components contained in the excitation signal.

In the signal generation side, memory registers containing the binary multitone signal are sampled by a clock. The result is fed to a Digital-to-Analog Converter (DAC) that transforms the binary signal to pre-defined voltage levels. A Low Pass Filter (LPF) is used in order to get rid of the high frequency quantization noise. A current driver is then used to convert the multitone voltage to an excitation current.

In the response analysis side, an Instrumentation Amplifier (IA) is used to amplify the bio-modulated signal, an Analog-to-Digital Converter (ADC) is used to convert the amplified signal to the digital domain, then a Discrete Fourier Transform is computed in order to determine the phase and amplitude of the frequency components of interest.

C. Frequency sweep architectures

The signal to noise ratio (SNR) is a critical criterion in order to assess the quality of the excitation signal, and thus the measurement precision. The SNR cannot be improved by increasing the amplitude of the excitation signal, since it is limited by the safety requirements. In BIS applications where high precision is needed, frequency sweep architectures are used in order to have a maximal SNR, since the energy of the signal is concentrated over a single frequency point each time.

In the signal generation side, a Phase locked Loop (PLL) is generally used as a frequency multiplier to get the higher frequency. On the negative feedback path, successive divisions are then performed to get the frequency components of interest. The digital signal is then scaled and delayed multiple

times and the sum of the results is used to construct a sine wave of the targeted frequency. The scale coefficients and the phase shifts are computed so that the constructed signal has the lowest total harmonic distorsion possible [4]. A current driver is then used to convert the constructed sine wave voltage to an excitation current.

In the response analysis side, an IA is used to amplify the bio-modulated signal, then a lock-in amplifier followed by a low rate ADC feed the memory with bit words representing the real and the imaginary parts of the impedance.

D. Proposed hybrid architecture

When analyzing the two architectures previously detailed, an integrated circuit designer could notice that the constraints vary all over the α and β frequency ranges. In fact, at the α dispersion, the frequency sweep architecture has longer measurement times since we process the measurement chain entirely for each frequency. Moreover, for response analysis, the LPF establishment time is significant since it operates at low frequencies. At the β dispersion, the multitone architecture needs high sampling frequencies for the generation side in order to push further quantization noise. On the other hand, for the response analysis side, a high frequency ADC is needed which brings design complexity. Moreover, at the β frequency range, the bioimpedance magnitudes are low (tens of Ohms to several Ohms). Thus, the energy of each of the frequency components contained in the multitone excitation signal is low, making the specifications of the IA difficult to reach.

Fig. 1 shows the proposed architecture, it is a combination of the frequency sweep and the multitone architectures described in subsections II-B and II-C. It uses a finite state machine in order to choose the multitone generation and analysis for the frequencies of the α frequency region (10 Hz to 10 KHz), then it switches to the frequency sweep architecture for the β frequency region (10 kHz to 10 MHz) using the different multiplexers and demultiplexers. The current driver and the instrumentation amplifier are common blocks to the two architectures. Thus, their design process is critical since they

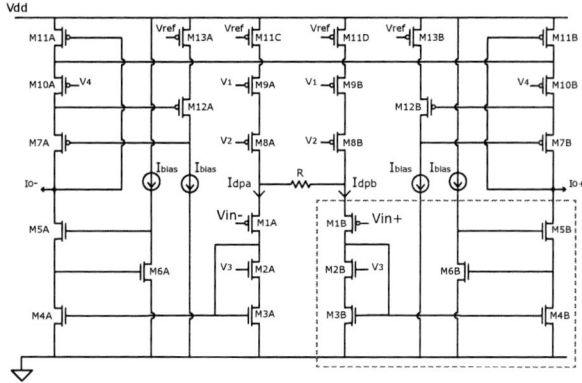

Fig. 2. Current driver architecture

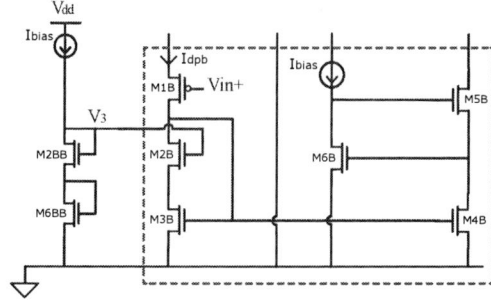

Fig. 3. Improved regulated cascode architecture

should be able to cover both the α and β frequency ranges. As an example, the current driver design and optimization is detailed in the next section.

III. CURRENT DRIVER DESIGN AND OPTIMIZATION

The current driver is a critical block of the presented architecture. Since α and β dispersions are associated with the most relevant biological aspects [3], the current should have a constant amplitude over these frequency ranges. This is achieved by having a large output impedance compared to the load. In fact, at lower frequencies, the sum of the electrodes-interface and tissue impedances have the highest magnitudes. Whereas at higher frequencies, stray capacitance represents the biggest challenge since it shunts the output resistance of the current driver which reduces its value.

A. Topology

Fig.2 shows the architecture of the current driver [5]. It is a symmetrical fully-differential Operational Transconductance Amplifier (OTA). The first stage of the OTA consists of a degenerated differential pair. A degeneration resistance was added to the M1 transistors in order to enhance the linearity of the transfer function and to maximize the range of input voltage. The current across the differential pair branches has been set in order to have a bandwidth of operation covering the α and β frequency dispersions.

The output stage is a critical part of the current driver structure. It needs to provide a high output impedance while enabling a maximum output swing. Using regulated cascode current mirrors as the output stage architecture provides the best compromise between a high output resistance and the lowest output voltage. Since the OTA is fully-differential, a common-mode feedback (CMFB) loop is implemented to stabilize the common-mode output voltage at half supply [6]. Due to low supply voltage specifications and in order to compensate for process variations, the output DC level is tunable using an independent reference voltage (Vref).

B. Output impedance increasing

The output stage of the current driver is critical. It has to provide an output current proportional to the input current at the high impedance output node. The output current amplitude needs to be constant regardless of loading. The important factors defining the performance of the output stage are thus output impedance, minimum output voltage and accuracy.

Bandwidth specifications imposed by the application require the use of short channel MOSFETs. A size reduction of the transistors leads to a decrease in the output resistance. Simple current mirror blocks or basic cascode structures are unable to provide an acceptable output resistance, especially at low frequencies where the load (electrode-tissue interface impedance) displays the highest impedance magnitudes, generating measurements errors in that frequency range. Therefore, improved output stage architectures for higher performances should be used.

Regulated cascode current mirror have the highest output impedance compared to simple or Wilson cascodes. It uses negative feedback in order to enhance the output impedance.

The improved regulated cascode shown in Fig.3 is the basic block of the output stage. In order to minimize loading effects, the transistor M_{6B} is used to enhance the output impedance by a $gm * ro$ factor compared to basic cascode structures. The output impedance of the improved regulated cascode can be written as :

$$Rout = ro_{M_4} * ro_{M_6} * gm_{M_6} * ro_{M_5} * gm_{M_5} \qquad (1)$$

The main drawback of the regulated cascode is the output minimum voltage, which is equal to $V_{th} + 2V_{eff}$. This could be improved to around $2V_{eff}$ by using the M_6 transistors in weak inversion. Although the matching of the M_6 transistors will be affected when operating in weak inversion, the loop gain of the structure is so important that the matching errors are negligible.

In order to enhance the accuracy of the current mirror, the drain to source voltages of M_{3B} and M_{4B} should be equalized. Since the V_{ds} of M_{4B} is imposed by the gate voltage of M_{6B}, the proposed solution consists in using in a mesh an M_{6BB} transistor of similar size as M_{6B} biased with the same current, together with an M_{2BB} transistor of similar V_{gs} as M_{2B} :

$$V_{ds_{M_{4B}}} = V_{gs_{M_{6B}}} \tag{2}$$

$$V_{gs_{M_{6B}}} = V_{gs_{M_{6BB}}} \tag{3}$$

$$V_{ds_{M_{3B}}} = V_{gs_{M_{2BB}}} - V_{gs_{M_{2B}}} + V_{gs_{M_{6BB}}} \tag{4}$$

$$V_{ds_{M_{3B}}} = V_{gs_{M_{6BB}}} = V_{ds_{M_{4B}}} \tag{5}$$

C. Results

The current driver was designed in a $0.18\,\mu m$ AMS CMOS process operating at 1.8 V power supply. The current driver is capable of providing currents up to $600\,\mu A$ peak to peak. The circuit design, simulations and layout were developed with Cadence suite using the toolkit provided by the foundry.

The Frequency response of the current driver transconductance is presented in Fig.4. It has a transconductance of $860\,\mu S$ up to 4 MHz decreasing to $854\,\mu S$ at 8 MHz. The cutting frequency is located at 67 MHz. The current driver was loaded with a $1\,k\Omega$ resistor with only its intrinsic stray capacitance.

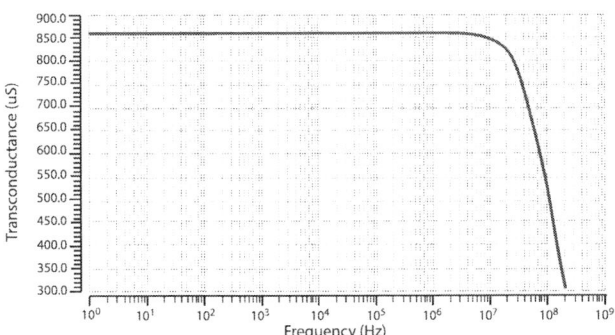

Fig. 4. Current driver bandwidth

Total Harmonic Distortion (THD) was simulated for a $400\,\mu A$ peak to peak output current. The simulation results have showed a THD below 0.3% at low frequencies increasing to 0.6% at 8 MHz.

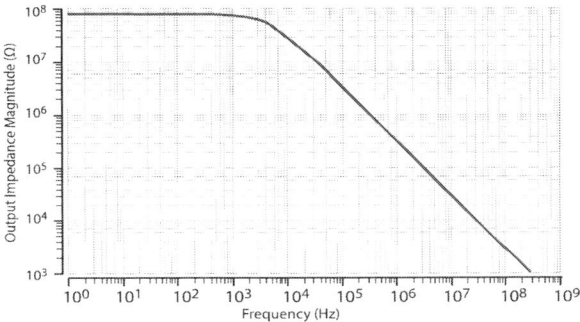

Fig. 5. Output impedance of the current driver

The frequency response of the output impedance of the current driver is presented in Fig.5. The output impedance is equal to $79\,M\Omega$ at DC up to 1 KHz. It decreases to $3.6\,M\Omega$ at 100 KHz, and to $324\,k\Omega$ at 1 MHz. The low frequency output impedance magnitudes are suitable for exploring the physiological processes occurring in the α region. At higher frequencies, the output impedance values are several orders of magnitude higher than the electrode-tissue impedance, since it is expected to decrease in the β region. A research presented in [7] shows a comparison of the impedance magnitudes of several Ag/AgCl commercial electrodes in the frequency range (10 Hz-1 MHz), The average electrode-tissue impedance magnitudes were higher than $10\,K\Omega$ for frequencies below 10 KHz, equal to $5\,K\Omega$ at 10 KHz and $337\,\Omega$ at 1 MHz. The current driver output impedance results compared to the electrode-tissue magnitudes presented in [7] show that it is capable of providing measurements with an error of 0.02% at 10 KHz and 0.1% at 1 MHz.

IV. CONCLUSION

A hybrid architecture for bioimpedance spectroscopy has been presented. This new architecture has been defined for a wide exploration of the electrical properties of the biological tissue. It combines the frequency sweep and multitone measurement techniques, thus providing fast measurement time and maximal precision over the α and β dispersions. Moreover, using the multitone measurement over the α dispersion and a frequency sweep over the β dispersion, enable the system architect to overcome the design challenges faced when using each technique separately. The current driver and the instrumentation amplifier are common blocks to the two measurement techniques, they have been designed in order to cover a frequency range of 10 Hz to 10 MHz. As an example, the current driver design and optimization using $0.18\,\mu m$ CMOS AMS process has been presented.

ACKNOWLEDGMENT

This research was part of the MERLIN-POPSTAR project funded by Ifremer and a Phd funded by Ifremer and labex Numev. We thank our colleagues from Ifremer/REM/RDT and RBE, CNRS, UM and IRD who provided insight and expertise that greatly assisted the research.

REFERENCES

[1] S. Grimnes and O. G. Martinsen, *Bioimpedance and Bioelectricity Basics*. Academic Press, 2014.

[2] H. P. Schwan, "Electrical properties of tissue and cell suspensions," *Advances in Biological and Medical Physics*, vol. 5, pp. 147–209, 1957.

[3] H. Dastjerdi, R. Soltanzadeh, and H. Rabbani, "Designing and implementing bioimpedance spectroscopy device by measuring impedance in a mouse tissue." *Journal of Medical Signals and Sensors*, pp. 187–194, 2013.

[4] S. David-Grignot, A. Lamlih, M. M. Belhaj, V. Kerzérho, F. Azaïs, F. Soulier, P. Freitas, T. Rouyer, S. Bonhommeau, and S. Bernard, "On-chip generation of sine-wave summing digital signals: an analytic study considering implementation constraints," *Journal of Electronic Testing*, vol. 34, no. 3, pp. 281–290, Jun 2018. [Online]. Available: https://doi.org/10.1007/s10836-018-5710-4

[5] A. Lamlih, P. Freitas, S. David-Grignot, J. Salles, V. Kerzérho, F. Soulier, S. Bernard, T. Rouyer, and S. Bonhommeau, "Wideband fully differential current driver with optimized output impedance for bioimpedance measurements," in *2018 IEEE International Symposium on Circuits and Systems (ISCAS)*, May 2018, pp. 1–5.

[6] W. M. C. Sansen, *Analog Design Essentials*. Springer, 2006.

[7] M. Rahal, J. M. Khor, A. Demosthenous, A. Tizzard, and R. Bayford, "A comparison study of electrodes for neonate electrical impedance tomography," *Physiological Measurement*, vol. 30, no. 6, p. S73, 2009.

Design Understanding: From Logic to Specification*

Goerschwin Fey[†]   Tara Ghasempouri[‡]   Swen Jacobs[§]   Gianluca Martino[†]   Jaan Raik[‡]   Heinz Riener[¶]

[†]*Hamburg University of Technology*   [‡]*Tallinn University of Technology*
Hamburg, Germany   Tallinn, Estonia
[§]*CISPA and Saarland University*   [¶]*École Polytechnique Fédérale de Lausanne*
Saarbrücken, Germany   Lausanne, Switzerland

Abstract—**We present an outline of the field of Design Understanding and summarize state-of-the-art research in deriving human-understandable knowledge in form of logic properties from an unknown design.**

Index Terms—**Design understanding, temporal logics, specification, verification, synthesis, assertions, properties**

I. INTRODUCTION

Design understanding is the process of reconstructing human-understandable knowledge from an unknown design. This problem arises frequently in practice, e.g., when a part of a design fails or behaves unexpectedly and needs to be debugged, but the initial designer of this part left the design team. In such a situation, the documentation of the design, which is still under development, is often not available, still incomplete, or deviates from the actually implemented behavior. Automated design understanding tools can then be used to assist a human in gaining an understanding of the design and implementing the required changes. Overall, automated design understanding has a great potential to improve fault localization, reduce time-to-market, and improve product quality.

However, design understanding is also a fuzzy process and until today no commonly accepted formalism, notation, and methodology for extracting design knowledge from an unknown implementation exist. We argue that *properties* written in a formal language—the standard formalism in formal verification—are not only useful to describe the input-output semantics of a system, but also capable of describing the interior behavior of a design in a human-understandable manner. Particular temporal logics, that describe relationships between signals of a circuit implementation over time, are attractive. We envision that a design understanding tool derives knowledge in form of temporal-logic formulæ from an unknown design. Such a tool can be useful in various situations:

1) *Reverse engineering*: When the intent of a design is not known or its Register-Transfer Level (RTL) specification got lost over the years, making sense of a gate level description (particularly after heavy optimization) is often impossible for humans. Automated design understanding tools have the potential to isolate a few core properties that give an idea of the overall usage of the design.

2) *Debugging*: Finding and fixing bugs in a design is often a complicated and time-consuming task. Automatically generated properties can assist in spotting faults more quickly, e.g., when the reported properties slightly deviate from expectations.

3) *Verification*: For certain designs, equivalence checking requires considerable effort. In these cases, checking simple automatically generated properties may still be possible due to property-specific abstraction mechanisms. The property-checks can then be used as a lightweight approach to increase the confidence in the correctness of the design or to pinpoint functional differences.

This paper summarizes a special session covering state-of-the-art research in Design Understanding with a focus on deriving human-understandable, logic properties from an unknown implementation in a concise and straight-forward way.

II. SYNTAX-GUIDED PROPERTY ENUMERATION

Automated assertion generation approaches [3], [17] for RTL derive properties of a design from simulation data. A formal verification engine guarantees that the assertions hold on the design. The derived assertions consequently depend on the quality and quantity of the simulation data used. Moreover, the generated assertions often capture cycle-accurate signal manipulations implemented in the RTL code, which tend to be long and not easy to understand for humans.

We propose syntax-guided property enumeration [13], a technique that derives temporal-logic formulæ of bounded length directly from a design. The idea is inspired by the recent success of syntax-guided synthesis (SyGuS) [1]. The SyGuS problem is, given a specification Γ in form of a logic formula with an uninterpreted symbol F and a context-free grammar G, to find a syntactic expression $\varphi \in G$ such that $\Gamma[F/\varphi]$ holds for all possible assignments to the free variables in Γ when F is replaced by φ.

Syntax-guided property enumeration [13] reformulates the SyGuS problem in the context of temporal logics and model checking. For a given hardware design S, a list $\varphi_1, \varphi_2, \ldots, \varphi_n$ of temporal-logic formulæ is enumerated by unwinding a grammar G. Each formula is model-checked on the design. Satisfied formulæ are kept and reported to a user, failing formulæ are used to prune the search space. A termination criterion, e.g., in form of a time limit or an upper bound on the maximum length of a formula, is required to guarantee termination. The

---

*This research was supported by H2020-ERC-2014-ADG 669354 CyberCare and by the German Research Foundation (DFG) under the project ASDPS (JA 2357/2-1).

978-1-5386-4757-8/18 $31.00 © 2018 IEEE

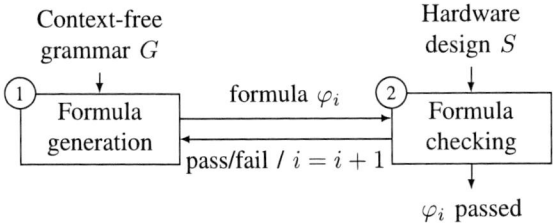

Fig. 1. High-level overview of syntax-guided property enumeration

$$
\begin{aligned}
S &::= \mathsf{AG}(F) \\
V &::= signal \\
F &::= \mathsf{false} \mid V \mid (\neg F) \mid (F \wedge F) \mid \\
    &\quad\ \ \mathsf{AX}(F) \mid \mathsf{AF}(F) \mid \mathsf{AG}(F) \mid (F)\,\mathsf{EU}(F)
\end{aligned}
$$

Fig. 2. CTL grammar for invariant generation with start symbol S

overall formula generation process is visualized in Fig. 1 and consists of two steps: 1) in the formula generation step, a new temporal-logic formula φ_i is generated from the grammar. 2) In the formula checking step, the formula φ_i is model-checked on S. The verdict of model checking is reported to the user and provided to the formula generation step to enable learning from previous results.

We have implemented a proof-of-concept of syntax-guided property enumeration in C++ using the enumeration library behemoth[1] and the model checker IImc [9]. Our implementation takes as input a gate level circuit design in the AIGER format, a grammar that describes a fragment of *Computation-Tree Logic* (CTL), a bound on the maximal number of logic operators in a formula, and a time limit in seconds. The implementation enumerates all formulæ by length and thus guarantees that shortest and best understandable properties are generated—assuming that shorter formulæ are easier understood by humans.

In an experiment, we have generated assertions (or invariants) using our implementation with the gammar in Fig. 2, a limitation to at most 4 logic operators per formula, and an overall time limit of 15 minutes for generating invariants. As signals, we consider all primary outputs and latch outputs of a design. For evaluation, we use six benchmarks provided with the model checker IImc. All experiments have been conducted on a Linux workstation with an Intel Core i7-7820HQ, which supports up to 8 parallel threads, and 16 GB RAM.

Fig. 3 shows the number of generated invariants and the number of invariants successfully verified by the model checker.

In a second experiment, we have demonstrated that syntax-guided property enumeration, due to the independence of the individual model checking problems, qualifies for multi-threaded implementation. As shown in Fig. 4, it is possible to exploit the parallelism capabilities of modern processors to speed-up invariant generation.

III. ASSERTION REWRITING TO IMPROVE READABILITY

Assertions describe the behavior of a system. They can be used to verify the consistency between design intent, the actual implementation, and the specification [6]. This work

[1]behemoth, https://github.com/hriener/behemoth

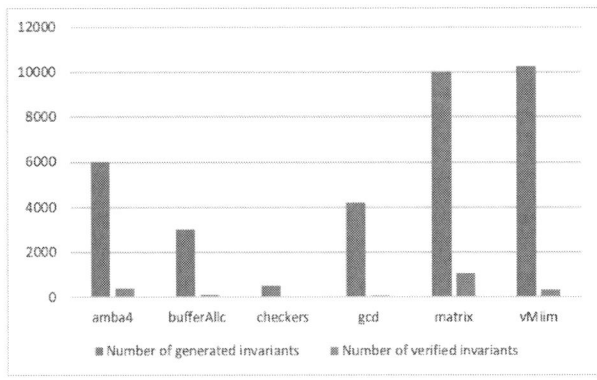

Fig. 3. Single-threaded invariant generation (time limit: 15 minutes)

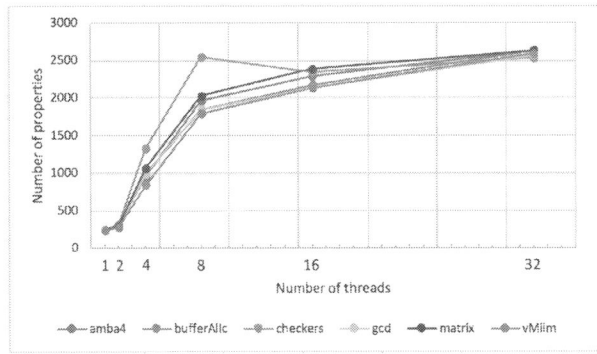

Fig. 4. Multi-threaded invariant generation (time limit: 90s)

proposes a methodology to estimate the quality of an assertion by analyzing its propositions and to rewrite the assertion to improve readability for humans. The quality estimation is based on a metric called Q evaluated with respect to a given test set of simulation traces. The metric Q is a linear combination of three metrics adapted from data mining: 1) The *support* refers to the number of occurrences of an assertion during simulation. 2) The *Correlation Coefficient* (CC) [16] refers to the dependency between antecedent and consequent of an assertion. 3) The *strength* [7] represents an assertion with a low occurrence but highly correlated to the other assertions which may cover corner cases.

The proposed methodology utilizes the three metrics to determine redundant and vague propositions and then generates a set of new assertions, which is more readable and shorter without loss of accuracy with respect to the original assertion when code coverage and fault coverage analyses are taken into account. Previous works have developed quality estimation techniques for assertions based on data-mining metrics; however, none of them have broken assertions down into their propositions to exercise them. For instance, in [10], an approach has been proposed that estimates the quality based only on the number of propositions. In [8], assertion quality is estimated based on their frequencies and correlation during simulation. The work does not consider assertions with a low number of frequencies, which may cover corner cases of a design.

Fig. 5 shows the flow of the proposed methodology. In Step 1, assertion A is broken down into its propositions. Consider the

978-1-5386-4757-8/18 $31.00 © 2018 IEEE

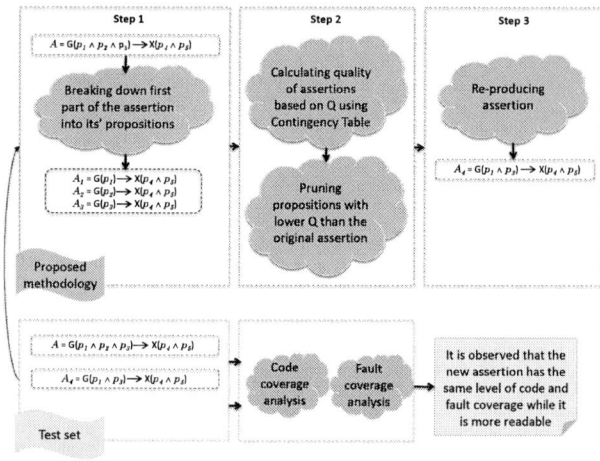

Fig. 5. Overview of the proposed methodology

assertion $A = \mathsf{G}(p_1 \wedge p_2 \wedge p_3) \to \mathsf{X}(p_4 \wedge p_5)$ in *Linear Temporal Logic* (LTL), which states it always happens that p_4 and p_5 are satisfied one simulation instant later than p_1, p_2 and p_3 become true. The propositions $p_1, p_2, \ldots, p_5$ are composed of invariants over the signals $v_1, v_2, \ldots$, such as $p_1 : (v_1 = \mathsf{true}) \wedge (v_2 > v_3)$, $p_2 : v_4 = \mathsf{false}$, $p_3 : (v_4 = \mathsf{false}) \vee (v_1 = \mathsf{true})$, etc. The assertion A is first broken down into three assertions A_1, A_2, and A_3 as shown in Fig. 5. In the next step, Step 2, the quality of assertion A and the three new assertions A_1, A_2, and A_3 are calculated with the help of the *Contingency Table* (CT). The CT represents the relation between antecedent φ and consequent ψ for each assertion A for form $A = (\varphi \to \psi)$ and counts the occurrences of φ and ψ.

Looking at (1), f_{11} represents the number of times where φ and ψ are true during simulation. f_{10} represents the number of times where φ is true but ψ is false. f_{01} is the dual of f_{10}, i.e., it is the number of times where φ is false and ψ is true during simulation. f_{00} is the number of times an assertion is not true through the simulation. f_{1X} is the sum of f_{11} and f_{10}, f_{0X} is the sum of f_{01} and f_{00} and so on. For more detail in calculating metric Q based on the CT please refer to [8] and [7]. The formula of Q is equal to:

$$Q(A) = \frac{f_{11}}{f_{XX}} + \frac{f_{11}f_{XX} - f_{1X}f_{X1}}{\sqrt{f_{1X}f_{0X}f_{X1}f_{X0}}} + \sqrt{\frac{f_{11}^2}{|f_{X1} - f_{X0}| \cdot |f_{1X} - f_{0X}|}} \quad (1)$$

At this step the quality of all the assertions are estimated. The assertions with the lower degree of quality than the original one are discarded. In Step 3 the antecedents of all the remained assertions are recomposed together to reproduce a more readable and shorter assertion with respect to the original one. We supported the effectiveness of the proposed methodology with code coverage and fault coverage analysis.

We broke 6 assertions from an LBDR design (one components of a router) down into 28 new assertions: 5 of these new assertions had a lower degree of Q with respect to the original assertion and were pruned. The final assertion set was reproduced and exercised with code and fault analysis. 331 faults were injected to the design. The original assertions and the final set detected the same number of faults i.e., 301, while 100% of the code were covered.

Experimental results show that the proposed methodology allows us to prune vague and useless propositions in assertions without loss of accuracy when fault and code coverage analyses are taken into account.

IV. UNDERSTANDING FORMAL SPECIFICATIONS

Using formal languages to specify properties that a system should or does satisfy has undeniable benefits: a fixed, formal semantics of the statements, as well as (the possibility of) formal proofs that the statement indeed holds. However, when formal languages are used to describe the *intended* behavior of a design, say in formal verification or synthesis, it is often difficult to write specifications that accurately reflect the intent of the designer. We argue that this is because of a gap between the formal semantics and the perceived meaning by a human designer, which must therefore also be taken into account in design understanding.

For example, consider the linear temporal logic specification

$$(\mathsf{G}\,\mathsf{F}\,r \wedge \mathsf{G}\,(r \to \mathsf{X}\,(\neg r\,\mathsf{W}\,g))) \to \mathsf{G}\,(r \to \mathsf{X}\,g)\,,$$

where r stands for a request that is issued by the environment, and g for a grant action triggered by the system under design. Intuitively, the formula states that the environment should never completely stop sending requests, and that after every request it waits until it was granted before sending a new request. Under this assumption, the system should grant every request one step after it appeared. Now, which systems do satisfy this specification? On the one hand, every system that grants a request one step after it appeared. But on the other hand, this specification is also satisfied by a system that *never grants a request*, because this forces the environment to stop sending requests, which means that the first part of the assumption will be violated.

While an expert in temporal logics may detect such pitfalls, one must take into account that this is a very simple example. Actual specifications of a system, in particular if they are automatically generated by a process for design understanding, may be much more complex.

In the following we consider some common pitfalls in writing specifications for verification and synthesis, try to predict what this might mean for design understanding, and offer a few directions that may be worth investigating in the future.

The example above highlights a problem that is very common in formal synthesis (and to some extent also verification): virtually no system runs on its own, but instead we have to consider its behavior in communication with other systems, or as one of many components of the same system. To obtain scalable formal methods for analysis and design, we need to be able to decompose systems into their components, which in turn requires that the specification of any component also contains the necessary *assumptions* on other components.

However, writing good formal assumptions is highly non-trivial. Two reasons make it particularly challenging:

1) By the standard interpretation of temporal logics like LTL, assumptions are interpreted in a *worst-case* manner. That is, we can only assume exactly what is specified, and in all other cases we have to expect utmost hostility from the other components. On the one hand, this is the only sound way to interpret assumptions. On the other hand, such an interpretation is overly pessimistic if we consider a system of components that are really intended to co-operate. Further, the worst-case interpretation requires specifications to be overly specific, which makes it harder for the other components of the system to satisfy them, and harder for a human to understand and write correctly.

2) Again by the standard interpretation of LTL, a specification of the form $\varphi \to \psi$ is satisfied if the assumption φ does not hold. As we have seen above, such a specification can of course be satisfied by fulfilling the guarantee, but in some cases the system can also force the environment to violate the assumptions in order to satisfy the specification.

Assume-guarantee reasoning [11], [14] and vacuity-detection [12] provide frameworks to handle these problems in verification. In synthesis, Bloem et al. [2] have considered these problems, among others, when handling assumptions in synthesis. When using automated methods for automatically *generating a specification* for a given system, we may well run into the dual problem: while the specification states exactly what the system does, the human designer may interpret it in a wrong way because of somewhat counter-intuitive corner-cases in the semantics of temporal logics.

For the problems above in particular, a possible solution would be to highlight such potential corner-cases, or otherwise increase the awareness of their existence. In general however, while we cannot solve the problem completely, we advocate to minimize it by guiding the automatic generation of formal properties towards statements that are less likely to be misinterpreted.

One obvious way to make formal statements understandable is to keep them short or simple, informally spoken. Formally, this may mean different things. Let us again look at the dual problem of formal synthesis. There, one tries to obtain systems that are human-understandable in a number of different ways:

- by starting from predefined sketches [15],
- by supplying a predefined grammar [1],
- by bounding their size [5], or
- by bounding more complex metrics of the implementation, such as the number of cycles in the state machine [4].

All of these give rise to similar approaches that could be taken when generating a specification from a given system. These approaches are:

- using predefined templates of properties, for each of which the intuitive meaning is clear or can be supplied to the designer,

- using a fixed grammar that is limited in a way as to make the resulting formulas clear and easy to understand,
- bounding the size of formulas with respect to standard metrics such as length, number of different subformulas, nesting depth of temporal operators, etc., or
- bounding the formulas under more complex metrics that correlate with human understanding of the property. This may include a small representation by an accepting automaton, the decomposition of the overall property into several (largely) independent and easy to understand properties, or, vice versa, the generalization of several related properties into one property that subsumes them.

The proposed property mining approaches described in the previous sections already show the benefits of some of these considerations.

V. Conclusion

In this paper, we have provided an overview of state-of-the-art research in the field of Design Understanding focusing on techniques to derive readable properties in temporal logic from an unknown hardware design.

References

[1] R. Alur, R. Bodík, E. Dallal, D. Fisman, P. Garg, G. Juniwal, H. Kress-Gazit, P. Madhusudan, M. M. K. Martin, M. Raghothaman, S. Saha, S. A. Seshia, R. Singh, A. Solar-Lezama, E. Torlak, and A. Udupa. Syntax-guided synthesis. In *Dependable Software Systems Engineering*, pages 1–25. IOS Press, 2015.

[2] R. Bloem, R. Ehlers, S. Jacobs, and R. Könighofer. How to handle assumptions in synthesis. In *SYNT*, volume 157 of *EPTCS*, pages 34–50, 2014.

[3] A. DeOrio, A. Bauserman, V. Bertacco, and B. Isaksen. Inferno: Streamlining verification with inferred semantics. *TCAD*, 28(5):728–741, 2009.

[4] B. Finkbeiner and F. Klein. Bounded cycle synthesis. In *CAV*, volume 9779 of *LNCS*, pages 118–135, 2016.

[5] B. Finkbeiner and S. Schewe. Bounded synthesis. *STTT*, 15(5-6):519–539, 2013.

[6] H. Foster, A. Krolnik, and D. Lacey. *Assertion-based design (2. ed.)*. Kluwer, 2004.

[7] T. Ghasempouri, S. Payandeh Azad, B. Niazmand, and J. Raik. An automatic approach to evaluate assertions' quality based on data-mining metrics. In *ITC-Asia*, 2018.

[8] T. Ghasempouri and G. Pravadelli. On the estimation of assertion interestingness. In *VLSI-SoC*, pages 325–330, 2015.

[9] Z. Hassan, A. R. Bradley, and F. Somenzi. Incremental, inductive CTL model checking. In *CAV*, volume 7358 of *LNCS*, pages 532–547, 2012.

[10] S. Hertz, D. Sheridan, and S. Vasudevan. Mining hardware assertions with guidance from static analysis. *TCAD*, 32(6):952–965, 2013.

[11] C. B. Jones. Software development based on formal methods. In *WSFA*, pages 153–172, 1986.

[12] O. Kupferman and M. Y. Vardi. Vacuity detection in temporal model checking. *STTT*, 4(2):224–233, 2003.

[13] G. Martino, H. Riener, and G. Fey. Coverage-guided CTL property enumeration for understanding models of reactive systems. In *IWLS*, 2018.

[14] A. Pnueli. In transition from global to modular temporal reasoning about programs. In Krzysztof R. Apt, editor, *Logics and Models of Concurrent Systems*, pages 123–144, Berlin, Heidelberg, 1985. Springer Berlin Heidelberg.

[15] A. Solar-Lezama. The sketching approach to program synthesis. In *APLAS*, volume 5904 of *LNCS*, pages 4–13, 2009.

[16] P.-N. Tan, V. Kumar, and J. Srivastava. Selecting the right interestingness measure for association patterns. In *KDD*, pages 32–41, 2002.

[17] S. Vasudevan, D. Sheridan, S. J. Patel, D. Tcheng, W. Tuohy, and D. R. Johnson. GoldMine: Automatic assertion generation using data mining and static analysis. In *DATE*, pages 626–629, 2010.

978-1-5386-4757-8/18 $31.00 © 2018 IEEE

Neuromorphic Computing - From Robust Hardware Architectures to Testing Strategies

Lorena Anghel[1], Denys Ly[3], Giorgio Di Natale[1], Benoit Miramond[2], Elena Ioana Vatajelu[1], Elisa Vianello[3]

[1] Univ. Grenoble Alpes, CNRS, Grenoble INP*, TIMA, 38000 Grenoble, France
[2] Univ. Côte d'Azur, LEAT, CNRS UMR 7248, Campus SophiaTech, 06903 Sophia Antipolis, France
[3] Univ. Grenoble Alpes, CEA, LETI 38000 Grenoble, France

Abstract— **This paper provides an overview of the challenges faced by hardware implemented Spiking Neural Networks, from device to circuit design, reliability and test. We present a comprehensive description of the state-of-the-art neuromorphic architectures inspired by brain computation, with special emphasis on Spiking Neural Networks (SNNs), together with emerging technologies that have enabled such systems, namely Phase Change and Metal Oxide Resistive Memories. Finally, we discuss the main challenges faced by hardware implementations of SNNs, their reliability and post-fabrication test issues.**

I. INTRODUCTION

Hardware implementation of neural networks is a hot research topic and is now considered as strategic for several large hardware-oriented companies such as Nvidia, IBM, Intel, as well as software-oriented companies such as Amazon, Facebook, Microsoft. The recent interest around deep neural networks for pattern recognition and classification has put a new spotlight on neuromorphic computing that brings brain modelling closer to data analysis. Top projects in neuromorphic engineering have led to powerful brain-inspired chips able to simulate numerous spiking neurons working in non-von Neumann computer architectures. These technologies need to fit embedded systems or Internet-of-Things (IoT) requirements therefore, their energy consumption is critical and needs to be minimized. Heterogeneous integration between CMOS and emergent technologies is seen as an opportunity to accomplish such goal. Indeed, emerging technologies have the potential of providing many benefits, such as energy efficiency, high integration density, CMOS-compatibility, reconfigurability, non-volatility, and open the path towards novel computational structures and approaches, for the traditional Von-Neumann architectures and beyond. Among the emerging technologies, memory technologies such as Resistive Memories (ReRAMs), Phase-Change Memories (PCMs), or spintronic based memories (STT-MRAMs) are triggering intense interdisciplinary activity, having driven the research community towards revisiting the existing computing and storage paradigms, providing hardware solutions for neuromorphic computing. Considering the large number of neurons and synapses required to perform efficient learning and classification, design teams face several obstacles: efficient storage of the synaptic weights, access to parameters in real time, reliable and testable design of hybrid, analog-digital- Non-Volatile heterogeneous architectures. Therefore, it is important to jointly consider device physics, circuit robust design, testability, and architecture constraints altogether.

The purpose of this paper is to provide a comprehensive overview of the bio-inspired hardware-implemented neuromorphic architectures, with special emphasis on their design, reliability and test. The paper is organized as follows.

The second section presents a comprehensive description of state-of-the-art neuromorphic architectures inspired by brain computation. The third section describes the latest achievements in Phase Change Memory and Metal Oxide Resistive Memory emerging technologies used as artificial synapses for Spiking Neural Networks (SNNs). The fourth section summarizes the main constraints of hardware implemented SNNs, possible faults and reliability issues, together with main challenges faced by post-fabrication test.

II. NEUROMORPHIC ARCHITECTURES: A BRIDGE BETWEEN MACHINE LEARNING AND BRAIN-INSPIRED COMPUTING

Artificial neural networks, inspired by the tremendous capabilities of the biological brain, have been constantly revisited, regularly opening new fields of application (computer vision, machine learning, robotics, artificial intelligence, etc.). Several electronic substrates are currently studied that offer interesting characteristics to achieve, on the one hand, the promises of energy efficiency of the biological model and, on the other hand, the technological maturity and the programming capacities expected by the applications.

The first challenge for neuromorphic hardware design is related to their power efficiency during the inference phase of the neural network in the application domains related to embedded systems. Learning is then supposed to be done offline and is not considered in the problem definition. In this context, many hardware accelerators have been proposed recently (such as Many-core, FPGA and SIMD) and they are now fully integrated in Machine Learning (ML) frameworks and available as commercial solutions. However, hardware Spiking Neural Networks (SNN) have shown to bring an improvement in efficiency compared to classical activity-based Artificial Neural Networks (ANN) while ensuring comparable classification performances [1]. SNN-based hardware remains an intense subject of interdisciplinary research since it explores the behavior and the great abilities of the brain. Although some promising solutions already exist [2], they only transpose the classical ML approaches into another coding scheme. The paradigm switch from ANN to SNN would really be reached when the properties of bio-inspired neurons will be revealed.

Indeed, intrinsic properties of bio-inspired neural networks exhibit unsupervised and distributed learning compared to ML approaches. But, taking advantage of these properties requires a better understanding of the global behavior of the brain through interdisciplinary research from neurosciences to electronics. This paradigm shift begins with the questioning of the back-propagation algorithm which shows every day its limits in terms of energy consumption and scalability. This centralized method performing the computation of

---

*Institute of Engineering Univ. Grenoble Alpes

classification error seems unlikely to be implemented by the brain, which clearly shows the interest and benefit of local, distributed and unsupervised computing.

The related bio-inspired learning rule is now known as STDP (Spike Based Dependent Plasticity) and is applied on each synapse independently of the global state of the network. In return, the synapse must be doted of computation capabilities. One of the most promising solutions is to use resistive materials, as the ones described in section III.

In addition, the change of learning rule also raises the question of the feedforward topology itself as we find it today in current deep networks. Actually, STDP is a kind of competitive learning process, where each neuron learns and represents a prototypical input. The competition between neurons is ensured by lateral inhibition, once again more biologically probable. It is implemented with a WTA (Winner-Takes-All) all-to-all connectivity in order to ensure the convergence of the learning and the good representation of the input data [3]. Such a topology clearly changes the hardware implementation but also brings redundancy in the representation of classes of data which in return is also more reliable face to hardware faults or variability observed in the most recent design process (discussed in section IV).

Finally, such unsupervised learning changes the current way of thinking about the use of learning methods by addressing the issue of on-line learning, or even long-term learning, naturally present in biological systems. It paves the way to new application domains where the system can adapt continuously to its environment.

Addressing jointly all the specificities of bio-inspired learning could provide an answer to the problem of excessive consumption of current off-line learning strategies. The main challenge of neuromorphic hardware is related to the control of the learning process itself. It remains still difficult to adapt it to different applications, furthermore when considering unsupervised learning. This also raises many exciting questions and challenges in the field of microelectronic design such as: joint design with heterogeneous technologies, analog/digital interfaces, programmability of neuromorphic circuits and materials, dynamic management of their consumption, interoperability with digital applications, and finally the verification and reliability of these systems.

III. PHASE CHANGE MEMORY AND METAL OXIDE RESISTIVE MEMORY AS ARTIFICIAL SYNAPSES IN SPIKING NEURAL NETWORKS

Spike based computational mechanism and architectural co-localization of processing and memory are two important features to design and fabricate hardware SNNs. This requires the integration of hardware neuron circuits must with specialized circuits modelling synapses. Synapses require to exhibit plasticity, that is modulation in their efficacy, and to support online learning algorithms, that manifest in changes in their conductivity. Phase Change Memory (PCM) and Metal Oxide Resistive Memory (OxRAM) can be used as synaptic elements thanks to their tunable conductivity, compatibility with advanced CMOS fabrication process, low power consumption, non-volatility and scalability. Two main approaches to emulate synaptic conductance modulation have been successfully demonstrated: a) the analog approach, where multiple resistance states to emulate long-term potentiation and depression (cumulative decrease and increase of resistance, LTP and LTD) are used; b) the binary approach, where only two distinct resistance states (Low Resistance State, LRS, and High Resistance State, HRS) per device associated with a probabilistic programming scheme are adopted. In the latter case, the conductance modulation is achieved by designing a single synapse as a composition of n multiple binary cells operating in parallel [4].

PCM devices can be used as an analog memory. However, they show a strong asymmetry between the SET and RESET process: the SET process is extremely gradual, very similar to synaptic potentiation, while the RESET process is abrupt. In [5] a 2-PCM synapse that recreates artificial symmetry between SET and REST by employing two devices per synapse has been proposed. This strategy works well, but it requires long and energy-hungry refresh operations, where all PCM devices are being reprogrammed. A narrow heater bottom electrode-based PCM associated with an initialization electrical pulse followed by a sequence of identical fast programming pulses is presented in [6] to implement a bi-directional synapse. Gradual long-term potentiation and depression are achieved by applying a long train of identical short SET and REST pulses (< 50 ns). The results are illustrated in Figure 1. Under these conditions, the created amorphous region does not cover the entire area of the bottom electrode (i.e. the heater), achieving a gradual resistance modulation. This strategy has been demonstrated to achieve unsupervised learning using STDP, on the character recognition application. An average classification rate of about 76% has been demonstrated by means of system level simulations. In agreement with experimental results, 200 SET and 30 RESET levels have been simulated. To improve the performance, the number of depression levels has to be increased. A classification rate of 82% is achieved for 100 depression levels (i.e., 100 RESET).

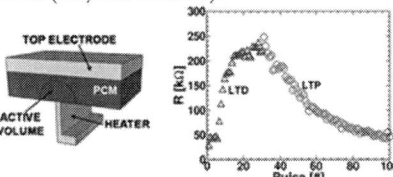

Fig. 1. Phase Change Memory (a) Schematic drawing of the PCM wall storage element. (b) Depression (LTD) and potentiation (LTP) characteristics. The memory is initialized by applying a 20 ns pulse of 1.25 V to set the cell at 30 kΩ. A train of identical 20 ns amorphizing pulses of 1.6 V (LTD) and 20 ns crystallization pulses of 1.25 V (LTP) are applied.

OxRAMs can be programmed consuming little power and can be integrated simply with advanced CMOS technologies. However, these devices suffer from high cycle-to-cycle and device-to-device conductance variability, presenting considerable challenge for standard memory applications. Figure 2 shows the cumulative distribution of HRS and LRS for a TiN/HfO$_2$/Ti/TiN stack. The measurements have been performed on a 4 kbit array. In OxRAMs, both directions of programming are non-cumulative. Therefore, an alternative method to implement stochastic STDP has been proposed [7]. The proposed learning scheme has been demonstrated on a Fully Connected Neural Network for automobile tracking. It has been demonstrated that the network can draw benefit from conductance variability since it increases the range of synaptic weights available during learning [8].

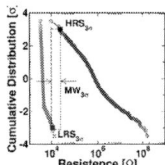

Fig. 2. Cumulative distribution of LRS and HRS measured on 4 kbit array (Vset = 2 V, Vreset = 2.5 V, Icc=200µA).

IV. ON THE DESIGN, ROBUSTNESS AND TEST OF SPIKING NEURAL NERWORKS

In order to get the maximum efficiency out of hardware implemented SNNs, functional modules (neuron-synapse) have to be designed in such a way that their input/output characteristics provide the learning and processing capability required by application. Additionally, the network connectivity has to allow for high integration with strong and reliable reconfiguration and adaptation characteristics.

Significant benefits can be gained by adopting RRAMs for neuromorphic computation as explained in the previous section. RRAM with bidirectional and continuous conductance tuning capability is considered as a natural electrically-controlled synaptic device. There is a considerable amount of work dedicated to the development of new synaptic-compliant devices, device modeling and algorithmic validation of such devices placed in the context of SNNs. Little work has been dedicated to the physical design of a full SNN and the identification of possible faults and reliability issues and no post-fabrication test solutions have been proposed so far. For a robust and efficient hardware implementation of SNNs we have to jointly-consider the characteristics of the SNN itself (connectivity, neuronal activation function, learning rule and synaptic update) and the characteristic of the devices used to implement it (CMOS ON/OFF current and threshold voltage, conductivity modulation and current-compliance of the RRAM, etc.).

Here we focus on a fully-connected SNN, that learns using the Spike Timing Dependent Plasticity (STDP) method with lateral inhibition, with integrate-and-fire neuron and resistive synapses (see Fig. 3). Fig. 3a) sketches the connectivity between two consecutive layers of such a network. Here N represents a spiking neuron, S represents a synapse and test wrapper is the circuitry not-pertaining to the SNN but necessary to perform the post-fabrication test. The synapse and its control circuit are illustrated in Fig. 3b). Several works propose using the resistive-based synaptic devices in a crossbar array (without access device) to assure minimum area footprint. However, such implementation suffers from large leakage currents in the half accessed cells (gravely detrimental to the learning process) and from inefficient forming process (when RRAM devices are used for synapse) and un-controlled current compliance (can cause synapse deterioration or inefficiency). For this reason, in the presented design, an access transistor is used for the access to each synapse. This transistor should isolate the synapse from the rest of the network when there is no activity on either of its connected neurons, and it should allow the passage signal coming from either of the neurons. This behavior is guaranteed by OR-ing enable signals generated by the neurons. These signals are generated every time a neuron spikes. The PPre is the presynaptic spike, comes from an input neuron and caries the input information, while PPost is the postsynaptic spike, comes from an output neuron and caries the learning control. The generation of these pulses is done by an integrate-and-fire neuron. One neuron receives information from all neurons in the previous layer of the network, modulated by the corresponding synaptic weight. This information is accumulated until it reaches a certain level, at which point the neuron sends signal towards the next layer. The functional structure of a neuron is illustrated in Fig. 3 c). The accumulation function is performed by an integrator structure, represented in the figure by the operational amplifier and capacitor C. The output of this amplifier is compared against a threshold Vth, and once the threshold is achieved, a pulse generator is activated, who's main functionality is to generate the postsynaptic spike (the PPost pulse) feedback to adjust the synaptic weight, and feedforward, as presynaptic spike (PPre pulse), to send the information to the next network layer. In addition to this information-carrying signal, the pulse generator has to provide the enable signals to control (i) the switch S1 and activate the neuron's refractory period, (ii) the switch S2 and activate the synaptic weight modulation (learning), or (iii) to allow the passage of the modulated presynaptic spike (W*PPre, with W being the synaptic weight). In addition, the pulse generator gives the enable signals controlling the OR gate in the synapse control and the disabled signals inhibiting neighboring neurons (lateral inhibition) - not shown in the figure. Once the circuit is designed and its parameters optimized by simulation, it will be fabricated.

Manufacturing process of integrated circuits in general is not totally controlled. There are many effects that could lead to defective circuits, including dust, spot defects on the silicon wafer, process variability, and assemblage faults. Manufacturing testing is the process able to ensure quality and reliability of integrated circuits. The main approach used to test integrated circuits is based on: (i) the generation of test input vectors able to target possible faults; (ii) the application of those vectors to the circuit; and (iii) the comparison of the responses provided by the circuit with the expected ones pre-calculated by a simulator. While many efficient solutions exist for testing traditional designs (analog, digital or mixed-signal), to the best of our knowledge, there is no work so far dealing with the post-fabrication testing of a hardware-implemented SNN. As described previously, circuits implementing SNNs have some major differences compared to classical circuits. Indeed, they resort to the combination of devices with both deterministic and stochastic behaviors and they include both digital and analog elements. A test strategy suitable for SNNs should be able to test the correct operation of both neurons and synapses, it should be economical (in area, power and performance overhead), it should not depend on the training data nor on the context the network would be used. This last condition is important for general-purpose SNNs, which could be used in different scenarios. Indeed, it is possible for SNNs with online learning to be used for pattern classification within different contexts, as long as the problem space is equal or smaller than the input-neuron space.

In order to identify and implement an efficient test strategy, it is necessary to understand the possible faulty behavior of the circuit under test. This is traditionally accomplished by fault modeling. The faults that can be found in resistive memory arrays can be classified into two categories: soft faults and hard faults [9-10]. Soft faults are caused by different cycle-to-cycle or device-to-device variations that appear during the fabrication, but also in-field during read/write operations [11]. Hard faults are provoked by fabrication steps or they can be caused by the forming process or by continuous stress; they are more difficult to be prevented. One typical type of hard fault occurs when the resistance of a resistive memory cell will no longer change; this category includes stuck-at-0 (SA0) and stuck-at-1 (SA1) faults caused by fabrication techniques and limited endurance. In this case, the faulty device is stuck at high resistance or low resistance state, and these situations are occurring with a quite high probability. It is reported that 63% of a storage array based on memristor are fault free in a 4Mb resistive RAM, with about 10% of the cell being of Stuck-At type. [12]. In [13] the authors showed that 10% of broken RRAM cells will lead to substantial degradation of the accuracy and overall performance of a convolutional neural network. These studies were performed for resistive devices used in memory arrays, where the resistive states are seen as binary. However, limited research has been dedicated to analyze and model the faults of a resistive synapse, where the resistive device is either used as an analog device (multiple resistance states are used to emulate the synaptic weights) or as a binary device but with a probabilistic programming scheme. Therefore, fault models and test strategies have to be devised according to the specificity of the synaptic array. For the analog synapse implementation, a classical analog test could be used, while for the binary implementation with probabilistic programing an entirely new test strategy should be devised. The main challenge here is to identify a strategy to test for the randomness of the resistive device programing. This test has to be low cost and have no impact on the network behavior at runtime. A possibility is to start use similar strategy as the ones used for the testing quality of true random number generators (TRNGs). However, since in this case the entropy is not of major concern, low cost test, such as NIST [14], could be used. Regarding the neuron, it is in most cases an analog device, and classical analog test can be used. The main challenge here is to find a way to apply the test vectors and read the circuit response.

Though dependability analysis and design-for-dependability are common sense when dealing with traditional (Von Neumann) computing architectures, they are not so common when dealing with spiking neuromorphic structures. It was considered that neural networks are inherently resilient as they imply algorithmic resilience, regularization, over-parameterized data. However, when the network is implemented in hardware, the fault-tolerance property degrades due to embedded systems restrictions: such as strong limited size of the network (i.e., number of hidden layers and number of neurons in each layer) and smaller neuron connectivity (i.e., number of synapses), size reduction of weights. Due to the fact that we target minimum size networks for maximum (possible) complexity and low power, hardware imperfections become relevant. Papers [15-16] show preliminary results of technology dependability threats impact on neuron and synaptic functionality, as well as the quantification of their effects on the operation efficiency of the functional reduced size synapses and also spiking neural networks. Evaluations have been performed on typical 2-layers of fully-connected spiking neurons with lateral inhibition. STDP is used as unsupervised learning for the recognition of static images and the recognition error can drop

by more than 35% with a reduced set of training data. In larger synapse storage capabilities, the drop of error recognition is still neighboring 20%, and can be brought to 10% only if the training set is increased by at least 6 times, which in turn will generate similar increase in the power envelope.

REFERENCES

[1] L. Khacef, N. Abderrahmane, B. Miramond, *"Confronting machine-learning with neuroscience for neuromorphic architectures design,"* International Joint Conference on Neural Networks (IJCNN), 2018

[2] L.A. Camuñas-Mesa, et al.,*"A Configurable Event-Driven Convolutional Node with Rate Saturation Mechanism for Modular ConvNet Systems Implementation,"* Frontiers in Neuroscience, 2018

[3] R. Kreiser, T. Moraitis, Y. Sandamirskaya and G. Indiveri, *"On-chip unsupervised learning in winner-take-all networks of spiking neurons,"* IEEE Biomedical Circuits and Systems Conference (BioCAS), 2017

[4] D. Garbin, et al., *"Hfo2-based oxram devices as synapses for convolutional neural networks,"* IEEE Transactions on Electron Devices, vol. 62, no. 8, pp. 2494–2501, 2015.

[5] M. Suri, et al. *"Phase change memory as synapse for ultra-dense neuromorphic systems: Application to complex visual pattern extraction,"* 2011 International Electron Devices Meeting, 4.4.1 - 4.4.4.

[6] S. La Barbera, et al., *"Narrow Heater Bottom Electrode-based Phase Change Memory as a Bidirectional Artificial Synapse,"* to appear on Advanced Electronic Materials.

[7] M. Suri, et al. *"CBRAM devices as binary synapses for low-power stochastic neuromorphic systems: Auditory (cochlea) and visual (retina) cognitive processing applications,"* 2012 IEDM.

[8] D. R. B. Ly; et al., *"Role of synaptic variability in spike-based neuromorphic circuits with unsupervised learning"*, ISCAS 2018.

[9] (Degrae2015) R.Degraeveetal.,*"Causes and consequences of thestochastic aspect of filamentary RRAM,"* Microelectronic Engineering, vol. 147, pp. 171–175, 2015

[10] E. I. Vatajelu, P. Prinetto, M. Taouil, S. Hamdioui, *"Challenges and Solutions in Emerging Memory Testing"*, IEEE TETC, 2017

[11] L.Xia, et al.,*"Technological exploration of RRAM crossbar arrayf or matrix-vector multiplication,"* Journal of Computer Science and Technology, vol. 31, 2016.

[12] C.-Y.Chen et al.,*"RRAM defect modeling and failure analysis based on march test and a novel squeeze-search scheme,"* IEEE Trans. on Computers, vol. 64, Jan 2015.

[13] L. Xia et al., *"Fault-Tolerant Training with On-Line Fault Detection for RRAM-Based Neural Computing Systems,"* DAC, 2017

[14] A. Rukhin, et al., *"A Statistical Test Suite for Random and Pseudorandom Number Generators for Cryptographic Applications, NIST,"* Special Publication 800-22, 2010

[15] E. I. Vatajelu, L. Anghel, Reliability Analysis of MTJ-based Functional Module for Neuromorphic Computing, IEEE IOLTS 2017.

[16] E. I. Vatajelu, L. Anghel, Fully-connected single-layer STT-MTJ-based spiking neural network under process variability, 2017 IEEE/ACM NANOARCH, 2017

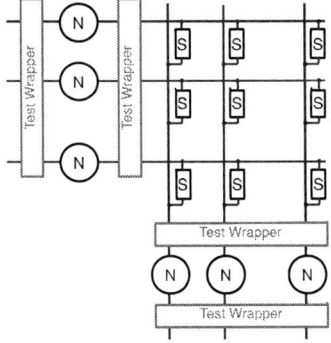

(a) Two-Layer Fully Connected SNN

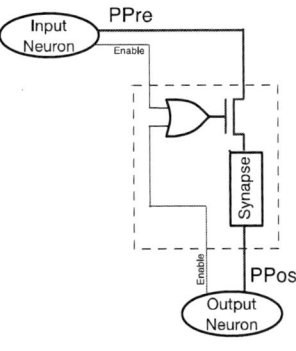

(b) Synapse and control circuit

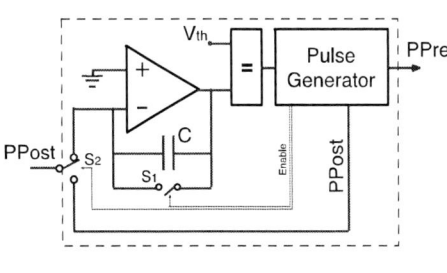

(c) Integrate and Fire Neuron

Fig. 3. Design of the Spiking Neural Network

978-1-5386-4757-8/18 $31.00 © 2018 IEEE

Prospects for energy-efficient edge computing with integrated HfO_2-based ferroelectric devices

Ian O'Connor, Mayeul Cantan,
Cédric Marchand, Bertrand Vilquin
Lyon Institute of Nanotechnology
University of Lyon – Ecole Centrale de Lyon – CNRS
Ecully, France
ian.oconnor@ec-lyon.fr

Stefan Slesazeck[1], Evelyn T. Breyer[1],
Halid Mulaosmanovic[1], Thomas Mikolajick[1,2]
[1]NaMLab GmbH, [2]Chair of Nanoel. Materials, TU Dresden
Dresden, Germany
stefan.slesazeck@namlab.com

Bastien Giraud, Jean-Philippe Noël
Univ. Grenoble Alpes, CEA, LETI, MINATEC Campus,
Grenoble, France
bastien.giraud@cea.fr

Adrian Ionescu, Igor Stolichnov
Nanolab
Ecole Polytechnique Fédérale de Lausanne (EPFL)
Lausanne, Switzerland
adrian.ionescu@epfl.ch

Abstract— **Edge computing requires highly energy efficient microprocessor units with embedded non-volatile memories to process data at IoT sensor nodes. Ferroelectric non-volatile memory devices are fast, low power and high endurance, and could greatly enhance energy-efficiency and allow flexibility for finer grain logic and memory. This paper will describe the basics of ferroelectric devices for both hysteretic (non-volatile memory) and negative capacitance (steep slope switch) devices, and then project how these can be used in low-power logic cell architectures and fine-grain logic-in-memory (LiM) circuits.**

Keywords—ferroelectric devices, non-volatile memory, steep slope switch, low-power logic, logic-in-memory

I. INTRODUCTION

Data size and functionality requirements for computing are increasing, according to the expectation that hardware performance will continue to improve, irrespective of the actual implementation. This is particularly true for emerging distributed computing paradigms for the Internet of Things, such as Edge Computing and Fog Computing, which are placing extraordinarily stringent constraints on computing hardware performance. Such paradigms are necessary to guarantee low-latency, secure and contextualized computation on inhomogeneous sensory data, as close as possible to the data source. This usually implies that energy sources are limited and consequently, that hardware energy efficiency must be maximized. Indeed, optimal usage of the constrained resources such as memory, bandwidth, processor, and most importantly power of IoT devices is necessary for sustainable and long-life IoT deployments.

Low-power microcontroller units (MCU) with embedded non-volatile memory (NVM) are typically to be found at the heart of today's IoT devices, hierarchically placed between the sensor nodes and the host microprocessor and given the task of pre-computing the data to reduce heavy loading of the host processor. In IoT applications, most of the power is consumed while the MCU is inactive, so NVM is used to realize a "normally-off" MCU, thus drastically cutting power consumption. The contents of the CPU are stored in the NVM and subsequently CPU power is shut down to zero power consumption sleep mode before the data are restored during MCU wake-up period.

While a centralized data transfer from the CPU to the NVM and vice-versa is a straightforward approach, more efficient approaches are highly desirable. In distributed memory concepts, NVM elements can be embedded in an advanced CMOS platform and distributed close to the logic circuits to store contents locally and essentially make logic circuits non-volatile (e.g. NV-registers, NV-SRAM, NV Code memory). This allows inactive logic circuits to be shut down, thus optimizing power savings, minimizing memory cycling and increasing reliability. The distributed memory concept is also in line with more advanced "fine grain" logic-in-memory (LiM) concepts with tighter integration of logic and memory and aiming to reduce latency and energy cost during data transfer. However, storing and re-storing CPU content costs energy, which imposes constraints on the NVM characteristics to be used. At present, the standard NVM used in MCUs is eFlash because it is high density, manufacturable and low cost. However, it suffers from low write speed, high power requirements, low endurance and vulnerability to radiation.

Therefore, a new, more robust NVM with higher speeds, lower power and high endurance is required to replace eFlash in (normally-off) MCUs to reduce the energy spent during storage and retrieval of CPU content and to allow flexible LiM designs with further improved energy efficiency. A number of NVM candidates with high speed/low power characteristics have emerged (STTRAM, ReRAM, and FeRAM) and several working prototypes have been demonstrated but there is no clear winner at present. FeRAM has the highest endurance of all candidates, low energy per bit and power consumption which could make it a good candidate to replace Flash in embedded applications. However, current embedded FeRAM devices with perovskite materials have serious problems with regard to memory cell scaling, compatibility with Si processing, manufacturability and cost that inhibit development as a mainstream NVM solution. New FE materials to overcome the shortcomings of present day FeRAM are needed.

In this paper, we explore the use of new FE HfO_2-based materials to develop a competitive and versatile FeRAM technology for NVM solutions. The structure of the paper is as follows: in section II, we describe device operation of hysteretic and steep-slope devices based on HfO_2 ferroelectric FETs (FeFETs). Section III gives examples of the use of such devices to build non-volatile flip-flops and reconfigurable logic gates. Finally, section IV discusses projections to new computing architectures based on such devices.

II. MATERIALS AND DEVICE OPERATION

A. HfO_2 as a ferroelectric material

High-k HfO_2 is key for modern nanoelectronics since it is compatible with silicon technology and, thanks to its high

978-1-5386-4757-8/18 $31.00 © 2018 IEEE

dielectric permittivity, has allowed downscaling without the prohibitive leakage currents associated with the traditional gate oxide. Used as a high-k gate oxide, HfO_2 is amorphous. In crystalline form, HfO_2 is generally centrosymmetric with a high temperature tetragonal phase and room temperature monoclinic phase. The tetragonal phase can be realized by doping HfO_2 with certain dopants and actually has the best high-k performance. However, under certain conditions of stoichiometry, doping and/or strain the polymorphism can be extended to a non-centrosymmetric orthorhombic phase in doped HfO_2 [1], and this has led to the recent discovery [2] of ferroelectricity in HfO_2 (Fig. 1). This material may thus impact embedded memory by allowing scaling of FeRAM cells [3] to increase storage capacity.

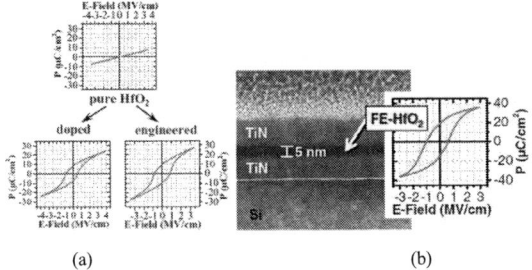

(a)   (b)

Fig. 1. (a) Ferroelectric HfO_2 may be obtained by doping and by strain engineering (b) 5nm HZO capacitor

In terms of silicon compatibility, ferroelectric HfO_2 can be integrated into transistor gates for 1T FeFET memory cells [4] with non-destructive read, integrated in the front end of line (FEOL) with CMOS. This opens the possibility of realizing new, fine-grained LiM designs [5] to enable the merging of logic and memory and to significantly improve energy efficiency of processing and storage units.

Finally, ferroelectric HfO_2 integrated in the gate of transistor could also produce negative capacitance FETs (NCFETs) [6] functioning as low power steep slope switches to further boost low power/high performance operation of LiM circuits.

In summary, the functionality and versatility of ferroelectric HfO_2 could have a significant impact on embedded NVM solutions, tightly integrated with logic to increase the energy efficiency of computation.

B. Ferroelectric transistor (FeFET) operation

From a structural point of view, a ferroelectric transistor (FeFET) is simply an extension of a regular bulk or FDSOI MOSFET with an additional layer of ferroelectric material inside the gate stack (Fig. 2) [8]. This leads to a functionality based on the inclusion of a ferroelectric capacitance located between the external gate and the "internal" gate, which actually controls the state of the FET channel. The potential simplicity of the process has led to speculation that every transistor in a standard CMOS process could be transformed into a non-volatile memory or steep-slope device.

FeFETs operate in two different modes: a non-volatile mode, which requires hysteretic operation, and a steep switching mode, which can be hysteretic or non-hysteretic. The ratio between the ferroelectric capacitance and the dielectric capacitance determines the FeFET operation mode.

1) Non-volatile mode: this mode leverages the hysteretic polarization vs. voltage characteristic of the ferroelectric

material (P versus V_{FE}) as already shown in Fig. 1. When this material is placed in the gate stack, it forms a capacitance in series with the gate of a transistor. From the point of view of the overall device, the I_{DS}-V_{GS} transfer characteristic then also becomes hysteretic (Fig. 3(a)). For V_{GS}=0V and a centered ferroelectric hysteresis, the FeFET demonstrates bistable states corresponding to positive and negative polarization retention in the ferroelectric layer. Hence the FeFET channel resistance is either in a high-resistance state (HRS) for low I_{DS}, or a low-resistance state (LRS) for high I_{DS}. For an n-type FeFET, HRS is achieved for P<0 and LRS for P>0, while the conditions are opposite for a p-type FeFET. The ferroelectric capacitance must be small with respect to the dielectric capacitance as the series capacitance leads to a reduced hysteretic window of P vs. V_{GS}. This usually means that the ferroelectric layer has to be relatively thick in order to achieve non-volatile FeFET operation.

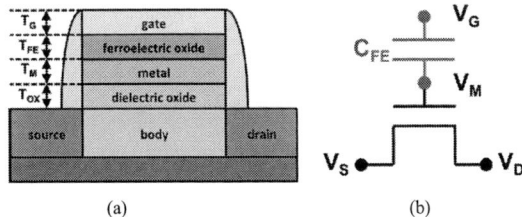

(a)   (b)

Fig. 2. (a) Ferroelectric HfO_2 device gate stack (b) equivalent circuit schematic

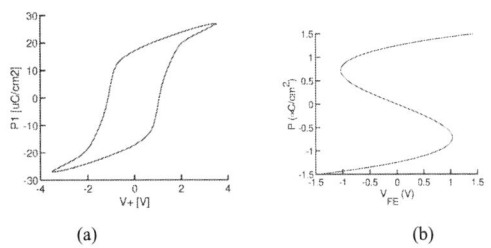

(a)   (b)

Fig. 3. Polarization versus voltage characteristics (a) of a non-volatile FeFET (b) of an NCFET

2) Steep switching mode: this mode relies on a direct non-hysteretic transition between positive polarization for negative voltage and negative polarization for positive voltage, originating from the "S-shaped" P vs. V_{GS} curve described by Landau-Ginzburg-Devonshire theory, without considering domain formation. This particular region of the curve displays *negative* differential capacitance (C=δQ/δV), which would be physically unstable in a standalone device, but can be associated in series with the dielectric capacitance in a stable way if the overall capacitance is positive. This condition typically requires a ferroelectric capacitance that is larger than the dielectric capacitance, implying that the ferroelectric layer is relatively thin. On the other hand, the higher ferroelectric capacitance means a weaker negative capacitance effect. Therefore, both capacitances have to remain comparable. The voltage division across the ferroelectric and dielectric capacitances gives

$$\frac{V_M}{V_G} = \frac{C_{FE}}{C_{FE} + C_{MOS}}$$

where V_M and V_G represent respectively the intermediate and gate potentials as shown in Fig. 2(b), and C_{FE} and C_{MOS} represent the ferroelectric and dielectric capacitances respectively. Usual voltage division would mean that $V_M<V_G$; but with a negative ferroelectric capacitance value, $V_M>V_G$ and is actually *amplified*. As V_M is the surface potential controlling the FET channel, the drain current I_{DS} is enhanced and from the point of view of the gate voltage V_G, the transistor can achieve "steep switching" with a subthreshold slope below the Boltzmann limit of 60mV/dec. However the use of an internal metal as shown in Fig. 2(a) can impair the stabilization of the negative capacitance state due to leakage currents and domain formation [10][11]. Therefore, experimental devices should avoid using such an internal metal layer. Such devices (commonly termed "negative capacitance FETs" or NCFETs) have clear advantages for low-power logic operation, and the intrinsic compatibility with CMOS coupled with the potential for non-volatile devices makes the case for a ferroelectric technology platform all the more convincing.

III. FeFET BASED LOGIC AND MEMORY CIRCUITS

In this section, we examine non-volatile memory cells and reconfigurable logic cells based on ferroelectric devices.

A. Non-volatile memory

An example of a non-volatile memory cell is the Black and Das non-volatile flip-flop depicted in Fig. 4. This structure combines the operation of a regular SRAM cell for normal operation, and two FeFETs for non-volatile storage, which can be written to independently. When the circuit is powered, the values stored in the ferroelectric capacitors can be written to the SRAM through the *sense* transistor. If the FeFETs are in different resistance states, the imbalance will determine the rest state the SRAM will return to, as the FeFET with the lowest resistance will pull its branch down. Such an approach reduces the number of write cycles, and allows the SRAM to retain voltage compatibility with traditional structures. Moreover, this device remains operational when HRS and LRS are relatively close.

B. Reconfigurable logic

As well as pure non-volatile memory applications, FeFETs have also been integrated inside logic gates themselves [9][13]. Two types are presented here: a 1-FeFET reconfigurable NAND/NOR logic gate and a 2-FeFET X(N)OR logic gate. In both cases, the ferroelectric material is HfO_2. To achieve sequential logic functionality, one logic input is stored in the polarization state of the FeFET by applying a write pulse to the gate terminal, whereas the second logic input is subsequently applied as readout voltage to the gate terminal (see inputs A and B in Fig. 5(a) and (b)).

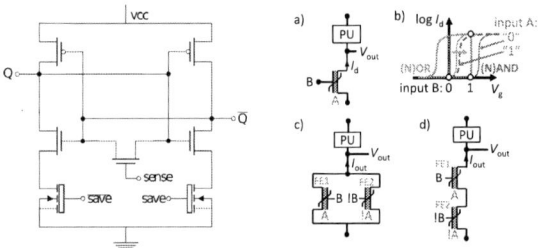

Fig. 4. Non-volatile FeFET based latch

Fig. 5. a) and b) 1-FeFET NAND/NOR schematic and basic functionality c) and d) 2-FeFET X(N)OR logic gates

Input A denotes the internal polarization state of the FeFET and can be either logic 0 (high resistance state – solid line) or 1 (low resistance state – dashed line). The second input – input B – is applied as a readout voltage to the gate terminal. Note that the applied voltage for logic input B is small enough not to switch the polarization state. Therefore, these gates constitute a sequential logic that immediately links the memory and the logic functionality of the FeFET.

Electrical measurement results prove the feasibility of the NAND/NOR logic gate concept for 22nm FD-SOI FeFETs. By applying a back bias voltage to the FeFET, the I_dV_g characteristics of the FeFET can be shifted along the V_g axis. As a result, the logic functionality switches between NAND and NOR behavior and electrical reconfigurability is introduced (see Fig. 5(b) and Fig. 6(a)). As a next step, two FeFETs of NAND or NOR functionality are connected in parallel or in series, respectively. If the inputs of the second FeFET are the logical complements of the inputs of the first FeFET (Fig. 5(c) and (d)), the resulting structure exhibits X(N)OR functionality (Fig. 6b). A direct integration into existing AND/NAND memory arrays is possible [13].

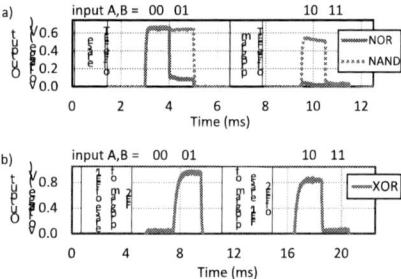

Fig. 6. Transient voltage measurements of (a) the proposed 1-FeFET logic NAND/NOR and (b) the 2-FeFET logic XOR gates

IV. PROSPECTS FOR ENERGY-EFFICIENT COMPUTING

Integrated ferroelectric devices allow data storage close to logic circuits to reduce energy cost of data transfer, allow smart gating for "normally-off" computing and open the way to novel energy-efficient computing paradigms such as logic-in-memory (LiM).

A. Normally-off computing

Normally-off (N-Off) computing uses a non-volatile memory array to immediately store the context of logic (data at logic nodes) during power-down, which enables significant reduction of activity duty cycles and energy consumption in IoT nodes due to leakage currents (Fig. 7(a)). An extension of this model also uses dedicated non-volatile memory close to data sources (e.g. sensors) to store and accumulate acquired data without expensive wakeup cycles for IoT node processors (Fig. 7(b)). The priority criteria are:

- For store (at power-down): 1. Low write power, 2. Low write time, 3. High-density
- For restore (at power-up): 1. Low read power, 2. Low read time 3. High-density
- For accumulation (off): 1. Low write power, 2. Low write time, 3. Low read power, 4. Low read time, 5. High density, 6. High endurance, 7. Low leakage

B. Logic-in-memory (LiM)

LiM represents a logically enhanced memory array that can be programmed to realize arithmetic and logic operations in an endurance-aware way. Several concepts and terms exist to identify means of associating logic with memory, such as LiM (logic in memory), IMC (in-memory computing) and PiM (processing in memory). In order to define clearly the contribution of HfO$_2$ FeFETs in this spectrum, we firstly make explicit our perception of these terms.

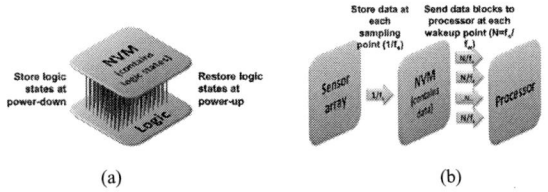

Fig. 7. Normally-off computing to: (a) store/restore logic states (b) accumulate data at sensor nodes

1) PiM - Processing in Memory: move PROCESSING functionality INto modified MEMORY. Here, memory is enhanced with a lightweight processor (complete with control unit and instruction register / decoder) located close to memory macros. The main processor offloads complete portions of tasks onto the lightweight processor which has direct and fast access to local data as they are physically situated inside the memory, and is capable of carrying out a set of simple but frequently used operations.

2) IMC - In Memory Computing: INside modified MEMORY add elementary COMPUTING functionality. It can be useful to enhance memory to enable it to do some computation locally. In this case, instead of loading operands and storing results from and to the memory, the processor can ask the enhanced memory to carry out elementary operations on operands before sending the result for subsequent (more complex) calculations. This alleviates load on processor-memory data communication but requires a richer processor-memory control communication.

3) LiM - Logic in Memory: do LOGIC operations IN existing MEMORY resources. *Coarse-grain LiM* requires a full (non-volatile) memory matrix that can be integrated sufficiently close to the processor to be used with the same latency as a L1 cache. The memory is used to contain known results of a frequently used operation (multiplication is a simple example). A key characteristic is that no additional logic is required inside the memory plane, although some additional interpretation of results (e.g. interpolation, normalisation) may be necessary in the logic plane. *Fine-grain LiM* is different and more prospective. No clear use model has yet emerged for this, but the key characteristic is that a (non-volatile) memory element (or small network of elements) can be connected to a small logic switch network (in different planes) to achieve novel functions. A non-volatile flip-flop is an example using the level of intimacy between logic and memory required for fine-grain LiM [15].

For FeFET-based coarse-grain LiM (Fig. 8(a)), the machine can be programmed using the state of NV devices to realize complex logical and arithmetic functions in an endurance-aware way, which can then be read using word lines to address function content. The priority criteria in this scenario are:

- For programming offline: 1. Low write power, 2. Low write time, 3. High-endurance
- For operation online: 1. Low read time, 2. Low read power, 3. High-density.

For FeFET-based fine-grain LiM (Fig. 8(b)), device-level association of logic and memory can be used for long-term variable assignments in mathematical accelerator functions such as function coefficients (e.g. filters) and table content.

Both arithmetic and logic functions can also be associated with sequencing. The priority criteria in this scenario are:
- For NV switches: 1. High endurance, 2. High density, 3. Low write time, 4. Low write power
- For logic switches: 1. Low leakage, 2. High speed, 3. High density, 4. Low voltage

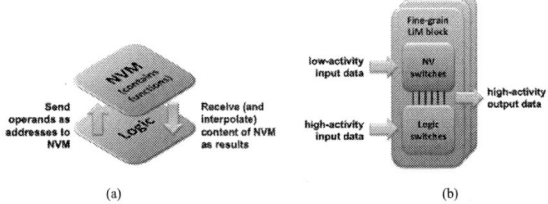

Fig. 8. LiM with non-volatile memory: (a) coarse-grain (b) fine-grain

V. CONCLUSION

Thanks to its proven compatibility with silicon, ferroelectric HfO$_2$ can be integrated in FET gate stacks to form non-volatile memory cells and low-power logic. In this paper, we have described the underlying physics and circuit operation and projected how normally-off computing and logic-in-memory can benefit. Future work will focus on experimental demonstration of large-scale HfO$_2$ FeFET memory arrays integrated with advanced CMOS platforms.

ACKNOWLEDGMENT

This project has received funding from the European Union's Horizon 2020 research and innovation programme under grant agreement No 780302 "3εFerro".

REFERENCES

[1] Böschke et al. Ferroelectricity in hafnium oxide thin films, Appl. Phys. Lett. 99, 102903 (2011)

[2] J. Müller, et al., "Ferroelectricity in simple binary ZrO2 and HfO2", NanoLett. 12, 4318 (12)

[3] J. Müller et al., "Ferroelectric Hafnium Oxide Based Materials and Devices: Assessment of Current Status and Future Prospects", ECS Journal of Solid State Science and Technology, 4 (5) N30-N35 (15)

[4] M. Trentzsch et al., "A 28 nm HKMG super low power embedded NVM technology based of ferroelectric FETs" IEDM 2016, 294-297

[5] X. Yin et al., "Exploiting Ferroelectric FETs for Low-Power Non-Volatile Logic-in-Memory Circuits" ICCAD'16, November 07–10, 2016, Austin, TX, USA

[6] S. Salahuddin, S. Datta, NanoLett. 8 , 405 (08)

[7] M. Hofmann et al., "Direct observation of negative capacitance in polycrystalline ferroelectric HfO2" Adv. Funct. Mater. 2016

[8] A. Aziz et al., "Computing with Ferroelectric FETs: Devices, Models, Systems, and Applications", DATE 2018

[9] M. Hoffmann et al., "Direct observation of negative capacitance in polycrystalline ferroelectric HfO2" Adv. Funct. Mater. 2016

[10] A. I. Khan et al. IEEE Trans. Electr. Dev. 63 (11), 4416-4422 (2016)

[11] M. Hoffmann et al. Nanoscale, 2018,10, 10891-10899 (DOI: 10.1039/C8NR02752H).

[12] E. T. Breyer, H. Mulaosmanovic, T. Mikolajick and S. Slesazeck, "Reconfigurable NAND/NOR logic gates in 28 nm HKMG and 22 nm FD-SOI FeFET technology" IEDM 2017, pp. 28.5.1-28.5.4

[13] E. T. Breyer, H. Mulaosmanovic, S. Slesazeck, and T. Mikolajick, "Demonstration of versatile nonvolatile logic gates in 28nm HKMG FeFET technology" ISCAS 2018, pp. 1-5

[14] M. Trentzsch et al., A 28nm HKMG super low power embedded NVM technology based on ferroelectric FETs IEDM 2016, p. 294

[15] N. Jovanovic, et al. "Design Considerations for Reliable OxRAM-based Non-Volatile Flip-Flops in 28nm FDSOI Technology". International Symposium on Circuits and Systems (ISCAS), IEEE, 2016.

Multi-bit nonvolatile flip-flop based on NAND-like spin transfer torque MRAM

Erya Deng[1,3], Zhaohao Wang[1,2,3], Wang Kang[1,3], Shaoqian Wei[4], Weisheng Zhao*[1,2,3]

[1]Beijing Advanced Innovation Center for Big Data and Brain Computing, Fert Beijing Research Institute, Beihang University
[2] Beihang-Geortek Joint Microelectronics Institute, Qingdao Research Institute, Beihang University
[3]School of Microelectronics, Beihang University
[4]School of Electronic and Information Engineering, Beihang University
Beijing, China
deng.erya@163.com, {zhaohao.wang, wang.kang, Shaoqian_Wei, weisheng.zhao}@buaa.edu.cn

Abstract—Nonvolatile flip-flops (NVFFs) integrating emerging spintronics devices such as magnetic tunnel junction (MTJ) are under intensive investigation. They allow computing systems to be powered-off during the standby state, hence high static power issue of conventional CMOS technology can be addressed. MTJ based on spin transfer torque (STT) effect provide non-volatility, good endurance and 3D integration with CMOS based circuits. However, it suffers from relative long switching delay, high switching power and asymmetric switching issues. In this work, we first present a multi-bit NVFF using NAND-like spintronics (NANS-SPIN) devices which are written by STT and spin orbit torque (SOT) currents. It shows advantages in terms of power consumption, area overhead and write voltage. Then, functionality and performance of the proposed NVFF will be simulated and validated.

Keywords—flip-flop, magnetic tunnel junction, spin-transfer torque, spin orbit torque

I. INTRODUCTION

As the basic units in digital logic circuits, flip-flops (FFs) have been widely used in electronic circuits. However, conventional FFs suffer from power issues due to the increasing leakage current as the technology node continuously scales down [1]. In recent years, nonvolatile FFs (NVFFs) that integrate nonvolatile memory (NVM) devices are widely investigated [2-7]. In the NVFFs, data are first stored into the NVMs before the system is powered off and then restored when the system is powered on. Therefore, the static power consumption of NVFFs can be significantly reduced by shutting down the supply power in standby mode while maintaining data.

NVFFs using spin transfer torque magnetic tunnel junction (STT-MTJ) are of great interest thanks to its attributes of non-volatility, infinite endurance, fast read/write speed and compatible resistance value with CMOS transistors [7, 8]. Their area overhead is significantly reduced since MTJs are distributed over the circuit plane by using the 3D integration technology. Nevertheless, the STT-based nonvolatile devices suffer from asymmetry writing because the critical current for switching MTJ from antiparallel to parallel state (AP-P) is smaller than that from parallel to antiparallel state (P-AP) [9, 10]. Besides, STT switching suffers from low write speed, high write power consumption and large write

This work was supported by the National Natural Science Foundation of China (No. 61704005, No. 61501013).

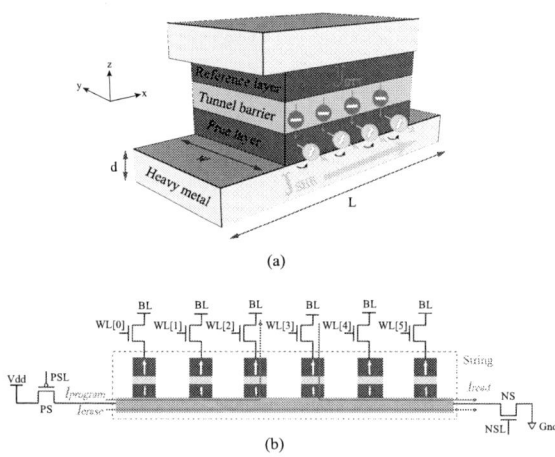

(a)

(b)

Fig. 1. (a) Three-terminal MTJ device structure. The free layer of the MTJ is connected with a heavy metal strip (β-W). STT write current (I_{STT}) and SHE write current (I_{SHE}) are applied to achieve the magnetization switching of the free layer. (b) Structure of the NAND-SPIN string. N MTJs are integrated in the same string and share the same selection transistors, i.e., PS and NS.

circuit area due to an intrinsic incubation delay at the initial process. In order to save data in several nanoseconds, large write current is required. Even though performance can be improved by enlarging the transistor size or reducing the critical write current, the area overhead of the write circuit and the current passing through the MTJ are increased, leading to higher risk of MTJ barrier breakdown [11, 12].

Spin orbit torque (SOT) provides a way to solve the long switching delay and high power consumption issues [12-18]. Spin-Hall-assisted STT switching was proposed to perform high-speed write operation for a perpendicular MTJ (pMTJ) [12, 19]. External magnetic field is replaced by an STT write current. Three-terminal MTJ device is shown in Fig. 1(a). It is mainly composed of two ferromagnetic (FM) layers, i.e., free layer (FL) and reference layer (RL), separated by an oxide barrier. The magnetization of RL is fixed whereas that of FL is free to be changed. MTJ presents either parallel (P) state or antiparallel (AP) state, corresponding to low resistance (R_P or R_L) or high resistance (R_{AP} or R_H). FL is attached at the top of a heavy metal strip (β-W). A current passing through the heavy metal strip injects a spin current into FL,

978-1-5386-4757-8/18 $31.00 © 2018 IEEE

exerting a SOT to assist the conventional STT [20]. In this way, STT write current can be largely reduced while keeping relative high write speed. However, three-terminal MTJ device has density problem as two access transistors are needed to control the current direction. In order to address this issue, NAND-like spin transfer torque (NAND-SPIN) memory was proposed in [9]. As illustrated in Fig. 1(b), a NAND-SPIN string integrates N MTJs whose FLs are attached to the same heavy metal strip. RF of each MTJ is connected with a transistor to select the MTJ bit-cell. A PMOS transistor and a NMOS transistor, i.e., PS and NS in Fig. 1(b), are used as the selection transistors of the string.

In this work, we propose a multi-bit NVFF based on NAND-SPIN memory. It combines STT and SOT write mechanism to erase and then program the storage MTJs. Thanks to the high spin injection of spin Hall effect (SHE), the required write current is much lower than that of the STT-based MTJ. Therefore, write circuit is much simpler when compared with previous works on STT-based NVFF. The circuit structures and operations are investigated, and the performance is evaluated at the 40 nm technology node.

The rest of this paper is organized as follows. Section II introduces the NAND-SPIN memory model that will be used in our design. In Section III, we illustrate the schematic and operation of the proposed multi-bit NVFF. In Section IV, simulation and discussion are performed to validate its functionality and performance. Finally, Section V concludes this paper.

II. NAND-SPIN MEMORY MODEL

The magnetization dynamics in FL of MTJ is described by a modified Landau-Lifshitz-Gilbert (LLG) equation [12], as

$$\frac{\partial \mathbf{m}}{\partial t} = -\gamma \mu_0 \mathbf{m} \times \mathbf{H}_{eff} + \alpha \mathbf{m} \times \frac{\partial \mathbf{m}}{\partial t} - \xi \varepsilon J_{STT} \mathbf{m} \times (\mathbf{m} \times \mathbf{m}_r) \quad (1)$$
$$- \xi \eta J_{SHE} \mathbf{m} \times (\mathbf{m} \times \boldsymbol{\sigma}_{SHE})$$

$$\varepsilon = \frac{P\Lambda^2}{(\Lambda^2 + 1) + (\Lambda^2 - 1)(\mathbf{m} \cdot \mathbf{m}_r)} \quad (2)$$

where $\mathbf{m}$ and $\mathbf{m}_r$ are unit vectors along the magnetization

TABLE I
SIMULATION PARAMETERS

Description	Default value
Oxide barrier thickness	0.85 nm
Free layer thickness	1 nm
MTJ surface	40 nm × 40 nm
Heavy metal width	40 nm
Heavy metal thickness	3 nm
Distance between two MTJs	60 nm
Resistance-area product	10 Ω·μm²
TMR ratio with V_{bias}=0	150%
MTJ Thermal stability factor	32
Supply voltage	1 V
Damping constant	0.02
Spin Hall angle	0.3
Saturation magnetization	1×10^6 A/m
Spin polarization	0.62
Heavy metal resistivity	200 μΩ·cm

orientation of the free layer and reference layer, respectively. J_{STT} and J_{SHE} are STT and SHE write current densities, respectively. $\boldsymbol{\sigma}_{SHE}$ is the polarization direction of spin current induced by SHE. $\mathbf{H}_{eff}$ is the effective field. ε, Λ and P are the STT efficiency and asymmetry factor, spin polarization, respectively [21]. More details about other coefficients can be found in [12].

A spice-compatible model for NAND-SPIN device has been developed with Verilog-A language. It provides a feasible interface between MTJ signals and CMOS circuits. Table I shows the critical parameters used in the following simulations.

III. PROPOSED MULTI-BIT NAND-SPIN NVFF

As shown in Fig. 2, the proposed multi-bit NAND-SPIN NVFF is composed of a pre-charge sense amplifier (PCSA) to evaluate logic results on output nodes OUT and OUTB [22]. Two transistors N_4 and N_5 are placed between MTJ cells and PCSA to effectively separate the read and write operation. Four storage MTJs (M_0-M_3) are placed on the same side while one reference MTJ (M_{ref}) is connected with the right branch of the circuit. During one operation phase, NMOS transistors N_6-N_9 are used to select one MTJ to be read or write. We can use a decoder to control the

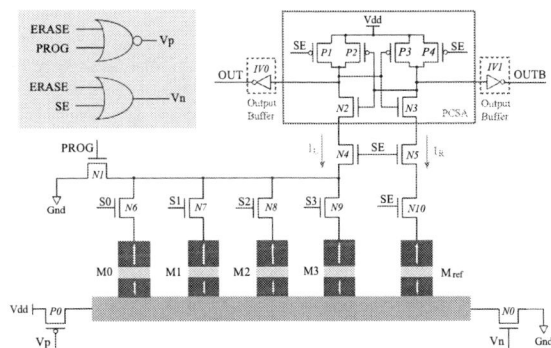

Fig. 2. Schematic of the proposed multi-bit NAND-SPIN NVFF. Pre-charge sense amplifier (PCSA) is used to read 1-bit data stored in MTJs. The write circuit is composed of four transistors, where one NMOS transistor (N0) is also the discharge transistor during the sensing operation.

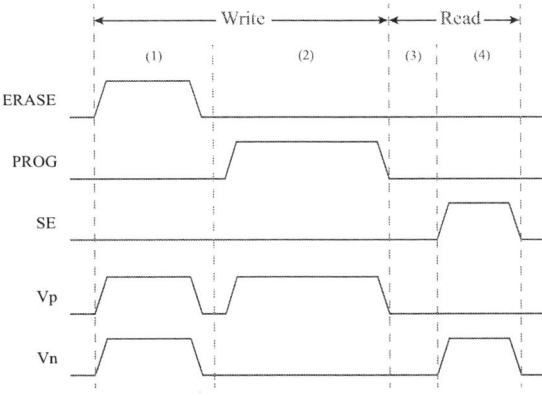

Fig. 3. Read/Write timing diagram of the multi-bit NAND-SPIN NVFF. (1)-(4) represent four operation phases, i.e., erase, program, pre-charge and discharge phases, respectively.

978-1-5386-4757-8/18 $31.00 © 2018 IEEE

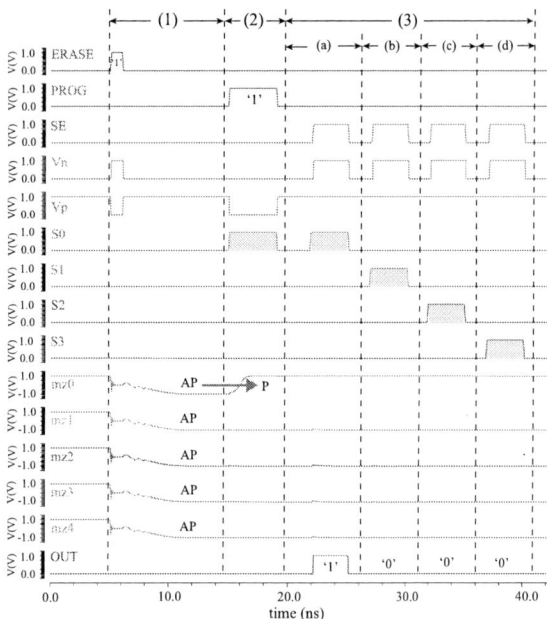

Fig. 4. Transient simulation of the multi-bit NAND-SPIN NVFF. (1), (2) and (3) represent the erase, program and read operations, respectively. (a)-(d) represent the read phases of four storage MTJs. m_z is the z-component magnetization of MTJ.

selection signals S_0-S_3, thus realize switching among four contexts. Write circuit part is composed of a PMOS transistor (P_0) and two NMOS transistors (N_0 and N_1), which control the direction of erase and program currents. Different from the FFs which use two MTJs that are always in complementary states to store 1-bit data, this structure needs less write power consumption because only one MTJ is switched. Moreover, the overall circuit area can be largely reduced because several MTJs share the same read and write circuits.

Fig. 3 illustrates the basic timing diagram of the multi-bit NAND-SPIN NVFF including read and write operations.

- The write operation includes two phases, i.e., erase phase (period (1) in in Fig. 3) and program phase (period (2) in in Fig. 3). During the erase phase, transistors P_0 and N_0 are open by forcing signal ERASE to high voltage. A current I_{erase} passes through the heavy metal strip and generates SOT to switch all MTJs to AP state. During the program phase, signal PROG is forced to high voltage. NMOS transistor N_0 is deactivated while N_1 is activated such that a current $I_{program}$ passes through the selected MTJ and switches it to P state. It should be noted that signal SE is always set at low voltage. Two transistors N_4 and N_5 are closed to separate the write circuit from the read one.

- The read operation is composed of two phases. During the pre-charge phase (period (3) in in Fig. 3), SE is set at low voltage. Outputs OUT and OUTB are pulled down to the ground through two inverters IV_0 and IV_1. During the discharge phase (period (4) in in Fig. 3), SE is at high voltage and the MTJs are connected to the pre-charge sense

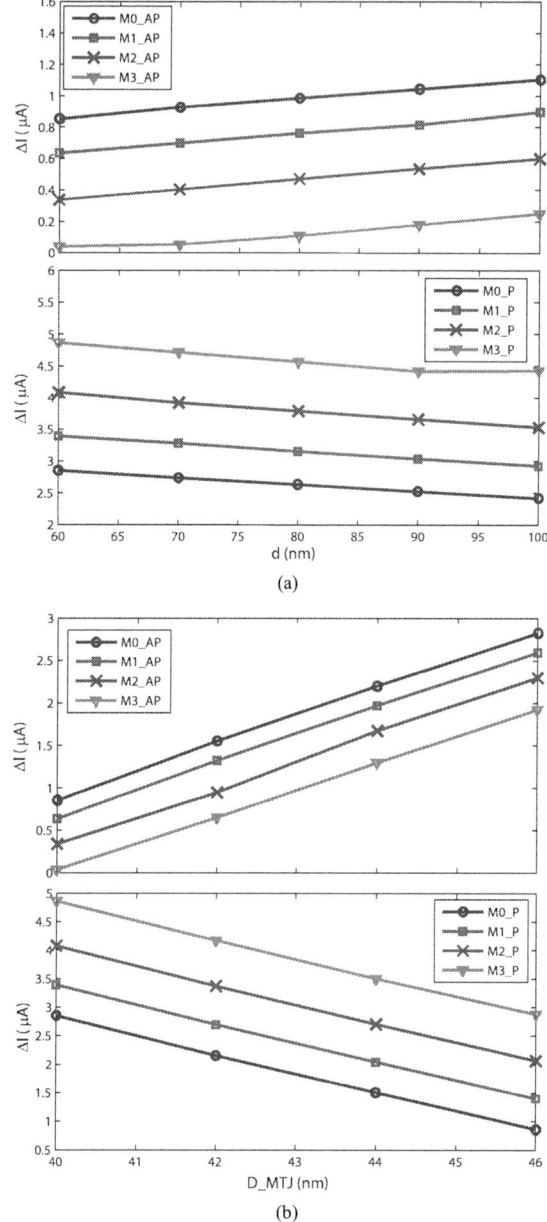

Fig. 5. Sense marge with respect to (a) the distance between the reference MTJ (M_{ref}) and the closest storage MTJ (M_3) (b) the size of M_{ref}

amplifier. P_0 and N_1 are both deactivated whereas N_0 is activated. In this operation, N_0 acts as a discharge transistor, enabling read current passing through the MTJs to the ground. Due to the resistance difference between two branches, I_L and I_R are different. The branch with lower resistance will be pulled down to the ground while the other branch will be pulled up to V_{dd} by the PMOS transistors, P_2 or P_3.

IV. FUNCTIONAL SIMULATION AND DISCUSSION

A. Transient simulation and performance analysis

Transient simulation is performed by using the CMOS 40 nm design kit and a spice-compatible NAND-SPIN MTJ model. The transistor width of write circuit (P_0, N_0 and N_1) is 480 nm while that of other transistors is at minimum size (120 nm) for high-speed switching. The distance between two MTJs is 60 nm. Other simulation parameters are listed in Table I.

Fig. 4 shows the timing diagram of the proposed multi-bit NAND-SPIN NVFF, which can be divided into three parts: 1) In the erase phase, signal ERASE is high. All the MTJs including four storage MTJs and one reference MTJ are erased to AP state by SOT current with a ultra-high speed (~0.5 ns). 2) In the program phase, S_0 = '1', only one MTJ cell (M_0) is selected to be written by opening the transistor N_6. After a delay of 1.6 ns, M_0 is switched from AP to P state whereas the state of other MTJs (M_{1-3} and M_{ref}) remain unchanged. 3) Data stored in the storage MTJs are sequential read in the third phase. The expected results, OUT = '1' when reading the first data stored in M_0 and OUT = '0' when reading other data stored in M_1-M_3, are obtained. The read time is lower than 200 ps, leading to high frequency operation. It should be noted that the read current passing through the heavy metal is too small to disturb the magnetization of the FL.

B. Reliability optimization

Simulation results show that there are more sensing errors when the storage MTJ is at AP state due to smaller read current margin (ΔI) between two branches. In this part, we propose several optimization methods to improve the reliability of the multi-bit NAND-SPIN NVFF. Fig. 5(a) shows that longer distance (d) between M_{ref} the M_3 leads to larger ΔI when MTJ is at AP state, because of an additional resistance added to the left branch. Fig. 5(b) shows that larger ΔI can be obtained by increasing the M_{ref} size (D_MTJ). The resistance of M_{ref} is decreased, thus the resistance difference between two branches is increased when MTJ is at AP state. We can also optimize the reliability by increasing the transistor size or the tunnel magnetoresistance (TMR) ratio which were described in the previous work [23].

V. CONCLUSION

We proposed a multi-bit NVFF based on NAND-like spintronics memory, which exploited SHE for fast and low-power switching. This structure showed fast erasing (~0.5 ns), programming (~1.6 ns) and sensing (<200 ps) operations. The asymmetric switching of conventional STT writing mechanism can be overcome since only switching from AP to P state was performed, which also save switching energy and simplify the write circuit. Moreover, multiple contexts share the same read and write circuits, resulting in high density and fast switching.

REFERENCES

[1] N. S. Kim et al., "Leakage current: Moore's law meets the static power," Computer, vol. 36, no.12, pp. 68-75, Dec. 2003.

[2] A. Lee et al., "A ReRAM-Based Nonvolatile Flip-Flop With Self-Write-Termination Scheme for Frequent-OFF Fast-Wake-Up Nonvolatile Processors," IEEE J. Solid-State Circuits, vol. 52, no. 8, pp. 2194-2207, Aug. 2017.

[3] J. Zheng, Z. Zeng and Y. Zhu, "Memristor-based nonvolatile synchronous flip-flop circuits," in Proc. International Conference on Information Science and Technology, 2017, pp. 504-508.

[4] W. Kang, W. Lv, Y. Zhang and W. Zhao, "Low Store Power, High Speed, High Density, Nonvolatile SRAM Design with Spin Hall Effect-Driven Magnetic Tunnel Junctions," IEEE Trans. Nanotechn., vol. 16, no. 1, pp. 148-154, Dec. 2016.

[5] M. Wang et al., "Current-induced magnetization switching in atom-thick tungsten engineered perpendicular magnetic tunnel junctions with large tunnel magnetoresistance," Nature Comm., 2018.

[6] S. Peng et al, "Giant interfacial perpendicular magnetic anisotropy in MgO/CoFe/capping layer structure," Appl. Phys. Lett., vol. 110, no. 7, 2017, doi: 10.1063/1.4976517.

[7] J. M. Portal et al., "An Overview of Non-Volatile Flip-Flops Based on Emerging Memory Technologies," J. Electron. Sci. Technol., vol. 12, no. 2, pp. 173-181, Jun. 2014.

[8] W. Kang, L. Zhang, J. O. Klein, Y. Zhang, D. R. Ravolosona and W. Zhao, "Reconfigurable Codesign of STT-MRAM under Process Variations in Deeply Scaled Technology," IEEE Trans. Electron Devices, vol. 62, no. 6, pp. 1769-1777, Mar. 2015.

[9] Z. Wang et al., "High-Density NAND-Like Spin Transfer Torque Memory With Spin Orbit Torque Erase Operation," IEEE Electron Device Lett., vol. 39, no. 3, pp. 343-346, Mar. 2018.

[10] C. J. Lin et al., "45nm low power CMOS logic compatible embedded STT MRAM utilizing a reverse-connection 1T/1MTJ cell," in Proc. IEEE Int. Electron Dev. Meeting, 2009, pp. 11.6.1–11.6.4.

[11] E. Deng, Z. Wang, J. O. Klein, G. Prenat, B. Dieny and W. Zhao, "High-Frequency Low-Power Magnetic Full-Adder Based on Magnetic Tunnel Junction With Spin-Hall Assistance," IEEE Trans. Magn., vol. 51, no. 11, pp. 1-4, Nov. 2015.

[12] Z. Wang et al., "Perpendicular-anisotropy magnetic tunnel junction switched by spin-Hall-assisted spin-transfer torque," J. Phys. D: Appl. Phys., vol. 48, no. 6, 2015.

[13] L. Liu et al., "Spin-Torque Switching with the Giant Spin Hall Effect of Tantalum," Science, vol. 336, pp. 555–558, May 2012.

[14] C. -F. Pai et al., "Spin transfer torque devices utilizing the giant spin Hall effect of tungsten," Appl. Phys. Lett., vol. 101, no. 12, Sept. 2012.

[15] A. van den Brink, S. Cosemans, S. Cornelissen, et al., "Spin-Hall-assisted magnetic random access memory," Appl. Phys. Lett., vol. 104, no. 1, pp. 012403, Jan. 2014.

[16] L. Liu, C. F. Pai, Y. Li, H. W. Tseng, D. C. Ralph and R. A. Buhrman, "Spin-torque switching with the giant spin Hall effect of tantalum," Science, vol. 336, no. 6081, pp. 555-558, May 2012.

[17] M. Yamanouchi et al., "Three terminal magnetic tunnel junction utilizing the spin Hall effect of iridium-doped copper," Appl. Phys. Lett., vol. 102, no. 21, pp. 212408, 2013.

[18] E. Eken et al., "Spin-hall assisted STT-RAM design and discussion," in Proc. ACM/IEEE International Workshop on System Level Interconnect Prediction, 2016, pp. 1-4.

[19] A. van den Brink, S. Cosemans, S. Cornelissen, et al., "Spin-Hall-assisted magnetic random access memory," Appl. Phys. Lett., vol. 104, no. 1, pp. 012403, Jan. 2014.

[20] J. E. Hirsch, "Spin Hall Effect," Phys. Rev. Lett., vol. 83, no. 9, pp. 1834–1837, Aug. 1999.

[21] L. Chang et al., "PRESCOTT: Preset-based cross-point architecture for spin-orbit-torque magnetic random access memory," in Proc. IEEE/ACM International Conference on Computer-Aided Design (ICCAD), 2017, pp. 245-252.

[22] W. Zhao, C. Chappert, V. Javerliac and J. -P. Noziere, "High speed, high stability and low power sensing amplifier for MTJ/CMOS hybrid logic circuits," IEEE Trans. Magn., vol. 45, no. 10, pp. 3784–3787, Oct. 2009.

[23] E. Deng, W. Kang, Y. Zhang, J. O. Klein, C. Chappert and W. Zhao, "Design Optimization and Analysis of Multicontext STT-MTJ/CMOS Logic Circuits," IEEE Trans. Nanotechn., vol. 14, no. 1, pp. 169-177, Jan. 2015.

978-1-5386-4757-8/18 $31.00 © 2018 IEEE

From Spintronic Devices to Hybrid CMOS/Magnetic System On Chip

Sophiane Senni, Frederic Ouattara, Jad Modad, Kaan Sevin, Guillaume Patrigeon,
Pascal Benoit, Pascal Nouet, Lionel Torres
LIRMM, University of Montpellier, CNRS, Montpellier, France
Email: firstname.lastname@lirmm.fr
François Duhem, Gregory Di Pendina, Guillaume Prenat
Univ. Grenoble Alpes, CEA, CNRS, INAC-SPINTEC, F-38000 Grenoble, France
Email: firstname.lastname@cea.fr

Abstract—"Beyond CMOS" is today one of the major research directions in semiconductor industries to address current integrated circuit issues. Many alternative technologies are currently under investigation to deal with the scaling limits of CMOS technology. This paper presents the design of a full system on chip based on a hybrid CMOS/Magnetic process. Spin-transfer-torque magnetic tunnel junctions are used to design different functions such as logic, memory, security and analog IP blocks.

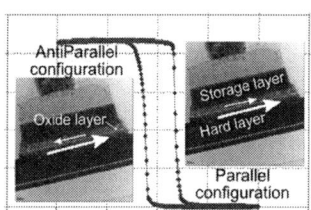

Fig. 1. Magnetic tunnel junction

I. INTRODUCTION

The number of connected devices around the world is growing fast with a forecast of 30 billion connected objects by 2020 [1]. The key requirements are ultra-low power, high processing capabilities, fast/dense storage, wireless communication, heterogeneous integration, and autonomy. Although it has been serving the industry well for several decades, CMOS technology has more and more obstacles to continue scaling into the future [2]. The main issue of the scaling limit of CMOS is the energy efficiency which is essential for battery-powered smart systems. New technology directions are being explored to continue providing denser, cheaper, faster and low power integrated circuits. This paper focuses on the spintronic device option. Unlike CMOS, they have the benefit of being non-volatile. This non-volatility can be integrated inside the logic devices to enable new computing paradigms and provide better energy efficiency. In this work, 200-nm spin-transfer-torque magnetic tunnel junctions (STT-MTJ) are used as elementary blocks in addition to 180-nm CMOS transistors to design a full system on chip (SoC) including multiple functions such as logic, memory, security and analog intellectual property (IP) blocks. The rest of the paper is organized as follows: Section II presents the STT-MTJ device and different STT-MTJ based IP blocks. Section III describes in details the architecture of the hybrid CMOS/magnetic SoC designed in this work. Section IV discusses the considered application scenarios for power analysis. Conclusions are finally given in Section V.

II. STT-MTJ DEVICE AND RELATED IP BLOCKS

A full design flow has been established for hybrid CMOS/Magnetic circuits based on STT-MTJ devices. On the one hand, physical and behavioral compact models have been created for circuit simulators such as SPICE. On the other hand, a process design kit (PDK) compatible with the standard design tools has also been developed to design IP blocks and chip-level systems based on a CMOS/Magnetic process. This section gives the basics of the STT-MTJ device and the related compact models before presenting different STT-MTJ based IP blocks used in the design of the SoC.

A. STT-MTJ device

1) Basics: Magnetic tunnel junctions (MTJ) are nanostructures composed of two ferromagnetic layers separated by a thin insulator (Figure 1). The resistance of the stack depends on the relative orientation of the two layers. A parallel (anti-parallel) state of the MTJ causes a low (high) resistance value which can be characterized as a logic zero (one). The tunnel magneto resistance (TMR) ratio gives the relative variation of resistance between the two states, which is typically around 150% to 200%. While the magnetization of one of the two layers is pinned to one direction (reference layer), the magnetic orientation of the other one (storage layer) can be flipped by applying a sufficient current through the MTJ, with a hysteretic behavior. The magnetization switching is driven by the spin-transfer-torque (STT) effect [3] where the current gets spin polarized by the reference layer and transfers the magnetic moment to the storage layer. Reading the magnetic state consists in measuring the resistance of the device.

978-1-5386-4757-8/18 $31.00 © 2018 IEEE

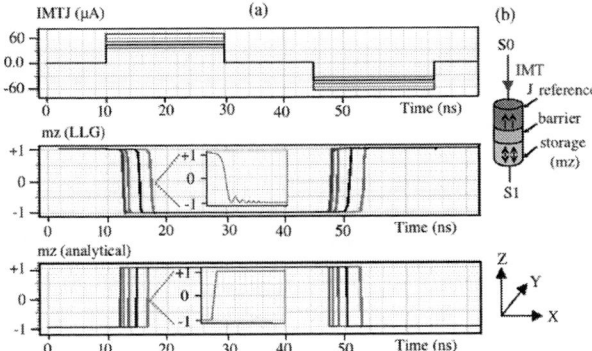

Fig. 2. Magnetization switching simulation with physical (middle) and behavioral (bottom) compact models

2) *Compact models:* Two main approaches have been considered to model the magnetization switching of the MTJ : physical [4] and behavioral [5]. The former uses the Landau-Liftshitz-Gilbert equation which gives an accurate dynamic evolution of the magnetic state driven by the STT effect, with a precessional behavior. However, this model is relatively slow and not adapted to the simulation of complex circuits such as memory arrays. The latter does not model the magnetization dynamics and needs a calibration of the parameters of the MTJ to fit the actual behavior of the device. As a result, this model allows much faster simulations. Regarding the resistance of the MTJ, both models take into account the dependence of the resistance upon the magnetic state and the polarization voltage, the dependence of the transport and magnetic parameters upon the temperature and the heating of the MTJ due to Joule effect.

Figure 2 shows the magnetization switching behavior of the STT-MTJ using the physical and the behavioral compact models.

B. STT-MTJ based IP blocks

1) Non-volatile Flip-Flop: Thanks to the PDK developed in this work for CMOS/Magnetic design, a non-volatile flip-flop (NVFF) is proposed as an extra standard cell (Figure 3) to give the possibility to make more complex circuits (e.g. processor) non-volatile. The NVFF is based on a standard CMOS flip-flop with additional MTJs together with their read/write circuits. The layout of the NVFF is shown in Figure 4. This cell has been intensively characterized for timing and power through electrical simulations including worst case corners and Monte-Carlo analysis. Furthermore, all necessary files are available to use the NVFF in each step of the standard design flow including synthesis, digital simulations and physical implementation.

2) STT-MRAM: The STT-MTJ device has been used as a memory cell to design a full magnetic random access memory (MRAM). A memory compiler approach was used to easily provide memories of different sizes (up to 16kB) according to the specifications. Figure 5 and 6 respectively show the memory architecture of a 16kB STT-MRAM and

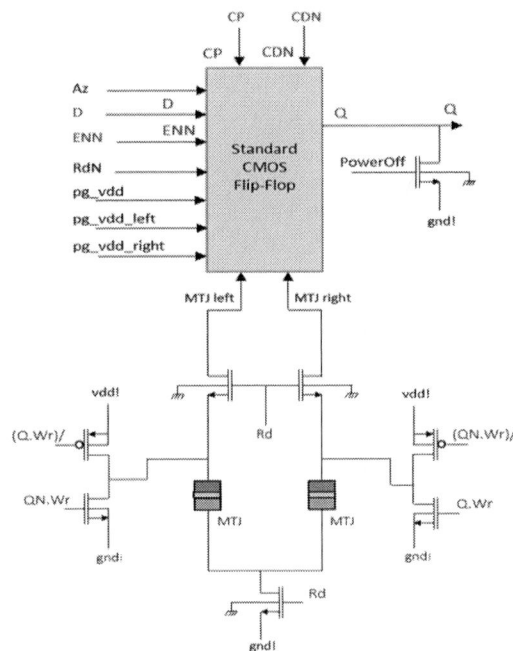

Fig. 3. STT-MTJ based non-volatile flip-flop architecture

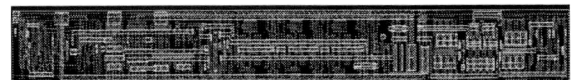

Fig. 4. Layout of the Non-Volatile Flip-Flop

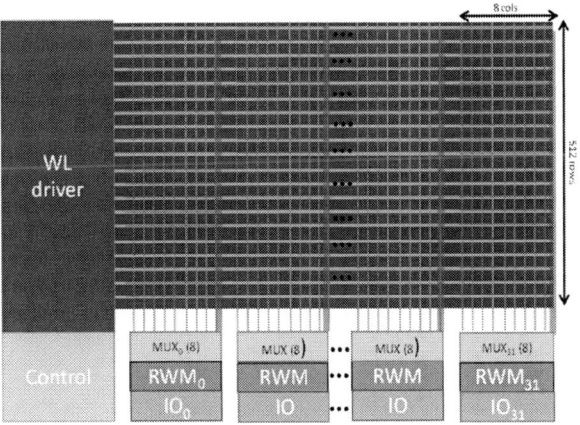

Fig. 5. STT-MRAM architecture

the corresponding layout. This memory is designed with a synchronous single-port SRAM-like interface and uses a 32-bit word organization. Electrical simulations were also performed to characterize this memory circuit for timing and power in order to use it in the digital design flow.

3) True Random Number Generator: This work includes the design of a bulding block for security applications. The

978-1-5386-4757-8/18 $31.00 © 2018 IEEE   189

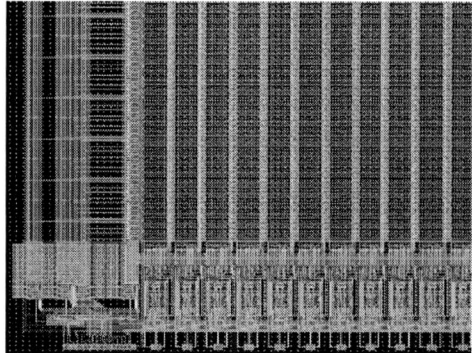

Fig. 6. STT-MRAM layout

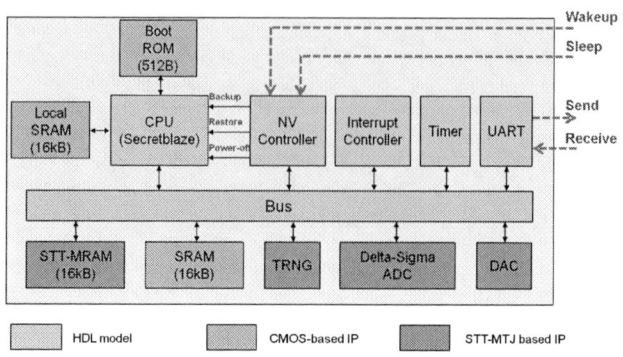

Fig. 7. SoC architecture

developed IP is a true random number generator (TRNG) based on STT-MTJ to generate random cryptographic keys for high secure data encryption. TRNGs use non-deterministic physical events to generate random numbers with a very high entropy and zero correlation. This STT-MTJ based TRNG uses the stochastic nature of the STT switching as a source of randomness to generate 1-bit random numbers by applying a write current with 50% of success rate.

4) Analog-to-Digital and Digital-to-Analog Converters: Two additional IP blocks have been considered for processing analog signals. On the one hand, a 1-bit delta-sigma analog function is proposed where the active bridge is used in a feedback structure to convert a relative variation of resistance into a digital word [6]. A frequency divider, whose configuration is stored in STT-MTJs, is implemented to control the sampling rate and therefore the signal bandwidth. On the other hand, a current output digital-to-analog converter (DAC) has been designed. The design has a segmented architecture with a R2R Ladder and a thermometer topology using STT-MTJs as resistors.

III. Hybrid CMOS/Magnetic system on chip

A. Architecture

Based on the building blocks presented in Section II-B, a full SoC has been designed to evaluate the co-integration of digital and analog IP blocks using STT-MTJ devices. The overall architecture is depicted in Figure 7. The design corresponds to a microcontroller including a processor core, memories and peripheral devices.

1) Non-volatile processor: The microcontroller is based on the Secretblaze [7], a 32-bit 5-stage pipeline RISC processor compatible with Xilinx's MicroBlaze instruction set architecture. The Wishbone bus interface is used to allow the connection of external memories and peripheral devices.

The Secretblaze core has been made fully non-volatile by using the non-volatile flip-flop (NVFF) presented in Section II-B1. The hybrid architecture of the NVFF gives the possibility to backup/restore the processor state into/from the STT-MTJs while the CMOS part is used during normal operation for high performance. This non-volatility inside the microarchitecture is a valuable benefit for energy efficiency

since aggressive low-power mode is possible with near zero leakage power. In addition, the wakeup time is significantly reduced as the system state is preserved after a shutdown.

2) Memory organization: The memory subsystem consists of a 512B read-only memory (ROM) for the boot process and three 16kB random access memories (RAM) to store instructions and data. Several RAMs have been implemented for comparison purposes. During normal operation, only one RAM is actually used to execute the application. On the one hand, a local SRAM directly connected to the processor is available with a dual-port interface to allow simultaneous read/write operations. On the other hand, a SRAM and the STT-MRAM presented in Section II-B2 are accessible via the Wishbone bus interface. The objective is to compare the SRAM-based microcontroller with the STT-MRAM based microcontroller on the same platform.

3) Peripheral devices: A set of peripherals has been integrated into the SoC including digital IP cores (interrupt controller, timer, UART) and the analog IP blocks presented in Section II-B3 and II-B4 (TRNG, Delta-Sigma ADC, DAC).

Moreover, a specific controller ("*NV Controller*" in Figure 7) has been developed to manage the backup/recovery of the system state. Before entering into sleep mode, the *NV controller* backs up all the NVFFs of the processor. To avoid electrical integrity issues due to the high write current during this operation, the NVFFs have been arranged in several clusters. Thus, a sequential backup is performed. To wake up the system, the NVFFs are also restored in a sequential way.

B. Operation

After the power is switched on, a normal boot consists in loading the binary code of the application via the UART interface and copying it into the local SRAM, or the SRAM or the STT-MRAM. Then, the processor fetches the first instruction to be executed. Alternatively, the system can restart from a known state in the case of returning from sleep mode.

Regarding the power modes management, the *NV controller* is called for placing the system in sleep mode either by software or by an external signal ("*Sleep*" in Figure 7). Returning to active mode is possible either by an external signal ("*Wakeup*" in Figure 7) or further to an event from the interrupt

978-1-5386-4757-8/18 $31.00 © 2018 IEEE

controller. Above scenarios have been validated through post-layout simulations considering several applications.

C. Layout and features

A full layout of the SoC is shown in Figure 8 and some features are given in Table I. This non-volatile microcontroller can run at $20MHz$ with a backup/recovery time of about $4\mu s$. The energy costs are respectively $437nJ$ and $98.7nJ$. It is worth noting that the current implementation of the SoC aims at giving a proof of concept instead of an optimized design. For instance, this work has made all the flip-flops of the processor non-volatile while this is not mandatory to preserve the system state after a shutdown. Thus, optimizations are still possible for the backup/recovery time and energy. Moreover, considering a more advanced technology node would further improve the overall performances.

IV. APPLICATION SCENARIOS

Typical application scenarios of the internet of things (IoT) are under development to analyze the power consumption of the SoC. IoT objects essentially do three actions: sensing, processing, sending. The system first captures information from the external environment thanks to a sensor followed by an analog to digital conversion. Second, the measured data are stored in memory and minor or more complex computations

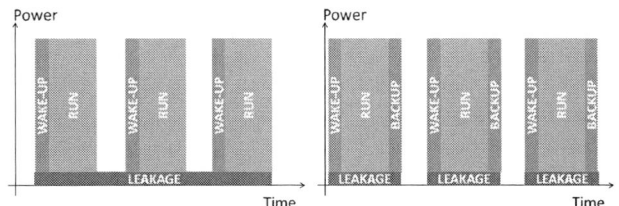

Fig. 9. Current solution (left) vs normally-off computing (right)

are performed. Finally, data are sent to a data center wirelessly. Once the wireless transmission ends, the system can go into sleep mode and wake-up later to restart the overall process.

The objective is to demonstrate how the normally-off computing (Figure 9) will allow better energy efficiency for a set of a real applications. Preliminary experiments show that staying in sleep mode for several milliseconds is sufficient to optimize the power consumption.

V. CONCLUSION

This paper presented the design of a full system on chip based on a hybrid CMOS/Magnetic process, from the STT-MTJ device to the design of the entire chip by way of STT-MTJ based IP blocks. A non-volatile microcontroller has been developped to enable new computing paradigms and provide better energy efficiency for beyond CMOS circuits. Future work will consider more advanced technology nodes to evaluate the potential of spintronic technologies.

ACKNOWLEDGMENT

This work has received funding from the European Union's Horizon 2020 research and innovation programme under grant agreement No 687973 - GREAT (heteroGeneous integRated magnetic tEchnology using multifunctional standardized sTack (MSS)).

REFERENCES

[1] B. Ahlgren, M. Hidell, and E. C.-H. Ngai, "Internet of things for smart cities: Interoperability and open data," *IEEE Internet Computing*, no. 6, pp. 52–56, 2016.

[2] M. T. Bohr and I. A. Young, "Cmos scaling trends and beyond," *IEEE Micro*, vol. 37, no. 6, pp. 20–29, 2017.

[3] G. Fuchs, N. Emley, I. Krivorotov, P. Braganca, E. Ryan, S. Kiselev, J. Sankey, D. Ralph, R. Buhrman, and J. Katine, "Spin-transfer effects in nanoscale magnetic tunnel junctions," *Applied Physics Letters*, vol. 85, no. 7, pp. 1205–1207, 2004.

[4] W. Guo, G. Prenat, V. Javerliac, M. El Baraji, N. De Mestier, C. Baraduc, and B. Dieny, "Spice modelling of magnetic tunnel junctions written by spin-transfer torque," *Journal of Physics D: Applied Physics*, vol. 43, no. 21, p. 215001, 2010.

[5] K. Jabeur, F. Bernard-Granger, G. Di Pendina, G. Prenat, and B. Dieny, "Comparison of verilog-a compact modelling strategies for spintronic devices," *Electronics letters*, vol. 50, no. 19, pp. 1353–1355, 2014.

[6] S. Hacine, T. El Khach, F. Mailly, L. Latorre, and P. Nouet, "A micropower high-resolution $\sigma\delta$ cmos temperature sensor," in *Sensors, 2011 IEEE*. IEEE, 2011, pp. 1530–1533.

[7] L. Barthe, L. V. Cargnini, P. Benoit, and L. Torres, "The secretblaze: A configurable and cost-effective open-source soft-core processor," in *Parallel and Distributed Processing Workshops and Phd Forum (IPDPSW), 2011 IEEE International Symposium on*. IEEE, 2011, pp. 310–313.

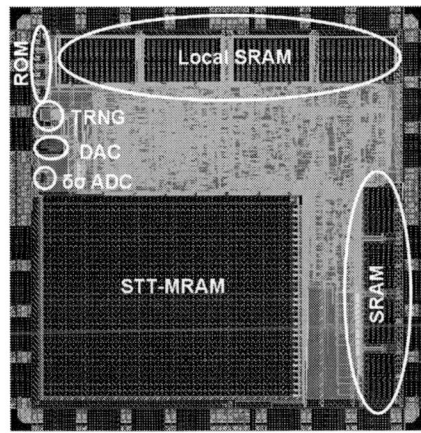

Fig. 8. SoC layout

TABLE I
SoC FEATURES

Die Area	$23mm^2$
Process	$180nm$ CMOS / $200nm$ STT-MTJ
Supply Voltage	1.8V Core / 3.3V IO
Frequency	$20MHz$
Number of NVFFs	2126
Backup time	$4.15\mu s$
Recovery time	$4.15\mu s$
Backup energy	$437nJ$
Recovery energy	$98.7nJ$

Reliable ReRAM-based Logic Operations for Computing in Memory

Mathieu Moreau[1], Eloi Muhr[1], Marc Bocquet[1], Hassen Aziza[1], Jean-Michel Portal[1]
Bastien Giraud[2], Jean-Philippe Noël[2]

[1]Aix-Marseille Univ., Université de Toulon, CNRS, IM2NP, Marseille, France
[2] CEA-Leti, Univ. Grenoble Alpes, Grenoble, France
mathieu.moreau@univ-amu.fr

Abstract — The development of non-conventional Von-Neumann architectures becomes essential for breakthrough computing in Internet of Things (IoT) devices. The main objective for IoT application is to lower as much as possible the power consumption to promote autonomy. The key to solve this challenge is to reduce the data transfer between memory and computing unit. As emerging non-volatile memories and especially resistive switching technologies (ReRAM) can today be co-integrated with CMOS on hybrid process, we propose in this paper to develop bitwise logic operations inside and close to the memory array. Using two transistors – one ReRAM (2T1R) memory cell architecture with differential approach to enhanced read reliability, we can perform logic operations without impacting the global memory architecture. Thanks to parallel data sensing, the structure enables fast computation of any bitwise logic operations (ID, AND, OR, XOR in their natural or complementary form) with high reliability, promoting the computing in memory (CiM) concept.

Keywords — ReRAM, memory array, Computing in Memory (CiM), Processing in Memory (PiM), Logic in Memory (LiM)

I. INTRODUCTION

Data exchange between core and memory has been identified as the major bottleneck to reduce power consumption and enhance performances in Internet of Things (IoT) applications based on embedded memory. To overcome this limitation, the computing in memory concept (CiM), also known as processing in memory (PiM) or logic in memory (LIM) concepts, has raised [1]. Some CiM/PiM/LiM architectures, based on resistive switching technology (ReRAM), have been proposed at architectural level like PLiM [2], MAGIC [3], Imply [4] and PRIME [5] taking their roots in works proposed at device level [6-8]. These architectural proposals tend to fully exploit the opportunity offered by the high-density crossbar architecture targeting High Performance Computing (HPC) applications. On the other hand, for IoT application and CiM/PiM/LiM concept assessment at circuit level, proposed studies rely on two transistors – one ReRAM (2T1R) architectures. As proposed in [9], reliability of the operations, especially during read process, is of high concern. In this paper, operations are performed thanks to comparison of the read current versus OR/AND reference current while considering cell variability.

In this context, the aim of our paper is to propose a ReRAM architecture to perform either bitwise logic operations or regular memory operation. Bitwise logic operations are based on a differential approach with self-referencing as proposed in [10] to improve read operation reliability. Moreover, the use of this read scheme allows to enrich the operations since ID, AND, OR, XOR operations

can be computed in their natural or complementary form, thanks to our differential read scheme.

The rest of the paper is organized as follows: section II introduces briefly the ReRAM technology of which the compact model used for simulation was validated. Section III focuses on the bit-cell and array description together with the enriched pre-charge sense amplifier architecture. Section IV presents the principle of operations and section V validated the concept through a XOR computation example with theoretical development and simulations. Finally, Section VI draws some concluding remarks.

II. ReRAM TECHNOLOGY AND COMPACT MODEL OVERVIEW

For this study, a ReRAM technology, based on HfO_2 is considered as a part of the elementary bit-cell. The ReRAM stack is composed of a 5 nm thick HfO_2 resistive switching layer embedded in-between a TiN/Ti Top Electrode (TE) and a TiN Bottom Electrode (BE) as shown in Fig. 1(a) [11].

ReRAM modeling is described in the work presented in [12]. This approach relies on electric field-induced creation/destruction of oxygen vacancies within the switching layer. The model enables continuous accounting for FORMING, SET and RESET operations into a single master equation in which the resistance is controlled by the radius of a conductive filament.

After model card extraction, the model satisfactorily matches quasi-static and dynamic experimental data measured on actual HfO2-based memory elements [11]. Moreover, to account for the variability of ReRAM technology, two corners cases were simulated. They include the two extreme behaviors observed experimentally: one favoring the SET mechanism and slowing the RESET, and the other being the exact opposite.

Fig. 1 shows the quasi-static behavior of ReRAM devices and the good modeling correlation. The corners encompass the full range of features that ensures to consider the worst cases of FORMING, SET and RESET.

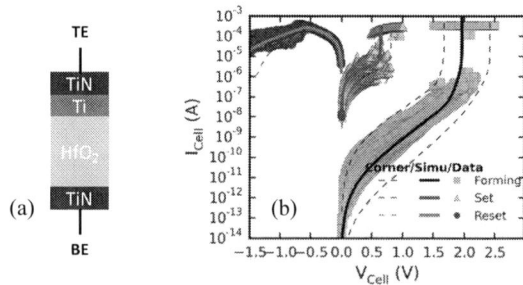

Fig. 1. (a) ReRAM stack (unrepresentative scale). (b) Experimental [11] and simulated (including corners definition) [12] I(V) characteristics for FORMING, SET, and RESET.

III. ARRAY ARCHITECTURE FOR LOGIC OPERATION

In this section, the elementary bit-cell and the array structure are presented. Based on differential data sensing approach, the array is built to perform several tasks including conventional memorization operations and bitwise logic functions. Doing so, the set of Logic operation can be computed in natural form (ID, AND, OR and XOR) or complementary form (NOT, NAND, NOR and XNOR).

A. 2T1R elementary cell

The elementary cell is composed of two NMOS transistors and one ReRAM to define a 2T1R bit-cell, as depicted Fig. 2. This structure is connected to five access lines: Set-Line (SL) linked to the top electrode (TE) of the ReRAM, the classical Bit-Line (BL) and Word-Line (WL) to address the ReRAM and BL' and WL' to program the ReRAM.

B. Array organization

The memory array organization is described in Fig. 2. The 2T1R elementary cells are connected by pair for a given column "i". Left cells of the column for a given row "j" store an X value [X being either High Resistive State (HRS) or Low Resistive State (LRS)] whereas right cells of the column for a given row "j" store the opposite state $\overline{X}$ ($\overline{X}$ being either LRS or HRS) in memory mode. It is important to note that the BL and $\overline{BL}$ (respectively BL', $\overline{BL'}$, SL and $\overline{SL}$) signals hold complementary values in normal memory operation. However, when performing Logic operation, programming operation of variable in the memory array may need to be non-complementary as it will be depicted in the next section, requiring a specific programming and reading signal activation.

As illustrated in Fig. 3, read operations are based on differential data sensing between BL_i and $\overline{BL_i}$ for standard memory or logic operation computation.

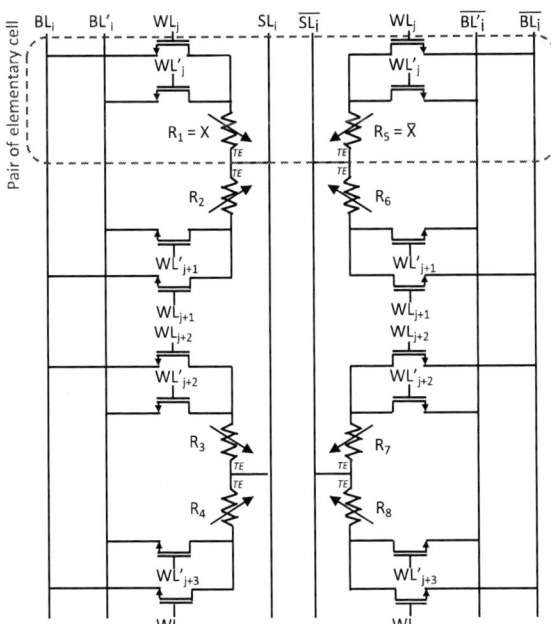

Fig. 2. Schematic of a given column (i) and 4 successive rows (j, j+1, j+2, j+3) as a basic element of the array organization highlighting the differential approach promoted to enhance read reliability.

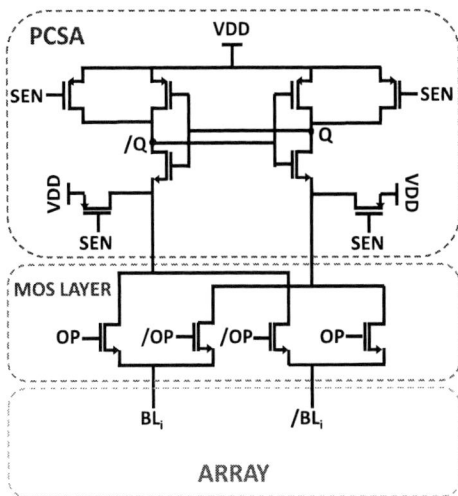

Fig. 3. Sensing scheme illustration based on PCSA [9] including a MOS layer to control data or function state (natural or complementary).

The sensing circuit is composed of two parts: a pre-charge sense amplifier (PCSA) which shows high performance in terms of reliability, power and speed [10] and a MOS layer to disconnect the array from the PCSA during programming and retention phases. The MOS layer also controls whether the data or logic functions are inverted or natural using a XOR scheme.

The following section shows how logical operations are performed inside the memory array.

IV. MEMORY ARRAY OPERATIONS

For programming operation, each memory cell is used in a classical way, as illustrated in Fig. 4. For the addressed elementary bit-cell (*WL'* is activated to turn on NMOS transistors), *SL* and *BL'* lines are used to perform the FORMING/SET operations (SET or FORMING voltage is applied to the *TE* and *BL'* is grounded). As a consequence, the ReRAM switches to a low resistive state (LRS), which represents a logic '1'. For RESET operation, *SL* is grounded and the reset voltage is applied on *BL'* to ensure a negative bias between *TE* and *BE* and get a high resistive state (HRS), which represents a logic '0'.

It is worth to note that complementary values are programmed in pair of elementary cells (Fig. 2) in normal memory mode of operation, and read through a classical differential approach for enhance the reliability, inhibiting the MOS layer (*OP* = '1' in Fig. 3 for ID function). However, an alternative mode is to use *OP* = '0' to perform the NOT logical operations.

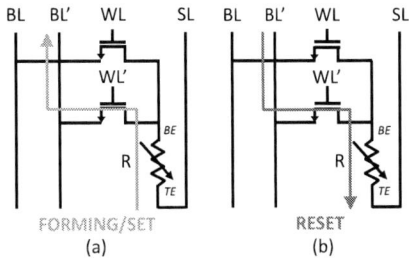

Fig. 4. 2T1R cell schematic with different paths to perform (a) FORMING/SET and (b) RESET operations.

978-1-5386-4757-8/18 $31.00 © 2018 IEEE   193

TABLE I. ReRAM STATE FOR AND, OR AND XOR LOGICAL OPERATIONS

Logical functions	ReRAM states							
	R_1	R_2	R_3	R_4	R_5	R_6	R_7	R_8
OR	$\overline{A}$	$\overline{B}$	HRS	HRS	A	LRS	B	LRS
AND	$\overline{A}$	LRS	$\overline{B}$	LRS	A	B	HRS	HRS
XOR	A	$\overline{A}$	$\overline{B}$	B	A	$\overline{A}$	B	$\overline{B}$

In contrast, as represented in Table I, considering four rows and the associated ReRAMs, resistance values may not be complementary in each pair of elementary cells depending on the operation to compute between two variables A and B. The computation operation in this mode is evaluated during the read process, here also using a differential read but with a specific biasing scheme. Indeed, SL and $\overline{SL}$ are in high impedance; the group of four rows, where the computation takes place, is activated simultaneously (WL_j, WL_{j+2}, WL'_{j+1} and WL'_{j+3} are biased to V_{DD}). The read process is then classically performed in two phases. During the first phases (SEN = '0'), BL and $\overline{BL}$ are both pre-charged to V_{DD} through the MOS layer whereas BL' and $\overline{BL'}$ are disconnected to avoid any extra consumption. During the second phase (SEN = '1'), BL and $\overline{BL}$ are discharged through BL' and $\overline{BL'}$ that are grounded. Doing so, the overall resistance path on BL and $\overline{BL}$ are compared to perform the computation in a reliable manner.

V. XOR OPERATION: THEORETICAL STUDY AND SIMULATION

Fig. 5 presents the simplified scheme of the schematic presented Fig. 2 for a XOR operation during the evaluation process (read phases). Transistors have been removed for sake of clarity since they are biased so as to minimize their contribution in the resistive path versus the ReRAM contribution. From this schematic, we can express the logical equation of both branches connected to $\overline{BL}$ (linked to $\overline{Q}$ output in the left branch, see Fig. 3) and $\overline{BL}$ (linked to Q output in the right branch, see Fig. 3) in the following way [(1) and (2)], which confirms from a Logic point of view the XOR operation:

$$\overline{Q}=A.(\overline{A}+B)+\overline{B}.(B+\overline{A})$$
$$\overline{Q}=A.\overline{A}+A.B+\overline{B}.B+\overline{B}.\overline{A}$$
$$\overline{Q}=A.B+\overline{B}.\overline{A} \tag{1}$$

$$Q=A.(\overline{A}+\overline{B})+B.(\overline{B}+\overline{A})$$
$$Q=A.\overline{A}+A.\overline{B}+\overline{B}.B+B.\overline{A}$$
$$Q=A.\overline{B}+B.\overline{A} \tag{2}$$

These equations are established by considering all path between BL and BL' (respectively $\overline{BL}$ and $\overline{BL'}$), where serial resistances are considered as AND and parallel resistances are considered as OR.

To assess the proper functionality of the sensing process, here also considering a XOR operation, the resistances of both branches (function of ReRAM resistance values: R_{HRS} and R_{LRS}), connected to BL ($\overline{Q}$ branch) and $\overline{BL}$ (Q branch), have to be considered depending on the associated A and B logical state. Theoretical calculation of the global resistance values in each branch for all A and B states is proposed in Table II, and considering a ratio n between R_{HRS} and R_{LRS} (R_{HRS} = $n \times R_{LRS}$) in Table III. These values are calculated following the equations below for the resistance in the branch connected to BL (3) and to $\overline{BL}$ (4):

$$R_{BL}= \frac{(R_A \times R_{\overline{B}})}{(R_A+R_{\overline{B}})} + \frac{(R_{\overline{A}} \times R_B)}{(R_{\overline{A}}+R_B)} \tag{3}$$

$$R_{\overline{BL}}= \frac{(R_A \times R_B)}{(R_A+R_B)} + \frac{(R_{\overline{A}} \times R_{\overline{B}})}{(R_{\overline{A}}+R_{\overline{B}})} \tag{4}$$

To ensure the right result of logic operations, the resistance conditions between both branches is indicated in Table II and III. For example, the result of XOR between A = '0' and B = '0' must equal to Q = '0' and $\overline{Q}$ = '1' which is the case if the resistance in the left branch (R_{BL}) is lower than the resistance in the right branch ($R_{\overline{BL}}$). From Table II and III, it appears that the resistance balance between BL and $\overline{BL}$ is always respected theoretically for a ratio $n > 1$. To validate this analysis, simulation of the read operation has been performed for a R_{LRS} value of 15 kΩ, evaluated from [9] (worst case LRS resistance value), and compared to theoretical value as depicted Fig. 6. From the simulation, it appears that considering an LRS value of 15 kΩ and depending on the read time, the ratio n for a successful read is just below 2. This result clearly shows an improvement versus single ended solution in term of read reliability, where ratio n > 10 is a classical minimal requirement.

TABLE II. BL AND $\overline{BL}$ RESISTANCE VALUES VERSUS A AND B STATE FOR XOR OPERATION.

XOR		Initial expressions		Conditions
A	B	R_{BL}	$R_{\overline{BL}}$	
0	0	$\frac{2R_{HRS}R_{LRS}}{R_{HRS}+R_{LRS}}$	$\frac{R_{HRS}}{2}+\frac{R_{LRS}}{2}$	$R_{BL} < R_{\overline{BL}}$
0	1	$\frac{R_{HRS}}{2}+\frac{R_{LRS}}{2}$	$\frac{2R_{HRS}R_{LRS}}{R_{HRS}+R_{LRS}}$	$R_{BL} > R_{\overline{BL}}$
1	0	$\frac{R_{HRS}}{2}+\frac{R_{LRS}}{2}$	$\frac{2R_{HRS}R_{LRS}}{R_{HRS}+R_{LRS}}$	$R_{BL} > R_{\overline{BL}}$
1	1	$\frac{2R_{HRS}R_{LRS}}{R_{HRS}+R_{LRS}}$	$\frac{R_{HRS}}{2}+\frac{R_{LRS}}{2}$	$R_{BL} < R_{\overline{BL}}$

TABLE III. CONDITIONS ON ARRAY RESISTANCE VALUES FOR XOR OPERATION CONSIDERING RESISTANCE RATIO

XOR		Simplified expressions (R_{HRS}/R_{LRS} = n)		Conditions
A	B	R_{BL}	$R_{\overline{BL}}$	
0	0	$\frac{2n}{n+1}R_{LRS}$	$0.5(n+1)R_{LRS}$	$R_{BL} < R_{\overline{BL}}$
0	1	$0.5(n+1)R_{LRS}$	$\frac{2n}{n+1}R_{LRS}$	$R_{BL} > R_{\overline{BL}}$
1	0	$0.5(n+1)R_{LRS}$	$\frac{2n}{n+1}R_{LRS}$	$R_{BL} > R_{\overline{BL}}$
1	1	$\frac{2n}{n+1}R_{LRS}$	$0.5(n+1)R_{LRS}$	$R_{BL} < R_{\overline{BL}}$

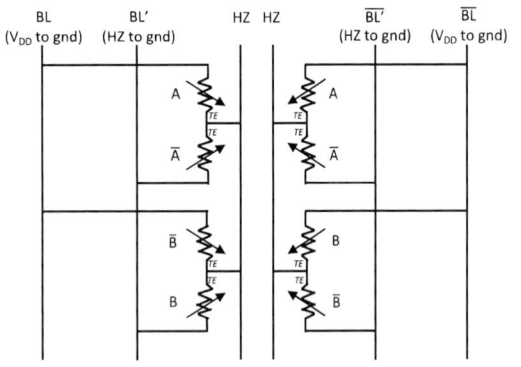

Fig. 5. Simplified scheme of the column schematic presented in Fig. 2 in the case of a XOR operation (two-stage read phase, e.g. BL = V_{DD} first and next is grounded), with the corresponding ReRAM values as identified in Table I.

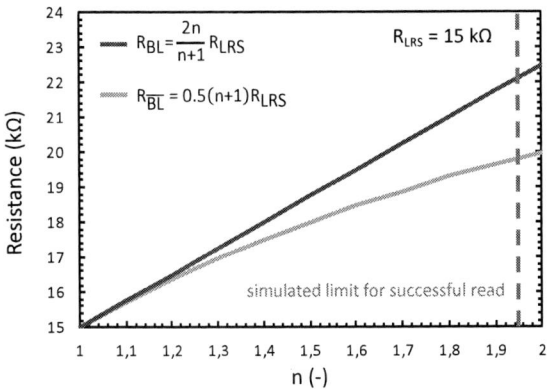

Fig. 6. Evolution of the resistance in *BL* (blue line) and $\overline{BL}$ (green line) versus the R_{HRS}/R_{LRS} ratio (n) with a R_{LRS} value set to 15 kΩ.

To fully validate our solution, the proposed array is designed with a 130nm CMOS design kit from STMicroelectronics associated with ReRAM technology from CEA-Leti. The full solution is simulated using our compact model (see section I) for the complete set of *A* and *B* value in XOR and XNOR configurations (depending on *OP* value). Only transient simulation results of read phases are reported Fig. 7. These simulation results fully validate our concept and allow extracting read time (510 ps for *Q* stabilization after *SEN* rises in the worst case). Programming scheme is classical thus performance evaluation is out of the scope of this paper and is not reported here.

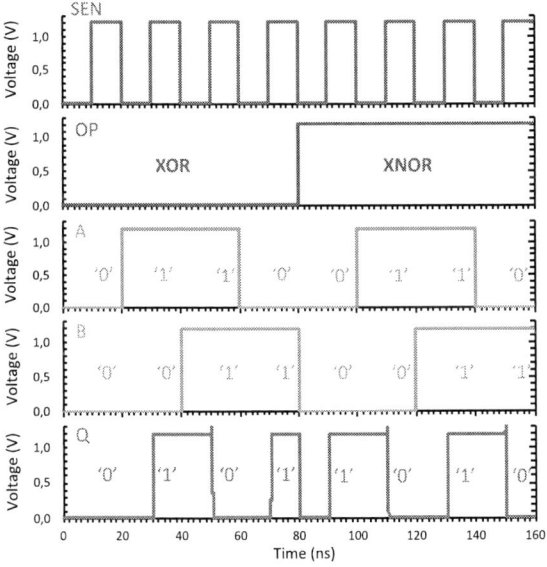

Fig. 7. Transient simulation for XOR computation using CMOS 130nm STMicroelectronics design kit and our ReRAM compact model during read phases.

VI. CONCLUSION

CiM/PiM/LiM concepts are seen as a major feature to reduce consumption in IoT chip by avoiding intense data transfer between memories and processing core. Moreover, ReRAM technology is fully compatible with CMOS process and offers a clear opportunity to deploy this concept, but still suffers from large variability. In this context, we propose a CiM approach that allows computing any logical operations (ID, AND, OR, XOR) and their complementary counterpart between two variables stored in a ReRAM array. Theoretical study as well as simulation results show that our solution is reliable for ratio above 2 between LRS and HRS resistance values compared to a minimal ratio of 10 for single ended solution.

ACKNOWLEDGMENT

This work was supported in part by the ANR funding agency through MACS inter-Carnot project (Carnot STAR and Leti).

REFERENCES

[1] H.-S. P. Wong and S. Salahuddin, "Memory leads the way to better computing", Nature Nanotechnology, vol. 10, pp. 191-194, March 2015.

[2] P. E. Gaillardon, L. Amaru, A. Siemon, E. Linn, R. Waiser, A. Chattopadhyay, and G. De Micheli, "The Programmable Logic-in-Memory (PLiM) computer," 2016 Design, Automation & Test in Europe Conference & Exhibition (DATE), Dresden, 2016, pp. 427-432.

[3] S. Kvatinsky, D. Belousov, S. Liman, G. Satat, N. Wald, E. G. Friedman, A. Kolodny, and U. C. Weiser, "Magicmemristor-aided logic," IEEE Transactions on Circuits and Systems II: Express Briefs (TCASII), vol. 61, pp. 895–899, November 2014.

[4] S. Kvatinsky, G. Satat, N. Wald, E. G. Friedman, A. Kolodny, and U. C. Weiser, "Memristor-based material implication (IMPLY) logic: design principles and methodologies," IEEE Transactions on Very Large Scale Integration Systems (TVLSI), vol. 22, pp. 2054–2066, October 2013.

[5] P. Chi, S. Li, C. Xu, T. Zhang, J. Zhao, Y. Liu, Y. Wang, and Y. Xie, "PRIME: a novel processing-in-memory architecture for neural network", 2016 International Symposium on Computer Architecture (ISCA), Seoul, 2016, pp. 27-39.

[6] C. Kügeler, R. Rosezin, E. Linn, R. Bruchhaus, and R. Waser, "Materials, technologies, and circuit concepts for nanocrossbar-based bipolar RRAM," Appl. Phys. A - Mater. Sci. Process., vol. 102, pp. 791-809, March 2011.

[7] E. Linn, R. Rosezin, C. Kügeler, and R. Waser, "Complementary resistive switches for passive nanocrossbar memories," Nature Materials, vol. 9, pp. 403-406, April 2010.

[8] J. Borghetti, G. S. Snider, P. J. Kuekes, J. J. Yang, D. R. Stewart, and R. S. Williams, "Memristive switches enable stateful logic operations via material implication," Nature, vol. 464, pp. 873-876, April 2010.

[9] W. H. Chen et al., "A 16Mb dual-mode ReRAM macro with sub-14ns computing-in-memory and memory functions enabled by self-write termination scheme," 2017 IEEE International Electron Devices Meeting (IEDM), San Francisco, CA, 2017, pp. 28.2.1-28.2.4.

[10] W. S. Zhao, M. Moreau, E. Deng, Y. Zhang, J.-M. Portal, J.-O. Klein, M. Bocquet, H. Aziza, D. Deleruyelle, C. Muller, D. Querlioz, N. Ben Romdhane, D. Ravelosona, and C. Chappert, "Synchronous Non-Volatile Logic Gate Design Based on Resistive Switching Memories," IEEE Transactions on Circuits and Systems I (TCASI), vol. 61, pp. 443–454, February 2014.

[11] E. Vianello, O. Thomas, G. Molas, O. Turkyilmaz, N. Jovanović, D. Garbin, G. Palma, M. Alayan, C. Nguyen, J. Coignus, B. Giraud, T. Benoist, M. Reyboz, A. Toffoli, C. Charpin, F. Clermidy and L. Perniola, "Resistive Memories for Ultra-Low-Power embedded computing design," 2014 IEEE International Electron Devices Meeting (IEDM), San Francisco, CA, pp. 6.3.1-6.3.4, 2014.

[12] M. Bocquet, D. Deleruyelle, H. Aziza, C. Muller, J.-M. Portal, T. Cabout and E. Jalaguier, "Robust compact model for bipolar oxide-based resistive switching memories," IEEE Transactions on Electron Devices (TED), vol. 61, pp. 674 - 681, March 2014.

An ultra-low power active diode using a hysteresis common gate comparator for low-voltage and low-power energy harvesting systems

Kaori Matsumoto, Tetsuya Hirose, Hiroki Asano, Yuto Tsuji,
Yuichiro Nakazawa, Nobutaka Kuroki, and Masahiro Numa
Department of Electrical and Electronic Engineering, Kobe University
1-1 Rokkodai, Nada, Kobe 657-8501, Japan
E-mail: matsumoto@eedept.kobe-u.ac.jp, hirose@eedept.kobe-u.ac.jp

Abstract—**This paper proposes an ultra-low power active diode using a hysteresis common gate comparator for low-voltage and low-power energy harvesting systems. The proposed active diode consists of a MOS switch and hysteresis common gate comparator, which eliminates unwanted ripple and noise voltages. The hysteresis comparator controls the MOS switch to turn ON or OFF, depending on the input and output voltages. The hysteresis voltages of the comparator can be controlled by the current flowing in the comparator. Simulation results demonstrated that the hysteresis comparator has a -27 and 25 mV hysteresis voltages and the active diode using the hysteresis comparator eliminates unwanted ripple voltage.**

I. Introduction

Energy harvesting has attracted attention to realize battery-less and maintenance-free IoT devices [1]–[12]. To realize such devices, highly efficient power management systems are strongly required because the output voltages of the small harvesters are basically weak and are easily lost depending on their power generation environment.

Fig. 1 shows a block diagram of a conventional power management system. The system consists of an energy harvester, voltage boost converter (VBC), series regulator, active diode, and supercapacitor. The VBC boosts the harvester's output voltage because it is too low to fully charge the supercapacitor (e.g., the output voltage of a single photovoltaic cell is around 0.5 V). The boosted voltage is regulated to an optimum voltage (e.g., 2.4 V) by the series regulator. The active diode transfers the charge to the supercapacitor with a quite low voltage difference [13]–[17]. When the output voltage of the energy harvester is lost, the active diode will detect the input voltage reduction and cut off the current path to prevent the reverse current.

To achieve highly efficient power management system, we have to develop the system with ultra-low power. As a VBC, switched-capacitor (SC) VBCs are widely used because it shows high efficiency in low current applications and can be integrated on a chip. The power dissipation of the series regulator and a comparator used in the active diode can be re-

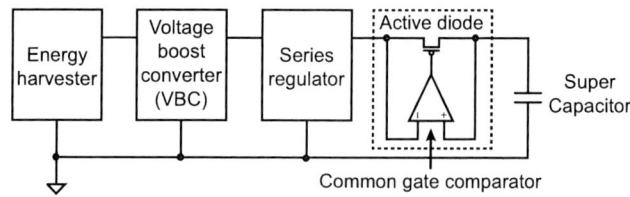

Fig. 1. Conventional power management system.

duced by using a nano-ampere bias current [18]–[22] and large feedback resistors. However, in some cases, the active diode cannot operate correctly because the comparator has a finite offset voltage and monitors two large time constant voltage nodes. In the worst case, both nodes decrease simultaneously without turning the pMOS switch off.

To solve the problem, we modified the power management system architecture and developed a hysteresis common-gate (CG) comparator and active diode using the CG comparator for low-voltage and low-power energy harvesting systems. In our proposed power management system, the hysteresis CG comparator monitors the output voltage of the VBC and stored voltage of the supercapacitor. The hysteresis CG comparator is required because the output voltage of the VBC has a certain ripple voltage. This paper is organized as follows: Section II briefly summarizes our proposed power management system. Section III explains our proposed hysteresis CG comparator. Section IV discusses the simulation results. Section V concludes the paper.

II. Proposed Power Management System

We modified the power management system architecture. Fig. 2 shows the modified system. The inverting terminal of the CG comparator is connected to the output of the VBC. The active diode can easily detect the power generation condition of the harvester because the output voltage of the VBC directly depends on the harvester's condition. However, the output voltage of the SC VBC has a ripple voltage. Fig. 3 shows waveforms of input and output voltages with and without hysteresis. The ripple voltage will oscillate the comparator's

978-1-5386-4757-8/18 $31.00 © 2018 IEEE

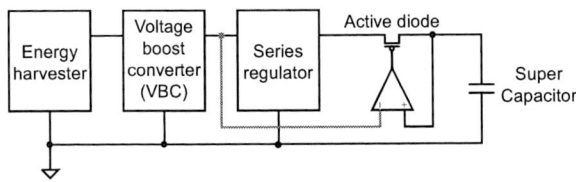

Fig. 2. Proposed power management system.

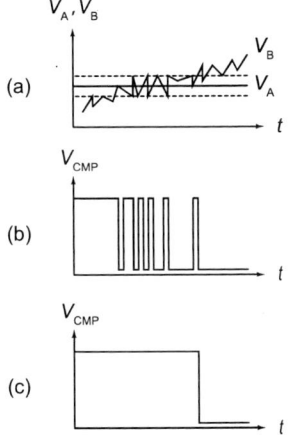

Fig. 3. Waveforms of (a) input voltages, (b) output voltage of the comparator without hysteresis, and (c) output voltage of the comparator with hysteresis.

output when we use a comparator without hysteresis (Fig. 3(b)). To eliminate the oscillation (Fig. 3(c)), we use hysteresis comparators. There are several hysteresis comparators that are based on the differential pair circuit. However, there are few reports on the common gate comparator with hysteresis. Therefore, we develop a hysteresis common gate comparator for the active diode.

III. HYSTERESIS COMMON GATE COMPARATOR

A. Conventional Common-Gate Comparator

Fig. 4(a) shows a simplified core circuit of the conventional CG comparator. The circuit consists of two current sources and two pMOS transistors. Input voltages V_A and V_B are connected to the source terminals of the pMOS transistors. Fig. 4(b) illustrates waveforms of the output current I_X and output voltage V_X when V_A is set to a certain voltage and V_B is changed from Low to High. When V_B is lower than V_A, V_X is high because current flowing in MP2 is larger than I_B. On the other hand, when V_B is higher than V_A, V_X is low because current flowing in MP2 is lower than I_B. Therefore, current flowing in MP2 will determine the voltage of V_X. This can be analyzed as follows.

When the source-drain voltage V_{SD} of a PMOS transistor is 0.1 V or more, the subthreshold drain current I can be expressed as

$$I = K I_0 \exp\left(\frac{V_{SG} - |V_{THP}|}{\eta V_T}\right), \tag{1}$$

where $K(= W/L)$ is the aspect ratio of the transistor, $I_0(= \mu C_{ox} V_T^2 (\eta - 1))$ is the process dependent parameter, μ is the carrier mobility, C_{ox} is the gate-oxide capacitance, $V_T(= k_B T/q)$ is the thermal voltage, k_B is the Boltzmann constant, T is the absolute temperature, q is the elementary charge, η is the subthreshold slope factor, V_{SG} is the source-source voltage, and $|V_{TH}|$ is the threshold voltage of a pMOS transistor. From Eq. (1), the source-gate voltage V_{SG1} of MP1 is expressed as

$$V_{SG1} = |V_{THP}| + \eta V_T \ln\left(\frac{I_B}{K_1 I_0}\right), \tag{2}$$

where I_B is the bias current and K_1 is the aspect ratio of MP1. From Eq. (2), V_{G1} can be expressed as

$$\begin{aligned} V_{G1} &= V_B - V_{SG1} \\ &= V_B - |V_{THP}| - \eta V_T \ln\left(\frac{I_B}{K_1 I_0}\right). \end{aligned} \tag{3}$$

The source-gate voltage V_{SG2} of MP2 can be expressed as

$$V_{SG2} = V_A - V_{G1}. \tag{4}$$

Therefore, from Eqs. (1) and (3), the current I_2 can be expressed as

$$I_2 = \frac{K_2}{K_1} I_B \exp\left(\frac{V_A - V_B}{\eta V_T}\right), \tag{5}$$

where K_2 is the aspect ratio of M2. When $K_1 = K_2$, Eq. (5) can be expressed as

$$I_2 = I_B \exp\left(\frac{V_A - V_B}{\eta V_T}\right). \tag{6}$$

Therefore, the current I_X flowing into V_X can be expressed as

$$\begin{aligned} I_X &= I_2 - I_B \\ &= I_B \left\{\exp\left(\frac{V_A - V_B}{\eta V_T}\right) - 1\right\}. \end{aligned} \tag{7}$$

From Eq. (7), when $V_B < V_A$, I_X becomes positive and V_X increases. On the other hand, when $V_B > V_A$, I_X becomes negative and V_X decreases. Fig. 4(b) illustrates I_X and V_X as a function of V_B. When $V_B = V_A$, the output voltage V_X changes from High to Low.

B. Proposed Hysteresis Common-Gate Comparator

As discussed in the previous section, we can control the switching voltage of the CG comparator by changing the bias current I_B connected to V_X. Fig. 5 shows a simplified schematic of our proposed hysteresis CG comparator. A switch (SW) is used to control the bias current.

When SW is OFF, I_3 in Fig. 5 becomes $I_B/2$. From Eq. (7), I_X can be expressed as

$$\begin{aligned} I_X &= I_2 - I_3 = I_2 - \frac{1}{2} I_B \\ &= I_B \left\{\exp\left(\frac{V_A - V_B}{\eta V_T}\right) - \frac{1}{2}\right\}. \end{aligned} \tag{8}$$

978-1-5386-4757-8/18 $31.00 © 2018 IEEE   197

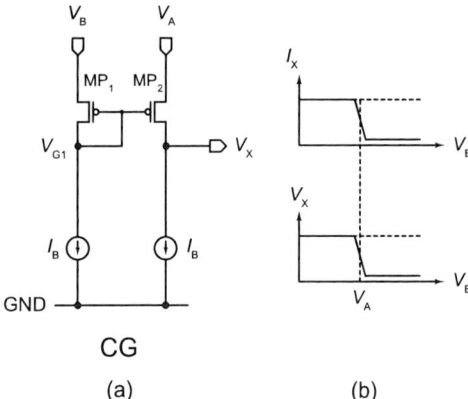

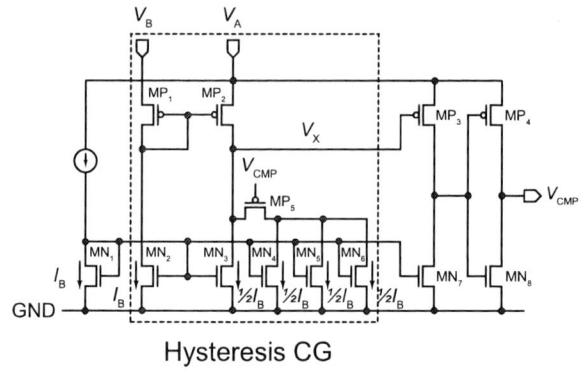

Fig. 4. (a) Schematic of the simplified CG comparator and (b) illustrations of the output current and voltage (I_X and V_X).

Fig. 6. Schematic of our proposed hysteresis CG comparator.

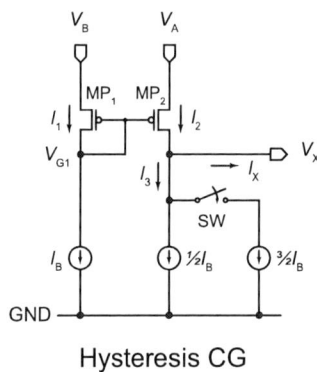

Fig. 5. Simplified schematic of our hysteresis CG comparator.

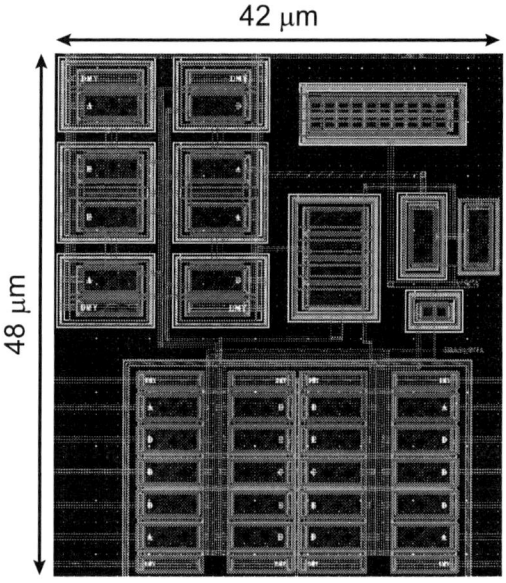

Fig. 7. Chip layout.

Therefore, from Eq. (8), I_X becomes positive, when

$$V_B < V_A + \eta V_T \ln 2. \qquad (9)$$

The switching voltage becomes higher than V_A by $\eta V_T \ln 2$. It is 27 mV, when we set η and V_T at room temperature to 1.5 and 26 mV, respectively.

On the other hand, when SW is ON, current I_3 in Fig. 5 becomes $2I_B$. Therefore, from Eq. (7), I_X can be expressed as

$$\begin{aligned} I_X &= I_2 - I_3 = I_2 - 2I_B \\ &= I_B \left\{ \exp\left(\frac{V_A - V_B}{\eta V_T} \right) - 2 \right\}. \end{aligned} \qquad (10)$$

Therefore, from Eq. (10), I_X becomes positive, when

$$V_B < V_A - \eta V_T \ln 2. \qquad (11)$$

The switching voltage becomes lower than V_A by $\eta V_T \ln 2$. It is -27 mV, when we set η and V_T at room temperature to 1.5 and 26 mV, respectively.

Therefore, when the input voltage V_B changes from Low to High, a positive hysteresis voltage can be obtained by turning SW OFF. On the other hand, when the input voltage V_B changes from High to Low, a negative hysteresis voltage

can be obtained by turning SW ON. Fig. 6 shows a complete schematic of our proposed hysteresis CG comparator. A common source amplifier and inverter are added to enhance the voltage gain and a nano-ampere current source is used to achieve ultra-low power dissipation [18]–[22]. The transistor sizes of MN1:MN2:MN3-6 are set to 2:2:1. The pMOS transistor switch MP5 is used to control the bias current. The gate of the MP5 is connected to the output of the comparator.

IV. RESULTS

We evaluated the performance of our hysteresis CG comparator using SPICE with a set of 65-nm standard CMOS process parameters. Fig. 7 shows a physical design layout of our comparator. The area occupied 0.0020 mm^2. Fig. 8 shows a schematic of the simulated hysteresis comparator. Tab. I lists the transistor sizes ($M \times (W/L)$, M: multiplier, W: channel width, and L: channel length) of the comparator.

TABLE I
TRANSISTOR SIZES

Transistor	Size
MP$_1$, MP$_2$	4×(2 μm/5 μm)
MP$_3$	4×(1 μm/5 μm)
MP$_4$	1×(1 μm/5 μm)
MP$_5$	2×(1 μm/1 μm)
MP$_6$	20×(1 μm/1 μm)
MN$_1$, MN$_2$, MN$_7$	4×(2 μm/5 μm)
MN$_3$, MN$_4$, MN$_5$, MN$_6$	2×(2 μm/5 μm)
MN$_8$	4×(1 μm/5 μm)

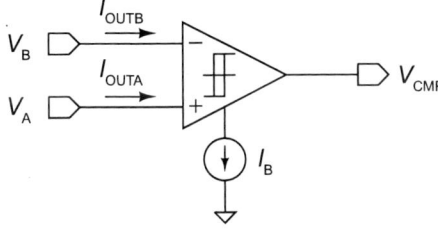

Fig. 8. Block diagram of hysteresis comparator.

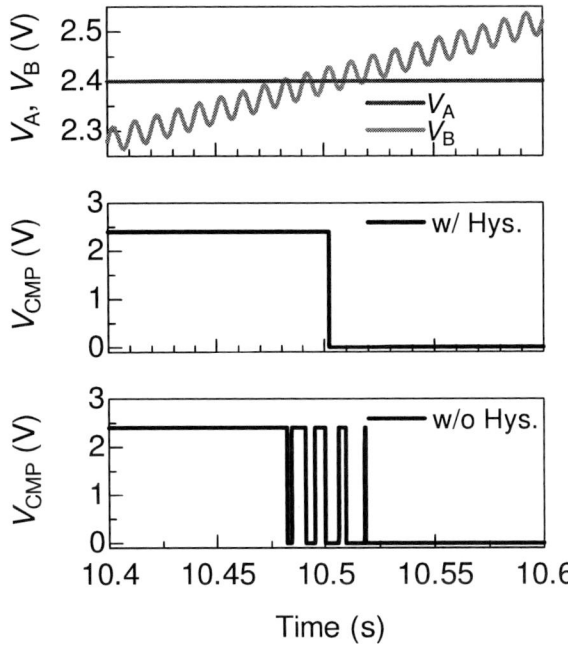

Fig. 10. Simulated transient waveforms.

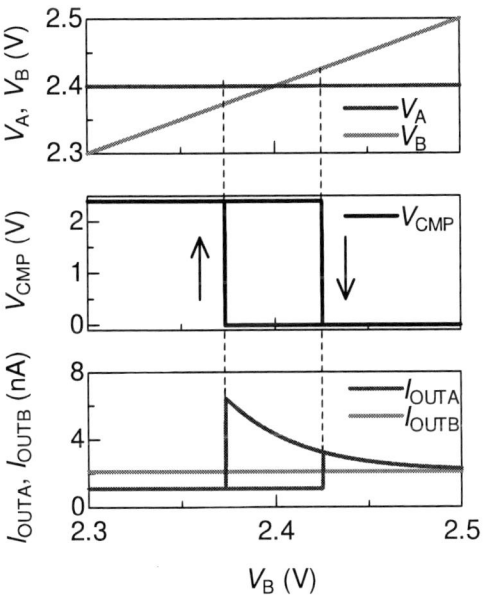

Fig. 9. Simulated transfer function.

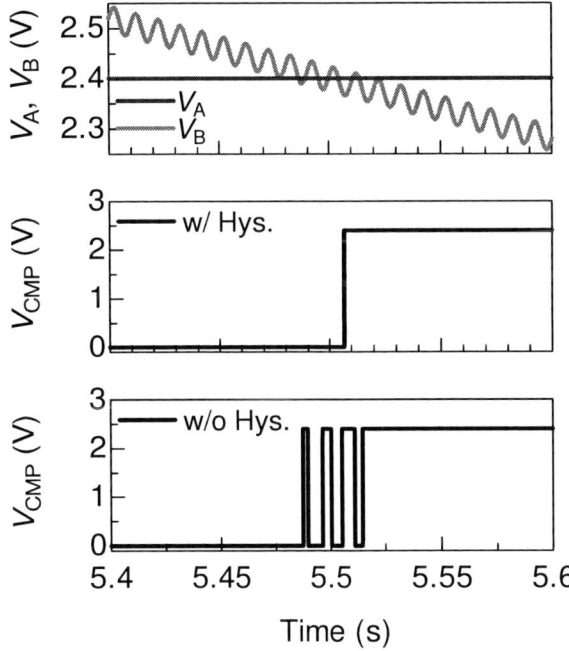

Fig. 11. Simulated transient waveforms.

Fig. 9 shows the simulated V_A, V_B, V_{CMP}, I_{OUTA}, and I_{OUTB}, as a function of V_B. The V_A and I_B were set to 2.4 V and 2 nA, respectively. The V_B increased from 2.3 to 2.5 V, and then decreased from 2.5 to 2.3 V. As shown in Fig. 9, we confirmed that V_{CMP} changes with hysteresis voltages. The positive and negative hysteresis voltages were 25 and −27 mV, respectively, which are almost the same as the calculated results.

Figs. 10 and 11 show the simulated transient waveforms of the comparator. The V_B was changed from low to high (Fig. 10), and high to low (Fig. 11) with a ripple voltage. The frequency and amplitude of the ripple voltage were set to 100 Hz and 50 mV, respectively. The output voltage of the comparator without hysteresis oscillated as shown on the bottom in Figs. 10 and 11. We confirmed that our proposed

978-1-5386-4757-8/18 $31.00 © 2018 IEEE   199

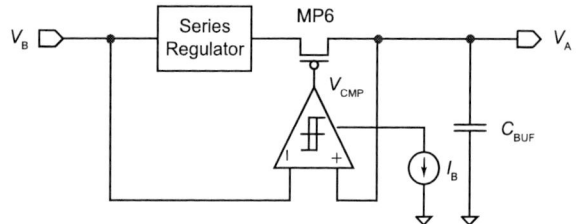

Fig. 12. Test bench of our active diode.

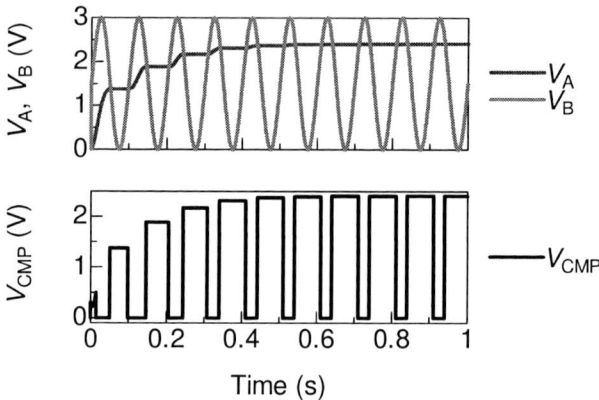

Fig. 13. Simulated transient waveforms.

comparator operates correctly.

Fig. 12 shows a test bench circuit of the active diode with our proposed hysteresis comparator. The I_B and C_{BUF} were set to 2.0 nA and 47 μF, respectively. The frequency and peak-to-peak voltage of V_B were set to 10 Hz and 3 V, respectively. The output voltage of the series regulator was set to 2.4 V. Fig. 13 shows the simulated waveforms of the active diode. The simulated V_A increased gradually and settled to 2.4 V.

V. CONCLUSION

This paper presented a hysteresis common gate comparator and active diode using the comparator for low-voltage and low-power energy harvesting systems. The proposed active diode consists of a MOS switch and hysteresis comparator, which eliminates unwanted ripple and noise voltages. The hysteresis comparator controls the MOS switch to turn ON or OFF, depending on the input and output voltages. The hysteresis voltage of the comparator can be controlled by the current flowing in the common gate comparator. Simulation results showed that the hysteresis comparator has a -27 and 25 mV hysteresis voltages and the active diode using the hysteresis comparator eliminates unwanted ripple voltage.

ACKNOWLEDGMENT

This work was based on results obtained from a project commissioned by the New Energy and Industrial Technology Development Organization (NEDO) of Japan, and was conducted through a dedicated licensing program provided by the VLSI Design and Education Center (VDEC) at the University of Tokyo with the cooperation of Cadence Design Systems, Inc. and Mentor, a Siemens business. This work was also partially supported by JSPS KAKENHI Grant Number JP15H01679 and Canon Foundation.

REFERENCES

[1] L. D. Xu et al., "Internet of things in industries: A survey," *IEEE Trans. Ind. Inform.*, vol. 10, no. 4, pp. 2233-2243, 2014.

[2] D. Blaauw et al., "IoT design space challenges: Circuits and systems," in *Symp. VLSI Technol. Dig. Tech. Papers*, 2014, pp. 1-2.

[3] S. Chen et al., "A vision of IoT: applications, challenges,and opportunities with china perspective," *IEEE Internet of Things J.*, vol. 1, no. 4, 2014

[4] R. J. M. Vullers et al., "Energy harvesting for autonomous wireless sensor networks," *IEEE Solid State Circuits Mag.*,vol. 2, pp.29-38, 2010.

[5] V. Raghunathan and P. H. Chou, "Design and power management of energy harvesting embedded systems," in *Proc. Int. Symp. Low Power Electron. Design (ISLPED)*, 2006, pp. 369-374.

[6] A. P. Chandrakasan et al., "Next Generation Micro-power Systems," *IEEE Symp. VLSI Circuits*, pp.2-5, 2008.

[7] T. Hirose et al., "Watch-dog circuit for quality guarantee with subthreshold MOSFET current," *IEICE Trans. Electron.*, vol. E87-C, no. 11, pp. 1910-1914, 2004.

[8] K. Ueno et al., "CMOS smart sensor for monitoring the quality of perishables," *IEEE J. Solid-State Circuits*, vol. 42, no, 4, pp. 798-803, 2007.

[9] H. Asano et al., "A 1.66-nW/kHz, 32.7-kHz, 99.5ppm/$^\circ$C, fully integrated current-mode RC oscillator for real-time clock applications with PVT stability," in *Proc. Eur. Solid-State Circuits Conf. (ESSCIRC)*, 2016, pp. 149-152.

[10] T. Ozaki et al., "Fully-Integrated High-Conversion-Ratio Dual-Output Voltage Boost Converter with MPPT for Low-Voltage Energy Harvesting," *IEEE J. Solid-State Circuits*, vol. 51, no. 10, pp. 2398-2407, 2016.

[11] T. Ozaki et al., "A highly efficient switched-capacitor voltage boost converter with nano-watt MPPT controller for low-voltage energy harvesting," *IEICE Trans. Fundam. Electron. Commun. Comput. Sci.*, vol. E99-A, no. 12, pp. 2491-2499, 2016.

[12] Y. Tsuji et al., "A 0.1-0.6 V Input Range Voltage Boost Converter with Low-Leakage Driver for Low-Voltage Energy Harvesting," in *Proc. Int. Conf. Electron. Circuits Syst. (ICECS)*, 2017, pp. 502-505.

[13] Y. Rao et al., "An input-powered vibrational energy harvesting interface circuit with zero standby Power," *IEEE Trans. Power electronics*, vol. 26, no. 12, pp. 3524-3533, 2011.

[14] P. F. Becker et al., "Efficient energy harvesting with electromagnetic energy transducers using active low-voltage rectification and maximum power point tracking," *IEEE J. Solid-State Circuits*, vol. 47, no. 6, pp. 1369-1380, 2012.

[15] Q. Li et al., "A wide input amplitude range, highly efficient rectifier for low power energy harvesting systems," *Nonlinear Theory and Its Applications, IEICE*, vol. 5, no. 4, pp. 499-511, 2014.

[16] E. E. Aktakka et al., "A micro inertial energy harvesting platform with self-supplied power management circuit for autonomous wireless sensor nodes," *IEEE J. Solid-State Circuits*, vol. 49, no. 9, pp. 2017-2029, 2014.

[17] D. A. Sanchez et al., "A parallel-SSHI rectifier for piezoelectric energy harvesting of periodic and shock excitations," *IEEE J. Solid-State Circuits*, vol. 51, no. 12, pp. 2867-2879, 2016.

[18] T. Hirose et al., "Ultralow-power current reference circuit with low temperature dependence," *IEICE Trans. Electron., vol. E88-C*, no. 6, pp. 1142-1147, 2005.

[19] T. Hirose et al., "Temperature-compensated CMOS current reference circuit for ultralow-power subthreshold LSIs," *IEICE Electron. Express*, vol. 5, no. 6, pp. 204-210, 2008.

[20] K. Ueno et al., "1-uW, 600-ppm/$^\circ$C current reference circuit consisting of sub-threshold CMOS circuits," *IEEE Trans. Circuits Syst. II, Express Briefs*, vol. 57, issue 9, pp. 681 - 685, 2010.

[21] T. Hirose et al., "A nano-ampere current reference circuit and its temperature dependence control by using temperature characteristics of carrier mobilities" in *Proc. Eur. Solid-State Circuits Conf. (ESSCIRC)*, 2010, pp. 114-117.

[22] Y. Osaki et al., "Temperature-compensated nano-ampere current reference circuit with subthreshold metal-oxide-semiconductor field effect transistor resistor ladder," *Jpn. J. Appl. Phys.*, vol. 50, no. 4, pp. 04DE08, 2011.

A Bandwidth-Aware Authentication Scheme for Packet-Integrity Attack Detection on Trojan Infected NoC

Mubashir Hussain and Hui Guo
School of Computer Science and Engineering
The University of New South Wales, Australia
{m.hussain, h.guo}@unsw.edu.au

Abstract—Bandwidth is a fundamental issue in the NoC (Network on Chip) design and increasing use of the third-party NoC IPs adds the security as another design concern. The third-party NoC may have hardware Trojans that can break the confidentiality and integrity of the data transferred over the NoC, thereafter exposing the system to varied attacks. Basically, encryption can be applied to protect the data confidentiality; For the data integrity, authentication is often used. A general authentication approach for network communication is using a tag to identify the data; The tag is attached to the data at the source and the data is verified against the tag at the destination. For the effectiveness of the authentication, a large tag size is desired. But large tags, when transferred over the NoC, will consume considerable network bandwidth, potentially conflicting with the system performance requirement. In this paper, we propose a progressive packet data authentication scheme, where a packet is progressively authenticated against a set of small tag segments such that the authentication overhead is low yet the effective tag size is large. The experiments in our case study of 2D mesh NoC demonstrate that our design can achieve savings about 36% on bandwidth and 56% on implementation area cost as compared to the baseline design with the traditional non-progressive authentication scheme.

Index Terms—Network-on-Chip, Packet integrity, Data Authentication, Hardware Trojan

I. INTRODUCTION

With the advance of the semiconductor technology, many intellectual property (IP) cores can be integrated on a single chip. A typical example is the MPSoC (Multi-Processor System-on-Chip) design. MPSoC contains multiple processors, memory elements, power management blocks and I/O interfaces. To meet the high bandwidth needs of the MPSoCs, the router based scalable communication architecture, network-on-chip (NoC), is often preferred.

To reduce the cost and the time-to-market of MPSoCs, designers often use off-the-shelf components, for example, using NoC (Network-on-Chip) IP from the third-party vendors. However, concerns about the security of the third-party components become increasingly salient. The vendor can modify the hardware components in the design to leak sensitive information and/or tamper the system data for various attacks.

In this paper, we target an MPSoC system that uses the third-party NoC IP. We assume that the NoC may contain a hardware Trojan (a malicious hardware component embedded in the NoC) in some router and the Trojan can be activated and de-activated during the system operation. Once activated, the Trojan can observe the passing packet and attempt to modify it for varied attack purposes, such as changing the packet destination to divert the packet to a different node, altering the packet data of an online money transaction.

For the integrity verification of a packet, a common way is to tag the packet data at the source node; both the data and its tag are then transferred over the network to the destination, as illustrated in Figure 1. At the destination, the packet data is authenticated against the tag. If the data is altered, the tag becomes invalid.

The authentication calculates the tag value of the received data and compares it with the tag in the packet. If they are different, the packet is deemed tempered. However, the attack can pass the authentication and go undetected if the tag of the altered data collides with the tag of the original data (**tag collision**). Therefore, it is important to design tags that have no or low tag collisions, which entails a big tag size. However big tags will consume considerable packet space, greatly degrading the network performance.

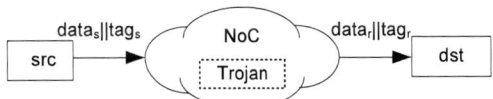

Fig. 1. Traditional Authentication Scheme for Data Integrity Checking: tag generated at the source (src), data/tag transferred over the network, and data authenticated at the destination (dst)

In this paper, we aim for a packet authentication design that is low in bandwidth overhead yet highly effective in verification of the packet data transferred over the third party NoC. Our main contributions are

- We proposed a novel progressive packet authentication scheme, where a packet is tagged at the source but progressively authenticated during its transmission. For a given packet, it has varied tag segments for different network nodes; When the packet comes to a node, the packet is verified against the specific tag segment. Each tag segment is small (hence incurs a low authentication cost), but the effective tag used for the overall packet authentication is formed by a chain of tag segments and can be sufficiently large.

- We undertook a case study on the 2D-mesh NoC. For the random distributed packets, we identified the most effective attack position for a Trojan. For the Trojan in this position, we investigated and verified the effectiveness of the proposed progressive authentication design.

The rest of the paper is organized as follows. Section II reviews existing work related to data authentication. Our progressive packet authentication scheme is proposed in III. The case study based on the 2D-mesh NoC is presented in Section IV. The paper is concluded in Section V.

978-1-5386-4757-8/18 $31.00 © 2018 IEEE

II. RELATED WORK

The data integrity checking has been well studied, especially for the messages transferred over the network. The common approach for message authentication is to use a MAC (short for message authentication code, also called tag in the data authentication), to identify a message.

Most MAC schemes divide messages into blocks and the MAC generation is made of the operations on blocks. Depending on how those operations are structured between blocks, MAC schemes can be divided into iterated MAC [21] [2] [16][14][3][11] and parallel MAC [1][4][17].

In the iterated MAC design, the operations on blocks are performed in sequence. The result of one block relies on the output of previous blocks. In the parallel MAC scheme, all blocks of a message are processed in parallel.

In terms of data authentication, designs can be cryptographic hash function based (such as CRC [15]), Added Redundancy Explicit Authentication ((AREA) based[7], and MAC (Message Authentication Code) scheme based.

In [22], Yan et al. proposed an authentication and encryption design (**AE**) to protect the confidentiality and integrity of memory data with Galois/Counter Mode (GCM) operations proposed in [13].In [18] Rogers and Milenkovic proposed another AE design aiming to protect both code and data in the memory. A PMAC [4] scheme is used for memory data authentication.

Hong et. al in [10] proposed a tag generation design that is based on a sequence of randomization operations. The design aims for low tag storage consumption and on-chip area cost, and can be optimized for a given tag size.

There are also approaches proposed for the packet error detection on the NoC.

Yu et al. in [24] investigated the data integrity attacks by hardware Trojans on the NoC links, where a Trojan on a link can flip the bit value of the link. Single error correction double error correction code is used and flits with double-bit errors will be retransmitted. When retransmitted, the flits are reshuffled so that the Trojan location can be identified.

Boraten et al. [6] also addressed the data integrity attacks by hardware Trojans.They proposed to use different error detection codes (CRC or algebraic manipulation and detection) for the different level of security required by the individual applications.

Frey et al. [8] proposed a dynamic flit permutation and the ECC (Error Correcting Code) techniques to protect against NoC bandwidth depletion attacks within a router, where the ECC code for a flit is generated and scrambled into the flit at the input port of NoC router. At the output port, the flit is verified using ECC.

However, those approaches are not effective in that if an altered packet goes undetected by the authentication on the next transmission node, it will never be detected. In addition, the large error-code adds the computing burden to authentication on each node, hence degrading the network performance.

Instead of using a big tag and attaching the tag directly to the data, our design uses small tags in the authentication and a packet is progressively authenticated by a sequence of nodes on its transmission path, which is described in the next section.

III. PROGRESSIVE AUTHENTICATION SCHEME

Threat Model: We target a system that uses the third-party NoC IP, as shown in Figure 2. Each node in the NoC can connect to a designer's processing unit (PU). We assume the hardware and software of the PU are trusted, but the NoC is untrusted. Some node in the NoC may have a hardware Trojan. The Trojan can observe and modify the passing packets.

Here we focus on the detection of tampered packets. Our progressive authentication scheme is elaborated below.

For a new packet, the PU of the source node generates a set of tag segments. Each tag segment corresponds to a specific node and the tag value is created by a cryptographic function[1] controlled by the key that is local to the node.

To detect whether a packet has been tampered, we attach a secure authentication unit (AU) to each node (see Figure 2)[2]. For an incoming packet, the authentication unit calculates the tag segment based on the received packet and compares it with the original one generated by the PU of the packet source node; if they are the same, the packet is passed to the next node; otherwise, the packet is dropped.

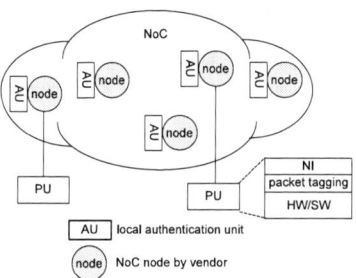

Fig. 2. System Overview

In our design, each node has a dedicated key for the local tag generation. Tags and keys are stored in a centralized repository that holds two tables, as shown in Figure 3. Both tables are only accessible via a secure channel to PUs and AUs. At the source of a packet, all local tag segments for the nodes on the packet transmission path are generated with the related keys provided by the key table (Figure 3 (a)). Those tag segments are then saved in the tag table, as depicted in Figure 3(b). The AU of a node can have a local copy of its key to avoid repeat access of the key table.

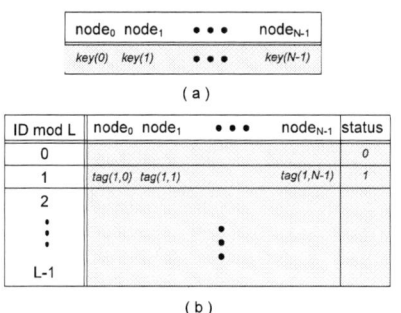

Fig. 3. Centralized Repository (a) key table (b) tag table

The tag table only needs to hold tags for packets that are alive in the NoC. For a NoC with N nodes, assume there are maximally L alive packets[3], the tag table contains L entries

---

[1]An existing MAC/Tag generation design can be used. In our case study to be presented in Section IV, we use a MAC-based tag generation.

[2]NoC is often a soft IP, as assumed in the existing work [24][8][6]. It is possible to attach a component to the network node.

[3]L can be determined by profiling.

and each entry holds N tag segments plus a status bit. A packet is labeled with an ID. We allow for all packets of the same mod ID (mod to L) to share one entry in the table. For a packet with ID i, its related tag segments are saved in entry (i (mod L)). Packet i and packet $i + L$ should be temporally far away from each other. When packet $i + L$ is generated by a PU, packet i is expected to be off the NoC and the table entry is free to packet $i + L$. However, it is possible that the profiling may have missed cases where the alive packet number is larger than L. Under these circumstances, some entries in the tag table will not be available to new packets; the new packets should, therefore, be stalled at the their source nodes until the tag entries are free.

We use the status bit (located in the last column of the tag table) to control the use of a tag entry in the table. The bit serves two functions: one, indicating whether the tag entry is available for a new packet, and two, showing whether authentications on the path nodes are required for the related packet. If the bit is 1, the tag segments in the entry are valid and the packet should be authenticated during the packet transmission. If the status bit is 0, then to the source of a new packet that is mapped to the entry, this entry is free to use, and to the network nodes there is no need to authenticate the related packet.

When authenticating packet i at node j, the AU on the node first calculates the tag for the packet, and then compares the tag value with $tag(i \pmod{L}, j)$ in the tag table. If they are the same, the node passes the packet to the next node; otherwise, the packet is deemed tampered and will be discarded.

The related tag table entry will be reset and free to the next packet (with the same mod ID value) when the current packet is off the network (either dropped or received at the destination).

The packet tagging at the source and the authentication on a path node are summarized in Algorithm 1 and Algorithm 2, respectively.

Algorithm 1: Tag Segments Generation at the PU of Source Node for a Packet

input : Key Table, T_{key}; packet, D
output: tag segments for D, saved in Tag Table, T_{tag}

1 $i = genPacketID(D)$;
2 $P = getTransmissionPath(D)$;
3 **for** *each node, j, on P* **do**
4      $key(j) = getKey(j, T_{key})$;
5      $tag(i, j) = generateTag(key(j), D)$;
6      $T_{tag} \leftarrow saveTag(tag(i, j))$;
7 **end for**

Algorithm 2: Authentication on Node, i, for a Packet

input : Tag Table, T_{tag}, packet D, node, j
output: Authentication outcome

1 $i = readPacketID(D)$;
2 $tag_{ori} = readTag(i, j, T_{tag})$;
3 $tag_{cal} = calculateTag(key(j), D)$;
4 **if** $tag_{cal} = tag_{ori}$ **then**
5      $passPacket(D)$;
6 **else**
7      $dropPacket(D)$;
8 **end if**

For a packet to be sent to the NoC, Algorithm 1 first generates an ID and finds the transmission path[4] (Lines 1 & 2). For each node on the path, it then generates a tag segment based on the key available in the key table and saves the tag

---

[4]If a dynamic routing algorithm is used, all nodes may need to be considered as a potential part of the path.

---

segment in the tag table (Lines 4-6). The tag generation can be an existing design such as CBC [21]. An example will be given in Section IV-B.

On a node along the packet path, in order to authenticate a packet, Algorithm 2 reads the packet ID from the packet and the related tag segment from the tag table (Lines 1 & 2), and then uses the same tag generation function as adopted by the source PU to calculate the tag segment based on the received data (Line 3). If the calculated tag segment is the same as the original one in the tag table, the packet transmission continues; otherwise, the transmission is stopped (Lines 4-8).

Algorithm 1 is implemented in PUs that need to send data to other PUs and require packet tagging and Algorithm 2 is implemented in each of the AUs for packet authentication.

As can be seen, our design may look like the traditional hop-to-hop detection in that packets are checked at each node along their transmission paths. But in the traditional hop-to-hop design, a packet, if passed an authentication, it will be re-tagged based on the current packet value and the new tag is used by the authentication in the next hop; Therefore, if a modified packet cannot be immediately identified in the next hop, the subsequent detections are just in vain but only to waste energy. With our design, on the other hand, a packet will undergo a sequence of effective authentications; If a modified packet escaped the authentication on one node (due to tag collision), it may be caught in the subsequent authentications on the downstream nodes. In addition, the progressive authentication also offers the potential of early detection of a tampered packet. Once detected, the tampered packet is dropped, hence reducing unnecessary network traffic.

IV. CASE STUDY: ON 2D-MESH NOC

To obtain an insight of the effectiveness of the progressive packet authentication, we perform a study on the NoC with a 2D-mesh topology, popular in MPSoCs. We assume that the XY dimensional-ordered routing is used in the packet transfer. For a packet transferred from source (x_s, y_s) to destination (x_d, y_d), it first flows horizontally to (x_s, y_d) then vertically to (x_d, y_d).

The location of the Trojan may affect its ability to capture a target packet for attack. For the 2D-mesh with the fixed routing scheme, some packets may never pass through the Trojan node. Those packets are immune from the Trojan attack. Furthermore, the location of the Trojan also affects the effective tag size used for a packet authentication.

To make our evaluation well-founded and effective, we target the Trojan that holds the maximal advantages in terms of attack probability and attack success probability, which is discussed below.

A. Advantageous Trojan Locations

Without targeting specific-applications, we assume packet sources and destinations are randomly distributed over the whole network. We also assume that the Trojan is located at (x_t, y_t) in the mesh that has m rows and n columns, and we call the row and column that the Trojan is located, the **Trojan row** and **Trojan column**, respectively.

1) Trojan Locations for High Attack Probability

Intuitively, a Trojan can gain maximal attack advantage when it is located in the center of the network, which can be verified below.

We divide the mesh into four sections A11, A12, A21, A22, as shown in Figure 4. Given the XY-routing, a packet, from source (x_s, y_s) to the destination (x_d, y_d), that can be attacked

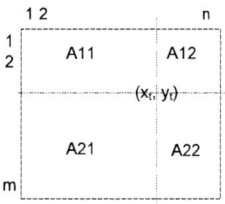

Fig. 4. Trojan in Mesh for High Attack Probability

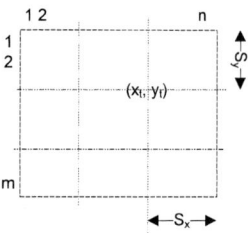

Fig. 5. Trojan in Mesh for High Attack Success Probability

should pass the Trojan node. Therefore, the following two types of packets are vulnerable:

- The packet source is on the Trojan row (i.e. $x_s = x_t$) and the source and destination are separated by the Trojan column;
- The packet destination is on the Trojan column (i.e. $y_d = y_t$) and the source and destination are separated by the Trojan row.

For a Trojan at the location (x_t, y_t), the number of the packets from the left Trojan row to the right section is $(y_t - 1) * (n - y_t + 1) * m$, and that from the right Trojan row to the left section is $(n - y_t) * y_t * m$.

Similarly, the number of the packets from the bottom section to the top Trojan column is $(m - x_t) * n * x_t$, and from the top section to the bottom Trojan column is $(x_t - 1) * n * (m - x_t)$.

The total number of possible packets that can be attacked by the Trojan is

$$
\begin{aligned}
P &= (y_t - 1) * (n - y_t + 1) * m + (n - y_t) * y * m + \\
&\quad (m - x_t) * n * x_t + (x_t - 1) * n * (m - x_t)
\end{aligned} \tag{1}
$$

From[5]

$$
\frac{dP}{dy_t} = 2n - 4y_t + 1 = 0 \tag{2}
$$

and

$$
\frac{dP}{dx_t} = 2m - 4x_t + 1 = 0, \tag{3}
$$

we can obtain that when $x_t = (m/2 + 1/4)$ and $y_t = (n/2 + 1/4)$, the value of P reaches to the maximum. Namely, when the Trojan is positioned at the center of the mesh network, it has the highest probability to intercept packets for an attack.

2) Trojan Locations for High Attack Success Probability

The success probability is closely related to the effective tag size, which is determined by the Trojan path length to the packet destination. To calculate such a path length, we divide the mesh vertically into three sections: two sections of the same size symmetrical to the Trojan column and the rest non-symmetric section, as shown in Figure 5. The width of the symmetric section is $S_x = min(x_t - 1, n - x_t)$ or 0 if the Trojan is on the either of two edge columns. Similarly, we can divide the mesh horizontally, where $S_y = min(y_t - 1, m - x_t)$ or 0 if the Trojan is on the edge rows.

The total path length of the Trojan to all nodes on the Trojan row in a symmetric section is $\sum_{i=0}^{S_x} i$, and for the non-symmetric section, it is $S_x * (m - 2S_x - 1) + \sum_{i=0}^{m-2S_x-1} i$.

The total path length of the Trojan to all nodes on the same row is

$$
\begin{aligned}
Lx &= 2 * \sum_{i=0}^{S_x} i + Sx(m - 2S_x - 1) + \sum_{i=0}^{m-2S_x-1} i \\
&= Sx^2 - (m - 1)Sx + \frac{m(m - 1)}{2}
\end{aligned} \tag{4}
$$

Given the path length for the Trojan to all nodes on the same row, we can easily find the length to the nodes on other rows and the total path length of the Trojan to all nodes is:

$$
\begin{aligned}
L &= Lx * n + m * (2 * \sum_{i=0}^{Sy} i + Sy * (n - 2Sy - 1) + \sum_{i=0}^{n-2Sy-1} i) \\
&= n * (Sx^2 - (m - 1)Sx + \frac{m(m - 1)}{2}) + \\
&\quad m * (Sy^2 - (n - 1)Sy + \frac{n(n - 1)}{2})
\end{aligned} \tag{5}
$$

With $dL/dS_x = 0$ and $dL/dS_y = 0$, we can then obtain that when $S_x = (m - 1)/2$ and $Sy = (n - 1)/2$ (namely, the mesh center), L is minimal.

As can be seen from above derivations, **when a Trojan is placed at the center of the mesh[6], it gains the highest chance and success probability to attack a packet.**

B. Experiments

We ran a set of experiments to investigate the effectiveness of our progressive authentication strategy, as given below.

1) Verification of Advantageous Trojan Position

To verify that the center location offers the advantage to Trojan for the highest possible attack, we built a simulation platform written in C to simulate the packet flows over the 2D NoCs of varied sizes (from 3x3 to 10x10). In each simulation, the packets' sources and destinations are uniformly distributed over the whole NoC, and for the Trojan sitting in a different location, the effective authentication path length of all packets are obtained.

For demonstration, Table I shows a small set of experiment results – on the 3x3 mesh. Each row in the table lists the number of attacked packets (NoA) among all packets (Row 1) transferred over the network and the average authentication path length (APL) per attacked packet for a given Trojan location (Column 1) under varied number of random packets (from 100 to 500000), where the authentication path of an attached packet is the path from the Trojan node to the destination node of the packet. As can be seen from the table, at the center location (2,2), the Trojan can intercept a maximal number of packets and the related attacks will go authentication with a minimal tag size. The similar results can be observed for other sizes of meshes in our experiments.

The attack path lengths for different mesh sizes are summarized and shown in Figure 6.

The average effective tag size with the 8-bit tag segment is given in Table II. As can be seen, the effective tag size

---

[5]To reduce the complexity of the derivation, we treat all discrete functions as continuous.

[6]The center of the mesh can be one node, or a group of nodes, depending on the parity of the node number in each mesh dimension.

978-1-5386-4757-8/18 $31.00 © 2018 IEEE

TABLE I
PASSING TROJAN PACKET NUMBER AND AVERAGE ATTACK PATH LENGTH 3X3 MESH UNDER DIFFERENT TEST SIZES

Trojan Location	100		1000		5000		10000		50000		100000		500000	
	NoA	APL	NoA	APL	NoA	APL	NoA	APL	NoA	APL	NoA	APL	NoA	APL
(1,1)	27	2	283	2	1517	2	3033	2	15306	2	30336	2	152186	2
(1,2)	25	1	405	1	1878	1	3902	1	19086	1	38086	1	191275	1
(1,3)	33	2	342	2	1452	2	3000	2	15203	2	30386	2	151462	2
(2,1)	41	2	384	1	1942	1	3854	1	19103	1	38871	1	192158	1
(2,2)	51	1	472	1	2348	1	4609	1	23139	1	46343	1	232339	1
(2,3)	46	2	406	1	1919	1	3812	1	19042	1	38331	1	191304	1
(3,1)	35	2	326	2	1555	2	2989	2	15210	2	30531	2	151774	2
(3,2)	42	2	378	1	1923	1	3793	1	19284	1	38405	1	192484	1
(3,3)	24	2	284	2	1496	2	2963	2	15217	2	30426	2	151945	2

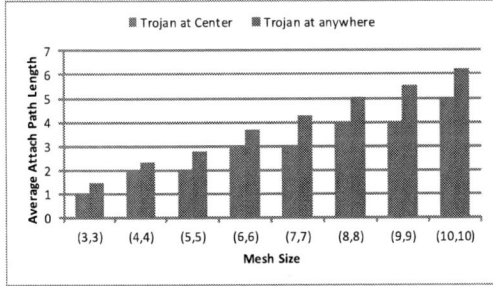

Fig. 6. Average Attack Path Length

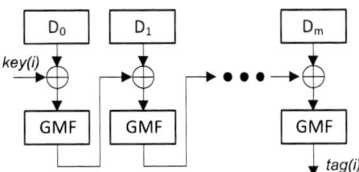

Fig. 7. Tag Segment Generation

TABLE II
EFFECTIVE TAG SIZE

Mesh Size	Trojan at center		Trojan at anywhere	
	Attack Path Length	Effective Tag Size	Attack Path Length	Effective Tag Size
(3,3)	1	8	1.4	12
(4,4)	2	16	2.3	19
(5,5)	2	16	2.8	23
(6,6)	3	24	3.7	30
(7,7)	3	24	4.3	35
(8,8)	4	32	5.0	41
(9,9)	4	32	5.5	45
(10,10)	5	40	6.2	50

segment generation is controlled by the key dedicated to the node, as explained in Section III.

To see the bandwidth savings from our design, we also ran application traces collected by Netrace 1.0 [9] on a cycle-accurate NoC simulator, Booksim 2.0 [12].

Table III shows the overhead savings on the area, packet space, and bandwidth of our progressive authentication as compared to the traditional non-progressive authentication for a set of applications. The progressive authentication does not carry tags in the packets and therefore saving packet space by 25% (the ratio of the tag size over the packet size), because of which, when the progressive authentication is applied on each applications, about 36% overhead on bandwidth (i.e, the extra number of packets) and 11.66% on packet latency, incurred by the traditional design can be saved. Furthermore, due to the small-tag based authentication used on the network nodes, the cost of the tag table and key table in the progressive authentication is mitigated. The area overhead is about 56% that of the traditional design.

TABLE III
SAVINGS (%)

Application	Area	Packet Space	Bandwidth	Packet Latency
blackscholes	0.55	0.25	36.02	11.59
bodytrack	0.57	0.25	36.05	11.67
canneal	0.57	0.25	36.05	11.68
dedup	0.56	0.25	36.08	11.63
fluidanimate	0.55	0.25	36.05	11.68
swaptions	0.58	0.25	36.05	11.62
vips	0.54	0.25	36.12	11.72
x264	0.54	0.25	36.05	11.67
AVG	0.56	0.25	36.06	11.66

increases with the network size and varies with the Trojan location. For the 8x8 mesh and 8-bit tag segment, 32-bit tags are effectively used in the authentication when the Trojan is located in the network center, and with the randomly located Trojan, tags of average 41 bits are used.

2) Overhead Savings

For the case of Trojan located in the center on the 8x8-mesh network, we further evaluated the costs of our detection design with the Synopsys Design Compiler. The NoC in our experiments uses 3-stage pipelined routers, and the matrix arbitration and wormhole switching. The packet size is 128 bits.

For tag segment generation, we adopted a block-based CBC approach, similar to [21], as shown in Figure 7.

With this design, the packet is divided into blocks with the block size equal to the tag segment size. Each block is randomized with a GFM (Galois-field Multiplication) function. The operations are chained into all blocks and the tag segment value is produced from the last block. For each node, the tag

3) Detection Rate

To evaluate the security effectiveness of our approach, we implemented three types of Trojans [8], [23], [6], [5], [19], [20] that focus on different packet fields for different attack purposes. The first type of Trojans ($Tro1$) modify the packet addresses to leak information or to degrade the system performance. The second type of Trojans ($Tro2$) alter the packet ID, type and control bits to create denial-of-services. The third type of Trojans ($Tro3$) only modify the data bits to change the packet payload.

For each Trojan type, we run the simulation for 10 million attack packets. The detection rates of both the traditional non-progressive authentication designs (of different tag sizes, 8-32 bits) and our progressive authentication design (8 bits for tag segment) on the three types of Trojans are given in Table IV. As can be seen from Table IV, for $Tro1$, when 8-bit tag is used, the detection rate is just 57.35%, due to the high collision rate of the small tags. The detection rate increases with the tag size. When the tag size is 32 bits, the detection rate reaches to 98.54%. Similarly, for $Tro2$ and $Tro3$, the detection rates are related to the tag size and reach to 99.89% and 99.9%, respectively, when 32-bits tags are used. However, with our authentication approach, even if the 8-bit tag segment is used in authentication, the overall tag size is effectively large and the detection rates for the three Trojans are all above 99.99%, as shown in the last column of the table.

TABLE IV
DETECTION RATES (%) OF TRADITIONAL AUTHENTICATION VS PROGRESSIVE AUTHENTICATION

Application	Non-Progressive Scheme				Progressive Scheme
	8	16	24	32	
Tro1	57.35	68.46	92.33	98.54	99.9989
Tro2	57.67	68.88	93.02	99.89	99.9999
Tro3	57.55	68.91	95.81	99.99	99.9899
AVG	57.52	68.75	93.72	99.47	99.9962

V. CONCLUSION

In this paper, we presented a progressive authentication scheme for the packet data authentication on the third-party NoC, where a packet is tagged at the source for all path nodes it may pass. When the packet travels to a node, it is authenticated against the tag for this node. Our design has two special features: one, the tag for a packet is not carried by the packet, hence avoiding the bandwidth overhead; two, for a given packet, the authentication tag is different from node to node. If a tampered packet went undetected on one node, it may be caught in the next node. Therefore, the effectiveness of the detection is greatly improved.

Our experiment on an 8x8 mesh NoC shows that, given a small 8-bit tag segment used in the authentication unit, an average of 32-bit tag is effectively used for highly advantageous Trojan and the effective tag size can grow when the Trojan is randomly located in the network and the network size is increased. According to our experiments on a set of applications, for the similar attack detection rate (over 99%), the progressive authentication design can save about 36% bandwidth overhead and this saving is achieved at 56% less area overhead as compared to the traditional non-progressive authentication.

REFERENCES

[1] M Bellare, R Guerin, and P Rogaway. XOR MACs: New methods for message authentication using finite pseudorandom functions. In *Advances in Cryptology - CRYPTO '95. 15th Annual International Cryptology Conference. Proceedings*, pages 15–28, Berlin, Germany, 1995.

[2] A Berendschot, Jean-Paul Boly, Antoon Bosselaers, J Brandt, David Chaum, Ivan Damgård, Peter de Rooij, Markus Dichtl, Walter Fumy, Cees Jansen, et al. Integrity primitives for secure information systems. final report of race integrity primitives evaluation (ripe-race 1040). *Lecture Notes in Computer Science*, 1007, 1995.

[3] J Black and P Rogaway. CBC MACs for arbitrary-length messages: the three-key constructions. In *Advances in Cryptology - CRYPTO 2000. 20th Annual International Cryptology Conference. Proceedings (Lecture Notes in Computer Science Vol.1880)*, pages 197–215, Berlin, Germany, 2000.

[4] J Black and P Rogaway. A block-cipher mode of operation for parallelizable message authentication. In *Advances in Cryptology - EUROCRYPT 2002. International Conference on the Theory and Applications of Cryptographic Techniques. Proceedings (Lecture Notes in Computer Science Vol.2332)*, pages 384–397, 2002.

[5] Travis Boraten and Avinash Kodi. Mitigation of hardware trojan based denial-of-service attack for secure nocs. *Journal of Parallel and Distributed Computing*, 111:24–38, 2018.

[6] Travis Boraten and Avinash Karanth Kodi. Packet security with path sensitization for nocs. In *2016 Design, Automation & Test in Europe Conference & Exhibition (DATE)*, pages 1136–1139. IEEE, 2016.

[7] Reouven Elbaz, Lionel Torres, Gilles Sassatelli, Pierre Guillemin, Michel Bardouillet, and Albert Martinez. Block-Level Added Redundancy Explicit Authentication for Parallelized Encryption and Integrity Checking of processor-memory transactions. *Transactions on Computational Science X. Special Issue on Security in Computing, Part I*, 6340(PART 1):231–260, 2010.

[8] Jonathan Frey and Qiaoyan Yu. A hardened network-on-chip design using runtime hardware trojan mitigation methods. *Integration, the VLSI Journal*, 56:15–31, 2017.

[9] Joel Hestness, Boris Grot, and Stephen W Keckler. Netrace: dependency-driven trace-based network-on-chip simulation. In *Proceedings of the Third International Workshop on Network on Chip Architectures*, pages 31–36. ACM, 2010.

[10] Mei Hong, Hui Guo, and Sharon X Hu. A cost-effective tag design for memory data authentication in embedded systems. In *Proceedings of the 2012 International Conference on Compilers, Architectures and Synthesis for Embedded Systems (CASES 2012)*, pages 17–26, 2012.

[11] T Iwata and K Kurosawa. OMAC: one-key CBC MAC. In *Fast Software Encryption. 10th International Workshop, FSE 2003. Revised Papers (Lecture Notes in Comput. Sci. Vol.2887)*, pages 129 — 53, 2003.

[12] Nan Jiang, James Balfour, Daniel U Becker, Brian Towles, William J Dally, George Michelogiannakis, and John Kim. A detailed and flexible cycle-accurate network-on-chip simulator. In *Performance Analysis of Systems and Software (ISPASS), 2013 IEEE International Symposium on*, pages 86–96. IEEE, 2013.

[13] D A McGrew and J Viega. The security and performance of the Galois/counter mode (GCM) of operation. In *Progress in Cryptology - INDOCRYPT 2004. 5th International Conference on Cryptology in India. Proceedings (Lecture Notes in Computer Science Vol.3348)*, pages 343 – 55, Berlin, Germany, 2004.

[14] NIST. Recommendation for block cipher modes of operation: The cmac mode for authentication. http://csrc.nist.gov/publications/nistpubs/800-38B/SP_800-38B.pdf.

[15] William Wesley Peterson and Daniel T Brown. Cyclic codes for error detection. *Proceedings of the IRE*, 49(1):228–235, 1961.

[16] F Petrank and C Rackoff. CBC MAC for real-time data sources. *J. Cryptol. (USA)*, 13(3):315 – 38, 2000.

[17] P Rogaway. Efficient instantiations of tweakable blockciphers and refinements to modes OCB and PMAC. In *Advances in Cryptology-ASIACRYPT 2004. 10th International Conference on the Theory and Application of Cryptology and Information Security. Proceedings (Lecture Notes in Computer Science Vol.3329)*, pages 16–31, Berlin, Germany, 2004.

[18] A Rogers and A Milenkovic. Security extensions for integrity and confidentiality in embedded processors. *Microprocess. Microsyst. (Netherlands)*, 33(5-6):398–414, 2009.

[19] Hassan Salmani, Mohammad Tehranipoor, and Ramesh Karri. On design vulnerability analysis and trust benchmarks development. In *Computer Design (ICCD), 2013 IEEE 31st International Conference on*, pages 471–474. IEEE, 2013.

[20] Bicky Shakya, Tony He, Hassan Salmani, Domenic Forte, Swarup Bhunia, and Mark Tehranipoor. Benchmarking of hardware trojans and maliciously affected circuits. *Journal of Hardware and Systems Security*, 1(1):85–102, 2017.

[21] ANSI X9.9. American national standard for financial institution message authentication. 1981.

[22] Chenyu Yan, B Rogers, D Englender, D Solihin, and M Prvulovic. Improving cost, performance, and security of memory encryption and authentication. In *Proceedings. 33rd International Symposium on Computer Architecture*, 2006.

[23] Qiaoyan Yu, Jaya Dofe, Yuejun Zhang, and Jonathan Frey. Hardware hardening approaches using camouflaging, encryption, and obfuscation. In *Hardware IP Security and Trust*, pages 135–163. Springer, 2017.

[24] Qiaoyan Yu and Jonathan Frey. Exploiting error control approaches for hardware trojans on network-on-chip links. In *2013 IEEE International Symposium on Defect and Fault Tolerance in VLSI and Nanotechnology Systems (DFTS)*, pages 266–271. IEEE, 2013.

Upgrading QoSinNoC: Efficient Routing for Mixed-Criticality Applications and Power Analysis

Serhiy Avramenko[1], Siavoosh Payandeh Azad[2], Behrad Niazmand[2],
Massimo Violante[1], Jaan Raik[2], Maksim Jenihhin[2]

[1]DAUIN, Politecnico di Torino, Italy, {serhiy.avramenko | massimo.violante}@polito.it
[2]Computer Systems, Tallinn University of Technology, Estonia, {siavoosh | bniazmand | jaan | maksim}@ati.ttu.ee

Abstract—Multi-processor system-on-chip (MPSoC) devices are a well known replacement of single-core devices. Some industries, like avionic, are particularly sensitive to the weight and power consumption. Such industries would really take advantage of reducing the number of computers by using MPSoCs. However, the usage of such devices in safety critical domain is currently more than limited. The main issue, which hinders the MPSoC usage in safety critical field, is the presence of on-chip resources shared by the cores. The interconnection itself is the most evident these shared resources. This aspect is an issue as it undermines the safety aspect of the system by providing non functional dependencies, especially from the timing point of view. The certification process is crucial in safety critical field and the system complexity makes the whole certification process harder and even unfeasible. The certification requires to prove that the system exhibits a precise set of guarantees to the safety-critical tasks. While the complexity was an issue for the well known bus-based MPSoCs, it becomes even more critical for the emerging network-on-chip (NoC) interconnection model. The QoSinNoC framework has been created to analyze a set of simple NoC architectures and the related techniques to enable their usage in the scope of mixed criticality. The main contribution of this work is an alternative routing algorithm which allows a better system utilization without any hardware modifications to the NoC architectures considered by QoSinNoC. Furthermore the framework has been upgraded with the power estimation feature which allows a better design space exploration.

Keywords—*mixed-criticality, network-on-chip, quality-of-service.*

I. INTRODUCTION

The Network-on-Chip (NoC) interconnection, has been developed to overcome the limitations of the well known bus- and crossbar-based counterparts in those fields where it is convenient to integrate a considerable amount of cores on a single chip. The main advantage of NoC usage is the reduction of the amount of wiring and thus temperature, delay, and power consumption, when a considerable amount to cores and peripherals are integrated on a single chip. NoC interconnection thus allows to integrate on a single device even thousands of cores, which could be processors (CPUs), DMA or DDR, hardware accelerators, IPs, etc. In this work we will refer to the devices integrating multiple cores on a single chip as Multi-Processor System-on-Chip (MPSoC), independently from the interconnection type they use. The usage of the NoC technology is already in place for some fields, as for instance networking and internet-of-things (IoT). However the usage of such technology struggles to get into the safety-critical domain due to strict insurances required by the mandatory certification process.

Well known members of such domain are aerospace and avionic industries. The electronic systems directly involved in the operations of an airborne vehicle, *avionics*, are required to carry on hard real time tasks, at least for a subset of their functionalities. The related certification is subject to strict standards and has to be issued by a dedicated certification agency. The safety critical certification is responsible for most of the design effort, both in terms of time and money. Such a high cost leads avionic stakeholders to be reluctant in adoption of new technologies.

To be precise, the MPSoC devices are actually used by the avionic industry quite often, however their particular usage de facto makes them single-core devices. In fact, for avionics, MPSoC's processing cores are usually all shut down but one during execution, so that the system is actually used as a single-core processor [1][2]. The main motivation of such a solution, is that for safety critical applications, a worst-case execution time (WCET) has to be estimated with specific and strict precision. The estimated value is required to assure that the scheduled hard real time tasks are able to met their deadlines under any circumstance. An MPSoC has several sources of non-determinism due to the presence of shared resources, with interconnection the most evident of them. Several solutions to perform WCET analysis has been developed [3]. However such solutions does not take into account the nature of the interconnections, as the non-determinism can be caused even without a direct competition for a shared resource [4].

As NoC interconnection is more complex with respect to the bus and crossbar counterparts, its adoption in the avionic field depends on availability of adequate techniques to ensure a set of specific requirements. QoSinNoC framework [5] was developed to propose a set of low complexity architectures and techniques to allow their usage in the context of safety critical systems. The focus of the work has been put on the systems running applications with different levels of requirements in terms of safety, namely mixed criticality (MC) systems. This paper will present an alternative routing technique to be flacked to the one used by QoSinNoC. Power estimation of the considered architectures will be presented as well.

The rest of this paper is organized as follows: Section II provides an overview of the state-of-the-art for MC on NoC-based many-core systems; Section III gives some additional details on MC in avionic field and describes the MC use-case we are considering; Section IV explains the proposed solution to tackle the drawbacks of the techniques used by QoSinNoC; Section V describes some of relevant aspects in scope of the techniques used by QoSinNoC; Sections VI and present the experimental results, finally VII draw some conclusions.

978-1-5386-4757-8/18 $31.00 © 2018 IEEE

II. RELATED WORK

The Æthereal network [6] was the first effort to implement MC on NoC-based system, considering quality-of-service (QoC) concept. The Æthereal network implemented a router microarchitecture able to differentiate among Best-Effort (BE) and a Guaranteed-Throughput (GT) traffic. The SuperGT NoC [7] extends the concept of the Æthereal NoC, introducing a third type of traffic, the SuperGT, and a priority inversion mechanism.

A different approach was taken in [8], where a modified wormhole protocol is proposed to address the mixed-criticality scenario. The Wormhole NoC Protocol for Mixed-Criticality Systems (WPMCS) adds a criticality concept to the scheduling scheme, thus, allowing schedulability of mixed-criticality systems on wormhole networks. In this protocol, a router can be either in a Low-Criticality (LO) or High-Criticality (HI) state and the mechanism used to signal a state transition can be subject to interference by LO traffic. Despite some improvements [9][10] this approach implies increased complexity in the router's arbiter, which should be able to read the slack counter and increase priority of the packet.

Approaches based on run-time dynamic scheduling like the one proposed in [11] can support mixed-criticality applications, however dynamic solutions cannot comply with the verification requirements of high level Safety-Integrity Levels (SIL) as defined in IEC-61508 – or their equivalent in other safety standards.

The QoSinNoC framework, we proposed in [5], considers a set of simple architectures and explores some techniques to enforce the MC usage of such architectures. The approach used by QoSinNoC is the one of traffic isolation done at HW level.

III. MIXED-CRITICALITY APPLICATION USE CASE FOR AVIONICS

For the avionic industry the safety aspect was the most important one. With time, multiple standards was developed in order to describe how the system should be designed and implemented in order to ensure specific and stringent levels of safety [12][13]. Originally the *federated architecture* approach was used, which is a "one function – one computer" approach. This approach allows to assure that a faulty application is not able to affect other applications running onboard of an aircraft. As the number and the complexity of functions to be implemented was keeping growing, the maintenance complexity and the growing cost of such approach led to the development and standardization of Integrated Modular Avionics (IMA) architecture approach [14]. The latter allows multiple functions, also having different levels of criticality to be deployed on single computer. The base idea, also for IMA approach, is to separate functions as it was for the federated architecture. This partitioning concept [15] has to be ensured in order to target a system certification. However this standard only considers the single-core devices and the usage of MPSoC is still avoided.

QoSinNoC framework focuses on a particular MC scenario, which is made by exactly one critical application and one or more non-critical applications. A further characteristic of the considered MC scenario is the absence of shared resources between the applications. The considered MC scenario could match a number of realistic scenarios. For instance the critical application could be a control application which would need a computing core to implement the control function and a ADC/DAC node to interface with sensors and actuators. The non-critical application could be for instance the data logging application, which is in charge of compressing data and sending it to the ground station. This kind of applications is quite common for instance for Unmanned Aerial Vehicles (UAVs). The certification requires to insure that under unexpected behavior non-critical application is not able to interfere with the critical application.

IV. UPGRADING QoSinNoC: ROUTING TECHNIQUE

This section will describe the architectures considered by QoSinNoC as well the techniques used by the latter to allow a MC to be deployed on such architectures. As the final goal is to achieve a certifiable system, the architectures considered by QoSinNoC are fairly simple, having 2D mesh topology and are using worm-hole routing. All the considered architectures are based on single physical network, as the presence of virtual networks would add complexity to the system. Each node of the considered NoC-based system is structured as a processing element connected to a router and the routers are connected by two unidirectional links. The nature of the processing elements is flexible as we can consider a heterogeneous system, where each processing element can be either a computing element (e.g., processor) or a passive slave (e.g., memory). The same low-complexity concept holds for the proposed techniques to enforce MCA deployment as well.

This section will first describe the basic idea of the proposed approach. Then the architectures will be described as well as the implementation of the proposed technique for each of them. The architectures considered by QoSinNoC are described in detail in the work where the framework was first introduced [5]. Still some details has to be recalled in order to explain the routing technique proposed in this work.

A. QoSinNoC mixed criticality enforcment technique

Both the routing technique proposed in the original work [5] (hereinafter called either *original routing technique* or *Rout. A*) and the one proposed in this work (hereinafter called *Rout. B*) allows to implement two levels of the quality of service (QoS) in order to allow MC deployment on the considered NoC architectures. The notation used in this work for the two QoS levels is GS (guaranteed service) and BE (best effort), respectively for safety-critical and non-safety-critical traffic. The proposed notation slightly differs from the one used in the original work [5] as originally the QoS level related to safety-critical traffic was called GL (guaranteed latency) to put emphasis on the guaranteed response time, mandatory for the hard real time applications. However also in the original work, the related QoS level (called GS in this work) was guaranteeing both latency and throughput. Thus, there is no difference between GL of the original work and the GS of this work apart from the name. As the original routing technique, the one proposed in this work guarantees latency and throughput to the critical traffic by ensuring the isolation between the applications.

Both Rout. A and Rout. B are based on traffic isolation, preventing non-critical traffic from using the links used by critical traffic as well as all the other internal to the router resources used by critical traffic (e.g., buffers). If any non-critical node were able to interfere with critical traffic, creating congestion on the path used by the latter, this would mean temporal dependency between the applications. This dependency is exactly what we have to ensure will not happen under any condition. The main cause which could lead to such interference is the presence of bugs inside the non-critical application. In fact, differently from the critical applications, the non-critical ones have a much loose safety requirements and thus cannot be considered as virtually bug-free applications.

Both Rout. A and Rout. B, should allow the deployment of more levels of criticality, although this aspect is outside the scope of this work. The original routing technique will not be described in detail, although some details are necessary to allow a comparison between the two. As will be described in detail in section VII, the proposed technique represents an alternative to the original technique rather than its replacement.

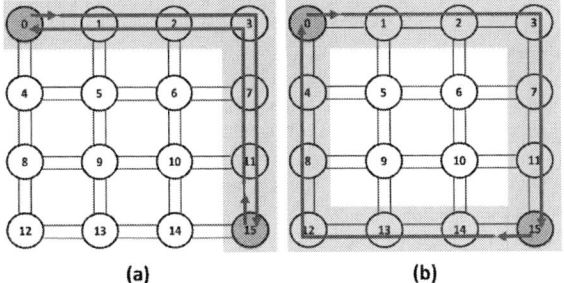

(a) **(b)**

Figure 1. Scenario 1 (SC_1): two critical nodes – node 0 and node 15; shaded area: critical region; (a) original routing technique - *Rout. A*; (b) proposed routing technique - *Rout. B*.

In order to explain the proposed technique (Rout. B), two placement scenarios are considered, SC_1 and SC_2. Both the placement scenarios considers critical application using exactly two nodes and the non-critical application to use all the other nodes. Both the considered scenarios refer to the MC scenario described in section III. The only difference between SC_1 and SC_2 is the placement of the second critical node, which is node **15** for SC_1 and node **14** for SC_2. Different placement scenarios can be justified by the heterogeneous system where some nodes (DDR memory, hardware accelerators) are bound to a specific node, defined at the design phase. A different placement can also be justified by thermal considerations, regardless of whether the considered system is homogeneous or heterogeneous.

The idea behind the proposed technique is to exploit the bidirectional link in order to overcome the dead nodes phenomena induced by the proposed technique. This side effect, already observed in [5] have two different causes, which depends on the NoC architecture. The first cause, experienced by B_LBDR architecture (described in section IV.B), is the absence of any means to prevent a non-critical node located inside a critical channel from injecting its (non-critical) traffic inside the critical channel. The second cause of dead nodes side effect, experienced by the other two architectures (described in sections IV.C and IV.D), is due to unavailability of physical links for non-critical nodes, as the links are used by critical traffic.

The Rout. B technique we are proposing in this work targets the second cause of dead nodes phenomena, while it does not solve the one experienced by B_LBDR architecture. Figure 1.a shows how the node **3** has all the links used by critical traffic.

This will mean that the node is not able to use the NoC and thus it is isolated. The proposed routing technique is shown in Figure 1.b, and it is the well known XY routing. This approach will allow node **3** to exploit its output west link and its input south link. This means that the concept of *critical region* (also used in the original work), defined as the region traversed by the critical traffic, has been completely overcame. In fact, the proposed solution not only allows the non-critical traffic to traverse the critical region but even to flow inside it, as far as the links used by critical traffic are not used by non-critical traffic. For the considered architectures, the condition that the links are not used by the non-critical traffic also imply that there are no internal to the router resource used by critical traffic which is also used by non-critical traffic.

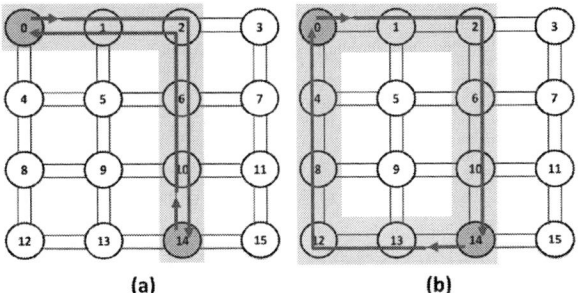

(a) **(b)**

Figure 2. Scenario 2 (SC_2): two critical nodes – node 0 and node 14; shaded area: critical region; (a) original routing technique - *Rout. A*; (b) proposed routing technique – *Rout. B*.

Both SC_1 and SC_2 only consider two critical nodes per critical application. However it is important to notice that the proposed approach does not limit the number of critical nodes as the base idea is to prevent the non-critical traffic to interfere with the critical traffic, and the latter is scheduled in order assure the traffic generated by a critical node will not create congestion to the traffic generate by any other critical node. The aforementioned assumption comes from the best practice of race condition avoidance and WCET estimation of real time scheduling, as the interconnection itself is a shared resource. It is worth to mention that this considerations hold only when the critical application are considered, as the main issue of non-critical applications is the impossibility to assure that these are virtually bug free. This in turn makes it impossible to assume non-critical applications will behave as for scheduling derived for them.

However, as the proposed solution degrades the allowed connectivity between the nodes, it should be clarified that some rules have to be followed in order to avoid further side effects when more than two nodes per application are considered.

First of all, to have the critical region as small as possible can be considered as the general rule. When more than two nodes per critical application are deployed, the rule is not to modify the "two nodes morphology" of the region – L-shape for Rout. A, and *hollow rectangle* in case of Rout. B (see Figure 1 and Figure 2). For Rout. A this simply means that, placed the two most distant nodes, the other have to be placed inside the critical region. For the Rout. B some further considerations are required. Once the two most distant nodes are placed, if the further node is placed in the corner position, it can only communicate to the nodes of the two sides it belongs to as well as to the nodes

978-1-5386-4757-8/18 $31.00 © 2018 IEEE

located in other corner positions. In case the node is placed not in the corner position, it should only communicate with the nodes of the critical region's side it belongs to. The exception to this rule is the case when the critical region does not encircle any non-critical node, in that case there are no restrictions apart from placing the nodes inside the critical region. An example of the last condition is, considering Figure 2.b, a critical node can be added in position **6** but it can only communicate to node **14** (and eventually to node **2** and **10**, if also these nodes will be placed to the same application). If these rules are not followed the MC enforcement will be still valid, however there will easily be a sever connectivity reduction as the critical region will enlarge. As follows from the aforementioned rules, the Rout. B have a drawback to require more constrains for the placement of critical nodes, with respect to Rout. A. I.e., Rout. B reduces placement freedom for the critical nodes.

A natural upper bound to the number of critical nodes is represented by interconnection not to be able to support a large number of nodes under the throughput of latency point of view. However this is rather a limit for the NoC than for the proposed approach.

B. Baseline LBDR

Logic-Based Distributed Routing (LBDR) [16] architecture is characterized by allowing multiple routing algorithms while resulting much more scalable with respect to solutions that rely on usage of routing tables. In detail, this kind of architecture is derived from turn modes analysis [17] to create deadlock-free routing by blocking some of the turns for the packets. The considered baseline LBDR architecture (B_LBDR) allows to implement all the nonadaptive (e.g., the well known XY) and semi-adaptive routings, and thus it provides the means to overcome the connectivity limitations that are introduced by the proposed MC enforcement techniques. To allow some minimal means to enforce the proposed techniques for MCAs deployment, the considered architecture also has 4 connectivity bits per router, which allows to block the usage of the links by the router. These bits thus allow to create completely separated regions, with hardware isolation between them. The routing solution adopted in [5] is shown in Figure 1 (a) and Figure 2 (a). It can be observed how the critical traffic is using XY and YX. For the non-critical traffic, QoSinNoC infers the best, under the reachability (discussed in detail in section V) point of view, considering all nonadaptive and semi-adaptive deadlock-free turn model routing algorithms. However this solution require the non-critical nodes, to be powered down. This is due the absence of any means to prevent a non-critical node from injecting traffic inside the critical region. Thus, this would create congestion and timing interference between the two levels of criticality.

C. Extended LBDR

Extended LBDR architecture (Ext_LBDR) was considered in order to overcome the dead nodes issue experienced by LBRD architecture. As the name states, this architecture is derived the LBDR architecture described in section IV.B. In addition to the turns of LBDR, this architecture also allows to code the straight paths (i.e. W2E, S2N, etc.) and the paths from/to Local (L) port (i.e. N2L, L2N, etc.). From the previous study [5], under many aspects, this architecture is the most suitable for the MC deployment.

D. Routing Table-based

The routing tables based architecture (Table-based) was considered in order to have an architecture which allows non-minimal routing, which in turn is able to tackle the connectivity reduction provoked by the proposed technique. The considered architecture implements a routing table for each router's input port. The table stores the output direction(s) for each destination in the network. Unfortunately this topology does not scale with the growing size of the NoC [18]. The MC enforcement, for this architecture, is implemented by simply not supporting any path (for non-critical traffic) which would create contention to the critical traffic. The region separation is implemented by just non supporting any path which would violate the partitioning.

V. REACHABILITY

Reachability is one of the most important metrics for the techniques of MC enforcement we are considering in QoSinNoC. This metric takes into account the logical connectivity limitations induced by the particular approach we are using. This connectivity reduction is not necessarily physical, as the particular routing algorithm inferred by the proposed approach can prevent some nodes to communicate to each other although a physical connection exists between them. *Reachability* of a node **j** is defined as number of nodes to which node **j** is able to (successfully) send packets to. The reachability figures depend on both the considered architecture and the considered placement scenario. In this work we are considering the following two slightly different definitions of reachability. *Nett reachability* is computed only taking into account the non-isolated nodes. *Rough reachability* is computed also taking into account the dead nodes. Only non-critical nodes are considered for both nett and rough reachability.

A. Reachability optimization

In order to exploit potentialities of LBDR-based and Table-based architectures, QoSinNoC framework optimizes the reachability for the particular considered placement scenario. The particular constrains on the nodes that should be able to communicate with each other are also considered by the framework.

B. Deadlock and livelock

The routing algorithm that is used for the non-critical traffic in this work is devised using the approach proposed in [18]. Using the tool, all uniform turn models that guarantee deadlock- and live-lock freeness are extracted from the whole search space and it is also possible to select the one that provides maximum reachability among the nodes in the network. Proof of deadlock-freeness is also provided in [18], which is based on the concept of proving the absence of cycles in routing graph (RG). By traversing the RG, it is possible to find the paths from each source to destination. Both in the case of using Ext_LBDR and also Table-based, since it is assured that no cycles exist in the RG corresponding to the chosen turn model, therefore, even non-minimal paths would not lead to any deadlock ([18]).

978-1-5386-4757-8/18 $31.00 © 2018 IEEE   210

VI. POWER ESTIMATION

The total power consumption (in mW) consists of static and dynamic power, where the latter consists of internal and switching power. The power results are obtained via Synopsys Design Compiler using AMS 180 nm CMOS technology library. In order to obtain the power results, first each architecture has been simulated under each scenario using a specific packet injection rate (pir). Simulation traces are collected in form of VCD files. To make a fair comparison, the same scenarios for both architectures are considered (SC_1 and SC_2). The switching activity of the components and signals are then stored in SAIF format (switching activity interchange format) which is generated by ModelSim. Further, during the synthesis process, the annotated switching activities (the SAIF files) are fed into the synthesis tool (Synopsys Design Compiler) in order to calculate the final static and dynamic power consumption (in mW).

VII. EXPERIMENTAL RESULTS

QoSinNoC framework has been used to evaluate the proposed routing technique. A MC scenario made of two applications, one safety critical and one non-safety-critical has been considered. For the described MC scenario two placement scenarios have been evaluated, as the results are highly dependent on the placement. The two placement scenarios and the considered routing techniques are described in Section IV. SC_1 is the worst case scenario as it has the maximum Manhattan distance for the critical nodes. SC_2 is a quasi-worst case scenario, having the property to create two non-critical regions isolated by the critical one. As explained in Section IV, the proposed approach (for both Rout. A and Rout. B) is not limited to critical application using only two critical nodes. Instead an arbitrary number of critical nodes are supported as far as the critical region topology is not modified. However, to simulate a scenario having more than 2 critical nodes will not add much to the general discussion as the scheduling of a critical application should be designed (best practice) considering the access time for each shared resource in order to avoid contention. As interconnection is a shared resource itself, this means that only congestion-free figures make sense for the critical nodes.

TABLE 1. ARCHITECTURE FEATURE

Architectures	B_LBDR	Ext_LBDR	Table-Based
Scalability	+	+	-
Non-Minimal Path Support	-	-	+
Area (μm^2) (4x4 Network)	1.49×10^6	1.54×10^6	2.12×10^6
Area Overhead (%)	0 %	2.9 %	41.9 %
Critical Path Delay (ns)	9.1	9.0	9.0
Critical Path Delay Overhead (%)	0 %	-0.9 %	-0.8 %

Both the considered scenarios was simulated injecting 5000 packets per node. Each node always sent 8-flit packets. The destination of non-critical nodes has been computed randomly to achieve a random uniform traffic pattern. Table I compares the placement-independent characteristics of each architecture. Table 2 is related to the SC_1 and compares the original routing technique and proposed routing technique respectively called *Rout. A* and *Rout. B*. Table 3 is related to the SC_2 and performs the same comparison done by Table 2. In Figure 3 are reported the non-critical traffic latency comparison between the Rout. A

and Rout. B approaches for the two considered scenarios. Table 4 presents the figures related to power consumption of the considered architectures. Table 1 was already commented in [5], so it will not be commented – it is included here to provide the global picture.

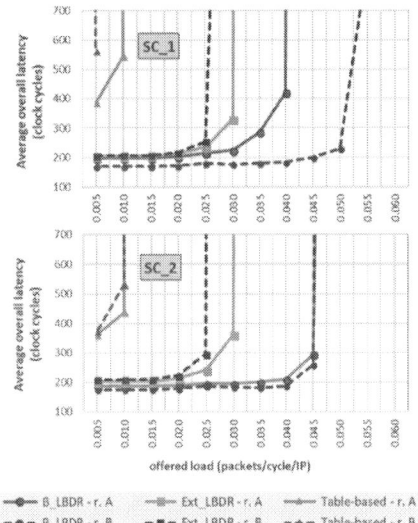

Figure 3. Average latency for non-critical nodes for the two considered scenarios (SC_1 and SC_2).

Before commenting the figures reported in the Table 2, Table 3 and Figure 3 it is worth to explain some aspects of the reported metrics. First of all, the number of dead nodes can easily be considered as the most important metric. The reachability metric could be considered the second important, as the throughput and latency are dependent on such metric. Often high reachability means that also distant nodes can communicate with each other, which increase the congestion and thus makes throughput and latency be worse. However it should be clear that this will not be a ceteris paribus comparison (as different reachability imply different traffic patterns), so the throughput and latency figures have to be considered accordingly.

From Table 2 and Table 3 it can be observed how the B_LBDR suffers a doubling of isolated nodes. Thus, we can conclude that the new routing approach proposed in this paper is not suitable for B_LBDR architecture. Apart from B_LBDR, considering the SC_1 (Table 2), the proposed Rout. B allows to solve the issue of dead nodes. Thus Rout. B is better with respect to Rout. A for this placement scenario. Considering the CS_2 (Table 3), where there are no dead nodes for both Rout. A and Rout. B, we can observe how Rout. B provides a reachability metric improvement for Ext_LBDR, while Table-based has better reachability with Rout. A. From Table 2 and Table 3 it could be also observed how the proposed routing approach does not change the figures related to critical nodes. From the Figure 3 it can be observed how Rout. B exhibits worse latency figures, however, as explained earlier, this is explained by more active nodes and better reachability.

The total power consumption for the 4x4 NoC when using baseline B_LBDR, Ext_LBDR and Table-based architectures under two scenarios SC_1 and SC_2 is illustrated in Table 4. As it can be observed in Table 4, under SC_1, Ext_LBDR has the least dynamic power, while table-based routing has the highest dynamic power, and baseline B_LBDR is close to Ext_LBDR. Due to higher area overhead of the routing table entries per each input port, the table-based approach imposes higher dynamic power results as well, having more logic with switching activity. Similar results can be observed for SC_2.

TABLE 2. PERFORMANCE COMPARISON: SC_1

Architectures		B_LBDR	Ext_LBDR	Table-Based
Non-Critical Nodes Metrics				
Number of dead nodes	Rout. A	5	1	1
	Rout. B	10	0	0
Average Rough Reachability	Rout. A	5.1	7.7	11.1
	Rout. B	0.9	7.5	11.9
Average Nett Reachability	Rout. A	8.0	8.3	12.0
	Rout. B	3	7.5	11.9
Saturation Throughput (flits/clk cycle/system)	Rout. A	3.0	3.4	1.1
	Rout. B	1.9	3.2	0.8
Critical Nodes Metrics				
Congestion-free Latency (clk cycles)	Rout. A	30.5	30.5	30.5
	Rout. B	30.5	30.5	30.5
Saturation Throughput (flits/clk cycle/node)	Rout. A	0.37	0.37	0.37
	Rout. B	0.37	0.37	0.37

TABLE 3. PERFORMANCE COMPARISON: SC_2

Architectures		B_LBDR	Ext_LBDR	Table-Based
Non-Critical Nodes Metrics				
Number of dead nodes	Rout. A	4	0	0
	Rout. B	8	0	0
Average Rough Reachability	Rout. A	3.0	7.6	13.0
	Rout. B	1.0	7.9	12.5
Average Nett Reachability	Rout. A	4.2	7.6	13.0
	Rout. B	2.3	7.9	12.5
Saturation Throughput (flits/clk cycle/system)	Rout. A	3.7	3.5	1.2
	Rout. B	2.3	2.9	1.2
Critical Nodes Metrics				
Congestion-free Latency (clk cycles)	Rout. A	27.5	27.5	27.5
	Rout. B	27.5	27.5	27.5
Saturation Throughput (flits/clk cycle/node)	Rout. A	0.37	0.37	0.37
	Rout. B	0.37	0.37	0.37

TABLE 4. POWER CONSUMPTION FOR ROUT. B: DETAILS

Architectures	B_LBDR		Ext_LBDR		Table-Based	
	SC1	SC2	SC1	SC2	SC1	SC2
Static power consumption (mW) (leakage)	0.003	0.003	0.003	0.003	0.004	0.004
Internal power consumption (mW)	48.9	48.7	44.9	44.9	61.1	60.0
Switching power consumption (mW)	3.5	3.5	6.3	5.9	3.8	2.8
Dynamic power consumption (mW) (internal + switching)	52.4	52.2	51.2	50.8	64.9	62.8
Dynamic power consumption overhead (%)	0%		-2.3%	-2.8%	23.8%	20.3%

VIII. CONCLUSIONS

This work presents some improvements to QoSinNoC, a framework created to facilitate the usage of NoC technology by the industries operating in safety critical domain. As the simplicity is crucial to obtain a certification, the purpose of the framework is to evaluate the performance of a set of simple NoC architectures, when MC enforce ment techniques are applied.

The improvements proposed in this work target both the framework itself and the techniques used to enforce MC. In detail, we propose a routing technique exploiting the bidirectional physical link. The technique has been implemented and compared to the one originally used by QoSinNoC. We considered a particular case of a MCA: one critical application and one non-critical application. For the critical application only two nodes have been simulated, although more critical nodes can

be supported with no overhead as far as the critical region's morphology is not modified. The experimental results show how the proposed routing approach allows to solve the dead nodes side effect for some of the considered architectures. The proposed routing approach is also providing an alternative routing, as the one originally used by QoSinNoC performs better in some cases. Furthermore, the power consumption estimation figures has been reported to better explore the design space.

ACKNOWLEDGMENT

The work has been supported by EU H2020 TWINN TUTORIAL, EU H2020 RIA IMMORTAL and Estonian institutional research grant IUT19-1 and the European Regional Development Fund.

REFERENCES

[1] X. Jean, M. Gatti, G. Berthon, and M. Fumey, "MULCORS - Use of Multicore Processors in airborne systems," Final study report EASA.2011/6, 2012.

[2] CAST, "Position Paper CAST-32, Multi-core Processors," 2014.

[3] A. Burns and R. I. Davis, Mixed Criticality Systems - A Review, 7th Ed., University of York, 2016.

[4] Z. Shi and A. Burns, "Schedulability analysis and task mapping for real-time on-chip communication," Real-Time Syst., vol. 46, 2010, pp. 360–385.

[5] S. Avramenko, S. P. Azad, S. Esposito, B. Niazmand, M. Violante, J. Raik, M. Jenihhin, "QoSinNoC: Analysis of QoS-Aware NoC Architectures for Mixed-Criticality Applications," in 21st IEEE International Symposium on Design and Diagnostics of Electronic Circuits and Systems, DDECS 18, 2018.

[6] K. Goossens, J. Dielissen, and A. Radulescu, "Æthereal Network on Chip:Concepts, Architectures, and Implementations," IEEE Des. Test Comput., vol. 22, no. 5, pp. 414–421, 2005.

[7] T. Marescaux and H. Corporaal, "Introducing the SuperGT network-on-chip," in Design Automation Conference, pp. 116–121, 2007.

[8] A. Burns et al. "A wormhole NoC protocol for mixed criticality systems," in 2014 IEEE Real-Time Systems Symposium, 2014.

[9] L. S. Indrusiak, J. Harbin, and A. Burns, "Average and Worst-Case Latency Improvements in Mixed-Criticality Wormhole Networks-on-Chip," in Euromicro Conference on Real-Time Systems, pp. 47–56, 2015.

[10] S. Tobuschat and R. Ernst, "Efficient latency guarantees for mixed-criticality networks-on-chip," in Proceedings of the IEEE Real-Time and Embedded Technology and Applications Symposium, RTAS, 2017.

[11] A. Kostrzewa, S. Saidi, and R. Ernst, "Dynamic Control for Mixed-Critical," in Real-Time Systems, pp. 317–326, 2015.

[12] RTCA Inc., "DO-254 Design Assurance Guidance for Airborne Electronic Hardware", Issue date 4/19/2000.

[13] RTCA Inc., "DO-178c Software Considerations in Airborne Systems and Equipment Certification", Issue date 12/13/2011.

[14] P. J. Prisaznuk, "Integrated modular avionics," Aerosp. Electron. Conf. 1992. NAECON 1992., Proc. IEEE 1992 Natl., pp. 39–45.

[15] J. Rushby, "Partitioning in Avionics Architectures: Requirements, Mechanisms, and Assurance," NASA Langley Research Center, NASA CR-1999-209347, 1999.

[16] S. R. Mocholi, "Cost Effective Routing Implementations for On-Chip Networks," Universidad Politécnica de Valencia, 2010.

[17] C.J. Glass, L.M. Ni, "The Turn Model for Adaptive Routing", in 19th Annual International Symposium on Computer Architecture, 1992.

[18] S. P. Azad et al., "Comprehensive performance and robustness analysis of 2D turn models for network-on-chips," 2017 IEEE International Symposium on Circuits and Systems (ISCAS), Baltimore, MD, 2017, pp. 1-4.

Testability of Switching Lattices in the Stuck at Fault Model

Anna Bernasconi
Dipartimento di Informatica
Università di Pisa, Italy
anna.bernasconi@unipi.it

Valentina Ciriani   Luca Frontini
Dipartimento di Informatica
Università degli Studi di Milano, Italy
{valentina.ciriani, luca.frontini}@unimi.it

Abstract—Switching lattices are two-dimensional arrays of four-terminal switches proposed in a seminal paper by Akers in 1972 to implement Boolean functions. Recently, with the advent of a variety of emerging nanoscale technologies based on regular arrays of switches, synthesis methods targeting lattices of multi-terminal switches have found a renewed interest. In this paper, the testability under the stuck-at-fault model (SAFM) of switching lattices is analyzed, and properties of fully testable lattices are identified and discussed. Experimental results are given to analyze the testability of lattices synthesized with different methods.

Index Terms—Switching lattices; testability; logic synthesis.

I. INTRODUCTION

A switching lattice is a two-dimensional lattice of four-terminal switches linked to the four neighbors of a lattice cell, so that these are either all connected, or disconnected. A Boolean function can be implemented by a lattice associating each four-terminal switch to a Boolean literal, so that if the literal takes the value 1 the corresponding switch is ON and connected to its four neighbors, otherwise it is not connected. The function evaluates to 1 if and only if there exists a connected path between two opposing edges of the lattice, e.g., the top and the bottom edges (see Figure 1 for an example). The synthesis problem on a lattice consists in finding an assignment of literals to switches in order to implement a given target function with a lattice of minimal size.

The idea of using regular two-dimensional arrays of switches to implement Boolean functions dates back to a seminal paper by Akers in 1972 [2], but has found a renewed interest recently, thanks to the development of a variety of nanoscale technologies. Synthesis algorithms targeting lattices of multi-terminal switches have been designed [3], [5], [13], [14], and methods based on function decomposition techniques have been exploited to mitigate the cost of implementing switching lattices [8], [9], [10]. Moreover, several studies on fault tolerance for nano-crossbar arrays have been published recently [4], [15], [16], [17].

Besides synthesis and fault tolerance, testability is a major aspect of the design process. While detailed studies on testability have been performed for standard two-level and three-level networks (see for instance [1], [6], [7], [11], [12],

[18]), to the best of our knowledge, the testability of switching lattices has not been considered so far. Therefore, in this paper, we study redundancies of lattices under a static fault model: the stuck-at-fault model (SAFM). In particular, we prove that under the SAFM, switching lattices minimized with respect to the number of literals controlling the switches are free of redundancies by construction. Whereas, it can be shown by counter examples that lattices minimized with respect to the number of switches, i.e. minimized with respect to the size, are not in general fully testable. We also identify the properties that make a switching lattice fully testable in the SAFM, and show how these properties resemble the properties that guarantee the full testability of the SOP forms in the SAFM, i.e., the primality of the products and the irredundancy of the cover. Finally, we propose a method for identifying redundant cells in a lattice. We conclude the paper reporting experimental results regarding the testability of lattices synthesized with two different methods [5], [13].

II. PRELIMINARIES

A. Fault Models (FMs)

The standard stuck-at faults model (shortly, SAFM) is well-known and used throughout the industry for many years [1], [11]. In SAFM it is assumed that a defect causes a basic cell input or output to be fixed to either 0 or 1, i.e., signal lines can assume constant values independent of the inputs.

Definition 1: A *stuck-at fault* with fault location v is a tuple $(v[i], \epsilon)$ or $([i]v, \epsilon)$. $v[i]$ ($[i]v$) denotes the i-th input (output) pin of v, $\epsilon \in \{0, 1\}$ is the fixed constant value.

In the following we simply speak of stuck-at-0 (SA0) and stuck-at-1 (SA1) faults. Now, let C be any combinational logic circuit over a fixed library.

Definition 2: An input t to C is a *test* for a fault f, iff the primary output values of C on applying t in presence of f are different from the output values of C in the fault free case.

A fault is *testable*, iff there exists a test for this fault. The goal of any test pattern generation process is a *complete* test set for the circuit under test, i.e., a test set that contains a test for each testable fault. The construction of complete test sets requires the determination of the faults which are not testable (= *redundant*), even though it is easy to see that in general the detection of redundancies is *coNP-complete*. Redundancies have further unpleasant properties: they may

978-1-5386-4757-8/18 $31.00 © 2018 IEEE

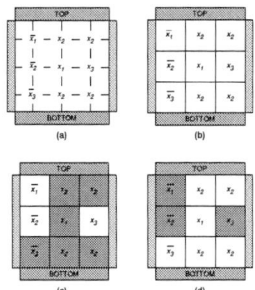

Fig. 1. A four terminal switching network implementing the function $f = \overline{x}_1\overline{x}_2\overline{x}_3 + x_1x_2 + x_2x_3$ (a); its corresponding lattice form (b); the lattice evaluated on the assignments 1,1,0 (c) and 0, 0, 1 (d), with grey and white squares representing ON and OFF switches, respectively.

invalidate the completeness of the test set and often correspond to locations of the circuit where area is wasted [11]. For this, synthesis procedures which result in non-redundant circuits are desirable. A node v in C is called *fully testable*, if there does not exist a redundant fault with fault location v. If all nodes in C are fully testable, then C is called *fully testable*.

Finally, we recall that the investigations with respect to the SAFM are usually based on the single fault assumption, i.e., one assumes that there is at most one fault in the circuit.

B. Switching Lattices

A switching lattice is a two-dimensional array of four-terminal switches linked to the four neightbours of a lattice cell, so that these are either all connected (when the switch is ON), or disconnected (when the switch is OFF). A Boolean function can be implemented by a lattice in terms of connectivity across it:

- each four-terminal switch is controlled by a literal;
- each switch may be also labelled with the constant 0, or 1;
- if the literal takes the value 1, the corresponding switch is connected to its four neightbours, else it is not connected;
- the function evaluates to 1 if and only if there exists a connected path between two opposing edges of the lattice, e.g., the top and the bottom edges;
- input assignments that leave the edges unconnected correspond to output 0.

For instance, the 3×3 network of switches in Figure 1 (a) corresponds to the lattice form depicted in Figure 1 (b), which implements the function $f = \overline{x}_1\overline{x}_2\overline{x}_3 + x_1x_2 + x_2x_3$. If we assign the values 1, 1, 0 to the variables x_1, x_2, x_3, respectively, we obtain paths of gray square connecting the top and the bottom edges of the lattices (Figure 1 (c)), indeed on this assignment f evaluates to 1. On the contrary, the assignment $x_1 = 0, x_2 = 0, x_3 = 1$, on which f evaluates to 0, does not produce any path from the top to the bottom edge (Figure 1 (d)).

The synthesis problem on a lattice consists in finding an assignment of literals to switches in order to implement a given target function with a lattice of minimal size. The size is measured in terms of the number of switches in the lattice.

A switching lattice can similarly be equipped with left edge to right edge connectivity, so that a single lattice can implement two different functions. This fact is exploited in [5] where the authors propose a synthesis method for switching lattices simultaneously implementing a function f according to the connectivity between the top and the bottom plates, and its dual function[1] f^D according to the connectivity between the left and the right plates. In [13], the authors have proposed a different approach to the synthesis of minimal-sized lattices, which is formulated as a satisfiability problem in quantified Boolean logic and solved by quantified Boolean formula solvers. This method uses the previous algorithm to find an upper bound on the dimensions of the lattice. It then searches for successively better implementations until either an optimal solution is found, or a preset time limit has been exhausted. Experimental results show how this alternative method can decrease lattice sizes considerably. In this approach the use of fixed inputs (i.e., constant values 0 and 1) is allowed.

III. LATTICES: DEFINITIONS AND PROPERTIES

In this section we introduce some definitions and present some properties of switching lattices that will be exploited in Section IV for the analysis of their testability.

Let the first row of a lattice be the *top row*, the last row be the *bottom row*, and any other row be an *internal row*. Two cells in a lattice are *adjacent* if they are in the same column and in two adjacent rows or in the same row and in two adjacent columns. Hereafter, in a lattice we denote *path* any list $l_1, l_2, \ldots, l_{m-1}, l_m$ of literals such that l_i and l_{i+1} (for all $1 \leq i < m$) are contained in adjacent cells and: 1) l_1 is contained in a cell in the top row, 2) l_m is contained in a cell in the bottom row, and 3) all the other literals (i.e., $l_2, \ldots, l_{m-1}$) are contained in cells of the internal rows. Note that paths in lattices may contain more occurrences of the same literal.

Definition 3: A path in a lattice is *unsatisfiable* (resp., *satisfiable*) if contains (resp., does not contain) both a variable x and its complement $\overline{x}$.

Definition 4: The *product associated to a satisfiable path* is the conjunction of all literals of the path, without repetitions. The *product associated to an unsatisfiable path* is 0.

For example, in the lattice in Figure 1 (b) the path x_2, x_1, x_2 is satisfiable and the path $\overline{x}_1, \overline{x}_2, x_1, x_2$ is unsatisfiable. The associated products are x_1x_2 and 0, respectively.

With a slight abuse of notation, we consider the products associated to all paths in a lattice L as implicants of the function f_L implemented by L. Indeed, f_L evaluates to 1 on the set of minterms covered by these products. In this framework, the set of minterms covered by an implicant can be empty: this happens whenever a path is unsatisfiable, as, in this case, the associated product evaluates to 0.

Definition 5: An *accepting path* for a minterm v in a lattice is a satisfiable path whose associated product covers v.

We now introduce the concept of *primality of a path* in a lattice which is strictly related to the concept of prime implicant in a SOP:

---

[1]The dual of a Boolean function f depending on n binary variables is the function f^D such that $f(x_1, x_2, \ldots, x_n) = \overline{f^D}(\overline{x}_1, \overline{x}_2, \ldots, \overline{x}_n)$.

978-1-5386-4757-8/18 $31.00 © 2018 IEEE

Definition 6: A path $l_1, \ldots, l_i, \ldots, l_m$ in a lattice L is *prime* w.r.t. a literal l_i ($1 \le i \le m$), if the product associated to the sequence of literals obtained removing l_i from the path is not an implicant of the function implemented by L.

The primality of a path with respect to a literal l implies that:

(1) the path cannot contain other occurrences of l, since the corresponding product would not change if we remove one occurrence of l from the path, leaving the others;

(2) the path cannot contain pairs $x, \overline{x}$, with $x \ne l$, since the removal of l would leave the associated product unchanged, and equal to 0;

(3) the path might contain cells associated to $\overline{l}$ (in this case the path is unsatisfiable, and becomes satisfiable after the removal of l).

For instance, in the lattice in Figure 1 (b):

- the satisfiable path x_2, x_3, x_2 is prime w.r.t. x_3, as the product x_2 obtained removing x_3 from the path is not an implicant of $f = \overline{x}_1 \overline{x}_2 \overline{x}_3 + x_1 x_2 + x_2 x_3$;
- the satisfiable path x_2, x_3, x_2 is not prime w.r.t. x_2, as the removal of one occurrence of x_2 leaves the associated product $x_2 x_3$ unchanged;
- the unsatisfiable path $x_2, x_1, \overline{x}_2, \overline{x}_3$ is prime w.r.t. x_2, as the product $x_1 \overline{x}_2 \overline{x}_3$ is not an implicant of f;
- the unsatisfiable path $\overline{x}_1, \overline{x}_2, x_1, x_2$ is never prime, as the removal of any of its literal leaves the associated product unchanged, and equal to 0.

We finally focus on the single cells in a lattice, and we introduce a property that can be associated to the irredundancy of a SOP. Let c be a cell in a switching lattice L that implements a function f_L.

Definition 7: The cell c is *essential* in L if there exists at least a minterm v in the on-set of f_L whose accepting paths always contain c.

For instance, in the lattice in Figure 1 (b) all cells on the leftmost column are essential, as they form the only accepting path for the on-set minterm 000; while the top-left cell c in the lattice in Figure 2 (b) is not essential, since for any on-set minterm of the function implemented by the lattice there exists an accepting path that does not include c.

Observe that setting to the constant value 0 a literal in a cell c is equivalent to removing all paths that include c from the lattice L: indeed, the 0 in c disconnects, i.e., makes unsatisfiable, all paths going through c. If, in addition, the cell c is essential for a minterm v, then all accepting paths for v are removed from L. Thus, the function implemented by the lattice changes, at least on v. In this sense, we can associate the notion of essential cell to that of irredundant product: if we remove an irredundant product from a SOP, the function represented by the expression changes, and if we remove from the lattice L all paths that include an essential cell, the function implemented by L changes. We conclude this section with the following two propositions that characterize how the function implemented by a lattice may change setting one cell to a constant value, i.e., forcing a stuck-at-fault in the cell. Let L be a switching lattice, and let f_L be the function implemented by L. Now, consider the lattice $L^{c \leftarrow 1}$ obtained replacing a literal in a cell c of L with the constant 1.

Proposition 1: The on-set of the function $f_{L^{c \leftarrow 1}}$ implemented by $L^{c \leftarrow 1}$ is a superset of the on-set of f_L, i.e., $f_L^{on} \subseteq f_{L^{c \leftarrow 1}}^{on}$.

Proof. We show that any satisfiable path in L remains satisfiable in $L^{c \leftarrow 1}$. Let $p = l_1, l_2, \ldots, l_{m-1}, l_m$ be a satisfiable path in L. If this path does not include the cell c, than the corresponding path in $L^{c \leftarrow 1}$ is composed by exactly the same literals of p, and it is satisfiable. If p includes the cell c, than the corresponding path in $L^{c \leftarrow 1}$ is obtained replacing one of the literals of p with the constant 1, and it is satisfiable. ∎

Consider now the lattice $L^{c \leftarrow 0}$ obtained replacing a literal in a cell c of L with the constant 0. We have:

Proposition 2: The on-set of the function $f_{L^{c \leftarrow 0}}$ implemented by $L^{c \leftarrow 0}$ is a subset of the on-set of f_L, i.e., $f_{L^{c \leftarrow 0}}^{on} \subseteq f_L^{on}$.

Proof. We show that any satisfiable path in $L^{c \leftarrow 0}$ is satisfiable in L. Let $p = l_1, l_2, \ldots, l_{m-1}, l_m$ be a satisfiable path in $L^{c \leftarrow 0}$. The thesis immediately follows since the satisfiable path p cannot include the cell c (that has value 0), thus the corresponding path in L is composed by exactly the same literals of p, and is satisfiable. ∎

IV. TESTABILITY IN THE SAFM

In this section we analyze the properties that make a switching lattice fully testable in the SAFM. As we will see, these properties resemble the two properties that guarantee the full testability of the SOP forms in the SAFM: the primality of the products, that ensures the testability of AND gates, and the irredundancy of the cover, that guarantees the testability of the OR gate. In our analysis we will consider stuck-at-faults at the literals that control the four-terminal switches in the lattice. We first introduce the notion of irredundant literal in a lattice.

Definition 8: A literal in a lattice's switch is *0-irredundant* (resp., *1-irredundant*) if it cannot be substituted by the constant 0 (resp., 1) without changing the function computed by the lattice.

For instance, the literal x_1 in the top-left cell of the lattice in Figure 2 (b) is not 0-irredundant, while the literal x_4 in the center-right cell of the lattice in Figure 2 (a) is not 1-irredundant. The notion of irredundancy can be extended to the whole lattice as follows:

Definition 9: A lattice is *0-irredundant* (resp., *1-irredundant*) if any literal contained in it is 0-irredundant (resp., 1-irredundant).

Definition 10: A lattice is *irredundant* if it is 0-irredundant and 1-irredundant.

Observe that 0-irredundant literals guarantee the testability of SA0 faults in the corresponding cell of the lattice, while 1-irredundant literals guarantee the testability of SA1 faults. Indeed, since the function implemented by the lattice changes if we set a literal in a cell c to the constant 1 or to the constant 0, the fault in c can be tested on the minterm on which the function changes. Thus, we have

Proposition 3: An irredundant lattice is fully testable with respect to the SAFM.

The two lattices in Figure 2 (c) and (d) are irredundant, and thus fully testable in the SAFM.

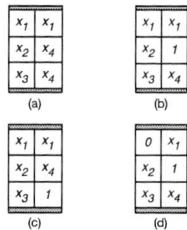

Fig. 2. Four different minimum size switching lattices implementing the function $f = x_1 x_2 x_3 + x_1 x_4$. (a) A 0-irredundant, but not 1-irredundant, lattice; (a) a 1-irredundant, but not 0-irredundant, lattice; (c) a fully testable lattice; (d) a fully testable lattice with a minimum number of literals.

We now investigate the relations between minimality of a switching lattice and full testability in the SAFM. We first consider the minimality with respect to the number of switches, i.e., the area of the lattice. Under the SAFM, it can be shown by counter examples that a lattice of minimum size for a given target function is not in general fully testable. Consider for instance the lattice in Figure 2 (a) that implements the function $f = x_1 x_2 x_3 + x_1 x_4$: this lattice is of minimum size, but it is not fully testable, because it is not 1-irredundant. On the contrary, lattices minimized with respect to the number of literals in the switches are free of redundancies.

Theorem 1: A switching lattice L with a minimum number of literals is fully testable in the SAFM.

Proof. Suppose that L is not fully testable. This means that there exists at least one redundant cell in the lattice whose literal can be replaced by a constant value, 0 or 1, without changing the implemented function. If replace the literal in this cell with a constant value, we get a new lattice for the same function, with a smaller number of literals, in contradiction with the minimality of the number of literals in L. ∎

The lattice in Figure 2 (d) contains a minimum number of literals (4) and is therefore fully testable. Note that the minimality of the number of literals is not necessary for the full testability. For example, the lattice in Figure 2 (c) is fully testable, but not minimum with respect to the number of literals, as it contains 5 literals instead of 4.

Finally, we prove in the following theorems that the structural properties of switching lattices that guarantee their full testability in the SAFM are the primality of the paths (for the stuck-at-1 faults) and the essentiality of the cells (for the stuck-at-0 faults).

Theorem 2: A SA1 in a lattice cell c with literal l is testable if and only if there exists a path p that contains the cell c and is prime with respect to l.

Proof. Let L be a switching lattice, and let f_L be the function implemented by L.

(If part). Consider a path p in L that contains the cell c and is prime w.r.t. l. Recall that the primality of p w.r.t. l implies that p cannot contain other occurrences of l, p might contain cells associated to $\bar{l}$, and p cannot contain pairs $x, \bar{x}$, with $x \neq l$. We must show that the SA1 in c can be propagated to

the output of the lattice L in order to be tested, i.e., we must prove that the function implemented by the faulty lattice differs from the original function f_L on at least one minterm. This immediately follows from the primality of the path p. Indeed, if we substitute with the constant 1 the unique occurrence of l in p, the resulting product, which cannot be empty because of the properties of prime paths (see Section III), is not an implicant of f_L. Therefore, there exists at least one minterm in the off-set of f_L covered by this new product. On this minterm the faulty lattice computes 1, instead of 0, since the path p, with l replaced by 1, becomes an accepting path.

(Only-if part). Now suppose that the SA1 in the cell c, with literal l, is testable. As we are injecting a constant value 1 into the lattice, we know, by Proposition 1, that the faulty lattice is always correct on the on-set of f_L. Thus, the testability of the SA1 implies the existence of an off-set minterm v of f_L on which the faulty lattice computes 1, while the original lattice computes 0. Since whenever the literal l gets the value 1, correct and faulty lattices are identical and have the same behaviour, l must be 0 on v. Let us now denote with p an accepting path for v in the faulty lattice. p is satisfiable and contains the cell c with the constant 1, while it cannot contain cells with literal l since l is 0 on v. Consider the corresponding path q in the original lattice L. p and q are identical in all cells but c, which is labelled by 1 in p and by l in q. The product of all literals in q is an implicant of f_L, possibly empty if it contains $\bar{l}$. If we remove l, the resulting product is not an implicant since it covers the off-set minterm v. Therefore path q is prime w.r.t. l. ∎

Theorem 3: A SA0 in a lattice cell c is testable if and only the cell c is essential.

Proof. Let L be a switching lattice, and let f_L be the function implemented by L.

(If part). We show that the function implemented by the lattice with a SA0 in the essential cell c differs from f_L on at least one minterm. As we are injecting the value 0 into one cell of L, Proposition 2 implies that the faulty lattice is always correct on the off-set of f_L. Thus, to prove the testability of the SA0, we must show that there is an on-set minterm of f_L on which the faulty lattice computes 0, while the original lattice computes 1. Let l be the literal associated to cell c. Since c is essential, there exists an on-set minterm v for which all accepting paths include c. On all these paths, l must be 1. Now, if we substitute with the constant 0 the literal l in c, we disconnect all paths going through c, and in particular all accepting paths for v. This the faulty lattice computes 0 instead of 1 on v, and the fault can be tested.

(Only-if part). We prove by contraposition that if c is not essential, then a SA0 in c cannot be tested. So, suppose that cell c, with literal l is not essential. By Proposition 2, we know that the faulty lattice, with a SA0 in c, is always correct on the off-set of f_L. Thus, it is enough to show that the faulty lattice is correct also on the on-set of f_L. Let v be a minterm of f_L. Since c is not essential, there exists an accepting path p for v not including c. Thus, the SA0 in c has no effect on p, and the faulty lattice correctly computes 1 on v. ∎

V. ALGORITHMS FOR IRREDUNDANCY TEST

In this section we describe a strategy for identifying the *redundant*, i.e., non irredundant, cells in a lattice. Recall that an irredundant lattice is fully testable under the SAFM. Moreover, recall that in our analysis we only consider stuck-at-faults in non-constant cells, i.e., only stuck-at-faults at the literals that control the switches. We first describe a methodology to test whether a given cell in a lattice is 0-irredundant (resp., 1 irredundant). Algorithm 1 shows the strategy based on Theorem 3: a cell c, in the lattice L, is 0-irredundant if and only if c is essential, i.e., there exists at least one miniterm v in the on-set of f_L whose accepting paths always contains c.

Algorithm 1: Algorithm for the testing of the 0-irredundancy of a cell c.

0-irredundant (cell c)
INPUT: A cell c (containing the literal l) in a lattice L
OUTPUT: true if c is 0-irredundant in L, **false** otherwise
forall sub-path p_T from a top cell of L to c ($c \notin p_T$)
  if (p_T contains $\bar{l}$) discard p_T;
  if (p_T contains x and $\bar{x}$) discard p_T;
  else
    forall sub-path p_B from c to a bottom cell of L ($c \notin p_B$)
      if (p_B contains $\bar{l}$) discard p_B;
      if ($p_T p_B$ contains x and $\bar{x}$) discard p_B;
      else forall miniterm m of the product associated to p_T, l, p_B
        if m is not in the on-set of $L^{c \leftarrow 0}$ **return true** ;
return false;

The algorithm starts from the given cell c (containing the literal l) and considers any sub-path p_T in the lattice from one top cell to c (where c is non included in p_T). If p_T contains $\bar{l}$ the path is discarded since the SA0 in c will not change the output of any miniterm computed by L through p_T. Moreover, if p_T contains a variable x and its complement $\bar{x}$ any path containing as a prefix p_T outputs 0. Thus, the output is not affected by the SA0 in c and p_T can be discarded. Any other sub-path p_T is consider in combination with a sub-path p_B from c to a bottom cell of L. Again, we can discard some of the sub-paths p_B in a similar way. In order to check if c is essential, we are left to consider any miniterm m of the product p associated to p_T, l, p_B and test if m is in the on-set of $L^{c \leftarrow 0}$. We can omit to test the miniterms in the product associated to $p_T, \bar{l}, p_B$ since we have $c = 0$ in $L^{c \leftarrow 0}$. If we find a miniterm of p that outputs 0 in $L^{c \leftarrow 0}$, the fault is testable.

Algorithm 2 shows the strategy based on Theorem 2: a cell c (containing the literal l), in the lattice L, is 1-irredundant if and only if there exists a path p that contains the cell c and is prime with respect to l, i.e., the product associated to the sequence of literals obtained removing l from the path is not an implicant of the function implemented by L.

Algorithm 2: Algorithm for the testing of the 1-irredundancy of a cell c.

1-irredundant (cell c)
INPUT: A cell c (containing the literal l) in a lattice L
OUTPUT: true if c is 1-irredundant in L, **false** otherwise
forall sub-path p_T from a top cell of L to c ($c \notin p_T$)
  if (p_T contains l) discard p_T;
  if (p_T contains x and $\bar{x}$) discard p_T;
  else
    forall sub-path p_B from c to a bottom cell of L ($c \notin p_B$)

if (p_B contains l) discard p_B;
if ($p_T p_B$ contains x and $\bar{x}$) discard p_B;
else forall miniterm m of the product associated to $p_T, \bar{l}, p_B$
  if m is not in the on-set of L **return true** ;
return false;

The first part of the algorithm is similar to Algorithm 1. In order to check if there exists a path p that contains the cell c and is prime with respect to l, we are left to consider any miniterm m of the product p associated to $p_T, \bar{l}, p_B$ and test if m is in the on-set of L. It is easy to see that a lattice L can be modeled with an undirected graph $G_L = (V, E)$ where each vertex v in V corresponds to a cell c in L and it is labeled with the literal in c. The edge (v, w) is in E iff the vertices v and w correspond to adjacent cells in L. A vertex corresponding to a cell in the top (resp., bottom) row is a *top* (resp., *bottom*) *vertex*. Both the algorithms consider sub-paths from top cells to c and from c to bottom cells. They can be easily implemented by DFS visits in the graph G_L from c to top (resp., bottom) vertices. The final test in both algorithms can be implemented by OBDD based methods, as classically performed in SOP forms for primality and irredundancy test. More precisely, we need to build the OBDD representing the target function f computed by L, and the OBDD containing all products associated to a satisfiable path through c. The two algorithms have then time complexity polynomial in the size of the OBDDs and the graph G_L. An alternative strategy is a simulation in the given lattice L and in the transformed ones $L^{c \leftarrow 0}$ and $L^{c \leftarrow 1}$. The irredundancy of the overall lattice is simply tested by applying the 0-irredundant and 1-irredundant tests on all the non-constant cells of the lattice.

VI. EXPERIMENTAL RESULTS

In this section we discuss the experiments aimed at evaluating the testability of switching lattices synthesized with the recent methods presented in [5], [13]. These experiments are based on the fault injection in lattices by substituting a literal controlling a single cell with a SA1 or a SA0. The fault injection procedure is repeated for each cell of the lattice.

The defect simulations have been run on a machine with two AMD Opteron 4274HE for a total of 16 CPUs at 2.5 GHz and 128 GByte of main memory, running Linux CentOS 7. The benchmarks functions are expressed in PLA form and are taken from a subset of LGSynth93 [19]. A total of about 580 functions were considered, and for each function each output is implemented as a separate Boolean function. The software used for simulations is written in C++. We used ESPRESSO to implement the method described in [5], and a collection of Python scripts for computing minimum-area lattices by transformation to a series of SAT problems, to simulate the results reported in [13].

Table I reports a sample of benchmark functions, referring to lattices synthesized as described in [5] and [13]. The benchmarks synthesized with [13] method were stopped after ten minutes of each SAT execution. The first column in the table reports the name and the number of the considered output of each function. The following columns report, for each synthesis method, the dimension ($r \times s$, and area) of the lattice, and the percentages of 0-redundant and 1-redundant

TABLE I

A SAMPLE OF BENCHMARK FUNCTIONS SYNTHESIZED WITH [5] AND [13] APPROACHES AND THEIR PERCENTAGES OF 0-REDUNDANT AND 1-REDUNDANT CELLS

			[5]				[13]	
name	col × row	area	$(R_0 /$ area$)\%$	$(R_1 /$ area$)\%$	col × row	area	$(R_0 /$ area$)\%$	$(R_1 /$ area$)\%$
addm4 (6)	10×11	110	49%	79%	6×4	24	0%	0%
b11 (3)	3×6	18	22%	56%	3×4	12	8%	8%
b7 (27)	2×5	10	0%	30%	3×3	9	22%	0%
bench (3)	4×6	24	8%	58%	4×3	12	8%	0%
dc2 (1)	7×12	84	40%	62%	6×4	24	4%	13%
ex5 (34)	10×4	40	8%	53%	6×4	24	0%	8%
exps (32)	2×7	14	43%	29%	2×5	10	10%	0%
m3 (3)	5×4	20	10%	55%	5×3	15	7%	7%
m3 (4)	8×6	48	27%	42%	7×3	21	0%	0%
max128 (23)	11×12	132	33%	82%	–	–	–	–
newtag (0)	8×4	32	13%	69%	6×3	18	0%	0%
newxcpla1 (18)	10×7	70	44%	71%	3×7	9	0%	0%
p3 (10)	6×10	60	10%	67%	4×5	20	0%	15%
p82 (13)	5×7	35	29%	34%	3×5	15	0%	0%
rd53 (1)	10×10	100	18%	80%	–	–	–	–
risc (21)	2×5	10	20%	20%	2×4	8	13%	0%
root (1)	8×8	64	36%	73%	6×4	24	8%	8%
sex (4)	3×5	15	40%	27%	3×4	12	17%	17%
tms (0)	4×11	44	32%	41%	3×6	18	0%	0%

TABLE II
OVERALL RESULTS OF THE SIMULATIONS

Synthesis Method	Average area	$(R_0/$area$)\%$	$(R_1/$area$)\%$
[5]	30	20%	29%
[13]	15	4.5%	4.5%

cells. Table II describes the overall results for the benchmarks we considered, and it shows the average values for lattice area and percentages of 0-redundant and 1-redundant cells.

We can note that the percentage of cells that are redundant is higher in the [5] synthesis method. This is due to the more constrained structure of the lattices. Indeed, the method proposed in [5] computes a lattice that implements both the target function and its dual, and is in general less compact than the corresponding lattice given by [13].

VII. CONCLUSION

In this paper we have analyzed the testability of switching lattices under the SAFM, and characterized the properties of fully testable lattices. We have also proposed an algorithm to detect redundancies. Future work includes the design of a method to transform non testable lattices into testable ones, by replacing some literals with a constant value, without changing the implemented function.

VIII. ACKNOWLEDGMENTS

This project has received funding from the European Union's Horizon 2020 research and innovation programme under the Marie Skłodowska-Curie grant agreement No 691178.

REFERENCES

[1] M. Abramovici and M. Breuer, *Digital Systems Testing and Testable Design*. Wiley-IEEE Press, 1994.

[2] S. B. Akers, "A Rectangular Logic Array," *IEEE Trans. Comput.*, vol. 21, no. 8, pp. 848–857, Aug. 1972.

[3] D. Alexandrescu, M. Altun, L. Anghel, A. Bernasconi, V. Ciriani, L. Frontini, and M. B. Tahoori, "Logic synthesis and testing techniques for switching nano-crossbar arrays," *Microprocessors and Microsystems - Embedded Hardware Design*, vol. 54, pp. 14–25, 2017.

[4] M. Altun, V. Ciriani, and M. B. Tahoori, "Computing with nano-crossbar arrays: Logic synthesis and fault tolerance," in *Design, Automation & Test in Europe Conference & Exhibition, DATE 2017, Lausanne, Switzerland, March 27-31, 2017*, 2017, pp. 278–281.

[5] M. Altun and M. D. Riedel, "Logic Synthesis for Switching Lattices," *IEEE Trans. Computers*, vol. 61, no. 11, pp. 1588–1600, 2012.

[6] A. Bernasconi, V. Ciriani, R. Drechsler, and T. Villa, "Logic Minimization and Testability of 2-SPP Networks," *IEEE Trans. on CAD of Integrated Circuits and Systems*, vol. 27, no. 7, pp. 1190–1202, 2008.

[7] A. Bernasconi, V. Ciriani, G. Trucco, and T. Villa, "Logic Minimization and Testability of 2SPP-P-Circuits," in *Euromicro Conference on Digital Systems Design: Architectures, Methods and Tools (DSD)*, 2009.

[8] A. Bernasconi, V. Ciriani, L. Frontini, V. Liberali, G. Trucco, and T. Villa, "Enhancing logic synthesis of switching lattices by generalized shannon decomposition methods," *Microprocessors and Microsystems - Embedded Hardware Design*, vol. 56, pp. 193–203, 2018.

[9] A. Bernasconi, V. Ciriani, L. Frontini, and G. Trucco, "Synthesis on switching lattices of dimension-reducible boolean functions," in *IFIP/IEEE International Conference on Very Large Scale Integration (VLSI-SoC)*, 2016.

[10] ——, "Composition of Switching Lattices and Autosymmetric Boolean Function Synthesis," in *Euromicro Conference on Digital System Design, DSD 2017, Vienna, Austria, August 30 - Sept. 1, 2017*, 2017, pp. 137–144.

[11] M. Breuer and A. Friedman, *Diagnosis & Reliable Design of Digital Systems*. Computer Science Press, 1976.

[12] V. Ciriani, A. Bernasconi, and R. Drechsler, "Testability of SPP Three-Level Logic Networks in Static Fault Models," *IEEE Transactions on Computer Aided Design of Integrated Circuits and Systems*, vol. 25, no. 10, pp. 12 241–2248, 2006.

[13] G. Gange, H. Søndergaard, and P. J. Stuckey, "Synthesizing Optimal Switching Lattices," *ACM Trans. Design Autom. Electr. Syst.*, vol. 20, no. 1, pp. 6:1–6:14, 2014.

[14] M. C. Morgul and M. Altun, "Synthesis and optimization of switching nanoarrays," in *Design and Diagnostics of Electronic Circuits and Systems (DDECS), 2015 IEEE International Symposium on*. IEEE, 2015, pp. 161–164.

[15] O. Tunali and M. Altun, "Permanent and transient fault tolerance for reconfigurable nano-crossbar arrays," *IEEE Trans. on CAD of Integrated Circuits and Systems*, vol. 36, no. 5, pp. 747–760, 2017.

[16] ——, "A survey of fault-tolerance algorithms for reconfigurable nano-crossbar arrays," *ACM Comput. Surv.*, vol. 50, no. 6, pp. 79:1–79:35, 2017.

[17] ——, "Logic synthesis and defect tolerance for memristive crossbar arrays," in *2018 Design, Automation & Test in Europe Conference & Exhibition, DATE 2018, Dresden, Germany, March 19-23, 2018*, 2018, pp. 425–430.

[18] T. Williams and K. Parker, "Design for Testability - A Survey," *IEEE Transactions on Computers*, vol. 31, no. 1, pp. 2–15, 1982.

[19] S. Yang, "Logic Synthesis and Optimization Benchmarks User Guide Version 3.0," Microelectronic Center, User Guide, 1991.

978-1-5386-4757-8/18 $31.00 © 2018 IEEE

An MCTS-based Framework for Synthesis of Approximate Circuits

Muhammad Awais, Hassan Ghasemzadeh Mohammadi, Marco Platzner

Department of Computer Science
Paderborn University, Paderborn, Germany
Email: {mawais,hgm,platzner}@mail.upb.de

Abstract—Approximate computing has become a very popular design strategy that exploits error resilient computations to achieve higher performance and energy efficiency. Automated synthesis of approximate circuits is performed via functional approximation, in which various parts of the target circuit are extensively examined with a library of approximate components/transformations to trade off the functional accuracy and computational budget (i.e., power). However, as the number of possible approximate transformations increases, traditional search techniques suffer from a combinatorial explosion due to the large branching factor. In this work, we present a comprehensive framework for automated synthesis of approximate circuits from either structural or behavioral descriptions. We adapt the *Monte Carlo Tree Search* (MCTS), as a stochastic search technique, to deal with the large design space exploration, which enables a broader range of potential possible approximations through lightweight random simulations. The proposed framework is able to recognize the design Pareto set even with low computational budgets. Experimental results highlight the capabilities of the proposed synthesis framework by resulting in up to 61.69 % energy saving while maintaining the predefined quality constraints.

Index Terms—Approximate computing, High-level synthesis, Accuracy, Monte-Carlo tree search, Circuit simulation.

I. INTRODUCTION

The quick proliferation of highly complex embedded systems, has increased the demand for their energy and performance efficient implementation. A wide range of embedded applications in several domains of *Digital Signal Processing* (DSP), wireless communication, *Machine Learning* (ML), and data mining, can still produce acceptable outputs in the presence of inexactness and approximation in their underlying computations [1]. The imprecision is permitted owing to the various sources of error resiliency such as lack of ground-truth data for the exact computations, lack of human perception on the quality degradation, and the existence of lossy algorithms over the redundant input datasets [2]. The inherent error-tolerance of these applications can be explicitly utilized for ingenious hardware implementations, that leads to great reduction of both power consumption and silicon footprint of the design.

In the recent years, an increasing amount of work has been performed in the field of approximate computing ranging from high-level programming languages to low-level transistor logic circuits [3]–[7]. Approximate circuits can be generated by two main approaches. The first approach utilizes controlled quality degradation, i.e., by over-clocking or voltage over-scaling, to obtain imprecise circuit instances. The second one exploits functional approximation to bring down the area and the energy cost of a given circuit while maintaining a quality threshold. Functional approximation is achieved either by altering the selected circuit components using possible slightly modified functions or by applying Boolean optimizations for relaxing the functionality of circuit components. Both strategies rely on massive search during modification or optimization process required to reduce circuit costs.

Generating approximate circuits efficiently, for various domains of applications, is a challenging problem for designers due to the exponential growth in the search space. Additionally, the number of possible approximations that each operand, operation, or circuit block can take changes as the multi-objective approximation process progresses. *Monte Carlo Tree Search* (MCTS) has exposed promising outcomes in several *Artificial Intelligence* (AI) problems with extremely large search space. Since it has been introduced, MCTS has been a successful technique for a number of AI applications such as computer Go [8] or automated narrative generation [9]. Motivated by this, we have adopted MCTS to provide an efficient synthesis framework in the domain of approximate circuits.

In this paper, we introduce a comprehensive framework for automated synthesis of the approximate circuits which utilizes MCTS, as an intelligent search technique, to efficiently cover the design search space with a lightweight computational budget. An input circuit, either in structural or behavioral descriptions, is fed to the framework and then is parsed to determine its potential operands and operations. The candidates along with their potential approximations are then used by MCTS to generate the optimal design Pareto set in terms of accuracy, power, and area. The framework is quite general and also can be integrated with other circuit synthesis frameworks for any further optimizations. The framework is evaluated by five Verilog benchmarks from different application domains: FIR and IIR filters usually used in signal processing, a configurable convolution filter commonly used in image processing, an 8-point FFT filter typically exploited in speech processing and finally a *Discrete Cosine Transform* (DCT) block widely used in compression algorithms. The main contributions of this paper are as follows:

- The proposed framework presents a new success for efficiently solving the problem of design space exploration for automated approximate circuit synthesis even with a tight computational budget.
- The underlying search technique is modified to prune all the combinations that are not promising for the target design in terms of quality. Using this, it avoids the combinatorial explosion of inadequate possible approximations.
- We validate the benefit of using the proposed framework in circuit energy and area for five widely used benchmark circuits in various embedded applications. We also compare the obtained results with that of other approximate circuit synthesis tools to illustrate the effectiveness of the proposed framework.

The remainder of the paper is organized as follows. The background of circuit synthesis in the domain of approximate computing is presented in Section II. The details of the proposed framework for synthesis of approximate circuits are demonstrated in Section III. Section IV provides the experimental results, and finally Section V concludes the paper along with the future directions.

II. RELATED WORK

Approximate computing circuits have been broadly investigated at different levels of design abstraction. In this section, we mainly focus

978-1-5386-4757-8/18 $31.00 © 2018 IEEE

on the previous research dealing with the functional approximation in the domain of approximate circuits synthesis. The shared characteristic of previous studies is the use of various search techniques to either explore the best possible approximate transformations needed for the target circuit or examine various approximate transformations via Boolean optimization while maintaining a quality constraint. As a method for automated approximate circuit synthesis, SALSA [10] encodes error constraints as quality functions, thereby utilizing don't care optimizations iteratively to generate approximate circuits. SASIMI [11] achieves functional approximation by determining pairs of circuit nodes that have similar functionalities and substituting them with each other. In a more recent work [12] SCALS framework iteratively exploits three simple transformations (e.g., adding extra logic gates among the circuit nodes, randomly changing the type of logic gates, and flipping signal polarities) on the sub-networks extracted from the original circuit. The candidate nodes for approximation are selected randomly, and *Markov Chain Monte Carlo* (MCMC) sampling is used to accept or reject the recently applied transformation.

Chandrasekharan et al. [13] propose approximation techniques for Boolean functions represented as *AND Inverter Graphs* (AIGs). The technique called AIG re-writing first identifies the critical paths of the circuit and then iteratively selects and replaces the cuts on the critical path. The cut selection scheme assumes that starting approximation transformations with a smaller size cut can result in better approximations hence it sorts the cuts in an increasing order before selection. The error constraint is then formally verified through a SAT solver. In [14], *Binary Decision Diagrams* (BDDs) are exploited for the approximate circuit synthesis in which a set of rounding transformations are used to simplify the Boolean functions. Circuit Carving [15] aims to prune the maximum part of an exact design represented as a *Direct Acyclic Graph* (DAG) by finding a cut that has the maximum number of gates. The approximate design is achieved by setting the weights of the outgoing edges of the identified cut to a constant value. All the aforementioned techniques are able to only leverage combinational circuits. The search methods used by these techniques can be mainly categorized as either greedy or random search except the Circuit Carving that employs binary search to find cuts in the graph. However, it faces the combinatorial explosion in case of large size circuits not only for the search but also for the weight labeling of the nodes in the DAG.

To address sequential approximate circuits, ASLAN [16] extends SALSA by introducing a virtual quality constraint checking to measure the precision loss of the multi-cycle circuits. Since the quality function encodes the error constraints, identifying the type of approximation, that leads to the highest power and area efficiency, needs a tremendous exploration over the large space of possible transformations. ASLAN uses a gradient-based search technique which needs a large simulation budget for reasonably large designs, and may get stuck in the local optimums for highly non-linear quality constraints of circuits.

The problem of efficient design space exploration for synthesis of approximate circuits is investigated by a few number of works [17] [18] [19]. In [17], ABACUS is propsed as a framework for generating a Pareto set of the target design with different accuracy, area and power trade-offs. The proposed framework employs a greedy best-first search algorithm by which the best available approximation is selected for a circuit component in each search step. This, however, strictly confines the search space in early steps of the algorithm and therefore the framework becomes remarkably prone to overlook the global optimum. Nepal et al. [18] try to improve the efficiency of the previous exploration algorithm by selecting multiple optimal transformations in each search step via a genetic algorithm-based technique

called NSGA-II. However, the iterative nature of ABACUS may cause a set of approximations, belonging to the former generations, dominate the global search space and hence limit the selection of other promising approximations that may appear in later generations. Furthermore, IDEA [19] proposes a synthesis technique utilizing a branch-and-bound algorithm. In this manner, IDEA iterates over the possible approximations of each node in the design and performs a depth-first search until no more approximations are possible and then backtracks. The performance of this technique to generate an optimal approximate design is limited since the large portion of the search budget is dedicated to identify the best possible approximation of the firstly visited nodes of the design.

A number of other works deal with the approximate circuit synthesis problem at other levels of abstraction [20] [23] [24]. In these works, the error analysis is usually limited to a subset of the design search space. Therefore, they often lead to a marginal saving on the circuit budget.

In the following, we address the problem of design space exploration for synthesis of approximate circuits through an efficient stochastic search technique that intelligently eliminates undesired approximations even with low computational budget.

III. THE MCTS-BASED FRAMEWORK FOR SYNTHESIS OF APPROXIMATE CIRCUITS

This section explains the proposed framework for Synthesis of Approximate Circuits. First, we describe the workflow of the proposed framework. This is followed by a brief review of the *Monte Carlo Tree Search* (MCTS) algorithm and the explanation of its modifications to be used in the proposed framework.

A. Overview of the proposed framework

The overall flow of the MCTS-based approximate circuit synthesis framework is illustrated in Figure 1. The framework is fed with the behavioral or structural description of the exact circuit. The framework exploits an HDL parser [25] in order to identify the target operations and operands for approximation. As a result, the obtained parsing tree is later modified by the search engine to generate a set of approximate circuits. In our framework, we use the approximate transformations such as precision reduction for arithmetic operations and memory elements, broken carry chain (e.g., ACA [26]) for additions and multiplications, operator relaxation (e.g., partially or completely replacing additions with OR or XOR operations) and loop unrolling (which uncovers the possibilities for further approximations for behavioral descriptions).

The framework starts with extracted components being configured as the root node of the search tree. A new approximate design variant is generated by applying an available approximate transformation. This design variant is evaluated with a batch of test vectors selected randomly from the training dataset. At the end of each test, the framework evaluates the quality of the obtained approximate circuit through a statistical hypothesis testing using the student's T-test [27]. For a user-defined error boundary such as Er and a confidence level such as Cl, the problem is formulated as the following:

$$\begin{cases} H_0 : & training\,error > Er \\ H_1 : & training\,error < Er \end{cases} \qquad (1)$$

where H_0 and H_1 are the null and alternative hypotheses respectively. To be sure that the error of the approximation remains within the boundary, the null hypothesis is evaluated by using a batch from training set. By comparing the probability value for each batch regarding to the Cl value, we can evaluate that the null hypothesis remains true or not. In each iteration, we update the result of the

978-1-5386-4757-8/18 $31.00 © 2018 IEEE

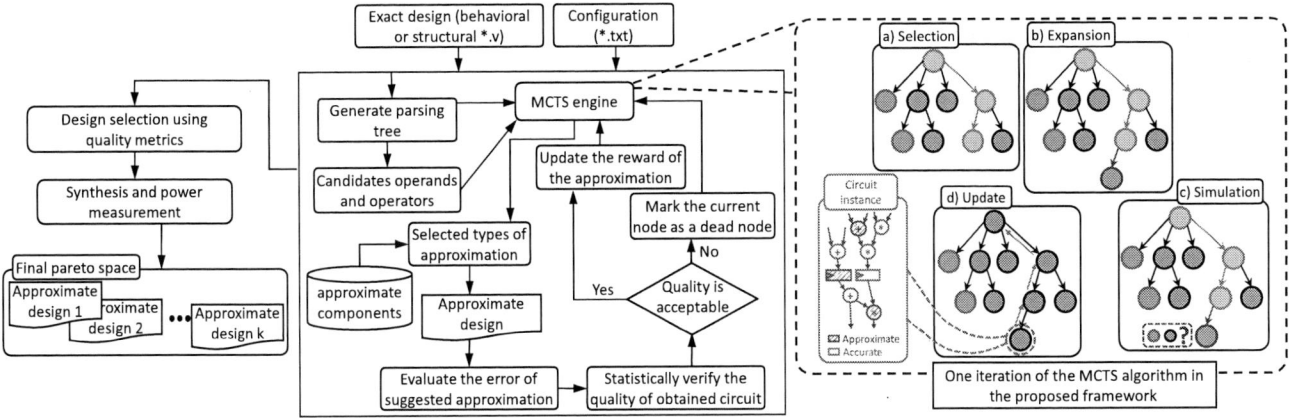

Fig. 1. Overview of the proposed framework.

search tree based on the conclusion of hypotheses over each batch. If the design fulfills the quality criteria, it is added as a node in the search tree. The node keeps the circuit configuration of the design variant, the error value and a number of MCTS specific data fields that are: reward value, number of visits, and pointers to the parent and children nodes. The search engine iteratively continues to build the tree until the computational budget assigned to it is exhausted. The result of the search engine is the set of approximate variants with various circuit characteristics. The set of approximate variants is further evaluated using additional quality metrics to find a set of promising approximate designs. This is done to avoid overfitting of approximate design to the input data. This process is then followed by circuit synthesis, determining area, delay and power consumption. The suitable design candidates are presented as the set of Pareto points to the designer.

In the following subsection we provide the details of the adapted MCTS-based framework and discuss the modifications to improve the performance of MCTS in approximate circuit synthesis domain.

B. MCTS and its adaptation to the circuit synthesis problem

MCTS is a popular learning-based stochastic search algorithm that uses random lightweight simulations to intelligently build a search tree. In the recent past, MCTS has received an ample amount of attraction due to its success in the domain of games and a wide variety of other search-based problems such as story generation and planing [8], [9]. The ability of MCTS to provide efficient design space exploration and a balanced tree selection policy makes it an obvious choice for the synthesis problem which has a very large branching factor.

However, the conventional MCTS algorithm [8] needs substantial modifications in order to be applied to the circuit synthesis problem due to a number of reasons. First, the simulation time of a circuit instance in comparison to the games (e.g., Go) is very large. A playout in Go for example, completes in a few microseconds where as the cycle-accurate simulation for even a small circuit could take up to a few seconds. Also, unlike a definitive win or loss reward, the reward value for a circuit instance is based on circuit characteristics(area or power) or quality(error). Finally, the design space may contain certain approximation choices that can lead to large errors and therefore further approximations from those choices onward will not provide any benefit. This is not the case in the games where some initial bad moves can still lead to a win. The modifications done to the conventional MCTS to use it for circuit synthesis problem are discussed in detail at the end of this section.

The detailed flow of the proposed synthesis framework is represented in Algorithm 1. The algorithm initiates with the description of the original circuit and builds a search tree in which the branches represent different approximation paths and the leafs specify approximate circuit instances. In each iteration, a circuit instance is selected in the search tree (or added if it is the first iteration). One of the exact operators in this circuit instance is replaced with the approximate operator to produce an approximate version of the circuit. The new approximate circuit then is added to the search tree (the blue node in Figure 1b).

Algorithm 1: The MCTS-based synthesis framework

Input: O= Verilog description of the circuit M= Computation budget
$\quad \epsilon_{max}$ = pre-defined quality threshold
Output: S={A Pareto set of approximate circuits A_i}

1   $K \leftarrow$ Circuit components;
2   Initialize search tree $(root.config \leftarrow K)$;
3   **while** $M > 0$ **do**
4     $node \leftarrow root$;
5   **while** $(f(S, L)==0$ OR $node!=leaf)$ **do**
6       $node \leftarrow UCTSelect(node.children)$;
7   **end**
8   **if** *(node is not fully approximated AND node is not dead)* **then**
9       $a \leftarrow$ approximation chosen with weighted random heuristic;
10   add $node_{new}$ as a child of the $node$;
11      $node_{new}.config \leftarrow$ update with a;
12   **else**
13   backpropagate 0 as reward; // Leaf node reached
14   **continue**;
15   **end**
16     $\epsilon \leftarrow$ Simulate $node_{new}$ and compute the error (ϵ);
17   **if** $(\epsilon > \epsilon_{max})$ **then**
18   mark $node_{new}$ as a dead node;
19   **else**
20   calculate the reward $R(\epsilon)$;
21      $S_{ActiveNodes} \leftarrow S_{ActiveNodes} \cup node_{new}$;
22   **end**
23   backpropagate the reward $R(\epsilon)$;
24     $M \leftarrow M - 1$;
25   **end**
26   $S \leftarrow$ Pareto points from $S_{ActiveNodes}$;
27   **return** S

The next step is circuit simulation in which a random playout has to be run. In the game domain, the random playout means applying actions in random sequences to reach a terminal state (the state which finishes the game). The playout results in a reward value which in turn is an indication whether the playout resulted in a win, loss or draw situation. For a circuit instance though, this means the circuit instance undergoes a validation with a random test set and then the reward value is estimated on the basis of its circuit characteristics. The circuit instance is added to the search tree if the error is under the maximum allowed error ϵ_{max}(the green nodes in Figure 1); otherwise it is marked as a dead node (the red nodes in Figure 1) in the search tree (Algorithm 1- l. 18).

Finally, in the update step the error value (ϵ) is then updated for the newly added circuit instance in the search tree (Figure 1d). This will complete one iteration of MCTS and depending upon the computational budget, the search tree continues to grow with each iteration. The algorithm terminates when the computational budget assigned to it is exhausted.

To have a control on the number of branches in the search tree, we use a decision function (explained in the following subsection) that determines the number of branches to add at each exact operator location by looking at already available branches and the total number of simulations done at that time. This makes sure that the computational budget is fairly utilized by allowing a reasonable number of branches in the search tree (Algorithm 1- l. 5). The weighted heuristic for pool components can guide the expansion step to select promising approximate operators (Algorithm 1- l. 9). In addition, node pruning is used to cut the branches that violate the error constraint ϵ (Algorithm 1- l. 13).

In the following, we briefly explain the modifications in our MCTS search engine for the synthesis of approximate circuits.

1) Selection and expansion policy: The selection policy in the MCTS plays an important role in building the search tree. The *Upper Confidence Bound applied to Trees* (UCT) is the most common formula to select the child node in the search tree [22]. It selects the child node i that has the maximum UCT value:

$$UCT_i = \frac{W_i}{V_i} + C\sqrt{\frac{\ln(V_N)}{V_i}} \qquad (2)$$

where W_i represents the reward value for the current node, V_i shows the number of visits for the current node, V_N shows the number of visits for the parent node, and C is the exploration constant which takes a small positive value [22].

The UCT formula has an important feature that it keeps a balance between selecting nodes that proved promising in the previous iterations and the nodes which have not been explored yet. This property of UCT suits very well to the design space exploration of circuit synthesis since it is equally important to explore the new paths in the MCTS to find new approximation possibilities (exploration) as well as to expand the previously explored paths (exploitation) for further approximations.

The expansion phase of the basic MCTS as described earlier, adds one or more children nodes to the search tree. While this is suitable for the domain of games, the computational complexity of circuit simulation are considerably larger for the circuit synthesis and it is not optimal to add multiple children nodes in each iteration. To balance the number of children nodes with the available computational budget in our domain, we introduce a heuristic decision function $f(S, L)$ to decide whether a new child node should be added to the search tree. The decision function has a Boolean output and is defined as:

$$f(S, L) = \begin{cases} 1 & \text{if } \frac{S^\alpha}{(L+1)} > n_c \\ 0 & \text{otherwise} \end{cases} \qquad (3)$$

where S represents the total number of simulations done (number of nodes in the tree) so far, L represents the depth of the tree at the current node, $\alpha \in [0, 1]$ is the expansion parameter, and n_c represents the number of children nodes for the current node. The parameter α is able to manage the computational budget in an efficient way to balance the number of nodes in the search tree. The impact of the

α and the exploration parameter C on the quality of the results is further discussed in section IV.

The framework also utilizes weighted random heuristic to intelligently select the approximate transformations. All the transformations have weights computed using equation 4 which determines their probabilities for the selection during the search process.

$$W = \gamma \times Area + \lambda \times Power\ Consumption \qquad (4)$$

where γ and λ are the coefficients can be set according to a designer's preference, and the power consumption is measured using a uniformly distributed random input workload.

2) Reward function: In MCTS for games, the random playout results in win, loss or a draw state which is then quantified as a discrete value of 1, -1, or 0. This representation cannot be directly applied to the synthesis problem at hand since the reward of a node in this case describes the precision of its computation. So, we need a function to map the error value to the equivalent reward value. This means that the higher values for measured error (ϵ) will have small rewards and the lower error values will result in large reward values. In this way, the reward value can be obtained by applying a strictly decreasing function over this interval since we only care about the error value in the interval $[0, \epsilon_{max}]$. We will use quadratic function to calculate rewards in this work (i.e., $R(\epsilon) = (\epsilon - \epsilon_{max})^2$).

IV. Result and Discussion

To evaluate the proposed framework, we applied it to five benchmark circuits selected from several domains of applications. Table I summarizes the main features of these test benches and their related quality metrics as well. These benchmarks include an eight-tap Gaussian low-pass FIR filter, an 8-tap IIR filter, a configurable convolutional filter (for 256×256 RGB colored images), an 8-point FFT filter, and an eight-stage pipelined *Discrete Cosine Transform* (DCT) block that all are coded in Verilog. Each circuit is evaluated by a dataset which is used both during synthesis process for quality evaluation and after synthesis for design validation purposes. The datasets are generated through uniform distribution of random numbers over the whole input range. The quality evaluation is achieved via random sampling, described in Section III, by using a batch size of 2000 samples for training data in each epoch. This will guarantee that the obtained approximate variants do not overfit to a specific dataset. We certify the validation results through statistical student's T-test with a Cl of 95% to provide the statistical guarantee about the quality of generated circuits.

Table I also shows the bit-width and the sizes of benchmarks in terms of lines of code. The number of nodes for approximation refers to the total number of exact operators which are to be targeted for the approximation and the pool size refers to the number of available approximate operators with different energy/area trade-offs.

We implemented the proposed MCTS core in C++. The core wrapped by several Python-based modules altogether make a custom tool for the synthesis of approximate circuits. The modules evaluate the benchmark circuits to identify the candidate nodes for approximation, provide approximate instances of the target benchmarks and also connect the core to a standard CAD tool flow which includes *Mentor Graphics QuestaSim* and *Synopsys Design Compiler*. Moreover, a 22 nm industrial technology library was used in our experiments. In our framework, only the computational parts of the data-path are used for approximation, and the control parts are left unchanged.

Figure 2 represents the energy savings of the proposed framework over the five benchmark circuits and compares its results with those of three other competitors (ABACUS, ABACUS+NSGA-II, and IDEA discussed in Section II). All techniques were given the same computational effort in terms of the number of iterations. A

978-1-5386-4757-8/18 $31.00 © 2018 IEEE

TABLE I
BENCHMARK CIRCUITS AND THEIR CHARACTERISTICS

Benchmark circuits	Application domain	Data bit-width	# of lines (Verilog)	# of nodes for approximation	Pool size (app. components)	Area (μm^2)	Power(mW)	Quality metric
DCT block	Data compresion	22	346	100	44	17088.92	6.62	MSE ($< 5\%$)
FIR filter	Signal processing	64	148	25	69	15309.82	6.71	MSE($< 5\%$)
Convolution filter	Computer vision	32	476	29	69	16192.98	8.73	PSNR ($\geqslant 25$)
IIR filter	Signal processing	64	167	45	69	30093.63	16.83	MSE($< 5\%$)
8-Point FFT	Speech processing	32	485	200	56	94944.74	19.47	MSE($< 5\%$)

total of 10,000 iterations were run for the FIR and IIR designs and 1,000 iterations were run for convolution filter, FFT filter, and DCT block. The average run time per iteration was 14, 15, 518, 203 and 172 seconds for FIR, IIR, convolution filter, FFT, and DCT respectively on a 2.66 GHz Intel® Xeon® X5650 processor with 47 GB of RAM. The values of λ and γ are selected as 0.5 each in all experiments. The hyper parameters C, and α were respectively selected as $\{2, 1.5, 0.5, 0.5, 0.5\}$ and $\{0.35, 0.35, 0.20, 0.15, 0.15\}$ for FIR, IIR, convolution filter, FFT and DCT benchmarks. This hyper-parameter assignment allows the circuits to reach the maximum number of approximations under the predefined quality constraint. The parameter C provides a balance between exploration and exploitation of the search tree and can be experimentally found and tuned to get a trade-off between finding new search paths or to go deeper in the existing paths [8]. The parameter α controls the number of children nodes at each level. Higher values of alpha would allow the addition of more children nodes resulting in more approximate candidates to choose from in case the computational budget is higher. Lower values however, can be set to achieve the same approximation level under a tight computational budget. For example, we observed that a higher value of α such as 0.45 would force the MCTS to add large number of branches to the search tree and the MCTS will not be able to go deeper in the search tree. On the other hand, a smaller value (such as $\alpha = 0.05$) would make the MCTS go deeper quickly in the search tree but reduce the chance of adding more approximation branches.

The hyper parameters of the other techniques were also selected to achieve the best results in terms of both the maximum number of possible approximate transformations and energy/area savings. For example, we used 25 generations using ABACUS for the FIR circuit, which were then followed by 40 approximate transformations in each generation. The results then were selected among the obtained variants with best energy and area saving. Our proposed framework presents better energy savings than others by reaching 32.08%, 61.69%, 11.46%, 24.97%, and 9.65% savings for DCT, convolution filter, FIR, FFT, and IIR respectively. Figure 4 also represents the area saving of each technique over the five benchmarks. Our technique outperforms the others by 5.51%, 5.09%, 1.99%, 5.34%, and 1.25% area reduction for DCT, convolution filter, FIR, FFT, and IIR respectively.

Figure 5 also shows two representative samples from our training set (first row) and testing set (second row) used for evaluating the approximated convolution filter. Here, the filter was configured to generate blurry images. Figure 5 (a), and (d) represent the original images given to the filter. Figure 5 (b), and (e) represent the obtained outputs from the original hardware filter. Both cases represent a high PSNR value (> 60 dB) related to tiny truncations of input data since the filter exploits a 32-bit fixed point (24 bit for fractional parts) number format. Figure 5 (c), as a member of training set, just shows the PSNR value of 42.01 dB for the obtained filter with 29 approximations. Figure 5 (f) also depicts a sample of the test set which results in the PSNR value of 25.32 dB.

We also experimentally evaluated the impact of the framework

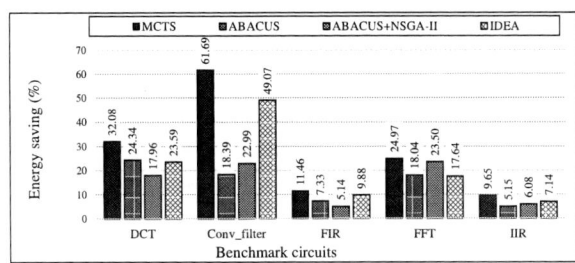

Fig. 2. Energy savings comparison obtained for various benchmarks using the proposed framework

hyper-parameters on the quality (i.e., error) and the characteristics (i.e., energy consumption and area) of generated approximate circuits. For this purpose, we performed experiments with different values for expansion parameter (α) and exploration constant (C). For each pair of (α, C), we performed approximate circuit synthesis for FIR benchmark with the budget of 10,000 iterations. Among the generated circuits, the one with higher number of approximate transforms, and more power/area saving in the design space was selected as the outcome of the framework. Figure 3 shows the impact of these parameters on the generated circuits. A larger value of the α (selected form $[0, 1]$) causes the search engine to spend more budget to find a better approximate transform for each circuit node. Therefore, the search budget may exhaust before visiting a large portion of nodes that could possibly be approximated. Besides, the high value of the C parameter forces the framework to increase the number of searches for finding the optimal transformation of the nodes visited few times so far. For α=0.45 only 17 transformations among 25 possibilities can be reached even with the minimum C value of 0.5 (Figure 3 (a)). That is why the design space is dedicated to more precise circuits with lower power and area savings (Figure 3 (b) and (c)). Alternatively, a smaller value for the parameter α forces the framework to find approximate transformations for higher number of circuit nodes with just a small budget. Consequently, more power and area savings can be obtained at the cost of a larger precision loss. The designer then can trade off among quality, optimization budget, and circuit characteristics by tuning these parameters.

V. CONCLUSION AND FUTURE WORK

Approximate computing is getting an increasing amount of attention due to the great benefits it provides in terms of computation budget in a wide range of error resilient applications. However, the combinatorial explosion of the possible approximations remains a challenge for efficient exploration of design space. In this paper, an MCTS-based framework for synthesis of approximate circuits is proposed to overcome the challenge of efficient search space exploration. We propose modifications to the main steps of MCTS to tailor it with the synthesis of approximate circuits along with node pruning in the search tree. The framework is then evaluated with five verilog-based benchmarks. The results, in principle show that MCTS is a potentially promising method for synthesis and provide

978-1-5386-4757-8/18 $31.00 © 2018 IEEE

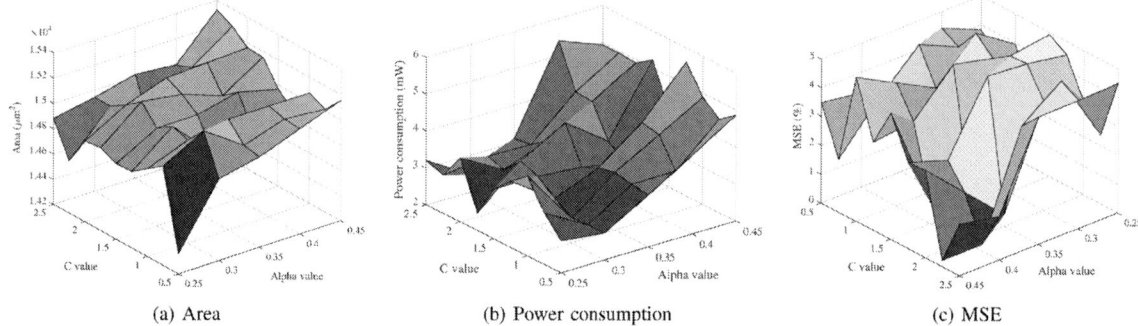

(a) Area   (b) Power consumption   (c) MSE

Fig. 3. Impact of hyper-parameters (α and C) on precision, power consumption and the area of FIR benchmark circuit

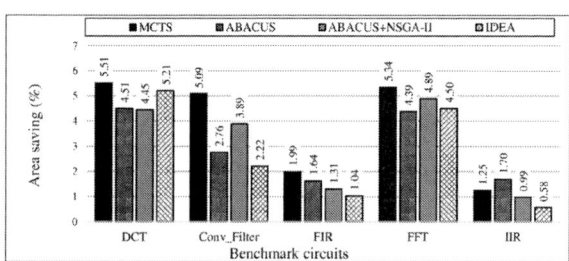

Fig. 4. Area savings comparison for various benchmarks using the proposed framework

reasonably good results with limited computational budget. It also reveals the efficiency of the proposed framework by resulting in average energy reduction of 28.99% and average area saving of 3.83% for the popular benchmark circuits.

In future, we are going to enhance this work by adding heuristics to the MCTS-based framework to make the search more intelligent. This includes the use of error propagation information to predict more suitable approximate operators for subsequent iterations to yield better approximations. Further, the dynamic adaptation of the MCTS parameters such as the expansion parameter (α) and the exploration parameter (C) can be done to change the way the MCTS builds the tree. This could be performed by considering the characteristics of the circuits and designer expectations.

REFERENCES

[1] X. Qiang, T. Mytkowicz, and N.S. Kim, "Approximate computing: A survey." IEEE Design & Test 33(1):8-22, 2016.

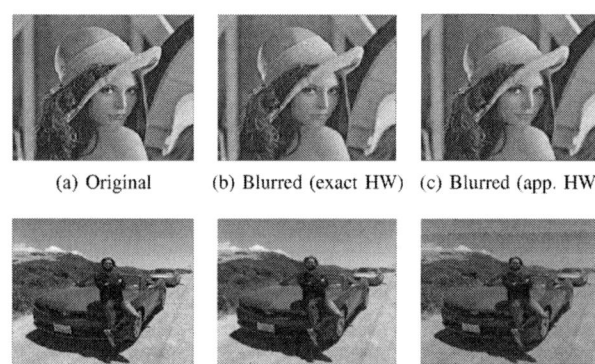

(a) Original   (b) Blurred (exact HW)   (c) Blurred (app. HW)

(d) Original   (e) Blurred (exact HW)   (f) Blurred (app. HW)

Fig. 5. Samples of images from the training (a-c) and testing (d-f) sets for the convolution filter benchmark

[2] S. Mittal, "A survey of techniques for approximate computing." ACM Computing Surveys (CSUR) 48(4): 62, 2016.

[3] H. Jie, and M. Orshansky, "Approximate computing: An emerging paradigm for energy-efficient design." IEEE ETS Tech. Dig., 2013.

[4] V. Gupta, et al., "IMPACT: imprecise adders for low-power approximate computing." IEEE ISLPED Tech. Dig., 2011.

[5] V. Rangharajan, et al., "MACACO: Modeling and analysis of circuits for approximate computing." IEEE ICCAD Tech. Dig., 2011.

[6] M. Thierry, et al., "SNNAP: Approximate computing on programmable socs via neural acceleration." IEEE HPCA Tech. Dig., 2015.

[7] A. Mercat, et al., "Exploiting computation skip to reduce energy consumption by approximate computing, an HEVC encoder case study." IEEE DATE Dig., 2017.

[8] C. B. Browne, et al., "A survey of monte carlo tree search methods." IEEE Trans. on Computational Intelligence and AI in games 4(1):1-43, 2012.

[9] B. Kartal, J. Koenig, and S. J. Guy. "User-driven narrative variation in large story domains using monte carlo tree search." AAMAS Tech. Dig., 2014.

[10] S. Venkataramani, et al., "SALSA: systematic logic synthesis of approximate circuits." 49th DAC Tech. Dig., ACM, 2012.

[11] S. Venkataramani, K. Roy, and A. Raghunathan. "Substitute-and-simplify: A unified design paradigm for approximate and quality configurable circuits." DATE Tech. Dig., 2013.

[12] G. Liu and Z. Zhang "Statistically Certied Approximate Logic Synthesis" ICCAD Tech. Dig., 2017.

[13] A. Chandrasekharan et al. "Approximation-aware Rewriting of AIGs for Error Tolerant Applications" ICCAD Tech. Dig., 2016.

[14] M. Soeken, et al., "BDD minimization for approximate computing." IEEE ASP-DAC Tech. Dig., 2016.

[15] I. Scarabottolo et al. "Circuit Carving: A Methodology for the Design of Approximate Hardware" DATE Tech. Dig., 2018.

[16] A. Ranjan, et al., "ASLAN: Synthesis of approximate sequential circuits." DATE Tech. Dig., 2014.

[17] K. Nepal, et al., "Abacus: A technique for automated behavioral synthesis of approximate computing circuits." DATE Tech. Dig., 2014.

[18] K. Nepal, et al., "Automated high-level generation of low-power approximate computing circuits." IEEE TETC PP(99):1-1, 2016.

[19] M. Barbareschi, F. Iannucci, and A. Mazzeo, "Automatic design space exploration of approximate algorithms for big data applications." WAINA Tech. Dig., 2016.

[20] A. Yazdanbakhsh, et al., "Axilog: Language support for approximate hardware design." DATE Tech. Dig., 2015.

[21] M. Magnuson, "Monte Carlo Tree Search and Its Applications." Scholarly Horizons: University of Minnesota, Morris Undergraduate Journal 2(2):4, 2015.

[22] L. Kocsis, and C. Szepesvri. "Bandit based monte-carlo planning." ECML. 6: 282-293, 2006.

[23] V. Gupta, et al., "IMPACT: imprecise adders for low-power approximate computing." ISLPED Tech. Dig. 2011.

[24] P. Kulkarni, et al.. "Trading accuracy for power in a multiplier architecture." Journal of Low Power Electronics 7.4 (2011): 490-501.

[25] S. Takamaeda-Yamazaki, "Pyverilog: A Python-based Hardware Design Processing Toolkit for Verilog HDL." ARC Tech. Dig. 2015.

[26] A. B. Kahng and S. Kang, "Accuracy-configurable adder for approximate arithmetic designs." DAC Tech. Dig. 2012.

[27] B. J. Winer, "Statistical principles in experimental design." (1962).

FLUTE-EM: Electromigration-Optimized Net Topology Considering Currents and Mechanical Stress

Steve Bigalke and Jens Lienig
Institute of Electromechanical and Electronic Design
Dresden University of Technology
✉ steve.bigalke@outlook.com, ✉ jens@ieee.org

Abstract—**The future reliability of integrated circuits is endangered by ever shrinking feature sizes and the resulting rise in electromigration (EM) damage. In order to guarantee reliability in future circuits, new approaches are needed in physical synthesis. These approaches must prioritize reliability constraints, such as EM-induced stress reduction during net-topology generation. In line with these insights, our rectilinear Steiner tree is optimized for currents and mechanical stress. We thus aim for optimized EM robustness rather than minimal wire length in the generated net topology. Our results imply a mechanical stress reduction in most cases of more than 50%, thereby significantly abating EM vulnerability. In addition, we show that reservoirs can further reduce the absolute mechanical stress level, and we present an equation for directly calculating the optimal reservoir length.**

Index Terms—**Reliability, Electromigration, Net Topology, Stress, Reservoir**

I. INTRODUCTION

In the physical synthesis of very large scale integration (VLSI) designs, producing a net topology with minimal wire length (WL) plays an important role in layout synthesis. It is used in the placement and routing steps to estimate routing congestions, interconnect parasitics or routing paths. The rectilinear Steiner minimum tree (RSMT) is a WL-optimal net topology and the fast lookup table estimation (FLUTE) from Chu [1] is probably the library that is most widely used. to determine the RSMT.

Since shrinking feature sizes drastically compromise reliability, we may need to focus our design efforts in future on enhancing reliability instead of minimizing WL. Electromigration (EM) is becoming one of the main cause of chip failures, as downscaling not only increases the EM effect but also lowers EM thresholds. Reliability will therefore be more often endangered by EM damage and might become the downscaling bottleneck [2]. The International Technology Roadmap for Semiconductors (ITRS) predicts that there will soon be no viable solutions available for EM damage [3]. Consequently, a shift from a traditional (post-layout) EM verification towards a robust (pro-active) EM-aware design is needed [4], along with new approaches such as load-aware redundant vias [5] or the presented generation of EM-robust net topologies.

EM is a material migration process caused by collisions between flowing electrons and atoms. Therefore, the main driving force behind EM is current density, which causes atomic dislocation. This dislocation depends not only on current

density, but also on the length of an interconnect. It is physically quantifiable by the hydrostatic stress, which we refer to as stress from now on. This stress is not only a consequence of EM but can also serve as its indicator. It is therefore more accurate to identify an EM-robust solution by a relatively low stress than by the widely used current density metric.

In our approach, we aim to reduce the stress by producing an EM-robust net topology (FLUTE-EM) instead of a WL-optimal one, like FLUTE [1]. Our net topology optimizes the connection between the pins in order to enhance the chip reliability in global routing. The subsequent detailed routing can further improve our results, e.g., by vias. Figure 1 illustrates how we reduced the stress of an RSMT net topology by more than 50%. The cost of this reduction is a 60% longer WL.

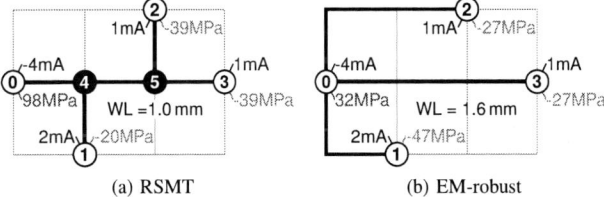

(a) RSMT   (b) EM-robust

Fig. 1. Reduction of EM-induced stress (red and blue) between the RSMT in (a) and an EM-robust net topology in (b). Pin 0 is a source of $-4\,\text{mA}$ and pins 1 to 3 are sinks of $2\,\text{mA}$, $1\,\text{mA}$ and $1\,\text{mA}$, respectively. Points 4 and 5 are Steiner points. The calculation technique is described in Section II-B.

The first studies on EM-aware net topologies were published in the field of analog design. In 2003, Lienig et al. presented an approach to plan wires (net topologies) in order to reduce the current connection area - representing pin distances, wire widths and currents [6]. Xue et al. solved the Lienig example with a simulated annealing approach, and reduced the area [7]. In 2008, Yan et al. presented an approach for building a rectilinear Steiner tree to reduce the current-driven wire widths. Lin et al. proposed an integer linear programming (ILP) approach to perform the EM-aware wire planning with consideration of obstacles [8]. In 2010, Jiang et al. solved the same problem with a multi-source multi-sink flow network formulation [9]. Tsai et al. took the channel width between the obstacles into consideration [10]. In 2014, Martins et al. minimized the total wiring area with regard to EM and IR-drop constraints for an extended multiport example [11]. Approaches for increasing the EM robustness of digital layouts by addressing detailed routing solutions based on fixed net topologies have been published by Zhang et al. [12] and Paris et al. [13].

978-1-5386-4757-8/18 $31.00 © 2018 IEEE   225

All of these studies considered mainly the current or wire width to counteract EM but none of them took the EM-induced stress - produced by material migration - into consideration. Our work is the first to design an EM-aware net topology by considering currents and mechanical stress. We also propose an equation for calculating a reservoir length (this is a passive interconnect structure with no current flow that can further reduce stress).

II. THEORY

A. Electromigration in a Single-Branch Interconnect

Electromigration is a material migration caused by the momentum exchange between flowing electrons and fixed atoms. Because of their collisions with electrons, atoms break out of their lattice positions and migrate in the direction of the electron flow. Due to the depletion of atoms at the cathode and their aggregation at the anode, tensile and compressive stresses build up forming voids and hillocks, respectively. A resulting stress gradient is formed producing stress migration (SM) that compensates EM (Fig. 2).

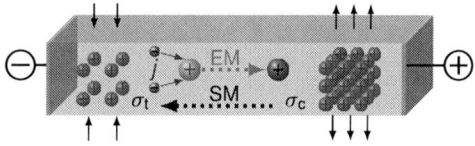

Fig. 2. Electrons (blue) collide with atoms (red), causing the atoms to migrate in the direction of the electron flow. This depletes atoms at the cathode and accumulates them at the anode. These concentration changes introduce tensile ($\sigma_t > 0$) and compressive stresses ($\sigma_c < 0$), respectively.

The stress build-up within an interconnect with blocking boundaries at both ends can be described by Korhonen's equation [14] as

$$\frac{\partial \sigma}{\partial t} = \frac{\partial}{\partial l}\left[\frac{DB}{kT}\left(eZ^*\rho j + \Omega\frac{\partial \sigma}{\partial l}\right)\right], \quad (1)$$

with the hydrostatic stress σ, time t, length l, diffusivity D, bulk module B, Boltzmann's constant k, temperature T, electric charge e, effective charge number Z^*, resistivity ρ, current density j and atomic volume Ω. Equation (1) confirms that the main EM driving force is the current density, and that EM compensation by SM is a function of the length.

Figure 3 plots the stress development described by Eq. (1) in principle. It shows that EM and SM reach a steady state determining the final maximum and minimum stresses within an interconnect. If these steady state stresses are higher than a critical technological stress threshold (σ_{crit}), EM damage can occur in the form of voids or hillocks.

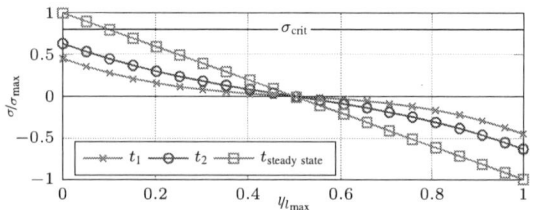

Fig. 3. Basic one-dimensional spatial stress development over time in an interconnect under EM and SM.

B. Electromigration in Multi-Branch Interconnects

Analog and digital nets consist mainly of multi-branch interconnects. Therefore, one needs to consider the entire net topology. To calculate the stress for multi-branch interconnect problems, we use the voltage-based EM immortality check from Sun et al. [15]. Here, Eq. (1) is applied for multi-branch interconnects to evaluate the overall EM load for any net type. Since our approach produces EM-robust net topologies, we assume that all branches have the same width. This allows us to shorten Sun's approach.

To investigate the cause of the different stress results between Fig. 1a and b, we use the equations from Sun's method [15] for Fig. 1a:

$$V_0 = 0 \qquad\qquad L_0 = l_{04} \qquad\qquad \sigma_0 = \beta V_g \quad (2)$$
$$V_1 = i_{04}l_{04} + i_{41}l_{41} \qquad L_1 = l_{41} \qquad \sigma_1 = \beta(V_g - V_1) \quad (3)$$
$$V_2 = i_{04}l_{04} + i_{45}l_{45} + j_{52}l_{52} \quad L_2 = l_{52} \qquad \sigma_2 = \beta(V_g - V_2) \quad (4)$$
$$V_3 = i_{04}l_{04} + i_{45}l_{45} + i_{53}l_{53} \quad L_3 = l_{53} \qquad \sigma_3 = \beta(V_g - V_3) \quad (5)$$
$$V_4 = i_{04}l_{04} \qquad\qquad L_4 = l_{04} + l_{41} + l_{45} \quad \sigma_4 = \beta(V_g - V_4) \quad (6)$$
$$V_5 = i_{04}l_{04} + i_{45}l_{45} \qquad L_5 = l_{45} + l_{52} + l_{53} \quad \sigma_5 = \beta(V_g - V_5) \quad (7)$$

$$V_g = \frac{\sum_{i=0}^{5} V_i L_i}{\sum_{i=0}^{5} L_i} = \frac{V_0 L_0 + V_1 L_1 + V_2 L_2 + V_3 L_3 + V_4 L_4 + V_5 L_5}{L_0 + L_1 + L_2 + L_3 + L_4 + L_5} \quad (8)$$

where V_i represents the equivalent voltage at node i, L_i the length at node i, i_{ij} the current in the edge ij (flowing from node i to node j) and V_g the equivalent virtual voltage. In general, upper cases refer to pins or Steiner points (vertices) and lower cases to connections (edges). From now on, we draw the connections as fly lines in our figures but the calculations are always based on the Manhattan length. With Eqs. (2) to (8), the given currents I in Fig. 1a, an assumed area A of $25\,\mu m^2$ ($j = I/A$), an equal length of $l = l_{ij} = 200\,\mu m$ and a factor β of $2460\,V\,s\,m^{-2}$ ($\beta = eZ^*\rho/\Omega$), one can obtain the annotated stress values at each node i in Fig. 1a.

To investigate the EM influence of the Steiner point locations in Fig. 1a, we consider lengths l_{04} and l_{45} as independent parameters ($0 < l_{04}, l_{45} < l$). This allows us to calculate the stress as a function of both lengths, as depicted in Fig. 4.

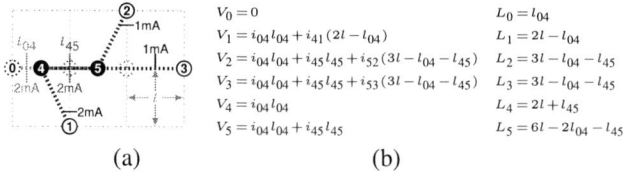

$$V_0 = 0 \qquad\qquad L_0 = l_{04}$$
$$V_1 = i_{04}l_{04} + i_{41}(2l - l_{04}) \qquad L_1 = 2l - l_{04}$$
$$V_2 = i_{04}l_{04} + i_{45}l_{45} + i_{52}(3l - l_{04} - l_{45}) \quad L_2 = 3l - l_{04} - l_{45}$$
$$V_3 = i_{04}l_{04} + i_{45}l_{45} + i_{53}(3l - l_{04} - l_{45}) \quad L_3 = 3l - l_{04} - l_{45}$$
$$V_4 = i_{04}l_{04} \qquad\qquad L_4 = 2l + l_{45}$$
$$V_5 = i_{04}l_{04} + i_{45}l_{45} \qquad L_5 = 6l - 2l_{04} - l_{45}$$

(a)   (b)

Fig. 4. (a) The Steiner point lengths l_{04} (red) and l_{45} (blue) can vary from 0 to $200\,\mu m$. In (b), we consider both as independent parameters.

The curves in Fig. 5a and Fig. 5b show the maximum stress at pin 0 and the corresponding total wire length. Clearly, the stress and the wire length develop in opposite directions. In other words, the greater the wire length, the lower the stress.

Given that Steiner points reduce the wire length, we come to the following conclusion:

Axiom 1: Steiner points increase the EM-induced stress within a net because they accumulate currents from different branches and reduce the wire length. Therefore, Steiner points worsen the effects of EM.

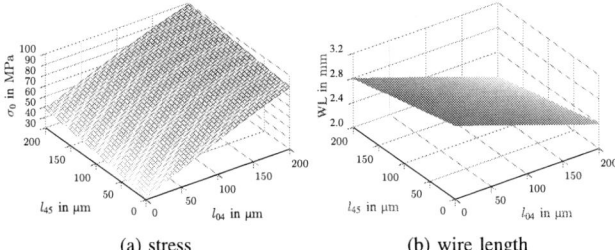

(a) stress (b) wire length

Fig. 5. Stress and wire length curves depending on the location (lengths l_{04} and l_{45}) of the Steiner points 4 and 5 in Fig. 4a. The stress is inversely proportional to the wire length.

In the case of a net with one source and any number of sinks (typical in digital designs), the EM net topology with the lowest stress is the clique net model, because it realizes the longest wire length and prevents an accumulation of different currents within one branch. However, if multiple sinks and sources with different currents are present (typical in analog nets), a robust EM net topology is difficult to obtain. To overcome this problem, we propose the following brute force, and iterative approaches for low-degree and high-degree nets, respectively.

III. BRUTE FORCE APPROACH FOR LOW-DEGREE NETS

The chicken-and-egg problem here is that without a net topology, one cannot calculate the currents within the net and without the currents, you cannot find an EM-robust net topology. To solve this dilemma, we first randomly select a topology from all possible topologies. Currents and stress values can then be calculated based on the selected topology and pin locations within the net. To find a robust EM net topology, we try all possible net topologies without Steiner points. To do this, we iterate through all possible spanning trees to calculate the occurring stresses. Based on Axiom 1 from the previous section, we expect a robust EM net topology to be without Steiner points. To speed-up our time-consuming brute force approach, we store all valid spanning trees in a look-up table together with their calculation matrices for three to nine pins.

A. Spanning Tree Generation

Our spanning tree generation is a straight forward trial-and-error approach, because the time to create the look-up table is a one-time cost and therefore of minor interest. Having a net with a number of pins p, the number of possible edges e in a spanning tree is equal to the number of edges in a full connected graph given by

$$e = \binom{p}{2} = \frac{p(p-1)}{2}. \tag{9}$$

Knowing that we need exactly $p-1$ edges to obtain a valid spanning tree leaves us with the number of possible trees t as

$$t = \binom{e}{p-1}. \tag{10}$$

Not all of these possible trees are valid spanning trees. We, therefore, have to check that all pins are continuously connected to each other without a loop. All resulting valid spanning trees for a four-pin net are shown in Fig. 6.

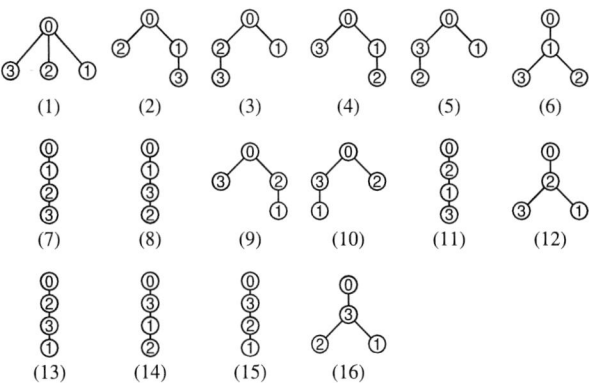

Fig. 6. All valid spanning trees for a four-pin net; they are all tested for their EM robustness.

B. Direction of Current and Current Calculation

Once the net topology has been picked, one can calculate the current for each edge in the tree by solving Kirchhoff's equations. Since the direction of current is very important for EM, we add a direction to the edges in the tree representation from Fig. 6, as shown in Fig. 7. A positive or negative current value sign signifies respectively that the current flows in the same direction as, or in the opposite direction to, the edge direction.

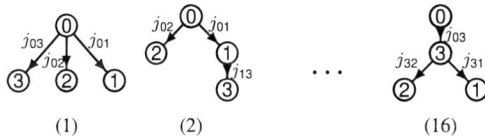

Fig. 7. The edge direction is added to the tree view because the direction of current is an important EM contributing factor. The root is set to the zero node by definition.

C. Stress calculation

As already mentioned, the stress is then calculated according to the technique presented in [15] by solving Eqs. (2) to (8) to obtain the stress values at each node. We use the currents, Manhattan lengths between pins, area and beta factor as input values for our approach to estimate the stress within each net topology based on the range between the maximum and minimum stress.

D. Look-Up Table Entries

We show what the look-up entries look like using Fig. 7(1) as an example. The inputs for our algorithm are the X_i and Y_i locations and the currents I_i at each pin:

$$X_i = \begin{bmatrix} X_0 \\ X_1 \\ X_2 \\ X_3 \end{bmatrix}, \qquad Y_i = \begin{bmatrix} Y_0 \\ Y_1 \\ Y_2 \\ Y_3 \end{bmatrix}, \qquad I_i = \begin{bmatrix} I_0 \\ I_1 \\ I_2 \\ I_3 \end{bmatrix}. \tag{11}$$

In the look-up table, the matrices m_s and m_t contain the connections between the source and target nodes for each edge, line by line. The matrix m_{ij} serves to calculate the edge currents based on the pin currents. Matrices m_L and m_V are needed to calculate the node lengths L_i and equivalent voltages V_i based on the edge currents i_{ij} and lengths l_{ij}, viz.:

$$m_s = \begin{bmatrix} 1 & 0 & 0 & 0 \\ 1 & 0 & 0 & 0 \\ 1 & 0 & 0 & 0 \end{bmatrix}, \quad m_t = \begin{bmatrix} 0 & 1 & 0 & 0 \\ 0 & 0 & 1 & 0 \\ 0 & 0 & 0 & 1 \end{bmatrix}, \quad m_{ij} = \begin{bmatrix} 0 & 1 & 0 & 0 \\ 0 & 0 & 1 & 0 \\ 0 & 0 & 0 & 1 \end{bmatrix},$$

$$m_L = \begin{bmatrix} 1 & 1 & 1 \\ 1 & 0 & 0 \\ 0 & 1 & 0 \\ 0 & 0 & 1 \end{bmatrix}, \quad m_V = \begin{bmatrix} 0 & 0 & 0 \\ 1 & 0 & 0 \\ 0 & 1 & 0 \\ 0 & 0 & 1 \end{bmatrix}.$$

Edge lengths and currents, as well as the node length, equivalent voltages and stresses are calculated as follows:

$$l_{ij} = |m_s X_i - m_t X_i| + |m_s Y_i - m_t Y_i|, \tag{12}$$

$$i_{ij} = m_{ij} I_i, \tag{13}$$

$$L_i = m_L l_{ij}, \tag{14}$$

$$V_i = m_V(i_{ij} \circ l_{ij}), \quad (\circ... \text{ element-wise multiplication}) \tag{15}$$

$$\sigma_i = \beta(V_g - V_i). \tag{16}$$

IV. ITERATIVE APPROACH FOR HIGH-DEGREE NETS

Since the brute force algorithm is very time-consuming for nets with more than nine pins (computing time increases from seconds to minutes), we developed an iterative approach that attempts to reduce the stress with the RSMT solution step by step. It can also achieve a compromise between the RSMT and EM-optimal net topology.

Our iterative approach checks all edges in the initial RSMT solution, and reconnects one edge at a time until it finds the edge that reduces stress the most. This means, we remove one edge after another and build up a set of source and target pins as possible reconnection pins for the currently removed edge. The source and target sets contain all reachable pins found by a depth first search from the source and the target pin of the currently removed edge. Here, we exclude Steiner points, as we expect a robuster EM solution to be without Steiner points based on Axiom 1 above. Removing an edge can cause Steiner points to be obsolete, so we remove Steiner points with less than three connections.

To find the best reconnection pins for an edge, we permute all source and sink pins with each other and calculate the resulting stress based on the reconnection. As we go through all permutations, when we find a more EM robust solution than the previous one, we save the reconnection and continue with the next edge. At the end of the loop, we select the best reconnection edge offering the greatest stress reduction, and mark this new edge as fixed. Then, we continue with the next run, which rechecks all unfixed edges to further improve the stress results until the results can no longer be improved. Searching through edge by edge may seem computationally demanding, but it is still faster than the brute-force (see results in Section V). We must search step by step, as reconnecting an edge affects the other edges.

We select a good result by comparing the stress range between the maximum and minimum occurring stress (a narrow range indicates a more balanced solution). In some cases, the stress range is equal to the previous value. In this case, we take the wire length into consideration as well, and select the

TABLE I
STEPS IN OUR ITERATIVE APPROACH FOR THE EXAMPLE OF FIG. 1. RED EDGES ARE FIXED SINCE THEY REDUCE THE STRESS MOST.

Current Edge	Source Pins	Target Pins	Better Edge
(5,2)	[3,1,0]	[2]	(0,2)
(5,3)	[2,1,0]	[3]	(0,3)
(4,1)	[0,2,3]	[1]	(0,1)
(0,4)	[0]	[1,2,3]	–
(4,5)	[1,0]	[2,3]	–
(4,2)	[3,0,1]	[2]	(0,2)
(4,3)	[2,0,1]	[3]	(0,3)
(0,4)	[0,1]	[2,3]	–
(0,2)	[0,3,1]	[2]	(0,2)

reconnection if the wire length is shorter than the previously selected length.

Table I lists the edges with source and target vertices for each step for the example in Fig. 1a. In this simple example, our iterative approach finds the brute-force solution.

V. EXPERIMENTAL RESULTS

A. Brute-Force Approach

To demonstrate the stress reduction achieved with our brute-force approach from Section III, we apply it to two different types of nets. The first example is a typical signal net with alternating currents in a digital design from Fig. 1a. This time, we include the stress values for the charge and discharge net phases, meaning that all pins change from source to sink, and vice versa. Figure 8 clearly shows that the alternating current cause the maximum tensile stress to become the minimum compressive stress, and vice versa.

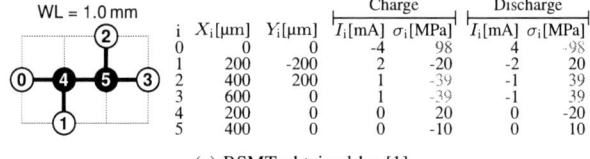

(a) RSMT obtained by [1]

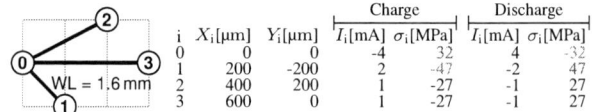

(b) The EM net topology obtained with our brute-force approach

Fig. 8. The absolute maximum stress in the RSMT net topology in (a) is almost twice as high as the stress in our EM net topology in (b). The lower stress achieved by our brute-force method comes with a 60% increase in wire length.

The second example is a net taken from [6]; it represents a typical net in an analog design with multiple sources and sinks as well as direct currents. In this example, we compare the stress values for the RSMT net topology, the net topology from [9] (latest study without obstacles) and our net topology.

978-1-5386-4757-8/18 $31.00 © 2018 IEEE   228

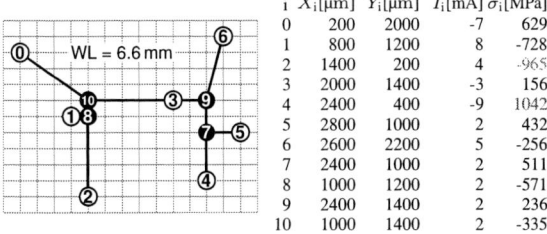

i	$X_i[\mu m]$	$Y_i[\mu m]$	$I_i[mA]$	$\sigma_i[MPa]$
0	200	2000	-7	629
1	800	1200	8	-728
2	1400	200	4	-965
3	2000	1400	-3	156
4	2400	400	-9	1042
5	2800	1000	2	432
6	2600	2200	5	-256
7	2400	1000	2	511
8	1000	1200	2	-571
9	2400	1400	2	236
10	1000	1400	2	-335

(a) RSMT obtained by [1]

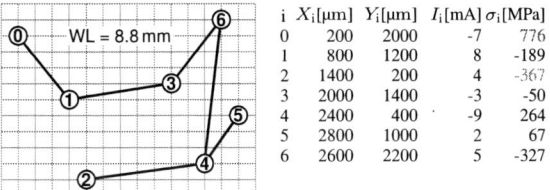

i	$X_i[\mu m]$	$Y_i[\mu m]$	$I_i[mA]$	$\sigma_i[MPa]$
0	200	2000	-7	776
1	800	1200	8	-189
2	1400	200	4	-367
3	2000	1400	-3	-50
4	2400	400	-9	264
5	2800	1000	2	67
6	2600	2200	5	-327

(b) Net topology from [9]

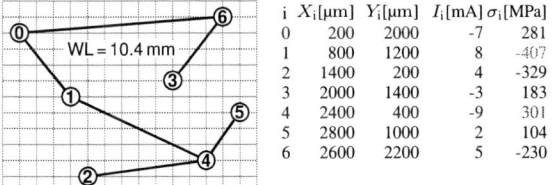

i	$X_i[\mu m]$	$Y_i[\mu m]$	$I_i[mA]$	$\sigma_i[MPa]$
0	200	2000	-7	281
1	800	1200	8	-407
2	1400	200	4	-329
3	2000	1400	-3	183
4	2400	400	-9	301
5	2800	1000	2	104
6	2600	2200	5	-230

(c) Our EM net topology obtained by our brute force approach

Fig. 9. Our net topology in (c) reduces the absolute maximum stress by 61% compared to (a) and by 48% compared to (b). In order to achieve this EM robustness, we had to invest 58% more wire length than in (a) and 18% than in (b).

B. Iterative Approach

To demonstrate the performance of our iterative approach, we solve the same 7-pin-net example from [6], as above, because we can compare it to the best EM solution found by our brute-force method. We improve the RSMT in the first step by reconnecting edge (10,3) to (2,4). In the second step, we reconnect edge (0,10) to (0,1), and we remove the Steiner points 8 and 10, as they are no longer needed.

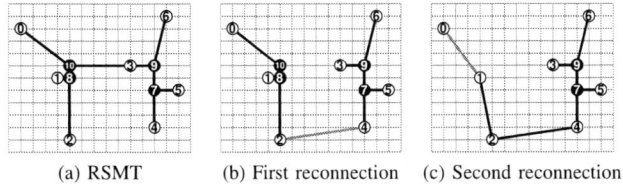

(a) RSMT   (b) First reconnection   (c) Second reconnection

Fig. 10. Improving the RSMT solution iteratively: (a) replace edge (10,3) with (2,4); (b) reconnect edge (0,10) to (0,1); and (c) remove the Steiner points 8 and 10.

The result of our iterative approach seems to be a good compromise between wire-length increase and stress reduction. The approach reduces the stress by 50% with a rise of only 6% in wire length compared to the RSMT.

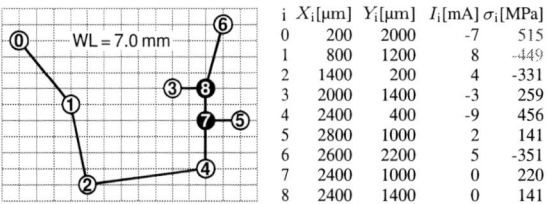

i	$X_i[\mu m]$	$Y_i[\mu m]$	$I_i[mA]$	$\sigma_i[MPa]$
0	200	2000	-7	515
1	800	1200	8	-449
2	1400	200	4	-331
3	2000	1400	-3	259
4	2400	400	-9	456
5	2800	1000	2	141
6	2600	2200	5	-351
7	2400	1000	0	220
8	2400	1400	0	141

Fig. 11. Net topology generated by our iterative approach for the example from [6]. The absolute stress is about 50% lower than with the RSMT. There is only a 6% increase in wire length. Comparing these values with the best EM net topology found by our brute force approach, which enables a 61% stress reduction with a 58% increase in wire length, shows that this approach offers a good trade-off between stress reduction and increased wire length.

VI. ADDITIONAL STRESS IMPROVEMENTS WITH RESERVOIRS

In the case of an unbalanced stress distribution, as in Fig. 1b, where there is a marked difference in the absolute minimum compressive, and maximum tensile, stresses, the two absolute values can be equalized with a reservoir. The balance between absolute compressive and tensile stresses is expressed by

$$|\sigma_{min}| = |\sigma_{max}|. \tag{17}$$

Since the minimum tensile stress is always negative and results from the maximum equivalent voltage (and the positive maximum compressive stress from the minimum equivalent voltage), one can resolve the previous equation as follows

$$-\beta(V_g - V_{max}) = \beta(V_g - V_{min}). \tag{18}$$

Given now that the equivalent virtual voltage V_g depends on the node voltage V_r and reservoir length L_r, the approach to balance the stresses is

$$V_{max} + V_{min} = 2V_g(V_r, L_r). \tag{19}$$

The equivalent virtual voltage $V_g(V_r, L_r)$ expanded by the node voltage and reservoir length is given by

$$V_g(V_r, L_r) = \left(\frac{\sum_{i=0}^{n} V_i L_i + 2V_r L_r}{\sum_{i=0}^{n} L_i + 2L_r} \right), \tag{20}$$

where the reservoir node voltage V_r is equal to the voltage of the node V_i to which the reservoir is connected, since no current flows through the reservoir.

Substituting Eq. (20) into Eq. (19) and resolving the equation according to the reservoir length, yields to

$$L_r = \frac{(V_{max} + V_{min}) \sum_{i=0}^{n} L_i - 2 \sum_{i=0}^{n} V_i L_i}{-2(V_{max} + V_{min} - 2V_r)}, \tag{21}$$

where V_{max} and V_{min} are the nodes in the net with the respective maximum and minimum equivalent voltages.

If the absolute minimum stress is greater than the maximum stress, the reservoir should be connected to any sink, otherwise to any source. The minimum reservoir lengths can be attained if the reservoir is connected as follows

$$\min L_r = \begin{cases} V_r = V_{min}, & \text{if } |\sigma_{max}| > |\sigma_{min}|, \\ V_r = V_{max}, & \text{otherwise.} \end{cases} \tag{22}$$

978-1-5386-4757-8/18 $31.00 © 2018 IEEE

If one were to consider the width of a connection, Eq. (21) would contain the node areas A_i instead of the node lengths L_i. Different void or hillock stress thresholds or residual tensile stress from the manufacturing process could be considered by lowering the appropriate stress value in Eq. (17).

Analog and digital signal nets alike can benefit from reservoirs because they equalize the absolute minimum tensile stress and maximum compressive stress until the absolute values are equal. In order to prevent any EM damage, it is advisable to keep the absolute maximum stress as low as possible.

Figure 12 visualizes the shortest possible reservoirs for balancing the stress distributions in the examples from Fig. 1a and b, as well as from Fig. 9c. Reservoirs can be located anywhere as long as they realize the calculated length.

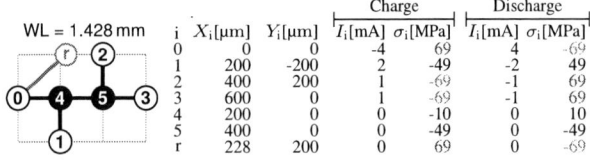

(a) RSMT obtained by [1] with reservoir

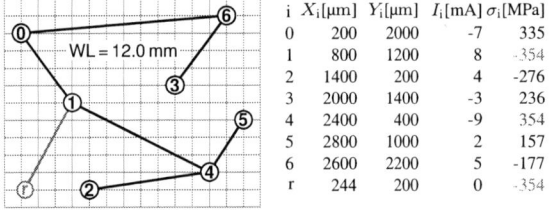

(b) Our EM net topology with reservoir for the example from Fig. 1

(c) Our EM net topology with reservoir for the example from [6]

Fig. 12. In (a), we improve the stress values in the RSMT solution by 30% with a reservoir, and increase the wire length by 43% over the solutions in Section V. In (b) and (c), we further lower the absolute maximum respective stress by 20% and 15% compared to our brute force solution in Section V with wire-length increases of 19% and 15%, respectively.

VII. IMPLEMENTATION

Our implementation of the brute-force, iterative and reservoir approaches will be publicly available as open source code under the name EMTO. It is implemented in C++ using the boost and FLUTE libraries. We provide an interface for inputting the X and Y locations as well as the current I for each pin. Our implementation calculates each node's connection given by the predecessor node, the stress value and a reservoir length for each pin based on the theory outlined in this paper. It can be used to determine an EM-robust net topology for any net. It is advisable to harden the nets with the highest EM-induced stresses, and to keep the wire length as short as possible.

Table II contains sample runtimes for our brute force (best result) and iterative (good result) approach on a single core of an Intel Xeon E5-2620 at 2.40 GHz. Runtimes for the brute-force approach in digital nets are clearly very fast. This is because only the clique net topology needs to be calculated.

TABLE II
RUNTIMES FOR ONE SOURCE AND MULTIPLE SINKS (DIGITAL NET) AS WELL AS MULTIPLE SOURCES AND SINKS (ANALOG NET).

No. Pins	Digital Net		Analog Net	
	Brute Force	Iterative	Brute Force	Iterative
3	30μs	700μs	60μs	400μs
9	40μs	17ms	10s	15ms
35	200μs	12s	-	15s

VIII. SUMMARY AND CONCLUSION

While FLUTE [1] and the RSMT are well established, it is time to enhance these important approaches with reliability requirements. Given that EM damage is on the rise with every new technology node, an approach like ours that increases the reliability by generating EM-robuster net topologies is absolutely needed. To the best of our knowledge, we are the first to consider stress in the pursuit of EM-robust net topologies. Our net topologies can more than halve the stress in most cases, and thus significantly harden the layout against EM by investing more routing resources. We also show that reservoirs can further lower EM-induced stress, and we are the first to provide an equation for calculating the optimal reservoir length.

ACKNOWLEDGMENT

We cordially thank Chris Chu, author of FLUTE [1], for his permission to use the name "FLUTE-EM" in the title of our work.

REFERENCES

[1] C. Chu and Y.-C. Wong, "FLUTE: Fast lookup table based rectilinear steiner minimal tree algorithm for vlsi design," *IEEE Trans. on Computer-Aided Design*, vol. 27, no. 1, pp. 70–83, Jan. 2008.

[2] J. Lienig and M. Thiele, "The pressing need for electromigration-aware physical design," in *Proc. of the ACM 2018 Int. Symposium on Physical Design (ISPD '18)*, 2018, pp. 144–151.

[3] International Technology Roadmap for Semiconductors 2.0 (ITRS 2.0), "More moore," http://www.itrs2.net/itrs-reports.html, 2015 edn (2016).

[4] S. Bigalke, T. Casper, S. Schöps, and J. Lienig, "Increasing em robustness of placement and routing solutions based on layout-driven discretization," in *14th Conf. on PhD Research in Microelectronics and Electronics (PRIME)*, Jul. 2018, pp. 89–92.

[5] S. Bigalke and J. Lienig, "Load-aware redundant via insertion for electromigration avoidance," in *Proc. of the 2016 ACM Int. Symposium on Physical Design (ISPD'16)*, Apr. 2016, pp. 99–106.

[6] J. Lienig and G. Jerke, "Current-driven wire planning for electromigration avoidance in analog circuits," in *Proc. of the Asia and South Pacific Design Automation Conference (ASP-DAC)*, Jan. 2003, pp. 783–788.

[7] B. Xue and X. He, "Electromigration avoidance aware net splitting algorithm in analog circuits," in *2006 Int. Conf. on Communications, Circuits and Systems*, vol. 4, Jun. 2006, pp. 2805–2808.

[8] C.-C. Lin, H.-H. Huang, H.-A. Chien, and T.-M. Hsieh, "Obstacle-avoiding electromigration aware wire planning for analog circuits," in *Proc. of the 2009 12th Int. Symposium on Integrated Circuits (ISICAS)*, Dec. 2009, pp. 651–654.

[9] I. H.-R. Jiang, H.-Y. Chang, and C.-L. Chang, "Optimal wiring topology for electromigration avoidance considering multiple layers and obstacles," in *Proc. of the ACM 2010 Int. Symp. on Physical Design (ISPD'10)*, 2010, pp. 177–184.

[10] Y. C. Tsai, T. H. Li, T. C. Chen, and C. W. Yeh, "Electromigration- and obstacle-avoiding routing tree construction," in *2013 Int. Symposium on VLSI Design, Automation, and Test (VLSI-DAT)*, Apr. 2013, pp. 1–4.

[11] R. Martins, N. Loureno, A. Canelas, and N. Horta, "Electromigration-aware and IR-drop avoidance routing in analog multiport terminal structures," in *2014 Design, Automation Test in Europe Conference Exhibition (DATE)*, 2014, pp. 1–6.

[12] J. Zhang and H. Yao, "FaSEA: Fast single-trunk detailed router for electromigration avoidance," in *2013 Int. Conf. on Communications, Circuits and Systems (ICCCAS)*, 2013, pp. 395–398.

[13] L. de Paris, G. Posser, and R. Reis, "Electromigration aware circuits by using special signal non-default routing rules," in *2016 IEEE Int. Symposium on Circuits and Systems (ISCAS)*, May 2016, pp. 2795–2798.

[14] M. A. Korhonen, P. Børgesen, K. N. Tu, and C.-Y. Li, "Stress evolution due to electromigration in confined metal lines," *Journal of Applied Physics*, vol. 73, no. 8, pp. 3790–3799, Apr. 1993.

[15] Z. Sun, E. Demircan, M. D. Shroff, T. Kim, X. Huang, and S. X. D. Tan, "Voltage-based electromigration immortality check for general multi-branch interconnects," in *2016 IEEE/ACM Int. Conf. on Computer-Aided Design (ICCAD)*, Nov. 2016, pp. 1–7.

Meta-model Based Automation of Properties for Pre-Silicon Verification

Keerthikumara Devarajegowda[1,2] and Wolfgang Ecker[1,3]

Infineon Technologies AG[1] - University of Kaiserslautern[2] - Technical University of Munich[3]

Email:Keerthikumara.Devarajegowda@infineon.com, Wolfgang.Ecker@infineon.com

Abstract—In the last decade, several hardware generation languages (HGLs: chisel, metartl, spinalhdl, coreir and more) that focus on generation of RTL code have been proposed. These languages rise the level of abstraction from RTL description to RTL generation and utilize high-level languages such as Python or Scala for describing the generation intent. As a result, they are guiding the overall productivity and chip complexity on the rising trend. On the other front, pre-silicon verification is an equally important aspect of the design process and consumes more than 50% of the overall development time. As a consequence of increased chip complexity, the existing verification gap becomes wider. Therefore it neutralizes the productivity gain achieved from RTL generation or other productivity improvement techniques.

In this paper, we describe a Python based generation language and framework for hardware properties in order to increase formal verification productivity. The described approach follows the model driven architecture (MDA) vision of OMG. The MDA approach includes transformations, that make the approach platform independent: Different property languages as well as different simulation and formal verification tools are supported. Applied to formal hardware verification, the approach empowers comparable productivity increase as HGLs in RTL designs. In addition, it is an ideal partner for HGLs, since it enables to generate a property set for each HGL generated RTL.

The applicability of our approach for real-life industrial designs is demonstrated by generating properties for a RiscV CPU core and also for peripheral devices of a CPU system. The correctness of the generated properties are validated by verifying these designs with a formal verification tool.

Index Terms—Property generation, Models-of-Property (MoP), Functional verification, Productivity, Formal specifications

I. Introduction

Semi-conductor industry is strongly driven by consumer demands that include rich functionality, safety, high performance and robustness of the products. Time-to-market of these products plays a significant role for the manufacturing company's revenue. By-product of this situation are extremely complex chips, coupled with short development cycles. To tackle the problem, companies are constantly increasing their head count and R&D expenditures [14]. However, with ever increasing chip complexity and the trend of billion IoT (Internet-of-Things) devices in near future, this approach may not be sustainable. As a counter measure for design productivity loss, hardware generation and modeling languages have been proposed, that improve the design productivity by a factor of 3x or more as reported [4]. In addition, High level Synthesis

(HLS), early modeling and optimization techniques are also adding to design productivity. As a result, hardware systems are increasingly designed in short cycles and are becoming even more complex, thereby widening the verification gap.

Effective approaches for pre-silicon verification must be easy to build, find early design bugs, improve observability of the designs and significantly reduce the design cycle [21], [7]. Assertion[1] Based Verification (ABV) has all the required traits and uses properties to capture the design intent. Although the usage of properties for design exercise has known benefits, writing large number of assertions, maintaining them, and frequent changes to specification documents are making it a process requiring plenty of additional manual effort. Further, manual definition of properties from informal specifications is time tedious and error-prone. Insufficient and incorrect properties may hide design bugs that may prove expensive, when the bugs are uncovered later in the product cycle [12].

The discussion until now clearly emphasizes a well-defined, efficient, easy to construct, and easy to maintain property automation framework. We propose a novel property generation flow that is inspired from Object Management Group's (OMG) Model-Driven-Architecture (MDA) principle for code generation[18]. The main motivation behind following MDA flow are the flexibility, scalability and re-usability. An important aspect of the generation flow is to address multiple targets and at the same time, abstracting away from different artifacts of the target languages. To achieve the same we construct an intermediate model called Models-of-Property (MoP), by extracting specification details from abstract models called Model-of-Things (MoT). MoP is a platform independent way of describing abstract temporal traces.

As already mentioned, a major advantage of the approach is its ability to generate properties in multiple target languages (SVA - SystemVerilog Assertions, PSL - Property Specification Language, ITL - Interval Language). For instance, some property languages only support assertions for RTL level and not for transaction-level verification. With our approach, properties can be generated for RTL and transaction-level verification simultaneously and in different languages. Also, use of Python as generation language allows to employ rich

---

[1]Please note that *assertion* is a property that is evaluated with the verification directive *assert*. However, it is a common practice to use assertion and property, interchangeably. Other types of directives are *cover* and *assume*.

978-1-5386-4757-8/18 $31.00 © 2018 IEEE

features of the language to address several limitations posed by property specification languages. Another key aspect of our approach is that the properties are generated for both functional verification and RTL generation validation. The former targets generating properties from specification for verifying functional correctness of the RTL. In latter case, we use our generation flow to automatically derive properties to validate the generated RTL views and formal models.

We contemplate key challenges that must be addressed by the property generation flow in Sec. II. In Sec. III, we elaborate on the generation flow and outline steps taken to tackle key challenges. We demonstrate the effectiveness of our approach by considering a real-life design as a test vehicle in Sec. IV. We provide a brief summary of a selection of the existing approaches and compare them with our approach in Sec. V. A brief summary of the work in Sec. VI concludes the paper.

II. FACTORS INFLUENCING PROPERTY GENERATION

The level of automation that can be achieved relies on several factors such as target languages, availability of formal specifications, underlying automation framework, target verification techniques and many more. For example, the degree of expressiveness of the specifications and the level of automation that can be achieved have a direct trade-off [10]. Therefore, a crucial phase is to analyze the challenges and limitations to define an efficient generation flow considering all key challenges. In this section, we provide a brief summary of the key challenges addressed by our generation framework.

1) Informal specifications: A common practice in the industry is to outline requirements and specifications of a design in an informal natural language (ex: English, German).

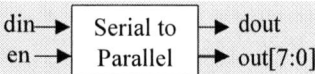

Input "*din*" is serial-in data line and "*dout*" is serial out data line and emits the same stream, delayed by 8 clock cycles. "*out[7:0]*" is parallel output. If "*en*" is "*high*", "*out*" is the last 8 bits input at "din". If "*en*" is low, "*out*" is all zeros.

Fig. 1: Informal specifications for a serial to parallel converter

Informal specifications are considered as the main source of design bugs, since context based languages are vulnerable to different interpretations. Further, ambiguous and incompletely defined behaviors only add to the known issue. Consider an example in Fig. 1, in which a specification for a serial to parallel converter are drawn. As can be seen, specifications for the given system is quite detailed. However, attempting to build properties from these specification lead to following questions:

- Is current bit included in the *last 8 bits input*?
- What is the value of "*out*" for first 7 clock-cycles?

From the simple example, it can be seen that informal specifications are vague and incomplete. Thus, informal specifications must be translated to formal ones to provide an abstract and complete view of the system.

2) Effort for formalizing specifications and building generators: The generation flow should be able to quantify the advantages with regard to productivity and achieving more quality. Even for simpler designs, translating informal specification to formal specification may take considerable effort and time. Thus, the effort needed for transforming the specifications into formal models and describing the property trace decides the effectiveness of generation approach. Also, the degree of expressiveness in formal specifications must be controlled to avoid huge effort needed for building generators [10].

3) Grey-box approach: A significant benefit of assertions is that they can be used across all verification techniques such as simulation, formal verification and emulation. A verification environment can be built to verify the Design Under Verification (DUV) following different approaches such as black-box, grey-box or white-box [21]. In the black-box approach, the verification environment is unaware of the internal signal details of the DUV and typically deployed to verify top level behaviors. Properties for such an approach need only top level port details. However, this approach is not ideal for formal verification as the technique requires design details to avoid false negatives and also to speed-up the property proof. In the white-box approach, the verification environment has complete access to internal details of the DUV. This approach may dilute the verification quality if the verification environment is close to design implementation and further changes in the design requires significant amount of rework. The grey-box approach meets both approaches in the middle and provides controlled insights into the DUV. The grey-box approach is suitable for property evaluation with all techniques and hence, we consider controlled design information for defining properties.

4) Static and dynamic generation techniques: In general, property generation can be classified as static or dynamic technique. Static techniques involve generating properties by statically analyzing RTL description or specifications. In dynamic techniques, properties are generated from simulation traces by using data mining or machine learning algorithms. Depending on the objective of the generated properties (bug-hunting, filling coverage holes, effective verification), an effective flow needs to define the technique being adapted for the generation.

5) 4-eyes principle: According to a well established paradigm, RTL design and verification of the same must be carried out by at-least two engineers. When both hardware design and properties (or test-cases) are generated in an automated manner using the same flow, the possibility of a bug being hidden by the flow cannot be ruled out. Consequently, the design and property generation flows must be decoupled from one another. In other words, the generation framework must satisfy 4 eyes principle. This helps to avoid the possibility of bug escapes due to inherent flaws in the generation framework.

III. PROPERTY GENERATION

Our property generation flow is built on-top of an existing meta-model based automation framework. A brief summary

978-1-5386-4757-8/18 $31.00 © 2018 IEEE

of the same is outlined in III-A. We elaborate the model-driven approach taken for property generation in III-B. Later, we depict how our flow is also used for automatic validation of the generated views of RTL in III-C.

A. Meta-model based automation framework

An in-house automation framework uses meta-models to automate the code generation [5], [6]. Use of meta-models allows to develop the generators not only for one model but, for all valid instances of the meta-model. This approach enables higher degree of productivity gain as reported in [6]. Meta-models can be represented either graphically using UML class diagrams or in textual format. In principle, meta-models are used to define things, their attributes and relations between them. Further, an advanced feature of the framework is a common meta-meta-model, which allows to combine different meta-models and to generate meta-models from EBNF notations [16].

For a given hardware function to be implemented, a specific meta-model is constructed to represent different components and their relations. Once a meta-model is created/generated, an infrastructure is derived in order to facilitate flexible and easy maintaining of code generators. In detail, for a given meta-model, automatic APIs are provided together with model structures to combine, modify and generate target code. Further, a graphical tool is generated to manually enter the meta-model instances (models). Target code transformations are coded in Python and Mako templates and are largely assisted by the generated infrastructure. Python and Mako templates are used as automation languages due to their simple use, dynamic (type)-checks and ad-hoc on the fly extendability. However, developing generators in other object-oriented languages can also be realized using *plugin & extension* mechanisms available in the framework.

B. Model-driven property generation flow

The approach taken for automating properties is inspired by the MDA vision for code generation from OMG. The core idea of MDA is to perform a series of model-to-model transformations, in which each model represents a higher level of abstraction than the next target model. We are currently upgrading the existing automation flow to adapt MDA principle and successfully developed the flows for both hardware/RTL and properties. For further read on MDA and our adaption for RTL generation, please refer to [4]. The approach for the generation of properties is depicted in Fig. 2. The flow is conceptually divided into 3 layers as Model-of-Things (MoT) layer, Models-of-Property (MoP) layer and View layer.

MoT layer: From Sec. II-1, informal specifications are one of the key challenges for automating properties. To address the same, we capture informal specifications in formal models as shown in upper part of Fig. 2. We first create a meta-model called Meta-Model-of-Things (MMoT). This meta-model captures high-level attributes, but does not include details of the micro-architecture. An instance of MMoT is MoT which in turn is a valid abstract formal model of the hardware

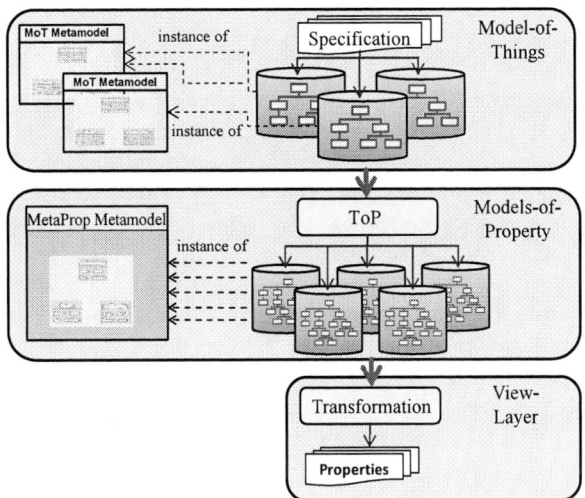

Fig. 2: Model-Driven Property Generation Flow.

system. In short, MoTs provide information about "what?" and "when?" with respect to the intended hardware function and hide implementation details ("How?"). MMoT are graphically represented using UML class diagrams and deriving MoT from MMoT is largely assisted by the framework. Thus the efforts needed for translating informal specification to formal models are substantially lowered by utilizing the underlying automation framework.

MoP layer: The next layer in our flow is Models-of-Property layer and is shown in central part of Fig. 2. This intermediate and also the core model of our approach is built by extracting information from MoT and DUV. As shown in the figure, transformations, which are Python programs, are used to extract the required details and are termed as Templates-of-Property (ToP). In addition to abstract requirements from MoT, micro-architecture details such as clock-cycle delays between events[2] and port names are needed for describing temporal traces[3]. From the extracted details, ToPs define MoP which holds the abstract temporal trace capturing a required hardware behavior.

Each MoP is an abstract and platform independent way of

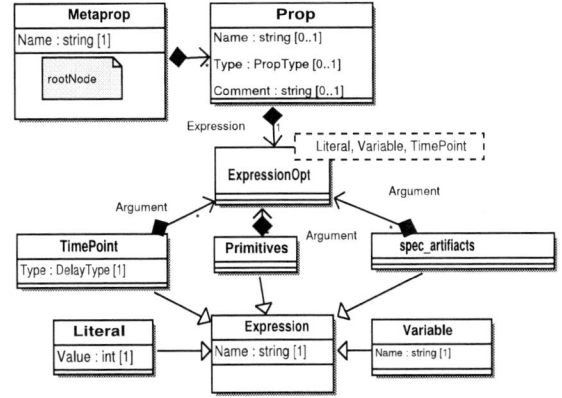

Fig. 3: A simplified meta-model of Model-of-Property

describing a property trace. MoP is defined by a meta-model

---

[2]Events are transitions from one concrete state to another
[3]Temporal traces are boolean expressions spanning multiple clock cycles

978-1-5386-4757-8/18 $31.00 © 2018 IEEE

called MetaProp and is shown in Fig. 3. The objective of the meta-model of MoP is to define the structure and elements of a property trace. Through MoP, an expected hardware behavior is modeled as a sequence of expressions spanning multiple clock-cycles. Hence, MetaProp is conceived on the notion of a temporal expression tree. The expression tree consists of primitive operators, delay operator (time variable) and certain pre-defined special constructs.

Rootnode of the meta-model has a composite relation to property class (*Prop*) with multiplicity *many*. Each *Prop* has a *Name*, a *Property Type*, an optional description and an *Expression*. The *Property Type* categorizes a property as an assert, assume or cover. *Expression* class is the base class for all elements that constitute the trace. *Literal*, *Variable*, *TimePoint*, *Primitives* and *spec_artifacts* are the elements of an expression. *TimePoint*, *Primitives* and *spec_artifacts* classes in turn accept expression as an argument to build the expression tree. In order to keep the meta-model simple and compact, class *Primitive* acts as a placeholder for all primitive operations (ex: Logical-AND, Bitwise-OR, NOT, etc.). All primitive operators are components of a separate meta-model called *MDA_Expressions* and are merged with the main meta-model MetaProp. Similarly, *spec_artifacts* class acts as a placeholder for special system tasks (ex:*PAST*, are used to simplify the definition of traces in ToP) and are combined with MetaProp to complete the meta-model of MoP.

ToPs are used to define the MoP and are largely aided by the APIs that are generated by the framework for the MetaProp meta-model. An example of ToP is shown in Fig. 5. Since ToPs are Python programs, the complete feature set of the language is utilized for flexible, efficient and compact MoP definitions. Due to the feature rich and flexible nature of Python, different coding styles, different data structures, the aspect oriented programming paradigm and the functional programming paradigm can be used, to effectively define property models.

View layer: The final and least abstract layer of our approach is the view layer shown in lower part of Fig. 2. MoPs, which are automatically generated from ToP definitions are transformed by Templates-of-View (ToV) into properties in one or more target languages. The ToVs are a combination of Python and Mako template programs. Python programs are used to map primitive and special task operators to their pre-defined structure in a specific target language. Using Python functions allows effective and flexible mapping of each operator to its corresponding syntax and semantics in the specification language. Mako templates are only used to define the structure of the file that contains the properties along with necessary constructs such as verification directives and binding to the design module. Currently, our flow supports both SVA and ITL target languages.

C. RTL generation validation

In preceding paragraphs, we outlined the property generation flow from abstract specifications. However, due to the

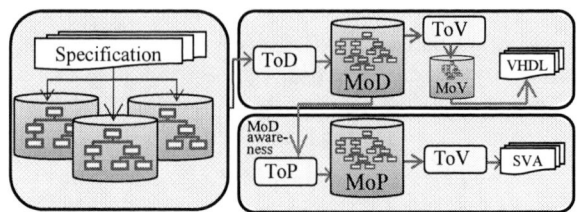

Fig. 4: Automatic validation of RTL generated views.

flexibility and adaptability of the flow, the approach is also employed for automatic validation of the generated RTL views. The flow employed for RTL code generation is depicted in the upper part of Fig. 4. In brief, Templates-of-Design (ToDs) extract abstract specification information from MoTs and define Model-of-Design (MoD), which holds the micro-architectural details of the intended hardware system. The MoD consists of a set of highly flexible hierarchical components and their interconnections. Each component is precisely defined by a Mealy machine [15]. Finally, the MoD is transformed into Model-of-View (MoV), which is an abstract syntax tree of the target language to realize the target RTL code (VHDL, Verilog) [4].

A significant aspect of any hardware modeling and generation is to validate the correctness and conformity of the generated RTL code as perceived by the designer while developing the high-level abstractions. In other words, RTL generation validation is not the functional verification of the design, instead validation of the design generation process itself. To achieve the same, a set of automatic properties describing the intended design behavior are generated from MoD for each sub-component, as shown in lower part of Fig. 4. Since each component (also composition of components) in MoD is defined by a Mealy machine with clear semantics, ToPs are setup to automatically generate property models (MoPs). Developing ToPs for validation of MoD is one time effort and are re-used for any MoD thereafter. Finally, a generated RTL files and properties are evaluated either in a formal tool or in a simulator.

IV. APPLICATION

To demonstrate the applicability and advantages of our model-driven flow, we generated properties for verifying different instances of a CPU core. In addition, peripheral devices such as timer, register interface and connectivity modules are formally verified by using properties generated from the proposed flow. We describe the steps followed for generating properties for different instances of processor core in the following subsections.

A. MetaRISC: A meta-model to symbolize RISC ISA

In general, MMoT is formulated considering various alternatives of the intended hardware system. On the same notion, a meta-model called MetaRISC is created to formalize the abstract elements and behavior of a generic RISC processor. This meta-model for RISC ISA consists of an instruction encoding, instruction behavior, and architectural state objects (ex: CSRs, Program Counter, GPRs) as major elements of a processor system. We considered a RiscV core supporting

978-1-5386-4757-8/18 $31.00 © 2018 IEEE   234

RV32I Base Integer Instruction Set [1]. We created an instance of MetaRISC meta-model for RiscV 32I Base Integer Instruction Set. Similarly, different MoT instances can be created that support different extensions of the RiscV ISA from the MetaRISC meta-model. RTL code of the RiscV processor core is generated following the model-driven flow outlined in [4].

B. Decoding MoT and describing ToPs

The next step is to describe property models and this step is the substantial part of our property generation flow. MoTs are fed into ToPs through an *extension* mechanism available in the automation framework. In ToPs, MoTs (MetaRISC instances) are decoded to extract the supported instructions, their encoding and corresponding behavior (dataflow strings). Additionally, RTL high-level implementation details such as port names and clock-cycle delays are also fed into the ToPs (grey-box approach). For each instruction encoding type (ex: jump, store, load, branch, arith, etc.), a property model (MoP) is described. This MoP is an abstract model of all instructions that belong to the corresponding encoding type. This property model is then used to generate properties with required operations (or sequence of operations) needed to be performed by each instruction.

```
def get_alu_reg_prop(stages):
  self.stages = stages
  dec_delay=1
  (ex_delay=1) if (stages==5 or stages==4) else (ex_delay=0)
  (mem_delay=1) if (stages==5) else (mem_delay=0)
  ##antecedent
  antecedent= LAND(CHANGED(inst), EQ(pc, HWPLUS(pc,4)),
                   EQ(SLICE(inst,6,0),    R_TYPE),
                   EQ(SLICE(inst,14,12), R_TYPE_ADD),
                   EQ(SLICE(inst,31,25), FUNCT_7_ADD))
  ##consequent
  consequent= LAND(##execute phase
                   DELAY(dec_delay),
                   EQ(alu_in1, RS1_DATA),
                   EQ(alu_in2, RS2_DATA),
                   EQ(alu_res, getOP(alu_in1,alu_in2,inst)),
                   ##mem-access phase
                   DELAY(ex_delay),
                   EQ(MEM_WR_EN, 0), EQ(MEM_RD_EN, 0),
                   ##write back phase
                   DELAY(mem_delay),
                   EQ(REG_EN,1),EQ(REG_DATA,PAST(alu_res,2)),
                   EQ(REG_ADDR, PAST(SLICE(inst,11,7),3)))
  ##return MoP
  return IMPLY(antecedent, consequent)
```

Fig. 5: Code snippet of ToP describing MoP (simplified)

Describing property traces in Python allows to utilize language features to efficiently setup the program for multiple hardware alternatives. Fig. 5 shows an example for describing a property trace for capturing *"ADD"* instruction (encoding: arithmetic register type) behavior in ToP. As can be seen from the Fig. 5, the same ToP is used to generate properties for 3-stage, 4-stage and 5-stage CPU cores. The re-usability factor of ToP is high as they are built for all valid instances of the MetaRISC meta-model. In addition, another benefit of using the generation flow is in describing simple property models. For each instruction, we generate multiple properties checking specific behaviors. For example, same MoP is used to generate properties to verify the core behavior with and without for-

warding scenario. This makes the generated properties simpler and more efficient for formal analysis.

TABLE I: LoC gain and re-usability of ToPs

Design module	Instances	ToP	SVA	ITL
CPU Core	3	1.5k	4.2k[*]	3.7k[*]
Timer	Any	750	2.5k[*]	2.2k[*]
Connectivity module	Any	278	2.7k[*]	1.8k[*]
Register interface	Any	320	1.2k[*]	1.1k[*]

[*]The numbers are shown for one specific instance

In Table I, Lines of Code (LoC) in ToP and the corresponding generated LoC in SVA and ITL are tabulated. As seen from the table, ToP for CPU core is built in such a way that the same ToP is used to generate properties for 3 different instances (3/4/5-stage pipelines). Similar to CPU cores, RTL code for *timer*, *connectivity module* and *register interface* are also generated following the model-driven flow described in III-C. A timer is a common unit in a processor system and is generally employed as a watchdog timer, event counter, for raising interrupts at specific time interval and many more. Connectivity module is a design unit that is used to allow configurable connections between different modules. Register interface is used to allow interactions between the software program and peripheral devices through control and status registers (CSRs) and is common to all peripheral devices. ToPs for timer, connectivity module and register interface are built in such a way that the properties are generated for any instance of the corresponding RTL blocks. The LoC gain when coding properties in ToP as compared to SVA or ITL is at-least 3x or more. In special cases the LoC gain is 10x or more and shows the efficiency of the generation flow. This shows that our generation approach saves effort already in the first application and gains more benefit with additional use.

V. RELATED WORK

Several academic and industrial solutions exists for the generation of properties. A brief summary of selected approaches is listed in the following.

Commercial formal tools offer pre-packaged verification solutions in the form of so called formal apps. A majority of these apps automate properties from meta-data structures. They are mostly limited to ip-xact, csv or excel sheets [2], [11], [8]. Formal apps (with exceptions) are restricted to specific verification tasks such as SoC connectivity or register file verification and hence, do not cover major aspects of property generation. These cases are already covered by our approach as we utilize an automation framework that also supports meta-data description formats such as xml, xmi, ip-xact, csv, xls and many more.

A number of approaches exist that generate/extract properties from the data collected during simulation runs [20], [17], [13], [3], [9]. Although these approaches differ in their implementation mechanism, they all use simulation test cases to generate and dump data, required for automatic extraction of assertions.

In [20], [17], the authors use data mining, static analysis of the RTL and machine learning algorithms to generate candidate assertions. These candidate assertions may also include incorrect assertions and hence, need revision in a formal tool to eliminate the spurious assertions. Approaches in [9], [3], [13] use dynamic analysis to generate assertions from simulation runs. Even though the approaches that utilize a simulation database for generating property traces are more automatic in nature, there are several significant limitations of these mechanisms. Building extensive/exhaustive set of test cases for simulation runs is not realistic (due to large and complex designs). Thus, the generated assertions are incomplete and insufficient for an effective validation of the design. Moreover, simulation test cases are generally constructed to verify high-level design behaviors. The properties generated from such traces may not be suitable for effective formal verification.

A static approach is proposed [22], in which the assertions are generated by adopting an RTL ATPG algorithm. A set of assertions are generated for each output signal and only a minimal set required for covering the complete design space are considered. Even though the approach can be used to obtain 100% design space coverage, potential bug escapes are possible as the assertions are not generated from specifications.

Although several solutions exist, a structured approach for generating properties from abstract property models is not known to our knowledge. Many solutions adapt a dynamic flow using simulation traces or static analysis of RTL code, focusing on coverage metrics or bug hunting. Hence, these solutions are limited to specific verification problems and do not present an approach for generation of properties from specifications. In [19], an analogous approach is presented in which abstract properties are created from high-level system descriptions. The generated abstract properties are used as a reference for an intuitive RTL design process. These properties are later refined with implementation details for effective formal analysis of the design. Due to the flexibility of our generation framework, we envision a similar design approach, where the generated properties are used as sound formal specification models to automate the RTL code.

VI. SUMMARY

We presented a structured approach for the automation of properties for pre-silicon verification. The proposed flow raises the level of abstraction from property specification to property generation. The novelty of our approach is the adaptation of MDA principle for code generation and flexibility in generating properties for both functional verification and automatic validation of generated RTL code. The abstract property models are platform independent and support properties in multiple target languages (SVA, ITL, PSL). Moreover, capturing high-level requirements in meta-models, setting up ToPs for meta-model instances, introduces an efficient and promising approach for lowering the verification barrier. The same ToP is re-used to generate properties for different micro-architectures of the same design specification (ex: 5-stage, 4-

stage, single-stage pipelines of RiscV core). The results so far show that describing property traces in ToP is compact and LOC gain (ToP - generated properties) is at-least 3x and in special cases 10x or more. Since the approach uses model-to-model transformations to generate properties, the flow can be easily leveraged for requirement based development flows.

REFERENCES

[1] Krste Asanović and David A. Patterson. Instruction sets should be free: The case for risc-v. Technical Report UCB/EECS-2014-146, EECS Department, University of California, Berkeley, Aug 2014.

[2] Cadence Design Systems, Inc. JasperGold Apps User Guide. https://www.cadence.com/. [Online:accessed 05-Apr-18].

[3] A. Danese, T. Ghasempouri, and G. Pravadelli. Automatic extraction of assertions from execution traces of behavioural models. In *2015 Design, Automation Test in Europe Conference Exhibition (DATE)*, pages 67–72.

[4] W. Ecker and J. Schreiner. Metamodeling. In Soonhoi Ha and Jürgen Teich, editors, *Handbook of Hardware/Software Codesign*, chapter 10, pages 266–290. Springer, 2017.

[5] Wolfgang Ecker, Michael Velten, Leily Zafari, and Ajay Goyal. Metamodeling and code generation -the infineon approach. In Wolfgang Mueller and Wolfgang Ecker, editors, *MeCoES - Metamodelling and Code Generation for Embedded Systems: Workshop with ESWEEK*.

[6] Wolfgang Ecker, Michael Velten, Leily Zafari, and Ajay Goyal. The metamodeling approach to system level synthesis. In Gerhard Fettweis and Wolfgang Nebel, editors, *DATE*, pages 1–2. European Design and Automation Association, 2014.

[7] Harry Foster. Applied assertion-based verification: An industry perspective. *Found. Trends Electron. Des. Autom.*, 3(1):1–95, January 2009.

[8] Mentor Graphics. Questa Property Generation. https://www.mentor.com/products/fv/questa-property-generation. [Online:accessed 05-Apr-18].

[9] Sudheendra Hangal, Sridhar Narayanan, Naveen Chandra, and Sandeep Chakravorty. Iodine: a tool to automatically infer dynamic invariants for hardware designs. In *Proceedings. 42nd Design Automation Conference, 2005.*, pages 775–778, June 2005.

[10] Thomas Kropf. *Introduction to Formal Hardware Verification: Methods and Tools for Designing Correct Circuits and Systems*. Springer-Verlag New York, Inc., Secaucus, NJ, USA, 1st edition, 1999.

[11] Onespin Solutions. Onespin User Manual. https://www.onespin.com/products/dv-verify-apps/. [Online:accessed 05-Apr-18].

[12] D. Price. Pentium FDIV flaw-lessons learned. *IEEE Micro*, Apr 1995.

[13] Frank Rogin, Thomas Klotz, Görschwin Fey, Rolf Drechsler, and Steffen Rülke. Automatic generation of complex properties for hardware designs. In *Proceedings of the Conference on Design, Automation and Test in Europe*, DATE '08, 2008.

[14] Ron Collett and Dorian Pyle. What happens when chip-design complexity outpaces development productivity? {https://www.mckinsey.com/~/media/McKinsey/dotcom/client_service/Semiconductors/Issue%203%20Autumn%202013/PDFs/4_ChipDesign.ashx}.

[15] J. Schreiner, R. Findenig, and W. Ecker. Design Centric Modeling of Digital Hardware. In *IEEE International High Level Design Validation and Test Workshop, HLDVT 2016*, pages 46–52, 2016.

[16] J. Schreiner, F. Willgerodt, and W. Ecker. A new approach for generating view generators. In *Design and Verification Conference - US*, Feb 2017.

[17] D. Sheridan, L. Liu, H. Kim, and S. Vasudevan. A coverage guided mining approach for automatic generation of succinct assertions. In *2014 27th International Conference on VLSI Design and 2014 13th International Conference on Embedded Systems*, pages 68–73, Jan 2014.

[18] F. Truyen. The fast Guide to Model Driven Architecture.

[19] J. Urdahl, S. Udupi, T. Ludwig, D. Stoffel, and W. Kunz. Properties First? A New Design Methodology for Hardware, and Its Perspectives in Safety Analysis. In *Proceedings of the 35th International Conference on Computer-Aided Design*, ICCAD '16, pages 84:1–84:8. ACM, 2016.

[20] S. Vasudevan, D. Sheridan, S. Patel, D. Tcheng, B. Tuohy, and D. Johnson. Goldmine: Automatic assertion generation using data mining and static analysis. In *DATE 2010*, pages 626–629, March 2010.

[21] Bruce Wille, John C. Gross, and Wolfgang Roesner. *Comprehensive Functional Verification*. Morgan Kaufmann Publishers, 2005.

[22] T. Zhang, D. Saab, and J. A. Abraham. Automatic assertion generation for simulation, formal verification and emulation. In *2017 IEEE Computer Society Annual Symposium on VLSI (ISVLSI)*, pages 471–476.

Cyber-Physical Systems Integration in a Production Line Simulator

Stefano Centomo, Marco Panato, Franco Fummi
Department of Computer Science – University of Verona (Italy)
name.surname@univr.it

Abstract—**Digital Twin represents a simulated model of a production line which allows making analyses of future states concerning the real factory. More in details, these analyses are related to the variability of production quality, prediction of the maintenance cycle, the accurate estimation of energy consumption and other extra-functional properties of the system. This is the core of what is so called Industry 4.0. Every single node of the manufacturing process needs to be modelled as a Cyber-physical system to be able to make the mentioned analyses. However, Manufacturing simulators represent these systems with a high level of abstraction, making impossible precise analyses. In the state of the art, some solutions try to solve the problem connecting multiple domain-specific simulators, to preserve details but requiring complex co-simulation environments. This paper presents a methodology for the integration of Cyber-Physical Systems in production line simulators, avoiding these issues. The proposed solution is based on a new promising technology: the Function Mockup Interface (FMI). This standard defines an interface to exports models as blocks called Functional Mockup Units (FMUs). These FMUs can be easily integrated together composing heterogeneous systems. The methodology is composed of two steps: 1) Exporting Digital and Physical systems as different FMUs 2) Integration of the FMUs into a production line simulator. An example is used to validate our solution which clearly shows the limitations of the production line simulator without the integration of CPSs. This paper aims at producing Cyber-Physical Production Systems (CPPS) to make more accurate simulations, and hence more accurate analysis of the production line.**

I. INTRODUCTION

Industry 4.0 represents the fourth industrial revolution and its goal is to evolve the actual factories into smart factory systems [1]. A smart factory is a factory that can make analysis and rational decisions to optimize and maximize the entire production. To be able to do that the smart factory requires a simulable model of the factory, usually called *Digital Twin*. A *Digital Twin* is a combination of Cyber-Physical System(CPS), where each CPS represents a process of the real factory. Nowadays, there are several tools developed by different stakeholders, that allow to model a production line [2]. All these modeling tools use Model-based design approaches, thus giving to modelers intuitive and easy-to-use environments. Unfortunately, these tools do not provide mechanisms to simulate Cyber-Physical Systems making impossible to model the production line processes with more details. Because of this limitation, the resulting simulation is not accurate enough in order to make precise analysis and planning optimal strategies. For instance, [3], couples real equipments with a production line simulator. The approach is promising but it does not consider the models of the physical processes, thus making the solution not usable to make accurate analysis. Mosterman

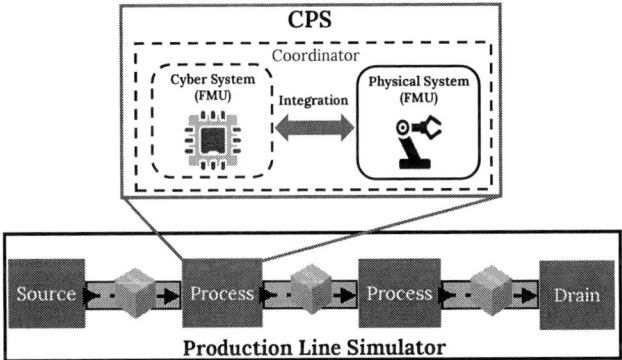

Figure 1: Overview of the CPS integration in a production line simulator

et al. [4] proposed an example of a simple logistic production line, modeling the entire process by using Simulink. This is a complex solution, but not easily reusable in different scenario. Some others [5], [6] try to fill this weakness by using co-simulation techniques. Co-simulation approches require complex environments and they are extremely error prone in the connection of components and time consuming. For this reason, in this work we present a methodology to integrate Cyber-Physical Systems in a Production Line Simulator but avoiding Co-simulation issues (see Figure 1). The proposed methodology starts from the modeling phase of the CPS by using specific domain languages and tools. For the digital system this methodology starts from VHDL or Verilog models at Register Transfer Level (RTL). On the contrary, the physical system is modeled by using OpenModelica tool [7], a state-of-the-art tool for this field. Both of the systems are then exported by using the Functional Mock-up Interface (FMI) technology [8], as Functional Mock-up Units (FMUs). Finally, the methodology relies on a coordinator written in C language, that manages the FMUs representing the two systems on the production line side. This work adopts Siemens Plant Simulation [9], that provides proprietary C-Interface. With this interface, it is possible to import C dynamic libraries (.dll) inside Plant Simulation. The proposed coordinator, and the FMUs, are then compiled as a C dynamic library and integrated in the simulator. The resulting dynamic Library contains a CPS, which represents a process of the production line. To enable the CPS integration, there are some issues to face with:

- FMU Generation for homogeneous and standardize communication;
- FMU Coordination;

978-1-5386-4757-8/18 $31.00 © 2018 IEEE

Listing 1: Sketch of Simtalk method that use an external library

```
1  openConsole
2
3  var lib_path := to_str("../library.dll")
4  var lib_ref : loadLibrary(lib_path)
5
6  var op1 : integer := 2
7  var op2 : integer := 3
8
9  var result := callLibrary(lib_ref, "add", op1, op2)
10
11 end
```

Listing 2: Simple C function implemented using SimTalk C-Interface

```
1  #include "cwinfunc.h" // C-Interface library
2
3  extern "C" __declspec(dllexport)
4  void add(UF_Value *ret, UF_Value *arg){
5
6    int op1 = arg[0].value.integer;
7    int op2 = arg[1].value.integer;
8
9    ret->type = UF_INTEGER;
10   ret->value.integer = op1 + op2;
11 }
```

- Integration of the CPS in the simulation environment;
- Timing scale differences among plant, cyber and physical models.

The paper examines and solves all such issues, allowing the designer to easily integrate CPS in a plant simulator, thus enabling precise analyses about the production line in order to optimize all the manufacturing processes.

The experimental results show the benefits of the integration of CPS in a plant simulator, in terms of simulation accuracy. The paper is organized as follow: Section II presents the necessary background about Production Line Simulators and about the FMI standard. Section III explains the integration methodology of the CPSs in Siemens Plant Simulation. Section IV shows experimental results related to the proposed methodology, while Section V reports conclusions and possible future works.

II. PLANT SIMULATION AND INTEGRATION ALTERNATIVES

This Section presents two main background concepts: an overview of production line simulators, particularly Siemens Plant Simulation, and a brief description of the FMI standard. The two concepts are necessary to understand the methodology explained in section III.

A. Production Line Simulators

In these years several providers proposed different tools to model Manufacturing processes [2], [10]. Report [10] summaries periodically all tools by proposing comparisons on their main characteristics. The last report showed the main differences between tools in terms of provided functionalities, usability, multiprocessing execution, costs and other characteristics. These data help in the selection of the most suited plant simulator with respect to the parameters to be evaluated. A production line is mainly the composition of a chain of production processes. These processes require handling information like geometric properties of the products, the processing time, the energy consumption, the failure rate, *etc.*. The simulation of production systems, with this information, allows making decisions in order to optimize the entire production line in terms of cost, quality and productivity. These simulators have some common principles such as:

- *Layout Planning*: Represents the geometrical structure of the production line. All of these simulators have a library of components (*i.e.*, generic processes, assembly stations, buffers, *etc.*) which is possible to model the factory, by considering physical constraints.

- *Material Flow/Fluid Simulation*: Represents the transportation of the products from a process to the others. This is made possible with components like line transporters or pipe, depending if the product of the production line is solid or fluid.

- *Process Simulation*: Represents the physical transformation made by the processes to products.

All of these simulators use the Discrete-event Model of Computation. When a product enters in one of the production process chain, an event is triggered and that specific process can execute the relative action.

B. Siemens Plant Simulation: SimTalk C-Interface

Plant Simulation [9] is the widely used production line simulator developed by Siemens. It offers an internal programming language called *SimTalk*. This language gives the entire control over the simulation in order to customize the production line models. *SimTalk* allows to define methods that can be used inside every available object of the Plant Simulation library. In particular, it is possible to couple *SimTalk* methods with occurring events (*i.e.* Entrance or Exit of a product from a process) in every object of the production line model. Furthermore, it is also possible to import dynamic libraries written in C, with a proprietary interface, called *C-Interface*.

SimTalk provides a `callLibrary` method that permits to call a function of an external dynamic library that is implemented using the *C-Interface* (Listing 1, line 9). Listing 2 shows a sketch of a function written in C implemented by using *SimTalk C-Interface*. The function has two parameters that represent the input and the output values, in order to respect the *C-interface* structure (Listing 2, line 4). The UF_VALUE is a structure defined by *SimTalk C-Interface* and it has to be used as the datatype of the function parameters. The UF_VALUE structure is composed of two fields: `type` and `value`. The `type` field represents the datatype of the parameters (Listing 2, line 9), while the `value` field represents the value of the parameters (Listing 2, line 10). *SimTalk C-Interface* allows the functions to have multiple inputs but only one output value. [2], [10] show comparisons between different simulators. From the survey [10] and the simulator comparison [2], we decided to adopt Siemens Plant Simulation for the purpose of this work because of *SimTalk* utilities. In particular the *C-Interface* is a very usable modeling environment that allows to extend the tool by creating models not natively supported.

978-1-5386-4757-8/18 $31.00 © 2018 IEEE   238

Listing 3: Sketch of SimTalk method that uses CPS library functions

```
1  openConsole
2
3  var libPath := to_str("../CPS.dll")
4  var CPSLib : loadLibrary(libPath)
5
6  //get properties from the product
7  var prodType := product.prodType
8
9
10 //simulate the CPS with the product properties
11 callLibrary(CPSLib, "simulateCPS",prodType)
12
13
14 //Get the properties
15 var time:=callLibrary(CPSLib,"getTime")
16 var energy:=callLibrary(CPSLib,"getEnery")
17 var prodProperty:=callLibrary(CPSLib,"getProdProp")
18
19
20 //Set the properties in Plant Simulation Objects
21 SingleProcess.time:=time
22 SingleProcess.energy:=energy
23 //Set the new properties of the product
24 product.property:=prodProperty
25 end
```

Listing 4: Sketch of an FMU coordinator using FMISupport library with *SimTalk C-Interface*

```
1  #include "cwinfunc.h" //C-Interface library
2  #include "FMISupport.h" //FMUs Support library
3
4  extern "C" __declspec(dllexport)
5  void simulateCPS(UF_Value *ret, UF_Value *arg)
6  {
7
8    Product prod=getProductProperties(arg);
9    bool eventFlag=false;
10
11   time = 0;
12   while(!eventFlag)
13   {
14
15     //FMU Simulation
16     fmuDoStep(cyberFMU,time,step);
17     fmuDoStep(physicalFMU,time,step);
18
19     //Data Exchange between FMUs and product
20     eventFlag=dataExchange(cyberFMU,physicalFMU,prod);
21
22     time=time+step;
23
24   }
25 }
```

C. Functional Mockup Interface (FMI)

In the state-of-art there are some solutions that try to uniform the interfaces and functionalities of the models. For instance, IP-XACT [11] is a standard for the definition of of Hardware model interface, but it does not consider their functionalities. Other solutions, like S-functions provided by Mathworks, define both interface and functionality of the models, but they are proprietary and are not supported by other tools.

The Functional Mockup Interface [8] is a relatively new standard that defines an interface that allows to export and imports models from different tools. The main goal of the standard is the simulation of heterogeneous systems. A component which implements the interface is called FMU. The standard defines two different interfaces: *Model Exchange* and *Co-Simulation*. An FMU may implement one or the other. In a Model Exchange FMU the simulation environment has to provides an external numeric solver. With *Co-Simulation* FMU the numeric solver is provided inside the model. This work considers only *Co-Simulation* Interface being more portable and easier to use. An FMU is composed of two parts: an XML file and a dynamic library. The XML file contains the information about the interface of the FMU, that will be visible to the simulation environment [8]. Each port of the interface has different properties like *name*, *causality* (*e.g.*, input, output, parameter, *etc.*), a *type* and a *value reference*. The value reference of a port represents a numeric identificator that must be unique. The dynamic library must implement the functionality through a set of functions defined by the standard. The most relevant functions are:

- `fmi2SetupExperiment`: initializes the initial condition of the FMU.

- `fmi2Set`: sets the value of an input port of the FMU.

- `fmi2Get`: gets the value of an output port of the FMU.

- `fmi2DoStep`: executes the model contained in the FMU.

III. INTEGRATION METHODOLOGY

The integration relies on the FMI standard [8], and *SimTalk C-Interface*. The Cyber and the Physical systems are modeled and exported as FMUs. Then, a coordination algorithm with the *SimTalk C-Interface* is implemented, in the view of enabling the integration with the production line simulator.

A. Cyber System: Modelling and FMU Generation

Cyber Systems are designed by using HDL languages (VHDL & Verilog) usually at Register Transfer Level (RTL). The entire Cyber System needs to be exported as a portable FMU, to allow the integration with the Physical System.

Thus, a first step is required to translate the HDL model into a C/C++ equivalent model, in order to simply wrap it with the FMI interface. In the years, the translation issue has been addressed by different works. For instance, [12] considered only the translation from VHDL to C. Verilog is not supported. In [13], VHDL and Verilog are both considered, but no other manipulation tools are provided.

This methodology relies on HIFSuite framework [14]. HIF-Suite provides a set of manipulation tools for VHDL and Verilog models and also offers a set of APIs that allows to extend its functionalities. Some of these manipulations allow generating semantic equivalent models written in C++ [15]. After the manipulation and generation of C++ code, HIFSuite can also generate the XML file and the C wrapper, necessary for the composition of the FMU [16], [17]. As explain in Section II an FMU is a zip file that contains an XML file and a dynamic library. Then, the resulting C++ and C code is compiled in order to obtain the dynamic library. Finally, to generate the FMU, the XML and the dynamic library are zipped together. The used abstraction [15] changes the model of computation of the Cyber System from event-driven to cycle-accurate or even transaction-level, but preserving the initial behavior. This step is fundamental for the correct FMU integration with the other FMUs.

B. Physical System: Modelling and FMU Generation

Physical Systems can be modeled by using two different solutions, depending on the complexity of the system to be represented.

$$\begin{cases} \dot{x}(t) = Ax(t) + Bu(t) \\ y(t) = Cx(t) + Du(t) \end{cases} \quad (1)$$

The first solution is used for Linear Time-Invariant (LTI) systems represented by a set of equations (see Eq. 1) and four matrices (A,B,C,D). The first equation represents the evolution of the internal states of the system. The second represents the output function, that depends on the input and the actual state of the system. To easily model LTI systems, we provided a C template, which allows to define these four matrices. The template uses a fixed step integration as numeric solver for the LTI system. The template is then wrapped with the FMI interface and compiled as dynamic library, to obtain an FMU of the system.

The second solution is used when dealing with more complex dynamic systems. In this cases a modeling framework like OpenModelica [7] can be used. OpenModelica is an open-source Model-based design environment. It offers a library of components that allows to model the physical system as the composition of these blocks, by using the Model-based design paradigm. Figure 2 shows a simple OpenModelica model that represents a second-order dynamical system, with a PID, secondOrder and feedback block. The model has one input port, "u", and one output port, "y". Furthermore, OpenModelica exports the system as an FMU. OpenModelica exposes the input and the output ports in the XML file of the FMU. The obtained Physical System uses Synchronous Data Flow Model of Computation (MoC) and it needs to be synchronized with Cycle-accurate MoC of the Cyber System [17].

C. Cyber-Physical System: Coordination and Integration

The two FMUs, representing respectively the Cyber and the Physical system, required to be connected and simulated together to obtain a CPS behavior. In order to do that an FMU coordinator is needed. For this methodology, we have developed a C library called *FMISupport* that allows managing Co-Simulation FMUs. The coordinator has to be integrated into Plant Simulation using the *SimTalk C-Interface*, to allow precise analysis concerning the production process. Listing 4 sketches a basic coordinator algorithm that uses the *FMISupport* library and *SimTalk C-Interface*. The function `simulate_CPS` is defined with the C-Interface signature and it represents the simulation function of the CPS (Listing 4, line 5). This function is called by a *SimTalk* method (Listing 3, line 11) associated to the Entrance event of a product in a process called *SingleProcess*. The properties of this product (*i.e.* type of the product, geometric properties, *etc.*) are passed to the `simulateCPS` function as parameter, and saved in a local structure (Listing 4, line 5-8). Then, the two FMUs are simulated for the same amount of time, using the `fmuDoStep` function provided by the *FMISupport* library (Listing 4, lines 16-17). After the simulation step of the line, data are exchanged between the FMUs and the product (Listing 4, line 20). The simulation stops when a certain event is reached. This event represents, for example, the end of the physical process. The time of the entire process could depend on the

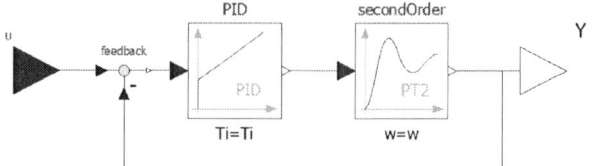

Figure 2: Example of an OpenModelica Model

product properties of the line. For instance, if the process has to deal with different types of product, it could affect the processing time. Furthermore, the processing time could also affect the energy consumption of the process. *SimTalk C-interface* does not define a structure to return more than one value for each function. Because of this limitation related to *SimTalk C-interface*, it is needed to define a get function for each desired property. For that reason, some functions are defined in the Coordinator to retrieve the information related to the product and the process. Regarding the process, this methodology focuses on two aspects: the processing time and the energy consumption. In Listing 3 it is possible to see the *SimTalk* method that uses the CPS Library implemented in Listing 4. The *SimTalk* method retrieves the product properties (Listing 3, line 7) and then calls the `simulateCPS` with the product properties as parameter (Listing 3, line 11). After the execution of the `simulateCPS`, the *SimTalk* method retrieves the processing time, the energy consumption and the new product properties (Listing 3, lines 15-17). Finally, the *SimTalk* method propagates the properties to the product and to the *SingleProcess* (Listing 3, lines 20-24).

IV. METHODOLOGY APPLICATION

The proposed methodology has been applied to a three-process production line. Metal sheets cross the line that must bend sheets to a desired angle. Involved processes are the following:

- *Sensing*: it analyzes sheets to discover their bend angle which is written over an applied barcode;
- *Bending*: it reads angles from barcodes, then it computes the bending actions and it controls the bending machine;
- *Checking*: it ensures that sheets have been correctly bent; thresholds are defined to separate correct, acceptable and damaged sheets.

This example has been chosen because it contains the three-steps sequence common to many production lines: generation, production and validation of items. Plant Simulation models this production line with different nodes, as shown in Figure 3:

- *Source* node provides metal sheets to the line;
- *Drain* node receives manufactured metal sheets;
- *SingleProcess* nodes represent the three processes above.

Plant Simulation provides *SimTalk* to describe production line functionalities: this leads to a high-level simulation which is very fast but not much detailed. The proposed methodology integrates CPSs into *SingleProcess* nodes to get simulations enriched with time, energy and mechanical wear. This is done by using the *SimTalk C-interface* which permits to call customized C-functions directly from the simulator.

978-1-5386-4757-8/18 $31.00 © 2018 IEEE

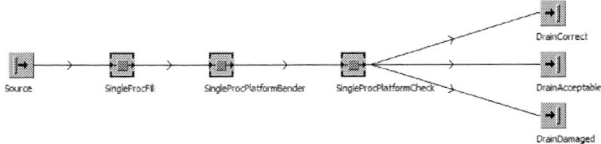

Figure 3: Overview of proposed production line in Plant Simulation

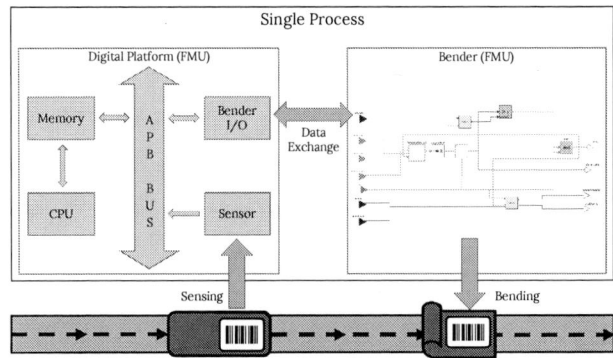

Figure 4: Overview of bending machine CPS

A. Bending machine CPS

This CPS consists of a digital virtual platform and an analog model of a bending machine. The digital platform is composed of M6502 CPU, RAM and ROM memory, bus, sensor and bender I/O interfaces as shown in Figure 4. CPU can access bus peripherals by MMIO. For The bending machine there are two models obtained by using two solutions proposed in Section III-B. The first model is a partial model of the bending machine and it is described using the C template, defining the values of the four matrices A,B,C,D (Section III-B, Eq. 1). This model allows to have faster simulation but does not have accurate details about the energy consumption of the equipment. The second model is more detailed than the first it is described using OpenModelica tool [7], integrating information concerning the energy consumption. These models are characterized by different bending speeds according to bending direction, different power consumption, different bending times, increasing bending error due to bender wear. Two versions of this CPS are provided:

- *Partial platform*: it is composed of the digital platform and a partial bending machine model, described as LTI system.

- *Complete platform*: it is composed of the digital platform and the complete bending machine model, described using OpenModelica;

The CPU reads the angle from the previously applied barcode, then computes the bending action as delta angle and bend direction. The bending action is provided to the bending machine through the bus. When the bending process completes, metal sheet can advance to the next node. This behavior is shown in figure 4. The models of the digital platform are written in VHDL and Verilog. The entire digital platform is automatically abstracted into their C++ cycle-accurate level representation by state-of-the-art [17] and then wrapped in order to obtain the Digital Platform FMU. The CPU Software is cross-compiled stored inside the Digital Platform FMU. The two models of the Bending machine, partial and complete, are exported as FMUs using the C template and the OpenModelica framework. The Digital platform FMU and one of the Bender FMUs, are manually interconnected with a coordinator(see Listing 4). The FMUs and the coordinator are wrapped using the *SimTalk C-interface* exposing the functions needed to simulate the CPS (`simulateCPS`) and retrieve the properties (*i.e.*,`getTime` and `getEnergy`). Finally, the entire code is compiled into a dynamic Library ready to be loaded from the simulator (see Section III-C). A model composed of only the bending machine analog model is not provided. **This is due to the digital platform simulation time that is negligible when compared with the one required to simulate the bending machine model.**

Table I: Properties evaluated by the different simulation line versions

Version	Functionality	Time	Energy	Mechanical Wear
SimTalk	✓	≈		
Partial Platform	✓	✓	≈	≈
Complete Platform	✓	✓	✓	✓

B. Alternatives Taxonomy

Table I summarizes all aspects supported by the different modeling alternatives proposed in this paper. It compares the standard Plant Simulation production line model with CPS-enriched ones. The *SimTalk* approach can only simulate bending functionalities and it provides an estimation of the simulated time. This is the highest level of abstraction because a bending machine simulable model is not used. With a CPS model, more information can be retrieved. *Partial Platform* grants detailed information about CPU processing time. Partial bending machine model used in Partial Platform provides only an estimation of energy and mechanical wear but it can give the precise bending time. With *Complete Platform* instead, it is possible to get detailed information about all properties. This is possible thanks to the refined model.

C. Simulation speed

Simulation data are analyzed from different points of view to demonstrate the usability of this work. First adopted method is to measure the simulation times and the required times to simulate the bending of a single metal sheet: it is important that simulated line can advance its time at least as the real factory does.

Table II reports the simulation time respect to the total number of processed metal sheet. This paper proposes two different coordinators for the *Complete Platform*: the *Basic* and the *Optimized Coordinator*. The difference between *Basic* and *Optimized Coordinator* concerns the step size used to simulate an entire operation of the bending machine FMU. The *Basic Coordinator* always uses a step size equal to the clock period of the digital FMU. The *Optimized Coordinator* adopts longer steps during bendings, that decouple the simulation of the FMUs and reduce the number of synchronization point. The *Complete Platform* with the *Optimized Coordinator* achieves up to 374x speed-up, respect the *Basic Coordinator*, with 1000 bending operations. The real bending machine requires on average ~ 1 second at every new sheet (~ 1.8

978-1-5386-4757-8/18 $31.00 © 2018 IEEE 241

Table II: Execution Time comparison of different approaches.

Metal Sheets	SimTalk		Partial Platform		Complete Platform			
					Basic Coordinator		Optimized Coordinator	
	Total Time	Time per Sheet	Total Time	Time per Sheet	Total Time	Time per Sheet	Total Time	Time per Sheet
10	0.01	0.0001	18.17	1.8165	48.33	4.8334	1.40	0.1404
100	0.01	0.0001	202.35	2.0235	499.89	4.9989	2.85	0.0285
500	0.03	0.0001	1021.89	2.0438	2562.85	5.1257	9.06	0.0181
1000	0.06	0.0001	2290.06	2.2901	6364.07	6.3641	17.04	0.0170

Table III: Simulation times and percentage of errors of the proposed approaches, respect to the real bending machine.

Metal Sheets	Partial Platform		Compl. Plat. Basic		Compl. Plat. Opt.	
	Sim Err.	Time/Sheet	Sim Err.	Time/Sheet	Sim Err.	Time/Sheet
10	0.00%	1.8165	0.00%	4.8334	0.00%	0.1404
100	33.33%	2.0235	0.00%	4.9989	25.00%	0.0285
500	42.86%	2.0438	0.00%	5.1257	33.93%	0.0181
1000	40.30%	2.2901	0.00%	6.3641	19.40%	0.0170

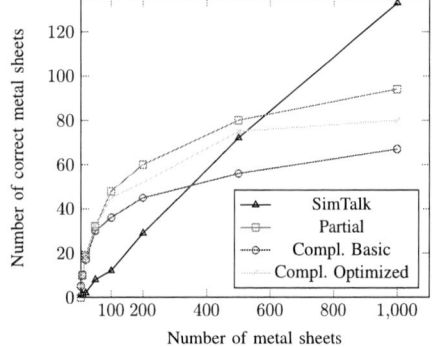

Figure 5: Number of correct metal sheets

seconds in the worst case). This leads to about 3 seconds per metal sheet considering also times required to transport sheets between nodes. In table II, *Complete Platform* with *Basic Coordinator* requires more than 3 seconds to simulates a single bending operation. This result does not met real-time requirements in order to plan strategies on the real factory. In conclusion, *SimTalk*, *Partial platform* and *Complete platform* with *Optimized Coordinator* are solutions that can be adopted to advance in parallel to real production line.

D. Simulation errors

The second analysis focuses on the measurements of simulation results: an abstracted model leads to less detailed simulations. Figure 5 shows the number of correct metal sheets produced by the proposed approaches. *Complete Platform* with *Basic Coordinator* represents the correct trend: it is the most detailed simulation, but requires a lot of time and can not be used in a real-time context. *SimTalk* version grows almost linearly because it does not take care of mechanical wear. *Partial Platform* leads to smaller error thanks to its approximated wear model. *Complete Platform* with *Optimized Coordinator* leads to an error smaller than *Partial Platform* one. It is due to different approximations done by analog numeric solvers when changing step size. Thus, *Complete Platform* with *Optimized Coordinator* represents an optimal trade-off solution between simulation time and the percentage of errors.

V. CONCLUSIONS

We presented a methodology for the integration of Cyber-Physical Systems in a production line simulator. This work

represents an efficient alternative to co-simulation methodologies for the modeling of Digital Twin concerning Industry 4.0.

The methodology has been implemented and applied to a common use case scenario in manufacturing process. The results obtained from the simulation clearly show the benefits from the integration of Cyber-physical systems in terms of accuracy. **CPS integration allows to estimate properties of the production line that could not be estimated elsewhere**. The simulation time required to compute the CPS is 100 times faster then the real processing time. These promising results enable the possibility to make precise analysis in order to plan a strategy to optimize the entire production line. Future work will focus on improving the simulation speed proposing a new coordination of the FMUs and mixing models with different levels of abstraction.

REFERENCES

[1] R. Drath and A. Horch, "Industrie 4.0: Hit or hype? [industry forum]," *IEEE Industrial Electronics Magazine*, vol. 8, no. 2, pp. 56–58, jun 2014.

[2] D. Mourtzis, M. Doukas, and D. Bernidaki, "Simulation in manufacturing: Review and challenges," *Procedia CIRP*, vol. 25, pp. 213–229, 2014.

[3] J. Vachalek *et al.*, "The digital twin of an industrial production line within the industry 4.0 concept," in *2017 21st International Conference on Process Control (PC)*. IEEE, jun 2017.

[4] P. J. Mosterman and J. Zander, "Industry 4.0 as a cyber-physical system study," *Software & Systems Modeling*, vol. 15, no. 1, pp. 17–29, oct 2015. [Online]. Available: https://doi.org/10.1007/s10270-015-0493-x

[5] D. Pfeifer, J. Valvano, and A. Gerstlauer, "SimConnect and SimTalk for distributed cyber-physical system simulation," *SIMULATION*, vol. 89, no. 10, pp. 1254–1271, mar 2013.

[6] T. Peter and S. Wenzel, "Coupled simulation of energy and material flow using plant simulation and MATLAB simulink," *SNE Simulation Notes Europe*, vol. 27, no. 2, pp. 105–113, jun 2017.

[7] P. Fritzson *et al.*, "OpenModelica - a free open-source environment for system modeling, simulation, and teaching," in *2006 IEEE Conference on Computer-Aided Control Systems Design*. IEEE, oct 2006.

[8] T. Blochwitz *et al.*, "Functional mockup interface 2.0: The standard for tool independent exchange of simulation models," in *Proc. of MODELICA Conference 2012*, 2012, pp. 173–184.

[9] Siemens, "Tecnomatrix, Plant Simulation."

[10] "Simulation software survey," 2017. [Online]. Available: https://www.informs.org/ORMS-Today/OR-MS-Today-Software-Surveys/Simulation-Software-Survey

[11] IEEE, "IP-XACT, Standard Structure for Packaging, Integrating, and Reusing IP within Tool Flows."

[12] OSTATIC, "VHDLC."

[13] Veripool, "Verilator."

[14] N. Bombieri *et al.*, "Hifsuite: tools for hdl code conversion and manipulation," *EURASIP Journal on Embedded Systems*, vol. 2010, no. 1, pp. 1–20, 2010.

[15] S. Vinco, V. Guarnieri, and F. Fummi, "Code Manipulation for Virtual Platform Integration," *IEEE Transactions on Computers*, vol. 65, no. 9, pp. 2694–2708, 2016.

[16] S. Centomo, M. Lora, A. Portaluri, F. Stefanni, and F. Fummi, "Automatic Generation of Cycle-Accurate Simulink Blocks from HDL IPs," in *Proc. of ECSI/IEEE Forum on Specification & Design Languages 2017 (FDL 17)*, 2017, pp. 1–8.

[17] M. Lora, S. Centomo, D. Quaglia, and F. Fummi, "Automatic Integration of Cycle-accurate Descriptions with Continuous-time Models for Cyber-Physical Virtual Platforms," in *Proc. of ACM/IEEE DATE 2018*, 2018, pp. 1–6.

Key Architectural Optimizations for Hardware Efficient JPEG-LS Encoder

Yakup Murat MERT
TÜBİTAK İLTAREN
Ankara, TURKEY
murat.mert@tubitak.gov.tr

Abstract—**This paper presents the set of optimizations for the efficient implementation of lossless JPEG-LS encoder. Given approaches are not only based on digital design techniques but also emerged from scrutinizing the regularity of the algorithm. Although it is well known for the low complexity, former designs have significant divergences hence varying performances. In this context, critical design issues of JPEG-LS encoder and trade-offs are discussed in detailed benchmark. FPGA implementation results suggest that encoder with the proposed optimizations outperforms the state-of-art counterparts in terms of hardware cost and throughput.**

Keywords—JPEG-LS, LOCO-I, FPGA, run mode, lossless compression.

I. INTRODUCTION

The need for the image compression gained significance with increase in the resolution and the dimension of the captured images. This is mainly due to the constraints over the processing and storage systems as well as the transmission costs. Compression can be achieved with some compromise on the outcome quality while in some applications information loss on the data cannot be tolerated at all. Hence, performance of a lossless compression algorithm is decisive reason of selection for those systems. Other factors are resource requirements and encoding/decoding speed of the algorithm.

JPEG-LS is a lossless and near-lossless compression standard for continuous-tone still images based on LOCO-I algorithm [1][2]. Despite to its low complexity, lossless compression performance on the various fields such as natural images [3], medical images [4] or even video sequences without additional temporal process [5] is outstanding compare to other compression standards. There are also recent studies that aims to improve its video compression capability using additional interframe processes [6][7][8]. This implies an inclination on utilization of this simple, yet effective algorithm in broader application areas. Consequently, an efficient hardware implementation of the JPEG-LS encoder is a crucial task regarding the embedded host systems.

JPEG-LS encoder has two main operational modes, regular mode and run mode respectively. It has been implemented several times by discarding the run mode [10][13][14][17], or even sometimes modifying the regular mode itself [16][18] for the sake of ease. First ever implementation involving the run mode is presented by [15]. However, this design used multiple clock domains and alternative datapaths with varying level of

pipeline stages for different modes which exposes significant hardware cost. A recent study reported a fully synchronous implementation that also supports the run mode [11]. But, the dichotomy based run mode parameter update scheme in that design lead to inevitable modification on the mode selection mechanism. Fully synchronous design that is also fully compatible with the standard is first time showcased by [9] which is the base design of this work. In this paper, final optimizations on the run mode as well as the other novel techniques to address the bottlenecks of the regular mode are introduced. Besides, the design points for the operational requirements are discussed. Synthesis results indicate a clear improvement on both throughput and hardware usage.

II. OVERVIEW OF JPEG-LS IMAGE COMPRESSION STANDARD

Encoding process described by the JPEG-LS standard consists of two main parts. First one is the context modeler which uses neighboring pixels, a, b, c and d, as input to generate specific context for the current pixel, x. Context is calculated by quantizing the local gradient vector $\{Rd-Rb, Rb-Rc, Rc-Ra\}$ then mapping to a subset addressed by Q. If at least one element of the vector is not zero, encoder will switch to the regular mode. Otherwise, encoder will execute run-length coding in the run mode.

When encoder is in the regular mode, it predicts the value of the current pixel based on the median predictor given as $median(Ra, Rb, Ra+Rb-Rc)$. Then, prediction error is computed which is the deviation of prediction from the real value of the current pixel. JPEG-LS encoder uses four statistical parameters for each context stored as tables which are the accumulator of the magnitude of the prediction errors **A**, bias values for the prediction error **B**, prediction error correction coefficient **C**, and frequency of occurrence counter **N**. Computed prediction error is corrected with **C**[Q] then encoded with Golomb encoder using an encoding parameter, **k**, which is calculated using **A**[Q] and **N**[Q] parameters. Golomb encoder output size is bounded by a parameter **LIMIT** in case of an outlier. All

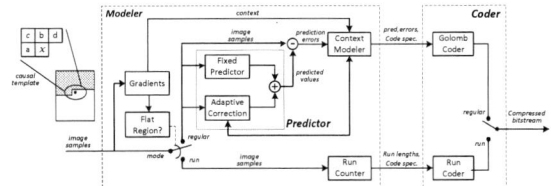

Fig. 1. Block diagram of the JPEG-LS encoder [3].

978-1-5386-4757-8/18 $31.00 © 2018 IEEE
243

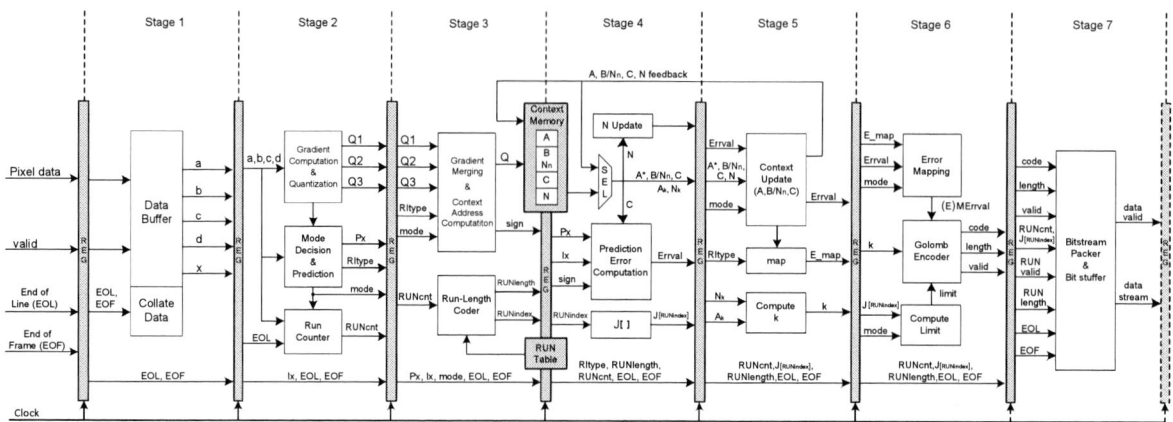

Fig. 2. Optimized fully pipelined JPEG-LS encoder architecture.

parameters are updated after completion of encoding process for each pixel. When occurrence counter **N** reaches to its pre-defined upper limit, named as **RESET**, all variables of that context is halved to avoid overflow. Block diagram of JPEG-LS encoder is given in Fig. 1.

III. PROPOSED ENCODER DESIGN AND OPTIMIZATIONS

Optimized pipelined architecture of the designed JPEG-LS encoder is illustrated in Fig. 2. Some minor operations are not shown in the figure in order to simplify it. The design consists of seven pipeline stages in which the first five stages are dedicated for the context modeling while last stages are for the Golomb coder and the stream packing respectively. It is fully synchronous and can fetch one pixel at every clock cycle. There is an internal data buffer that accommodates one row of the image. The encoder has a simple interface which utilizes two signals for input stream control purpose. End-of-line, EOL, signal accompanies to the last pixel of the current line while end-of-frame, EOF, flag is triggered at the frame end in a similar way. Compressed and bit-stuffed data is emitted with 32-bit length packets accompanying the valid flag.

JPEG-LS encoder fetches the pixels of the image in raster-scan order and buffers the neighboring pixels of the current one. First step of the encoding is computing the local gradients according to causal-template which determines the mode that encoder will execute and the context addresses for the regular mode. In the standard, local gradient quantization calls for comparison with the nine quantization levels [1]. This is due to the three threshold levels (T1, T2 and T3) and their negative values as well as the near-lossless compression parameter (NEAR) and its negative correspondent. This process can be simplified substantially by working on the absolute values of the local gradients as described in Fig 3a. Note that, for lossless compression, sign of the local gradients will not be involved in this scheme since the quantization outcomes already have the same sign with them. Consequently, magnitudes of the gradient quantization will be non-negative and range between 0 and 4.

Having the local gradients quantized ($Q1$, $Q2$ and $Q3$), next step is computing the context addresses. Although the standard does not explicitly describe the way of computing the regular mode context addresses, a simple computation method can be deduced from the standard's quantization scheme as in (1). At first sight, the expression seems to require multiplication operations. However, multiplication by 9 can be achieved by three bit left shifting the input and then summing with input itself. Keep in mind that optimized gradient quantization scheme eliminated the negative levels and quantized gradients can now be expressed with only three bits. If the multiplicand is in three-bit width as in this case, sum operation will also be eliminated and multiplication by 9 will reduce to the concatenation of input into six-bits. Multiplication by 81 is realized in a similar manner, so that one logical-OR operation and concatenation suffice. Finally, context addresses for the regular mode is determined based on the *sign* parameters as given in figure Fig. 3b. This approach renders context address computation possible without need for complex circuitry or a look-up table [10][12].

$$81 \cdot Q1 + 9 \cdot Q2 + Q3 \qquad (1)$$

Computed context addresses map the prediction correction coefficients and variables that store the prediction error accumulators and their counters. These variables should be updated after the prediction error computation scheme for each pixel. From the implementation point of view, two challenges arise at this point. One of them is the cost of the correction on the prediction error. This challenge is solved with a decision tree based implementation [9] which helps to reduce the number of consecutive arithmetic operations hence delay. Other challenge is the data hazard on the variables table due to the pipelined implementation. In order to target this issue, the data forwarding technique is adopted rather than bias estimated

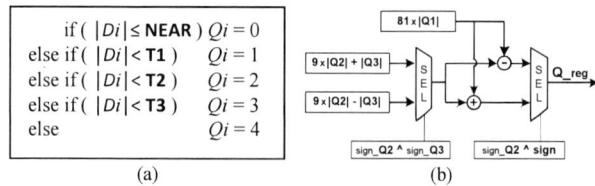

Fig. 3. a) Simplified local gradient quantization scheme b) Proposed context address computation circuitry.

978-1-5386-4757-8/18 $31.00 © 2018 IEEE   244

based approaches [11][14] or update on the different clock domains [10][15][17]. When the data hazard is detected, recently updated variables are fed to the previous pipeline stage input. In order to optimize the delay on the feedback, context counter, **N**, update process is delegated to the pipeline stage 4 as it doesn't have dependence on another parameter. Remaining variables, on the other hand, should wait for the most recent **N** and the prediction error for being updated in the subsequent pipeline stage (stage 5). Described data forwarding and variable update scheme is illustrated in Fig. 4.

Encoder switches to the run mode by means of checking the local gradients if it encounters to a flat region in the image [1]. Thus, run scan counter, *RUNcnt,* is implemented in the same pipeline stage with the mode detection unit which immediately triggers the run counting after the mode decision. Run-length coding employs two main parameters. First one is the *RUNindex* which holds the state of the run mode order and its map on J[ ] table. The *RUNindex* is updated according to its value before the run mode and the *RUNcnt* [1]. In order realize this computation, a normalization scheme is applied which removes the dependency on the initial value of the *RUNindex*. For this purpose, an addressable table named *RUNTable* is defined which accommodates the normalization coefficients for the *RUNcnt* as described in Fig. 5a. The coefficient fetched from the table is named as the *RUNadd* and its sum with the *RUNcnt* is called the Normalized *RUNcnt*. Consequently, as soon as the run mode ends, *RUNindex* can be determined using single lookup table from the Normalized *RUNcnt*, since normalization process produced a quasi-state in which initial value of the *RUNindex* equals to zero. Thanks to this optimization, update of the *RUNindex* is completed in a single pipeline stage rather than multiple dichotomy steps [11].

When the run mode ends, next step is the encoding of the run scan count and the run parameters update. Run-length coding is performed within two steps. The standard describes the first step as encoding of the run segments of length *rg*. Pseudo-code of the procedure is given in Fig. 5b and encoder should emit one bit logical "1" every time this iteration runs [1]. This encoding scheme is implemented by means of subtracting the updated *RUNindex* according to the *RUNcnt* and the registered *RUNindex* value before the run scan. Result of this subtraction is a newly defined parameter, *RUNlength*, which basically carries the information that the number of 1s to be appended to the output stream. The *RUNcnt* is also updated with the same normalization scheme before the *RUNindex*'s final update process. This middle state is denoted as the *rg_RUNindex* in Fig. 6 for the sake of distinguishing.

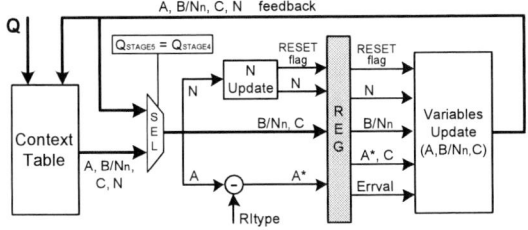

Fig. 4. Data forwarding scheme for the variables table update.

$$RUNadd[i] = \begin{cases} i = 0, & 0 \\ i > 0, & \sum_{k=0}^{i-1} 2^{J[k]} \end{cases}$$

(a)

```
while (RUNcnt ≥ (1 << J[RUNindex]) )
    RUNcnt = RUNcnt − (1 << J[RUNindex])
    if (RUNindex < 31)
        RUNindex = RUNindex + 1
```

(b)

Fig. 5. (a) *RUNTable* normalization content, (b) Encoding of run segments of length *rg* [1].

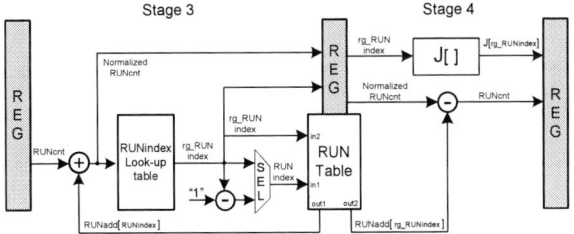

Fig. 6. Optimized run mode encoding scheme.

Other step is the encoding of the run segments of length less than *rg*. This step is executed only if the run mode is interrupted other than the EOL and consists of emitting the residual *RUNcnt* in J[*RUNindex*] bits delimited by one bit "0" from the encoded stream of the run segments of length *rg*. Table J[ ] indicates order of run-length codes. Then the final update of the *RUNindex* is executed and if the *RUNcnt* is still greater than zero, logical "1" will be appended to the stream as an indicator [1]. Bit insertion procedure is implemented easily with utilization of a proper flag which affects the stream length and convenient bit location in the stream. Parameters of both run-length coding steps are computed and updated in two pipeline stages using *RUNindex* lookup table and 32 entry dual-port *RUNTable*. Due to its regularity, J[ ] is realized as a combinational circuit rather than another ROM. Up-to-date parameters are transferred all the way to the final stage for the bitstream generation.

When run-length coding is interrupted other than EOL, following sample will be encoded. This procedure is named as run interruption sample encoding [1]. Although substantial resemblance with the regular mode, this encoding does not employ prediction correction scheme for prediction error and has a dedicated counter, **Nn**, for negative prediction errors. **A** table update is also different from the regular mode and run interruption sample encoding has its own context addresses (365 and 366). Absence of the prediction error correction can be fulfilled without any change on the structure designed for regular mode given in Fig.4. For this purpose, error correction table, **C**, returns zero for the context addresses of the run interruption sample encoding. Difference on the **A** table update procedure is implemented based on its identity described in

978-1-5386-4757-8/18 $31.00 © 2018 IEEE   245

TABLE I.   FEATURES OF THE JPEG-LS VARIABLES

Variable Properties for 8 bit JPEG-LS Encoder				
Table Name	**Value Bounds**	**Optimal Bits to Represent**	**Table Size***	**Notes**
A	$[0, (\textbf{RESET}-1) \cdot \textbf{RANGE}/2]$	$bpp - 1 + \lceil \log_2(\textbf{RESET}) \rceil$	367 x 13	Error accumulator will be halved at N equals RESET before being stored.
B	$[1-\textbf{RESET}, 0]$	$\lceil \log_2(\textbf{RESET}) \rceil$	365 x 6	Updated values are either zero or negative. Sign bit need not to be stored.
C	$[\textbf{MIN\_C}, \textbf{MAX\_C}]$	$\lceil \log_2(\max(|\textbf{MIN\_C}|,|\textbf{MAX\_C}|)) \rceil$	365 x 8	--
N	$[1, \textbf{RESET}]$	$\lceil \log_2(\textbf{RESET}) \rceil$	367 x 6	N counter starts from 1. Stored RESET can be substituted with 0 if RESET is power of 2.
Nn	$[0, \textbf{RESET}-1]$	$\lceil \log_2(\textbf{RESET}) \rceil$	2 x 6	Nn counter starts from 0. Its count will be halved at N equals RESET before being stored.

* RESET equals to 64, RANGE equals to 256, MIN_C and MAXC equal to -128 and 127 [1].

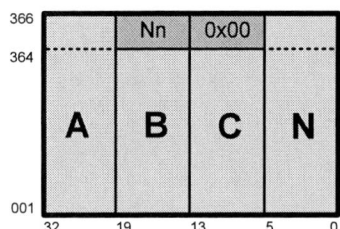

Fig. 7.   Memory map of the variables table.

Fig. 9 [9]. This update scheme is almost same as the regular mode except an arithmetic subtraction of run interruption coding index, *RItype*, flag from A[*Q*]. Therefore, all values fetched from the **A** table is subtracted by *RItype* (denoted as **A***) before any further process. *RItype* is always kept zero during regular mode to avoid erroneous updates. By this way, unlike former designs [11][15], variable update and prediction error computation in the run interruption sample encoding can share the same data path with the regular mode.

Since the variable table access and update is the crucial part of the design, it is important to optimize this for both hardware efficiency and performance without violating the standard. Some designs implemented each variables table separately [15][16][18] and some grouped them as two blocks; **A** and **N** tables are one group while **B** and **C** are another group [9] since they won't be involved in the run interruption sample encoding. In this design, all variables are integrated into single memory block. Details about the variable tables and optimizations are summarized in Table I. to clarify the approach. The key part is determining the value bounds of the given parameter. Among them, range of the **C** table is predefined thus its bit-size is constant. **N** counter is determined by the **RESET** parameter which halves the content of the all variables when N equals to it [1]. However, since it is initialized with "1", bit size of the **N** table can be arranged conveniently if **RESET** is power of two as in this case. Values of the variable **B** range from 1-**N** to 0, stating that it is either zero or negative. Consequently, sign of the **B**[*Q*] can be determined by an equality check and need not to be stored. In this picture, **B** table's context addresses at 365 and 366 will be unused while for **C** table they will be filled with zero since run interruption sample encoding procedure does not involve prediction correction. Negative error counter,

Nn, of the run interruption sample encoding is similar to **N** counter. Due to the relation between **N** and **B** variables, **Nn** counters can be accommodated at the empty space of the **B** table rather than standalone register based implementation. Note that both **Nn** and **B** variables depend on the magnitude of the **RESET** implying that change on this parameter will not lead to any mismatch. 366x33 block memory is organized for the variables as given in Fig. 7. By this way, there will be no wasted memory space. Besides, hazard detection and handling units will not be multiplied for separate tables [9].

In order to perform the Golomb encoding, prediction errors are mapped to non-negative values. This mapping process is straightforward for the regular mode while for the run interruption sample encoding, it is a bit tricky. In the latter case, computation of mapped prediction error, *EMErrval*, employs two additional flags *map* and *RItype* respectively. The flag *RItype* is computed in pipeline stage 2 as it uses causal template for input. On the other hand, *map* computation needs updated variables and prediction error itself which calls for five comparisons and two equality checks in total [1]. This scheme is simplified with the logical identity given in Fig. 8 which consists of two comparisons and two equality checks. Then, *map* and *RItype* flags are concatenated conveniently to generate possible values (0, -1 and -2) in order to complete the error mapping with single subtraction operation.

Final operation of the encoder is the packing of the variable length encoded data. Minimum packet size is determined by the **LIMIT** parameter which is 32 for 8-bit encoders [1]. The task is quite trivial for Golomb encoder products. However, for the products of the run mode and run interruption sample encoding, a convenient multiplexing scheme is needed. Since the run-

$$map = \{ (k == 0) \,\&\&\, [(2 \cdot Nn[Q] < N[Q]) \sim^\wedge (Errval > 0)] \,\&\&\, (Errval \mathrel{!=} 0) \}$$

Fig. 8.   Identity of the *map* parameter for run interruption sample encoding.

Run Interruption[i]	$EMErrval = 2 \cdot \mathrm{abs}(Errval) - RItype - map$
	$A[Q] = A[Q] + ((EMErrval + 1 - RItype) >> 1)$
Equivalent[ii]	$A[Q] = A[Q] + \mathrm{abs}(Errval) - RItype$
Regular Mode[iii]	$A[Q] = A[Q] + \mathrm{abs}(Errval)$

Fig. 9.   *i)* A[*Q*] update scheme for run interruption sample encoding [1]   ii) equivalent of the procedure iii) A[*Q*] update scheme for regular mode [1]

length coding consists of two steps, its outcome is emitted in two parts. As long as the run mode prevails, the pixels will not be coded by Golomb encoder which provides the time slot for the emission of the run segments of length rg. For this, computed *RUNlength* is transferred all the way to the data packing unit and stream of 1s produced at this stage right before the packing, then stream of 1s are emitted. If following sample will be coded by the run interruption sample encoding scenario, allowed Golomb code length is reduced as in (2) [1]. By this way, run segments of length less than rg can be concatenated and emitted with the run interruption sample encoding product in subsequent clock cycle. A very critical point arises if run mode is interrupted by EOL since the first sample of the following row might be coded in regular mode. In such case there will not be a room for the code of the run segment. However, regularity of the JPEG-LS removes the problem by nature. Run segments of length less than rg is not needed to be emitted if run mode is interrupted by EOL, only single bit '1' will suffice as an indicator [1]. Implementation for this case is nothing but the EOL flag itself at the end of 1s stream and as a stream length coefficient. Designed encoder is equipped with the circuitry for multiplexing all these cases conveniently without alternative pipeline stages [15]. In the end, all the encoding products are reduced to a data stream and a marker for its length. This stream is collated in a buffer of 2·**LIMIT**-1 bits length and most significant **LIMIT** bits of the buffer are shifted down to output as soon as filled.

$$glimit = \textbf{LIMIT} - \text{J[RUNIndex]} - 1 \qquad (2)$$

Packed stream should fulfill the bit-stuffing policy in which consecutive eight 1s should be followed by a one bit "0". Therefore, variable length codes should also be packed into variable length streams and padded with zero before the emission if necessary. As the 8-bit encoder's minimum packet size is 32, there are two possible zero-padding locations. If one location is padded, remainder bits will be shifted to the right which may lead to another zero-padding case. So, six different substrings of 8-bit are evaluated in brute force fashion in order to find the candidate bit locations. Final zero-padding decision is achieved by taking the priority of the upper nibbles into account. After the timing analysis, it is concluded that latency of this procedure can be hid behind the latency of the mode switching multiplexing prior to the data packing. Hence, these entities are implemented in the same pipeline stage to eliminate a redundant stage [9].

When the encoder reaches to the end-of-frame, EOF, internal states will return to its initial value and residuals in the data packing unit will be flushed. This task is realized by emitting one dummy stream of zeros in length of **LIMIT**-1 to the bit-stream packer right after coding the last pixel. As a result, final packet will be padded with the convenient number of zeros before emission like the standard describes. EOF flag also sets the pointer in the packing unit back to the head of the packing buffer and resets the buffer registers of the causal-template. Similar mechanism can be employed for continuous stream compression with mid-point terminations which might be desirable where input image is divided into tiles. In such case, variables table space should be doubled. While encoder employs one space for storing, a dedicated circuitry sets the

other memory space with the initial values and waits the arrival of EOF to toggle.

There are also two noteworthy optimizations in the architectural level. One of them is the implementation of mode detection, local gradient quantization and edge-detection prediction in the same pipeline stage unlike former designs [10][11][12][13][16][17]. These operations use causal-template pixels as input and have independent computation processes hence run in parallel. Dispersing them into multiple pipeline stages bring about no advantage in terms of clock frequency. Another critical decision for hardware efficiency is locating the Error mapping and *glimit* computation in the stage 6 with the Golomb encoder (Fig. 2). This optimization is about avoiding premature calculation of the final parameters that consume more resource than their inputs [11][14][18].

IV. IMPLEMENTATION DETAILS AND DISCUSSIONS

Designed JPEG-LS encoder is described using Verilog-HDL and implemented targeting the XC7VX485-2 Xilinx FPGA. Post place-and-route results given in Table II indicate that implemented encoder's throughput can exceed 200 MSample/sec while employing 544 registers. Allocated input buffer allows the compression of images up to 16 Kpixel/line resolution which calls for 16Kx8 memory space. This buffer utilizes two block memories while 366x33 variables table utilize one dual-port memory. Only FPGA primitive needed by the design is the memory blocks and no DSP unit is involved. Base encoder design has been verified on the VC707 FPGA board using high swath width, 8-bit monochrome satellite images. Optimizations described in this paper are verified with simulation results using the same satellite images.

The design is compared with the two most recent JPEG-LS encoder implementations listed in Table III. For fair comparison, this design is synthesized targeting the same FPGAs with the references. The work given in [10] implemented only the regular mode while other reference, [9], is fully synchronous and supports both regular mode and run mode. The two references already outperformed many recent Xilinx FPGA realizations mentioned in this context [11][14][16]. Implementation summary suggests that, optimized encoder has lower hardware utilization which saved nearly 100 registers compare to [9]. Besides, encoding throughput is improved up to 10% compare to regular mode only implementation [10].

Among the all mentioned optimizations, primary impact on the hardware cost reduction is stemmed from merging the bit-stream packer and the byte-stuffer in the same pipeline stage. This optimization is more crucial for the encoders that will compress higher dynamic range images like 16-bits. In such case, **LIMIT** parameter would be 64 which is equal to the size of the minimum bit-stream and without this optimization as

TABLE II.   IMPLEMENTATION DETAILS OF THE JPEG-LS ENCODER

Xilinx Design				
Target device	*Frequency (MHz)*	*Slice FlipFlops*	*Slice LUTS*	*Slice*
XC7VX485-2	207,8	544	1373	567

978-1-5386-4757-8/18 $31.00 © 2018 IEEE

TABLE III.   COMPARISON WITH THE FORMER JPEG-LS ENCODER IMPLEMENTATIONS

Design	Target device	Frequency (MHz)	Slice FlipFlops	Slice LUTS	Slice
[9] / **This work**	XC5VLX330	140.5 / **149.6**	645 / **546**	1711 / **1542**	649 / **542**
[9] / **This work**	XC7VX485	196.1 / **207.8**	642 / **544**	1554 / **1373**	632 / **567**
[10]* / **This work**	XC4VLX15	120 / **132.8**	- / **551**	- / **2085**	2596 / **1221**

* Regular mode only.

much registers would be utilized redundantly. On the other hand, variables table optimization is the key of the high clock frequency as the data-forwarding constitutes the critical path of the design. Specifically, embedding **Nn** into the variable-table memory not only eliminated the wasted memory space but also removed its data-forwarding circuitry with another path.

V.   CONCLUSION

This study summarizes the design challenges and proposed solutions for the efficient implementation of the JPEG-LS encoder that is fully compatible with the standard. Given optimizations rendered significant reduction on the hardware footprint possible while improving the encoding throughput as high as 207.8 MSample/s for VX7 FPGA.

Data forwarding in the variables table constitutes the bottle-neck of the overall encoding process in JPEG-LS. Determining the optimal bit representations and merging both regular mode and run interruption sample encoding variables into single table alleviated the critical path on the feedback. Other optimizations on the costly procedures such as context address computation, update of the run-length coding and run-interruption sample encoding parameters and bit-stream packing contributed to reduce the encoder hardware considerably.

REFERENCES

[1]   Information Technology -Lossless and near-lossless compression of continuous-tone still images- Baseline, International Telecommunication Union (ITU-T Recommendation T.87). ISO/IEC 14495-1(1998).

[2]   M.J. Weinberger, G. Seroussi and G. Sapiro, "LOCO-I: A Low Complexity, Context-Based, Lossless Image Compression Algorithm", Proc. of the 1996 Data Compression Conference (DCC'96), Snowbird, Utah. pp. 140-149, (1996).

[3]   M. Weinberger, G. Sapiro and G. Seroussi, "The LOCO-I lossless image compression algorithm: Principle and standardization into JPEG-LS", IEEE Trans. Image Process. vol.9, no. 8, pp.1309-1324 (2000).

[4]   G. Schaefer, R. Starosolski and S.Y. Zhu, "An evaluation of lossless compression algorithms for medical infrared images", Engineering in Medicine and Biology Society, 2005. IEEE-EMBS 2005. 27th Annual International Conference of the, pp. 1673-1676. IEEE, 2006.

[5]   Q. Cai, L. Song, G. Li and N. Ling, "Lossy and Lossless Intra Coding Performance Evaluation: HEVC, H.264/AVC, JPEG2000 and JPEG-LS", In Signal and Information Processing Association Annual Summit and Conference (APSIPA ASC), Asia-Pacific, pp.1-9, IEEE, 2012.

[6]   Z. Wang, D. Chand, S. Simon and T. Richter, "Memory Efficient Lossless Compression of Image Sequences with JPEG-LS and Temporal Prediction", Picture Coding Symposium (PCS), pp. 305-308, IEEE, 2012.

[7]   A. Ukhanova, A. Sergeev and S. Forchhammer, "Extending JPEG-LS for low-complexity scalable video coding", Image Processing: Algorithms and Systems IX, vol. 7870, p. 78701D. International Society for Optics and Photonics, 2011.

[8]   J. Park, J.W. Kim, J. Jeong and B. Choi, "Lossless Image Coding Based on Inter-color Prediction for Ultra High Definition Image", Pacific-Rim Symposium on Image and Video Technology, pp. 1-12, Springer, Berlin, Heidelberg, 2011.

[9]   Y.M. Mert, "FPGA based JPEG-LS Encoder for Onboard Real-Time Lossless Image Compression", Satellite Data Compression, Communications and Processing XI, vol. 9501, p.950106. International Society for Optics and Photonics, 2015.

[10]   B.S. Kim, S. Baek, D.S. Kim and D.J Chung, "A high performance fully pipeline JPEG-LS encoder for lossless compression", IEICE Electronics Express 10, no.12 (2013): 20130348-20130348.

[11]   W. Wei, J. Lei and Y. Li, "Onboard Optimized Hardware Implementation of JPEG-LS Encoder Based on FPGA", Satellite Data Compression, Communications, and Processing VIII, vol. 8514, p. 851406. International Society for Optics and Photonics, 2012.

[12]   N.Y. Kang, L. Jie, L. Yunsong, S. Changhe & W. Xianyum "VLSI design of the JPEG-LS near lossless image encoder", Journal of Xidian University, 2016, pp 81-86.

[13]   L.J. Kau and S.W. Lin "High Performance Architecture for the Encoder of JPEG-LS on SOPC Platform", Signal Processing Systems (SiPS), 2013 IEEE Workshop on, pp. 141-146, IEEE, 2013.

[14]   M. Papadonikolakis, A.P. Kakarountas and C. E. Goutis, "Efficient high-performance implementation of JPEG-LS encoder", Journal of Real-Time Image Processing, no. 4, pp. 303-310 (2008).

[15]   X. Li, X. Chen, X. Xie, G. Li, L. Zhang, C. Zhang and Z. Wang, "A Low Power, Fully Pipelined JPEG-LS Encoder for Lossless Image Compression", International Conference on Multimedia & Expo, pp. 1906-1909, IEEE 2007.

[16]   Z. Wang, T. Zhang, L.Yan and C. Gong, "A high Performance Fully Pipelined Architecture for Lossless Compression of Satellite Image", Multimedia Technology (ICMT), 2010 International Conference on, pp.1-4. IEEE, 2010.

[17]   P. Merlino and A. Abramo, "A Fully Pipelined Architecture for the LOCO-I Compression Algorithm", Very Large Scale Integration (VLSI) Systems, IEEE Transactions on 17, no 7, pp. 967-971 (2009).

[18]   H. Daryanavard, O. Abbasi and R. Talebi, "FPGA Implementation of JPEG-LS Compression Algorithm for Real Time Applications", Electrical Engineering (ICEE), 19th Iranian Conference on pp. 1-4. IEEE, 2011.

A graph-based approach for mobile localization exploiting real and virtual landmarks

Florenc Demrozi*, Kevin Costa*, Federico Tramarin[†] and Graziano Pravadelli*

*Department of Computer Science, University of Verona, Italy,
Email: name.surname@univr.it
[†]Department of Information Engineering, University of Padua, Italy,
Email: federico.tramarin@unipd.it

Abstract—In the last decade, several localization systems have been proposed, some of them achieving an accuracy in the order of decimeters. Nonetheless, a broad range of these methods may reveal unattractive for low-cost and resource-bounded mobile application scenarios. Indeed, significant constraints stem from costs and availability of the required technological infrastructure for the target environment. In addition, resources necessary to train and run the localization algorithm may pose stringent requirements in terms of time, power and computational. In the aforementioned context, this paper presents an effective radio–based positioning algorithm able to localize the user by exploiting a graph-based representation of the environment. The graph is created by analyzing the received signal level and the movement directions of a mobile device with respect to a set of real and virtual landmarks. The comparison with a commercial localization system proves the effectiveness and accuracy of the proposed approach.

I. INTRODUCTION

A localization system is a technique for estimating the position of people and objects within a specific environment without any knowledge of previous locations. They can be roughly categorized either in radio–based (i.e. exploiting Radio Frequency (RF) [1], Radio Frequency Identification (RFID) [2], ZigBee or Wireless Local Area Network (WLAN) [3]) or non radio–based (i.e. based on light impulse [4], sound [5], vibration [6], or magnetic field [7], [8] techniques). In general, localization systems exploit the knowledge of several sensor measurements, for instance the Received Signal Strength Indicator (RSSI) relevant to beacons sent by Wi-Fi Access Points (APs) to estimate the distance from known anchor points [9].

In last decades, they gained increased popularity in several areas, exploiting different technologies on both hardware and software sides, to support applications such as navigation, localization and monitoring, in the fields of security, management, health–care, etc. Even more, in the Internet of Things (IoT) context, localization systems constitute one of the fundamental components [10]. Despite this, it is worth highlighting that existing solutions are often subjected to several design and implementation issues [11], [12], mainly due to environmental constraints, involved devices and signal features.

On the one hand, in outdoor environment, where a common choice is to use Global Positioning System (GPS) or other satellite–based systems, the presence of obstacles and buildings may hinder received signals, and in addition such techniques require specific, complex and quite expensive components. Analogously, in the case of indoor radio–based localization, we have to consider that signal propagation depends on walls thickness, obstacles, number of people, etc., which may cause issues like shadow fading and multipath propagation [13].

On the other hand, localization in indoor environment (where GPS system lacks accuracy or may even become unavailable) is still a relevant open point and there is not a general consensus on the solutions to be deployed in this scenario. Indeed, many indoor positioning systems have been proposed during time, which differentiate on the basis of the employed technologies. A typical result is that, to obtain a reasonable accuracy, localization is performed exploiting quite complex architectures, which in turn impair either on the cost and availability of the technological infrastructure for the target environment, and/or on the time, power and computational resources necessary to run (and in some case, to train) the localization algorithm.

A significant example comes from the fingerprinting technique, which is becoming a very popular positioning method for indoor localization where it may achieve, under well specific circumstances, an accuracy in the order of decimeters [12], [7]. In fingerprinting, the environment is subdivided in sub–areas, and the user is able to univocally identify the sub–area to which she/he belongs exploiting specific algorithms based on different metrics (RSSI from WiFi and Bluetooth sources, Magnetic Field value, etc.). Nevertheless, despite the potential of fingerprinting techniques (for instance IndoorAtlas [7]) we should also take into account some relevant issues in terms of human intervention due to the initial *learning* phase, computational effort required by the algorithm for the *localization* phase, and unavoidable impairments affecting the input metrics, given the dependence of radio signals on external factors.

The drawbacks highlighted above make several existing solutions unattractive for low-cost and resource-bounded application scenarios [12]. The goal of this paper is hence to overcome some of the aforementioned limitations, in particular avoiding the initial learning phase, typical of fingerprinting, and introducing a computational efficient algorithm.

To this aim, the paper presents a radio-based positioning algorithm that localizes the user by exploiting a graph-based representation of the environment with respect to a set of real and virtual radio signal emitters (*landmarks*). The most significant characteristics of the proposed approach are:

- reuse of existing infrastructures (for instance, the APs of an existing WLAN installation);
- lightweight computation on mobile devices;

978-1-5386-4757-8/18 $31.00 © 2018 IEEE

- introduction of the *virtual landmark concept;*
- ease of extension to other protocols (e.g. AltBeacon).

The paper is organized as follows. Section II introduces some preliminary concepts and definitions, that will be then exploited in Section III to describe the proposed methodology. The outcomes obtained through a significant experimental campaign are then presented in Section IV. Subsequently, we briefly discuss about both some current limitations of the proposed approach and future work in Section V. Finally, Section VI reports some concluding remarks.

II. PRELIMINARIES

This section briefly introduces some preliminary definitions required for the description of the proposed approach.

A. Technologies and Devices

Wireless communication technologies are highly pervasive and surround us in everyday life, the most widespread being WiFi and Bluetooth systems based on the IEEE 802.11 and IEEE 802.15.1 communication standards, respectively. The System on a Chip (SoC) of common mobile devices nowadays always include both technologies, making them highly available even in low–cost and resource–bounded scenarios. Although these technologies lead to very different communication systems, both can be effectively exploited for indoor localization purposes. Indeed, IEEE 802.11 infrastructure networks (the most widespread ones) are managed by an AP, a specific network device that periodically transmit *beacon* frames to signal the presence of the AP itself, and that contain information about the network. Analogously, in Personal Area Networks (PAN) based on Bluetooth, it is possible to introduce devices called Beacon, that are special hardware components periodically broadcasting a Bluetooth frame in a well–defined format (for instance iBeacon, EddyStone or AltBeacon).

Independently of the adopted communication standard, these data frames allow a receiver to univocally recognize the sender, whose position is fixed and known *a priori*, and also to obtain a rough estimation of the distance between them, based on the knowledge of the received power of each data frame (e.g. RSSI).

B. Received Signal Strength Indicator (RSSI)

In the considered communication standards, and in many others as well, devices are required to provide a mean to express the perceived power relevant to a received data frame. A widespread metrics is provided by the *RSSI*, which in particular is an indication of the power level being received after antenna and cable loss. This value is often reported as an integer within a predefined range (typically an 8-bit field), where typically higher values of RSSI correspond to stronger received signal [11].

In different localization techniques, fingerprinting being one of the prominent examples, such measurement is largely exploited [12]. Nonetheless, we should also recall that RSSI presents some important issues, for instance due to inaccurate RSSI readings obtained by different transceivers [14], or to impairment of the radio signal [13], which has a high sensitivity to environmental phenomena, causing considerable signal variation hence impairing on the localization accuracy.

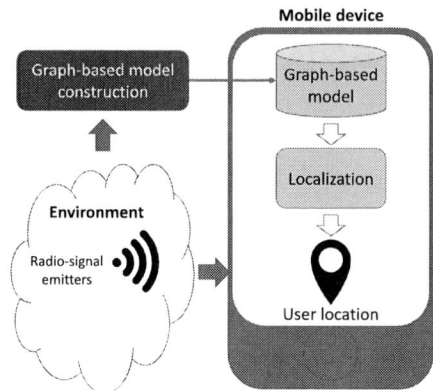

Fig. 1: Overview of the proposed approach.

III. METHODOLOGY

With reference to Figure 1, the proposed approach is composed of two phases:

i) *Off-line graph construction*: a graph model of the environment is constructed starting from the knowledge about the location of a set of landmarks;

ii) *On-line localization*: an application, running on the user's mobile device, loads and exploits the previously created graph model to compute the most probable location of the user within the environment, driving its decision through landmarks emitted signals and the user's direction.

A. Landmarks

In our approach, two kind of landmarks are considered: real landmarks and virtual landmarks.

We define a **real landmark** as a uniquely identifiable radio signal emitter (e.g., AP or BlueTooth Low Energy (BLE) Beacon) that is located in a known and fixed position. It identifies a *sub–area* of the environment where the RSSI values measured by a receiver are greater than a fixed threshold. Such a sub-area represents the **range of action** of the landmark. The **threshold** is set as:

$$T_{RSSI} = -(10 * n) * log(r) + TxPower \qquad [\text{dBm}] \qquad (1)$$

where r is the desired range of action for the landmark in meters, n is the path loss index (normally $n = 2$) and *TxPower* refers to the landmark transmission power, expressed in dBm [11]. When the RSSI values received by the user are above the threshold, we localize the user in the range of action associated to the corresponding landmark. Smaller are the ranges of action, more accurate is the localization of the user. However, the range of action of each real landmark cannot be arbitrarily reduced, as we need to avoid having uncovered sub-areas where the user cannot be localized. Each threshold is then set according to the number of real landmarks and their distribution in the environment to reduce the risk of black holes, where the user cannot be localized.

In order to further increase the accuracy of the localization approach, we then introduce the concept of **virtual landmark**. Since it does not physically exist, it does not emit any signal. Anyway it is associated to a specific sub–area between two real landmarks. In particular, we virtualize the presence of one

978-1-5386-4757-8/18 $31.00 © 2018 IEEE

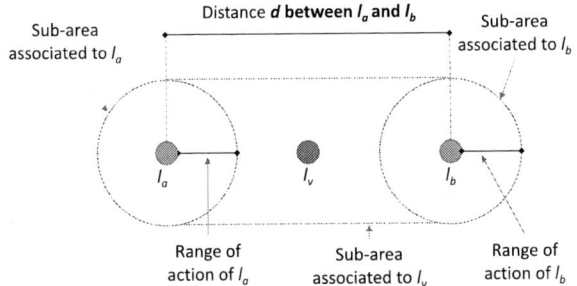

Fig. 2: Real and virtual landmarks.

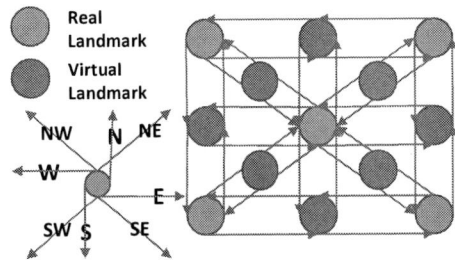

Fig. 3: Example of a graph model of the environment.

virtual landmark l_v between every couple of real landmarks (l_a, l_b), as shown in Fig. 2. The thresholds of l_a and l_b are set such that their ranges of action are 1/4 of the physical distance d between l_a and l_b, i.e., $range(l_a) = range(l_b) = d/4$.

When the signal received by the user in a given position is lower than the thresholds of the nearest real landmarks, we localize the user in the range of action of the nearest virtual landmark, according to his/her previous position with respect to the last visited real landmark, and his/her movement direction. For example, consider Fig. 2. If, at time t, the user is located in the sub-area of l_a, and it is moving towards east (according to the information gathered by the sensors on the mobile device), as soon as the RSSI received by the device is lower than the threshold of l_a, our approach locates the user in the sub–area represented by l_v.

The introduction of virtual landmarks and the sensing of the user's movement direction allow us increasing the number of *sub-areas* in the environment, thus improving the accuracy of the proposed localization approach.

B. Graph construction

Given the concept of real and virtual landmarks, the first phase of the localization approach creates a graph-based model of the environment according to the following definition.

A graph-based model is a tuple $G = (L, E, O, D, R, T)$ where:

- $L = \{\{L_r\} \cup \{L_v\}\}$ is the set of real and virtual landmarks;
- $E = \{(l_i, l_j) | l_i \in L_r \text{ and } l_j \in L_V\}$ is the set of edges between landmarks;
- $O : E \to C$ is an labeling function that assigns to each edge $e = (l_i, l_j) \in E$ a cardinal direction $c \in C = \{N, NE, E, SE, S, SE, W, NW\}$ representing that a user can reach l_j from l_i moving towards the direction c;
- $D : E \to \mathbb{R}$ is a labeling function that assigns to each edge $e \in E$ a length $d \in \mathbb{R}$ representing the distance between the landmarks connected by e;
- $R : L \to \mathbb{R}$ is a labelling function that assigns to each landmark $l \in L$ its range of action $r \in \mathbb{R}$;
- $T : L_r \to \mathbb{R}$ is a labelling function that assigns to each real landmark $l_r \in L_r$ its threshold $th \in \mathbb{R}$.

An example of a graph-based model is shown in Figure 3. Note that two neighbour landmarks are always connected by two edges, with opposite cardinal directions to represent that a user can move from one landmark to the other and vice versa.

Algorithm 1 explains the graph construction procedure. It takes in input the positions within the environment of real

---

Algorithm 1 Graph-based model generation

Input: *Real landmark positions (pos_L), and wall/obstacle positions (pos_W)*
Output: *Graph-based model of the environment (G)*

1: $G = create\_graph(pos_L, pos_W)$
2: $G.L_v = \emptyset$ // *set of virtual landmarks*
3: **for** $l_i \in G.L_r$ **do**
4:     $r_i = set\_range(l_i, neighbors(l_i))$
5:     $th_i = set\_threshold(l_i, r_i)$
6: **end for**
7: **for** $e_1, e_2 \in G.E$ connecting two real landmarks **do**
8:     $G.L_v = G.L_v \cup add\_v\_landmark(e_1, e_2)$
9:     $G.E = update(G.L_v)$
10: **end for**
11: **return** G

---

landmarks and walls/obstacles, and it returns the graph-based model according to the previous definition.

The procedure creates a graph composed of only real landmarks, according to their positions in the environment and taking care of constraints due to walls and obstacles (line 1). At this time, thresholds and ranges of action of real landmarks are not yet defined and the set of virtual landmarks is empty (line 2). Therefore, for each real landmark l_i, it sets its range of action and its threshold according to the distance between l_i and its neighbors (lines 3-5). In particular, the function *set_range* and *set_threshold* set the range of action and the threshold for l_i as follows:

1) select the nearest neighbour l_j of l_i according to the graph G;
2) set the range of action of l_i as $r = d/4$, where d is the distance between l_i and l_j;
3) given the range r set the threshold of l_i according to the equation 1 reported in Section III-A.

Then, for each couple of edges e_1, e_2 connecting two real landmarks l_i and l_j, respectively with cardinal direction from l_i to l_j and vice-versa, the graph is updated as follows to insert virtual landmarks:

1) add a virtual landmark l_v with range of action $r_v = (d - r_i - r_j)/2$, where d is the distance between l_i and l_j, and r_i, r_j represent the ranges of action, respectively, of l_i and l_j;
2) remove e_1 and add two new edges connecting l_v with l_i and l_j whose lengths are, respectively, $r_i + r_v$ and $r_j + r_v$.
3) set the cardinal direction of the two new edges equal to the cardinal direction of e_1.

978-1-5386-4757-8/18 $31.00 © 2018 IEEE

4) repeat actions 2 and 3 for e_2.

C. Localization

According to the graph-based model created by Algorithm 1, our system univocally localizes the user inside the sub-area corresponding to the range of action of the nearest landmark with respect to his/her actual position in the environment.

Algorithm 2 Localization

Input: *Graph-based model of the environment (G)*
Output: *Location of the user (location)*
1: *event = wait_event()*
2: **if** *event = Event* 1 **then**
3:      *l_r = get_real_landmark()*
4:      *visited_landmarks.push(l_r)*
5:      *location = l_r*
6: **else** // *Event 2*
7:      *last_l_r = visited_landmarks.pop()*
8:      *c = sense_movement_direction()*
9:      *location = get_virtual_landmark(G, last_l_r, c)*
10: **end if**

The localization works as shown in Algorithm 2. The background process *wait_event()* (line 1) activates whenever one of the following event happens:

- *Event 1:* the transceiver of the mobile device perceives a radio signal emitted by a real landmark l_r and the measured RSSI value is higher than the threshold defined for such a landmark; this means that the user is located in the range of action of a real landmark;
- *Event 2:* the transceiver of the mobile device stops perceiving a radio signal emitted by a landmark or the measured RSSI value becomes smaller than the threshold defined for such a landmark. This means the user entered in the range of action of a virtual landmark l_v.

When Event 1 happens, the real landmark l_r associated to the perceived signal is identified and stored into a list of visited landmarks (lines 3-4). Then, it is returned as the current location of the user (line 5). When Event 2 happens the algorithm extracts from the list the last real landmark l_r visited by the user (line 7) and it senses the movement cardinal direction c of the mobile device (line 8). Then, it makes a query to the graph model to retrieve the virtual landmark l_v adjacent to l_r such that l_v and l_r are connected by an edge whose cardinal direction is c (line 9). Such a virtual landmark is finally returned as current location of the user.

On the basis of the localization mechanism a shortest-path Dijkstra-based algorithm has been also implemented to provide a navigation system that guides the user between two arbitrary landmarks belonging to the same graph.

IV. EXPERIMENTAL CAMPAIGN

In this section the outcomes of a thorough and extensive experimental campaign are presented, based on the application of the proposed localization method. The experiments have been carried out both in indoor and outdoor environments. Furthermore, exploiting the same setups, we also performed a performance comparison between the outcomes obtained from our proposal and those obtained through the adoption

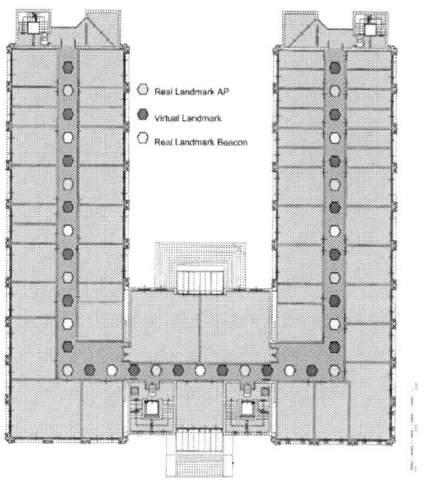

Fig. 4: Map of the *indoor* environment — *Id* = 3

Fig. 5: Map of the *outdoor* environment — *Id* = 5

of another popular localization platform, namely InDoorAtlas [7], as better detailed in the following.

A. Environment and devices characteristics

We performed our tests in two different environments, one for assessing indoor localization performance and another for assessing outdoor ones. The former indoor environment is located at the second floor of the Department of Computer Science building at the University of Verona, with a total area of $256 \, m^2$. Figure 4 provides the actual map of this indoor environment, where we had permission to perform experiment only in the corridors represented by the green area, for security and privacy concerns.

The outdoor experiments were performed in the parking pertaining to the same Department, whose map is reported in Figure 5 and which is extended over an area of $3000 \, m^2$. In this scenario, movements within the area were allowed to occur only in walkable roads.

To provide a thorough assessment, six different experiments have been carried out: four within the indoor environment (identified with *Id*=1..4) and two in the outdoor environment (*Id*=5..6).

To implement the real landmarks in the experimental campaign, we exploited both BLE beacon devices and IEEE 802.11 APs. For the sake of comprehensiveness, in each experiment the proposed localization technique has been tested with different sets (and combination) of real landmarks on the basis of the peculiarities of the environment where each test has been carried out. Table I presents the characteristics of the devices adopted to implement real landmarks.

978-1-5386-4757-8/18 $31.00 © 2018 IEEE   252

TABLE I: Real landmark characteristics

Type	Name	Producer	Tx Power	Protocol
BLE Beacon	Location Beacons	Estimote, Inc	-16 dBm	iBeacon/EddyStone
AP	Aironet 2602i	Cisco	-17 dBm	IEEE 802.11 beacons
Beacon Simulator	Android App	Android App	-16 dBm	iBeacon/EddyStone

TABLE II: Users characteristics

User ID	Height	SmartPhone	Operating System
1	1.70 m	Nexus 5	Android 6.0.1
2	1.75 m	Huawei P9 lite	Android 6.0.1
3	1.95 m	Samsung Galaxy S7 Edge	Android 7.0.1

In particular, tests in the indoor environment exploited any of the four possible combinations of BLE beacon devices and IEEE 802.11 beacon frames from APs of the existing WiFi network infrastructure. Conversely, in the outdoor area the application was tested at first using only BLE beacons devices, then adding a further set of simulated beacons obtained through a specific and purposely-developed mobile application (third row in Table I). Finally, for the sake of completeness, we have reported in Table II data relevant to the users (ID and Height) and mobile devices characteristics (Smartphone model and operating system).

In this experimental campaign we evaluated the performance of the proposed localization method on the basis of two metrics:

- *sub-area error*: this occurs when the localization algorithm fails to match the correct sub-area (i.e., landmark) with the real user location.
- *absolute accuracy*: distance in meters between the real and the estimated position.

B. Experimental Results

A resume of the obtained outcomes is reported in Table III, where we show experimental results, for both the indoor and the outdoor scenarios (Column 2), at varying the type of real landmarks (Column 4), their number (Column 5), the number of virtual landmarks (Column 6), and the range of action of the real landmark associated to the smallest sub–area (Column 7). For example, the row with $Id = 3$ represents the situation depicted in Fig. 4, where the indoor area was divided into 39 sub–areas: 19 of them were associated to real landmarks (in Fig. 4 APs and BLE Beacons are depicted in blue and yellow, respectively), while 20 sub–areas were associated to virtual landmarks (depicted in red). Conversely, the row with $Id = 5$ refers to the outdoor configuration reported in Fig. 5, with 12 real landmarks (green nodes in the figure), and 15 virtual landmarks (red nodes). As it can be observed looking at Column 8, our approach resulted in a zero error rate in terms of sub-area localization, i.e. the user was always located in the correct sub–area. Moreover, even if the user enters one of the offices (grey sub–areas in Fig. 4) the localization pointer remained correctly in the nearest sub-area.

In Column 9 the localization accuracy is reported, which is related to the range of action r of the adopted landmarks (in meters) and hence for each experiment varies between the range of the smallest and the largest sub–areas where the user was localized, and clearly improves with the increase of the number of real landmarks.

Columns 12 and 13 show instead, respectively, localization error and the absolute accuracy when the localization algorithm of Section III.C was used by considering only real landmarks. In this configuration, we observed several localization errors, primarily due to the user entering sub-areas where no known signal (or *signal* $\leq$ *threshold*) was perceived. In this case we have no meaningful information about user location and hence in the worst case, the error may tend to the number of sub-areas n_{sa} (column 12). Looking at the accuracy, in the best case it still depends on the range of action of the landmarks (r) and having maintained the same thresholds throughout the same experiment Id, the values are the same as in Column 7. In the worst case, instead, the accuracy deteriorates to $n_{sa} \times r$ meters (column 13).

This experiment hence definitely proves the benefits derived from the introduction of the virtual landmarks.

As a final note, during all the experiments the users followed different paths, performing changes of direction, accelerations and decelerations of the movements. Our system was able to localize the users correctly independently of the followed path, without variations in the sub-area errors and absolute accuracy. This proves that the proposed approach is not error-prone with respect to issues like the variation of the RSSI values and irregularities in the movement, which indeed may negatively affect solutions based on fingerprinting and sensor fusion.

C. Comparison with InDoorAtlas

We also carried out a comparison with InDoorAtlas [7]. This is a fingerprinting localization platform that is based on a preliminary learning phase of the environment. This phase consists on the collection of data (learning dataset) by different sources, such as magnetic field and RSSI levels, coming from selected positions called Reference Point (RP). At runtime, during the *online* phase, the localization algorithm continuously monitors the same types of sources, and returns the current location based on an estimation carried out by exploiting the learning dataset. Looking at Table III, Columns 10 and 11 report the localization error and the absolute accuracy for the InDoorAtlas solution. For a fair comparison, the number of fingerprints was set equal to the number of landmarks. Exploiting InDoorAtlas in the indoor environment, in the same conditions, we obtained up to 8 ($Id = 4$) sub-area localization errors, since this platform keeps evolving its estimation based on the direction and the acceleration of the user at the last predicted location. Comparing the localization absolute accuracy, it is worth observing that our approach shows the best accuracy in correspondence of the smallest landmark range, but the accuracy worsen when large virtual landmark range are employed. Conversely, InDoorAtlas in the best case resulted to be more accurate than our approach, while in the worst case it provides a lower accuracy, stemming from the limitations stated above.

We also performed a test with InDoorAtlas in the outdoor scenario, although we are aware of the fact that this platform was designed for indoor localization. Nonetheless, in the outdoor environment, it showed a localization error in 3 cases, compared with no error found with our method. The localization accuracy is instead in the range of one meter when the user is very close to the RPs, becoming greater than 8 meters in the other points.

It is worth highlighting that more data are collected during the learning phase of InDoorAtlas, more accurate the

978-1-5386-4757-8/18 $31.00 © 2018 IEEE

TABLE III: Experimental Results

		Experiment setup					Our approach		InDoorAtlas [7]		Without virtual landmarks	
Id	Environment	Area dimension m^2	Landmarks type	Number of real landmarks	Number of virtual landmarks	Minimum action range (m)	Error (sub-areas)	Accuracy (m) (min, max)	Error (sub-areas)	Accuracy (m) (min, max)	Error (sub-areas)	Accuracy(m) (min, max)
1	InDoor	256	Beacons	10	11	4	0	4,6	4	1,≥8	n_{sa}	4,$(n_{sa} \times r)$
2	InDoor	256	AP	10	11	4	0	4,6	4	1,≥8	n_{sa}	4,$(n_{sa} \times r)$
3	InDoor	256	Beacons+AP	19	20	4	0	4,6	4	1,≥8	n_{sa}	4,$(n_{sa} \times r)$
4	InDoor	256	Beacons	30	32	2	0	2,5	8	1,≥8	n_{sa}	2,$(n_{sa} \times r)$
5	OutDoor	3000	Beacons	12	15	10	0	10,50	3	1,≥8	n_{sa}	10,$(n_{sa} \times r)$
6	OutDoor	3000	Beacons	24	24	5	0	5,20	~	~	n_{sa}	5,$(n_{sa} \times r)$

~ = not performed for lack of beacon devices n_{sa} = defines the number of *sub-areas*

localization in the online phase is. However, it should also taken into account that this learning phase is particularly time-consuming and it drastically increases with the area dimension. Indeed, as a figure of merit, to perform the learning phase in experiment with *Id*=3 (with 39 landmarks) we spent 25 minutes, which became 40 minutes in the case of experiment *Id*=5 (with 27 landmarks) due to the larger area dimension and distance between landmarks. In addition, this platform bases the learning phase also on data collected from the accelerometer and gyroscope sensors of the mobile device, which are then used to calculate distances among different RPs and direction from one RP to another. This hence imposes an increased attention and caution during the learning phase.

D. Battery consumption

The final figure of merit we aim to provide is relevant to a qualitative estimation of the battery consumption. Our approach provided slightly better performance with respect to InDooAtlas in this regards. Indeed, using the proposed approach on the three devices reported in Table II, we found that the mean battery consumption, due to the designed application, was about 11% of the total consumption per hour. Conversely, using InDoorAtlas during the same tests, we found a mean consumption of about 13% of the total consumption per hour.

V. LIMITATIONS AND FUTURE WORK

The proposed localization method is based on perceived RSSI values to identify real landmarks, and on the last perceived real landmark and user direction to identify virtual landmarks. Some limitations exist with our current approach. At first, it is theoretically possible to receive two different beacons whose RSSI level is above the threshold. However, since the threshold is set considering the distance between two real landmarks according to Eq. 1 and Algorithm 1, this should happen with a quite low probability. We are refining the proposed algorithm in order to recognize and solve such situations. Secondly, we actually do not consider the movement of the users, and this in some cases can be a limitation of the approach. Furthermore, taking as example Fig. 3, if the user is located in the virtual landmark at the top of the graph model and moves directly to another virtual landmark, then our proposal may return a wrong location. This may be avoided, for example, exploiting the mobile device accelerometer by implementing a step counter. This further development of the application introduces a higher number of virtual landmarks, hence increasing the accuracy of the system.

VI. CONCLUSIONS

In this paper we proposed a localization and navigation approach based on real landmarks represented by the radio-signal emitters existing in the target environment. A graph-based model of such an environment is automatically created,

and virtual landmarks are added to increase the accuracy of the localization without requiring the introduction of further radio-signal emitters.

The presence of virtual landmarks allowed to reduce the range of actions of real landmarks making our approach less susceptible to errors due to outliers signal values, users body, or obstacles, unlike existing techniques base on signal analysis. A comparison was made between the proposed approach and a commercial localization platform. The results demonstrated that our approach provides advantages in both the accuracy in identifying the user location and the battery consumption of the mobile device used for the localization application.

REFERENCES

[1] P. Bahl and V. N. Padmanabhan, "Radar: An in-building rf-based user location and tracking system," in *INFOCOM 2000. Nineteenth Annual Joint Conference of the IEEE Computer and Communications Societies. Proceedings. IEEE*, vol. 2. Ieee, 2000, pp. 775–784.

[2] M. Bouet and A. L. Dos Santos, "Rfid tags: Positioning principles and localization techniques," in *Wireless Days, 2008. WD'08. 1st IFIP*. IEEE, 2008, pp. 1–5.

[3] S. Hilsenbeck, D. Bobkov, G. Schroth, R. Huitl, and E. Steinbach, "Graph-based data fusion of pedometer and wifi measurements for mobile indoor positioning," in *Proceedings of the 2014 ACM international joint conference on pervasive and ubiquitous computing*. ACM, 2014, pp. 147–158.

[4] Y.-S. Kuo, P. Pannuto, K.-J. Hsiao, and P. Dutta, "Luxapose: Indoor positioning with mobile phones and visible light," in *Proceedings of the 20th annual international conference on Mobile computing and networking*. ACM, 2014, pp. 447–458.

[5] R. Want, A. Hopper, V. Falcao, and J. Gibbons, "The active badge location system," *ACM Transactions on Information Systems (TOIS)*, vol. 10, no. 1, pp. 91–102, 1992.

[6] J. Niu, B. Lu, L. Cheng, Y. Gu, and L. Shu, "Ziloc: Energy efficient wifi fingerprint-based localization with low-power radio," in *Wireless Communications and Networking Conference (WCNC), 2013 IEEE*. IEEE, 2013, pp. 4558–4563.

[7] L. IndoorAtlas, "Ambient magnetic field-based indoor location technology: Bringing the compass to the next level," *IndoorAtlas Ltd*, 2012.

[8] H. Xie, T. Gu, X. Tao, H. Ye, and J. Lv, "Maloc: A practical magnetic fingerprinting approach to indoor localization using smartphones," in *Proceedings of the 2014 ACM International Joint Conference on Pervasive and Ubiquitous Computing*. ACM, 2014, pp. 243–253.

[9] G. Mao, B. Fidan, and B. D. Anderson, "Wireless sensor network localization techniques," *Computer networks*, vol. 51, no. 10, pp. 2529–2553, 2007.

[10] Z. Chen, F. Xia, T. Huang, F. Bu, and H. Wang, "A localization method for the internet of things," *The Journal of Supercomputing*, vol. 63, no. 3, pp. 657–674, 2013.

[11] A. Khalajmehrabadi, N. Gatsis, and D. Akopian, "Modern wlan fingerprinting indoor positioning methods and deployment challenges," *IEEE Communications Surveys & Tutorials*, vol. 19, no. 3, pp. 1974–2002, 2017.

[12] A. Khalajmehrabadi, N. Gatsis, D. J. Pack, and D. Akopian, "A joint indoor wlan localization and outlier detection scheme using lasso and elastic-net optimization techniques," *IEEE Transactions on Mobile Computing*, vol. 16, no. 8, pp. 2079–2092, 2017.

[13] J. Parsons and A. Turkmani, "Characterisation of mobile radio signals: model description," *IEE Proceedings I (Communications, Speech and Vision)*, vol. 138, no. 6, pp. 549–556, 1991.

[14] G. Lui, T. Gallagher, B. Li, A. G. Dempster, and C. Rizos, "Differences in rssi readings made by different wi-fi chipsets: A limitation of wlan localization," in *Localization and GNSS (ICL-GNSS), 2011 International Conference on*. IEEE, 2011, pp. 53–57.

Understanding the Design Space of Wavelength-Routed Optical NoC Topologies for Power-Performance Optimization

Mahdi Tala
Department of Engineering
University of Ferrara, Ferrara, Italy
mahdi.tala@unife.it

Davide Bertozzi
Department of Engineering
University of Ferrara, Ferrara, Italy
davide.bertozzi@unife.it

Abstract—Silicon photonics is the most promising emerging technology to deliver on- and off-chip communication performance and power that vastly exceed the capabilities of electronics. However, a significant abstraction gap does exist between novel devices and circuits and the higher-order switching structures that system designers need to instantiate. Currently, designers mostly rely on their intuition to bridge this abstraction gap. This paper lays the groundwork for a more rigorous and effective approach, by vertically-integrating the most advanced design methods and tools for topology synthesis and refinement in the context of a novel performance analysis framework. As a result, we can extract the highest aggregate bandwidth out of an optical network-on-chip topology, and provide an early-stage analysis of its static power, thus unveiling unexplored portions of the design space and interpreting its characteristics.

Index Terms—Emerging technologies; wavelength-routed optical NoC; design space exploration; design automation.

I. INTRODUCTION

There has been a recent surge of interest in optical networks-on-chip (ONoCs) as a key interconnect technology to keep up with the communication requirements of big data applications and to seamlessly scale on-chip communications to off-chip ones [1]. Dealing with optical data transmission on-chip makes designers and researchers face a whole new set of obstacles and challeges, well beyond the traditional concern about the maturity of silicon photonic devices. Examples include mapping flows of logic abstractions into new circuit primitives, new verification and testing challenges, different sources of design predictability gaps, variability-aware design, etc.

This context provides design automation with the opportunity to go beyond its traditional supportive role of a technology. In fact, it rather provides the means for an emerging technology to happen and to become competitive. Unfortunately, most contemporary design automation techniques are fine tuned for the silicon CMOS technology, and the extension of their paradigms and methodologies to emerging technologies such as silicon nanophotonic interconnects is admittedly still in the early stage. As a result, the gap between technology developers and system designers remains currently huge. There is an unmistakable evidence of such gap, consisting of a lack of knowledge of the design space, which forces designers to work out system-level solutions based on their intuition only.

This paper is a contribution to bridge this gap for a relevant family of ONoCs, namely wavelength-routed ones

(WRONoCs). They univocally associate the wavelength of an optical signal with a statically-defined lightpath across the optical transport medium. They are gaining momentum as all-optical solutions for on-chip communication, capable of relieving the latency and power overhead of electrically-assisted ONoCs dealing with optical contention. In fact, WRONoCs avoid any form of routing and arbitration by trading point-to-point channel bandwidth for enhanced communication concurrency [2]–[4].

The design automation gap for WRONoCs is revealed by the fact that topology solutions for this paradigm are typically compared in terms of optical resource and power requirements. There is no relative performance evaluation, since they are all supposed to achieve the same performance, namely contention-free all-to-all connectivity. This stems from the hidden assumption that a single bit is used for communication over an optical channel connecting a sender with a receiver. In practice, WRONoCs are typically built with microring resonators (MRRs), which have a periodic transmittance characteristic depending on their radius. While their periodic optical spectra raise tremendous opportunities for concurrent transmission and switching by using each resonant peak, the exploitation of this bit-level parallelism is hampered by misrouting concerns. In fact, whenever MRRs of different radii co-exist in a topology, some of their resonant peaks may end up in close proximity, thus causing an optical signal (or a significant fraction of its power) heading to a specific destination to be coupled on an optical path leading to a different destination.

Clearly, the enhanced parallelism of WRONoC topologies can be harnessed only through a suitable design automation support providing cross-layer visibility. Researchers have recently started to address this problem. First, they have laid the groundwork for the automated synthesis of generic topology design points [5]. Second, they have provided methods and tools for optical crosstalk analysis [6]. Third, they have related routing fault avoidance (hence WRONoC scalability and communication parallelism) to the proper selection of physical device parameters and to the uncertainty of the manufacturing process [7].

Although the field is far from consolidating, the key takeaway from current literature is that technology-aware performance and power estimations for early stage WRONoC topology analysis is becoming feasible. However, such estimations

978-1-5386-4757-8/18 $31.00 © 2018 IEEE

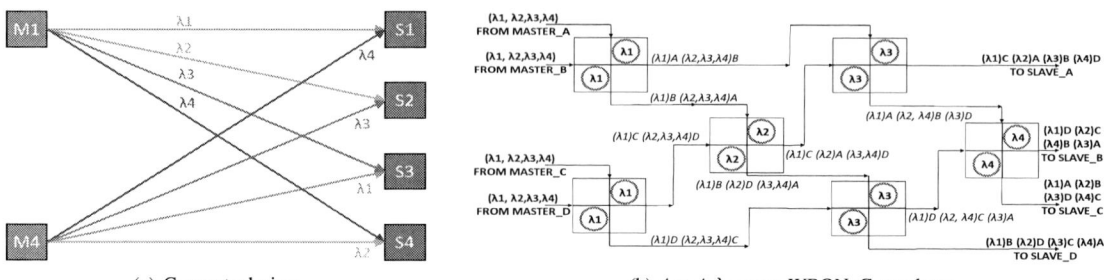

(a) Conceptual view.

(b) 4 × 4 λ-router WRONoC topology.

Fig. 1. Wavelength-selective routing.

are currently overly pessimistic, since they are based on conservative assumptions for the sake of problem simplification. For instance, the work in [7] assumes that an optical signal will always meet all MRR types on its way to destination, and ends up addressing the routing fault concern in the worst-case scenario. This is justified by the fact that some of the known topologies actually exhibit such a connectivity pattern. At the same time, the onset of methods and tools to populate the design space beyond the few design points available today [5] is revealing the existence of alternative scenarios, better suited to extract more communication parallelism. Unfortunately, these different tools have never been combined together in a consistent performance analysis framework of WRONoC topologies.

This is the challenge this work takes on in order to unveil new portions of the design space of WRONoC topologies, featuring higher-levels of on-chip communication bandwidth with respect to state-of-the-art methodologies. Another goal is to determine and interpret the distribution of these design points in the aggregate bandwidth-static power design space, thus coming out with guidelines for future automated synthesis tools in order to prune the design space towards the most promising directions.

From a methodology viewpoint, we couple an advanced framework for the generation of topology design points with a physical device parameter selection tool, and elaborate the outcome with a novel per-path analysis methodology that extracts the highest aggregate bandwidth out of it. At the same time, we set up a cost analysis framework, returning the cost of each refined design point in terms of number of laser sources and of their power requirements. The latter are derived based on an early-stage power model, processing the available pre-layout information during front-end topology synthesis. Thus, we establish a correlation with the synthesis process of VLSI digital circuits, where post-synthesis power estimations are given.

II. PREVIOUS WORK

WRONoC topologies exhibit different physical properties such as number of micro-ring resonators types and optical power losses.

The λ−Router [8] resembles multi-stage interconnection networks due to the cascaded stage organization. The GWOR topology [4] is built around a basic and symmetric 4x4 routing fabric, with generalization rules for scalability. The snake topology [9] exhibits a characteristic half-matrix connectivity

pattern. The topology presented in [10] consists of a multi-waveguide wavelength-routed optical ring. A hierarchical topology featuring regularity, vertex symmetry, and constant node degree is presented in [11]. Overall, only few topology design points for wavelength routing are today available from literature. An attempt to populate the design space more exhaustively has been reported in [5].

In any case, performance and scalability of WRONoC topologies are rarely investigated, since this requires cross-layer visibility and hence a suitable design automation support. The pioneer tool in [7] returns the maximum achievable parallelism by a WRONoC topology while overcoming misrouting issues and accounting for the uncertainty of the manufacturing process. However, provided performance figures are overly pessimistic, and no power estimates are provided. This paper aims at a more aggressive extraction of bit-level parallelism, and at relating it with the static power cost, thus making a wider design space available for topology optimization.

III. WAVELENGTH-ROUTED ONoCs

Wavelength-routed optical NoCs (WRONoCs) rely on the principle of wavelength-selective routing (Fig.1a), which statically associates an optical channel modulated onto a specific wavelength (*wavelength channel* from now on) to each source-destination pair. In particular, master $M1$ uses n wavelengths λ_1 to λ_n to reach slaves $S1$ to Sn, respectively. At the same time, each slave receives packets from masters $M1$ to Mn on different wavelengths. The intermediate topology (not shown in figure) should be synthesized in such a way that same-wavelength optical channels never overlap on the topology waveguides to avoid interference (hence loss of information). The distinctive property of wavelength-selective routing is that it delivers contention-free concurrent all-to-all connectivity without any form of resource arbitration.

A well-known 4 × 4 WRONoC topology is illustrated in Fig.1b, named λ-router [8]. Boxes represent 2x2 photonic switching elements, which progressively resolve wavelength-division multiplexed (WDM) signals from source nodes into their individual components (*drop function*), and recombine them into the output WDM signals for each receiver (*add function*).

IV. REFINED METHODOLOGY FOR THE EXTRACTION OF BIT-LEVEL PARALLELISM

A. Approach

Fig. 2 shows the guiding principle of our methodology for selecting the resonant wavelengths that each path can

978-1-5386-4757-8/18 $31.00 © 2018 IEEE

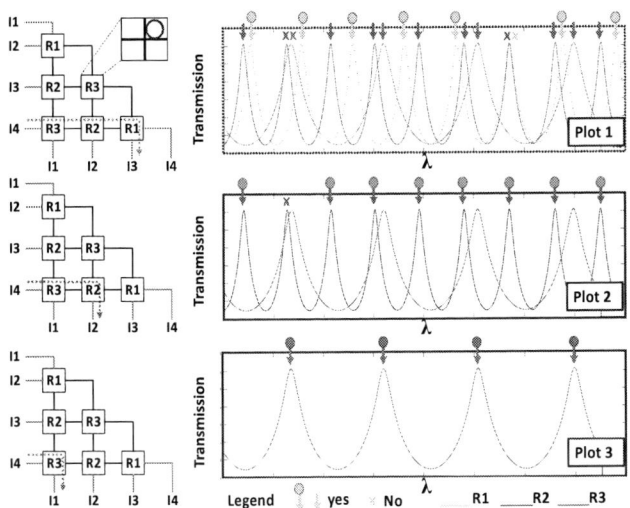

Fig. 2. Methodologies for the selection of useful resonant peaks in WRONoC Topologies

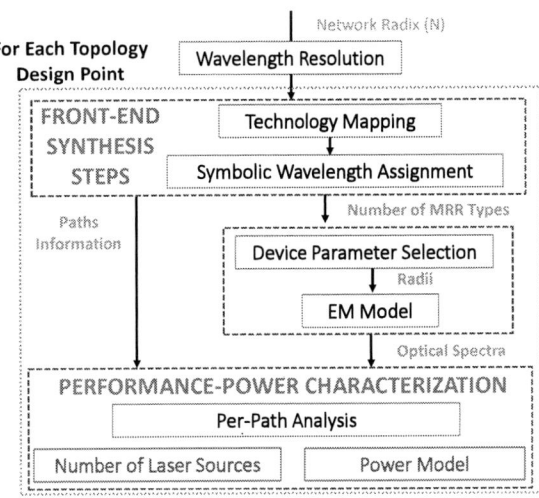

Fig. 3. Characterization methodology of the WRONoC design space.

use to achieve a certain bit-level parallelism. An example topology (snake) is considered in the figure, together with its assigned MRR types R1, R2 and R3. State-of-the-art analysis methodologies first plot the optical spectrum of all the MRR types together, then they choose the non overlapped peaks, which can be safely used for transmission since no routing fault will take place. A minimum distance is defined below which the peaks are considered to be overlapped. This is shown in plot 1, where the 3 optical spectrums of the 3 MRRs are displayed; the arrows with different colors correspond to the selected peaks (since the new methodology will need only the green arrows from this plot, we add a circle above them for distinction). This methodology ends up in pessimistic performance projections, since it relies on the worst-case assumption that each optical channel will always meet all MRR types on its way to destination. As a result, the bit-level parallelism achievable on each optical path is equal to $min(P1, P2, P3)$, where P1, P2 and P3 represent the number of selected/non-overlapped resonant peaks from the optical spectrum of R1, R2 and R3 respectively.

In practice, in many topologies optical paths are less constrained. Our methodology aims at exploiting this by leveraging a path-by-path analysis, trading analysis complexity for achievable performance.

For each path, we first identify the MRR types that an optical signal passes through, and the one which drops it. Then, we select the legal peaks of the latter, which will be used for transmission and will deliver the bit-level parallelism that can be achieved on that path. Legal peaks are those located outside the -3dB bandwidth of the resonant peaks of the other MRR types along the path. This way, the crosstalk concern is alleviated, and routing faults are avoided.

In the figure, we show 3 signals sent by Initiator 4; the first is for destination I3, the second for I2 and the last one for I1. The first one passes through R3 and R2, and is dropped by R3; the second one passes through only R3 before R2 drops it; finally, R1 immediately drops the signal. Thus, paths have different characteristics. In order to compute the parallelism for each of them, we show three plots (plot 1,2 and 3). In plot 1 we select the legal peaks from the optical spectrum of R3 subject to non-interference with R2 and R1, while in plot 2 we select the legal peaks from the optical spectrum of R2 subject to non-interference with R1. Finally, in plot 3 all the peaks in the optical spectrum of R1 are legal ones because R1 is the unique MRR type in the path.

By using the proposed methodology, we can achieve a bit-level parallelism of $P(I4 \rightarrow I3) = 7$ (plot 1, only green arrows), $P(I4 \rightarrow I2) = 8$ (plot 2, blue arrows) and $P(I4 \rightarrow I1) = 4$ (plot 3, red arrows), then the overall parallelism of the three paths amounts to 19. With state-of-the-art analysis methodologies, we would have selected from plot 1 $P(R1) = 7$, $P(R2) = 7$ and $P(R3) = 3$ resonant peaks. Then, we would have noticed that the minimum parallelism supported by all optical channels is three, and therefore the three paths under test would have given rise to an aggregate parallelism of $3 \times 3 = 9$. Even performing per-path analysis on the aggregate optical spectrum of plot 1 would have led to sub-optimal bandwidth exploitation (17 resonant peaks). Thus, our decoupled per-path analysis allows to extract better overall performance.

B. Proposed Methodology

In this section we present the complete methodology we followed to characterize the design space of $N \times N$ WRONoC topologies (see Fig. 3). We first fix the topology radix N that we target, then we proceed with the exploration as follows:

1) **Topology Synthesis.** In the first step, we feed the target radix N to a synthesis tool for WRONoC topologies, from our previous work in [5]. This tool implements a methodology that systematically synthesizes all the points of the $N \times N$ WRONoC topology design space by building upon the combination and aggregation of a basic filtering primitive: the 1x2 add-drop filter. The tool first expresses the needed filtering primitives to fulfill the connectivity requirements of the system at hand. Then, it combines them together to map the design on higher-order switching structures (e.g., 2x2 photonic switching elements, PSEs), and finally assigns symbolic resonant wavelengths to the MRRs inside the PSEs. The generation strategy is potentially exhaustive, although

978-1-5386-4757-8/18 $31.00 © 2018 IEEE

there are practical limits due to the large number of potential design points $(((N-1)!)^N)$.

At this stage, we have the following inputs for our later per-path analysis: MRR types each optical path in the topology includes, MRR types an optical signal passes through, dropping MRRs on each path, number of logical waveguide crossings (those inside PSEs) for early-stage power analysis.

2) **MRR selection.** Next, we feed the determined number of MRR types to an optimization tool, which we developed in our previous work [7]. This tool leverages a formal methodology to select WRONoC physical parameters while avoding the routing fault concern. It aims at maximizing the levels of connectivity and/or of bit-level parallelism that WRONoCs can achieve by properly selecting the radii of the assigned number of MRR types, as well as the exact value of the transmission wavelengths. Following state-of-the-art methodologies, the tool maximizes the common parallelism to all optical paths through the aggregate analysis of plot 1 in Fig. 2, hence providing conservative results. Nonetheless, we consider its output as an ideal starting point for our refined analysis, since it maximizes the availability of non-overlapped resonant peaks in a worst-case scenario (all MRR types on all optical paths, like in most paths of the $\lambda-$router). Our methodology then discards the provided transmission wavelengths and searches for new ones, taking the maximum advantage of the given set of MRR types and of their specified radii.

3) **Optical spectrum.** With the values of the MRR radii, an electromagnetic model generates the optical spectrum of the photonic switching elements, like those showed in Fig.1.

4) **Per-Path Analysis.** By leveraging topology-level connectivity information and physical device parameters, our per-path analyis methodology extracts the maximum bit-level parallelism that each path can achieve, as well as the real carrier wavelengths that each path should use. For this, the novel approach described in section IV-A is implemented. As a result, the aggregate topology bandwidth is computed by combining the aggregate bandwidth on each optical channel.

5) **Cost analysis.** We then compute the number of required laser sources to achieve that aggregate bandwidth, and consequently the early-stage total laser static power required for the topology under test. We assume that the ONoC is powered by an array of off-chip continuous-wave laser sources providing multiple optical carriers. For this reason, we assume that a Power Distribution Network (PDN) delivers the optical carriers to all the initiators. In practice, the laser sources are shared among all initiators, and all the identical wavelength carriers used in different paths in the topology come from the same laser source. This way, the number of needed laser sources is the sum of the distinct wavelengths used in the topology.

6) **Power Model.** Similar to pre-layout power estimations in digital VLSI design, we provide an early-stage analysis of the dominant power source in a WRONoC: laser static power. For this, we compute the insertion loss of

TABLE I
TECHNOLOGY PARAMETERS.

Photonic components	Device Information
Crossing	0.05 dB [13]
Splitter	0.2 dB [13]
Additional modulator array	5.5 dB [13]
MRR Drop	1 dB [14]
Receiver	1 dB [14]
Modulator	2.4 dB [13]
Coupling	1 dB [13]
Laser efficiency	20% [13]
Receiver sensitivity	-20 dBm [14]

each optical path in the topology based on the parameters in Table I. Then, we compute the optical power requirement on each path so that the receiver sensitivity is guaranteed. Finally, we introduce the optical power waste on the PDN due to equalization effects (see [12] for details). As a result, the total laser static power requirement for a topology design point is achieved. At this stage, where placement and routing have not been performed yet, our power model overlooks propagation losses as well as the additional waveguide crossings that arise during place&route. However, they are useful for power analysis with emphasis on relative accuracy.

V. EXPERIMENTAL RESULTS

We fixed the network radix to 8 and went through the flow in fig. 3. With a typical clustering factor of 16 cores per cluster, this correspond to a system with 128 cores. Each one of the 8 cluster gateways is both sender and receiver. The full enumeration of all points of the design space for N = 8 is computationally unaffordable. Therefore, we decided to randomly generate a sample of 40000 different topology design points. In order to avoid biasing the sample, the synthesis tool was instructed to make random choices whenever one was needed. For each design point, we characterize: the maximum aggregate bandwidth, the minimum number of laser sources to deliver it, an early-stage estimation of total laser static power.

In order to try out a different technology mapping, we notice that a 2x2 PSE works correctly by instantiating either two identical MRRs inside it (as assumed so far) or a single one (see Fig. 5). Therefore, we consider topologies built out of both physical devices.

The device parameter selection tool was given a range of 5 to 30 μm for MRR radii selection, with an incremental step of $0.25\mu m$. The returned MRRs' radii range between 25.5 μm and 29.25 μm, indicating that the tool prefers large MRRs because of their smaller free spectral range, while avoiding routing faults. This confirms that its output is a good starting point for our performance analysis methodology. Referring to [15], we estimated a conservative value of the 3dB-bandwidth of the MRRs to be roughly 0.15 nm, which we use to minimize crosstalk effects or routing faults.

A. Exploring the Design Space

Fig. 4 shows a sample of 40000 different design points for each technology mapping, with two different distributions. Thus, Fig. 4a shows the population distribution based on the

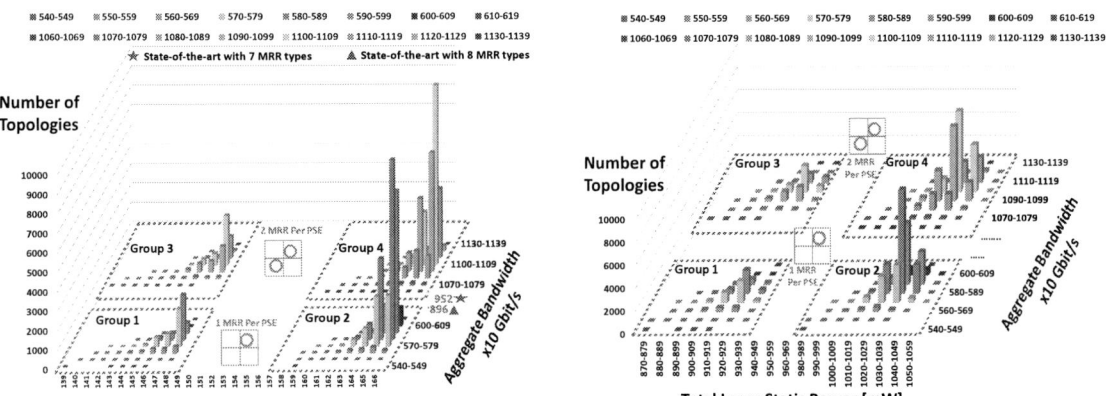

(a) Laser sources and aggregate bandwidth distribution.

(b) Total laser static power and aggregate bandwidth distribution.

Fig. 4. Sample of 40000 randomly-generated 8×8 WRONoC topologies.

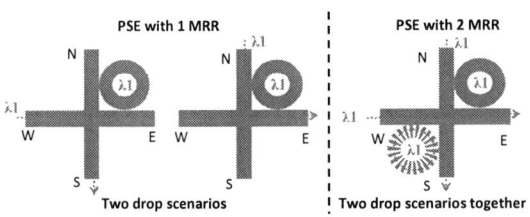

Fig. 5. Wavelength dropping scenarios in 1- vs. 2-MRR PSEs.

required number of laser sources and the aggregate total bandwidth, while Fig. 4b characterizes the static power-aggregate bandwidth trade-off.

As can be seen, the two technology mappings with 1 MRR per PSE (groups 1 and 2) and 2 MRRs per PSE (groups 3 and 4) are clearly differentiated: the latter case leads to an almost 2x increase in the aggregated bandwidth when using the same number of laser sources. This is due to the additional constraint a PSE with only 1 MRR introduces to guarantee proper operation of the network: two paths which are dropped by the same PSE cannot share the same wavelengths for sending data, in order to avoid inter-channel interference inside the MRR waveguide.

As a consequence, if the two paths have two resonant peaks that are legal for each of them, then they cannot use them at the same time. In this case, the wavelength channel is statically assigned to one of the two optical paths. Instead, the PSEs with 2 MRRs do not have this problem, because signal routing will take place separately and the paths can share the same carrier wavelengths (if any). These effects are pictorially illustrated in Fig.5. As a result, more aggregate bandwidth can be delivered with the same number of laser sources.

The reader should notice that despite the clear advantage, 2-MRR PSEs pose tighter fabrication constraints and post-fabrication tuning requirements. In fact, the MRRs should be identical, which is rarely the case if at all possible.

We can also notice that the laser source distribution in Fig. 4a and the total laser static power distribution in Fig. 4b are almost the same in both case studies. This means that there is a high correlation between the number of laser sources that a topology requires and the total laser static power it needs. In other words, by increasing the number of laser sources,

their individual output power requirement is not cut down significantly. The equalization effect of power requirements in the PDN sustains this trend.

Another important consideration from the figure is that in both case studies, the population is divided into two separated groups: one to the left (groups 1 and 3) and one to the right (group 2 and 4).

This is due to the wavelength assignment step during the topology generation process, where the used greedy algorithm always tries to reduce the number of MRR types used to synthesize a topology. For this reason, the groups to the left use only 7 MRR types, while the groups to the right use 8 of them. The groups to the right require a higher number of laser sources, because a new MRR type has been introduced, thus new resonant peaks are used and new laser sources are required.

Last but not least, we can see in Fig. 4a that in all groups the population is highly concentrated in the region where they have an intermediate aggregate bandwidth and a higher number of laser sources. In the future, automated topology synthesis tools should avoid this region when searching for the most promising design points matching predefined input requirements.

In fact, the best topologies in the sample are those that achieve the highest bandwidth with a lower number of laser sources (e.g., the design points which require 148 laser sources and 6100-6190 Gbit/sec or 11300-11390 Gbit/sec aggregate bandwidth range). The worst topologies are instead those that require the highest number of laser sources but achieve the lowest aggregate bandwidth.

Finally, we manually synthesized two relevant topologies in literature, i.e., the Snake [9] and the λ-Router [8], and applied the analysis methodology to them. We found that the Snake requires 148 laser sources and achieves an aggregate bandwidth of 5860 or 11100 Gbit/sec, by using 1 or 2 MRRs per PSE, respectively. The λ-Router requires 139 laser sources and achieves an aggregate bandwidth of 5480 or 10720 Gbit/sec depending on the use case. As a result, neither the Snake nor the λ-Router represent the best topologies, since the automatic exploration tool flow can generate dominating solutions.

978-1-5386-4757-8/18 $31.00 © 2018 IEEE

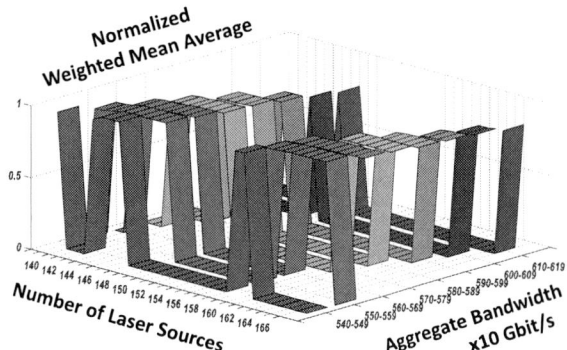

Fig. 6. Correlating topology distribution to its characteristics.

B. Comparison with state-of-the-art

By running the tool in [7], we get pessimistic performance projections for WRONoC topologies. We consider here only 2-MRR PSEs. Fig. 4a reports the achievable aggregate bandwidth with the previous methodology. Interestingly, the tool does not come up with per-topology figures, but provides a comprehensive estimate based on the number of MRR types in a topology, therefore we just have two values: one for 7 MRR types and one for 8. In both cases, the provided aggregate bandwidth is one order of magnitude lower than the differentiated performance figures our methodology can extract (groups 3 and 4).

C. Understanding the Design Space

In this section we aim at understanding the root cause for the distribution of the topologies that we have seen in Fig 4.

For this purpose, we defined the weighted mean of the number of MRR types per path in a topology as follows:

$$weighted\ mean = \frac{\sum_{n=1}^{N} NP_n \times n}{\sum_{n=1}^{N} NP_n}$$

where n is the number of MRR types, N is the network radix (which is also the maximum number of MRR types a path can cross), NP_n is the number of paths which have n MRR types. This mean increases when the number of paths with lots of MRR types in a topology increases, while the opposite holds when the number of paths with fewer MRR types increases. A topology which has a lower number of MRR types in its paths, i.e., lower weighted mean, achieves higher bandwidth; we anticipated this trend in section IV-A. In order to see how this metric changes throughout the design space, we chose the 1-MRR PSEs case study and we computed the weighted mean for each topology in the sample. Then, we calculated the arithmetic average of the weighted means of all topologies found in the same column, and normalized the obtained values to the greatest one. Resulting values are plotted in Fig. 6. We can clearly see that the average weighted mean decreases when the aggregated bandwidth increases, while it slightly increases when the number of laser sources increases. This means that a topology performs better when the corresponding weighted mean results low, which implies that it mainly exhibits a low number of MRR types on its paths.

VI. CONCLUSIONS

In this work, we synergistically cascade the most advanced design automation tools for wavelength routing in combination with a novel analysis methodology that extracts the maximum aggregate bandwidth out of a WRONoC topology. Thus, we shed light on the characteristics of a design space which is still largely unknown, and out-of-reach for state-of-the-art methodologies. We find that a higher number of MRR types in the topology paths leads to a higher number of laser sources as well as a higher total laser static power, while no improvements appear in the communication bandwidth. As a result, the design points which have lower weighted mean of the number of MRR types in their paths achieve higher communication bandwidth and lower number of laser sources, and consequently lower total laser static power. The next step consists of leveraging the knowledge provided by this paper to automatically synthesize a connectivity pattern that meets the requirements of the system at hand by consciously pruning the design space.

REFERENCES

[1] S. Werner, J. Navaridas, and M. Luj, "A survey on optical network-on-chip architectures," *ACM Comput. Surv.*, vol. 50, no. 6, pp. 89:1–89:37, 2017.

[2] M. Briere *et al.*, "System level assessment of an optical noc in an mpsoc platform," in *2007 Design, Automation Test in Europe Conference Exhibition*, 2007, pp. 1–6.

[3] S. Koohi, M. Abdollahi, and S. Hessabi, "All-optical wavelength-routed noc based on a novel hierarchical topology," in *Proceedings of the Fifth ACM/IEEE International Symposium*, 2011, pp. 97–104.

[4] X. Tan, M. Yang, L. Zhang, Y. Jiang, and J. Yang, "A generic optical router design for photonic network-on-chips," *Journal of Lightwave Technology*, vol. 30, no. 3, pp. 368–376, 2012.

[5] M. Tala, M. Castellari, M. Balboni, and D. Bertozzi, "Populating and exploring the design space of wavelength-routed optical network-on-chip topologies by leveraging the add-drop filtering primitive," in *2016 Tenth IEEE/ACM International Symposium on Networks-on-Chip (NOCS)*, 2016, pp. 1–8.

[6] M. Nikdast *et al.*, "Clap: a crosstalk and loss analysis platform for optical interconnects," in *2014 Eighth IEEE/ACM International Symposium on Networks-on-Chip (NoCS)*, 2014, pp. 172–173.

[7] A. Peano, L. Ramini, M. Gavanelli, M. Nonato, and D. Bertozzi, "Design technology for fault-free and maximally-parallel wavelength-routed optical networks-on-chip," in *2016 IEEE/ACM International Conference on Computer-Aided Design (ICCAD)*, 2016, pp. 1–8.

[8] A. Scandurra and I. O'Connor, "Scalable cmos-compatible photonic routing topologies for versatile networks on chip," 2011.

[9] L. Ramini, P. Grani, S. Bartolini, and D. Bertozzi, "Contrasting wavelength-routed optical noc topologies for power-efficient 3d-stacked multicore processors using physical-layer analysis," in *2013 Design, Automation Test in Europe Conference Exhibition (DATE)*, 2013, pp. 1589–1594.

[10] S. L. Beux *et al.*, "Optical ring network-on-chip (ornoc): Architecture and design methodology," in *2011 Design, Automation Test in Europe*, 2011, pp. 1–6.

[11] S. Koohi, M. Abdollahi, and S. Hessabi, "All-optical wavelength-routed noc based on a novel hierarchical topology," in *Proceedings of the Fifth ACM/IEEE International Symposium*, 2011, pp. 97–104.

[12] M. Ortín-Obón, M. Tala, L. Ramini, V. Viñals-Yufera, and D. Bertozzi, "Contrasting laser power requirements of wavelength-routed optical noc topologies subject to the floorplanning, placement, and routing constraints of a 3-d-stacked system," *IEEE Transactions on Very Large Scale Integration (VLSI) Systems*, vol. 25, no. 7, pp. 2081–2094, 2017.

[13] R. Hendry, D. Nikolova, S. Rumley, and K. Bergman, "Modeling and evaluation of chip-to-chip scale silicon photonic networks," in *2014 IEEE 22nd Annual Symposium on High-Performance Interconnects*, 2014, pp. 1–8.

[14] S. Beamer *et al.*, "Re-architecting dram memory systems with monolithically integrated silicon photonics," in *Proceedings of the 37th Annual International Symposium on Computer Architecture*, 2010, pp. 129–140.

[15] W. Bogaerts *et al.*, "Silicon microring resonators," *Laser & Photonics Reviews*, vol. 6, no. 1, pp. 47–73.

978-1-5386-4757-8/18 $31.00 © 2018 IEEE

Design of Latch based Configurable Ring Oscillator PUF Targeting Secure FPGA

Mahabub Hasan Mahalat*, Nikhil Ugale *, Rohit Shahare † and Bibhash Sen*
*Department of Computer Science & Engineering
†Department of Electronics & Communication Engineering
National Institute of Technology Durgapur, Durgapur, India
Email: {mahabubhasan.mahalat, unikhil76, rpshahare, bibhash.sen}@gmail.com

Abstract—Physically unclonable function (PUF) is one of the most advocated security primitives which extracts the uncontrollable intrinsic physical property of the fabrication process to generate secret bits for authentication, random number generation and key generation. Ring oscillator (RO) PUF is the widely adopted PUF design to implement in FPGA platform, but it is highly error prone to environmental noise (i.e. temperature and voltage). The configurable RO (CRO) PUF is advocated to resolve this issue without increasing the area overhead. This paper proposes an enhanced CRO framework which uses latch instead of inverter to build an RO. The use of dedicated latch (i.e. available in an FPGA) in place of inverter eliminates the restriction to use odd number of delay units (inverters) in an RO configuration. The proposed design efficiently utilizes the resources found in a configurable logic block (CLB) to increase the number of RO configurations while using the same area. Also, it provides the flexibility to include a latch in an RO configuration which in turns improve the reliability and the security as well. Experimental results on Xilinx Spartan 3E FPGA establish that the proposed design exhibits high stability despite varying environmental conditions without using any error correcting code or post-processing technique.

I. INTRODUCTION

Security is one of the most growing concerns of this digital era. An algorithmically secure cryptographic protocol relies on the underlying hardware to deliver security. In recent years, some active (i.e reverse engineering, hardware Trojan, counterfeiting etc.) as well as some passive (i.e side channel analysis) attacks have raised which targets the base hardware [1], [2]. Moreover, with the advancement of IoT, the devices which were isolated till now are also exposed to threats [3]. IoT devices have to work in resource constraint environment, hence conventional cryptographic algorithms are not feasible to use. In this scenario, a low cost, hardware-based security system is of urgent need.

Physically unclonable function (PUF), is a promising technology which provides a low cost and trustworthy security solution at hardware level. It derives an output called response to a particular input called challenge based on the intrinsic manufacturing variation of the physical body embedding it [4]. Practically, a PUF is infeasible to clone. Ideally, it is inherently tamper resistant which means, any attempt of tampering the device embedding PUF will break the PUF circuit or change the PUF behavior [5]. Different kinds of PUF designs are proposed in literature, e.g arbiter PUF, ring oscillator (RO) PUF, SRAM PUF, butterfly PUF etc. [4], [6]–[8].

Field programmable gate array (FPGA) provides a flexible hardware platform which can be configured by end users to implement any digital circuit. Nowadays, FPGAs play a very important role in embedded systems. But, cloning has become the most common security threats to FPGAs due to its binary bitstream file that can be easily copied by third party [9]. FPGA core industry is facing a lot of difficulties to control/monitor the use of IP cores configured into FPGAs by a particular end user. In this context, PUF can be applied to provide a low-cost solution for securing several FPGA based applications, like securing FPGA bitstream, authentication, IP core binding etc. [7], [10]. The RO PUF generates a response by comparing the frequency of two identical ring oscillators where the oscillator frequency depends on the intrinsic manufacturing variation [6]. The RO PUF, for its ease of implementation, is most well suited for FPGA platform [11]. An RO can be easily implemented as a hard macro on FPGA.

Besides its advantages, the RO PUF is mostly criticized due to its sensitivity to environmental variations (i.e temperature and voltage) [12]. Environmental variations may alter its response, which makes the RO PUF inefficient. Different techniques have been proposed to mitigate the effect of environmental variations on RO PUF, but most of them incur additional hardware like temperature sensor, additional RO pairs, or error correcting circuit [6], [13], [14]. On the other hand, Maiti et. al. [12] proposed Configurable RO (CRO) PUF, which efficiently used the resources in a configurable logic block (CLB) to generate reliable response despite environmental variations without using any additional hardware. The CRO generates 8 different RO configurations from a single CLB and chooses the most stable one which generates a reliable response (output). Thus, it mitigates the effect of environmental variations without increasing any hardware burden.

In this paper, we propose a new latch based CRO PUF framework, which efficiently utilizes hardware resources. Instead of inverter, the proposed design utilizes latch present in the CLB which results a significant improvement in CRO configurability. The proposed CRO has the flexibility to include or not to include a latch in a CRO configuration, which

978-1-5386-4757-8/18 $31.00 © 2018 IEEE

in turns also improve the reliability. Moreover, unlike the conventional CRO the proposed design does not restrict the RO configurations. In conventional CRO the number of inverters (delay elements) in a configuration must be odd, whereas, in the proposed CRO an RO configuration may contain odd as well as even number of latches. In the proposed design, latch has been considered as a delay element.

The main contributions of the proposed work are summarized below

- A latch based CRO (LCRO) PUF is proposed which provides the flexibility in selection of a particular delay unit with an increase in number of configurations.
- The proposed LCRO PUF eliminates the restriction in selection of RO configuration by replacing inverter with latch. The design exploits more intrinsic variations while using same area as the basic RO.
- Finally, LCRO PUF is implemented on Xilinx Spartan 3E FPGA. The experimental results depict, the implemented PUF is able to generate highly stable response despite environmental variation without using any error correcting code or post-processing technique.

The rest of the paper is organized as follows. Section II discussed the related works. The proposed PUF design is described in section III. Section IV contains the implementation methodologies. Experimental results and analysis are investigated in section V. Finally, section VI concludes the paper.

II. Related Works

The basic RO PUF proposed in [6], derives a single bit response by comparing the frequency of two symmetric ring oscillators but it is susceptible to voltage or temperature variation (i.e environmental variation). Various RO PUF designs have been investigated so far to mitigate the effect of environmental variations adding extra hardware [6], [13], [14].

The configurable RO (CRO) PUF (Fig. 1) has the ability to minimize the effect of environmental variations without imposing extra hardware burden. The first CRO PUF was introduced in [12]. It uses MUXes available in the CLB to generate more configuration to improve reliability. However, only 8 RO configurations can be produced using the CRO PUF proposed in [12]. An improved CRO design is proposed in [15], which can produce 256 RO configurations. However, both the designs lack flexibility to choose a particular delay element (inverter) [12], [15]. In [16], a highly flexible CRO (Fig. 2) design is proposed, which provides the flexibility to select a delay element. But, the flexible CRO PUF proposed in [16], put restrictions on selection of an RO configuration which reduces its configurability. All the CRO PUF designs investigated so far utilizes inverter as the major delay element. Recently, Zhang et. al. proposed a novel CRO PUF design which utilizes XOR as a delay element in place of inverter [17]. The proposed work targets to increase reliability and configurability of CRO PUF while utilizing latch as a delay element.

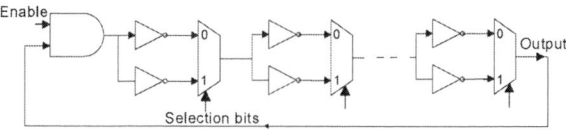

Fig. 1. A Configurable RO

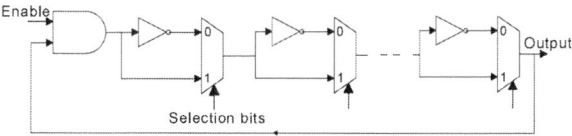

Fig. 2. A highly flexible RO

III. Latch based CRO (LCRO) PUF framework

A. Construction of latch based CRO (LCRO)

The proposed latch based CRO (LCRO) PUF replaces inverter with transparent latch. A conventional RO requires an odd number of inverters to oscillate. A single inverter is enough to provide the oscillation but the frequency will be very high. The additional inverters are required to increase the delay which reduces the RO frequency. Based on this observation, we have employed single inverter and obtained the additional delay by using transparent latches. In conventional RO design, the inverters are generally implemented by using lookup tables (LUTs). The proposed design deploys dedicated latch instead of the inverter, so the LUTs are free to hold other logic. This free LUTs are utilized to increase the number of configurations. Fig. 3 presents the logic diagram of the proposed RO. A single switch block in the design contains two transparent latches and three 2 : 1 MUXes. 'Inselect' line picks either the transparent latch or the direct path to drive the signal. It offers the flexibility to include or not to include the corresponding latch in an RO configuration. The flexible selection of a delay element, in this case latch, increases the frequency difference between a pair of RO configuration which eventually improves the reliability [16]. The 'Outselect' line expands the number of possible RO configurations by selecting either one of the two 'Inselect' MUXes. At the end of the switch blocks, an inverter is employed which performs the oscillation. The oscillator can be enabled or disabled using a MUX which uses 'Enable' as the select line. Each switch block contains 2 select lines, total $2n$ select lines are required for n switch blocks. So, 2^{2n} different RO configurations are possibles by using n such switch blocks. The proposed LCRO design takes full advantage of all the resources available in a CLB, the LCRO implementation is discussed in section IV.

B. RO comparison

A single bit PUF response can be extracted by comparing the frequency of two ROs. If the frequency of the first RO is greater than the second RO then the response is 1 otherwise the response is 0. As an LCRO with n switch blocks can be configured in 2^{2n} ways, so there are $2^{2n} \times 2^{2n}$ possibilities

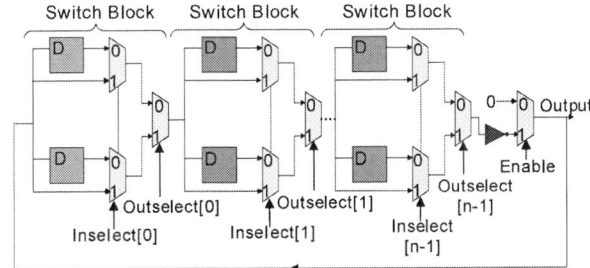

Fig. 3. Proposed latch based CRO

to select a pair of RO configuration from two LCROs. But, comparing two RO configurations with an unequal number of delay elements raises a security issue. The RO configuration with less number of delay elements will likely to be faster. Moreover, even comparing two RO configurations with an equal number of delay elements (i.e. latch) but with different select line may be biased due to systematic variation which caused by the unequal wire length.

So, in the proposed PUF design, we only consider the equivalent RO configurations for comparison. In this way, 2^{2n} pairs of RO configurations are possible from two LCROs with n switch blocks. Among 2^{2n} pairs, the pair with maximum frequency difference is finally selected to generate a reliable response. The intuition is that the response bit of the RO configuration pair with maximum frequency difference will likely to be not altered by environmental variation [6], [12]. Also, with a given reliability threshold, the proposed method can generate large number of responses by selecting the RO configuration pairs with frequency difference larger than the threshold. To minimize the measurement error, the LCROs under comparison are enabled and disabled simultaneously using a reference clock.

C. Selection of RO

Total $^nC_2$ pairs are possible from n *LCROs* (i.e considering a single configuration) but all of them are not independent [12], [18]. Moreover, random selection of RO pairs may raise security issues, as RO frequency also depends on the location of the RO in the FPGA [18]. Several strategies are proposed to avoid systematic variation during RO comparison [18]. We have adopted the chain like strategy for selecting an LCRO pair. This strategy only selects the neighboring ROs for comparison. For example, $LCRO_1$ compared with $LCRO_2$, $LCRO_2$ compared with $LCRO_3$, ... $LCRO_{n-1}$ compared with $LCRO_n$.

Using this strategy, $(n-1)$ independent LCRO pairs are possible from n LCROs. Another logical explanation to adopt this strategy is that theoretically, it may diminish the effect of environmental variations. Voltage or temperature variation will affect the neighboring ROs (i.e LCROs) in a similar way, so their frequency will likely to be changed in the same direction (i.e either increase or decrease). Thus, it will not hamper the frequency difference.

IV. IMPLEMENTATION OF LATCH BASED CRO (LCRO) PUF

The proposed PUF design has been implemented on Xilinx Spartan 3E (XC3S500E) FPGA using Spartan 3E Starter board. A Configurable logic block (CLB) of the aforementioned FPGA contains 4 slices, where each Slice contains two 4 input lookup tables (LUTs), two latches and two $2:1$ wide function MUXes (F5 MUX). An LCRO, with three switch blocks, is implemented inside a single CLB. The LCRO implementation is shown in Fig. 4.

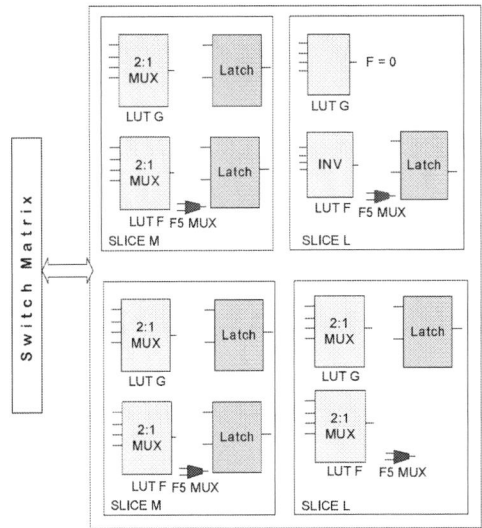

Fig. 4. The proposed LCRO inside a CLB

The 'Inselect' and 'Outselect' MUXes are implemented using LUTs and dedicated F5 MUXes respectively. The dedicated flip-flops are configured as transparent latches. A LUT and a dedicated F5 MUX is used to implement the inverter (INV) and the enable MUX respectively. A LUT is configured as constant 0, which is connected as an input to the 'enable' MUX (i.e F5 MUX). The implemented LCRO requires 6 select lines (i.e 3 'Inselect' and '3' 'Outselect' line) to select a configuration. So, total $2^6 = 64$ configurations are possible from a single LCRO. To make each LCRO symmetric, it has been implemented as a hard macro which preserves the wiring as well as the component. The hard macro only uses the CLB and the switch box connected with it. A hard macro of an LCRO is shown in Fig. 5. 100 similar LCROs are created by instantiating a hard macro. The LCROs are placed as a 10×10 matrix in the middle of the FPGA. $(100-1) = 99$ reliable response bits are finally extracted from 100 LCROs.

The generalized diagram of the proposed LCRO PUF is shown in Fig. 6. An LCRO pair is selected using 2 MUXes, which are controlled using $\lceil \log_2 n \rceil$ common select lines, where n is the number of LCROs. The LCROs $(0:n-2)$ are connected with the first MUX and the LCROs $(1:n-1)$ are connected with the second MUX. A select line always picks a pair of neighbor LCROs. The implemented LCRO PUF requires total 13 bits to select an LCRO configuration (i.e 7 bits

978-1-5386-4757-8/18 $31.00 © 2018 IEEE

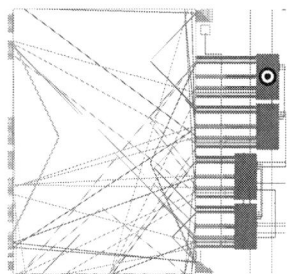

Fig. 5. Hard macro of an LCRO

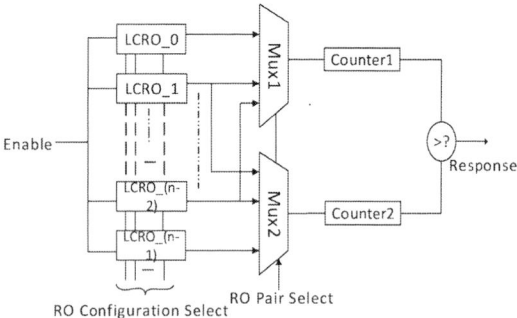

Fig. 6. Generic design of the proposed LCRO PUF

to select an LCRO pair and 6 bits to select a configuration). Two counters simultaneously measure the frequency of the selected LCRO configurations with respect to the on-board 50 MHz clock. Finally, a comparator compares the RO frequency to generate a single bit response.

It should be noted that hard macro creation is an essential step for implementation of RO PUF, otherwise the implemented ROs may not be exactly symmetric. But, after creating a hard macro the particular CLB cannot be used to hold any further logic. So, the proposed LCRO utilizes most of the resources in a CLB to improve the overall resource utilization. An LCRO uses 8 LUTs, 6 transparent latches and 4 wide MUXes from a CLB.

V. EXPERIMENTAL RESULT AND ANALYSIS

To check the quality factor, the same LCRO PUF design has been synthesized using Xilinx ISE 14.7 and implemented on 20 similar Spartan 3E starter boards. Additionally, in 5 boards measurement is taken in differential temperature and voltage environment. The Chip Scope Pro software available with Xilinx ISE is used to monitor RO frequency. The challenge (input) response (output) pairs and the frequency of RO configurations are collected by controlling the system with a finite state machine, which is implemented using the MicroBlaze MCS core. The frequency of the different configurations of an LCRO should be different otherwise this will not lead to any additional benefit. Fig. 7 represents the average frequency of all the 64 configurations of Board D487943 (serial no.). A significant change in frequency has been observed when the number of latches varies. It happens because a latch introduces

more delay compared to other elements in the RO. However, the average frequency varies in the range of 85 MHz to 180 MHz.

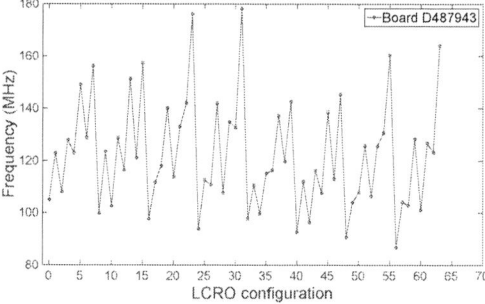

Fig. 7. Average frequency of the 64 RO configurations of board D487943

Any correlation between the selected stable (i.e. reliable) RO configurations may raise security issue. Theoretically, the stable RO configurations should be equally distributed over all the RO pairs. There are 64 possible RO configurations and 99 RO pairs. So, the 64 RO configurations will be distributed over the 99 pairs. Fig. 8 plots the count of selected stable RO configuration (i.e for convenience of visibility only 10 boards are shown). It can be noticed that, the selected stable RO configurations are almost random, which means observing the stable RO configurations selected in one board an adversary cannot accurately predict the stable configurations selected in other boards. The selected stable RO configurations are used

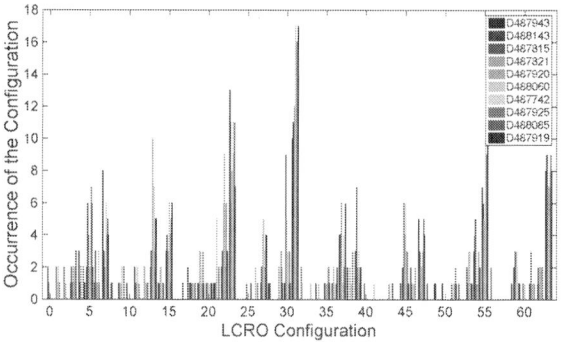

Fig. 8. Count of selected stable RO configurations on 10 boards

during the measurement of reliability, in any other quality measurement all the configurations are considered.

A. Bit aliasing of the response bits

Bit aliasing measures the entropy of PUF response across different PUF instances using same challenge. For i^{th} challenge in R different devices bit aliasing is measured as

$$Bit\ Aliasing = \frac{1}{R} \sum_{j=0}^{R-1} (^{j}r_i \times 100\%) \qquad (1)$$

978-1-5386-4757-8/18 $31.00 © 2018 IEEE

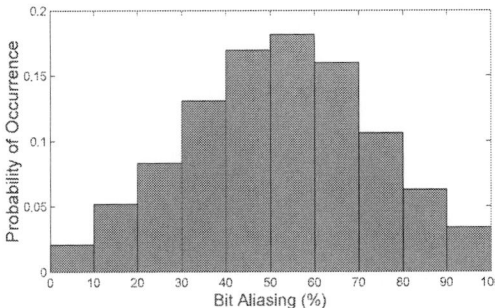

Fig. 9. Histogram of the percentage of bit aliasing

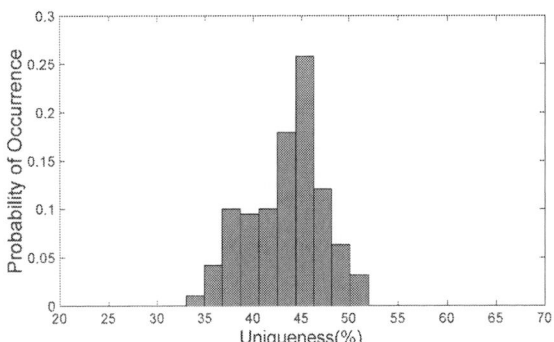

Fig. 11. Probability of the occurrence of Uniqueness(%)

where $^j r_i$ is the ith bit response of jth PUF instance. For a particular select line, ideally, bit aliasing should be 50%. Histogram of the percentage of bit aliasing is shown in Fig. 9. From the plot, it can be observed that the bit aliasing value tends to the ideal value of 50% with high probability. It signifies that the proposed design generates high-quality response with better security.

B. Uniformity of the response bits

Uniformity denotes the number of 1 in the PUF output. Ideally, the output should contain an equal number of 0 and 1. The uniformity is calculated as

$$Uniformity = \frac{1}{n} \sum_{i=1}^{n} (^p r_i \times 100\%) \quad (2)$$

Where $^p r_i$ is the ith response of PUF instance p, n is the number of responses. Ideally, uniformity should be 50%. The probability of the occurrences of uniformity is shown in Fig. 10. On average uniformity is 49.65%, which is very close to the ideal value of 50%.

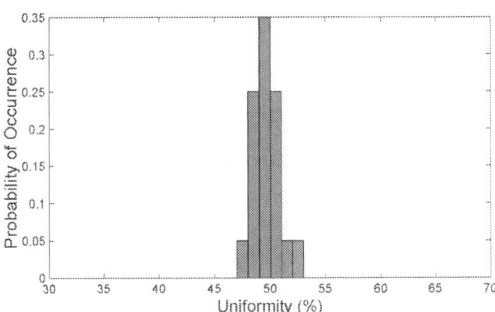

Fig. 10. Probability of the occurrence of Uniformity(%)

C. Uniqueness

It is the measurement of inter die variation. It indicates the ability of PUF to differentiate two devices. The same challenge is applied to two different devices and the hamming distance

between their response is calculated as uniqueness. For M PUF instances it is calculated as

$$Uniqueness = \frac{2}{M(M-1)} \sum_{i=1}^{M-1} \sum_{j=i+1}^{M} AHD(R_i, R_j) \times 100\% \quad (3)$$

Where $AHD(R_i, R_j)$ is the average hamming distance between the response of PUF instance R_i and R_j. Ideally, uniqueness should be 50%. On average the proposed design possesses a uniqueness of 43.40% with a standard deviation of 3.78. The minimum and maximum value of uniqueness are 33.08% and 51.70% respectively, which is sufficient to uniquely distinguish two PUF instances. The distribution of uniqueness is shown in Fig. 11.

D. Reliability analysis

Reliability is the most important factor to estimate PUF quality. It indicates the stability of the PUF responses across different measurements. Ideally, a PUF instance should always produce same response to the same challenge. Reliability is quantified as

$$Reliability = (1 - \frac{1}{M} \sum_{i=0}^{M-1} AHD(^k R_i, ^k R_C)) \times 100\% \quad (4)$$

Where $^k R_C$ is the response of kth PUF instance at nominal voltage and temperature. $^k R_i$ is response of ith measurement. M is the number of measurements in different environmental conditions and AHD is the average Hamming distance. The ideal value of the reliability is 100%.

The reliability of the proposed design is evaluated in normal as well as extreme operating conditions. In nominal operating condition (i.e $25°C$ and $1.2V$), repeated measurement of responses does not show any bit flip which means in the normal operating condition the proposed design possesses 100% reliability. In five boards, responses are extracted in different voltage and temperature level to estimate the effect of environmental variations on reliability. The temperature and voltage (i.e core voltage) ranges are $\{25°C, 35°C, 45°C, 65°C\}$ and $\{.8V, 1.0V, 1.2V, 1.4V, 1.6V\}$ respectively.

The implemented PUF produces 100% reliable responses despite temperature variations, so the results of temperature

978-1-5386-4757-8/18 $31.00 © 2018 IEEE

TABLE I
NUMBER OF BIT FLIPS DUE TO VOLTAGE VARIATION (T=25°C)

	.8V	1.0V	1.2V	1.4V	1.6V
Board D487943	0	0	0	0	0
Board D488143	1	1	0	0	0
Board D487815	0	0	0	0	0
Board D487821	2	2	0	2	2
Board D487920	0	0	0	0	0

variation are not listed. But, it has been observed that the voltage has a significant impact on RO frequency. The RO frequency increases or decreases as the core voltage increases or decreases. However, bit flip only happens, if the frequency change alters the sign of the frequency difference [6]. Number of bit flips (i.e out of 99 responses) in different voltage regions are shown in Table I. On average the reliability of the implemented PUF against voltage variation is $(100 - 0.51) = 99.49\%$, with a maximum and minimum value of 100% and 97.98% respectively .

Hardware overhead and number of reliable response bits are the well-known trade-offs for PUF characterization. The proposed PUF design generates almost 100% reliable response despite extreme environmental noise which eliminates the need for any additional resources for error correcting code or post-processing. On average, the proposed PUF produces almost a one bit reliable response from a single CLB. Moreover, the design generates 64 RO configurations from single CLB, so the number of response bits can also be increased by using a reliability threshold [15], [16]. The reliability comparison with the state of the art CRO PUFs is shown in Table II. The comparison results show that the proposed design has better reliability against environmental variation compare to the other designs. The proposed design utilizes latch instead of inverter which in turn allow to use the free LUTs for improving the configurability.

TABLE II
COMPARISON OF AVERAGE RELIABILITY

CRO design	Reliability	Delay unit
Basic CRO PUF [12]	99.14%	Inverter
Improved CRO PUF [15]	98.98%	Inverter
Highly flexible RO PUF [16]	NA	Inverter
Low cost CRO [19]	98.41%	Inverter
XOR based CRO [17]	97.72%	XOR
Proposed CRO PUF	99.49%	Latch

VI. CONCLUSION

In this paper a novel latch based CRO (LCRO) PUF is proposed. The design utilizes latches instead of inverters which generates more reliable response bits by exploring more hardware resources without increasing the area. Experimental results show that the proposed design is efficient in terms of all the PUF parameters. Moreover, it generates 100% reliable response in nominal condition as well as in the presence of thermal noise. It yields an average reliability of 99.49% despite

voltage variation which is significantly better than the previous CRO PUF designs. Also, the design is able to produce 64 different RO configurations from a single CLB (i.e CLB of Xilinx Spartan3e 3e FPGA).

REFERENCES

[1] S. P. Skorobogatov, "Semi-invasive attacks – a new approach to hardware security analysis," 2005.

[2] S. T. King, J. Tucek, A. Cozzie, C. Grier, W. Jiang, and Y. Zhou, "Designing and implementing malicious hardware," in *Proceedings of the 1st Usenix Workshop on Large-Scale Exploits and Emergent Threats*, ser. LEET'08. Berkeley, CA, USA: USENIX Association, 2008, pp. 5:1–5:8.

[3] T. Xu, J. B. Wendt, and M. Potkonjak, "Security of iot systems: Design challenges and opportunities," in *Proceedings of the 2014 IEEE/ACM International Conference on Computer-Aided Design*, ser. ICCAD '14. Piscataway, NJ, USA: IEEE Press, 2014, pp. 417–423.

[4] B. Gassend, D. Lim, D. Clarke, M. van Dijk, and S. Devadas, "Identification and authentication of integrated circuits: Research articles," *Concurr. Comput. : Pract. Exper.*, vol. 16, no. 11, pp. 1077–1098, Sep. 2004.

[5] M. Rostami, M. Majzoobi, F. Koushanfar, D. S. Wallach, and S. Devadas, "Robust and reverse-engineering resilient puf authentication and key-exchange by substring matching," *IEEE Transactions on Emerging Topics in Computing*, vol. 2, no. 1, pp. 37–49, March 2014.

[6] G. E. Suh and S. Devadas, "Physical unclonable functions for device authentication and secret key generation," in *2007 44th ACM/IEEE Design Automation Conference*, June 2007, pp. 9–14.

[7] J. Guajardo, S. S. Kumar, G.-J. Schrijen, and P. Tuyls, "Fpga intrinsic pufs and their use for ip protection," in *Cryptographic Hardware and Embedded Systems - CHES 2007*, P. Paillier and I. Verbauwhede, Eds. Berlin, Heidelberg: Springer Berlin Heidelberg, 2007, pp. 63–80.

[8] R. Maes, P. Tuyls, and I. Verbauwhede, "Intrinsic pufs from flip-flops on reconfigurable devices."

[9] J. Zhang and G. Qu, "A survey on security and trust of fpga-based systems," in *2014 International Conference on Field-Programmable Technology (FPT)*, Dec 2014, pp. 147–152.

[10] E. Öztürk, G. Hammouri, and B. Sunar, "Towards robust low cost authentication for pervasive devices," in *Proceedings of the 2008 Sixth Annual IEEE International Conference on Pervasive Computing and Communications*, ser. PERCOM '08. Washington, DC, USA: IEEE Computer Society, 2008, pp. 170–178.

[11] S. Morozov, A. Maiti, and P. Schaumont, "An analysis of delay based puf implementations on fpga," in *Reconfigurable Computing: Architectures, Tools and Applications*, P. Sirisuk, F. Morgan, T. El-Ghazawi, and H. Amano, Eds. Berlin, Heidelberg: Springer Berlin Heidelberg, 2010, pp. 382–387.

[12] A. Maiti and P. Schaumont, "Improved ring oscillator puf: An fpga-friendly secure primitive," *Journal of Cryptology*, vol. 24, no. 2, pp. 375–397, Apr 2011.

[13] C.-E. Yin and G. Qu, "Temperature-aware cooperative ring oscillator puf," in *Proceedings of the 2009 IEEE International Workshop on Hardware-Oriented Security and Trust*, ser. HST '09. Washington, DC, USA: IEEE Computer Society, 2009, pp. 36–42.

[14] M. D. Yu and S. Devadas, "Secure and robust error correction for physical unclonable functions," *IEEE Design Test of Computers*, vol. 27, no. 1, pp. 48–65, Jan 2010.

[15] X. Xin, J. P. Kaps, and K. Gaj, "A configurable ring-oscillator-based puf for xilinx fpgas," in *2011 14th Euromicro Conference on Digital System Design*, Aug 2011, pp. 651–657.

[16] M. Gao, K. Lai, and G. Qu, "A highly flexible ring oscillator puf," in *2014 51st ACM/EDAC/IEEE Design Automation Conference (DAC)*, June 2014, pp. 1–6.

[17] L. Zhang, C. Wang, W. Liu, M. O'Neill, and F. Lombardi, "Xor gate based low-cost configurable ro puf," in *2017 IEEE International Symposium on Circuits and Systems (ISCAS)*, May 2017, pp. 1–4.

[18] W. Liu, Y. Yu, C. Wang, Y. Cui, and M. O'Neill, "Ro puf design in fpgas with new comparison strategies," in *2015 IEEE International Symposium on Circuits and Systems (ISCAS)*, May 2015, pp. 77–80.

[19] Y. Cui, C. Wang, W. Liu, Y. Yu, M. O'Neill, and F. Lombardi, "Low-cost configurable ring oscillator puf with improved uniqueness," in *2016 IEEE International Symposium on Circuits and Systems (ISCAS)*, May 2016, pp. 558–561.

A low power keyword spotting algorithm for memory constrained embedded systems

Gionata Benelli
Information Engineering Dept.
University of Pisa
Pisa, Italy
g.benelli1@studenti.unipi.it

Gabriele Meoni
Information Engineering Dept.
University of Pisa
Pisa, Italy
gabriele.meoni@ing.unipi.it

Luca Fanucci
Information Engineering Dept.
University of Pisa
Pisa, Italy
luca.fanucci@unipi.it

Abstract—Nowadays Voice User Interfaces (VUIs) have become popular thanks to their easiness of use that makes them accessible to the elderly and people with disability. Nevertheless, their use in embedded systems for the realization of portable devices is limited by the computation complexity, the memory requirements and power consumption of the keyword spotting (KWS) algorithms, usually based on deep neural networks. In this paper we propose a new algorithm based on convolutional neural networks for the keyword spotting task, that offers a good trade-off among accuracy, power consumption and memory footprint. To select our proposed solution, we compared different neural network architectures to select the best trade-off of these metrics. For further improvements of these performances we implemented our solution on a dedicated hardware platform as Myriad 2 by Movidius. The use of this chip has reduced inference time and energy per inference by 50%.

Index Terms—keyword spotting, speech recognition, neural network, convolutional neural network, neural compute stick, low-power, memory footprint, machine learning

I. INTRODUCTION

With the advent of portable embedded devices, the capacity of modern Human Machine Interfaces (HMIs) to exploit voice, gestures and facial expressions have transformed the interaction with the user, making it more natural and intuitive [1]. In particular, Voice User Interfaces (VUIs) have been increasing in popularity thanks to their easiness of use and their accessibility to the elderly and people with disabilities [2]. Such HMIs recognize predefined keywords in the user speech that might be associated with specific commands. Such approach is known as Keyword Spotting (KWs) in literature. KWS algorithms are generally processed offline. Such approach guarantees lower latency in the processing and the possibility of using the interfaces without an internet connection [3]. In addition, as it happens for *Hey Siri* or *OK Google*, KWS is used to allow the online processing of the more complex speech sentences only when the correspondent keywords are detected, guarantees the user privacy by avoiding a continuous streaming of data on the web.

Modern KWS algorithms are realized through Deep Neural Networks (DNNs) which have outperformed the traditional Hidden Markov Model (HMM) - Gaussian Mixture Model

(GMM) approach thanks to the better hardware complexity / performance compromise offered [4]. In addition, they do not require any hypothesis on data correlation [5].

However, even if DNNs seem to be very promising, such algorithms are computationally intensive. For such reason, their implementation onboard embedded design requirements is tricky, owing to the low power and memory footprint constraints which might lead to limited accuracy and throughput performances [4]. In fact, for this kind of algorithms accuracy scales with the number of parameters and layers. For that reason, limiting the memory footprint might reduce the classification accuracy.

As a consequence, many works focuse on searching the best DNN model offering the best design trade-offs. Standard DNN based on Fully Connected (FC) layers are not able to exploit the nature of data so they usually require a higher number of parameters to reach acceptable accuracy with respect to other approaches. In [6] Convolutional Neural Networks (CNNs) are used. In [7] CNNs are used in a combined approach with Recurrent Neural Networks (RNNs). Thanks to the presence of an internal state, RNNs can exploit the correlations between consequent data, improving the accuracy with less parameters. In [4] Long Short Time Memories (LSTMs) are implemented thanks to their better ability of exploiting the correlation of data more distant in time.

Other works propose to use hardware accelerators for the implementation of DNNs with high improvements on the system timing performances and power consumption [8].

By using both the approaches, in this paper we presented a KWS algorithm based on an optimized CNN model. The latter has been selected by comparing different DNN models in terms of accuracy, number of parameters and number of required operations since they are good proxys of power consumption and latency [9].

All the models have been implemented onboard a Movidius Myriad 2 Visual Processing Unit (VPU) with high improvements on power consumption, throughput and latency performances.

The remainder work is structured as follows: section II shows the architecture of the KWS system and it compares the different usable DNN models; section III shows the results of such comparison. In section IV the system prototype using the

978-1-5386-4757-8/18 $31.00 © 2018 IEEE

Fig. 1: Architecture of the system and processing of the "OK" keyword

NCS for the DNNs inferences is described; results in terms of performances improvements are discussed in section V. Finally, section VI sums up our work.

II. MODEL DESCRIPTION

The system architecture is shown in Figure 1.

The 16 kHz sampled input audio stream is preliminarily divided into non-overlapped 32 ms long audio frames. The frame division is shown in figure 2. As explained by our previous work [10], a Voice Activity Detector (VAD) can be used in order to prevent the activation of the KWS algorithm in presence of frames free of voice. For this aim, the VAD processes and marks each frames as *voiced* or *silent*. This approach offers high advantages in term of reduction of power consumption. Even if the VAD labels each frame independently, the KWS algorithm will not process the input data until $N = 5$ consecutive frames are marked as voiced. When this happens, the input stream reasonably contains a keyword to detect. For this reason, the N voiced frames are stored in a memory buffer together with some of their successive frames. One frame precedent to the N voiced frames is also stored in the buffer to make the system more robust to VAD errors, which can be significant in presence of unvoiced consonants at the beginning of a word.

Since all the used command words always require less than 1s for their pronunciation, the buffer length has been sized to contain 1s of registration. In this way, an entire command word is contained in the buffer.

Some works use directly the stored data as input of the DNNs without any preprocessing. This approach is defined *end to end*. The basic idea of this method consists in leaving to the DNN the task to filter from the raw data the information necessary for the recognition of the words. The end to end technique permits to avoid data preprocessing with advantages in term of system simplification and reduction of the risk of eliminating information useful for the recognition.

However, this approach requires very large training datasets and the use of very deep neural networks, whose dimensions conflict with the low power and low memory footprint constraints. For this reason, complex design trade-offs are necessary, which generally produce solutions with limited accuracy performances [11].

For this reason, a *features extraction* preprocessing has been performed on the memorized data. It consists in the extraction of precise audio features which containing information usable for the command words recognition. In particular, the buffer samples are collected in 32 ms frames and the Mel Frequency Spectral Coefficients (MFSCs) are extracted for each frame by using the Hann window. Each frame is overlapped by 16 ms with the previous one. By collecting the MFSC coefficients of all the frames, the entire buffer spectrogram is calculated. The latter is used as input of a CNN. More details on the features extraction algorithm and on the CNN model selection can be respectively found in sections II-B and II-C.

Since the CNN processes the entire spectrogram at once, the whole keyword is searched in one step. This approach offers the advantage that no *Posterior Handling* is necessary to extract the information of consecutive CNN outputs [4].

A. Voice Activity Detector

The VAD has been modelled as a Bayes classifier which chooses if each frame belongs to the *Voice / Silence* class. For such aim, the VAD executes a *Maximum A Posteriori* estimation of the class $\tilde{C}$ according to the equation 1:

$$\tilde{C} = \underset{k \in \{1,2\}}{\operatorname{argmax}} \{ p(C_k) \prod_{i=1}^{2} p(x_i|C_k) \} \qquad (1)$$

where x_1 and x_2 are the used features, and C_k are the two possible *Voice / Silence* classes. The classifier works under the *naive* assumption of independence of the features, which in this works are calculated as shown in equation 2:

$$\begin{cases} x_1 = \frac{1}{N} \sum_{j=0}^{N-1} |sgn(x_j) - sgn(x_{j-1})| \\ x_2 = \frac{1}{N} log(\sum_{j=0}^{N-1} |x_j|^2) \end{cases} \qquad (2)$$

x_1 is the *zero-crossing frequency* normalized on the number of samples in a frame N [12]. $sgn(x_j)$ is the sign of the sample x_j. At the begin of each frame, x_{j-1} is set to zero. x_2 is the *Logarithmic Frame Energy* averaged on the frame length.

Fig. 2: Frame division of the input audio stream

978-1-5386-4757-8/18 $31.00 © 2018 IEEE

For both the features it has been assumed that $p(x_i|C_k)$ is a Gaussian density probability with statistics $(\eta_{x_i}, \sigma_{x_i})$. The latter has been estimated through a direct measurement of the ones of the used appropriate training set.

As regard the *a priori* probability $p(C_k)$, it has been assumed that $p(C_1) = p(C_2) = 0.5$.

B. Features Extraction

The features extraction process extracts information from the input stream that is useful to the keyword recognition. As previously explained, MFSC coefficients have been used in this work. Such features provide information about the spectral properties of the frame. In particular, they allow an estimation on the presence and the frequency position of the *formants* in the spectrum which allow to classify the pronounced phoneme in the analysed frame [13].

In order to calculate the MFSC coefficients, the Short Time Fourier Transform (STFT) calculation of the frame samples is performed. The square modulus of the resulting spectral samples is then calculated to get the short-term power spectrum. The latter is then convolved with the *Mel Frequency filter bank* to model the particular dependence with frequency of the human ear masking curve. Finally, the logarithm of the previous step output is performed to model the nonlinear amplitude response of the human ear.

Some works perform an additive step by extracting the *Discrete Cosine Transform* (DCT) of the MFSCs; the resulting features are defined Mel Frequency Cepstrum Coefficients (MFCCs). As explained in [13], the use of MFCCs is justified with a HMM - GMM approach, thanks to the decorrelation property of the DCT. However, features correlation does not seem to offer high advantages with DNNs. On the contrary, as measured by [14], the use of MFSCs offer better classification accuracy with a lower number of operations.

C. Convolutional Neural Network

In this section we give an overview of the different neural networks architecture explored in this work, including FC, CNN, Dephtwise Separable Convolution Neural Network (DCNN), and our proposal of Separable Convolution Neural Network (SCNN). All the architectures based on recursive link, like RNNs and LSTMs were excluded from this work since the chosen CNN hardware accelerator does not support this types of network.

The FC is a neural network made of a stack of fully-connected layers and non-linear activation layers linked in a feed-forward fashion. The input to the network is the flattened feature matrix, which feeds into a stack of d hidden fully-connected layers each with a different number of neurons. Each fully-connected layer is followed by a batch-normalization layer [15] and a Rectified Linear Unit (ReLU) based activation function. The output layer implements a *softmax* function activation to generate the output probabilities of the k keywords.

The major issue with FCs is their memory footprint, that grows quadratically with the number of nodes when a layer

is added. In addition, they are not good to exploit the input data structure to model time and spectral correlation.

On the other hand, CNNs take advantage of data structure and can model correlation inside input features by treating them like an image and by performing a 2-D convolution between them and a filter, whose coefficients have been learned during a training phase. Convolution permits to use a small set of parameters, the filter, to elaborate large amounts of input features. Convolutional layers are followed by batch-normalization and ReLU activation function. The output of each filter is a feature map.

A convolutional layer is composed by N filters, which gives an image with N-channels as an output. If the input to a convolutional layer is composed by a number of channels, the filters span all of them. For this reason, convolution layers can require millions of operations, if performed on data with multiple channels. No pooling is necessary in the intern layers thanks to the small input features matrix dimensions.

Convolutional layers help in reducing memory footprint but are computational expensive. For this reason, some new layers, like depthwise covolutional layers have been proposed [16]. DCNN first convolves each channel in the input feature map by a separate 2-D filter and then uses *pointwise* convolutions to combine the outputs of each input channel. For this reason, depthwise separable convolutions are more efficient both in number of parameters and operations, since the 2-D convolution is performed independently on each channel, enabling the development of deeper architectures. Each depthwise layer is followed by a batch-normalization and a ReLU based activation. Even in this case no pooling is performed between layers.

Following the example of DCNN, we proposed a separable convolution more suitable for audio processing. If we take a look at a spectrogram, we can clearly see that each dimension has a different physical meaning, since they represent different physical dimensions. For this reason, it is more appropriate to study a spectrogram by stacking two 1-D convolutions. In this way, we can more efficiently utilize the data structure and reduce the number of operations and parameters required. Figure 4 shows how the two 1-D convolutions process the spectrogram.

Each separable layer is followed by a batch-normalization

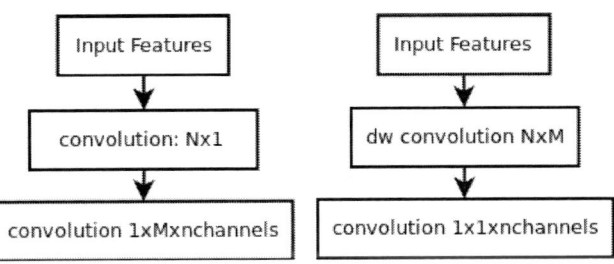

Fig. 3: Comparison between a separable Convolution(left) and a depthwise separable convolution(right)

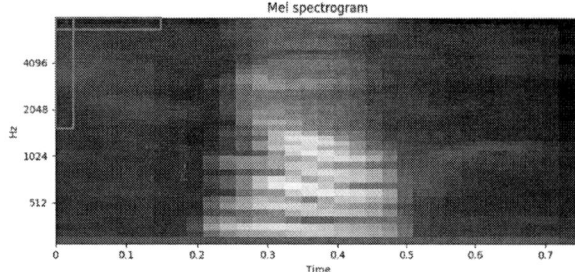

Fig. 4: Separable Convolution on Mel-Spectrogram, the different filters of the separable convolution are highlighted in red.

layer and a ReLU based activation. No pooling is performed between layers.

The output layer of each different model has been realized and tested by using two methods: in the first approach, global pooling is performed on the last layer to flatten the output of the network. The global pooling layer is then followed by a Fully-connected layer with batch-normalization and soft-max activation to finally get the k output probabilities. In the second method, the last layer is fed to a *pointwise* convolutional layer with k-channels, and it is followed by a global pooling and a soft-max layer in order to obtain the final output probabilities. For this work, we adopt a name convention which defines the model name as *X+FC* (e.g. CNN+FC,DCNN+FC or SCNN+FC) if the first method is used. Otherwise, models are defined as *Full-X* (e.g.Full-CNN, Full-DCNN, Full-SCNN).

III. MODEL EVALUATION RESULTS

We use the Google speech commands dataset for the neural network architecture exploration experiments. The dataset consists of 65 k 1-second long audio clips of 30 keywords, by thousands of different people, with each clip consisting of only one keyword. The neural network models are trained to classify the incoming audio into one of the 10 keywords - "Yes", "No", "Up", "Down", "Left", "Right", "On", "Off", "Stop", "Go", along with "silence" (i.e. no word spoken) and "unknown" word, which is the remaining 20 keywords from the dataset. The dataset is split into training, validation and test set in the ratio of 80:10:10 while making sure that the audio clips from the same person stays in the same set, *notice that keyword classes are balanced*. All models are trained in Keras [17], using Tensorflow [18] as a backend, with Mini Batch Stochastic Gradient Descent algorithm and cross-entropy loss function. With a batch size of 64, the models are trained for 100 iterations at most, starting with a learning rate of 0.001 which is divided by 10 every 3 iterations that validation loss has reached a plateau. We also monitored validation loss for Early Stopping [19], deciding to stop training when validation loss does not improve for 10 epochs. Such approach guarantees a reduction of the network *over-fitting*. Networks are evaluated on test set classification accuracy.

In this work we explored the design space by performing a Montecarlo method analysis for each architecture, to search the best combination of hyper-parameters like number of nodes and layers, with the constraint of requiring only a 100 000 parameters for our neural networks, since we want our application to be comparable with other works. For each architecture we selected the best trade-off among classification accuracy, number of parameters and number of operations to get the most suitable architecture for our application. We decided to compare these parameters since they have been shown to be good proxies for power consumption and inference time [9]. This way, by comparing these few parameters we got a rough estimation of performance independently from the platform. FC networks show the worst accuracy among all networks using the same number of parameters. Indeed, they produce the largest networks requiring up to 90 000 parameters to get an acceptable accuracy (70 %). However, they require the smallest number of operations among all networks, which makes them suitable for systems with real-time requirements. For the same number of parameters, DCNN+FC and Full-DCNN architectures show the best performance in terms of accuracy, followed by Full-SCNNs and SCNN+FCs. Full-DCNN is the most compact architecture but requires a fairly high amount of operations, whereas the SCNN+FCs have a lower number of operations maintaining an acceptable accuracy and size. At the end SCNNs and DCNNs are almost equivalent in terms of number of parameters, operations and classification accuracy, but the separable convolution is easier to implement in most framework and on most hardware platforms for DNN. This is particularly true when the convolution filter has different sizes on time and frequency axis, which is optimal for KWS [6].

TABLE I: Results of parameters optimization for the different neural network architectures

Model	Params	Millions Ops	Test Accuracy
FC	97K	0,08	70,84%
SCNN+FC	13K	2,2	86,20%
Full-SCNN	15K	4,93	87,77%
CNN+FC	12K	4,37	75,40%
Full-CNN	9K	3,30	68,85%
DCNN+FC	14K	2,45	88,63%
Full-DCNN	10K	5,94	88,50%

IV. PROTOTYPE DESCRIPTION

Finally, we realized a proof-of-concept prototype to implement our algorithm on. Since VAD and features extraction requires flexibility, we opted for a solution based on a microprocessor. For such reason, we used a Raspberry Pi 3B to implement this stage of our KWS system. The machine learning engine, instead, was implemented on a Hardware accelerator, to better fit the requirements for power-consumption and latency. The chosen hardware accelerator is a Myriad 2 by Movidius, a Vision Processing Unit with 2 Leon microprocessors and 12 Very-Long Instruction Word (VLIW) custom micro-controllers, called SHAVEs. This VPU is especially designed to accelerate the inference operation of

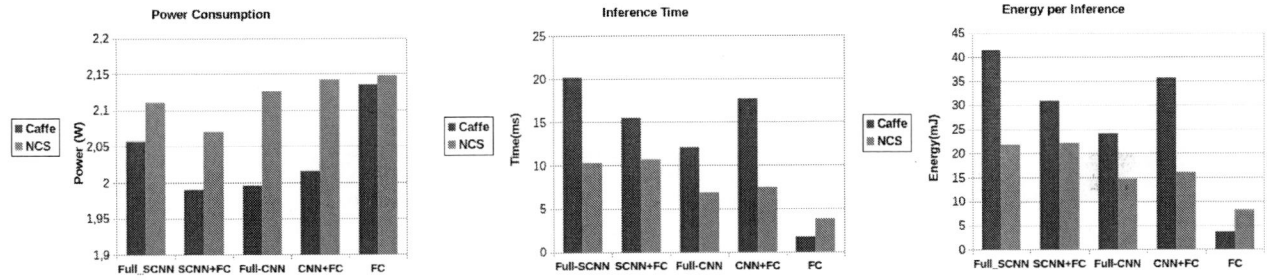

Fig. 5: Graphs showing the increase in performance thanks to the NCS

convolution neural network, but for our prototype we used the Neural Compute Stick(NCS), a USB stick that mounts the Myriad 2 and makes it easier to work with. Since the NCS does not support the dephtwise convolution with a rectangular kernel, DCNN+FC and Full-DCNN were not considered in this comparison.

To run model inference on the Raspberry Pi, we exported our proposed networks from Keras to Caffe [20], we tested the correctness of the conversion by comparing classification accuracy on test set before and after conversion, verifying a loss of accuracy below 0.4%. We also converted these models in a format compatible with NCS by using the tool included in the NC-SDK downloaded by the Intel Movidius official github. Evaluation was performed on a Raspberry Pi 3 Model B running Raspbian Stretch. On the Raspberry Pi, we ran the Caffe framework which imports a model and performs inference. To capture power measurements, the Raspberry Pi is plugged into a AVHzY CT2 USB tester, which has a USB port from which measurements can be automatically read. Power measurements are taken at a frequency of 1 Hz from an external laptop connected to the meter. The length of each experimental trial is sufficiently long (on the order of minutes) so that this resolution yields reasonably accurate measurements. During each experimental trial, a script on the Raspberry Pi iterates through all test samples for a fixed model, calling an Application Program Interface (API) served by the laptop to start and stop measurements after the inference. Each service call evaluates all testing dataset. We tested our system using a headless set-up and connecting to the board via Secure Shell (SSH) over cable protocol. For a system in idle mode, with no calculation executed, the power consumption resulted to be of 1.3 W in absence of the NCS and of 1.6 W when it is linked to the Raspberry PI.

V. EXPERIMENTAL RESULTS

From our results, we see that the use of the NCS has significantly enhanced performances. Indeed, although increasing slightly power consumption during inference time, the NCS reduced the latency up to 50% for some architectures. From our experiments it is clear that the introduction of FC layers does not reduce the inference time or power consumption if used in combination with NCS. This does not come as a surprise, since NCS has been developed for vision task, and

so for more complex Convolutional Neural Networks with respect to the ones proposed in this paper. For this reason, the impact of the FC layer on the overall performance is substantial while evaluating the performance on the device. In this way we obtained a KWS system with only 13 K parameters, which requires only 10 ms to perform inference. From this experiments, SCNNs outperform CNNs, giving a neural network with a higher accuracy with a smaller representation, but with a slightly longer inference time. Nevertheless, the latency introduced by SCNNs is still negligible in a standard KWS system used as a low-power voice wake up for other devices.

VI. CONCLUSIONS

In this paper we have shown a low-power architecture for keyword spotting based on Separable Convolution. In particular, we have shown that Separable Convolution has an accuracy comparable with the one of dephtwise convolution approach, but it is simpler to implement since it is supported by most frameworks and hardware platforms, including Myriad 2 by Movidius. At the end, our proposed solution gets an accuracy comparable with other neural networks architecture in literature [9] but requiring only a tenth of the parameters and a quarter of the execution time. The reduced computation time and the presence of the VAD can be used to enable and disable the KWS algorithm. The VPU utilization further reduces power consumption in a real application, increasing battery life.

REFERENCES

[1] A. Breen, H. H. Bui, R. Crouch, K. Farrell, F. Faubel, R. Gemello, W. F. Ganong III, T. Haulick, R. M. Kaplan, C. L. Ortiz *et al.*, "Voice in the user interface," *Interactive Displays: Natural Human-Interface Technologies*, pp. 107–163, 2014.

[2] F. Portet, M. Vacher, C. Golanski, C. Roux, and B. Meillon, "Design and evaluation of a smart home voice interface for the elderly: acceptability and objection aspects," *Personal and Ubiquitous Computing*, vol. 17, no. 1, pp. 127–144, 2013.

[3] I. McGraw, R. Prabhavalkar, R. Alvarez, M. G. Arenas, K. Rao, D. Rybach, O. Alsharif, H. Sak, A. Gruenstein, F. Beaufays *et al.*, "Personalized speech recognition on mobile devices," in *2016 IEEE International Conference on Acoustics, Speech and Signal Processing (ICASSP)*. IEEE, 2016, pp. 5955–5959.

[4] Y. Zhang, N. Suda, L. Lai, and V. Chandra, "Hello edge: Keyword spotting on microcontrollers," *arXiv preprint arXiv:1711.07128*, 2017.

[5] A. Graves, S. Fernández, F. Gomez, and J. Schmidhuber, "Connectionist temporal classification: labelling unsegmented sequence data with recurrent neural networks," in *Proceedings of the 23rd international conference on Machine learning*. ACM, 2006, pp. 369–376.

978-1-5386-4757-8/18 $31.00 © 2018 IEEE

[6] T. N. Sainath and C. Parada, "Convolutional neural networks for small-footprint keyword spotting," in *Sixteenth Annual Conference of the International Speech Communication Association*, 2015.

[7] S. O. Arik, M. Kliegl, R. Child, J. Hestness, A. Gibiansky, C. Fougner, R. Prenger, and A. Coates, "Convolutional recurrent neural networks for small-footprint keyword spotting," *arXiv preprint arXiv:1703.05390*, 2017.

[8] Y.-J. Lin and T. S. Chang, "Data and hardware efficient design for convolutional neural network," *IEEE Transactions on Circuits and Systems I: Regular Papers*, vol. 65, no. 5, pp. 1642–1651, 2018.

[9] R. Tang, W. Wang, Z. Tu, and J. Lin, "An experimental analysis of the power consumption of convolutional neural networks for keyword spotting," *arXiv preprint arXiv:1711.00333*, 2017.

[10] G. Meoni, L. Pilato, and L. Fanucci, "A low power voice activity detector for portable applications," in *14th Conference on Ph. D. Research in Microelectronics and Electronics (PRIME)*, Prague, Czech Republic, July 2018.

[11] R. Collobert, C. Puhrsch, and G. Synnaeve, "Wav2letter: an end-to-end convnet-based speech recognition system," *arXiv preprint arXiv:1609.03193*, 2016.

[12] H. Noguchi, T. Takagi, M. Yoshimoto, and H. Kawaguchi, "An ultra-low-power vad hardware implementation for intelligent ubiquitous sensor networks." in *SiPS*, 2009, pp. 214–219.

[13] A.-r. Mohamed, "Deep neural network acoustic models for asr," Ph.D. dissertation, 2014.

[14] L. Deng, J. Li, J.-T. Huang, K. Yao, D. Yu, F. Seide, M. Seltzer, G. Zweig, X. He, J. Williams *et al.*, "Recent advances in deep learning for speech research at microsoft," in *Acoustics, Speech and Signal Processing (ICASSP), 2013 IEEE International Conference on.* IEEE, 2013, pp. 8604–8608.

[15] S. Ioffe and C. Szegedy, "Batch normalization: Accelerating deep network training by reducing internal covariate shift," *arXiv preprint arXiv:1502.03167*, 2015.

[16] A. G. Howard, M. Zhu, B. Chen, D. Kalenichenko, W. Wang, T. Weyand, M. Andreetto, and H. Adam, "Mobilenets: Efficient convolutional neural networks for mobile vision applications," *arXiv preprint arXiv:1704.04861*, 2017.

[17] F. Chollet *et al.*, "Keras," https://keras.io, 2015.

[18] M. Abadi *et al.*, "TensorFlow: Large-scale machine learning on heterogeneous systems," 2015, software available from tensorflow.org. [Online]. Available: https://www.tensorflow.org/

[19] R. Caruana, S. Lawrence, and C. L. Giles, "Overfitting in neural nets: Backpropagation, conjugate gradient, and early stopping," in *Advances in neural information processing systems*, 2001, pp. 402–408.

[20] Y. Jia, E. Shelhamer, J. Donahue, S. Karayev, J. Long, R. Girshick, S. Guadarrama, and T. Darrell, "Caffe: Convolutional architecture for fast feature embedding," *arXiv preprint arXiv:1408.5093*, 2014.

978-1-5386-4757-8/18 $31.00 © 2018 IEEE

Author Index

Acquaviva, Andrea, 19, 31
Aldegheri, Stefano, 119
Anghel, Lorena, 176
Asada, Kunihiro, 55
Asano, Hiroki, 196
Austin, Todd, i
Avramenko, Serhiy, 207
Awais, Muhammad, 219
Azad, Siavoosh Payandeh, 207
Aziza, Hassen, 192

Barchi, Francesco, 19
Barros, Edna, 125
Belhaj, Mohamed-Moez, 168
Benelli, Gionata, 267
Benini, Luca, 43
Benoit, Pascal, 188
Bernard, Serge, 168
Bernasconi, Anna, 137, 213
Bertacco, Valeria, xi
Bertozzi, Davide, 255
Bhattacharjee, Debjyoti, 1
Bigalke, Steve, 225
Bocquet, Marc, 192
Boffa, Antonio, 137
Bombieri, Nicola, iii, 119
Bonhommeau, Sylvain, 168
Braun, Konstantin, 89
Brendler, Leonardo H., 71
Breyer, Evelyn T., 180

Calimera, Andrea, 113, 149
Cantan, Mayeul, 180
Cantoro, Riccardo, 59
Carbonara, Sara, 59
Centomo, Stefano, 237
Champac, Victor, 65, 77
Charles, Henri-Pierre, xiv
Chattopadhyay, Anupam, 1, 95
Chen, Suyuan, 83
Ciobanu, Catalin Bogdan, 143
Ciriani, Valentina, 213
Costa, Kevin, 249

De Micheli, Giovanni, x
De Oliveira Junior, Luiz Antonio, 125
Demrozi, Florenc, 249

Deng, Erya, 184
Devarajegowda, Keerthikumara, 231
Di Natale, Giorgio, 176
Di Pendina, Gregory, 188
Duhem, François, 188

Ecker, Wolfgang, 231
Elfadel, Ibrahim, 37
Enescu, Florian, 49

Fanucci, Luca, 267
Fey, Gorschwin, 172
Fioranelli, Francesco, 163
Floridia, Andrea, 59
Forero, Freddy, 65
Forno, Evelina, 31
Freitas, Philippe, 168
Fritzmann, Tim, 89
Frontini, Luca, 213
Fujita, Masahiro, iii
Fummi, Franco, 237

Gómez, Andrés Felipe, 65
Gaillardon, Pierre-Emmanuel, 107
Ghasempouri, Tara, 172
Giacomin, Edouard, 107
Giraud, Bastien, 180, 192
Guo, Hui, 201
Gupta, Utkarsh, 49

Hamdioui, Said, xiv
Haugou, Germain, 43
Hirose, Tetsuya, 196
Histace, Aymeric, 159
Huang, Juinn-Dar, 13
Hussain, Mubashir, 201

Iizuka, Tetsuya, 55
Ilioaea, Irina, 49
Ionescu, Adrian, 180

Jacob, Pierre, 159
Jacobs, Swen, 172
James, Rekha K., 25
Jenihhin, Maksim, 207
Jose, John, 25

Kalla, Priyank, 49

Kang, Wang, 184
Katkoori, Srinivas, 7
Keliris, Anastasis, 101
Kerzerho, Vincent, 168
Kobayashi, Yuki, 31
Koeppl, Heinz, xii
Konstantinou, Charalambos, 101
Kuroki, Nobutaka, 196

Lamlih, Achraf, 168
Le Kernec, Julien, 163
Lienig, Jens, 225
Liu, Chia-Hung, 13
Lorandel, Jordane, 163
Luccio, Fabrizio, 137
Ly, Denys, 176

Macii, Enrico, 19, 31
Mahalat, Mahabub Hasan, 261
Mandal, Swagata, 1
Maniatakos, Michail, 101
Manzato, Silvia, 119
Marchand, Cédric, 180
Maringer, Georg, 89
Martino, Gianluca, 172
Matsumoto, Kaori, 196
Meinhardt, Cristina, 71
Meoni, Gabriele, 267
Mert, Yakup Murat, 243
Mess, Jan-Gerd, 59
Mikolajick, Thomas, 180
Minervini, Francesco, 43
Miramond, Benoit, 176
Modad, Jad, 188
Mohammadi, Hassan Ghasemzadeh, 219
Moreau, Mathieu, 192
Muhr, Eloi, 192
Mulaosmanovic, Halid, 180
Muzaffar, Shahzad, 37

Nair, Mini, 25
Nakazawa, Yuichiro, 196
Nakura, Toru, 55
Niazmand, Behrad, 207
Noël, Jean-Philippe, 180, 192
Nouet, Pascal, 188
Numa, Masahiro, 196

O'Connor, Ian, xiv, 180
Ojima, Naoki, 55
Ouattara, Frederic, 188

Pagli, Linda, 137
Palesi, Maurizio, 25
Panato, Marco, 237
Patrigeon, Guillaume, 188
Peluso, Valentino, 113
Perez, Zahira, 77
Platzner, Marco, 219
Portal, Jean-Michel, 192
Pravadelli, Graziano, i, 249
Prenat, Guillaume, 188
Pudi, Vikramkumar, 95

Rabozzi, Marco, 143
Raik, Jaan, 172, 207
Ranjandish, Reza, 155
Rao, Vikas, 49
Reis, Ricardo, 71
Reorda, Matteo Sonza, 59
Riener, Heinz, 172
Romain, Olivier, 159, 163
Rouyer, Tristan, 168
Roy, Kaushik, 65
Ruospo, Annachiara, 43

Salles, Jérémie, 168
Sanchez, Ernesto, 43, 59
Santambrogio, Marco Domenico, 143
Sazos, Marios, 101
Schamberger, Thomas, 89
Schiavone, Pasquale Davide, 43
Schmid, Alexandre, 155
Sciuto, Donatella, 143
Sen, Bibhash, 261
Senni, Sophiane, 188
Sepúlveda, Johanna, 89
Sereno, Denis, 159
Sevin, Kaan, 188
Shahare, Rohit, 261
Sharma, Ashutosh, 131
Shrestha, Rahul, 131
Simon-Chane, Camille, 159
Sleeba, Simi Zerine, 25
Slesazeck, Stefan, 180
Souchaud, Marc, 159
Soulier, Fabien, 168
Srinath, Arpitha, 49
Stolichnov, Igor, 180
Stornaiuolo, Luca, 143
Stramondo, Giulio, 143

Tahoori, Mehdi, xiv

Tala, Mahdi, 255
Tavva, Yaswanth, 1
Tchuenté, Maurice, 159
Tenace, Valerio, 149
Torres, Lionel, xiv, 188
Tramarin, Federico, 249
Tsuji, Yuto, 196

Ugale, Nikhil, 261
Urgese, Gianvito, 19, 31

Varbanescu, Ana Lucia, 143
Vatajelu, Elena-Ioana, 176
Vemuri, Ranga, 83
Vianello, Elisa, 176
Villacorta, Hector, 77
Vilquin, Bertrand, 180
Violante, Massimo, 207

Wang, Zhaohao, 184
Wei, Shaoqian, 184
Wong, Ming Ming, 95

Yang, Shufan, 163
Yang, Wei-Hao, 13

Zaman, Md Adnan, 7
Zaruba, Florian, 43
Zhao, Weisheng, 184
Zimpeck, Alexandra L., 71
Zorian, Yervant, xiii

IEEE
445 Hoes Lane
Piscataway, NJ 08854-4141

ISBN 978-1-5386-4757-8